Nanomechanics for Coatings and Engineering Surfaces

Nanomechanics for Coatings and Engineering Surfaces

Test Methods, Development Strategies, Modeling Approaches, and Applications

Edited by

Ben Beake
Micro Materials Ltd., Wrexham, United Kingdom

Tomasz Liskiewicz
Department of Engineering, Manchester Metropolitan University, Manchester, United Kingdom

ELSEVIER

Elsevier
Radarweg 29, PO Box 211, 1000 AE Amsterdam, Netherlands
125 London Wall, London EC2Y 5AS, United Kingdom
50 Hampshire Street, 5th Floor, Cambridge, MA 02139, United States

Copyright © 2025 Elsevier Inc. All rights are reserved, including those for text and data mining, AI training, and similar technologies.

Publisher's note: Elsevier takes a neutral position with respect to territorial disputes or jurisdictional claims in its published content, including in maps and institutional affiliations.

No part of this publication may be reproduced or transmitted in any form or by any means, electronic or mechanical, including photocopying, recording, or any information storage and retrieval system, without permission in writing from the publisher. Details on how to seek permission, further information about the Publisher's permissions policies and our arrangements with organizations such as the Copyright Clearance Center and the Copyright Licensing Agency, can be found at our website: www.elsevier.com/permissions.

This book and the individual contributions contained in it are protected under copyright by the Publisher (other than as may be noted herein).

Notices
Knowledge and best practice in this field are constantly changing. As new research and experience broaden our understanding, changes in research methods, professional practices, or medical treatment may become necessary.

Practitioners and researchers must always rely on their own experience and knowledge in evaluating and using any information, methods, compounds, or experiments described herein. In using such information or methods they should be mindful of their own safety and the safety of others, including parties for whom they have a professional responsibility.

To the fullest extent of the law, neither the Publisher nor the authors, contributors, or editors, assume any liability for any injury and/or damage to persons or property as a matter of products liability, negligence or otherwise, or from any use or operation of any methods, products, instructions, or ideas contained in the material herein.

ISBN: 978-0-443-13334-3

For Information on all Elsevier publications
visit our website at https://www.elsevier.com/books-and-journals

Publisher: Matthew Deans
Acquisitions Editor: Dennis McGonagle
Editorial Project Manager: Emily Thomson
Production Project Manager: Sujithkumar Chandran
Cover Designer: Christian Bilbow

Typeset by MPS Limited, Chennai, India

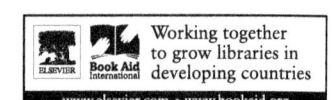

Contents

List of contributors	xiii
Preface	xvii

Section I Introduction

1. Development of nanomechanical testing — 3
Ben Beake and Tomasz Liskiewicz
 1.1 Historical background and development of methods for unloading curve analysis — 3
 1.2 Practical factors influencing the accuracy of nanoindentation data and reliability of the properties obtained — 6
 1.3 Indentation fracture — 14
 1.4 Indentation size effects — 16
 1.5 Relating hardness to yield stress: the constraint factor — 19
 1.6 Indentation energy — 19
 1.7 The continuous stiffness technique — 23
 1.8 High strain rate testing — 24
 1.9 High-speed mapping — 26
 1.10 Testing under environmentally relevant conditions — 27
 1.11 Summary — 27
 References — 28

2. Development of additional nanoscale wear test techniques — 37
Ben Beake and Tomasz Liskiewicz
 2.1 Introduction – single-asperity tribology — 37
 2.2 Scratch and wear testing — 40
 2.3 AFM scratch and wear – even smaller length scale/contact size — 48
 2.4 Experimental techniques for nano-/microscale fretting and reciprocating wear testing — 49
 2.5 Repetitive contact – nano and micro-impact tests — 51
 2.6 Conclusions — 55
 References — 56

3. Linking coating mechanical properties and wear resistance — 67
Ben Beake and Tomasz Liskiewicz
 3.1 Introduction — 67

3.2	Why coatings perform better in certain contact situations but not in others	71
3.3	Conclusions	96
	References	98

Section II Experimental methods and techniques

4. Nanoindentation—strategies for reliable measurements on coated systems   **113**
Thomas Chudoba

4.1	Preconditions for correct and reliable measurements	113
4.2	Application of a radial displacement correction	116
4.3	Combined calibration of area function and instrument compliance	118
4.4	The influence of a tip rounding	120
4.5	Theoretical considerations for the measurement of coatings	123
4.6	Modulus measurements by fully elastic indentations	127
4.7	The use of dynamic test methods for the measurement of hardness and modulus	132
4.8	Examples for the measurement of coatings	134
	References	139

5. Indentation energy-based analysis methods   **143**
Steve J. Bull

5.1	Introduction	143
5.2	Hardness and Young's modulus	143
5.3	Fracture toughness and fracture energy	148
5.4	Conclusions	156
	References	156

6. Nano-scratch testing   **159**
Ben Beake and Tomasz Liskiewicz

6.1	Introduction	159
6.2	Experimental considerations	160
6.3	General features	162
6.4	Multi-pass scratch testing	174
6.5	Influence of test temperature	179
6.6	Conclusions	180
	References	180

7. Reciprocating nano- and micro-scale wear testing   **187**
Ben Beake and Tomasz Liskiewicz

7.1	Introduction	187

7.2	Case studies on biomedical materials—Ti6Al4V, CoCrMo, and 316L stainless steel—influence of passivating oxide layer in reciprocating tests over different length scales	194
7.3	Case study on diamond-like carbon coatings on steel	202
7.4	Conclusions	209
	References	210

8. Nanowear by atomic force microscopy 217
Jiaxin Yu, Linmao Qian and Hongtu He

8.1	Single-asperity nanotribology by atomic force microscopy	217
8.2	Limitations and strengths of atomic force microscopy	223
8.3	Friction and wear of ultrathin films	225
8.4	Line and area scanning comparison	230
	References	232

9. Nano-impact testing 239
Xiangru Shi and Jian Chen

9.1	Introduction	239
9.2	Experimental setup of micro/nanoimpact testing	242
9.3	Multiple nano/microimpact testing	244
9.4	High-precision single nanoimpact testing	254
9.5	Conclusions	264
	References	264

10. Nanomechanical testing methods to understand the effects of residual stress on coating's performance 271
Marco Sebastiani, Edoardo Rossi and Saqib Rashid

	Abbreviations	271
10.1	Introduction	271
10.2	Recent advances in residual stress measurement at micron scale: the focused ion beam ring-core method	275
10.3	Micron-scale fracture toughness analysis: the pillar splitting method	287
10.4	Discussion and conclusions	295
	References	296

11. Multi-sensing approaches 301
Jan Tomastik, Adrian Harris, Vladimir Vishnyakov and Ben Beake

11.1	Introduction	301
11.2	Multi-sensing approach using simultaneous record of acoustic emission	301
11.3	Electrical contact resistance measurement	313
11.4	Friction measurement	317
11.5	Nanomechanical Raman spectroscopy	319
11.6	Sensors and actuators	320

11.7	Multi-sensing in high strain rate contact	321
11.8	Electroplastic effect	321
References		322

12. High-temperature testing 327
James S. K-L.Gibson

12.1	Introduction	327
12.2	Strategies for reliable high-temperature nanomechanical test measurements	327
12.3	High-temperature pillar compression and microcantilever bending	334
12.4	High-temperature scratch and impact/high strain rate testing	335
12.5	Summary and conclusions	336
References		336

Section III Modelling approaches

13. Analytical methods - applied 343
Troy vom Braucke, Norbert Schwarzer, Frank Papa and Julius Schwarzer

13.1	Introduction of analytical modeling of coated systems	343
13.2	Working through practical examples	346
13.3	Thought experiments	361
13.4	Outlook	364
References		367

14. Numerical simulation and finite element analysis 369
Roberto Martins Souza and Newton Kiyoshi Fukumasu

14.1	Introduction	369
14.2	Contact stresses	370
14.3	Thin film mechanics and substrate effects during indentation	371
14.4	Simulation of normal loads and lateral indenter displacement	380
14.5	Beyond indentation and scratch analyses for coating/substrate simulation	381
14.6	Concluding remarks	383
References		384

15. High-performance molecular dynamics simulations to investigate nanoindentation of advanced engineering materials 393
Saurav Goel and Pengfei Fan

Abbreviations		393
Nomenclatures		393
15.1	Ingredients of a trustworthy molecular dynamics simulation	395
15.2	Postprocessing of the molecular dynamics data to extract the nanomechanical properties	399

15.3	High-order postprocessing analysis of the molecular dynamics simulation files to correlate with experiments	411
15.4	Scale differences between molecular dynamics simulations and nanoindentation experiments	428
15.5	Concluding remarks	431
Acknowledgments		431
References		431

Section IV Coatings and engineering surfaces: design strategies and industrial applications

16. The significance of hardness and elastic modulus in the design of engineered surfaces — 441
Adrian Leyland and Allan Matthews

16.1	Introduction	441
16.2	The importance of hardness and elastic modulus in tribology and surface engineering	443
16.3	The Archard Wear Equation	447
16.4	A brief historical perspective on hardness, hardness-to-modulus ratio, and wear-rate determination	453
16.5	Practical use of hardness-to-modulus ratio based parameters	459
16.6	Appropriate use of hardness-to-modulus ratio as a design tool—fundamental considerations	463
16.7	A case study	464
16.8	Summarizing remarks	469
Acknowledgment		471
References		471

17. Thin protective coatings on silicon for microelectromechanical systems — 481
Ben Beake and Tomasz Liskiewicz

17.1	Introduction	481
17.2	Silicon (100)—phase transformation and cracking	482
17.3	Coatings to protect silicon – thin ta-C films	488
17.4	Conclusions	501
References		502

18. Diamond-like carbon coatings on steel for automotive applications — 509
Samuel James McMaster

18.1	Introduction to diamond-like carbon coatings	509
18.2	Performance of diamond-like carbons	511
18.3	Deposition of diamond-like carbon coatings	519
18.4	Mechanical and structural characterization	521
18.5	Scratch testing for adhesion testing of diamond-like carbons	532

	18.6	Fretting testing of diamond-like carbons	535
	18.7	Impact for fatigue determination	537
	18.8	Conclusion	542
	References		543

19. Machining of difficult materials — 557
Jose Luis Endrino, German Fox-Rabinovich and Ben Beake
- 19.1 Hard coatings for cutting processes — 557
- 19.2 Hard nitride coatings for machining ductile materials — 560
- 19.3 Self-organization processes during wear of cutting tools — 562
- 19.4 Influence of fracture resistance and high-temperature mechanical properties in self-adaptive behavior of TiAlCrSiYN-based coatings — 563
- 19.5 Summary — 566
- References — 567

20. Nanoindentation-based techniques for evaluating irradiated fuel and structural materials — 571
David Frazer and Peter Hosemann
- 20.1 Introduction — 571
- 20.2 Nanoindentation — 573
- 20.3 Other small-scale mechanical test techniques used today — 583
- 20.4 Future avenues for exploration — 588
- 20.5 Conclusions — 590
- References — 590

21. Aerospace coatings - surface engineering for operation in extreme environments — 599
John Rayment Nicholls, Christine Deborah Chalk and Luis Isern Arrom
- 21.1 Introduction — 599
- 21.2 Environmental protection coatings and bondcoats — 601
- 21.3 Thermal barrier coatings – design considerations — 604
- 21.4 Design tools for developing enhanced TBCs – probing sintering resistance, chemical resilience, and erosion resistance using nano- and microindentation — 614
- 21.5 Modeling erosion of thermal barrier coatings — 619
- 21.6 Conclusions — 621
- References — 622

22. Advanced solid lubricants — 627
Diana Berman
- 22.1 Introduction — 627
- 22.2 Mechanism of lubrication in traditional solid lubricants and approaches for application of solid lubricants — 628
- 22.3 Advanced solid lubricants — 630

	22.4 Conclusions	641
	Acknowledgments	641
	References	641
23.	**Nanomechanics of tribologically transformed surfaces**	**647**
	Guillaume Kermouche and Gaylord Guillonneau	
	23.1 Introduction	647
	23.2 About tribologically transformed surfaces	648
	23.3 About micromechanical tests used to characterize tribologically transformed surface mechanical properties	650
	23.4 Application to tribologically transformed surface resulting of manufacturing processes	653
	23.5 Application to glaze layers	658
	23.6 Thermal stability of tribologically transformed surface	661
	23.7 Upcoming developments	664
	23.8 Conclusion	668
	Acknowledgments	669
	References	669
24.	**Microtribology experiments for hardmetals**	**675**
	Mark G. Gee	
	24.1 Introduction	675
	24.2 Microtribology experiments	675
	24.3 Wear-corrosion synergy	680
	24.4 Electron back scattered diffraction analysis	682
	24.5 Summary and conclusions	688
	Acknowlegments	688
	References	689

Section V Summary

25.	**Trends and future directions**	**693**
	Ben Beake and Tomasz Liskiewicz	
	25.1 More extreme temperatures	693
	25.2 More complex experiments – multisensing, big data, and AI	693
	25.3 More complex experiments to simulate abrasion and erosion	694
	25.4 Mechanical property mapping	695
	25.5 Tribological and impact mapping	695
	25.6 Novel experiments	696
	25.7 A tool for surface optimization	696
	25.8 Summary	696

Index 697

List of contributors

Ben Beake Micro Materials Ltd., Wrexham, United Kingdom

Diana Berman Department of Materials Science and Engineering, University of North Texas, Denton, TX, United States

Steve J. Bull School of Engineering, Newcastle University, Newcastle upon Tyne, United Kingdom

Christine Deborah Chalk Coating Technology, Surface Engineering and Precision Centre, Cranfield University, Cranfield, United Kingdom

Jian Chen School of Materials Science and Engineering, Southeast University, Nanjing, Jiangsu Province, P.R. China

Thomas Chudoba ASMEC GmbH, Dresden, Germany

Jose Luis Endrino Universidad Loyola Andalucia, Dos Hermanas, Seville, Spain

Pengfei Fan Centre for Nanoscience and Nanotechnology, University of Bath, Bath, United Kingdom

German Fox-Rabinovich MMRI, McMaster University, Hamilton, ON, Canada

David Frazer General Atomics Electromagnetic Systems, Nuclear Materials & Technology Division, San Diego, CA, United States

Newton Kiyoshi Fukumasu Surface Phenomena Laboratory, Polytechnic School, Universidade de Sao Paulo, Sao Paulo, Brazil

Mark G. Gee Department of Materials and Mechanical Metrology, National Physical Laboratory, Teddington, United Kingdom

Saurav Goel School of Engineering, London South Bank University, London, United Kingdom; Department of Mechanical Engineering, University of Petroleum and Energy Studies, Dehradun, Uttarakhand, India

Gaylord Guillonneau Ecole Centrale de Lyon, CNRS, ENTPE, LTDS, UMR5513, Ecully, France

Adrian Harris Micro Materials Ltd., Wrexham, United Kingdom

Hongtu He Key Laboratory of Testing Technology for Manufacturing Process, Ministry of Education, Southwest University of Science and Technology, Mianyang, P.R. China

Peter Hosemann University of California Berkeley, Department of Nuclear Engineering, Berkeley, CA, United States

Luis Isern Arrom Coating Technology, Surface Engineering and Precision Centre, Cranfield University, Cranfield, United Kingdom

Samuel James McMaster Research Centre for Manufacturing and Materials (CMM), The Institute for Advanced Manufacturing and Engineering (AME), Coventry, United Kingdom; Pillarhouse International Ltd. Chelmsford, Essex, United Kingdom

Guillaume Kermouche Mines Saint-Etienne, LGF UMR5307 CNRS, Saint-Etienne, France

Adrian Leyland Department of Materials Science and Engineering, The University of Sheffield, Sheffield, United Kingdom

Tomasz Liskiewicz Department of Engineering, Manchester Metropolitan University, Manchester, United Kingdom

Allan Matthews Department of Materials, University of Manchester, Manchester, United Kingdom

John Rayment Nicholls Coating Technology, Surface Engineering and Precision Centre, Cranfield University, Cranfield, United Kingdom

Frank Papa GP Plasma, Medina, OH, United States

Linmao Qian Tribology Research Institute, The State Key Laboratory of Rail Vehicle System, School of Mechanical Engineering, Southwest Jiaotong University, Chengdu, P.R. China

Saqib Rashid Università degli Studi Roma Tre, Department of Civil, Computer Science and Aeronautical Technologies Engineering, Rome, Italy

Edoardo Rossi Università degli Studi Roma Tre, Department of Civil, Computer Science and Aeronautical Technologies Engineering, Rome, Italy; Consorzio Interuniversitario Nazionale per la Scienza e Tecnologia dei Materiali (INSTM), Florence, Italy

James S. K-L.Gibson Department of Materials, University of Oxford, Oxford, United Kingdom

Julius Schwarzer Siomec, Tankow, Ummanz, Germany

Norbert Schwarzer Siomec, Tankow, Ummanz, Germany

Marco Sebastiani Università degli Studi Roma Tre, Department of Civil, Computer Science and Aeronautical Technologies Engineering, Rome, Italy; Consorzio Interuniversitario Nazionale per la Scienza e Tecnologia dei Materiali (INSTM), Florence, Italy

Xiangru Shi School of Materials Science and Engineering, Hohai University, Nanjing, Jiangsu Province, P.R. China

Roberto Martins Souza Surface Phenomena Laboratory, Polytechnic School, Universidade de Sao Paulo, Sao Paulo, Brazil

Jan Tomastik Joint Laboratory of Optics of Palacký University and Institute of Physics AS CR, Palacký University in Olomouc, Olomouc, Czech Republic; Institute of Physics of the Czech Academy of Sciences, Joint Laboratory of Optics of Palacký University and Institute of Physics AS CR, Olomouc, Czech Republic

Vladimir Vishnyakov Inomorph Ltd., Huddersfield, United Kingdom

Troy vom Braucke GP Plasma, Medina, OH, United States

Jiaxin Yu Key Laboratory of Testing Technology for Manufacturing Process, Ministry of Education, Southwest University of Science and Technology, Mianyang, P.R. China

Preface

With advances in nanomechanical instrumentation and coatings being used increasingly to enable materials to operate in ever more challenging environments, it is timely to provide the reader with a one-stop-shop of best practice in nanoindentation measurements and related small-scale test methods and how to use the results to develop improved coatings and wear-resistant surfaces. As editors, we have gathered leading experts to contribute their insights, ensuring that this volume serves as an invaluable resource for researchers, engineers, and practitioners alike. As such, this book represents a collaborative effort by experts in the field to provide a comprehensive overview of the test methods, development strategies, modeling approaches, and applications that define this rapidly evolving domain.

The utilization of small-scale tests to design more durable and wear-resistant surfaces and materials is a central focus of the book. With nanomechanics and nanoscratch techniques gaining recognition through ISO and CEN standards, the contributing authors elaborate on current state-of-the-art practices for characterizing and developing resilient surfaces and coating systems. They delve into the criteria for a damage-tolerant surface, emphasizing the importance of reliable measurements of various metrics such as hardness, yield strength, adhesion, fracture toughness, and residual stress, as outlined in best practice guidelines. The authors underscore the significance of considering contact size in different applications and advocate for multiscale testing to enhance relevance across real-world scenarios. There is a recurring theme of employing nanomechanics for testing under extreme conditions of temperature and strain rate, particularly in industries such as aerospace, automotive, nuclear, cutting, and drilling. The experimental approach is complemented by modeling to identify and address weak points in design, thereby improving overall durability and performance.

We have tried to present the contents in a logical, easily assimilated manner. This book contains five sections written carefully to cover the fundamentals aspects as well as the key theoretical developments and experimental methodologies associated with nanomechanics. In Section I, we present the foundational aspects of nanomechanical testing. From the evolution of testing techniques to the intricacies of linking coating mechanical properties with wear resistance, this section lays the groundwork for understanding the complexities of nanomechanics. Section II, delves into the practical aspects of nanomechanical testing. Through chapters dedicated to nanoindentation, nanoscratch testing, reciprocating wear, AFM nanowear, nanoimpact and high-temperature testing, readers are equipped with the tools and methodologies necessary to explore coating behavior at the nanoscale. In Section III, we explore analytical and numerical simulations that complement experimental findings. From analytical methods to finite element analysis and atomistic simulations, this section offers insights into the theoretical frameworks underpinning nanomechanics. Section IV showcases the real-world implications of nanomechanics. With chapters devoted to thin protective coatings, automotive applications, aerospace coatings, nuclear materials applications, hard metals, solid lubricants, and tribologically transformed surfaces, readers gain a deeper understanding of how nanomechanics is shaping industries and driving innovation. Finally, in Section V, we reflect on current trends and future directions in nanomechanics. By examining emerging technologies and research directions, this section offers a glimpse into the exciting possibilities that lie ahead in this dynamic field.

This book is meant to be an active source of reference, which the reader can flip through to find useful guidance on specific topics when there is a need for some perspective or advice. There will be some degree of repetition between the chapters, which results from the fact that every chapter is a stand-alone contribution meant to provide a full understanding of a given topic. The reader is also advised to pay attention to the nomenclature used by the contributors, as this might differ between the chapters. We accepted this fact, as it spans from common practice in a field approached by scientists from different backgrounds, and did not try to manipulate the nomenclature as any attempt to create a unified glossary and list of symbols would be untrue to the field and could cause a degree of confusion.

As editors, we extend our sincere appreciation to all the contributors who have generously shared their expertise and insights. It is our hope that this book serves as a valuable resource for researchers, engineers, and practitioners seeking to unlock the potential of nanomechanics in coatings and engineered surfaces.

Ben Beake
Tomasz Liskiewicz

Section I

Introduction

Development of nanomechanical testing

Ben Beake[1] and Tomasz Liskiewicz[2]
[1]Micro Materials Ltd., Wrexham, United Kingdom, [2]Department of Engineering, Manchester Metropolitan University, Manchester, United Kingdom

1.1 Historical background and development of methods for unloading curve analysis

Nanomechanical testing has proved an important and highly effective technique to characterize the mechanical properties and performance of materials at small length scales. It has improved our fundamental understanding of mechanical properties and the importance of the nanoscale behavior of materials on their performance at larger length scales. Nanomechanical and nano/microtribological test techniques can be broadly divided into two main categories. The first of these are those designed for characterization of mechanical properties—primarily nanoindentation but also including other tests such as microcompression tests. The second category is primarily involved with simulating contact conditions—including nano- and microscratch testing, nano-/micro-impact and reciprocating nanowear, nanofretting tests—e.g., to aid the development of more wear-resistant or damage-tolerant materials or coating systems. In this chapter, we describe the development of the nanoindentation technique, from fundamentals to best practice to obtain reliable data. Developments in small-scale tribological techniques are described in the next chapter.

Indentation hardness measurements at macro- to nanoscale have been discussed by Fischer-Cripps (2011) and Broitman (2017). Although indentation was a well-established technique, the pace of its development and range of applications increased rapidly with the development of depth-sensing indentation (also termed nanoindentation or instrumented indentation testing) where the entire load − displacement curve is continuously recorded. Stilwell and Tabor investigated the elastic recovery of conical indentations into metallic materials in 1961 (Stilwell & Tabor, 1961). Borodich (2014) has noted that the first depth-sensing nanoindenter was developed in 1966 in the former Soviet Union by Kalei, and the first reported load − displacement curves were published in 1968 (Kalei, 1968). The introduction of a method for determination of the elastic modulus by analysis of the unloading curve by Bulychev and coworkers in 1975 was the next important step (Bulychev et al., 1975), as shown in Eq. (1.1), where E_r is the reduced indentation modulus, A_c is projected (or cross-sectional) area, S is the unloading stiffness, P is the load, and h is the displacement.

$$S = \frac{dP}{dh} = \frac{2}{\sqrt{\pi}} E_r \sqrt{A_c} \tag{1.1}$$

These developments were followed by research by Newey et al. (1982) and Pethica et al. (1983), which subsequently led to the development of the first commercial instrumentation for nanomechanical testing (the NanoTest and NanoIndenter, respectively). Their popularity was helped by the increasing need to characterize the mechanical properties of thin films at a length scale smaller than that possible in conventional microhardness testing and by the development of convenient methods of analyzing the load − displacement data. In 1984 Loubet and coworkers reported on unloading curve analysis of Vickers high-load indentation of MgO (Loubet et al., 1984). In 1986 Doerner and Nix proposed a method for determining hardness and modulus from low-load nanoindentation load − displacement data (Doerner & Nix, 1986). Their method was based on a flat punch approximation, in that contact area during the initial unloading remains constant, as it is for a flat punch, supported by experimental data suggesting that the initial unloading portions were linear. Pethica and Oliver (1987) noted that despite different stress distributions with a spherical tip and a flat-ended cylindrical punch, the surface stiffness was proportional to contact radius and elastic modulus in both cases. In 1992 the Doerner − Nix approach was refined by Oliver and Pharr (1992), who highlighted that a more detailed analysis of experimental data showed that the unloading curves only approximated to being linear, even on metals showing little elastic recovery. Since its publication in 1992, the original paper by Oliver and Pharr (2004) describing their power-law fitting to the unloading curve has been cited nearly 30,000 times! They showed that (1) the unloading portions are curved and (2) are better fitted by a power-law function of the type $P = \alpha(h-h_f)^m$ with exponents generally in the range 1.2−1.6 (i.e., $>m=1$ for flat punch). This approach produces a more accurate determination of the contact stiffness in the stiffness equation and, therefore, to elastic modulus and hardness, through contact depth, and projected contact area through the area function. The unloading stiffness is related to the contact depth through Eq. (1.2), where ε (Epsilon) is a constant that depends on the geometry of the indenter (often taken as 0.75). Fig. 1.1 shows a typical indentation curve on fused silica, a material showing high elastic recovery. A correction factor to Eq. (1.1) was introduced (Eq. 1.3), with a value typically taken as 1.034.

$$h_c = h_{max} - \varepsilon \frac{P_{max}}{S} \tag{1.2}$$

$$S = \beta \frac{2}{\sqrt{\pi}} E_r \sqrt{A_c} \tag{1.3}$$

The reduced modulus is defined as

$$1/E_r = (1-\nu_s^2)/E_s + (1-\nu_i^2)/E_i \tag{1.4}$$

Development of nanomechanical testing

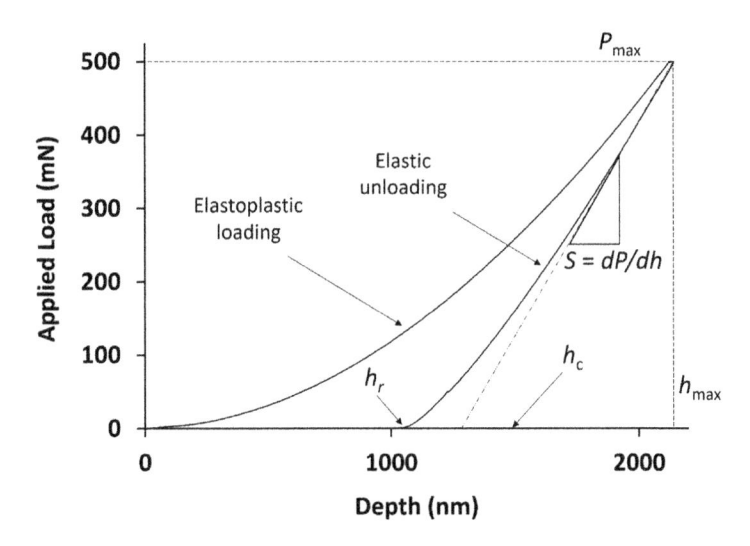

Figure 1.1 Nanoindentation curve 500 mN nanoindentation curve on fused silica with a Berkovich indenter.

where E_s and ν_s are the elastic modulus and Poisson's ratio of the sample, and E_i and ν_i are the corresponding values for the indenter. Eq. (1.4) shows that knowledge of the sample Poisson ratio is necessary to convert from the output of the test, E_r when E is required. E values obtained are not oversensitive to the exact value used in the conversion, and the Poisson ratio can usually be estimated within close limits (e.g., for hard coatings $\nu = 0.2-0.25$, for metallic substrates $\nu \sim 0.3$). Instead, ν values obtained are not overly sensitive to the exact value used in the conversion, and the Poisson ratio can usually be estimated within close limits (e.g., for hard coatings $\nu = 0.2-0.25$, for metallic substrates $\nu \sim 0.3$). Instead of E_r, the plane strain modulus, E^*, is also sometimes quoted ($E^* = E\,(1-\nu^2)$), for example, in ISO14577 (ISO 14577, 2015).

It was later realized that the correction (β) factor could be material-dependent, and later versions of ISO 14577 have proposed an improved correction procedure where β is set to 1. Chudoba and Jennett (2008) showed that data could be improved by variable epsilon and lateral (or radial) dilation correction. Implicit in the standard unloading curve analysis is the assumption that the projected contact area under load and the projected area when fully unloaded remain the same. Chudoba and Jennett showed that this is not the case, particularly for materials with high H/E ratios. How to correct this is explained in the current revision of ISO14577−1. Schwarzer and Pharr introduced the concept of an effective indenter shape, that is, the shape that produces the same normal surface displacements on a flat surface, which the actual indenter produces on the deformed surface produced by the indentation (Schwarzer, 2006). The effective indenter concept has been used to explain the power law exponents found and the epsilon factor. In these analysis methods, the contact is assumed to be frictionless, as adhesion work is only significant at very small length scales, as occurs in atomic force microscope — based nanoindentation tests with sharp probes and nN forces (Zhao et al., 2003). Provided the indentation size is large enough for tip rounding to be minimal, the hardness determined by nanoindentation can be converted to an equivalent Vickers hardness using HV (in kgf mm^{-2}) = 92.62 H (in GPa).

1.2 Practical factors influencing the accuracy of nanoindentation data and reliability of the properties obtained

Nanoindentation has been shown to be an effective test method for characterizing the mechanical properties of solid materials over a huge stiffness range from diamond to weakly cross-linked hydrogels (i.e., a stiffness range of a few kPa (Xu et al., 2022, 2023) to over 1 TPa (Fischer-Cripps, 2014). However, the precise experimental conditions can vary with the test requirements, and better results are obtained with optimized test conditions and analysis methods suitable for the material being tested and information needed.

1.2.1 Indenter choice

The first choice is the indenter material and geometry. Diamond is used almost exclusively for the indenter when testing at room temperature. The most popular indenter is the three-sided pyramidal Berkovich with face half-angle 65.27 degrees, which has the same projected area-to-depth ratio as the four-sided Vickers indenter more commonly used in microhardness testing. The Berkovich is preferable over the Vickers indenter because it is easier to polish the three sides of the diamond to an extremely sharp point, but there is inevitably an unwanted line of conjunction with the four-sided Vickers indenter. Berkovich's original idea was to use a three-sided pyramidal indenter (half angle 65.03 degrees) with the same actual area as the Vickers (68 degrees) but with the realization that the projected area is more important; the modified Berkovich geometry of 65.27 degrees was subsequent developed. In common practice, these "modified Berkovich" indenters are usually simply referred to as Berkovich indenters. They have an equivalent cone angle, which gives the same area-to-depth ratio of 70.3 degrees. Fig. 1.1 shows an example load — displacement curve on fused silica indented with a Berkovich indenter.

The choice of indenter geometry depends on the properties required. With the Berkovich indenters, full plasticity occurs at low load, and it is possible to measure the hardness and elastic modulus in a single test. Plasticity develops more gradually with spherical indenters so they can have advantages for measuring elastic modulus and can be used for indentation stress — strain measurements since they are not geometrically similar. Cube corner probes are preferable for fracture, and flat punch probes can be used for creep and contact fatigue.

1.2.2 ISO 14577—the international standard for depth-sensing indentation

Best practice and recommendations are summarized in ISO14577, which from its first publication in 2002 has been regularly reviewed and improved. Dub, Chaudhri, and Lim have been critical of the ISO standard stating that several key assumptions have not yet been proven experimentally (Chaudhri, 2018; Dub et al., 2010). Dub and coworkers found that the elastic anisotropy of copper single crystals was not observed in unloading curve analysis but was (prior to a sharp pop-in) present in loading curves when analyzed by Hertzian contact mechanics. It is, perhaps, helpful to separate potential intrinsic limitations of the method from inaccuracies coming from insufficiently accurate instrument calibration and/or experimental design, with the latter able to be fixed by better experimental methodology. Dub and coworkers reported strong load dependence in their measurements of the elastic modulus of copper samples when following the ISO14577 methodology, which were interpreted as having arisen from the limitations of the method (Dub et al., 2010). However, at least some of the experimental trends they observed might have been due to their chosen load history employed. Since their tests were performed with constant loading and unloading time, in conjunction with no stated hold at peak load for creep before unloading, possible anelastic effects could have adversely influenced the accuracy of the modulus measurements. Similarly, Chaudhri (2018) has noticed in the early work of Oliver and Pharr (1992) that the elastic moduli of several of the materials tested varied with load, citing, for example, tungsten. In this case, some of the load dependence in this work may come from a failure to account for the load dependence of frame stiffness in the test instrumentation since related instruments from the same manufacturer have subsequently been found to have lower stiffness at lower load (Avadanii et al., 2023; Li et al., 2013). This could account for the trend to increasing elastic modulus at low load for tungsten when analyzed with a constant frame stiffness, since tungsten is particularly sensitive (and hence, why it is used in calibration). In our experience, metallic materials such as copper and tungsten, while displaying a normal indentation size effect in hardness, show a constant elastic modulus throughout the entire load range when tested correctly.

The ISO method recommends an iterative approach to frame compliance ($C = 1/S$) and tip shape calibration. They are treated as follows to explain the key points, but in practice, it is possible, for example, to assume an approximate tip shape and then calibrate the compliance.

1.2.3 Frame compliance correction

In a nanoindentation test, the raw measured depth data contain contribution from the deformation of the sample and the elastic bending of the instrument frame. The total measured compliance is the sum of sample and frame compliances. Removal of the contribution of elastic bending of the nanoindentation instrument from the raw data is necessary before unloading curve analysis. Accurate determination of the frame compliance of the instrument is critical, especially for reliable modulus measurement. Although a higher frame stiffness is generally preferable, it is more important that the calibration to remove the contribution of the elastic bending from the $P-h$ data is reliable. There are various indirect ways to do this, which to an extent rely on the sensitivity of the reference material(s) to the instrument stiffness and to the accuracy of the indenter area function used. Alternatively, it is possible to make a rigid contact between the load application part of the instrument and the rest of the frame so that all the measured deformation results from the compliance of the frame with a zero contribution from sample compliance (Van Vliet et al., 2004). Although this method does not include additional compliances, which could occur if indenter and/or sample mounting is not rigid, in practice, good agreement between the two approaches has been found. The latter approach has the advantage of being able to conclusively show whether the instrument frame stiffness is constant across the load range or not. A key assumption of standard calibration methodologies is that the instrument frame (machine) stiffness does not vary across the entire load range. While this is usually the case, there have been reports that suggest that it is not necessarily true for some instruments, with Li et al. (2013) reporting variation with load and with indenter geometry, finding lower frame stiffness at lower load. Avadanii and coworkers adapted their calibration methodology to account for the variable compliance and variable tip radius in spherical indentation by using a third reference material (Avadanii et al., 2023). If the instrumentation does have a nonconstant frame stiffness with load, then analyzing the indentation data across the load range with a single value of compliance will produce errors. Accurate frame compliance correction becomes more important for deeper indentations since the proportion of deformation due to the elastic bending of the frame rather than the compliance of the contact progressively increases with increasing load (since the contribution from sample deformation increases more slowly). Kang and coworkers reported variable elastic modulus in spherical ($R = 250$ μm) microindentation data over the contact depth range $10-100$ μm on steels and aluminum (Kang et al., 2009). When the data were reanalyzed with a depth-dependent frame stiffness, it was possible to return constant elastic modulus across the entire range.

1.2.4 Reference materials

Fused silica has proven a very popular material for calibrating nanoindenters and indenter geometry. It has relatively little time-dependent behavior at room temperature, very low surface roughness, and is elastically isotropic with an elastic modulus that has been considered to be independent of depth. However, its mechanical properties are more complex than are commonly assumed. Bull and coworkers have shown that its near-surface properties may be reduced by hydration affecting the first 20 nm (Bull et al., 2016). Although normal glasses deform predominantly by shear flow in indentation, anomalous glasses such as fused silica have an open structure and densify. Fused silica can locally densify by up to 20% in indentation (Kermouche et al., 2008). Depending on the acuity of the indenter, there is a threshold load for fracture. This typically includes some cone cracking, in contrast to radial/lateral crack systems favored in normal glasses. ISO14577 suggests a threshold load of 75 mN for a Berkovich indenter, but the exact value will depend on indenter sharpness. Cracking is relatively minor for the Berkovich geometry compared with indenter geometries with more acute angles where discontinuities can clearly be seen in the loading curves, but cracks are clear in scanning electron microscopy images of indentations, and there will be an influence of these on the load − displacement curve even if an obvious pop-in is absent. It has one of the highest H/E values of any material and so is significantly more influenced by radial dilation than other materials (Chudoba & Jennett, 2008).

Although it is commonly used for frame compliance correction, its sensitivity to small errors in compliance is relatively small due to its high H/E. Stiffness is proportional to $(E/\sqrt{H})$ so that tungsten and aluminum are significantly more sensitive. Tungsten is essentially elastically isotropic (Zener ratio very close to 1), so polycrystalline tungsten is a popular choice. ISO14577 recommends the use of two materials with stiffness varying by at least a factor of 2 to improve reliability. Gan and Tin (2010) reported using three reference materials (fused silica, steel, and tungsten) to provide additional confidence in their calibration accuracy.

1.2.5 Area function determination

Indenter geometry is determined either by performing measurements into a reference material with constant elastic properties over a wide depth range or by direct measurement of the size of the indentation produced (e.g., by using an atomic force microscope). A function is generated that relates the projected contact area and the contact depth so that a constant reference value of elastic modulus is returned over the fit range. The relationship between area and depth for an ideal Berkovich indenter is $A_c = 24.56\ h^2$. Berkovich indenters are slightly rounded at the very apex, and area functions are improved by including more measurements at lower load in the fit. Although it is possible to determine higher-order polynomial fits, or spline fits between the projected area and the contact depth, in practice, many Berkovich indenters are well described (Bei et al., 2005) by the simple two-parameter relationship:

$$A_c = Ah + Bh^2 \tag{1.5}$$

This has a simple physical interpretation where the first term describes a spherical indenter in the limit $h_c \ll R$. Since $A_c = 2\pi R h_c - \pi h_c^2$, when $h_c \ll R$, this reduces to $A_c = 2\pi R h_c$ so that $R = A_c/2\pi$. The second term in Eq. (1.5) is the pyramidal indenter (24.56 for ideal Berkovich). These two terms have different dependencies on the contact depth with the first term dominating at small depths and the second term being more important at large depths. The second term in Eq. (1.5) is the pyramidal indenter (24.56 for ideal Berkovich). These two terms have different dependencies on the contact depth with the first term dominating at small depths and the second term being more important at large depths.

Higher-order polynomials should not be used outside of their fit range as they will diverge strongly. The simple two-parameter relationship has the advantage of being able to provide an estimate the shape for depths larger than those used in the fit, assuming that the face angles do not change with depth, so that measurements can be made on very soft materials. Alternatively, measurements on a softer reference sample such as aluminum can be used to extend the depth range directly. Aluminum is also a useful reference material for assessing the tip shape of spherical indenters, particularly with larger radii, for example, above 25 μm.

Lateral dilation is particularly important if reference materials have high H/E ratios (Chudoba & Jennett, 2008). The consequence of this is that, particularly on highly elastic materials, the tip shape determined from direct measurement of the indentation size can be different to that determined from unloading curve analysis until the lateral dilation correction is applied. ISO14577 recommends how to correct the effect of lateral dilation on an area function determined by indentation on a highly elastic material to obtain the true tip shape.

1.2.6 Surface roughness

Surfaces with high surface roughness are more challenging to test, so mitigating the influence of surface roughness is important. In principle, there can be a wavelength effect superimposed on the R_a roughness—that is, it is the height of the asperities and their spatial distribution—in that there are two limiting situations where the "wavelength" of the asperities is large compared with the size of the tip and vice versa. When the wavelength is large, the real contact area can be larger than that assumed for an ideally flat surface (similar to pile-up). More common is that the wavelength is lower than the tip size, and the real contact area is lower. While both effects can occur, it is more common that the effect of roughness is to lower the measured values. Pop-ins often occur on rough surfaces such as thermal barrier coatings. Johnson (1985) introduced a load-dependent surface roughness parameter, α defined by

$$\alpha = \frac{\sigma_s R}{a_0^2} \tag{1.6}$$

where σ_s is the maximum asperity height, a_0 is the equivalent contact radius under the same load on a smooth surface and R is the probe radius. Johnson determined that surface roughness was significant for $\alpha > 0.05$. For results to be completely free of surface roughness, the ISO standard recommends that the indentation depth be $\times 20$ greater than the R_a surface roughness. When testing at light load with spherical indenters, surface roughness can be significant (Fischer-Cripps, 2011). When testing the hardness of hard coatings with high surface roughness, measurements at larger loads will show smaller scatter due to roughness. The test load should be chosen to be large enough to mitigate roughness while also ensuring that the plastic stress field does not extend into the substrate. When surface roughness is high, then a larger number of measurements are necessary. For example, in arc-deposited PVD wear-resistant coatings designed for cutting tools, the surface roughness can be typically $R_a = 0.1$ μm or more, which typically requires 30–40 indentations. The use of a calibrated contact load, which can be increased for rougher samples, has also been shown to help mitigate the effects of high roughness.

Coating surfaces typically reflect the roughness of the substrates they are deposited on. For this reason, they are often deposited on smooth surfaces such as silicon wafers. For hard and stiff coatings, the elastic mismatch when deposited on silicon (with relatively low stiffness) leads to an underestimation in stiffness when measurements at 10% of the coating thickness are used. When deposited on metallic substrates such as hard metals or hardened steels, surface polishing before deposition is recommended. Polishing the surface after deposition can be more problematic since it can work-harden the coating although it may be necessary, for example, to remove metallic droplets in arc-deposited PVD coatings. More subtly, polishing can influence the dislocation density in the near-surface region and affect size effects in hardness and yielding. Bahr noted that electropolished tungsten samples showed pop-ins, but mechanically polished surfaces did not (Bahr et al., 1998). For mapping mechanical properties, it is recommended that the surfaces be as smooth as possible. Alternatively, imaging of the surface before testing allows the selection of smoother areas for testing.

1.2.7 Pile-up and sink-in

The unloading curve analysis, since it was derived from elastic solution contact (Sneddon, 1965), provides a good approximation of the projected contact area when the material sinks in. When cross-slip occurs and material piles up around the indent, as can occur in metallic materials that do not appreciably work-harden, the contact area will be underestimated, and the hardness and elastic modulus of the contact will be overestimated. Bolshakov and Pharr (1998) used a direct examination of contact profiles determined by finite element modeling with the contact area determined from unloading analysis. They were able to identify features in the indentation curve that could help assess the likelihood of pile-up occurring and estimated that when $h_f/h_{max} > 0.7$ and the material does not significantly work-harden, the contact area could be underestimated by up to 60%. Work hardening drives the plastic zone to greater depths into the material and decreases the pile-up since the work hardening during deformation constrains upward flow to the surface (Johnson, 1985). Gale et al. reported that in a copper alloy, more severely work-hardened regions showed greater pile-up than less damaged regions (Gale & Acheson, 2014). Provided that $h_f/h_{max} < 0.7$ or in all materials that only moderately work-harden, the unloading curve analysis could be expected to be reliable. A small effect of loading rate and hold time at peak load on the development of pile-up in polycrystalline copper, which shows strain rate sensitivity, has been shown (Chen et al., 2016).

Although not an ideal situation, reference materials can show some pile-up provided that the extent of pile-up does not change with depth over the load range being calibrated. For example, pile-up in tungsten can produce a small offset in the measured elastic modulus. In He$^+$ ion-implanted W-1% Re, there is extensive depth-dependent pile-up (Beck et al., 2017). He ion implantation is expected to reduce the stiffness of tungsten, but ~20% increase in stiffness was found. Using high-accuracy surface acoustic wave measurements of elastic modulus, it was possible to introduce a correction factor to account for this pile-up. Moharrami and Bull have compared the extent of pile-up in bulk materials and thin films on harder substrates (Moharrami & Bull, 2014). For bulk metals, they found that narrow and high pile-ups were found on copper but shorter and broader for Al, consistent with their different H/E ratios. They found that pile-up was considerably higher for the thin films. In general, the thin films have smaller grain sizes limiting grain mobility and promoting cross-slip and pile-up. When deposited on harder substrates, the effect was much more extreme since the harder substrate did not deform to the same extent as the softer coating.

A limitation of ISO14577 is that it does not contain methods for correcting the projected contact area for the effects of pile-up. Various methods such as (1) direct measurement of the projected contact area after the test, for example, by scanning probe microscopy; (2) measurement of contact resistance (only for metallic materials) (Kholkhujaev et al., 2023) can improve the accuracy of the projected area measurement and, therefore, the hardness and elastic modulus of the test material.

1.2.8 Zero point correction

Accurately determining the exact point of initial contact is critical for accurate measurements, particularly for small indentation depths. There are different strategies for doing this, based on whether precontact data are acquired or not. One approach does not require precontact data, with surface detected when a preset contact force (as low as possible) is reached. The depth sensor is set to zero and loading proceeds. Subsequent back-extrapolation to zero force to account for the elastic deformation occurring before this point provides a zero point correction. Since even sharp indenters always have some rounding at the tip, this can be done assuming Hertzian elastic loading, or more generally power law or linear extrapolation, as suggested in the ISO standard. Linear back-extrapolation can underestimate the zero point correction. An example of a Hertzian correction is given by Eq. (1.7).

$$F = C(h - h_0)^{1.5} \qquad (1.7)$$

This approach can be effective in dealing with surfaces with high roughness or in high-vibration environments. Alternatively, when monitoring precontact data, it is possible to apply a slope-based criterion where the contact is detected by a threshold value of the slope of the load − displacement response. When detecting contact on soft samples, the effectiveness of this will be dependent on the flexure stiffness of the instrument. Lower flexure stiffness improves sensitivity. Kaufman and Klapperich (2009) noted that inaccurate surface detection causes overestimation of mechanical properties in polymeric materials. More recently, Xu et al. used an instrument with low flexure stiffness to measure the elastic moduli of hydrogels with single-digit kPA stiffness (Xu et al., 2022, 2023).

1.2.9 Creep, time dependency, and thermal drift correction

Data quality can be adversely influenced by time dependency if this is not accounted for by experimental design, calibration, and/or suitable analysis methodology. The sources of time dependency include (1) materials-related effects (e.g., creep); (2) instrumental effects (thermal drift caused by expansion/contraction of the load frame), and it is important to not confuse the two.

The standard analysis of unloading curve data is based on elastic contact mechanics with no provision for time dependency. Metallic materials can show significant indentation creep. The experimental conditions (load vs time) can be designed so that this time-dependent behavior does not influence the accuracy of the elastic modulus measurements. Chudoba and Richter summarized the problems that can occur when indenting soft metals when the holding period at peak load is too short (Chudoba & Richter, 2001). In general, a reliable strategy to minimize/eliminate any influence from nonelastic behavior is to use the combination of (1) slow loading, (2) long hold at peak load, and (3) faster unloading (e.g., $\times 2$-$\times 5$ faster). Chen and coworkers showed that after a hold period of 100 s before unloading the elastic modulus of an as-rolled copper sample with predominantly (111) orientation was only 3% higher when loaded three orders of magnitude quicker—that is, the presence of the long hold was essentially sufficient to remove any rate dependence (Chen et al., 2016). ISO14577 has quantified this by stating that the hold period shall be long enough and/or the unloading time to be short enough that the following relationship holds:

$$\text{Unloading rate} > 10 \times \text{creep rate at maximum load/measured compliance at maximum load} \quad (1.8)$$

Feng and Ngan (2002) came essentially to the same conclusion and introduced a creep factor to assess the importance of creep over elasticity in a similar relationship recast in terms of unloading stiffness, with $C < 0.1$ minimizing any creep influence but huge discrepancies found for higher C, reaching apparent moduli of over 2000 GPa on Al and Cu when C was > 1.4.

If the creep rate is too large, as can happen at very high temperatures, or if the instrumentation has insufficient thermal stability to wait long enough for it to reduce sufficiently, then an alternative approach is to develop more complex analytical treatments that can handle the time dependency during unloading. Using the condition that the creep rate at the beginning of unloading is equal to that at the end of the hold period, Feng, Tang, and Ngan proposed (Feng & Ngan, 2002; Tang & Ngan, 2003) a viscoelastic correction:

$$\frac{1}{S_a} = \frac{1}{S} + \frac{\text{creep rate}}{\text{unloading rate}} \quad (1.9)$$

where S_a is the apparent stiffness, and S the true stiffness. The data should also not be influenced by any thermal or mechanical drift during the test duration. This is typically achieved by a suitable hold period at the initial contact load and/or after the removal of 90% of the peak load. It is important that when holding for a thermal drift correction that deformation due to creep or creep recovery is not erroneously thought to be due to thermal drift. An example of this is the continuing extensive creep recovery that occurs on all polymeric materials, which can be orders of magnitude greater than the actual thermal drift so mistakenly correcting data for it introduces error. A simple way to investigate whether a material is showing creep recovery is to perform measurements at different loads—creep recovery will increase with load, but thermal drift does not vary with load. Metallic materials can also show some creep recovery, although this is usually minimal in nanoindentation, it is noticeable at the microscale. Similarly, when the sample is very soft (e.g., solder), then the initial load for thermal drift should not be used as deformation during this period and is likely to be dominated by creep rather than drift.

1.3 Indentation fracture

Depending on the load, material, indenter geometry, and environment, the types of cracks that commonly occur in the indentation of brittle bulk materials are: (1) cone cracks, (2) Palmqvist radial cracks, (3) median cracks, (4) lateral cracks, and (5) half-penny radial cracks. Lawn and coworkers showed that for the half-penny crack configuration (i.e., well-developed radial/median crack system) induced by Vickers indentation that the K_c fracture toughness of the material could be estimated by Anstis et al. (1981) (Eq. 1.10).

$$K_c = \alpha (E/H)^{1/2} P/c^{3/2} \tag{1.10}$$

where P is the indentation load, c is the radial crack length from the indentation center to the crack tip, and α is a constant for a given indenter geometry. It was later found that this approach could be applied to other indenter geometries. The constant is often taken as 0.016 for Berkovich or Vickers and 0.032 for cube corner indenter geometries, although there is evidence that it also can be material-dependent, and limitations of the method for densifying materials such as fused silica and other silicate glasses have been highlighted (Yoshida, 2019). Jang and Pharr studied the application of this equation in indentation cracking in single crystal Si and Ge over a wide range of load and indenter geometry finding although there was some scatter, and indenter sensitivity reasonably constant fracture toughness could be obtained (Jang & Pharr, 2008). For most brittle materials, there is a cracking threshold, below which no radial cracking occurs. Switching to sharper indenter geometries such as cube corner triggers radial cracking. The sharper cube corner indenter displaces more than three times more material for a given load than the Berkovich and produces higher stresses beneath the indenter (Bruns et al., 2020; Jang & Pharr, 2008). In fused silica, this reduces the cracking threshold by three orders of magnitude, from $\sim 0.5-1.5$ N to around 1 mN. For this material, multiple crack systems are activated by high-load

Berkovich indentation. Radial cracking predominates with the cube corner geometry but is largely (~90%) accompanied by chipping (Bruns et al., 2020). Schiffmann has investigated the use of different models to determine fracture toughness with a cube corner indenter (Schiffmann, 2011).

The equation developed by Lawn and coworkers has been applied to >1 mm thick coatings where well-developed radial cracks are developed. When the crack system is not sufficiently developed, the method is invalid. This occurs when cracks are confined to the indentation impression (e.g., the radial cracks that run along the indenter edges in the hard coating on soft substrates). To eliminate the influence of substrate elastic − plastic deformation, smaller indentations are needed, but then the cracks are not well-developed compared with the indentation size. For <500 nm coatings, alternative energy-based models are needed. Chen and Bull have proposed a work of indentation versus displacement method that is able to separate the fracture energy dissipation from other deformation mechanisms (Chen & Bull, 2006) applying their method to determine the energy release rate and toughness of 240−400 nm thick solar control coating layers. To generate accurate fracture data, the tests were performed under displacement control.

Fracture toughness has also been determined in tests with simplified stress states by deforming micropillars and microcantilevers. Ast et al. (2019) have reviewed experimental approaches to microscale fracture toughness evaluation, focusing on the pillar splitting method for ceramic thin films sharp nanoindentation of microscale pillars, and the use of microcantilever bending for studying fracture mechanisms in brittle/semibrittle materials. In the pillar splitting method, once a crack nucleates under the indenter, it pops out to the pillar surface resulting in a displacement burst at the critical splitting load P_c. K_c can be determined by Eq. (1.11) with R the pillar radius and γ a calibration coefficient.

$$K_c = \gamma P_c / R^{3/2} \tag{1.11}$$

Ast and coworkers estimated, assuming that a maximum pillar diameter of 100 μm, which could be manufactured by FIB milling, that unstable crack propagation and pillar splitting with a cube corner indenter require K_c/H <0.8, confirming with an Ashby plot that the method is generally limited to ceramics and brittle glasses. Beirau et al. (2021) recently reported that the method was able to effectively reveal local fracture toughness differences in a sample of radiation-damaged zircon that had zones of higher crystallinity and zones with higher amorphization. They observed a direct correlation between the local mechanical properties and fracture toughness, with pillars FIB milled from regions with lower crystallinity having lower H/E and higher fracture toughness.

Although the test is limited to ceramic materials for fast fracture and toughness evaluation, microcompression of FIB machined pillars by flat punches has also been used as a small-scale uniaxial test, enabling stress − strain relationships to be determined at small length scales. For example, the technique has been used to study the stress − strain behavior of intermetallic inclusions in Al 7075 alloys showing that both Fe-bearing and Si-bearing inclusions possess higher strength than the

Al 7075 matrix (Singh et al., 2015). Microcompression testing also provides an opportunity to study plasticity and flow in macroscopically brittle materials (Korte-Kerzel, 2017) improving our understanding of the relationships between crystal structure and defect mobility in complex anisotropic crystals that are too brittle to test by other methods. FIB-manufactured micropillars have also been used in determining the residual stresses in thin coatings and their variation with depth (Renzelli et al., 2016) through incremental ion beam milling in combination with high-resolution SEM and digital image correlation, which has been termed the microring core method (FIB-DIC).

With the microcantilever geometry, there is the opportunity to produce a notch so that the crack location can be precisely controlled. Schaufler et al. used microcantilever tests to study the fracture strength and interfacial toughness of two a-C:H coatings produced with only Cr adhesion layer differences (Schaufler et al., 2012). The coating with enhanced adhesion in the Rockwell C adhesion test was found to have high interfacial fracture toughness. In microcantilever bend tests on less brittle materials, microscale plasticity can occur at the crack tip leading to stable crack growth, and fracture toughness has been assessed by a range of different approaches including J-integral and crack tip opening displacement methods.

1.4 Indentation size effects

1.4.1 Size effects in plasticity

The "indentation size effect," or ISE, commonly refers to the observation where the hardness tested with a geometrically self-similar indenter progressively increases as the depth is reduced, that is, "smaller is stronger." Nix and Gao (1998) proposed a model to explain this using strain gradient plasticity, where geometrically necessary dislocations (GNDs) are generated under the indenter due to strain gradients. Smaller indent sizes have increasing density of geometrically necessary dislocations (GNDs) and higher hardness through Taylor hardening. The theory predicts that the total dislocation density within the plastically deformed volume varies as $1/h_c$, so that $H^2 \propto 1/h_c$. Experimental hardness versus depth dependence was fitted by a two-parameter model, shown in Eq. (1.12), with H_0 (the hardness at infinite depth) and h^* (the characteristic length scale) being constants obtained by fitting.

$$H/H_0 = \sqrt{[1 + (h^*/h)]} \qquad (1.12)$$

It is more common to show this as a linear relationship between H^2 and $1/h$, which has been shown to be observed on a wide range of materials down to depths of around 150–350 nm. Below these depths, the relationship breaks down, and even though the hardness continues to increase with decreasing indentation depth, it does so much less rapidly with decreasing depth than predicted by the model. Several authors have proposed that tip rounding is important (Qiao et al., 2010; Qu et al., 2004) and that nonuniform distributions of GNDs are implicated. Wang and coworkers reported a bilinear behavior on single crystal tungsten, successfully modeling their data with nonuniform GND distribution (Wang et al., 2021). Ma and colleagues employed precession electron diffraction in TEM imaging to study the dislocation behavior under indentations at various depths in single crystal Nickel (Ma et al., 2021). They found a strongly depth-dependent behavior with progressive subgrain formation. At very small depths, deformation was controlled by dislocation source starvation, with mobile dislocations and low dislocation density in the plastically deformed volume. At larger depths, the ISE mechanism transitions, with entanglements and subgrain boundaries confining the dislocation movement to a small hemispherical volume below the indent leading to dislocation interaction hardening consistent with the Nix-Gao model. As ISEs can occur in materials that do not deform by dislocation-based plasticity, Milman and coworkers, noting that an empirical power law dependence was found ($P \propto h^m$), used a phenomenological approach to enable comparison between hardness measurements at different loads (Milman et al., 2011). To explain the indentation size effect in ceramics, Bull and coworkers proposed a model where rather than being continuous, plastic deformation occurs in discrete bands to progressively relieve stresses created by elastic flexure of the surface at the edges of the indentation (Bull et al., 1989). Alongside the size effect from the strain gradient, flow, and hence, hardness in metals can also show a concurrent strain rate effect (Liang & Pharr, 2022; Wang et al., 2021). Work on single crystal tungsten has shown that these appear to be additive, with the Nix − Gao model modified to include strain rate − dependent friction stress (Liang & Pharr, 2022). Although the presence of a native (passivating) oxide layer on most metallic materials will also exert some influence on their mechanical properties, it is usually too thin (e.g., 1.9 nm for 316 L stainless steel (Tardio et al., 2015)) that it does not noticeably affect the results.

Size effects in plasticity are also observed with spherical indenters. Jennett and coworkers have shown that the yield stress increases with the inverse cube root of indenter radius in relatively large grained fcc metals (Hou et al., 2008; Spary et al., 2006). Shim and coworkers found a strong radius dependence in Ni(100), which was interpreted in terms of changing the probe radius changing the size of the highly stressed zone relative to the dislocation spacing (Shim et al., 2008). Shim et al. considered that the increase in strength as the size of the contact decreases is a different type of indentation size effect to that commonly seen in hardness since the latter depends on the yielding and work-hardening behavior of the material and the former on the stress to initiate dislocation plasticity.

The size of the volume of material affected by an indentation contact has been estimated as $2.4a \times \pi a^2$, where a is the contact radius, and $2.4a$ is the depth of the primary indentation zone (Pathak & Kalidindi, 2015). Spherical indenters present an opportunity to systematically study responses at different length scales simply by varying the indenter radius (Pathak et al., 2017). When using spherical indenters, the extent of pile-up or sink-in in nonwork-hardening materials can change during an indentation as the indentation depth increases, with more pile-up being observed as the contact strain (a/R) and indentation depth increase (Taljat & Pharr, 2004). Spherical indenters also enable the onset of plasticity to be studied more conveniently than with sharper probes. Johnson has shown that for a flat surface in elastic/plastic contact with a rigid ball of radius R, the yield pressure (P_y) is given by Eq. (1.13) (Johnson, 1985).

$$P_y = 0.78R^2\left(H^3/E^2\right) \quad (1.13)$$

A sharp "pop-in" (displacement burst) can occur at the onset of plastic deformation in metallic materials. Up to pop-in (i.e., within the elastic regime), the mean pressure during loading can be determined by Hertzian analysis if the contact remains within the spherical part of the tip. Bei and coworkers studied the pop-in behavior of (110) Cr_3Si with a Berkovich indenter (Bei et al., 2005). For this material, they found that pop-in did not occur within the spherical part of the tip but over a depth range within the transition regime between the spherical and conical parts of the tip; consequently, the actual tip shape rather than that from the spherical radius should be used instead to calculate the contact pressure at yield (Bei et al., 2005). Studies have shown that before yield metallic materials can support much higher contact pressure (when the maximum shear stress under the indenter approaches the theoretical shear strength of the material) than their hardness determined after yield has occurred. Dub and coworkers reported mean pressure values at pop-in of 16.8 GPa for Cu(111) and 9.3 GPa for Cu(100) surfaces (Dub et al., 2010). Mean pressures of around 33 GPa have been reported on mechanically polished W(100) (Beake & Goel, 2018). In semiconductors, pop-ins may cause phase transformation events occurring once a critical pressure is reached. Pop-ins on Si usually occur when the contact reaches $\sim$ 12 GPa. Sapphire can also show well-defined pop-in behavior (Lu et al., 2007).

1.4.2 Contact size effects on deformation versus fracture

The deformation produced in an indentation test is a function of the radius of the indenter. More brittle deformation occurs with larger radii indenters (i.e., larger contact size), and more plasticity observed when using smaller radii indenters (i.e., small contact size) (Rhee et al., 2001). Rhee and coworkers introduced a brittleness index P_y/P_c, which was dependent on the radius of the indenter and reflected the competition between plasticity and fracture in the test. The relative proportions of elastic, plastic, and brittle deformation have a large influence on the subsequent development of wear in a tribological test, explaining why hardness alone often can be insufficient to predict even relative wear rate. Lawn and Marshall discussed the hardness/toughness ratio H/K on the competition between deformation and fracture (Lawn & Marshall, 1979). In particle technology, there is a size effect on fracture with the grinding limit marking the point at which further particle breakdown stops (Kendall, 1978). Similar size effects are observed in erosion (Hutchings, 1992).

1.4.3 Indentation size effects in coatings

In coated systems, the mechanical properties vary with penetration depth so that analysis methods may be required to determine the contribution of the individual layers to the composite response. Further complexity is introduced by the fact that the elastic and stress fields are sensitive to different volumes of material at any indentation depth. The elastic stress field is more diffuse or far field while the plastic stress yield is more localized. The methodology in determining accurate "coating-only" hardness and elastic modulus is summarized in ISO 14577−4 and explained in detail in a later chapter by Chudoba and Jennett.

More generally, the ratio of coating thickness to indenter radius (t/R) is an important parameter, which strongly influences the behavior of coated systems under contact loading. The location of initial yield relative to the coating − substrate interface and subsequent fatigue mechanisms in repetitive contact (including indentation fatigue, repetitive scratch testing, and impact fatigue) also vary with the t/R ratio (Michler & Blank, 2001) so to better understand the coating system behavior in tribo-contact, it can be very useful to perform repetitive tests with different tip radii to obtain data over a range of t/R. The contact size can be further controlled by performing tests at different applied load and therefore a/R.

1.5 Relating hardness to yield stress: the constraint factor

Linking hardness and yield stress is important for comparison to measurements with other techniques and for providing input yield stress for modeling. The hardness measured by nanoindentation is higher than the uniaxial yield stress in compression due to the confining pressure generated by the surrounding elastically strained material. Tabor introduced a constraint factor to describe the relationship between H and Y, $H = CY$, which was typically ~ 3 for ductile metals where deformation is fully plastic. Johnson determined that C depended on the plasticity index Y/E (Johnson, 1970). He proposed that there are effectively three deformation regimes. When the indentation is fully plastic Y/E is very low and C tends to the Tabor value, and when the elastic deformation is nonnegligible, as it is in elastoplastic contact the constraint factor gradually decreases. At $Y/E > 0.1$ the deformation is elastic and H/Y reduces further. Ghosh and coworkers noted that the constraint factor was also lower when using the cube corner indenter geometry, determining a theoretical value of 1.81 for ideal plastic materials (Ghosh et al., 2016).

1.6 Indentation energy

1.6.1 Relationship between H/E and plasticity index

In comparison to power-law fitting to unloading curves, the energy-based approaches have the potential benefit that an accurate (i.e., up-to-date) area function for the indenter geometry is not needed and may also be useful for situations where there is appreciable pile-up, which is not accounted for in the standard unloading

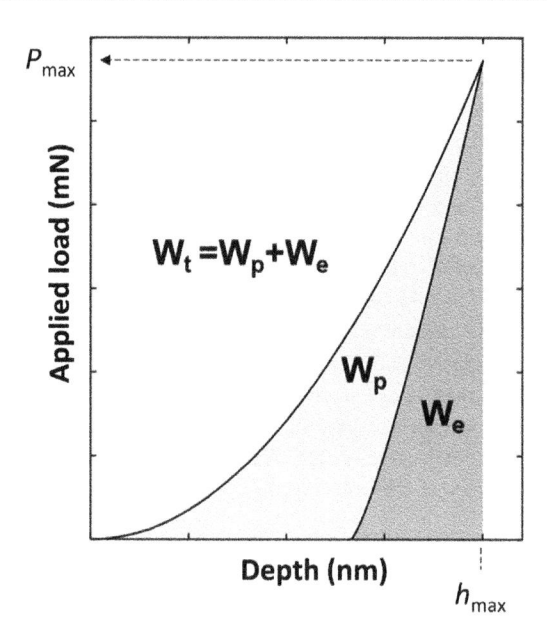

Figure 1.2 Indentation test elastic and plastic work in an indentation test.

analysis. Fig. 1.2 shows the elastic and plastic work done in an indentation test. Eq. (1.14) shows the general form of the relationship between the plasticity index and H/E_r for an indenter with a constant included angle, such as an ideally sharp pyramidal or conical indenter.

$$W_p/(W_p + W_e) \approx 1 - x(H/E_r) \tag{1.14}$$

where W_p and W_e are the plastic, and the elastic work done, $W_p/(W_p + W_e)$, is a dimensionless plasticity index, and x is a constant. Based on the analysis of indentation by a cone by Stilwell and Tabor (1961), the proportionality constant is related to the half-included angle of the indenter, through $x = \pi \tan \theta$. Cheng and Cheng have studied the interrelationships between (1) the indentation plasticity index (W_p/W_t) and H/E^*, (2) h_f/h_m and W_p/W_t, (3) W_p/W_t and h_f/h_m for a range of indenter geometries, sample work hardening, and Poisson's ratios by dimensional analysis and finite element (FE) methods (Cheng et al., 2002; Cheng & Cheng, 1998, 2004). As a rigid indenter was used for their simulation, the plane strain modulus, E^* was used ($E^* = E(1-\nu^2)$). H/E^* values for different work-hardening exponents and Poisson ratios at a cone angle of 70.3 degrees, that is, equivalent to a Berkovich indenter geometry, were compared with H/E_r experimental values with good agreement. A linear relationship between the plasticity index and H/E^* has recently been reported (Roa & Sirena, 2021). The shape of the experimental indentation curve is also influenced by the stiffness of the indenter material, which is taken to be infinitely rigid in many FE studies. As the diamond indenter is typically much stiffer than the materials being tested, the effect is very small in practice. However, it can be noticed in indentation curves on materials with high elastic modulus, such as tungsten, produced with indenters of much lower elastic modulus than diamond (e.g., sapphire, as is done at >400°C in high-temperature nanoindentation when diamond cannot be used due to graphitization), that is, when the indenter and sample elastic properties are much closer.

Across the range of behavior simulated (0.18 < W_p/W_t <1) by Cheng and Cheng, a linear best fit to the FE data gives $x \sim 5$ for the constant in Eq. (1.14). A more detailed examination of their original FE data clearly shows that the relationship only approximates to linearity. Experimental data suggest $x \sim 5$ for glasses and $\sim 6-7$ for metals. A near-to-linear relationship has also been reported for iron-oxide-based supercrystalline nanocomposites (Yan et al., 2022), and for metals and alloys, with a best fit proportionality constant of 6.25 for the latter (Yamamoto et al., 2021). Yang and coworkers reported an approximately linear relationship between W_p/W_t and H/E_r in their experimental data (Yang et al., 2008). They estimated the proportionality constant as 6.0 (analytically) or 5.8 (using FE) in close agreement with 5.7 determined from analysis of data by Cheng and Cheng with similar values summarized by Malzbender (2005). Yang and coworkers suggested that the unloading work may not be exactly equal to the elastic work, since their FE calculations suggested that the two properties were not equivalent if residual stresses were present, but in general, the work done in unloading can be assumed to be elastic.

Chen and Bull determined that x is not a constant, and it varies with material properties. They developed an analytical model, that incorporated the elastic stiffness and Poisson's ratio for the diamond indenter rather than assuming a rigid indenter, which enabled the direction simulation of the experimental relationship between W_p/W_t and H/E_r (Chen & Bull, 2009). Numerical simulations show that the divergence from linearity is more significant for H/E_r >0.1. Materials with higher H/E showed lower values of the apparent proportionality constant x, but differences in work hardening had little effect on the relationship (Chen & Bull, 2009).

It has often been assumed that when fracture does not occur, W_p/W_t of bulk material is constant when measured with a Berkovich indenter. However, it is dependent on tip geometry, reducing for very small contacts as full plasticity is not developed. Bull (2005) has shown that it increases slightly for deeper indentations on metallic materials (Bull, 2006). The increase may also be due to the influence of a very gradual change in indenter geometry, where any influence of tip rounding becomes negligible with increasing indentation depth. To investigate this, microindentation tests to 30 N were carried out on a sample of polycrystalline tungsten (Beake, 2022). A gradual increase in W_p/W_t was also observed with increasing indentation depth. However, there were no accompanying changes in the apparent proportionality constant x in Eq. (1.14). The relationship between H/E_r and W_p/W_t is also applicable to coating systems, provided that fracture/delamination does not appreciably influence energy dissipation (Malzbender & de With, 2002).

The relationship between H/E_r and h_r/h_m is also nonlinear, but a significant deviation occurs when H/E_r >0.1. Cheng, Li and Cheng found a one-to-one relationship between W_p/W_t and h_r/h_m that was linear when $h_r/h_m > 0.4$ or when $W_p/W_t > 0.2$ (Cheng et al., 2002). Chen and Bull noted that this requires $H/E_r < 0.1$ for a conical indenter equivalent to a Berkovich. A linear best fit gives $W_p/W_t = 1.27(h_r/h_m)$ -0.27 (Cheng et al., 2002). However, at high plasticity (low H/E), the relationship is close to $h_r/h_m = W_p/W_t$, as has been reported by Milman (2008). Experimental data on metallic materials have shown that within 0.01, the two indices h_r/h_m and W_p/W_t are equal. For nonmetallic materials with higher H/E, W_p/W_t is slightly lower than h_r/h_m, as predicted by the FE results of Chen and Bull.

To measure the hardness of a material (rather than simply the mean pressure generated under the indentation), it is necessary that a fully developed plastic zone is obtained. Fischer-Cripps (2014) has noted that when the combination of E^* and probe angle results in a mean contact pressure below the hardness, then the contact is fully elastic. Similarly, there may be a limit in H/E_r above which it should not be possible to induce plastic deformation in a material when using a Berkovich or a Vickers indenter with an equivalent cone angle of 70.3°, and more acute angle indenters such as a cube corner may be required if hardness is to be measured. For a Berkovich, $H/E_r \geq 0.179$ is consistent with the linear fit to $W_p/W_t = 0$ (Chudoba & Jennett, 2008). Nonlinearity in the relationships between W_p/W_t, H/E_r and h_r/h_m suggests that some plasticity may occur even at these high values.

1.6.2 Indentation work, hardness, and fracture

An alternative definition of hardness was provided by Stilwell and Tabor (1961).

$$H = W/V \tag{1.15}$$

where W is the plastic work done, and V is the deformed volume.

Tuck and coworkers proposed that the hardness could be determined from the work done in indentation without the need to measure the volume. They suggested that it was better to use the plastic work rather than the total work done and proposed Eq. (1.16) where hardness is related to the maximum load cubed divided by the plastic work squared (Tuck et al., 2001).

$$H = kP_m^3/9W_p^2 \tag{1.16}$$

where K is a constant related to indenter geometry and hardness definition, 0.0408 for a Berkovich using projected contact area, W_p is plastic work, and P_m is maximum load. They found that the approach was less sensitive to the effects of pile-up than unloading analysis and also found it effective for modeling hardness in coated systems, in terms of separating the contributions of coating and substrate. Jha and coworkers investigated the extent of correlation between the total indentation work approach and found that there was good agreement with unloading curve analysis for materials with $H < 1$ GPa, but divergence for harder materials (Jha et al., 2014).

These authors proposed a correction to the energy-based approach, which was dependent on the extent of elastic recovery, reporting improved agreement to hardness ~20 GPa. Energy-based approaches have also been used to study fracture, particularly for coatings that are too thin to be evaluated by alternative methods (Chen, 2012; Chen & Bull, 2007).

1.7 The continuous stiffness technique

The oscillation-based continuous stiffness technique is attractive since it offers the possibility to obtain mechanical property measurements as a function of penetration depth. However, the data obtained with the technique have been reported to be very sensitive to the experimental conditions, particularly for metallic materials with low hardness and high E/H ratios and at low indentation depths (Merle et al., 2017; Phani et al., 2021; Pharr et al., 2009; Siu & Ngan, 2013a). The technique has been critically investigated many times to identify the underlying issues with it and to recommend mitigation strategies (Phani et al., 2021; Pharr et al., 2009). The underlying issues include (1) contact detection accuracy, (2) intrinsic changes in strength due to the oscillation, and (3) plasticity errors.

An oscillation while approaching the surface to detect contact can introduce significant imprecision in the accuracy of the surface detection. Contact detection methods where the surface is defined as the point where contact stiffness is $\times 2$ the instrument spring stiffness have been criticized as being insufficiently reliable since it can lead to a false zero point several hundred nm away from the actual point (Moseon et al., 2008). Phani and coworkers reported finite element simulations showing that a $+/-$ 1 nm uncertainty in surface detection could dramatically affect the hardness and elastic modulus of fused silica, with even data at 100 nm depth showing $+/-$ 1% error asymptotically increasing for shallower depths (Phani et al., 2021). Their results appear to explain the large variability in mechanical properties reported with the technique for very shallow depths. For example, Liu et al. (2014) showed similar behavior on fused silica, reporting hardness between 5 and 20 GPa for depths just after contact. Detecting contact without oscillation has been recommended (Phani et al., 2021), which has been the standard approach in continuous stiffness-type measurements using other commercial instrumentation (NanoTest Vantage).

Maier et al. (2015) reported that long-term, depth (i.e., load)-dependent thermal drift strongly influenced fused silica creep data at room temperature (Durst & Maier, 2015; Maier et al., 2013) stating that a correction at the end of the test was insufficient to accurately correct for it. This could happen, for example, when the loading head of the instrument being used continually heats up with continued time. When drift is excessive compared with the timescale of the measurement, then Maier et al. (2015) proposed a neat solution for materials that exhibit constant elastic modulus versus depth. For these materials, it is possible to use dynamic stiffness measurements to correct the depth data.

Siu and Ngan reported that the oscillation in the continuous stiffness method artificially modifies the intrinsic strength of soft samples (Siu & Ngan, 2013b). With small indentations, the presence of oscillation increased the indentation depth. On aluminum, this was found to be due to an altered microstructure (observed by cross sections through indents) where subgrains form. Cordill and coworkers found that the oscillation alters the measured hardness of ductile metallic materials, especially at depths less than 200 nm, which they termed the "nanojackhammer effect" (Cordill et al., 2009). This change in the hardness is due to the added energy associated with the oscillation, which assists in dislocation nucleation.

Possible plasticity errors need to be considered when indenting soft and stiff materials. It was originally assumed that the unloading during the oscillation was sufficiently fast to remain elastic, but this was later shown not to be the case for materials with high E/H ratios (Merle et al., 2017). To mitigate the plasticity error, it is recommended to keep the parameter $(dF/dt)(1/F)/(f^* Fd_{max}/F)$ low (where f is frequency, $(1/F)\, dF/dt$ is strain rate, and Fd_{max}/F is the dynamic load fraction). The phase shift provides an indication of the plasticity error—around 5 degrees is usually acceptable. Phani et al. (2021) has recently recommended that a fixed oscillation load fraction of 0.2 be used, instead of the previously stated 2 nm fixed displacement amplitude. Although it is an improvement, data presented by Phani for fused silica and aluminum still show a wider scatter in both hardness and elastic modulus at low-to-moderate indentation depths (c. <80 nm for fused silica and <200 nm for Al) compared with data acquired without oscillation by multicycling or single cycle tests. There is a trade-off between absolute accuracy and data density.

There appears to be a maximum strain rate before artifacts occur, so it is generally preferable to perform tests at low strain rate. An apparent strong sensitivity to strain rate has been reported for materials such as fused silica, which exhibit little rate dependence at room temperature (Jia et al., 2017). At high strain rates, these authors noticed that there was an apparent size effect in hardness. Leitner and coworkers have suggested that higher hardness in dynamic indentation testing compared with quasistatic tests may be due to the changing strain rate during the hold segment at peak load in static tests (Leitner et al., 2017).

1.8 High strain rate testing

Understanding how materials deform under high strain rate (impact) loading is important for machining, forming, and aerospace applications. The strain rate in machining can approach $10^5\ s^{-1}$ when cutting at high speed (Zhang et al., 2020) and will only increase with the development of improved coating systems allowing cutting at even faster speeds. In aeroengines, composites are being increasingly used in load-bearing applications (e.g., as composite CFRP blades in the Rolls Royce UltraFan engine) to reduce weight, but it is vital to ensure adequate resistance to impact damage. High strain rate testing is conventionally achieved by techniques such as Split−Hopkinson bar testing where stress equilibrium can be an issue at high speeds. Considering the strain rate definition (velocity/length), higher strain rates may be accessed by reducing the length scale of the test. Testing at a small scale has the key benefits of providing highly spatially localized information and higher throughput than the macroscale tests, giving the potential, for example, to understand the role of microstructure on the high strain rate behavior.

Nanoindentation is a quasistatic technique where typically tests are performed at low strain rates in the range $10^{-4}-10^{-1}$ s^{-1}, but test instrumentation has been adapted to enable tests at higher strain rate. Phani and Oliver used a step load to obtain hardness at higher strain rates (Phani & Oliver, 2017). Guillonneau and coworkers described an in situ (in SEM) displacement-controlled indenter capable of performing indentation tests at constant strain rates at up to 1000 s^{-1} and micro-pillar compression tests to 100 s^{-1} (Guillonneau et al., 2018). Nano-impact testing provides an alternative approach where high strain rates can be achieved at lower speeds. The test works with electromagnetic actuation to retract a test probe a set distance above the sample surface (e.g., 10–50 μm), which is then rapidly accelerated to produce a high strain rate impact event. The impact energy and effective impact force can be controlled by varying the static load and/or the accelerating distance. For repetitive impact tests, once the probe has come to rest, it is retracted and reaccelerated to produce a set number of repetitive impacts at the same position on the surface. Being an impact test, the strain rate is not constant during a nano-impact test. The indenter geometry also indirectly influences the strain rate achieved in the test, with higher-aspect-ratio probes, such as cube corner indenters, generating higher strain rates than Berkovich or spherical indenters. The strain rate at contact in the nano-impact test can be extremely high, typically in the region of 10^4-10^5 s^{-1} (Faisal et al., 2014; Rueda-Ruiz et al., 2020; Zhang et al., 2021). This very high strain rate decreases on contact, but it remains high throughout the majority of the impact. As an illustration, the variation in strain rate with time when a cube corner diamond impacts alumina is shown in Fig. 1.3.

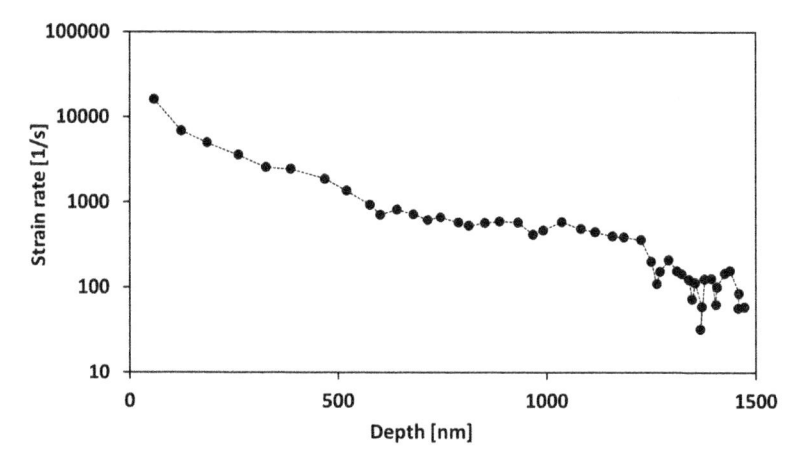

Figure 1.3 Strain rate during a nano-impact evolution of the strain rate during a nano-impact on alumina with a cube corner indenter.

The high strain rate in the nanoscale impact event induces high stresses, which lead to subsurface cracking and material removal on repetitive contact. This has been shown to be a highly efficient method for generating cracks in brittle materials and studying their damage tolerance under varying impact energy and/or test probe geometry. FIB sections through impact craters revealed subsurface intergranular cracking on alumina and subsurface transgranular cracking on MgO-partially stabilized zirconia (PSZ) (Beake et al., 2020). Acoustic emission (AE) monitoring during the tests showed that large displacement bursts with a sharp cube corner probe were associated with cracking. Damage progression was more complex with a 5 μm radius spherical indenter. Impacts could produce AE (cracking) without any depth change; the network of subsurface cracking revealed by FIB is consistent with an initiation process where cracks form subsurface, and several impacts are required before material removal. In multicycling indentation fatigue tests, the indenter remains in contact and the strain rate is lower. Since these repetitive high-cycle tests typically are longer than normal indentation tests, thermal drift becomes a more significant issue. Periodically pausing the cycling provides "hold periods," which can be used to correct for thermal drift more reliably than a single hold period at the end of a long test (Schmahl et al., 2023).

1.9 High-speed mapping

Fast mapping of the small-scale mechanical properties of heterogeneous materials has become possible with the advent of high-speed nanoindentation mapping, where each indent takes typically 1 s. Rossi and coworkers have reviewed recent advances in the technique and highlighted its applications in testing additively manufactured metals, advanced alloys, composites, thermal barrier coatings, and natural materials such as wood and minerals (Rossi et al., 2023). Key questions to address include (1) How close can the indents be spaced? (2) Is the influence of tip wear negligible? (3) How do the higher strain rate (approximate strain rate $\approx 5-10 \text{ s}^{-1}$) data compare with quasistatic results? (4) How is the influence of surface roughness and the presence of defects (e.g., pores) handled? (5) How are statistical deconvolution techniques used? How close the indentations can be placed depends on how sharp the indenter is, and how deep the indents are, which in turn depends on the quality of the surface, so improved results are obtained with extremely flat and smooth surfaces. ISO 14577 states that indents should be spaced (d) at least $d/h = 20$ apart. Phani and Oliver have suggested that this criterion is too strict and in practice $d/h = 10$ could be used (Phani & Oliver, 2019). This lower value of the normalized indent spacing means that an improved lateral resolution can be achieved in grid array mapping. Although in principle high-speed indentation maps can be made using load- or displacement-controlled approaches, generally it is better that all the indentations are of similar size, which can also be done conveniently by applying a "first condition" load termination, where the load ramp is terminated when either a set load or depth is reached. The possibility of chemical or mechanical wear resulting in gradual indenter blunting should be considered, so extremely sharp indenters may be more problematic, as these wear more rapidly. Rossi and coworkers highlighted that the results of high-speed nanoindentation also depend on the method of statistical deconvolution, and compared with ISO 14577, continuous stiffness technique, and high-speed mapping with 1D Gaussian, 2D Gaussian and K-means clustering (Rossi et al., 2023). To illustrate this, differences in the elastic modulus of phases in an Al-Cu eutectic alloy determined by the different methods were shown.

1.10 Testing under environmentally relevant conditions

Nanomechanical properties are usually determined by measurements under standard laboratory conditions with controlled temperature and humidity. The mechanical properties of materials should be characterized under environmentally relevant conditions (Kiener et al., 2023). When making measurements of nanomechanical properties for materials in high-temperature applications, it is important to perform the measurements at test temperatures that are more reflective of their operating conditions where the results are more relevant and the links between properties and performance and design of advanced materials systems for increasingly extreme environments are better understood (Beake, 2021; Beake & Harris, 2019). Best practice in high-temperature nanomechanical measurements relies on consideration of a combination of factors that make the measurements more challenging than at room temperature (Beake, 2021; Beake & Harris, 2019). It is particularly important to (1) ensure instrumental stability and thermal drift minimization, by making measurements in isothermal contact by actively heating the sample and indenter separately; (2) carefully consider the influence of test environment (temperature, oxygen level) on the chemical, oxidative, and mechanical stability of both the indenter and sample; (3) make suitable modifications to the experimental load − time profile and/or analysis procedures to mitigate the effect of often significantly greater indentation creep. Although it is also possible to anneal materials at the operating temperature and retest at room temperature, this approach cannot account for the thermally activated mechanisms occurring at temperature. Measurements have also been performed under subambient temperatures (Chen et al., 2010) and/or with varying levels of humidity (Altaf et al., 2011) or in liquid cells (Beake et al., 2008; Constantinides et al., 2008; Xu et al., 2022; Xu et al., 2023).

1.11 Summary

This chapter has introduced the background to the development of the nanoindentation technique and the unloading curve analysis, which has proved the most popular approach to determining mechanical properties at small contact sizes. The experimental factors that contribute to the accuracy of the determined mechanical properties are discussed and the best-practice recommendations in the international standard for depth-sensing indentation are explained. A limitation of the standard in addressing the influence of pile-up in ductile materials is highlighted. Indentation size effects in plasticity and fracture are investigated. Alternative approaches involving loading curve analysis, indentation energy, creep analysis, continuous stiffness measurements, tests with simplified stress states (micro-pillars and micro-cantilevers), high-speed mapping, and high strain-rate impact testing are also discussed.

References

Altaf, K., Ashcroft, I. A., & Hague, R. (2011). Modelling the effect of moisture on the depth sensing indentation response of a stereolithography polymer. *Computational Materials Science, 52,* 112–117.

Anstis, G. R., Chantikul, P., Lawn, B. R., & Marshall, D. B. (1981). A critical evaluation of indentation techniques for measuring fracture toughness: I, direct crack measurements. *Journal of the American Ceramic Society, 64*(9), 533–538. Available from https://doi.org/10.1111/j.1151-2916.1981.tb10320.x.

Ast, J., Ghidelli, M., Durst, K., Göken, M., Sebastiani, M., & Korsunsky, A. M. (2019). A review of experimental approaches to fracture toughness evaluation at the micro-scale. *Materials & Design, 173,* 107762.

Avadanii, D., Kareer, A., Hansen, L., & Wilkinson, A. (2023). Calibration and data-analysis routines for nanoindentation with spherical tips. *Journal of Materials Research, 38*(17), 4042–4056. Available from https://doi.org/10.1557/s43578-023-01041-6.

Bahr, D.F., Kramer, D.E., & Gerberich, W.W. (1998). Non-linear deformation mechanisms during nanoindentation. *Acta Mater 46,* 3605–3617.

Beake, B. D. (2021). Elevated temperature nanomechanics of coatings for high-temperature applications: A review. *Emergent Materials, 4*(6), 1531–1545. Available from https://doi.org/10.1007/s42247-021-00255-w, springer.com/journal/42247.

Beake, B. D. (2022). The influence of the H/E ratio on wear resistance of coating systems – Insights from small-scale testing. *Surface and Coatings Technology, 442,* 128272. Available from https://doi.org/10.1016/j.surfcoat.2022.128272.

Beake, B. D., Bell, G. A., & Bielinski, D. (2008). Influence of water on the nanoindentation creep response of nylon 6. *Journal of Applied Polymer Science, 107,* 577–582.

Beake, B. D., Ctvrtlik, R., Harris, A. J., Martin, A. S., Vaclavek, L., Manak, J., & Ranc, V. (2020). High frequency acoustic emission monitoring in nano-impact of alumina and partially stabilised zirconia. *Materials Science and Engineering: A, 780,* 139159. Available from https://doi.org/10.1016/j.msea.2020.139159, http://www.elsevier.com.

Beake, B. D., & Goel, S. (2018). Incipient plasticity in tungsten during nanoindentation: Dependence on surface roughness, probe radius and crystal orientation. *International Journal of Refractory Metals and Hard Materials, 75,* 63–69. Available from https://doi.org/10.1016/j.ijrmhm.2018.03.020, http://www.journals.elsevier.com/international-journal-of-refractory-metals-and-hard-materials/.

Beake, B. D., & Harris, A. J. (2019). Nanomechanics to 1000 °C for high temperature mechanical properties of bulk materials and hard coatings. *Vacuum, 159,* 17–28. Available from https://doi.org/10.1016/j.vacuum.2018.10.011.

Beck, C. E., Hofmann, F., Eliason, J. K., Maznev, A. A., Nelson, K. A., & Armstrong, D. E. J. (2017). Correcting for contact area changes in nanoindentation using surface acoustic waves. *Scripta Materialia*, *128*, 83−86. Available from https://doi.org/10.1016/j.scriptamat.2016.09.037, https://www.journals.elsevier.com/scripta-materialia.

Bei, H., George, E. P., Hay, J. L., & Pharr, G. M. (2005). Influence of indenter tip geometry on elastic deformation during nanoindentation. *Physical Review Letters*, *95*(4), 045501. Available from https://doi.org/10.1103/physrevlett.95.045501.

Beirau, T., Rossi, E., Sebastiani, M., Oliver, W. C., Pöllmann, H., & Ewing, R. C. (2021). Fracture toughness of radiation-damaged zircon studied by nanoindentation pillar-splitting. *Applied Physics Letters*, *119*(23), 231903. Available from https://doi.org/10.1063/5.0070597, http://scitation.aip.org/content/aip/journal/apl.

Bolshakov, A., & Pharr, G. M. (1998). Influences of pileup on the measurement of mechanical properties by load and depth sensing indentation techniques. *Journal of Materials Research*, *13*(4), 1049−1058. Available from https://doi.org/10.1557/jmr.1998.0146.

Borodich, F. M. (2014). The Hertz-type and adhesive contact problems for depth-sensing indentation. *Advances in Applied Mechanics*, *47*, 225−366. Available from https://doi.org/10.1016/B978-0-12-800130-1.00003-5, http://www.elsevier.com/wps/find/bookdescription.cws_home/704246/description#description.

Broitman, E. (2017). Indentation hardness measurements at macro-, micro-, and nanoscale: a critical review. *Tribology Letters*, *65*, 23.

Bruns, S., Petho, L., Minnert, C., Michler, J., & Durst, K. (2020). Fracture toughness determination of fused silica by cube corner indentation cracking and pillar splitting. *Materials & Design*, *186*, 108311. Available from https://doi.org/10.1016/j.matdes.2019.108311.

Bull, S. J. (2005). Nanoindentation of coatings. *Journal of Physics D: Applied Physics*, *38*, R393−R413.

Bull, S. J. (2006). Using work of indentation to predict erosion behavior in bulk materials and coatings. *Journal of Physics D: Applied Physics*, *39*(8), 1626−1634. Available from https://doi.org/10.1088/0022-3727/39/8/023.

Bull, S. J., Moharrami, N., Hainsworth, S. V., & Page, T. F. (2016). The origins of chemomechanical effects in the low-load indentation hardness and tribology of ceramic materials. *Journal of Materials Science*, *51*(1), 107−125. Available from https://doi.org/10.1007/s10853-015-9412-3.

Bull, S. J., Page, T. F., & Yoffe, E. H. (1989). An explanation of the indentation size effect in ceramics. *Philosophical Magazine Letters*, *59*, 281−288.

Bulychev, S. I., Alekhin, V. P., Shorshorov, M. K., Ternovskii, A. P., & Shnyrev, G. D. (1975). Determining young's modulus from the indentor penetration diagram. *Industrial Laboratory (USSR)*, *41*(9), 1409−1412.

Chaudhri, M. M. (2018). Some concerns about the current interpretation and analyses of indentation unloading $P - h$ curves highlighted with Young's modulus studies of single crystals of MgO (100). *Journal of Applied Physics*, *124*, 095107.

Chen, J. (2012). Indentation-based methods to assess fracture toughness for thin coatings. *Journal of Physics D: Applied Physics*, *45*(20), 203001. Available from https://doi.org/10.1088/0022-3727/45/20/203001.

Chen, J., Bell, G. A., Dong, H., Smith, J. F., & Beake, B. D. (2010). A study of low temperature mechanical properties and creep behaviour of polypropylene using a new sub-ambient temperature nanoindentation test platform. *Journal of Physics D: Applied Physics*, *43*(42), 425404. Available from https://doi.org/10.1088/0022-3727/43/42/425404, http://iopscience.iop.org/0022-3727/43/42/425404/pdf/0022-3727_43_42_425404.pdf.

Chen, J., & Bull, S. J. (2007). Indentation fracture and toughness assessment for thin optical coatings on glass. *Journal of Physics D: Applied Physics*, *40*(18), 5401−5417. Available from https://doi.org/10.1088/0022-3727/40/18/s01.

Chen, J., & Bull, S. J. (2006). Assessment of the toughness of thin coatings using nanoindentation under displacement control. *Thin Solid Films, 494*, 1−7.
Chen, J., & Bull, S. J. (2009). Relation between the ratio of elastic work to the total work of indentation and the ratio of hardness to Young's modulus for a perfect conical tip. *Journal of Materials Research, 24*(3), 590−598. Available from https://doi.org/10.1557/jmr.2009.0086.
Chen, J., Shen, Y., Liu, W., Beake, B. D., Shi, X., Wang, Z., Zhang, Y., & Guo, X. (2016). Effects of loading rate on development of pile-up during indentation creep of polycrystalline copper. *Materials Science and Engineering: A, 656*, 216−221. Available from https://doi.org/10.1016/j.msea.2016.01.042, http://www.elsevier.com.
Cheng, Y. T., & Cheng, C. M. (1998). Relationships between hardness, elastic modulus, and the work of indentation. *Applied Physics Letters, 73*(5), 614−616. Available from https://doi.org/10.1063/1.121873.
Cheng, Y.-T., & Cheng, C.-M. (2004). Scaling, dimensional analysis, and indentation measurements. *Materials Science and Engineering: R: Reports, 44*, 91−149.
Cheng, Y.-T., Li, Z., & Cheng, C.-M. (2002). Scaling relationships for indentation measurements. *Philosophical Magazine, 82*(10), 1821−1830.
Chudoba, T., & Jennett, N. M. (2008). Higher accuracy analysis of instrumented indentation data obtained with pointed indenters. *Journal of Physics D: Applied Physics, 41*(21), 215407. Available from https://doi.org/10.1088/0022-3727/41/21/215407.
Chudoba, T., & Richter, F. (2001). Investigation of creep behaviour under load during indentation experiments and its influence on hardness and modulus results. *Surface and Coatings Technology, 148*(2−3), 191−198. Available from https://doi.org/10.1016/s0257-8972(01)01340-8.
Constantinides, G., Kalcioglu, Z. I., McFarland, M., Smith, J. F., & Vliet, K. J. V. (2008). Probing mechanical properties of fully hydrated gels and biological tissues. *Journal of Biomechanics, 41*(15), 3285−3289. Available from https://doi.org/10.1016/j.jbiomech.2008.08.015.
Cordill, M. J., Lund, M. S., Parker, J., Leighton, C., Nair, A. K., Farkas, D., Moody, N. R., & Gerberich, W. W. (2009). The Nano-Jackhammer effect in probing near-surface mechanical properties. *International Journal of Plasticity, 25*(11), 2045−2058. Available from https://doi.org/10.1016/j.ijplas.2008.12.015.
Doerner, M. F., & Nix, W. D. (1986). A method for interpreting the data from depth-sensing indentation instruments. *Journal of Materials Research, 1*(4), 601−609. Available from https://doi.org/10.1557/jmr.1986.0601.
Dub, S. N., Lim, Y. Y., & Chaudhri, M. M. (2010). Nanohardness of high purity Cu(111) single crystals: The effect of indenter load and prior plastic strain. *Journal of Applied Physics, 107*, 043510.
Durst, K., & Maier, V. (2015). Dynamic nanoindentation testing for studying thermally activated processes from single to nanocrystalline metals. *Current Opinion in Solid State and Materials Science, 19*, 340−353.
Faisal, N. H., Ahmed, R., Goel, S., & Fu, Y. (2014). Influence of test methodology and probe geometry on nanoscale fatigue failure of diamond-like carbon film. *Surface and Coatings Technology, 242*, 42−53.
Feng, G., & Ngan, A. H. W. (2002). Effects of creep and thermal drift on modulus measurement using depth-sensing indentation. *Journal of Materials Research, 17*(3), 660−668. Available from https://doi.org/10.1557/jmr.2002.0094.
Fischer-Cripps, A. C. (2011). *Nanoindentation*. Springer.

Fischer-Cripps, A. C. (2014). Measurement of hardness of very hard materials. *Solid Mechanics and Its Applications*, *203*, 53−62. Available from https://doi.org/10.1007/978-94-007-6919-9_3, http://www.springer.com/series/6557.

Gale, J. D., & Acheson, A. (2014). The effect of work-hardening and pile-up on nanoindentation measurements. *Journal of Materials Science*, *49*, 5066−5075.

Gan, B., & Tin, S. (2010). Assessment of the effectiveness of transition metal solutes in hardening of Ni solid solutions. *Materials Science and Engineering: A*, *527*(26), 6809−6815. Available from https://doi.org/10.1016/j.msea.2010.06.071.

Ghosh, A., Arreguin-Zavala, J., Aydin, H., Goldbaum, D., Chromik, R., & Brochu, M. (2016). Investigating cube-corner indentation hardness and strength relationship under quasi-static and dynamic testing regimes. *Materials Science and Engineering: A*, *677*, 534−539. Available from https://doi.org/10.1016/j.msea.2016.08.067.

Guillonneau, G., Mieszala, M., Wehrs, J., Schwiedrzik, J., Grop, S., Frey, D., Philippe, L., Breguet, J. M., Michler, J., & Wheeler, J. M. (2018). Nanomechanical testing at high strain rates: New instrumentation for nanoindentation and microcompression. *Materials and Design*, *148*, 39−48. Available from https://doi.org/10.1016/j.matdes.2018.03.050.

Hou, X. D., Bushby, A. J., & Jennett, N. M. (2008). Study of the interaction between indentation size effect and Hall-Petch effect with spherical indenters on annealed polycrystalline copper. *Journal of Physics D: Applied Physics*, *41*(7), 074006.

Hutchings, I. M. (1992). Ductile-brittle transitions and wear maps for the erosion and abrasion of brittle materials. *Journal of Physics D: Applied Physics*, *25*(1A), A212−A221. Available from https://doi.org/10.1088/0022-3727/25/1a/033.

ISO 14577 (2015). Metallic materials—Instrumented indentation test for hardness and materials parameters.

Jang, J., & Pharr, G. M. (2008). Influence of indenter angle on cracking in Si and Ge during nanoindentation. *Acta Materialia*, *56*(16), 4458−4469. Available from https://doi.org/10.1016/j.actamat.2008.05.005.

Jha, K. K., Suksawang, N., & Agarwal, A. (2014). A new insight into the work-of-indentation approach used in the evaluation of material's hardness from nanoindentation measurement with Berkovich indenter. *Computational Materials Science*, *85*, 32−37.

Jia, Y. F., Cui, Y. Y., Xuan, F. Z., & Yang, F. (2017). Comparison between single loading-unloading indentation and continuous stiffness indentation. *RSC Advances*, *7*(57), 35655−35665. Available from https://doi.org/10.1039/c7ra06491h, http://pubs.rsc.org/en/journals/journalissues.

Johnson, K. L. (1970). The correlation of indentation experiments. *Journal of the Mechanics and Physics of Solids*, *18*(2), 115−126. Available from https://doi.org/10.1016/0022-5096(70)90029-3.

Johnson, K. L. (1985). *Contact mechanics*. Cambridge, UK: Cambridge University Press.

Kalei, G. N. (1968). Some results of microhardness test using the depth of impression. *Mashinovedenie*, *4*, 105−107.

Kang, S. K., Kim, J. Y., Kang, I., & Kwon, D. (2009). Effective indenter radius and frame compliance in instrumented indentation testing using a spherical indenter. *Journal of Materials Research*, *24*(9), 2965−2973. Available from https://doi.org/10.1557/jmr.2009.0358.

Kaufman, J. D., & Klapperich, C. M. (2009). Surface detection errors cause overestimation of the modulus in nanoindentation on soft materials. *Journal of the Mechanical Behavior of Biomedical Materials*, *2*(4), 312−317. Available from https://doi.org/10.1016/j.jmbbm.2008.08.004.

Kendall, K. (1978). The impossibility of comminuting small particles by compression. *Nature*, *272*(5655), 710−711. Available from https://doi.org/10.1038/272710a0.

Kermouche, G., Barthel, E., Vandembroucq, D., & Dubujet, P. H. (2008). Mechanical modelling of indentation-induced densification in amorphous silica. *Acta Materialia*, *56*(13), 3222−3228. Available from https://doi.org/10.1016/j.actamat.2008.03.010.

Kholkhujaev, J., Maculotti, G., Genta, G., & Galetto, M. (2023). Metrological comparison of available methods to correct edge-effect local plasticity in instrumented indentation test. *Materials*, *16*(12). Available from https://doi.org/10.3390/ma16124262.

Kiener, D., Wurmshuber, M., Alfreider, M., Schaffar, G. J. K., & Maier-Kiener, V. (2023). Recent advances in nanomechanical and in situ testing techniques: Towards extreme conditions. *Current Opinion in Solid State and Materials Science*, *27*(6), 101108. Available from https://doi.org/10.1016/j.cossms.2023.101108.

Korte-Kerzel, S. (2017). Microcompression of brittle and anisotropic crystals: Recent advances and current challenges in studying plasticity in hard materials. *MRS Communications*, *7*(2), 109−120. Available from https://doi.org/10.1557/mrc.2017.15.

Lawn, B. R., & Marshall, D. B. (1979). Hardness, toughness, and brittleness: An indentation analysis. *Journal of the American Ceramic Society*, *62*, 347.

Leitner, A., Maier-Kiener, V., & Kiener, D. (2017). Dynamic nanoindentation testing: Is there an influence on a material's hardness? *Materials Research Letters*, *7*, 486−493.

Li, W., Bei, H., Qu, J., & Gao, Y. (2013). Effects of machine stiffness on the loading-displacement curves during spherical nano-indentation. *Journal of Materials Research*, *28*, 1903−1914.

Liang, Z. Y., & Pharr, G. M. (2022). Decoupling indentation size and strain rate effects during nanoindentation: A case study in tungsten. *Journal of the Mechanics and Physics of Solids*, *165*, 104935. Available from https://doi.org/10.1016/j.jmps.2022.104935.

Liu, X., Wang, R., Ren, A., Jiang, J., Xu, C., Huang, P., Qian, W., Wu, Y., & Zhang, C. (2014). Evaluation of radiation hardening in ion-irradiated Fe based alloys by nanoindentation. *Journal of Nuclear Materials*, *444*(1-3), 1−6. Available from https://doi.org/10.1016/j.jnucmat.2013.09.026.

Loubet, J. L., Georges, J. M., Marchesini, O., & Meille, G. (1984). Vickers indentation curves of magnesium oxide (MgO). *Journal of Tribology*, *106*(1), 43−48. Available from https://doi.org/10.1115/1.3260865.

Lu, C., Mai, Y.-W., Tam, P. L., & Shen, Y. G. (2007). Nanoindentation-induced elastic−plastic transition and size effect in α-Al 2 O 3 (0001) α nanoindentation-induced elastic−plastic transition and size effect in α -Al 2 O 3 (0001). *Philosophical Magazine Letters*, *87*(6), 409−415. Available from https://doi.org/10.1080/09500830701203156.

Ma, X., Higgins, W., Liang, Z., Zhao, D., Pharr, G. M., & Xie, K. Y. (2021). Exploring the origins of the indentation size effect at submicron scales. *Proceedings of the National Academy of Sciences of the United States of America*, *118*(30), e2025657118. Available from https://doi.org/10.1073/pnas.2025657118, https://www.pnas.org/content/118/30/e2025657118.

Maier, V., Leitner, A., Pippan, R., & Kiener, D. (2015). Thermally activated deformation behaviour of ufg-Au: Environmental issues during long-term and high-temperature nanoindentation testing. *Journal of Materials*, *67*(2015), 2934−2944.

Maier, V., Merle, B., Göken, M., & Durst, K. (2013). An improved long-term nanoindentation creep testing approach for studying the local deformation processes in nanocrystalline metals at room and elevated temperatures. *Journal of Materials Research*, *28*(9), 1177−1188. Available from https://doi.org/10.1557/jmr.2013.39.

Malzbender, J. (2005). Comment on the determination of mechanical properties from the energy dissipated during indentation. *Journal of Materials Research*, *20*(5), 1090−1092. Available from https://doi.org/10.1557/jmr.2005.0162.

Malzbender, J., & de With, G. (2002). Indentation load−displacement curve, plastic deformation, and energy. *Journal of Materials Research*, *17*(2), 502−511. Available from https://doi.org/10.1557/jmr.2002.0070.

Merle, B., Maier-Kiener, V., & Pharr, G. M. (2017). Influence of modulus-to-hardness ratio and harmonic parameters on continuous stiffness measurement during nanoindentation. *Acta Materialia*, *134*(2017), 167−176.

Michler, J., & Blank, E. (2001). Analysis of coating fracture and substrate plasticity induced by spherical indentors: Diamond and diamond-like carbon layers on steel substrates. *Thin Solid Films*, *381*(1), 119−134. Available from https://doi.org/10.1016/s0040-6090(00)01340-7.

Milman, Y. V., Golubenko, A. A., & Dub, S. N. (2011). Indentation size effect in nanohardness. *Acta Materialia*, *59*, 7480−7487.

Milman, Y. V. (2008). Plasticity characteristic obtained by indentation. *Journal of Physics D: Applied Physics*, *41*(7), 074013. Available from https://doi.org/10.1088/0022-3727/41/7/074013.

Moharrami, N., & Bull, S. J. (2014). A comparison of nanoindentation pile-up in bulk materials and thin films. *Thin Solid Films*, *572*, 189−199. Available from https://doi.org/10.1016/j.tsf.2014.06.060.

Moseon, A. J., Basu, S., & Barsoum, M. W. (2008). Determination of the effective zero point for spherical nanoindentation. *Journal of Materials Research*, *23*, 204−209.

Newey, D., Wilkins, M. A., & Pollock, H. M. (1982). An ultra-low-load penetration hardness tester. *Journal of Physics E: Scientific Instruments*, *15*, 119.

Nix, W. D., & Gao, H. (1998). Indentation size effects in crystalline materials: A law for strain gradient plasticity. *Journal of the Mechanics and Physics of Solids*, *46*(3), 411−425. Available from https://doi.org/10.1016/S0022-5096(97)00086-0.

Oliver, W. C., & Pharr, G. M. (1992). An improved technique for determining hardness and elastic modulus using load and displacement sensing indentation experiments. *Journal of Materials Research*, *7*(6), 1564−1583. Available from https://doi.org/10.1557/jmr.1992.1564.

Oliver, W. C., & Pharr, G. M. (2004). Measurement of hardness and elastic modulus by instrumented indentation: Advances in understanding and refinements to methodology. *Journal of Materials Research*, *19*, 3−20.

Pathak, S., & Kalidindi, S. R. (2015). Spherical nanoindentation stress-strain curves. *Materials Science and Engineering R: Reports*, *91*, 1−36. Available from https://doi.org/10.1016/j.mser.2015.02.001.

Pathak, S., Kalidindi, S. R., Weaver, J. S., Wang, Y., Doerner, R. P., & Mara, N. A. (2017). Probing nanoscale damage gradients in ion-irradiated metals using spherical nanoindentation. *Scientific Reports*, *7*(1), 11918. Available from https://doi.org/10.1038/s41598-017-12071-6, http://www.nature.com/srep/index.html.

Pethica, J. B., & Oliver, W. C. (1987). Tip surface interactions in STM and AFM. *Physica Scripta*, *T19A*, 61−66. Available from https://doi.org/10.1088/0031-8949/1987/t19a/010.

Pethica, J. B., Hutchings, R., & Oliver, W. C. (1983). Hardness measurement at penetration depths as small as 20-nm. *Philosophical Magazine A*, *48*, 693-606.

Phani, P. S., & Oliver, W. C. (2017). Ultra high strain rate nanoindentation testing. *Materials*, *10*(6), 663. Available from https://doi.org/10.3390/ma10060663, http://www.mdpi.com/1996-1944/10/6/663/pdf.

Phani, P. S., & Oliver, W. C. (2019). A critical assessment of the effect of indentation spacing on the measurement of hardness and modulus using instrumented indentation testing. *Materials & Design, 164*, 107563.

Phani, P. S., Oliver, W. C., & Pharr, G. M. (2021). Measurement of hardness and elastic modulus by load and depth sensing indentation: Improvements to the technique based on continuous stiffness measurement. *Journal of Materials Research, 36*, 2137−2153.

Pharr, G. M., Strader, J. H., & Oliver, W. C. (2009). Critical issues in making small-depth mechanical property measurements by nanoindentation with continuous stiffness measurements. *Journal of Materials Research, 24*, 653−666.

Qiao, X. G., Starink, M. J., & Gao, N. (2010). The influence of indenter tip rounding on the indentation size effect. *Acta Materialia, 58*, 3690−3700.

Qu, S., Huang, Y., Nix, W. D., Jiang, H., Zhang, F., & Hwang, K. C. (2004). Indenter tip radius effect on the Nix−Gao relation in micro- and nanoindentation hardness experiments. *Journal of Materials Research, 19*(11), 3423−3434. Available from https://doi.org/10.1557/jmr.2004.0441.

Renzelli, M., Mughal, M. Z., Sebastiani, M., & Bemporad, E. (2016). Design, fabrication and characterization of multilayer Cr-CrN thin coatings with tailored residual stress profiles. *Materials and Design, 112*, 162−171. Available from https://doi.org/10.1016/j.matdes.2016.09.058.

Rhee, Y. W., Kim, H. W., Deng, Y., & Lawn, B. R. (2001). Brittle fracture versus quasi plasticity in ceramics: A simple predictive index. *Journal of the American Ceramic Society, 84*(3), 561−565. Available from https://doi.org/10.1111/j.1151-2916.2001.tb00698.x, http://www.blackwellpublishing.com/aims.asp?ref = 0002-7820.

Roa, S., & Sirena, M. (2021). A finite element analysis of conical indentation in elastic-plastic materials: On strain energy based strategies for an area-independent determination of solids mechanical properties. *International Journal of Mechanical Sciences, 207*, 106651. Available from https://doi.org/10.1016/j.ijmecsci.2021.106651.

Rossi, E., Wheeler, J. M., & Sebastiani, M. (2023). High-speed nanoindentation mapping: A review of recent advances and applications. *Current Opinion in Solid State and Materials Science, 27*, 101107.

Rueda-Ruiz, M., Beake, B. D., & Molina-Aldareguia, J. M. (2020). New instrumentation and analysis methodology for nano-impact testing. *Materials & Design, 192*, 108715.

Schaufler, J., Schmid, C., Durst, K., & Göken, M. (2012). Determination of the interfacial strength and fracture toughness of a-C:H coatings by in-situ microcantilever bending. *Thin Solid Films, 522*, 480−484. Available from https://doi.org/10.1016/j.tsf.2012.08.031.

Schiffmann, K. I. (2011). Determination of fracture toughness of bulk materials and thin films by nanoindentation: Comparison of different models. *Philosophical Magazine, 91* (7−9), 1163−1178. Available from https://doi.org/10.1080/14786435.2010.487984.

Schmahl, M., Müller, C., Meinke, R., Alves Alcantara, E. G., Hangen, U. D., & Fleck, C. (2023). Cyclic nanoindentation for local high cycle fatigue investigations: A methodological approach accounting for thermal drift. *Advanced Engineering Materials, 25*(10), 221676. Available from https://doi.org/10.1002/adem.202201676, http://onlinelibrary.wiley.com/journal/10.1002/(ISSN)1527-2648.

Schwarzer, N. (2006). Analysing nanoindentation unloading curves using Pharr's concept of the effective indenter shape. *Thin Solid Films, 494*(1−2), 168−172. Available from https://doi.org/10.1016/j.tsf.2005.08.253.

Shim, S., Bei, H., George, E. P., & Pharr, G. M. (2008). A different type of indentation size effect. *Scripta Materialia, 59*(10), 1095−1098. Available from https://doi.org/10.1016/j.scriptamat.2008.07.026.

Singh, S. S., Guo, E., Xie, H., & Chawla, N. (2015). Mechanical properties of intermetallic inclusions in Al 7075 alloys by micropillar compression. *Intermetallics, 62*, 69−75. Available from https://doi.org/10.1016/j.intermet.2015.03.008.

Siu, K. W., & Ngan, A. H. W. (2013a). The continuous stiffness measurement technique in nanoindentation intrinsically modifies the strength of the sample. *Philosophical Magazine, 93*, 449−467.

Siu, K. W., & Ngan, A. H. W. (2013b). Oscillation-induced softening in copper and molybdenum from nano- to micro-length scales. *Materials Science and Engineering: A, 572*, 56−64. Available from https://doi.org/10.1016/j.msea.2013.02.037.

Sneddon, I. N. (1965). The relation between load and penetration in the axisymmetric boussinesq problem for a punch of arbitrary profile. *International Journal of Engineering Science, 3*(1), 47−57. Available from https://doi.org/10.1016/0020-7225(65)90019-4.

Spary, I. D., Bushby, A. J., & Jennett, N. M. (2006). On the indentation size effect in spherical indentation. *Philosophical Magazine, 86*, 5581−5593.

Stilwell, N. A., & Tabor, D. (1961). Elastic recovery of conical indentations. *Proceedings of the Physical Society, 78*, 169−179.

Taljat, B., & Pharr, G. M. (2004). Development of pile-up during spherical indentation of elastic−plastic solids. *International Journal of Solids and Structures, 41*(14), 3891−3904. Available from https://doi.org/10.1016/j.ijsolstr.2004.02.033.

Tang, B., & Ngan, A. H. W. (2003). Accurate measurement of tip−sample contact size during nanoindentation of viscoelastic materials. *Journal of Materials Research, 18*(5), 1141−1148. Available from https://doi.org/10.1557/jmr.2003.0156.

Tardio, S., Abel, M. L., Carr, R. H., Castle, J. E., & Watts, J. F. (2015). Comparative study of the native oxide on 316L stainless steel by XPS and ToF-SIMS. *Journal of Vacuum Science and Technology A: Vacuum, Surfaces and Films, 33*(5), 05E122. Available from https://doi.org/10.1116/1.4927319, http://scitation.aip.org/content/avs/journal/jvsta.

Tuck, J. R., Korsunskyl, A. M., Bull, S. J., & Davidson, R. I. (2001). On the application of the work-of-indentation approach to depth-sensing indentation experiments in coated systems. *Surface and Coatings Technology, 137*(2-3), 217−224. Available from https://doi.org/10.1016/S0257-8972(00)01063-X.

Van Vliet, K. J., Prchlik, L., & Smith, J. F. (2004). Direct measurement of indentation frame compliance. *Journal of Materials Research, 19*, 325−331.

Wang, J., Volz, T., Weygand, S. M., & Schwaiger, R. (2021). The indentation size effect of single-crystalline tungsten revisited. *Journal of Materials Research, 36*(11), 2166−2175. Available from https://doi.org/10.1557/s43578-021-00221-6, https://www.springer.com/journal/43578.

Xu, D., Harvey, T., Begiristain, E., Domínguez, C., Sánchez-Abella, L., Browne, M., & Cook, R. B. (2022). Measuring the elastic modulus of soft biomaterials using nanoindentation. *Journal of the Mechanical Behavior of Biomedical Materials, 133*, 105329. Available from https://doi.org/10.1016/j.jmbbm.2022.105329.

Xu, D., Harvey, T., Martínez, J., Begiristain, E., Domínguez-Trujillo, C., Sánchez-Abella, L., Browne, M., & Cook, R. B. (2023). Mechanical and tribological characterisations of PEG-based hydrogel coatings on XLPE surfaces. *Wear, 522*, 204699.

Yamamoto, M., Tanaka, M., & Furukimi, O. (2021). Hardness-deformation energy relationship in metals and alloys: A comparative evaluation based on nanoindentation testing and thermodynamic consideration. *Materials, 14*, 7217.

Yan, C., Bor, B., Plunkett, A., Domènech, B., Schneider, G. A., & Giuntini, D. (2022). Nanoindentation of supercrystalline nanocomposites: Linear relationship between elastic modulus and hardness. *Journal of Materials, 74*(6), 2261−2276. Available

from https://doi.org/10.1007/s11837-022-05283-3, http://www.springer.com/materials/journal/11837.
Yang, R., Zhang, T., Jiang, P., & Bai, Y. (2008). Experimental verification and theoretical analysis of the relationships between hardness, elastic modulus, and the work of indentation. *Applied Physics Letters*, *92*(23). Available from https://doi.org/10.1063/1.2944138.
Yoshida, S. (2019). Indentation deformation and cracking in oxide glass—Toward understanding of crack nucleation. *Journal of Non-Crystalline Solids: X*, *1*, 100009. Available from https://doi.org/10.1016/j.nocx.2019.100009.
Zhang, H., Li, Z., He, W., Ma, C., Chen, J., Liao, B., & Li, Y. (2021). Damage mechanisms evolution of TiN/Ti multilayer films with different modulation periods in cyclic impact conditions. *Applied Surface Science*, *540*, 148366. Available from https://doi.org/10.1016/j.apsusc.2020.148366.
Zhang, K., Wang, K., Liu, Z., & Xu, X. (2020). Strain rate of metal deformation in the machining process from a fluid flow perspective. *Applied Sciences*, *10*(9), 3057. Available from https://doi.org/10.3390/app10093057.
Zhao, Y. P., Shi, X., & Li, W. J. (2003). Effect of work of adhesion on nanoindentation. *Reviews on Advanced Materials Science*, *5*, 348—353.

Development of additional nanoscale wear test techniques

Ben Beake[1] and Tomasz Liskiewicz[2]
[1]Micro Materials Ltd., Wrexham, United Kingdom, [2]Department of Engineering, Manchester Metropolitan University, Manchester, United Kingdom

2.1 Introduction – single-asperity tribology

Deformation and wear begin at the peaks of the asperities between surfaces in contact under load (Bhushan, 1996; Greenwood & Williamson, 1966; Sawyer & Wahl, 2008; Stoyanov & Chromik, 2017). In a macroscale tribological test with multiasperity contact, the actual contact pressures acting on these are not well known. Fig. 2.1 schematically shows how initially only the asperities are in contact when rough surfaces slide relative to each other. The contact zone is hidden, and the real area of contact is initially a small fraction of the apparent contact area and not known accurately. The contact pressure acting on the asperities is much larger than the nominal pressure (Bhushan, 1996; Dieterich & Kilgore, 1994; Sawyer & Wahl, 2008). This is relevant not only to dry sliding but also for components operating in lubricated conditions where isolated contact can occur under mixed or boundary lubrication. Numerical analysis has shown that maximum contact pressures between rough surfaces are significantly higher than those between smooth surfaces, consistent with the higher wear rates reported for DLC coatings deposited on rougher steel substrates in pin-on-disk tests (Jiang & Arnell, 2000) and lower critical loads in scratch tests.

An alternative approach to macroscale, multiasperity testing is to perform single-asperity tribological tests with much lower forces and sharper probes. The simplified contact conditions in these single asperity tests enable the onset of

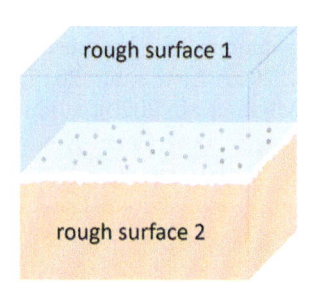

Figure 2.1 Rough surfaces in contact. In a tribo-contact initially, only the asperities are in contact, and the real area of contact is a small fraction of the apparent contact area.

wear, its correlation with friction, and the influence of the surface topography and mechanical properties of the contacting surfaces to be studied in detail (Jacobs et al., 2019; Sawyer & Wahl, 2008; Stoyanov & Chromik, 2017; Szlufarska et al., 2008). These single-asperity tests encompass a range of nano-/microtribological tests where the test probe is mechanically loaded and the deformation monitored at the level of an individual contact. The resultant overall deformation in a multiasperity contact is the sum of these individual contacts. One link between various single-asperity microtribological tests is that the contact size has an important role and contact mechanics is key. Szlufarska et al. (2008) stated that contact mechanics is critical in tribology, as it provides quantitative descriptions of contact area, elastic indentation, contact stiffness, and the stress and strain fields of a mechanically loaded asperity. An important trend in tribology research has been the continuing reduction in the testing length scale, aimed at improving our fundamental understanding of tribological contacts, which has led to the development of new coatings by identifying their property requirements at different length scales (Matthews et al., 2007). The capabilities of nanoindentation instruments have been extended beyond the characterization of mechanical properties to encompass a wide range of nano/microtribological test techniques involving single-asperity contacts between a diamond probe and the material under test that are primarily involved with simulating tribo-contact conditions. These tests include: (1) unidirectional sliding tests (i.e., scratch tests), (2) reciprocating contacts (including fretting, covering gross slip, and partial slip regimes), and (3) repetitive or cyclic impact tests. In combination with mechanical property information, these instrumented wear test methods provide unique insights into how mechanical properties influence friction and wear. When they can simulate contact conditions so that the major deformation mechanism(s) can be reproduced, they can be very effective tools in coating or alloy development screening campaigns. A significant benefit of this approach in comparison with multiasperity testing is the simplified geometry in the nano-/microscale tests. This allows modeling of the stresses involved and determination of the deformation mechanisms. The most common experimental configuration is the sphere-on-flat test geometry. Since the tests can be instrumented, retaining the depth-sensing capability of the nanomechanical test instrument, they are amenable to analysis and can provide valuable information on the relative importance of plastic deformation and fracture.

Various scanning probe techniques have been used including, in ascending contact size, TEMs (Jacobs et al., 2019; Liao et al., 2015; Liao & Marks, 2015, 2016; Mitchell & Shrotriya, 2007; Milne et al., 2020), atomic force microscopes (AFMs) (Beake et al., 2001, 2004; Colaço, 2009; Chen, Ji et al., 2011; Chen, Yang et al., 2011; Degiampietro & Colaço, 2007; Garabedian et al., 2019; Peng et al., 2009; Prioli et al., 2003; Schiffmann, 1998; Szlufarska et al., 2008; Varenberg et al., 2005; Yu, Dong et al., 2009; Yu, Qian et al., 2009; Yu et al., 2010; Yu et al., 2012; Yu, J.X., Qian et al., 2009; Liu et al., 2020), nanoindenters (Beake et al., 2010, 2012; Beake, Liskiewicz et al., 2011; Beake, Shi et al., 2011; Liskiewicz et al., 2010, 2013; Qian et al., 2007; Schiffmann & Hieke, 2003; Schiffmann, 2004, 2008; Stoyanov et al., 2010; Wilson, & Sullivan, 2008, 2009; Wilson et al., 2008, 2009), or microtribometers (Achanta et al., 2005, 2008; Drees et al., 2004; Gee & Gee, 2007; Gee et al., 2011) as test platforms. The sharpness of the test probes employed in these techniques has different advantages and disadvantages for probing different length scale contacts, which can be particularly critical for coatings. By altering the severity of the contact conditions, the contact size and length scale of tribo-contact are varied (so that small contact size reduces the length scale or "information depth" of the contact). By reducing the contact size (e.g., smaller probes, lower load), more "coating-dominated" properties are evaluated. Conversely, increasing the contact size results in the behavior of the entire coating system being evaluated. This can include, for example, its load-carrying capacity, substrate yield, the coating bending stresses thus generated, substrate ductility (and damage tolerance), and stress transfer over interfaces. The consequence of this is that the property requirements for wear resistance change with the contact size (Matthews et al., 2007).

Due to the requirement to cause failure within a single or a small number of cycles, the contact pressure is generally high in ramped load or repetitive scratch tests involving multiple passes over the same track, and in AFM experiments, they can be considered as "overload" tests or accelerated wear tests. When the test instrumentation has the necessary stability to perform longer tests, it becomes practical to perform higher cycle, lower-pressure fatigue-type tests, which may be closer to many practical contact situations.

By combining the results at different applied load and/or by using test probes of varying sharpness, it is possible to induce stresses at different locations within the coating-substrate system and evaluate the entire system not just the top layer. This can be important since tribological performance is usually strongly influenced by the extent of load support provided by the substrate under the test conditions. As an example, Wang et al. (2012) investigated the tribological behavior of graphite-like carbon coatings on titanium alloy or hard metal substrates with varying interlayer composition. They showed that, irrespective of the interlayer composition, the wear life of the carbon films was greatly increased when deposited on the harder substrate. Alternatively, by using the same test probe in a range of different test techniques (using multifunctional test instrumentation), it is possible to show the different requirements in different contact situations, which require more/less load support or fracture resistance to ensure low wear. This approach has been taken in studies of the influence of alloying on multilayer carbon films on hardened tool steel. The authors Beake, McMaster et al. (2021) were able to show that softer W-doped DLC films were susceptible to greater wear under sliding (scratch) or reciprocating conditions, but their reduced brittleness meant that they did not crack as readily under highly loaded impact as harder, undoped, DLC films did so that their performance under repetitive impact was dramatically improved. Small-scale tribological tests can also be performed at high temperatures to simulate contact conditions occurring, for example, through frictional heating in a cutting test. Although less well explored than tests at room temperature, the small-scale tests have been effective, for example, at revealing changing deformation mechanisms at elevated temperatures (Beake et al., 2017).

In the following sections, different test techniques (scratch, reciprocating, impact) are introduced, and the importance of the length scale of the contact on the results obtained (i.e., nano- vs micro- vs macroscale) is investigated.

2.2 Scratch and wear testing

The stress distribution in sliding contact is more complex than in indentation. In sliding, the shear stress distribution is asymmetric with maximum compressive stress in front of the probe and maximum tensile stress behind the probe. When the tensile stress exceeds the fracture strength of the coating, cracking occurs (Holmberg et al., 2007, 2008). This is illustrated schematically in Fig. 2.2.

Different probes are typically used for indentation and scratch testing of coated systems since sharp probes are needed for coating-only properties and blunter probes for locating high induced stresses deeper into the surface (e.g., in the vicinity of the interface). However, if the same probe is used for indentation and scratch testing, then the deformation in both tests can be directly compared. When sliding with spherical probes under nearly elastic conditions with low friction, the scratch deformation can be very similar to that under indentation (i.e., without sliding). This can be shown by removing instrument compliance from the scratch deformation as is done in indentation. For example, Beake and Liskiewicz showed that compliance and slope-corrected loading curves of a 0.5 μm ECR-CVD carbon film on Si indented and scratched by a 6.5 μm spherical probe diverged appreciably only when approaching 100 mN (Beake & Liskiewicz, 2017). Nevertheless, indentation and scratch stress fields increasingly diverge for (more asymmetric) higher friction contacts as occur on hard coatings at higher load and/or on more ductile materials. von Stebut and coworkers reported that the critical load for spallation of

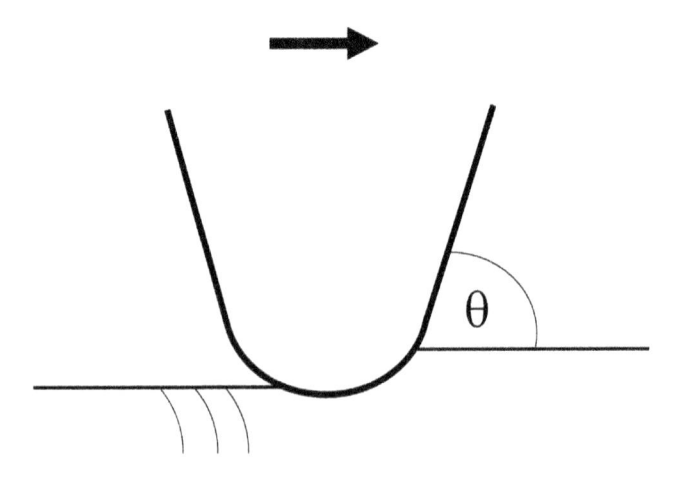

Figure 2.2 Stress distribution in sliding contact. Asymmetric stress distribution in the sliding contact.

TiN-coated high-speed steel was 45 N in a scratch test, but over four times higher (195 N) in indentation testing with the same probe.

The most common types of scratch test involve either (1) ramped (or progressive) load scratch tests where critical loads are recorded corresponding to yield or failure events or (2) repetitive constant load tests where the number of cycles to failure at a load lower than that causing failure in the ramped load test is determined. In ramped load nanoscratch tests, the contact pressures are often necessarily high due to the requirement to induce coating failure in a single cycle and the plowing component to the friction force can be relatively large resulting in higher tensile stresses at, or behind, the trailing edge of the sliding contact. By suitable choice of the applied load and probe radius, either coating-dominated wear properties or interfacial behavior can be studied in repetitive constant load nanoscratch tests. The applied load can be set to position the highest von Mises stresses close to the interface between coating and substrate and minimize substrate deformation as a precursor of coating debonding, with improved sensitivity to poor adhesion in comparison with progressive load scratch tests.

Since the scratch test can be considered as a model single-asperity contact where abrasive wear mechanisms can be reproduced, the technique has also found use in testing bulk materials, including ceramics (Petit et al., 2009), hard metals (Gee, 2001; Gee et al., 2017), and coatings (Bull, 1997), although Bull (1997) has cautioned that the test cannot make a quantitative prediction of abrasive wear rate. Gee et al. (2007) noted that the major deformation mechanisms in abrasive wear of hard metals, including plastic deformation, cutting, microfracture and delamination, can all be replicated in the scratch test. Zum Gahr (1987) proposed defining the material displaced by pileup and removed from the wear track to describe the relative fraction of cutting and plowing mechanisms, that is, cutting efficiency, in a scratch/abrasive contact.

In the next sections, the role of contact size is introduced and the deformation produced in macroscale scratch testing ($R = 200$ μm) compared with that in micro- and nanoscale scratch tests performed with sharper probe geometries. Macroscratch testing has been standardized by ASTM and more recently nano-/microscratch by CEN/TS (2021). The influence of coating thickness on the choice between nano- and microscale tests is discussed.

2.2.1 Macroscratch testing

The macroscale scratch test, in which a spheroconical diamond probe with a 200 μm end radius slides across the surface under an increasing normal load, was originally conceived as a test of coating adhesion strength, with higher critical loads directly corresponding to more adherent coatings (Steinmann et al., 1987). It has become clear subsequently that the mechanical properties of the coating and substrate influence on the deformation behavior in the test, and it is not controlled by the adhesion strength alone (ASTM C1624-05, 2015; Bull, 1991; Bull et al., 1988; Burnett & Rickerby, 1987; Diao et al., 1994; Heinke et al., 1995; Ichimura & Ishii, 2003; Kato, 1995; Larsson et al., 2000; Ollendorf & Schneider, 1999; Randall,

2019; Randall et al., 2001; Rezakhanlou et al., 1990; Steinmann et al., 1987; Valli et al., 1985; von Stebut et al., 1989; Xie & Hawthorne, 2001). The fundamental adhesion strength of the bond between the coating and the substrate is not directly measured in the scratch test. The critical load in the scratch test is dependent on a range of extrinsic and intrinsic factors in addition to the interfacial strength (Bull, 1997; Randall, 2019; Randall et al., 2001). The ASTM standard for the test explains that the test "provides a quantitative engineering measurement of the practical (extrinsic) adhesion strength and damage resistance of the coating-substrate system as a function of applied normal force" (ASTM C1624−05, 2015). Intrinsic test factors include the scratching speed, loading rate, radius of the test probe, probe wear, and machine stiffness, while extrinsic factors include the substrate properties (hardness, elastic modulus), coating properties (thickness, hardness, elastic modulus, residual stress), surface roughness, and friction coefficient (Bull, 1997; Randall, 2019; Randall et al., 2001). Major scratch deformation modes for hard coatings were listed by Bull (1991, 1997) as through-thickness cracking, chipping (cohesive failure), wedge spallation, and buckling failure, with only the latter two being related to adhesion failure (Bull, 1997).

The sensitivity of the progressive load scratch test to adhesion differences can be low. For example, to produce TiN coatings with a range of adhesion strengths, Ollendorf and Schneider varied the duration of presputtering Ar ion cleaning of the steel substrates they were deposited on (Ollendorf & Schneider, 1999). Contrary to their expectations, they found that the more adherent films they produced actually had lower critical loads for cracking and failure. The explanation for the unexpected inverse correlation is not immediately clear, but it may reflect mechanical differences in the films studied. Aldrich-Smith et al. (2004) reported a round-robin intercomparison study scratch testing 5 μm CrN on 304 stainless steel where the adhesion was controlled varying the thickness of a gold − palladium interlayer (0, 25, and 50 nm). Within the scatter of the data, it was not possible to discriminate between the different interlayer types. There was also considerable variability between laboratories most likely indicating differences in probe sharpness from the nominal 200 mm (since the critical load varies with probe radius).

The sharpness of the test probe also has a significant influence on the depth at which initial yielding occurs. In the conventional ASTM standardized macroscale scratch test with a 200 μm probe, the ratio of the coating thickness to the radius of the probe (t/R) is very small for thin coatings. The maximum von Mises stresses responsible for plasticity are located deep within the substrate at the critical load and typically significant substrate yielding occurs before any coating failure. The consequence of this is that the scratch test critical load is strongly correlated with the load-carrying capacity of the substrate. In scratch tests of TiN and CrN on steel substrates with varying hardness, the critical load for failure increased linearly with substrate hardness (Ichimura & Ishii, 2003). Failure behavior was dominated by plasticity and the cohesive strength of the coatings rather than their adhesion. In tests on DLC and TiN coatings on titanium substrates, the critical load was almost completely independent of the coating type, instead correlating with substrate hardness (Wang, Escudeiro et al., 2013; Wang, Zhang et al., 2013). Cross-sectional profiles through the scratch track at failure showed that the scratch depth was much greater than the coating thickness, confirming substrate plasticity. Finite element modeling showed that the increased load-carrying capacity was related to the stress generated in the substrate. Greater stresses were carried by the harder ultrafine-grained titanium substrate than with the softer coarse-grained Ti substrate, which reduced the strain at the coating − substrate interface. For each different coating type, there was a constant groove width at failure that was independent of substrate hardness implying a mechanism where coating failure is controlled by coating bending due to substrate deformation, with the critical level of coating bending varying with coating type. Wu et al. (2022) used cross-sectional analysis to investigate the failure mechanisms when scratching a model hard − soft coating system (9.5 µm thick TiN/600 nm Ti interlayer/304 L stainless steel substrate) with a 200 µm probe. They cautioned against using the scratch test critical loads as measures of adhesion strength since the cross sections showed that both L_{c2} (spallation) and L_{c3} (total peeling off) were not adhesion failures but were cohesive failure within the film and cohesive failure within the steel substrate, respectively. Additional load support to restrict coating bending can be provided by duplex coating systems, where the hard coating is deposited on a hardened substrate (e.g., by plasma nitriding). These duplex coating systems typically have shown larger scratch test critical loads than coatings deposited on nonhardened substrates (Batista et al., 2003; Bell et al., 1998; Leyland et al., 1991; Pujante et al., 2014). For example, higher critical loads were found for cohesive and adhesive failures on TiAlN and CrN coatings when deposited on plasma nitrided steel with a 30−60 µm diffusion zone than when deposited on H13 steel with 6 GPa hardness (Batista et al., 2003).

Coating mechanical properties and thickness and the surface roughness also influence the stresses induced in the scratch test and the resultant measured critical loads and deformation failure mechanisms. Holmberg and coworkers studied stress distributions and conditions for fracture in scratch tests of TiN and DLC-coated hardened steels employing 3D finite element methods (Holmberg et al., 2006a,b, 2007, 2008; Laukkanen et al., 2006). An increase in the thickness of CrN coatings on steel from 5 to 20 µm resulted in an improvement of the L_{c2} critical load from 21 to 55 N (Heinke et al., 1995). Conversely, Lin et al. (2023) reported lower critical loads for graphite-like-carbon/CrN coatings on a cold work die steel when the thickness of the carbon top layer was increased. Larsson et al. (2000) studied the influence of coating thickness on the critical loads of 1.6, 4.1, and 7.1 µm TiN coatings on high- speed steel. They found that the 4.1 µm coating showed the highest critical load for total coating removal, but it cracked at the lowest load in indentation. The minimum in load-carrying capability at the intermediate thickness could be due to thin coatings deflecting well without cracking while very thick coatings lower substrate deformation. The lower load for the thickest coating in the scratch test was explained by a change in mechanism. In a study of TiN on stainless steel, the critical load for coating detachment increased with thickness (Bull, 1997). Bull noted that this was a feature of the buckle failure mode and not an indication of higher adhesion. Larsson et al. (2000) found that the direction of scratching relative to the grinding grooves also has a significant effect on the critical load. The effect is surface roughness-dependent; the influence of scan direction being minimal when $R_a < 0.1$ μm, but a 50% reduction occurred when scanning across the grooves when R_a was 0.5 μm. In scratch testing, the test probe is almost invariably made of diamond so that it does not plastically deform during the test. An interesting recent exception was the use of a sharp conical aluminum pin (representing the work material) scratching against a cemented carbide (representing the tool) (Olsson & Cinca, 2024). During sliding, the tip flattens simulating a metal forming contact. Olsson and Cinca showed that submicron surface irregularities in the cemented carbide produced greater aluminum transfer and higher friction.

2.2.2 Nano- and microscratch

Sensitivity to coating and interfacial properties can be improved by performing tests at smaller contact size, which reduces the influence of substrate deformation on the overall coating system response. This can be achieved by reducing the test probe radius and using instrumentation with greater sensitivity at lower load to perform nano- or microscale scratch tests, which has been termed "dimensioning." Nanoscratch testing (e.g., with 5 μm or sharper probes) is more suitable for thinner coatings, and microscale scratch testing (e.g., with 25 μm probes) is more suitable for thicker coatings (e.g., >2 μm). The influence of probe radius is shown schematically in Fig. 2.3

In these tests, the coating and substrate mechanical properties, coating thickness, and test probe size all influence the initial yield location, which in turn affects how the interface is weakened and/or the crack evolution, which controls wear. Kato and coworkers have produced yield location maps from finite element modeling of sliding contacts (Diao et al., 1994; Kato, 1995). More recently, Schwarzer et al. (2011) developed commercial software to generate full 3D-simulated stress distributions in scratch tests of multilayer coating systems (Film Doctor, n.d.). The approach uses a physical-based analytical methodology to determine simulated stress distributions of von Mises, tensile, and shear stresses developing during the micro- or nanoscratch test. As important as the magnitude and location of stresses at yield is the subsequent cracking behavior/damage tolerance of the coating system. Akono and Ulm (2014) developed a linear elastic fracture mechanics model to obtain fracture

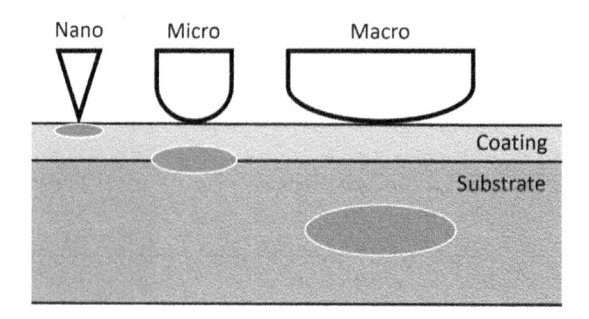

Figure 2.3 Position of maximum von Mises stress. Schematic representation of the influence of probe radius on position of maximum von Mises stress.

toughness from the applied load, probe geometry, and depth in macroscale scratch test. The correlation between the critical load in a scratch test and toughness has been studied extensively (Beake et al., 2006; Wang & Zhang, 2014; Wang, Escudeiro et al., 2013; Wang, Zhang et al., 2013; Zhang et al., 2015; 2015; Zhang, Bui et al., 2004; Zhang, Bui, Fu, et al., 2004; Zhang, Sun, et al., 2004). Voevodin and Zabinski (1998) proposed that the lower critical load in a scratch test could provide a measure of the fracture toughness. This was subsequently modified by Zhang and coworkers (Wang & Zhang, 2014; Wang, Escudeiro et al., 2013; Wang, Zhang et al., 2013; Zhang et al., 2004, 2015), who equated L_{c1} to crack initiation resistance and the increase to L_{c2}, the load for total failure of the coating (i.e., $L_{c2}-L_{c1}$) as a measure of crack propagation. Zhang et al. proposed a metric (later termed scratch crack propagation resistance or scratch toughness) that could represent the combined resistance to crack initiation and propagation as:

$$\text{Scratch toughness} = L_{c1}(L_{c2} - L_{c1}) \tag{2.1}$$

Although values of the scratch toughness are inevitably a function of the geometry of the scratch probe (increasing with probe radius) making them difficult to compare between studies when different probe radii have been used, they can provide a useful assessment of coating behavior and have been used effectively in several studies aimed at developing "hard yet tough" coatings (Wang & Zhang, 2014; Wang, Escudeiro et al., 2013; Wang, Zhang et al., 2013; Zhang et al., 2015; Zhang, Bui et al., 2004; Zhang, Bui, Fu, et al., 2004; Zhang, Sun, et al., 2004). For example, the relationship between scratch toughness and hardness of nc-TiN/a-Si$_3$N$_4$ produced by magnetron sputtering (Zhang, Sun, et al., 2004) or ion beam − assisted deposition (Beake et al., 2006) has been investigated and maximum scratch toughness found for coatings with hardness around 20 GPa in both cases, as illustrated by Fig. 2.4. In Fig. 2.4, the scratch toughness has been normalized to allow comparison

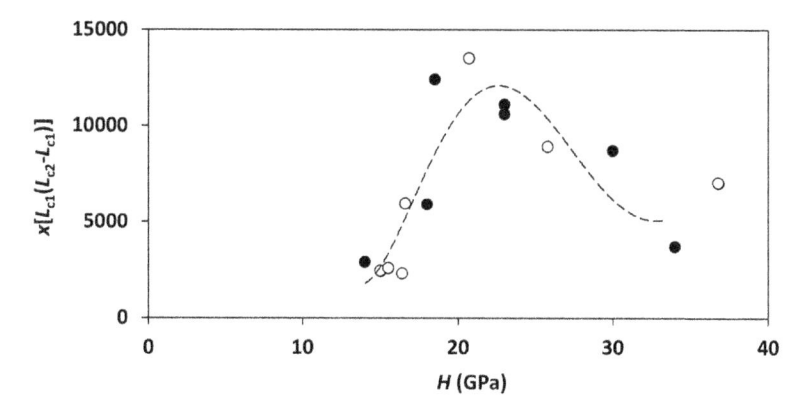

Figure 2.4 Impact of coating hardness on scratch toughness. Relationship between normalized scratch toughness and hardness of nc-TiN/a-Si$_3$N$_4$ films produced by different deposition methods (open circles = magnetron sputtering; filled circles = ion beam-assisted deposition).

of tests with different probe radii. Other studies have reported no maximum in scratch toughness at intermediate hardness but instead an inflexion point where coatings display adequate hardness and toughness (i.e., enhanced toughness without as much reduction in hardness) (Wang & Zhang, 2014; Wang, Escudeiro et al., 2013; Wang, Zhang et al., 2013).

Nano- and microscratch tests can simulate high contact pressure sliding/abrasive contacts involving coatings or bulk materials. For example, coated tools operate in tribologically extreme conditions with tangential loads in metal cutting. Since the components in highly loaded machining applications can operate in the region of their elastic limit, or above it, with peak stresses >4 GPa, (Bouzakis, Maliaris et al., 2012; Bouzakis, Michailidis et al., 2012), it is highly desirable to replicate these high stresses in laboratory tests aimed at assessing coating mechanical performance. Ahmed et al. reported a good correlation between the behavior of HIPed and as-cast Stellite 6 in nanoscratch tests and performance in ASTM G65 dry sand rubber wheel abrasion tests (Ahmed et al., 2014). In both tests, there was less wear for hot isostatically pressed Stellite 6 than for the conventionally cast Stellite 6. Other authors have reported a correlation between scratch resistance and crystallographic orientation (Pöhl et al., 2016).

2.2.3 Repetitive (multi-pass) scratch testing

Constant load, unidirectional multi-pass scratch testing was described by Bull and Rickerby (1989) and von Stebut et al. (1989) and has been shown to be an effective low cycle fatigue wear test (von Stebut, 2005). Gee (2001) compared the resistance of hard metals and ceramics with repetitive low load scratch tests. Although the ceramics were initially more scratch-resistant, the hard metals were more able to resist withstand repeated abrasion damage. Differences in the tribologically transformed structure (TTS) generated under the contact zone by continued sliding may be responsible. Efeoglu and Arnell (2000) performed multipass scratch tests with a 200 mm tip radius diamond on TiN coatings on steel deposited by closed-field unbalanced magnetron sputter ion plating following the frictional evolution with cycling. Under the test conditions, they reported that the coatings gradually wore away rather than failing cohesively or adhesively. Petit et al. (2009) performed repetitive ramped load scratch tests on monolithic ceramics and soda lime glass recording the load at which there was enhanced material removal rate through chipping for each cycle. They found that the materials studied showed a typical S−N fatigue-type behavior with a sharp decrease of critical load during the first few scratches followed by stabilization and more gradual decrease thereafter. Although the critical loads for single progressive load scratch tests ranked in order of their fracture toughness, under repetitive scratching, a different trend emerged with SiC showing a more dramatic reduction in critical load with cycling than the other materials.

It is normal for the repetitive scratches to be in the same track although there have been reports where one or more scratches are performed in parallel. Double-scratch experiments with a sharp Vickers indenter on BK7, a brittle glass, showed that at above a critical load, the damage caused by the first scratch caused increased material removal when the second scratch was ~60 μm or closer (Gu et al., 2011). Distributed arrays of parallel scratches are a related type of repetitive scratch test where damage from previous scratches influences subsequent scratches so that the accumulated damage builds up and the cycle-by-cycle development of an abraded region of the surface can be studied (Gant et al., 2017; Gee et al., 2017). Da Silva and de Mello (2009) have reported that the surface appearance after the test could closely resemble a worn surface after abrasion.

Repetitive constant load scratching in the same track has also proved effective in micro- and nano-scratch testing. In comparison with progressive load nanoscratch testing, repetitive constant load nano-scratch testing has the benefit that the applied load can be controlled to locate the maximum von Mises stress to be close to the coating − substrate interface and minimize substrate deformation as a precursor of coating debonding. In contrast to the gradual coating wear reported by Efeoglu and Arnell (2000), the failure in the nano-scratch test can be more extensive (cohesive and/or adhesive) and occur during a single or few scratches (Beake, Liskiewicz et al., 2011; Beake, Shi et al., 2011). Repetitive scratch tests can be more sensitive to adhesion differences than single-pass scratch tests.

Wear does not typically proceed at a constant rate in nano- or microscale sphere-on-flat tribological tests due to opposing effects from the changing contact geometry. The initially high contact pressure can rapidly decrease through deformation and wear increasing the contact depth and contact area. For a spherical probe, this increase in depth changes the angle between the probe and the sample in contact (the "attack angle")—which increases the contact strain, which can influence the predominant wear mechanism. This is illustrated in Fig. 2.5. Kato and coworkers observed that in repetitive scratch tests of steels with a 30 μm tungsten carbide indenter that increasing load (and hence contact strain) led to changes in the predominant deformation mode from plowing at low load to wedge formation and then finally to cutting at attack angles of >40 degrees (Hokkirigawa et al., 1998;

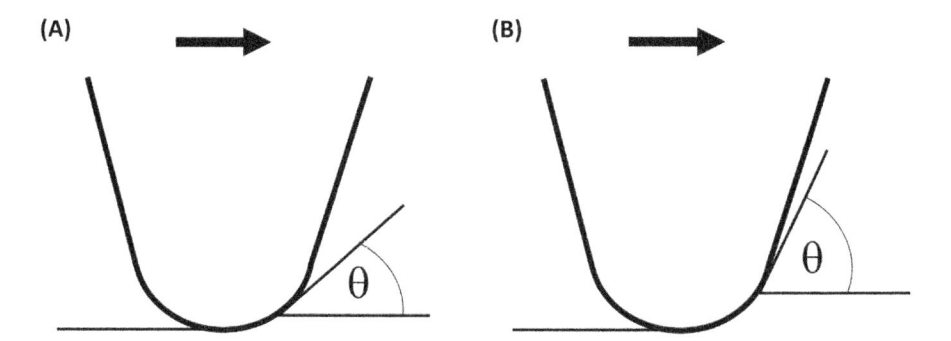

Figure 2.5 Probe attack angle for various tip penetration depths. Influence of probe depth on the attack angle and contact strain of a spheroconical probe; (A) lower attack angle at smaller depth (B) higher attack angle at greater depth.

Kitsunai et al., 1990). Coating systems are necessarily more complex. Kato (1995) noted that the predominant deformation mode changes with cycles.

In a repetitive sliding contact, the applied load controls the depth of the peak von Mises stresses. In general, the competition between plastic deformation and fracture dominated wear is dependent on the severity of the test, and in some cases, the relative ranking of coatings can change with wear load. An illustration of this is provided by an investigation into the nanoscratch behavior of 1 μm a-C PVD films deposited on silicon at -20 to -120 V substrate bias (Shi et al., 2008). Films deposited under higher bias were harder and more highly stressed. The progressive load nanoscratch test was relatively insensitive to this since films deposited with up to -100V bias all failed at a critical load of ~ 200 mN using a 4 μm diamond probe, although the critical load was under 100 mN when deposited at -120 V. Repetitive constant load scratch tests were performed at 50 mN (which was above L_y) and 150 mN (which was above L_{c1}). The contact was nearly elastic at 50 mN, with smaller residual wear depths found for harder a-C films deposited at high bias. Maximum von Mises stresses were located within the a-C coatings at 50 mN, so their mechanical properties dominated their wear resistance. At 150 mN, the developed von Mises stresses were located further below the contact and were largest close to the coating — substrate interface. The more severe loading conditions resulted in coating fracture and harder films deposited under high substrate bias performed poorly with extensive delamination outside of the scratch track within a few repetitive passes. There was a direct correlation between the substrate bias and the delamination area.

2.3 AFM scratch and wear — even smaller length scale/contact size

By further reducing the contact size, it is possible to study the fundamentals of friction and wear. TEM wear studies and molecular dynamics simulations of asperity-level contacts have been used to investigate fundamental wear mechanisms (Jacobs et al., 2019; Liao et al., 2015; Liao & Marks, 2015, 2016; Milne et al., 2020). Jacobs and Carpick (2013) used a TEM nanoindenter to perform studies of silicon wear against diamond, which was found by direct imaging to occur by atom-by-atom removal, which they described as atomic attrition, rather than plastic deformation or fracture. Sliding experiments performed in TEMs have been reviewed by Liao and Marks (2016). Liao et al. (2015) showed that the attack angle was critical in controlling the deformation behavior in nanoscale abrasive dry sliding of the fcc phase of CoCrMo against a silicon tip in vacuum in a TEM. Milne et al. (2020) found that after sliding, the adhesion between nanoscale silicon contacts increased. They hypothesized that passivating terminal species such as hydrogen or hydroxyl groups were temporarily removed by sliding. Combining small-radius probe tips, high resolution, low noise floor, and experimental convenience AFMs has been a popular choice for reciprocating single-asperity tribological tests (Degiampietro & Colaço, 2007; Szlufarska et al., 2008). A typical method is to image the surface under light load, perform a reciprocating wear test within a smaller region (e.g., a line scan or a "wear-box"), and then reimage the worn surface at the same size as the initial image so that the residual wear depth or volume can be measured. Degiampietro and Colaço (2007) performed wear tests of 2 μm × 2 μm regions of 316 L stainless steel with a 250 nm radius diamond tip at 2−20 μN. The worn volume increased linearly with the number of interactions between the surface and the abrasive tip and with the applied load but decreased with increasing scan velocity. The influence of the native oxide on wear has been studied. Mitchell and Shrotriya (2007) performed 70 pass wear experiments on CoCr with a 50 nm radius Si_3N_4 AFM probe. Wear tests at 2−3 nN (equivalent to 0.9 GPa contact pressure) were elastic, but above 25 nN ($\geq$2.3 GPa), the thin passive oxide could be damaged, resulting in wear. AFM abrasive wear studies have revealed bilinear behavior with a transition to a faster material removal rate occurring at a critical force. Celano et al. (2018) showed bilinear behavior in AFM wear experiments on semiconductors (Si, Ge, SiGe) vs. B-doped diamond. At low load, there was low-wear sliding, which transitioned to a higher wear rate sliding-plowing process as the load increased. They were able to fit this behavior by applying Archard's law at low load and a combination of Archard and DMT contact mechanics above the transition. In AFM wear tests on AISI 52100 steel with a diamond tip, Walker et al. (2021) found even more abrupt transition from a low wear region described as atom attrition to an elastoplastic plowing mechanism with much higher material removal and pile-up at the sides of the scratch. In addition to these abrasive wear situations where a harder probe slides against a softer material, transitions in wear mechanism have also been investigated for adhesive wear contacts where material is removed through sticking/tearing in contacts between bodies of comparable hardness. Aghababaei et al. (2016) employed atomistic simulations to show that there exists a critical junction size, which controls whether gradual smoothing or fracture-induced debris generation occurs. The predictions of their model were in agreement with AFM experiments. At the small scale of an AFM wear experiment, it can prove difficult to accurately pinpoint the onset of wear. When wear is dominated by plastic deformation rather than material removal measures of surface roughness, such as R_a, have been proposed as a more useful indicator of the progression of damage than the wear depth (Drees et al., 2004). AFM nanowear tests are limited by: (1) short track length and small sliding distances; (2) low sliding velocity; (3) high contact pressures and susceptibility to probe wear due to the small tip radius. Since AFM is a piezo-based technology, it does not have necessary stability for performing tests of extended duration (Mitchell & Shrotriya, 2007).

2.4 Experimental techniques for nano-/microscale fretting and reciprocating wear testing

Nano/microscale tribological tests performed with a nanomechanical test instrument typically have (1) larger probe radii (e.g., in the micron range, which minimizes the indenter wear that occurs with very sharp probes); (2) larger available force range; (3) longer sliding distance; (4) higher sliding velocity; and (5) better instrumental stability to run longer tests. Of particular benefit is the ability to perform reciprocating tests with a significantly larger number of test cycles than in AFM scratch or nanoindenter-based nanoscratch tests, enabling fatigue processes to be studied.

In nano-/microscale wear tests, the transitions between different wear regimes can be investigated by changing the experimental conditions (e.g., applied load, probe geometry, sliding distance) that combine to alter the contact size or length scale of the contact. The length scale of the contact can be considered as the "information depth" where the tribological response is dominated by the properties within this depth (where developed stresses can be high) and relatively unaffected by the mechanical properties below this (where developed stresses are much lower). By changing the track length, it is possible to test under gross slip or partial fretting conditions.

The behavior of solid lubricant coatings in reciprocating friction and wear tests at the mN force range has been investigated using nanoindenters and microtribometers. Nanoindenters have been modified by the addition of sample stages to enable the reciprocating motion over small and/or large track distances. The residual wear depth in these nano/microscale tests includes contributions from material removal and plastic deformation. Correlation between friction and the evolution of elastic, plastic deformation and wear in reciprocating testing of DLC and Si-doped DLC coatings on glass has been studied by Schiffmann (2004). By performing low single-cycle nanoindentation tests to the same loads—where the contribution of wear is minimal and the residual depth is due to plastic deformation—he was able to separate out the individual contributions of plastic deformation and material removal to the residual depth in the wear tests. His idea was subsequently used by Stoyanov and Chromik (2017) in tests on Ti-MoS$_2$. In these reciprocating tests, cracking may occur through low cycle fatigue at loads well below those needed to produce cracking in a single cycle test. The periodic loading can result in accumulation of plastic deformation and/or densification, which gradually increases the subsurface stress in the coating ultimately leading to microcracking (Schiffmann & Hieke, 2003; Schiffmann, 2004). To address reliability issues in microswitches, Stoyanov and coworkers investigated the microtribological properties of Au, Au-MoS$_2$ composite, and bilayer coatings with a sacrificial MoS$_2$ layer on top of the Au (Stoyanov et al., 2010, 2012). The composite and bilayer coatings showed improved behavior in the reciprocating tests than the monolayer Au. The lubrication strategy was for the MoS$_2$ layer to fail rapidly enabling electrical conductivity but leaving a thin MoS$_2$-based layer, which would promote tribofilm formation. Higher wear resistance for the Au-MoS$_2$ was attributed to a reduction in surface adhesion and differences in velocity accommodation modes (Stoyanov & Chromik, 2017). Wilson and coworkers employed an oscillating stage with a multilayer piezo-stack to generate reciprocating motion in a commercial nanoindenter (NanoTest system). The stability to run reciprocating (nanofretting) tests to over 200,000 cycles was shown (Wilson & Sullivan, 2009; Wilson et al., 2008). In reciprocating tests with a 150 μm radius ruby tip under 10−200 mN applied load and 2−14 μm displacement amplitude, the transition from fretting/partial slip to gross slip of Cr-doped amorphous carbon films was investigated by changing track length and applied load. The authors identified two distinct fretting wear regimes, with a W-shaped wear scar under low oscillation amplitude and a U-shaped wear scar produced at larger amplitudes. Deformation and failure mechanisms in nanofretting tests may differ from those in nanoscratch tests. For example, gradual changes in depth and friction were observed in nanofretting tests of thin ta-C films on silicon (Beake et al., 2013). In contrast, abrupt increases in depth and friction were observed at L_{c2} failure in ramped load scratch tests.

Reciprocating motion can also be generated in the same instrumentation using (1) SPM nanopositioning/imaging stage for track lengths below 100 μm (2) or by addition of a microtribology stage ("NanoTriboTest") for track lengths of up to 10 mm. The latter approach has been used to perform tests on biomedical alloys and sliding interconnector materials (Beake, Harris et al., 2021). By using electrically conductive metallic probes, it was possible to improve detection of the onset of wear and the subsequent failure mechanisms with a multisensing approach simultaneously monitoring friction and electrical contact resistance. Higher sliding speeds and larger sliding distances than used in other microtribological tests have enabled longer duration high or low contact pressure tests, for example, up to 35,000 cycles on an metallic interconnector. It has also been used to study the microscale friction and wear of DLC coatings in reciprocating tests over sliding distances that were large in comparison to those normally used in microscale testing. Friction coefficients between DLC coatings and diamond probes differed significantly between nanofretting and microtribological tests (Beake et al., 2013; Beake, McMaster et al., 2021). Higher friction in the nanoscale tests, particularly on the softest coating. was due to the greater plowing in the nanoscale test.

2.5 Repetitive contact – nano and micro-impact tests

Many materials undergo additional fatigue deformation mechanisms under cyclic loading that are not found for single-cycle tests (Kim et al., 1999; Ramírez et al., 2012). This is true even when the load is kept constant for an extended period in the latter to exclude the possibility of slow crack growth (Guiberteau et al., 1993). Cyclic impact tests are used as model tests for assessing coating durability under dynamic loading (Bantle & Matthews, 1995; Bouzakis et al., 2004; Yoon et al., 2004). Varying the impact energy and the geometry of the test probe controls the severity of an impact test on coating system and positions of peak impact-induced stresses relative to the coating−substrate interface.

Knotek et al. (1992) originally developed a ball-on-plate impact test to be used with large forces and mm-sized probes, which was effective at producing both adhesive and cohesive coating failures. The failure mechanisms of PVD coatings under repetitive impact from 3 mm radius hardened steel or tungsten carbide hard metal indenters were investigated by Bantle and Matthews (1995). Under the plastic strain from the deformation, a network of macrocracks in the coatings developed from the low yield stress steel substrate reducing the stress in the coating. The resultant crack growth was followed by coating delamination. Three different zones in the impact craters were identified, which are (1) a central zone of cohesive failure, (2) an intermediate zone of cohesive − adhesive failure, (3) a peripheral zone with circular cracks. Tests are usually performed at 90 degree angle. Gsellmann et al. (2020) recently reported an inclined impact testing method with the sample being inclined at 15 degree so that tests could be performed with 75 degree impact angle instead. In tests on a TiN coating on high-speed steel, near-interface cracking was observed.

In coating systems, impact fatigue mechanisms can vary with the ratio of coating thickness t to the indenter radius R, (t/R) (Michler & Blank, 2001). Macro, micro, and nano-impact tests with different contact sizes—controlled by applied load and indenter sharpness—can be used to obtain data over a range of t/R. The major differences between the experimental conditions in the nano-, micro-, and macroscale impact tests and the information that can be obtained in each are summarized in Table 2.1.

In macroscale tests of thin PVD coatings with cemented carbide or hardened steel spherical indenters with 1−3 mm end radius, the t/R values are of the order of ≈ 0.001. Under these conditions, the fatigue behavior can be strongly influenced by the substrate properties as the induced peak von Mises stresses that result in plastic deformation are found far from the interface and well into the substrate. With larger probe sizes, the results become progressively less sensitive

Table 2.1 Main features and differences between macro-, micro-, and nanoscale impact tests.

	Macro-impact	Micro-impact	Nano-impact
Depth-sensing test	N	Y	Y
Accurate cycles-to-failure	N	Y	Y
Test duration	Extended duration	5−10 min	5−10 min
Test probe material	WC-Co, hardened steel	Diamond	Diamond
Test probe radius	1−3 mm	5−100 μm	∼100 nm
Coating thickness ÷ test probe radius*	∼0.001	∼0.1	∼10
Sensitivity to coating mechanical properties	can be low	High	High
Sensitivity to adhesion	can be low	High	Medium
Automatic scheduling of multiple tests	N	Y	Y
Applied load	>> 100 N	0.1−5 N	0.001−0.2 N

Typical conditions. *for thin PVD coatings.

to coating properties and more strongly influenced by substrate hardness and toughness, as has also been reported in erosion testing under severe conditions (Bromark et al., 1995). These macroscale impact tests have some limitations. They are necessarily restricted to low t/R, and the exact number of cycles-to-failure is not recorded since the tests are usually not depth-sensing. Without a depth-sensing capability to monitor damage progression during the test, it is only possible to determine whether a coating has survived or failed after the test is ended. The failure criterion can be the failed area at a given number of cycles or the coating wear depth at the end of the test. Since the deformation substrate deformation may not be fully elastic, accurately deconvoluting the coating wear from substrate plasticity requires cross-sectioning, for example, by focused ion beam milling. Since the test probe has mm dimensions, the impact fatigue response of the coating system is averaged out over a similar sized area of the coating surface, making the test insensitive to point-to-point variations, for example, where the fatigue behavior varies across the sample surface.

An alternative approach to determining coating fatigue resistance is to perform nano- or microscale impact tests at higher t/R with sharper probes. Nano-impact testing utilizes the depth-sensing capability of a multifunctional nanomechanical test system (NanoTest system, Micro Materials Ltd.) to perform impact testing at strain rates that are several orders of magnitude higher than those in quasistatic indentation tests (Rueda-Ruiz et al., 2020; Somekawa & Schuh, 2012; Trelewicz & Schuh, 2008). These accelerated, high strain rate tests are of much shorter duration than macroscale tests and subject coatings to more severe conditions that replicate the high stresses generated in actual operating conditions. The high strain rate contact in nanoimpact tests can provide closer simulation of the performance of coatings systems under highly loaded intermittent contact and the evolution of wear under these conditions than tests at a lower strain rate. The importance of the strain rate on the fatigue failure of coatings has been highlighted by Bouzakis and coworkers, who have investigated the effect of strain rate on the fatigue failure of $Al_{60}Ti_{40}N$ coating on cemented carbide. These authors found that even a relatively modest increase in strain rate resulted in a lower fatigue endurance limit (Bouzakis, Maliaris et al., 2012).

Due to the smaller load and probe radius, the location of the generated peak von Mises stresses relative to the coating — substrate interface is completely different in the nano- and macroscale tests. Nanoimpact tests can be very sensitive to small differences in coating properties and have shown excellent correlation to coating performance in applications involving repetitive contact. In particular, there have been several studies on (Ti,Al)N-based PVD coatings on cemented carbide that have reported strong correlation between the wear of coated tools in high-speed machining applications and the fracture found in the nanoimpact test (Beake et al., 2007; Beake, Fox-Rabinovich et al., 2009; Beake, Goodes et al., 2009; Bouzakis et al., 2011, 2012). Sharp cube corner indenters (e.g., $R \sim 50-100$ nm) with high contact strain are commonly used to generate fracture rapidly even in hard and tough coatings. FIB cut cross sections through nanoimpact test craters on monolayered TiN and multilayered TiAlSiN PVD coatings on hardened steel showed that the chipping of the coatings did not extend to the interface, and there was no delamination/cracking found at the interface for either coating (Chen, Ji et al., 2011; Chen, Yang et al., 2011).

The small-scale tests provide more localized assessment of impact resistance so they are particularly well suited to testing the behavior of thin coatings and/or small volumes and in investigations of the influence of nano/microstructure on fatigue performance. Chen et al. (2012) found that the influence of thermal aging on the solid particle erosion testing of columnar EB-PVD thermal barrier coatings for aeroengines correlated with rapid nanoimpact tests. Worse coating performance was found after aging in both tests due to sintering between columns, with the correlation between them considered to be due to the similar contact footprints in both types of test. Since the exact number of cycles needed to cause coating failure is known precisely, it is easier than in macroscale tests to investigate the behavior of more closely related coatings (i.e., the technique has greater sensitivity). A related technique to nano-impact is nanoindentation fatigue (or cyclic indentation) where the load is continually cycled between two load levels while the probe stays in contact throughout the test (Islam et al., 2021). In this test, it is also possible to produce S-N curves with small radius probes and contact size, albeit at lower strain rate. Faisal et al. (2014) compared the behavior of 100 nm carbon films on silicon in the two types of test finding that both tests were effective ways to cause coating fatigue failure. The fatigue process was more efficient in the impact test, with fewer load cycles to failure.

The micro-impact test, employing higher loads (0.5−5 N) and larger probe sizes (e.g., $R \sim 20$ μm) than in nano-impact, enables tests to be performed at intermediate t/R between the nano- and macroscale tests. Due to the higher loads being accelerated over larger distances, the maximum impact energy that can be delivered in the microimpact test is $\times 100$ greater than in nanoimpact. Increasing the energy delivered per impact allows blunter indenter geometries to be used to induce coating failure within short experimental timescales. Spherical indenters have an intrinsic suitability for studying gradual fatigue processes (Guiberteau et al., 1993). With a suitable choice of microrange load, together with probe geometries similar to those used in microscratch testing, it is possible to position high stresses in the vicinity of the interface.

Both coating and substrate deformation can be important factors in the impact fatigue behavior under these conditions, and coatings can be subjected to high bending stresses. The coating behavior can show strong load sensitivity. This is illustrated for a graded a-C:H coating on M42 tool steel in Fig. 2.6. The on-load depth contains a significant elastic deformation contribution from the elastic bending, with clear overload failure observed at high load and a low-cycle fatigue behavior at lower load. To conveniently compare tests performed under different applied load, or for coatings deposited on substrates with different load-carrying capacity, the change in depth after the initial impact ($h-h_1$) can provide a more useful indication of the extent of coating failure than the final impact depth. Tracking the variation in the increase of depth with continued impacts allows the deformation behavior at different loads to be investigated in more detail (Fig. 2.7).

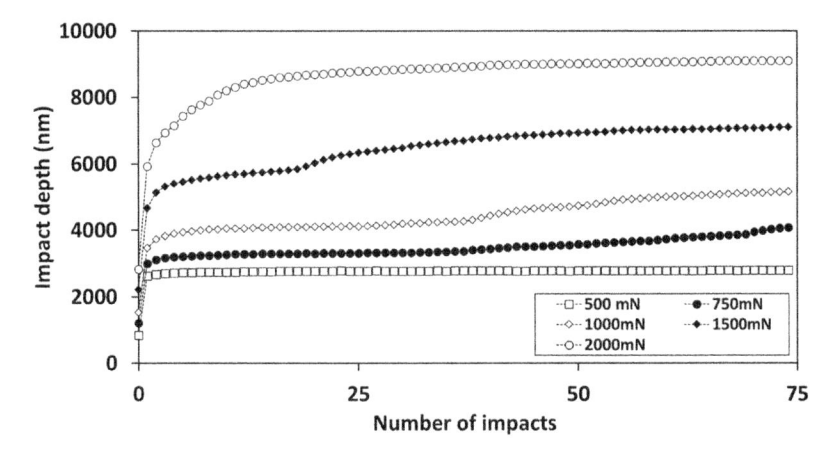

Figure 2.6 Impact depth as a function of the number of impacts. Impact behavior of a graded a-C:H coating on M42 tool steel: variation in depth with continued impact.

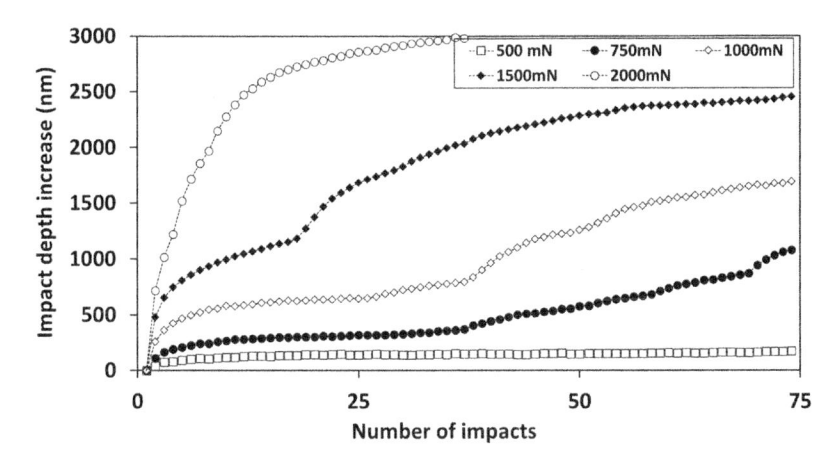

Figure 2.7 Impact depth increase as a function of the number of impacts. Impact behavior of a graded a-C:H coating on M42 tool steel: change in depth after the initial impact $(h-h_1)$ vs impact cycles.

2.6 Conclusions

The development of instrumented small-scale wear test methods that simulate contact conditions so that the major deformation mechanism can be reproduced has improved our fundamental understanding of wear processes. Scratch tests have been effective in simulating abrasive wear and impact tests in simulating erosion. The tests are typically much shorter than macroscale tests or field trials, and since the relative wear ranking can show excellent correlation to application performance, they can be effective screening coating tools. A major advantage in comparison to bulk-scale testing is the simplified contact geometry in the nano-/microscale tests, which allows modeling of the stresses involved and determination of the deformation mechanisms. Multisensing approaches where monitoring of multiple signals during the test (including wear depth, friction, acoustic emission, and electrical contact resistance) together with posttest microscopic analysis of the wear tracks can provide further information about the onset and progression of wear processes at a highly localized scale (as discussed in more detail in Chapter 11).

Predicting the performance of coating systems from their behavior in these small-scale wear tests is a more direct approach than attempting to predict relative ranking from mechanical properties obtained by nanoindentation. A wide load range available conveniently enables different sharpness indenters to be used. The contact size and hence length scale of the measurements can be tuned to interrogate either coating-dominated or coating – substrate system behavior. Small-scale impact tests have been particularly effective in predicting relative coating performance in highly loaded repetitive contacts such as cutting tests (especially milling operations) and erosion.

When designing coatings for high wear resistance, it is important to consider that the optimum coating mechanical properties can vary with the severity and type of mechanical contact. For example, the success or otherwise of doping DLC coatings with W has been found to vary with the type of mechanical contact. In highly loaded microscale scratch and impact tests, W-doped DLC performs well due to its higher crack resistance, but its relatively low hardness leads to higher wear in reciprocating sliding than was found on harder and more brittle undoped DLC coatings (Bai et al., 2021; Beake, McMaster et al., 2021).

References

Achanta, S., Drees, D., & Celis, J.-P. (2008). Friction from nano to macroforce scales analysed by single and multiple-asperity contact approaches. *Surface and Coatings Technology, 202*, 6127–6135.

Aghababaei, R., Warner, D. H., & Molinari, J.-F. (2016). Critical length scale controls adhesive wear mechanisms. *Nature Communications, 7*, 11816.

Ahmed, R., Ashraf, A., Elameen, M., Faisal, N. H., El-Sherik, A. M., Elakwah, Y. O., & Goosen, M. F. A. (2014). Single asperity nanoscratch behaviour of HIPed and cast Stellite 6 alloys. *Wear, 312*(1–2), 70–82. Available from https://doi.org/10.1016/j.wear.2014.02.006.

Akono, A. T., & Ulm, F. J. (2014). An improved technique for characterizing the fracture toughness via scratch test experiments. *Wear, 313*(1–2), 117–124. Available from https://doi.org/10.1016/j.wear.2014.02.015, https://www.journals.elsevier.com/wear.

Aldrich-Smith, G., Jennett, N.M., Housden, J. (2004). *A round robin to measure the adhesion of thin coatings, NPL Report DEPC-MPE 001.* National Physical Laboratory, Teddington, UK.

ASTM C1624-05. (2015). *Standard test method for adhesion strength and mechanical failure modes of ceramic coatings by quantitative single point scratch testing.*

Bai, M., Yang, L., Li, J., Luo, L., Sun, S., & Inkson, B. (2021). Mechanical and tribological properties of Si and W doped diamond like carbon (DLC) under dry reciprocating sliding conditions. *Wear, 484−485*, 204046.

Bantle, R., & Matthews, A. (1995). Investigation into the impact wear behaviour of ceramic coatings. *Surface and Coatings Technology, 74−75*(2), 857−868. Available from https://doi.org/10.1016/0257-8972(95)08314-6.

Batista, J. C. A., C. Godoy, C., & Matthews, A. (2003). Impact testing of duplex and non-duplex (Ti,Al)N and Cr−N PVD coatings. *Surface and Coatings Technology, 163−164*, 353−361.

Beake, B. D., Davies, M. I., Liskiewicz, T. W., Vishnyakov, V. M., & Goodes, S. R. (2013). Nano-scratch, nanoindentation and fretting tests of 5−80 nm ta-C films on Si(100. *Wear, 301*(1−2), 575−582. Available from https://doi.org/10.1016/j.wear.2013.01.073.

Beake, B. D., Endrino, J. L., Kimpton, C., Fox-Rabinovich, G. S., & Veldhuis, S. C. (2017). Elevated temperature repetitive micro-scratch testing of AlCrN, TiAlN and AlTiN PVD coatings. *International Journal of Refractory Metals and Hard Materials, 69*, 215−226. Available from https://doi.org/10.1016/j.ijrmhm.2017.08.017.

Beake, B. D., Fox-Rabinovich, G. S., Veldhuis, S. C., & Goodes, S. R. (2009). Coating optimisation for high-speed machining with advanced nanomechanical test methods. *Surface and Coatings Technology, 203*, 1919−1925.

Beake, B. D., Goodes, S. R., & Shi, B. (2009). Nanomechanical and nanotribological testing of ultra-thin carbon-based and MoST films for increased MEMS durability. *Journal of Physics D: Applied Physics, 42*, 065301.

Beake, B. D., Harris, A. J., Liskiewicz, T. W., Wagner, J., McMaster, S. J., Goodes, S. R., Neville, A., & Zhang, L. (2021). Friction and electrical contact resistance in reciprocating nano-scale wear testing of metallic materials. *Wear, 474−475*, 203886.

Beake, B. D., Leggett, G. J., & Shipway, P. H. (2001). Nanotribology of biaxially oriented poly(ethylene terephthalate) film. *Polymer, 42*, 7025.

Beake, B. D., & Liskiewicz, T. W. (2017). Nanomechanical characterization of carbon films. In D. A. Tiwcari, & S. Natarajan (Eds.), *Applied Nanoindentation inAdvanced Materials* (pp. 19−68). Wiley.

Beake, B. D., Liskiewicz, T. W., Pickford, N. J., & Smith, J. F. (2012). Accelerated nano-fretting testing of Si(100). *Tribology International, 46*, 114−118.

Beake, B. D., Liskiewicz, T. W., & Smith, J. F. (2011). Deformation of Si(100) in spherical contacts - Comparison of nano-fretting and nano-scratch tests with nano-indentation. *Surface and Coatings Technology, 206*, 1921−1926.

Beake, B. D., McMaster, S. J., Liskiewicz, T. W., & Neville, A. (2021). Influence of Si- and W- doping on micro-scale reciprocating wear and impact performance of DLC coatings on hardened steel. *Tribology International, 160*, 107063.

Beake, B. D., Shi, B., & Sullivan, J. L. (2011). Nanoscratch and nanowear testing of TiN coatings on M42 steel. *Tribology − Materials, Surfaces & Interfaces, 5*, 141.

Beake, B. D., Shipway, P. H., & Leggett, G. J. (2004). Influence of mechanical properties on the nanowear of uniaxially oriented poly(ethylene terephthalate) film. *Wear, 256*, 118.

Beake, B. D., Smith, J. F., Gray, A., Fox-Rabinovich, G. S., Veldhuis, S. C., & Endrino, J. L. (2007). Investigating the correlation between nano-impact fracture resistance and hardness/modulus ratio from nanoindentation at 25−500°C and the fracture resistance and lifetime of cutting tools with Ti1 − xAlxN (x = 0.5 and 0.67) PVD coatings in milling operations. *Surface and Coatings Technology, 201*(8), 4585−4593. Available from https://doi.org/10.1016/j.surfcoat.2006.09.118.

Beake, B. D., Vishnyakov, V. M., Valizadeh, R., & Colligon, J. S. (2006). Influence of mechanical properties on the nanoscratch behaviour of hard nanocomposite TiN/Si3N4 coatings on Si. *Journal of Physics D: Applied Physics, 39*(7), 1392–1397. Available from https://doi.org/10.1088/0022-3727/39/7/009.

Bell, T., Dong, H., & Sun, Y. (1998). Realising the potential of duplex surface engineering. *Tribology International, 31*, 127.

Bhushan, B. (1996). Contact mechanics of rough surfaces in tribology: Single asperity contact. *Applied Mechanics Reviews, 49*(5), 275–298. Available from https://doi.org/10.1115/1.3101928.

Bouzakis, K.-D., Klocke, F., Skordaris, G., Bouzakis, E., Gerardis, S., Katirtzoglou, G., & Makrimallakis, S. (2011). Influence of dry micro-blasting grain quality on wear behaviour of TiAlN coated tools. *Wear, 271*(5–6), 783–791. Available from https://doi.org/10.1016/j.wear.2011.03.010.

Bouzakis, K.-D., Maliaris, G., & Makrimallakis, S. (2012). Strain rate effect on the fatigue failure of thin PVD coatings: An investigation by novel impact tester with adjustable repetitive force. *International Journal of Fatigue, 44*, 89–97.

Bouzakis, K. D., Michailidis, N., Skordaris, G., Bouzakis, E., Biermann, D., & M'Saoubi, R. (2012). Cutting with coated tools: Coating technologies, characterization methods and performance optimization. *CIRP Annals − Manufacturing Technology, 61*(2), 703–723. Available from https://doi.org/10.1016/j.cirp.2012.05.006.

Bouzakis, K.-D., Siganos, A., Leyendecker, T., & Erkens, G. (2004). Thin hard coatings fracture propagation during the impact test. *Thin Solid Films, 460*(1–2), 181–189. Available from https://doi.org/10.1016/j.tsf.2004.02.009.

Bromark, M., Hedenqvist, P., & Hogmark, S. (1995). The influence of substrate material on the erosion resistance of TiN coated steels. *Wear, 186–187*, 189–194.

Bull, S. J. (1991). Failure modes in scratch adhesion testing. *Surface and Coatings Technology, 50*(1), 25–32. Available from https://doi.org/10.1016/0257-8972(91)90188-3.

Bull, S. J. (1997). Failure mode maps in the thin film scratch adhesion test. *Tribology International, 7*, 491–498.

Bull, S. J., & Rickerby, D. S. (1989). Multi-pass scratch testing as a model for abrasive wear. *Thin Solid Films, 181*(1–2), 545–553. Available from https://doi.org/10.1016/0040-6090(89)90523-3.

Bull, S. J., Rickerby, D. S., Matthews, A., Leyland, A., Pace, A. R., & Valli, J. (1988). The use of scratch adhesion testing for the determination of interfacial adhesion: The importance of frictional drag. *Surface and Coatings Technology, 36*(1–2), 503–517. Available from https://doi.org/10.1016/0257-8972(88)90178-8.

Burnett, P. J., & Rickerby, D. S. (1987). The relationship between hardness and scratch adhesion. *Thin Solid Films, 154*(1–2), 403–416. Available from https://doi.org/10.1016/0040-6090(87)90382-8.

Celano, U., Hsia, F. C., Vanhaeren, D., Paredis, K., Nordling, T. E. M., Buijnsters, J. G., Hantschel, T., & Vandervorst, W. (2018). Mesoscopic physical removal of material using sliding nano-diamond contacts. *Nature Publishing Group, Belgium Scientific Reports, 8*(1). Available from https://doi.org/10.1038/s41598-018-21171-w, http://www.nature.com/srep/index.html.

CEN/TS. (2021). 17629:2021 Nanotechnologies − Nano- and micro-scale scratch testing.

Chen, J., Beake, B. D., Wellman, R. G., Nicholls, J. R., & Dong, H. (2012). An investigation into the correlation between nano-impact resistance and erosion performance of EB-PVD thermal barrier coatings on thermal ageing. *Surface and Coatings Technology, 206*(23), 4992–4998. Available from https://doi.org/10.1016/j.surfcoat.2012.06.011.

Chen, J., Ji, R., Khan, R. H. U., Li, X., Beake, B. D., & Dong, H. (2011). Effects of mechanical properties and layer structure on the cyclic loading of TiN-based coatings. *Surface and Coatings Technology, 206*, 522−529.

Chen, L., Yang, M., Yu, J., Qian, L., & Zhou, Z. (2011). Nanofretting behaviours of ultrathin DLC coating on Si(100) substrate. *Wear, 271*, 1980−1986.

Colaço, R. (2009). An AFM study of single-contact abrasive wear: The Rabinowicz wear equation revisited. *Wear, 267*, 1772−1776.

Degiampietro, K., & Colaço, R. (2007). Nanoabrasive wear induced by an AFM diamond tip on stainless steel. *Wear, 263*, 1579−1584.

Diao, D. F., Kato, K., & Hayashi, K. (1994). The maximum tensile stress on a hard coating under sliding friction. *Tribology International, 27*, 267−272.

Dieterich, J. H., & Kilgore, B. D. (1994). Direct observation of frictional contacts: New insights for state-dependent properties. *Pure and Applied Geophysics PAGEOPH, 143*(1−3), 283−302. Available from https://doi.org/10.1007/BF00874332.

Drees, D., Celis, J.-P., & Achanta, S. (2004). Friction of thin coatings on three length scales under reciprocating sliding. *Surface and Coatings Technology, 188−189*, 511−518.

Efeoglu, I., & Arnell, R. D. (2000). Multi-pass sub-critical load testing of titanium nitride coatings. *Thin Solid Films, 377−378*, 346−353. Available from https://doi.org/10.1016/s0040-6090(00)01309-2.

Faisal, N. H., Ahmed, R., Goel, S., & Fu, Y. Q. (2014). Influence of test methodology and probe geometry on nanoscale fatigue failure of diamond-like carbon film. *Surface and Coatings Technology, 242*, 42−53. Available from https://doi.org/10.1016/j.surfcoat.2014.01.015.

Film Doctor. (n.d.) SIOMEC. http://www.siomec.de/FilmDoctor.

Gant, A. J., Nunn, J. W., Gee, M. G., Gorman, D., Gohil, D. D., & Orkney, L. P. (2017). New perspectives in hardmetal abrasion simulation. *Wear, 376−377*, 2−14. Available from https://doi.org/10.1016/j.wear.2017.01.038.

Garabedian, N. T., Khare, H. S., Carpick, R. W., & Burris, D. L. (2009). AFM at the macroscale: Methods to fabricate and calibrate probes, for millinewton force measurements. *Tribology Letters, 67*, 21.

Gee, M. G. (2001). Low load multiple scratch tests of ceramics and hard metals. *Wear, 250*(1−12), 264−281. Available from https://doi.org/10.1016/s0043-1648(01)00591-9.

Gee, M. G., Gant, A., & Roebuck, B. (2007). Wear mechanisms in abrasion and erosion of WC/Co and related hardmetals. *Wear, 263*, 137−148.

Gee, M. G., & Gee, A. D. (2007). A cost effective test system for micro-tribology experiments. *Wear, 263*(7−12), 1484−1491. Available from https://doi.org/10.1016/j.wear.2006.12.042.

Gee, M. G., Mingard, K., Nunn, J., Roebuck, B., & Gant, A. (2017). In situ scratch testing and abrasion simulation of WC/Co. *International Journal of Refractory Metals and Hard Materials, 62*, 192−201.

Gee, M. G., Nunn, J. W., Muniz-Piniella, A., & Orkney, L. P. (2011). Micro-tribology experiments on engineering coatings. *Wear, 271*(9−10), 2673−2680. Available from https://doi.org/10.1016/j.wear.2011.02.031.

Greenwood, J. A., & Williamson, J. P. (1966). Contact of nominally flat surfaces. *Proceedings of the Royal Society of London, 295*, 300−319.

Gsellmann, M., Klünsner, T. K. ü, Mitterer, C., Marsoner, S., Skordaris, G., Bouzakis, K., Leitner, H., & Ressel, G. (2020). Near-interface cracking in a TiN coated high speed steel due to combined shear and compression under cyclic impact loading. *Surface and Coatings Technology, 394*. Available from https://doi.org/10.1016/j.surfcoat.2020.125854.

Gu, W., Yao, Z., & Liang, X. (2011). Material removal of optical glass BK7 during single and double scratches. *Wear*, *270*, 241–246.

Guiberteau, F., Padture, N. P., Cai, H., & Lawn, B. R. (1993). Indentation fatigue. *Philosophical Magazine A: Physics of Condensed Matter, Structure, Defects and Mechanical Properties*, *68*(5), 1003–1016. Available from https://doi.org/10.1080/01418619308219382.

Heinke, W., Leyland, A., Matthews, A., Berg, G., Friedrich, C., & Broszeit, E. (1995). Evaluation of PVD nitride coatings, using impact, scratch and Rockwell-C adhesion tests. *Thin Solid Films*, *270*(1–2), 431–438. Available from https://doi.org/10.1016/0040-6090(95)06934-8.

Hokkirigawa, K., Kato, K., & Li, Z. Z. (1998). The effect of hardness on the transition of the abrasive wear mechanism of steels. *Wear*, *123*, 241–251.

Holmberg, K., Laukkanen, A., Ronkainen, H., Wallin, K., Varjus, S., & Koskinen, J. (2006a). Tribological contact analysis of a rigid ball sliding on a hard coated surface. Part I: Modelling stresses and strains. *Surface and Coatings Technology*, *200*(12–13), 3793–3809. Available from https://doi.org/10.1016/j.surfcoat.2005.03.040.

Holmberg, K., Laukkanen, A., Ronkainen, H., Wallin, K., Varjus, S., & Koskinen, J. (2006b). Tribological contact analysis of a rigid ball sliding on a hard coated surface, Part II: Material deformations, influence of coating thickness and Young's modulus. *Surface and Coatings Technology*, *200*, 3810–3823.

Holmberg, K., Ronkainen, H., Laukkanen, A., & Wallin, K. (2007). Friction and wear of coated surfaces—scales, modelling and simulation of tribomechanisms. *Surface and Coatings Technology*, *202*(4–7), 1034–1049. Available from https://doi.org/10.1016/j.surfcoat.2007.07.105.

Holmberg, K., Ronkainen, H., Laukkanen, A., Wallin, K., Erdemir, A., & Eryilmaz, O. (2008). Tribological analysis of TiN and DLC coated contacts by 3D FEM modelling and stress simulation. *Wear*, *264*, 877–884.

Ichimura, H., & Ishii, Y. (2003). Effects of indenter radius on the critical load in scratch testing. *Surface and Coatings Technology*, *165*(1), 1–7. Available from https://doi.org/10.1016/s0257-8972(02)00718-1.

Ichimura, H., & Ishii, Y. (2003). Effects of indenter radius on the critical load in scratch testing. *Surface and Coatings Technology*, *165*, 1–7.

Islam, M. M., Shakil, S. I., Shaheen, N. M., Bayati, P., & Haghshenas, M. (2021). An overview of microscale indentation fatigue: Composites, thin films, coatings, and ceramics. *Micron (Oxford, England: 1993)*, *148*. Available from https://doi.org/10.1016/j.micron.2021.103110.

Jacobs, T. D. B., & Carpick, R. W. (2013). Nanoscale wear as a stress-assisted chemical reaction. *Nature Nanotechnology*, *8*(2), 108–112. Available from https://doi.org/10.1038/nnano.2012.255, http://www.nature.com/nnano/index.html.

Jacobs, T. D. B., Greiner, C., Wahl, K. J., & Carpick, R. W. (2019). Insights into tribology from in situ nanoscale experiments. *MRS Bulletin*, *44*(6), 478–486. Available from https://doi.org/10.1557/mrs.2019.122, http://journals.cambridge.org/MRS.

Jiang, J., & Arnell, R. D. (2000). The effect of substrate surface roughness on the wear of DLC coatings. *Wear*, *239*, 1–9.

Kato, K. (1995). Microwear mechanisms of coatings. *Surface and Coatings Technology*, *76–77*(2), 469–474. Available from https://doi.org/10.1016/0257-8972(95)02570-7.

Kim, D. K., Jung, Y.-G., Peterson, I. M., & Lawn, B. R. (1999). Cyclic fatigue of intrinsically brittle ceramics in contact with spheres. *Acta Materialia*, *47*, 4711–4725.

Kitsunai, H., Kato, K., Hokkirigawa, K., & Inoue, H. (1990). The transitions between microscopic wear modes during repeated sliding friction observed by a scanning electron microscope tribosystem. *Wear, 135*(2), 237–249. Available from https://doi.org/10.1016/0043-1648(90)90028-9.

Knotek, O., Bosserhoff, B., Schrey, A., Leyendecker, T., Lemmer, O., & Esser, S. (1992). A new technique for testing the impact load of thin films: the coating impact test. *Surface and Coatings Technology, 54/55,* 102–107.

Larsson, M., Olsson, M., Hedenqvist, P., & Hogmark, S. (2000). Mechanisms of coating failure as demonstrated by scratch and indentation testing of TiN coated HSS. *Surface Engineering, 16*(5), 436–444. Available from https://doi.org/10.1179/026708400101517350.

Laukkanen, A., Holmberg, K., Koskinen, J., Ronkainen, H., Wallin, K., & Varjus, S. (2006). Tribological contact analysis of a rigid ball sliding on a hard coated surface, Part III: Fracture toughness calculation and influence of residual stresses. *Surface and Coatings Technology, 200*(12–13), 3824–3844. Available from https://doi.org/10.1016/j.surfcoat.2005.03.042.

Leyland, A., Fancey, A. S., & Matthews, A. (1991). Plasma nitriding in a low pressure triode discharge to provide improvements in adhesion and load support for wear resistant coatings. *Surface Engineering, 7,* 207–215.

Liao, Y., Hoffman, E., & Marks, L. D. (2015). Nanoscale abrasive wear of CoCrMo in in situ TEM sliding. *Tribology Letters, 57,* 28.

Liao, Y., & Marks, L. D. (2015). Direct observation of layer-by-layer wear. *Tribology Letters, 59*(3). Available from https://doi.org/10.1007/s11249-015-0567-5, http://www.springerlink.com/(snpxut45gxflnr45vb2gia45)/app/home/journal.asp?referrer = parent&backto = searchpublicationsresults,1,2.

Liao, Y., & Marks, L. D. (2016). *International Materials Reviews,* 1–17. Available from https://doi.org/10.1080/09506608.2016.1213942.

Lin, H., Dai, R., Shi, Y., Yuan, J., & Alfano, M. (2023). Effect of the GLC coating thickness on the mechanical and tribological properties of the CrN/GLC coatings. *Tribology Letters, 71*(3). Available from https://doi.org/10.1007/s11249-023-01768-7.

Liskiewicz, T. W., Beake, B. D., Schwarzer, N., & Davies, M. I. (2013). Short note on improved integration of mechanical testing in predictive wear models. *Surface and Coatings Technology, 237,* 212.

Liu, H., Zong, W., & Cheng, X. (2020). Load- and size effects of the diamond friction coefficient at the nanoscale. *Tribology International, 68,* 120.

Matthews, A., Franklin, S., & Holmberg, K. (2007). Tribological coatings: contact mechanisms and selection. *Journal of Physics D: Applied Physics, 40,* 5463–5475.

Michler, J., & Blank, E. (2001). Analysis of coating fracture and substrate plasticity induced by spherical indentors: diamond and diamond-like carbon layers on steel substrates. *Thin Solid Films, 381*(1), 119–134. Available from https://doi.org/10.1016/s0040-6090(00)01340-7.

Milne, Z.B., Bernal R.A., Carpick, R.W., (2020). *Sliding history-dependent adhesion of nanoscale silicon contacts revealed by in situ transmission electron microscopy,* Langmuir.

Mitchell, A., & Shrotriya, P. (2007). Onset of nanoscale wear of metallic implant materials: Influence of surface residual stresses and contact loads. *Wear, 263*(7–12), 1117–1123. Available from https://doi.org/10.1016/j.wear.2007.01.068.

Ollendorf, H., & Schneider, D. (1999). A comparative study of adhesion test methods for hard coatings. *Surface and Coatings Technology*, *113*(1−2), 86−102. Available from https://doi.org/10.1016/s0257-8972(98)00827-5.

Olsson, M., & Cinca, N. (2024). Mechanisms controlling friction and material transfer in sliding contacts between cemented carbide and aluminum during metal forming. *International Journal of Refractory Metals and Hard Materials*, *118*, 106481.

Peng, L., Lee, H., Teizer, W., & Liang, H. (2009). Nanowear of gold and silver against silicon. *Wear*, *267*(5−8), 1177−1180. Available from https://doi.org/10.1016/j.wear.2008.11.021.

Petit, F., Ott, C., & Cambier, F. (2009). Multiple scratch tests and surface-related fatigue properties of monolithic ceramics and soda lime glass. *Journal of the European Ceramic Society*, *29*, 1299−1307.

Pöhl, F., Hardes, C., & Theisen, W. (2016). Scratch behavior of soft metallic materials. *AIMS Materials Science*, *3*, 390−403.

Prioli, R., Chhoawlla, M., & Freire, F. L. (2003). Friction and wear at nanometre scale: A comparative study of hard carbon films. *Diamond and Related Materials*, *12*, 2195−2202.

Pujante, J., Vilaseca, M., Casellas, D., & Riera, M. D. (2014). High temperature scratch testing of hard PVD coatings deposited on surface treated tool steel. *Surface and Coatings Technology*, *254*, 352−357. Available from https://doi.org/10.1016/j.surfcoat.2014.06.040, http://www.journals.elsevier.com/surface-and-coatings-technology/.

Qian, L., Zhou, Z., Sun, Q., & Yan, W. (2007). Nanofretting behaviors of NiTi shape memory alloy. *Wear*, *263*(1−6), 501−507. Available from https://doi.org/10.1016/j.wear.2006.12.045.

Ramírez, G., Mestra, A., Casas, B., Valls, I., Martínez, R., Bueno, R., Góez, A., Mateo, A., & Llanes, L. (2012). Influence of substrate microstructure on the contact fatigue strength of coated cold-work tool steels. *Surface and Coatings Technology*, *206*, 3069−3081.

Randall, N. X. (2019). The current state-of-the-art in scratch testing of coated systems. *Surface and Coatings Technology*, *380*. Available from https://doi.org/10.1016/j.surfcoat.2019.125092.

Randall, N. X., Favaro, G., & Frankel, C. H. (2001). The effect of intrinsic parameters on the critical load as measured with the scratch test method. *Surface and Coatings Technology*, *137*, 146−151.

Rezakhanlou, R., Billard, A., Foos, M., Frantz, C., & Von Stebut, J. (1990). Influence of the intrinsic coating properties on the contact mechanical strength of perfectly adhering carbon-doped AISI 310 PVD films. *Surface and Coatings Technology*, *43−44*(1−3), 907−919. Available from https://doi.org/10.1016/0257-8972(90)90031-7.

Rueda-Ruiz, M., Beake, B. D., & Molina-Aldareguia, J. M. (2020). New instrumentation and analysis methodology for nano-impact testing. *Materials & Design*, *192*(108715), 12p.

Sawyer, G. W., & Wahl, K. J. (2008). Accessing inaccessible interfaces: In situ approaches to materials tribology. *MRS Bulletin*, *33*(12), 1145−1150. Available from https://doi.org/10.1557/mrs2008.244.

Schiffmann, K. I. (1998). Microfriction and macrofriction of metal containing amorphous hydrocarbon hard coatings determined by AFM and pin-on-disk tests. *Tribology Letters*, *5*(1), 109−116. Available from https://doi.org/10.1023/A:1019117002779, http://www.springerlink.com/(snpxut45gxflnr45vb2gia45)/app/home/journal.asp?referrer = parent&backto = searchpublicationsresults,1,2.

Schiffmann, K. I. (2004). Phenomena in microwear experiments on metal-free and metal-containing diamond-like carbon coatings: Friction, wear, fatigue and plastic deformation.

Surface and Coatings Technology, 177−178, 453−458. Available from https://doi.org/10.1016/j.surfcoat.2003.08.064, http://www.journals.elsevier.com/surface-and-coatings-technology/.

Schiffmann, K. I. (2008). Microtribological/mechanical testing in 0, 1 and 2 dimensions: A comparative study on different materials. Wear, 265, 1826−1836.

Schiffmann, K. I., & Hieke, A. (2003). Analysis of microwear experiments on thin DLC coatings: Friction, wear and plastic deformation. Wear, 254(5−6), 565−572. Available from https://doi.org/10.1016/S0043-1648(03)00188-1.

Schwarzer, N., Duong, Q.-H., Bierwisch, N., Favaro, G., Fuchs, M., Kempe, P., Widrig, B., & Ramm, J. (2011). Optimization of the scratch test for specific coating designs. Surface and Coatings Technology, 206(6), 1327−1335. Available from https://doi.org/10.1016/j.surfcoat.2011.08.051.

Shi, B., Sullivan, J. L., & Beake, B. D. (2008). An investigation into which factors control the nanotribological behaviour of thin sputtered carbon films. Journal of Physics D: Applied Physics, 41, 045303.

da Silva, W. M., & de Mello, J. D. B. (2009). Using parallel scratches to simulate abrasive wear. Wear, 267(11), 1987−1997. Available from https://doi.org/10.1016/j.wear.2009.06.005.

Somekawa, H., & Schuh, C. A. (2012). High-strain-rate nanoindentation behavior of fine-grained magnesium alloys. Journal of Materials Research, 27(9), 1295−1302. Available from https://doi.org/10.1557/jmr.2012.52.

von Stebut, J. (2005). Multi-mode scratch testing − A European standards, measurements and testing study. Surface and Coatings Technology, 200(1−4), 346−350. Available from https://doi.org/10.1016/j.surfcoat.2005.02.055.

von Stebut, J., Rezakhanlou, R., Anoun, K., Michel, H., & Gantois, M. (1989). Major damage mechanisms during scratch and wear testing of hard coatings on hard substrates. Thin Solid Films, 181(1−2), 555−564. Available from https://doi.org/10.1016/0040-6090(89)90524-5.

Steinmann, P. A., Tardy, Y., & Hintermann, H. E. (1987). Adhesion testing by the scratch test method: Influence of intrinsic and extrinsic parameters on the critical load. Thin Solid Films, 154, 333−349.

Stoyanov, P., & Chromik, R. R. (2017). Scaling effects on materials tribology: From macro to micro scale. Materials, 10(5). Available from https://doi.org/10.3390/ma10050550, http://www.mdpi.com/1996-1944/10/5/550/pdf.

Stoyanov, P., Chromik, R. R., Gupta, S., & Lince, J. R. (2010). Micro-scale sliding contacts on Au and Au-MoS$_2$ coatings. Surface and Coatings Technology, 205(5), 1449−1454. Available from https://doi.org/10.1016/j.surfcoat.2010.07.026.

Stoyanov, P., Gupta, S., Chromik, R. R., & Lince, J. R. (2012). Microtribological performance of Au−MoS$_2$ nanocomposite and Au/MoS$_2$ bilayer coatings. Tribology International, 52, 144−152.

Szlufarska, I., Chandross, M., & Carpick, R. W. (2008). Recent advances in single-asperity nanotribology. Journal of Physics D: Applied Physics, 41(123001), 39.

Trelewicz, J. R., & Schuh, C. A. (2008). The Hall-Petch breakdown at high strain rates: Optimizing nanocrystalline grain size for impact applications. Applied Physics Letters, 93(17). Available from https://doi.org/10.1063/1.3000655.

Valli, J., Mäkelä äMäkelä ä, U., Matthews, A., & Murawa, V. (1985). TiN coating adhesion studies using the scratch test method. Journal of Vacuum Science & Technology A: Vacuum, Surfaces, and Films, 3(6), 2411−2414. Available from https://doi.org/10.1116/1.572848.

Varenberg, M., Etsion, I., & Halperin, G. (2005). Nanoscale fretting wear study by scanning probe microscopy. *Tribology Letters*, *18*, 493.

Voevodin, A. A., & Zabinski, J. S. (1998). Load-adaptive crystalline-amorphous nanocomposites. *Journal of Materials Science*, *33*(2), 319–327. Available from https://doi.org/10.1023/A:1004307426887.

Walker, J., Umer, J., Mohammadpour, M., Theodossiades, S., Bewsher, S. R., Offner, G., Bansal, H., Leighton, M., Braunstingl, M., & Flesch, H. G. (2021). Asperity level characterization of abrasive wear using atomic force microscopy. *Proceedings of the Royal Society A: Mathematical, Physical and Engineering Sciences*, *477*(2250). Available from https://doi.org/10.1098/rspa.2021.0103, http://rspa.royalsocietypublishing.org/.

Wang, C. T., Escudeiro, A., Polcar, T., Cavaleiro, A., & Wood, R. J. K. (2013). Indentation and scratch testing of DLC-Zr coatings on ultrafine-grained titanium processed by high-pressure torsion. *Wear*, *306*, 304–310.

Wang, Y., Li, H., Liu, X., Wu, Y., Lv, Y., Fu, Y., Zhou, H., & Chen, J. (2012). Friction and wear properties of graphite-like carbon films deposited on different substrates with a different interlayer under high hertzian contact stress. *Tribology Letters*, *46*, 243–254.

Wang, Y. X., & Zhang, S. (2014). Toward hard yet tough ceramic coatings. *Surface and Coatings Technology*, *258*, 1–16. Available from https://doi.org/10.1016/j.surfcoat.2014.07.007, http://www.journals.elsevier.com/surface-and-coatings-technology/.

Wang, Y. X., Zhang, S., Lee, J.-W., Lew, W. S., & Li, B. (2013). Toughening effect of Ni on nc-CrAlN/a-SiNx hard nanocomposite. *Applied Surface Science*, *265*, 418–423.

Wilson, G.M., (2008). *An investigation of thin amorphous carbon-based sputtered coatings for MEMS and micro-engineering applications*, PhD Thesis. Aston University.

Wilson, G. M., Smith, J. F., & Sullivan, J. L. (2008). A nanotribological study of thin amorphous C and Cr doped amorphous C coatings. *Wear*, *265*(2008), 1633–1641.

Wilson, G. M., Smith, J. F., & Sullivan, J. L. (2009). A DOE nano-tribological study of thin amorphous carbon-based films. *Tribology International*, *42*, 220–228.

Wilson, G. M., & Sullivan, J. L. (2009). An investigation into the effect of film thickness on nanowear with amorphous carbon-based coatings. *Wear*, *266*(9–10), 1039–1043. Available from https://doi.org/10.1016/j.wear.2008.12.001.

Wu, G., Li, Y., Brittain, R., Lu, Z., & Yang, L. (2022). Understanding of fracture conditions and material response in a model TiN film/stainless steel substrate system – A cross-sectional scratch test study. *Surface and Coatings Technology*, *442*. Available from https://doi.org/10.1016/j.surfcoat.2022.128340.

Xie, Y., & Hawthorne, H. M. (2001). A model for compressive coating stresses in the scratch adhesion test. *Surface and Coatings Technology*, *141*(1), 15–25. Available from https://doi.org/10.1016/s0257-8972(01)01130-6.

Yoon, S. Y., Yoon, S. Y., Chung, W. S., & Kim, K. H. (2004). Impact-wear behaviors of TiN and Ti-Al-N coatings on AISI D2 steel and WC-Co substrates. *Surface and Coatings Technology*, *177–178*, 645–650. Available from https://doi.org/10.1016/j.surfcoat.2003.08.067, http://www.journals.elsevier.com/surface-and-coatings-technology/.

Yu, B., Dong, H., Qian, L., Chen, Y., Yu, J., & Zhou, Z. (2009). Friction-induced nanofabrication on monocrystalline silicon. *Nanotechnology*, *20*(465303), 8.

Yu, B., Li, X., Dong, H., Chen, Y., Qian, L., & Zhou, Z. (2012). Towards a deeper understanding of the formation of friction-induced hillocks on monocrystalline silicon. *Journal of Physics D: Applied Physics*, *45*(145301), 6.

Yu, J., Qian, L., Yu, B., & Zhou, Z. (2009). Nanofretting behavior of monocrystalline silicon (100) against SiO_2 microsphere in vacuum. *Tribology Letters*, *34*, 31–40.

Yu, J., Qian, L., Yu, B., & Zhou, Z. (2010). Effect of surface hydrophilicity on the nanofretting behavior of Si(100) in atmosphere and vacuum. *Journal of Applied Physics, 108*, 034314.

Yu, J. X., Qian, L. M., Yu, B. J., & Zhou, Z. R. (2009). Nanofretting behaviors of monocrystalline silicon (100) against diamond tips in atmosphere and vacuum. *Wear, 267*, 322–329.

Zhang, S., Bui, X. L., Fu, Y., Butler, D. L., & Du, H. (2004). Bias-graded deposition of diamond-like carbon for tribological applications. *Diamond and Related Materials, 13*, 867–871.

Zhang, S., Bui, X. L., Fu, Y., & Du, H. (2004). Development of carbon-based coating on extremely high toughness with good hardness. *International Journal of Nanoscience, 3*, 571–578.

Zhang, S., Sun, D., Fu, Y., & Du, H. (2004). Effect of sputtering target power on microstructure and mechanical properties of nanocomposite nc-TiN/a-SiNx thin films. *Thin Solid Films, 447–8*, 462–467.

Zhang, X., Beake, B. D., & Zhang, S. (2015). Toughness evaluation of thin hard coatings and films. In S. Zhang (Ed.), *Thin films and coatings: Toughening and toughness characterisation* (pp. 48–121). CRC Press, p. 53.

Zum Gahr, K.-H. (1987). *Microstructure and wear of materials*. Elsevier.

Linking coating mechanical properties and wear resistance

Ben Beake[1] and Tomasz Liskiewicz[2]
[1]Micro Materials Ltd., Wrexham, United Kingdom, [2]Department of Engineering, Manchester Metropolitan University, Manchester, United Kingdom

3.1 Introduction

A key aim of mechanical property measurements on coating systems is to predict their wear resistance in mechanical contact. Wear-resistant coatings are typically designed for high hardness since in the popular Archard wear model, the wear volume is inversely proportional to the hardness of the softer material in a sliding/adhesive contact (Archard, 1953). For several years, the development of superhard coatings was an important area of coating research (Veprek, 2008). For example, Veprek has claimed that nanocomposite coatings significantly harder than diamond can be produced. This claim has subsequently been critically examined, and the calibration protocol questioned (Fischer-Cripps et al., 2012). While "superhard" coatings with hardness ~60 GPa can be produced, it appears that claims of coatings with "ultrahardness" of 80–105 GPa are completely unfounded. Fischer-Cripps, Bull, and Schwarzer (Fischer-Cripps et al., 2012) used the analysis methodology by Veprek on industrial diamond, an ultrahard material of known properties, to show that it gave high hardness but underestimated elastic modulus by a factor of ca. 3 confirming that the calibration methodology adopted by Veprek was incorrect. For the nanocomposite coatings in question, having elastic modulus around 450 GPa, it should also not be possible to measure hardness >65 GPa with Berkovich or Vickers indenters due to the limit in mean contact pressure (Fischer-Cripps, 2014) that can be achieved or in measurable H/E. Brazhkin and Solozhenko have critically reviewed the experimental and computational evidence for superhard materials with elastic moduli and hardness significantly higher than diamond, finding them to be not be "scientifically reliable" (their words) (Brazhkin & Solozhenko, 2019).

However, notwithstanding the maximum hardness that can be achieved in superhard coatings, they are susceptible to brittle fracture, and several authors have shown that hardness alone is often not a particularly effective predictor of wear resistance (Baker et al., 2002; Bull, 1999; Leyland & Matthews, 2000, 2004; Rebholz et al., 1999). In many practical applications of coating systems, hardness should be combined with toughness to ensure adequate wear resistance (Baker et al., 2002; Bouzakis et al., 2012; Inspektor & Salvador, 2014; Leyland & Matthews, 2000, 2004; Musil & Jirout, 2007; Musil, 2006, 2012, 2015; Pei et al., 2005; Rebholz et al., 1999; Wang & Zhang, 2014, 2015; Wang et al., 2013; Zhang, Bui, Fu, Butler, et al., 2004;

Zhang, Bui, Fu, Du, et al., 2004; Zhang, Sun, et al., 2004). Ceramic thin film research has therefore moved away from focussing only on hardness and toward developing hard-yet-tough coatings that are crack-resistant and damage-tolerant under severe contact. Composite properties involving combinations of hardness and elastic modulus (e.g., H/E, H^3/E^2, H^2/E.) have all been shown to correlate with wear resistance under certain contact conditions. Leyland and Matthews proposed that the ratio of coating hardness (H) to its elastic modulus (E), H/E, could be used as an indicator of coating durability since it is a measure of the elastic-strain-to-break and coating resilience (Leyland & Matthews, 2000, 2004). The H/E ratio is strongly correlated with energy dissipation in mechanical contact. Plasticity indices have been introduced to provide an indication of the relative importance of elastic and plastic deformation in the overall deformation in contact. Greenwood and Williamson developed a plasticity index relating deformation in rough contacts to E_r/H multiplied by a geometric factor (Greenwood & Williamson, 1966) (Eq. 3.1).

$$\psi = (E_r/H)\sqrt{(\sigma/\beta)} \tag{3.1}$$

Where ψ = a plasticity index, E_r is the reduced modulus, σ = standard deviation of the height of the contacting asperities (i.e., surface roughness), and β = their average radius. An alternative formulation proposed later by Whitehouse and Archard (1970) is given by Eq. (3.2),

$$\psi = (E_r/H)(\sigma/\beta^*) \tag{3.2}$$

Where β^* is the correlation distance, the exponent of the correlation assuming that the surface profile is Gaussian (Palacio et al., 2006). In both treatments when $\psi \gg 1$, plastic deformation of asperities occurs even at minimal contact pressure, but deformation is largely or completely elastic when the plasticity index is much less than 1. These plasticity indices are inversely proportional to H/E. The ratio of hardness to elastic modulus in an indentation contact may be obtained from unloading curve analysis or alternatively from measuring the work of indentation (Cheng & Cheng, 1998). In the absence of fracture, there is an *approximately* linear relationship between H/E and the ratio of irreversible work to total work done in indentation (Cheng & Cheng, 1998, 2004, Cheng et al., 2002; Malzbender & de With, 2002; Malzbender, 2005) as shown in Eq. (3.3):

$$W_p/(W_p + W_e) \approx 1 - x(H/E_r) \tag{3.3}$$

In this equation, W_p and W_e are the plastic and the elastic work done, respectively, and $W_p/(W_p + W_e)$ is another dimensionless plasticity index, and x is a constant. When the sample Poisson's ratio is known (or can be estimated within limits), E_r can be converted to E using Eq. (3.4), so that H/E can be obtained.

$$1/E_r = (1 - \nu_s^2)/E_s + (1 - \nu_i^2)/E_i \tag{3.4}$$

Where E_s and ν_s are the elastic modulus and Poisson's ratio of the sample, and E_i and ν_i are the corresponding values for the indenter. Under a given contact pressure, higher H/E increases the likelihood of the deformation being predominantly elastic, which ultimately can have a strong influence on wear resistance in more tribologically complex loading situations, such as sliding/abrasion or impact/erosion. When coating fracture or delamination does not influence the dissipation significantly, the relationship between W_p/W_t and H/E_r in Eq. 3.3 is also applicable to coatings (Malzbender & de With, 2002).

Relationships between H and E also appear in equations for estimating fracture toughness, simulation of erosive wear, and in ductile grinding equations (Bousser et al., 2008, 2014; Hassani et al., 2008; Huang et al., 2020; Hutchings, 1992). H^3/E^2 is another parameter that can be obtained from nanoindentation, which has been studied in depth. Contact mechanics analysis shows that for a flat surface in elastic/plastic contact with a rigid ball of radius R, the yield pressure (P_y) is given by Eq. 3.5 (Johnson, 1985; Tsui et al., 1995). When H^3/E^2 is increased at constant contact pressure, then contact is more likely to be elastic. H^3/E^2 can be considered as a measure of resistance to plastic deformation or load-carrying capacity.

$$P_y = 0.78\, R^2 (H^3/E^2) \tag{3.5}$$

H/E and H^3/E^2 have been considered as potential indicators of coating fracture resistance. A correlation between H/E and surface cracking in indentation and bending of various oxide and nitride coatings has been reported by Musil and coworkers (Musil & Jirout, 2007, Musil, 2006, 2012, 2015). Bull found that the plasticity index (as defined in Eq. (3.1)), and therefore H/E, could characterize energy dissipated in damage processes during erosion. He related this to the volume of material removed by a single particle impact and hence to erosion rate (Bull, 2006).

Even if the nanoindentation test instrumentation has been correctly calibrated using the ISO 14577 methodology, there are additional requirements to be met so that truly comparable hardness and modulus measurements on coatings can be made so that reliable coating-only measurements of H/E and H^3/E^2 can be determined. This is especially important when coatings are of differing thickness or are deposited on a very soft and/or low-modulus substrate or are simply ultrathin. Although the measured hardness and elastic modulus from an indentation test can be used to provide estimate of coating H/E, it is important to ensure that full plasticity is achieved in the coating before substrate yield and understand the extent to which elastic modulus measurements are affected by the stiffness of the substrate.

Coating hardness should be determined with a sharp indenter (e.g., Berkovich) to induce plasticity within the coating before the substrate plastically deforms (i.e., the maximum Von Mises stress occurs within the coating causing plastic deformation while the stress in the substrate remains lower than its yield stress so the substrate only deforms elastically). The hardness of a hard coating on a soft substrate is commonly taken as the value at a relative indentation depth (RID) of 0.1 (the Bückle rule, indentation to less than 1/10 of the total film thickness to determine the true coating hardness independent of any substrate influence) or the value at a plateau in the measured hardness vs. RID, h_c/t_c, (h_c = contact depth, t_c = film thickness). With very thin coatings, it may not be possible for the developed mean pressure in the indentation region to reach the true hardness of the coating before plastic yield of the substrate occurs, and the measured hardness will be less than the true hardness, so a sharper probe (cube corner) would be required. It is possible to determine the hardness of soft coatings on harder substrates at significantly higher RID than 0.1 (Bull, 2005). An extensive plastic zone forms within the soft coating, but the substrate yield stress is not exceeded until much higher RIDs are reached.

Although not originally part of Bückle's idea, the 1/10 rule has since been widely applied to the measurement of coating elastic modulus as well. However, this is problematic since at any indentation depth, the elastic stress field extends much further than the plastic stress field (see Fig. 3.1). In the simplest treatment, the indenter bears down on an elastic coating connected to an elastic substrate so that reaction force is supported by a linear combination of two springs in series (Fischer-Cripps, 2006). The elastic modulus value at RID = 0.1 therefore includes an appreciable contribution from the substrate elasticity. This is particularly problematic where a large elastic mismatch exists between coating and substrate, for example, when a hard and stiff ceramic coating is deposited on a substrate with much lower stiffness such as silicon or glass. This mismatch results in H/E being significantly overestimated at an RID = 0.1. In indentation into a coating system, the relationship between elastic modulus and indentation depth is not quite linear (Bull, 2015; Lorenz et al., 2021; Saha & Nix, 2002; Zak et al., 2022). However, linear extrapolation over a limited depth range, as recommended in ISO14577-4, gives a more accurate measurement of the coating modulus than taking the value at an RID = 0.1. There is no "safe" RID at which the elastic response of a coated system is completely free of substrate influence (Chudoba, 2006; INDICOAT, 2001; ISO14577, 2015). However, measuring to RID = 0.03–0.05 has been proposed as a practical compromise between minimizing substrate contribution and minimizing uncertainties from surface roughness effects, which increase at lower RID, at least for cases where extreme elastic mismatch between coating and substrate is absent. Bull (2019) has suggested that the influence of the substrate can be apparent at RID = 0.01 while Zak and coworkers' studies of Mo and MoTa films on Si

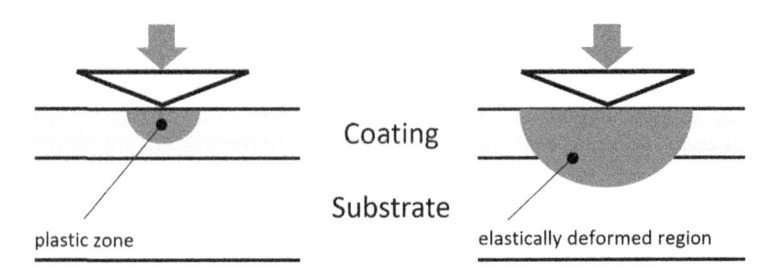

Figure 3.1 Plastic and elastic stress fields developed. A schematic representation of the plastic and elastic stress fields developed in indentation of a coating system.

(Zak et al., 2022) suggest that RID < 0.02 is needed. Finite element analysis of the same system showed that the elastic stress field extends 15 times deeper than the plastic stress field (Zak et al., 2022). Similarly, using an inaccurate coating modulus by performing a test at an RID of 0.1 results in an error in H/E and a larger error in H^3/E^2. As an illustration, a very high H^3/E^2 of 0.79 GPa was reported for an AlCrNbSiTiMoN high-entropy alloy nitride coating on Si(100) (Lo et al., 2020). In this case, the contribution from the relatively low stiffness substrate on the measured elastic modulus at an RID = 0.1 appears to have resulted in measured elastic modulus being lower than the real coating modulus so that H^3/E^2 was significantly in error.

3.2 Why coatings perform better in certain contact situations but not in others

Since, with sufficient care in experimental design, coating hardness and elastic modulus can be obtained in a single test technique (nanoindentation), the correlation between H/E or H^3/E^2 and wear resistance has been explored in detail, and H/E taken as a measure of wear resistance in many cases. Leyland and Matthews originally intended their H/E optimization approach primarily for coating deposition on relatively low-modulus engineering substrates such as steels and light alloys. They developed physical vapour deposited (PVD) coatings with high H/E by interstitial doping to create metallic nanocomposite coatings that were hard and wear-resistant even though they had lower elastic moduli than most ceramic coatings (Leyland & Matthews, 2000, 2004). The concept of H/E optimization has since been more widely adopted than perhaps originally intended. It is important to investigate the both the reasons behind its successes and its limitations to develop guidelines for when it could be used as a likely predictor of wear resistance and when a more cautious approach should be taken. An obvious case is where wear resistance is provided by tribo-film formation. Strauss et al. studied the influence of third bodies on friction and wear of PECVD nanocomposite Ti-Si-C-H coatings in pin-on-disk tests, finding wear resistance was dependent on formation of a stable transfer film, which was coating chemistry- and humidity-dependent (Strauss et al., 2011). Under their conditions, the tribo-film forming coatings showed increased wear resistance irrespective of their mechanical properties.

In the following sections of this chapter, we discuss possible reasons why coatings perform better in certain contact situations but not in others, focusing on (1) weakening effects/high stiffness limitations; (2) case studies investigating the influence of E_c/E_s in nanoscratch testing; (3) H^3/E^2, load-carrying capacity; (4) informing coating system design with nanomechanical measurements/breaking mutual exclusivity between hardness and toughness; (5) impact resistance in coating systems; (6) experimental simulation of contact conditions including high-temperature nanoindentation, microtribology, and impact.

3.2.1 Weakening effects/high stiffness limitations

It can be difficult to increase the hardness of ceramic coatings without an accompanying increase in their elastic modulus. The consequence of this is that (1) within a set of coatings, H/E may not vary much even if there is a large spread of hardness; (2) the high coating stiffness may have implications regarding fracture resistance. The dependence of the critical load for plastic deformation in an indentation contact on the yield stress of $2-3\,\mu m$ TiN layers on steel has been studied by Chudoba et al. (2002). The variation of the critical load for yielding with the probe radius showed a sharp discontinuity at the transition between plastic yield in coating (at low radius) and substrate (at higher radius). The TiN coating did not have a protective effect when indented by probes with large radii. This was due to the much higher stiffness of the coating than the substrate, which resulted in a stress concentration at the coating − substrate interface. This led to yield at lower load than the uncoated steel. While it is usually assumed that the deposition of a hard and stiff ceramic coating will mechanically protect the substrate, this result shows it is not always the case. Several other groups have also investigated weakening effects with thin hard coatings (Goltsberg et al., 2011; Goltsberg & Etsion, 2013; Huang et al., 2012, 2015a, 2015b; Komvopoulos, 1989; Sun et al., 1995) through modeling and experimental studies. Sun et al. performed finite element calculations showing that a $0.6\,\mu m$ TiN coating on high-speed steel reduced the load-carrying capacity in comparison to the uncoated substrate (Sun et al., 1995). Goltsberg, Etsion and coworkers provided an explanation in terms of higher substrate stresses due to the difficulty of the stiff coating to follow inward displacement of the more compliant substrate. The coating is stretched by the substrate, and in return, it restricts its deformation introducing additional stresses to cause earlier plastic yield (Huang et al., 2015a, 2015b).

Palacio et al. (2006) used Elastica software to determine the load for yield of $1\,\mu m$ coatings on hard (5 GPa yield stress) and soft (500 GPa yield stress) steel substrates. They reported that, except at very low radii, a fullerene-like CN_x coating with very high H/E ($H \sim 15$ GPa, $E \sim 60$ GPa) had higher load-carrying capacity on hardened steel than harder and stiffer coatings due to the reduction in contact stresses from its low elastic modulus. When deposited on soft steel, there were very sharp discontinuities between yield in the coating at low radius and in the substrate at higher radius.

3.2.2 Coating behavior in nano- and microscale scratch tests

The coating and substrate mechanical properties, coating thickness, and test probe sharpness all influence the location of yield in nano- and microscratch tests and subsequently affect how the interface is weakened and/or crack development, which controls wear. Kato and coworkers used finite element modeling to develop maps of yield location in sliding contacts (Diao et al., 1994; Kato, 1995). They showed that development of tensile stress occurring just behind the contact was influenced by high friction, the ratio of coating to substrate elastic modulus, and the severity of test conditions, which were varied through the dimensionless ratio t/a (t = coating thickness, a = contact half-width).

Under certain conditions, coating design to accommodate substrate strain can be a significantly more important factor in improving its wear resistance than having extremely high hardness. Close matching of strain in coating and substrate can be achieved by matching the elastic modulus of coating and substrate ($E_c/E_s \sim 1$). This minimizes the coating/substrate interfacial stress distribution under applied load and allows the coating to deflect with the substrate without cracking or debonding. As an illustration of this, the following section investigates what can happen when a large elastic mismatch between the coating and underlying substrate exists in a sliding contact.

3.2.2.1 The influence of elastic mismatch on coating behavior in nano-scratch testing – PVD thin films on Si

Studies investigating the behavior of thin PVD coatings on silicon in nano-scratch tests are described as follows. The end radius of the scratch probe was varied between 1 μm and 5 μm. The role of tensile stress developed at the coating side of the interface and changes in failure location in determining the observed behavior is highlighted.

The influence of the mechanical properties of TiFeN$_x$ nanocomposite films on their critical loads and failure mechanism in the nano-scratch test was investigated (Beake, 2021, 2022; Beake, Endrino, et al., 2017; Beake, Vishnyakov, et al., 2017). A set of 22 coatings with different compositions (18 (Ti, Fe)N$_x$, 2 FeN, and 2 TiN) but similar $\sim$1.3 μm thickness were deposited on Si(100) substrate with a $\sim$20 nm titanium interlayer to improve their adhesion using a dual ion beam system, with a 280–600 eV N$_2^+$ ion beam also used for ion assistance in some coatings. A worn Berkovich indenter (end radius $\sim$1 μm) scratching edge-forward was used for the nano-scratch tests. The large number of coatings allowed the interrelationships between their mechanical properties and the nano-scratch test critical load, friction, and deformation behavior over a wide range of coating hardness, H/E and H^3/E^2 to be investigated.

The deformation responses and critical loads were a strong function of the coating mechanical properties. Fig. 3.2 shows the variation in critical load with H^3/E^2 for ductile and brittle coatings. For films with the lowest hardness and titanium content (FeN and Fe(Ti)N, <15 GPa), ductile behavior was found. Increasing the titanium content in the film increased hardness to $\sim$20 GPa resulting in more brittle behavior with film delamination, which was accompanied by fluctuations in friction. For TiFeN films with higher hardness, there was localized brittle machining accompanied by delamination behind the contact zone at high load, which was not seen in the friction. For films with the highest H^3/E^2, there was brittle machining failure and/or adhesion failure at low load. The highest critical loads and scratch toughness were found for the TiFeN coatings with optimized mechanical properties ($H \sim$ 25 GPa and H^3/E^2 $\sim$0.13 GPa) rather than for the hardest coatings with the highest H^3/E^2.

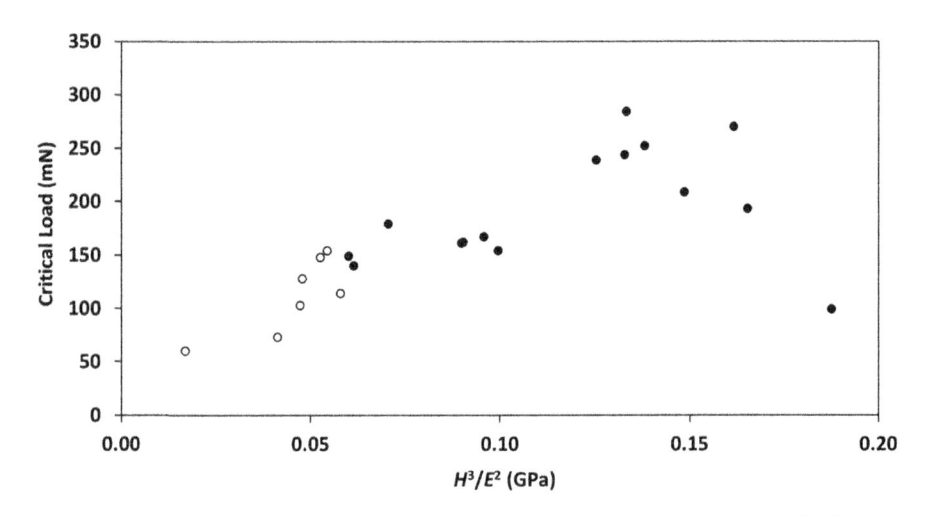

Figure 3.2 Variation in critical load for PVD coatings deposited on Si with H^3/E^2
Variation in critical load for ~ 1.3 μm PVD coatings deposited on Si (open circles = ductile failure; closed circles = brittle failure) with H^3/E^2.
Source: With permission from Beake, B. D. (2022). The influence of the H/E ratio on wear resistance of coating systems—Insights from small-scale testing. *Surface and Coatings Technology 442*, https://doi.org/10.1016/j.surfcoat.2022.128272. Copyright (2022).

The marked frictional oscillations that occur when failure was either in front or at the front/side of the moving scratch probe were not seen in failures originating behind the contact zone. Finite element modeling of the first principal stress by Holmberg and coworkers shows maximum tensile stress at the surface located around a half-contact width behind the sliding contact (Holmberg et al., 2007, Holmberg et al., 2008), in agreement with previous analytical results (Hamilton & Goodman, 1966). Its location at a small distance behind the contact rather than at the trailing edge of the contact can explain why failure can occur without frictional changes in some cases. Modeling by Diao and coworkers also showed a tensile maximum appearing just behind the trailing edge and increasing with elastic mismatch (Diao et al., 1994). Analytical modeling on the TiFeN$_x$ coatings has shown tensile stresses increasing rapidly with hardness (Beake, Endrino, et al., 2017; Beake, Vishnyakov, et al., 2017) and E_c/E_s, as predicted by Diao and Kato. Fig. 3.3 shows tensile stress as a function of E_c/E_s. Similarly, in indentation contact, Miranda and coworkers reported that damage at the interface was favored for $E_c/E_s > 1$ (Miranda et al., 2003).

The mechanical properties of the coatings influence the stresses developing in the high load sliding contact, which in turn control how the interface is weakened (by initial substrate or coating yield or both) and resultant failure mechanism. The "unloading" failure at low load with the hardest coatings can be understood as coating failure through a combination of substrate yielding and high tensile stress. There was a different failure mechanism for the softer coatings, through plastic

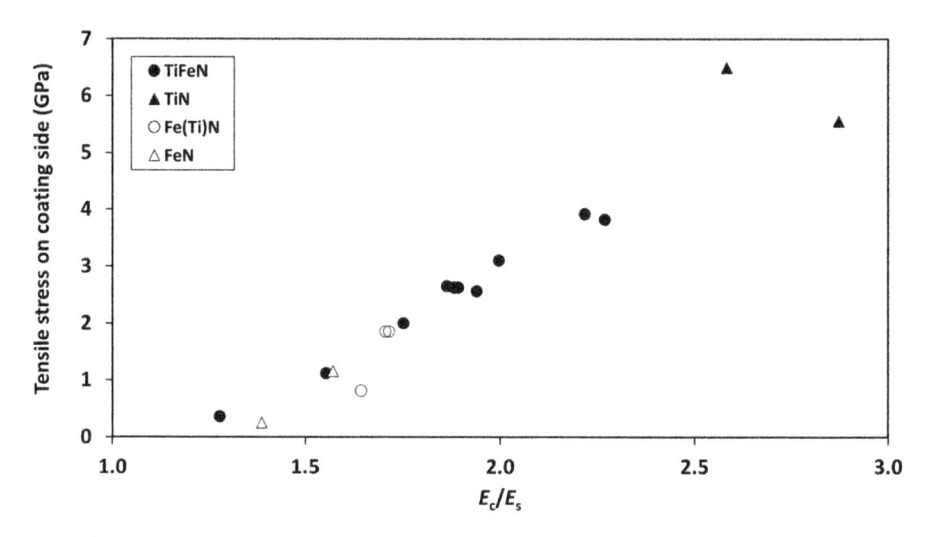

Figure 3.3 Variation in tensile stress in nitride coatings. Variation in tensile stress with E_c/E_s in nitride coatings.

yield initiating within the coating at lower tensile stress. For the softer coatings, the interface was weakened from the coating side, which did not result in the dramatic unloading failure observed when the interface was weakened first by substrate yield (Beake, Endrino, et al., 2017; Beake, Vishnyakov, et al., 2017).

This type of behavior is not restricted to TiFeN coatings. Similar behavior was reported for nanocomposite TiN/Si$_3$N$_4$ coatings on Si in nano-scratch tests with a spheroconical diamond probe of 3 μm end radius (Beake et al., 2006). The nc-TiN/Si$_3$N$_4$ coatings were produced by PVD with a wide range of mechanical properties (H = 13–30 GPa, E = 226–485 GPa). The main features observed in the nano-scratch tests as the load increased were (1) elastic-to-plastic transition (at L_y); (2) cohesive failure (edge or parallel cracking from L_{c1} critical load); (3) external transverse cracking; (4) "unloading" failure for the hardest films; and (5) total (compressive) film failure occurring in front of the probe. By using a spherical probe, it was possible to determine a clear transition from fully elastic behavior to plasticity, with higher critical loads found for higher H^3/E^2 films. The critical load for compressive failure in front of the sliding probe was generally higher for harder films. However, pronounced failure behind the probe with delamination extending over a large area outside the scratch track occurred at lower load for the films with highest H and H^3/E^2, being most extreme for a film with H = 29 GPa and H^3/E^2 = 0.183 GPa. For these hard TiSiN and TiFeN coatings deposited on silicon (a relatively low stiffness substrate) increasing the coating hardness, H/E and H^3/E^2 beyond an optimal point decreased the scratch test critical load. Increasing test temperature alters the mechanical properties of the coating and substrate and so may produce a similar effect in changing the location of yield and subsequent failure mechanism, as discussed later in this chapter.

A design strategy for high toughness could combine mechanical properties to give a high load threshold for the initiation of cracking with combined microstructural toughening mechanisms to minimize and retard the propagation of cracks. The improved behavior shown by the TiFeN coatings with $H \sim 25$ GPa is due to a combination of factors including: (1) a dense nanocomposite microstructure eliminating weak columnar boundaries, (2) avoiding the high tensile stresses that develop in the nanoscratch testing of the hardest and stiffest films, (3) sufficient hardness, (4) localized energy dissipative mechanisms (Beake, Endrino, et al., 2017; Beake, Vishnyakov, et al., 2017). Interestingly, TiFeMoN coatings produced with the same deposition apparatus and nanoscratch tested under similar conditions performed better than these TiFeN coatings (Beake et al., 2011). The TiFeMoN coatings exhibited more effective stress relief at high load by a combination of localized energy dissipative mechanisms including intergranular fracture, microcracking, and stick slip within the scratch track.

Tomastik et al. have studied the influence of nitrogen doping and annealing in air or vacuum at 700°C–1100°C on the mechanical properties and nanoscratch resistance of ~ 2.5 μm a-SiC$_x$N$_y$ coatings deposited on silicon (Tomastik et al., 2018). The a-SiC$_x$N$_y$ coatings were less hard but more durable than undoped a-SiC in nanoscratch tests with an $R = 5$ μm probe.

Fig. 3.4 shows that annealing and nitrogen-doping influence E_c/E_s and control L_{c1}. This is a result of higher tensile stresses being generated at the rear of the sliding contact due to the higher elastic mismatch with the silicon substrate. For a given value of E_c/E_s, annealing increases L_{c1}. For all the a-SiC and a-SiC$_x$N$_y$ compositions, H/E increased after annealing in air at 900°C and more significantly after annealing at 900°C in vacuum. However, there was little clear trend in H/E with nitrogen incorporation in the film, controlled by the N$_2$/Ar flow ratio, despite its

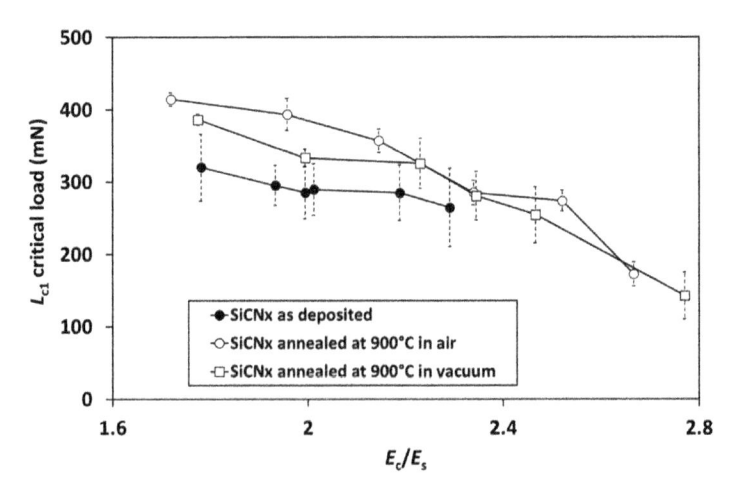

Figure 3.4 Variation in critical load for nitride coatings. Variation in critical load with E_c/E_s for as-deposited SiC$_x$N$_y$ coatings, and SiC$_x$N$_y$ coatings annealed in air or vacuum at 900°C.

strong influence on cracking. It can be concluded therefore that there was no obvious relationship between H/E and cracking resistance in these coatings. The 900°C vacuum-annealed samples showed lower durability, particularly for the films with lowest nitrogen incorporation and highest H/E (and H^3/E^2). However, there was a change in deformation behavior, with more brittle cracking and a reduction in delamination after annealing.

Apparent contact pressures in sliding contacts are often lower than those in the highly loaded scratch tests, but the surfaces in contact are not typically ideally smooth, particularly during the running in period. Coating behavior under highly loaded sliding contact is important as the actual contact pressures are significantly higher between rough surfaces than between ideal smooth surfaces (Jiang & Arnell, 1998).

3.2.2.2 Nano- and microscratch testing of diamond-like carbon coatings on hardened tool steel

The nanomechanical and nano/microtribological behaviors of Si-doped DLC and WC/C coatings on hardened steel have been compared with a typical hard hydrogenated DLC (Beake et al., 2015). WC/C coatings have been shown to perform well in a wide range of tribological conditions (Mutafov et al., 2014; Ramírez et al., 2015; Wänstrand et al., 1999). Si-DLC has low friction, but its wear resistance can be inferior to undoped DLC (Lanigan et al., 2016). The coatings were 3–4 μm thick with multilayer (Cr/W-C:H/a-C:H, Cr/W-C:H/Si-a-C:H, and CrN/a-C:H:W) architectures to improve their adhesion. The DLC and Si-DLC coatings were deposited by PECVD and WC/C was a commercial coating, Balinit C Star. In the PECVD coatings, the adhesion layer is a thin Cr, and then gradient layers were applied to adapt the elastic modulus of the softer substrate to the elastic modulus of the harder top coating. In WC/C, there was a CrN sublayer to provide load support.

The mechanical properties are shown in Table 3.1 and their behavior in nano- and microscratch tests to 500 mN and 5 N, respectively, using a spheroconical diamond probe with an end radius of 5 μm for the nanoscratch tests and 25 μm for the microscratch tests shown in Table 3.2.

In the scratch tests, the radius of the test probe has a significant influence on the yield behavior. With the 5 μm probe, there was a correlation between the coating properties and critical load for plastic yield, but this correlation was not observed in tests with the 25 μm probe. Fig. 3.5 shows simulated stress distributions. The

Table 3.1 Mechanical properties of diamond-like carbon coatings [coating-only values from nanoindentation (ISO14577 approach)].

	H (GPa)	E (GPa)	H/E	H^3/E^2 (GPa)
DLC	23.4	234	0.10	0.23
Si-DLC	16.8	151	0.11	0.21
WC/C	13.0	149	0.09	0.10

Table 3.2 Critical loads in nano- and micro-scratch tests (Peak load 500 mN in nano-scratch; 5 N in micro-scratch).

Coating	$R = 5$ μm nano-scratch			$R = 25$ μm micro-scratch		
	L_y (mN)	L_{c1} (mN)	L_{c2} (mN)	L_y (mN)	L_{c1} (mN)	L_{c2} (mN)
DLC	206 ± 5[a]	422 ± 4	>500	356 ± 9[a]	2179 ± 120	2612 ± 127
Si-DLC	110 ± 10[b]	445 ± 12	>500	383 ± 52[a]	1827 ± 111	2256 ± 112
WC/C	68 ± 4[b]	>500	>500	375 ± 49[a]	2830 ± 367	3695 ± 132

[a] Yield starts in substrate.
[b] Yield starts in coating.

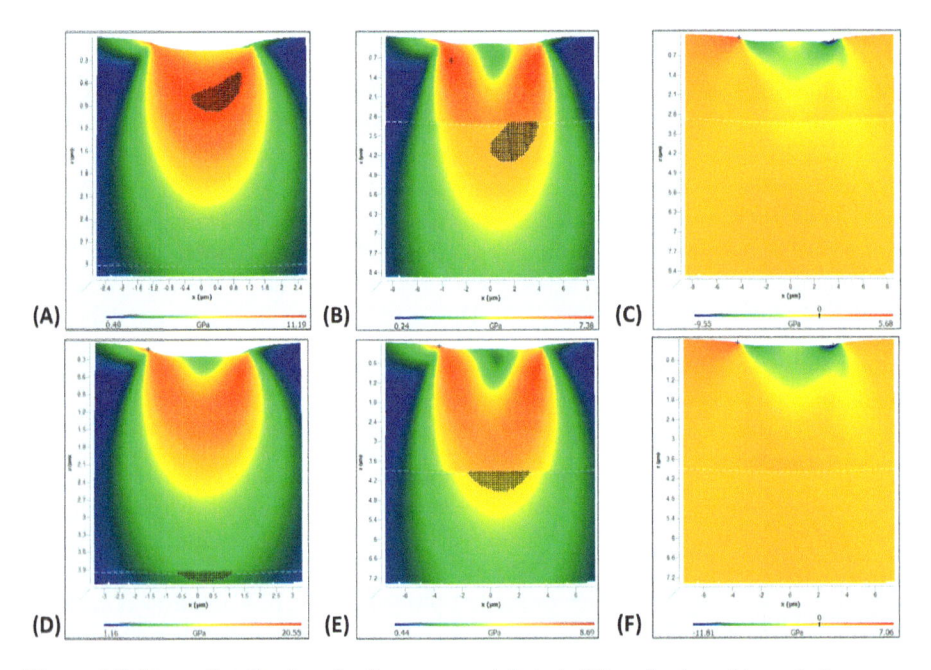

Figure 3.5 Stress distributions in the nanoscratch test. 2D projections (through the center-line of scratch) of simulated stress distributions in the nanoscratch test with $R = 5$ μm probe for a-C:H:W (in A–C) and a-C:H (in D–F). The scratch direction is left to right. a-C:H:W (A) Von Mises stresses at yield (B) Von Mises stresses at 500 mN with corresponding normal stresses in (C). a-C:H (D) Von Mises stresses at yield (E) Von Mises stresses at 422 mN (L_{c1}) with corresponding normal stresses in (F). In A,B,D,E the overstressed areas (where the magnitude of the von Mises stress is greater than the yield stress at that point) shown by hashed region.
Source: With permission from Beake, B. D. (2022). The influence of the H/E ratio on wear resistance of coating systems—Insights from small-scale testing. *Surface and Coatings Technology 442*, https://doi.org/10.1016/j.surfcoat.2022.128272. Copyright (2022).

simulated stress distributions were able to explain how the coating mechanical properties and scratch probe radius influence the behavior. The simulated distributions of the Von Mises, normal and shear stresses generated during the nano- and microscratch tests were obtained using the Scratch Stress Analyzer (SIO, Rugen, Germany), which uses a physical-based analytical methodology to determine simulated stress distributions utilizing George Pharr's effective indenter concept for the stress calculations (Schwarzer, 2006). The input parameters for the calculations were the hardness and elastic modulus of the coatings (taken as monolayered) and substrate, the coating thickness, the Poisson ratios of the coatings and substrate, the probe radius, material, applied load, and measured friction coefficient in the nanoscratch test. To convert from hardness to yield stress, it is necessary to also include H/Y values for coating and substrate. H/Y varies with elastic recovery and was taken as $H/Y = 1.2$ for coatings, $H/Y = 2.5$ for the steel substrate. The simulations show that when using the $R = 5$ μm probe, yield occurs initially within coating for Si-DLC and WC/C, but in the steel substrate for the DLC. There were correlations between the coating mechanical properties and crack resistance in nanoscratch tests and in reciprocating wear tests. Semicircular cracks at the rear of the contact were observed on DLC and Si-DLC, but cracking was not observed on the softer WC/C with lowest H/E (Beake et al., 2015). The more extensive plastic deformation for the softer WC/C coating reduced its susceptibility to cracking. With the $R = 25$ μm scratch probe, yield began in the substrate for all the coatings, consistent with the almost identical probe depths at that load. At yield, the maximum von Mises stresses were within the coating, but they remained below the coating yield stress.

Tensile stresses at the rear of the sliding contact were around 5–7 GPa at the critical load for cracking (L_{c1}) with the WC/C coating being able to sustain higher tensile stresses before cracking. On WC/C more extensive plastic flow in the substrate may have been beneficial in minimizing cracking since at L_{c2} there was substrate yield over a greater area on WC/C and less cracking around the scratch track than on the DLC and Si-DLC. Although there were marked differences in critical load between the three coatings, the interfacial shear stresses at failure were all around 2 GPa. The analytical simulations are consistent with the failure mechanism on all these DLC coatings being a combination of high tensile stress with plastic flow in the substrate adjacent to the coating – substrate interface. A similar mechanism was proposed for hard coatings on cemented carbide (Schwarzer et al., 2011). The WC/C coating had the lowest H^3/E^2 and was able to undergo more elastic deformation before the L_{c2} failure than the DLC and Si-DLC. This result appears to contradict the fact that higher coating H^3/E^2 improves its load spreading capability. In this case, the benefit of lower elastic modulus in reducing E_c/E_s and tensile stresses may also be important in highly loaded sliding contact, as was shown for PVD films on silicon in Section 3.2.2.1.

In designing for damage tolerance under highly loaded sliding, often the coating with the lowest elastic modulus performs best when deposited on a low-stiffness substrate. The lower E_c/E_s induces lower tensile stresses in the progressive load scratch test and results in a higher L_{c1}. Reduced brittle fracture in repetitive subcritical load scratch tests is then a consequence of being further below this cracking threshold so that microfatigue occurs more slowly. Similar behavior has been reported in microscratch testing where PVD AlTiN coatings with lower E_c/E_s outperformed harder AlCrN and TiAlN coatings in progressive load and repetitive subcritical load microscratch tests (Beake, Endrino, et al., 2017; Beake, Vishnyakov, et al., 2017) even though the carbide substrate has high stiffness.

Huang et al. (2015a, 2015b) studied the influence of substrate and coating mechanical properties on behavior in ball-on-flat tribological tests by depositing TiN and TiAlN on copper, high-speed steel, and WC-Co substrates. On the softer, lower-stiffness substrates, the wear was higher on the harder and stiffer TiAlN coating than TiN. Extensive cracking was observed, and the authors suggested that the higher elastic mismatch for TiAlN produced high tensile stresses leading to greater fracture. EBSD data for TiAlN and TiN on Cu showed smaller substrate distortion for TiN/Cu consistent with a weakening effect from higher E_c/E_s. On WC-Co, the trend was reversed as cracking was absent, and abrasion became the main wear mechanism (greater load support from the harder WC-Co). There was a marked trend with substrate hardness, with higher wear rates for both coatings on copper and lowest on WC-Co.

3.2.3 H^3/E^2, load-carrying capacity

H^3/E^2 is a measure of the resistance to plastic deformation or load-carrying capacity. At a given contact pressure, with higher H^3/E^2, the contact is more likely to be elastic, and more load is required to cause yield, than for a lower H^3/E^2 surface. For similar coatings, H^3/E^2 and H/E can be strongly correlated (Beake, 2022). The differences in H^3/E^2 are larger than H/E as shown in Fig. 3.6 for some PVD

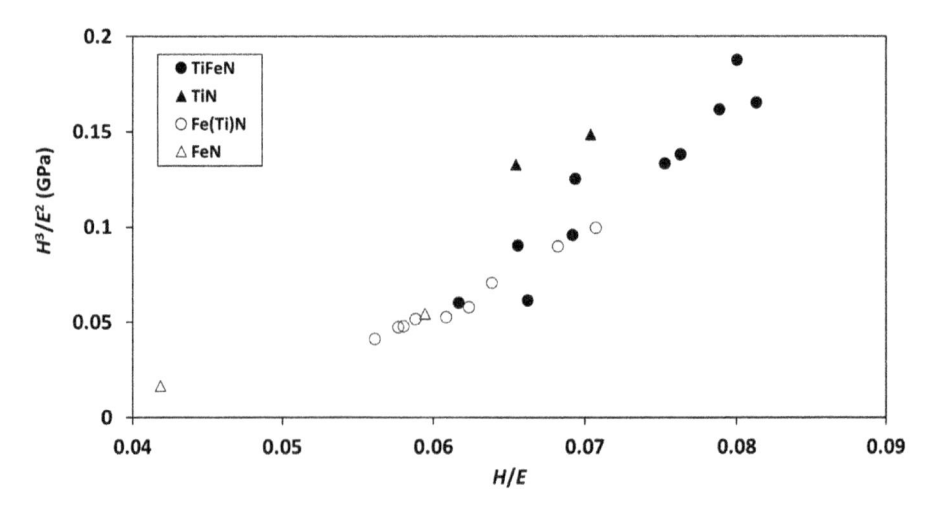

Figure 3.6 H^3/E^2 and H/E for PVD coatings. Relationship between H^3/E^2 and H/E for ~ 1.3 μm PVD coatings deposited on Si.

(Ti, Fe)N$_x$ coatings. In the example, an approximate doubling in H/E has resulted in nearly order of magnitude increase in H^3/E^2.

A correlation between H^3/E^2 and the onset for yield has been observed in nano-scratch tests on bulk materials with $R = 5$ μm diamond indenters. Higher critical loads were found for Ti6Al4V than CoCrMo (Beake & Liskiewicz, 2013), and for fused silica than silicon, despite having lower hardness in both cases. For coatings the extent of correlation is complicated by whether the yield is in the coating or at the interface/in the substrate. With a $R = 3$ μm probe, there was a very clear trend to increasing L_y with H^3/E^2 for TiSiN coatings on silicon, and for DLC coatings on steel with a $R = 5$ μm probe. However, in the latter case, little difference between the coatings was observed when tested with larger probe radius ($R = 25$ μm) since the yield position moved into the substrate ($\mu = 25$ μm) since the yield position moved into the substrate (Table 3.2). Rhee and coworkers indented bulk materials with 1–10 mm ball indenters (Rhee et al., 2001), plotting the dependence of P_y/R on H^3/E_r^2 (P_y was defined as the onset of quasiplasticity, since yield involved shear-driven subsurface microcracking in their samples). They reported an approximately linear correlation, except at very high H^3/E^2, where there were rapid increases in critical load for minimal changes in H^3/E^2. Resistance to particle erosion (Bousser et al., 2014) and cavitation erosion (Lavigne et al., 2022) can be strongly correlated with H^3/E^2. The erosion of TiN coatings by 50 μm Al$_2$O$_3$ particles has been simulated by Hassani et al., who reported lower erosion rate for TiN coatings with higher hardness, lower elastic modulus, and higher thickness (Hassani et al., 2008). Further simulations showed the volume removed through erosive wear correlated with both H/E and H^3/E^2. The simulations indicated that this combination of coating properties could lower the maximum tensile stresses developed under impact, which promote fracture.

In several cases, H^3/E^2 has also been taken as a direct measure of (i.e., a proxy for) coating fracture resistance (Xu et al., 2017). Bartosik et al. noted a correlation between fracture toughness determined from cantilever bending experiments and the H^3/E^2 ratio for annealed AlTiN coatings (Bartosik et al., 2017). Where a correlation between H^3/E^2 and resistance to cracking has been found, this may be a consequence of higher loads being required for fracture initiation since harder materials commonly can be more brittle. Chen et al. have cautioned that neither H/E nor H^3/E^2 accounts for the contribution of plasticity to fracture toughness citing examples where either no correlation or an inverse correlation between H^3/E^2 and toughness was found (Chen et al., 2019). For VC and W coatings, an inverse correlation between fracture toughness and H/E or H^3/E^2 was found, although increasing W in W/VC multilayers increased toughness without reducing hardness significantly (Shi et al., 2015).

In highly loaded contact requiring high resistance to (1) cracking and (2) plastic flow in several cases, the best wear resistance has been found for coatings with *intermediate*—rather than extremely high—values of hardness, H/E and H^3/E^2. For example, in erosion tests on 6 μm TiAlN coatings on Ti6Al4V by alumina, the maximum resistance to erosion was found for coatings with intermediate H^3/E^2 (Yang et al., 2004). Similarly, in a study of the resistance of CrN, AlCrN, and

AlTiN coatings to abrasion and impact, the AlCrN coating with medium H^3/E^2 among the three coatings showed the best resistance to abrasion and impact wear (Mo et al., 2013). Krella, in a review of the resistance of PVD coatings to solid particle erosion and cavitation erosion, has reported that there appears to be a limit to the ratio of H/E at which the best erosion resistance is obtained (Krella, 2020).

It has been proposed that to be crack-resistant, coatings should be as hard as possible with very high H/E and H^3/E^2 (Musil, 2006, 2012, 2015), although this is contrary to the usual antagonism between hardness and toughness. Improved crack resistance may be primarily a consequence of the higher load support as the critical load for plastic flow increases with H^3/E^2. A higher coating H^3/E^2 reduces substrate deformation, lowering bending stresses in the coating. In some cases, it may be this effect, rather than intrinsic structural toughening, which results in higher apparent toughness in the indentation test. Musil and coworkers have reported that to be crack-resistant in indentation, coatings require a combination of low E and H/E^* >0.1 ($E^* = E/(1-v^2)$) (Musil & Jirout, 2007, Musil, 2006, 2012). In the Al-Cu-O and nc-(γ-Al(Cu)O$_2$)/α-Al$_2$O$_3$ coatings, they studied that moderate hardness was combined with low stiffness. The lower modulus of the coatings reduces the modulus mismatch with the substrate, improving their crack resistance.

Increasing coating H^3/E^2 may result in higher brittleness, even at moderate hardness. For example, in a study of a-C, nc-TiC/a-C, and nc-TiN/a-C coatings on hardened steel, the a-C coating, which had the highest hardness (18 GPa), H/E (0.096) and H^3/E^2 (0.168 GPa) among them, spalled at low load in the scratch test (Zimowski et al., 2019). Conversely, the highest scratch resistance and wear resistance in highly loaded contact were found instead for the softest nanocomposite, the nc-TiC/a-C coating. Zawischa et al. investigated the influence of doping metallic and nonmetallic elements on the crack resistance of ta-C coatings on hardened steel (Zawischa et al., 2022). These authors found that there was a direct dependence between H^3/E^2 and crack resistance in these coating systems in indentation, scratch, and mandrel bending tests. Coatings with $H^3/E^2 > 0.2$ showed more brittle behavior with unstable crack propagation, but coatings with $H^3/E^2 < 0.2$ displayed more ductile behavior, with stable crack propagation and higher crack resistance.

3.2.4 Breaking mutual exclusivity between hardness and toughness

For nanomechanical measurements to reliably inform surface (i.e., coating system) design for wear resistance, it is important that the reasons behind and the extent of their correlation to the in-service performance of the coated surface are well understood. This includes what properties are needed for different applications, and the influence of surface roughness and contact size on the observed behavior. High wear resistance comes from combining high hardness and high toughness. However, in general, hardness and toughness are mutually exclusive, for example, as was shown in an extensive recent review of literature measurements of hardness and K_{1C} toughness of different grades of cemented carbides (Chychko et al., 2022).

The relative importance of hardness (i.e., resistance to plastic deformation and how this influences load support) and toughness (resistance to fracture) varies in different applications. Wear resistance can be improved by coating design to increase one property while only minimally decreasing the other. Ranade et al. (2012) explored a nanocomposite design strategy to produce Ti-TiB$_2$ coatings with significantly higher toughness but relatively less reduction in hardness (i.e., toughness increasing more rapidly than hardness decreasing) than nc-TiN/Si$_3$N$_4$. It was speculated that the improved toughness was due to activating dislocations in the Ti matrix caused by stresses developed by the formation of coherent Ti-TiB$_2$ interfaces.

When developing structure − properties − performance relationships, it is useful to consider that microstructural changes often are responsible for the changing coating mechanical properties. Hard coatings deposited by vapor deposition exhibit the widest variety of microstructures among materials in terms of grain size, crystallographic orientation, lattice defects, texture, surface morphology, and phase composition (Mayrhofer et al., 2006). Vapor deposition conditions produce metastable films with nonequilibrium density of structural defects offering considerable opportunities for microstructural optimization during deposition or by subsequent annealing. In coatings deposited with high ion energies high hardness—well above bulk materials—can be obtained utilizing strain hardening by high defect densities. The residual stress of the coating is important as it can dictate the maximum coating thickness without spallation (Bull, 1999). There can be an apparent linear relationship between residual stress and hardness, in single-phase coatings such as TiN (Bull, 1999; Mayrhofer et al., 2006). This is because the microstructural defects resulting in compressive residual stresses also act as obstacles to hinder dislocation movement. Grain refinement reduces the density of the dislocations responsible for plastic deformation, resulting in increased hardness by Hall − Petch strengthening.

Nanoscale control of the composition or phase modulation in hard coatings, such as nanocomposites or nanolaminates (superlattices), offers additional opportunities for strengthening, with hardness >40 GPa often reported through reduced dislocation mobility through strong grain boundaries. Superlattices—multilayers with coherently stacked alternating layers with fcc-fcc structure and nm periodicity—can combine high hardness with high toughness, breaking down the usual antagonism between hardness and toughness. Although Zhang et al. reviewed strategies for toughening ceramic coatings (Zhang et al., 2015), the specific cause of the "interface effect" toughness enhancement in superlattices was subsequently investigated by Hahn et al. (2016).

Multilayers of alternating metal/ceramic (i.e., compliant/stiff) layers have also shown enhanced toughness. Mishra and coworkers summarized the main changes to the crack driving force due to the multilayer architecture in metal − ceramic multilayered systems as (1) periodic stiffness variation, (2) crack tip blunting and bridging due to plasticity in the metal layer, (3) crack closure due to high compressive stresses in the hard layer, and (4) crack deflection along weak interfaces (Mishra et al., 2022). They reported that although a 50-layer Ti/TiN film had 82% higher fracture toughness than single layer TiN, the weak intercolumnar boundaries limited the advantages of crack shielding from alternating stiff-compliant layers. At larger scale, Meindlhumer et al. studied clamped microcantilevers composed of four alternating 5 μm CrN and Cr layers (Meindlhumer et al., 2021). For this system when a crack grew through a Cr layer to the adjacent CrN-Cr interface, it was arrested.

In practice, additional factors such as low friction may also be critical to ensure wear resistance. This solid lubrication can be achieved by a coating with intrinsically low friction such as an amorphous carbon film or by a coating that displays adaptive behavior—for example, by forming protective and lubricious tribo-films at elevated temperature (as discussed in more detail in Chapter 19). In this application, the mechanical properties of the surface are still important as they control the dynamics of the formation and regeneration of these tribo-films—high wear producing an environment where efficient film formation and regeneration are more difficult. Strauss et al. noted that in pin-on-disk tests of Ti-Si-C-H coatings sliding against a sapphire indenter the ability to form a stable, low friction, tribolayer was more important than coating mechanical properties (Strauss et al., 2011). These authors suggested that mechanical property − based optimization should be complemented by investigations on the third body (tribo-film) behavior.

Wear resistance can be limited by the "controlling defect" in the coating system. For monolayered PVD coatings, this can be the weak columnar boundaries, which act as sites for through-thickness cracking. Microstructural design to produce denser (e.g., nanocomposite) coatings or multilayers to remove the weak columnar boundaries often produces more crack-resistant coatings with improved mechanical properties, that is, provides a route where design can improve wear resistance by increasing both hardness and toughness.

3.2.5 Impact resistance

In many coatings, applications including engine components and discontinuous metal cutting operations resistance to repetitive impact are critical (Bouzakis et al., 2011; Knotek et al., 1992; Lawes et al., 2010). In an automotive engine system, diamond-like carbon (DLC) coatings are used to protect components in the power train such as tappets, pistons, piston rings, and fuel injectors, where they can be subjected to repetitive impacts in service (Lawes et al., 2010). High-performance machining operations such as milling or interrupted turning involve intermittent high-strain-rate contacts, which contact occur at high frequency, for example, 1 Hz−1 kHz. In a gas turbine engine, solid particle erosion results in damage to the turbine blades that lowers engine efficiency and can even be life-limiting in some conditions. Laboratory-based accelerated tests have been used in the development of advanced coatings for these applications. Additional deformation mechanisms (fatigue) can be present in repetitive (cyclic) impact tests that are not observed in single-cycle indentation tests (Kim et al., 1999), and hence, cyclic impact tests have been used as model tests for determining coating durability (Bantle & Matthews, 1995; Bouzakis et al., 2004; Yoon et al., 2004). In the most common test configuration, the impact occurs at 90 degree to the surface, but the sensitivity to interfacial toughness (adhesion) can be improved by performing inclined impact tests, which generate shear stresses at the interface (Bouzakis et al., 2007). Gsellmann et al. noted that a spectrum of normal and shear stresses are generated during cutting (Gsellmann et al., 2020).

Impact fatigue mechanisms in coating systems can vary with the ratio of coating thickness to the radius of the test probe (Michler & Blank, 2001). Tests can be carried out with different contact sizes through varying the applied load and probe sharpness (Bernoulli et al., 2015). In impact tests of PVD coatings with tungsten carbide or hardened steel ball indenters with 1−3 mm end radius typically, the t/R values can be as low as ~ 0.001. The peak von Mises stresses responsible for plasticity are within the substrate, and hence, the fatigue behavior of the system can be strongly affected by the properties of the substrate (Bantle & Matthews, 1995; Bouzakis et al., 2004; Bromark et al., 1995; Knotek et al., 1992; Yoon et al., 2004).

Improved impact resistance was reported for CrN when deposited on tool steel in comparison to when deposited on a hard metal substrate (Knotek et al., 1992). Yang et al. compared the performance of AlCrN coatings on tool steel and cemented carbide substrates in cyclic indentation tests finding that the harder cemented carbide provided better load support and resistance to crack initiation, but the tougher tool steel resisted crack propagation more effectively, which improved its damage tolerance (Yang et al., 2013). Yoon et al. noted that in impact tests on TiN and TiAlN coatings on tungsten carbide and tool steel with a 3 mm radius cemented carbide indenter, the influence of coating mechanical properties such as H^3/E^2 was dependent on the substrate (Yoon et al., 2004). When deposited on tool steel, there was more plastic deformation, but more brittle-dominated deformation when deposited on cemented carbide. On steel the TiAlN, with higher H^3/E^2, was more impact resistant than TiN. However, when deposited on cemented carbide, although initially more impact-resistant than TiN, with continued impact, TiAlN underwent rapid brittle failure leading to substrate exposure.

By decreasing the test probe radius, it is possible to perform impact tests at much lower load but at higher contact stress with higher coating bending stresses. These nano- and micro-scale impact tests at much higher t/R than in the macroscale tests provide a promising alternative approach to evaluating coating fatigue resistance. The accelerated impact tests can be performed in a much shorter period than macroscale tests since they probe coating system behavior under more severe conditions with higher coating bending. The resistance to impact fatigue of Al-rich PVD nitride coatings on cemented carbide and carbon coatings on hardened tool steels has been evaluated by microimpact testing (Beake et al., 2020; Beake, Bird, et al., 2019; Beake, Isern, et al., 2019; Beake, Isern, et al., 2021; Beake, McMaster, et al., 2021; Beake, Vishnyakov et al., 2021). A popular choice in micro-impact tests has been to use a spheroconical diamond probe with end radius ~ 20 μm, producing tests at $t/R \sim 0.15$ when testing 3 μm PVD coatings. This fills the gap in t/R between the nano- and macro-ranges, so that the test can be sensitive to coating and substrate together where stresses can be concentrated near interface(s) in the system (Beake et al., 2020; Beake, Bird, et al., 2019; Beake, Isern, et al., 2019; Beake,

Isern, et al., 2021; Beake, McMaster, et al., 2021; Beake, Vishnyakov et al., 2021; McMaster et al., 2020). By comparing the impact behavior of coating systems tested under the same conditions, it is possible to show the increasing contribution of the substrate properties (load support controlling coating bending, and its ductility influencing damage tolerance) to the coating system response at high load while maintaining a high sensitivity to the coating properties. As an example, Fig. 3.7 shows the behavior of Al-rich PVD nitride coatings on cemented carbide and carbon coatings on hardened tool steels. The PVD nitrides on cemented carbide were susceptible to impact-induced lateral fracturing as the test load increased. The highest load for fracture was found for a $Ti_{0.25}Al_{0.65}Cr_{0.1}N$ with slightly higher H^3/E^2 than the other nitrides (Beake, Bird, et al., 2019; Beake, Isern, et al., 2019). At

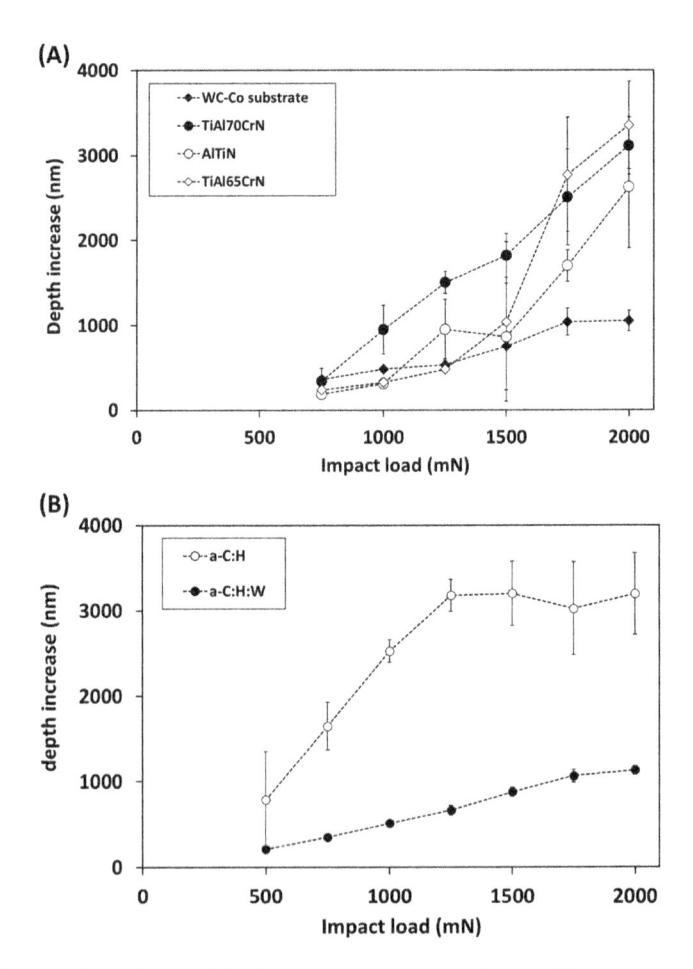

Figure 3.7 Load dependence of depth increases after initial impact (A) coatings on WC-Co and WC-Co substrate (B) diamond-like carbon coatings on hardened M2 tool steel. 75 impacts, accelerating distance = 40 μm; impact probe radii = 17−20 μm.

higher impact load, these fractures resulted in extensive substrate exposure and cracking of the hard WC crystals at the periphery of the impact crater, despite the greater load-carrying capacity of WC-Co.

The carbon coatings were subjected to higher bending strains on hardened tool steel, as evidenced by the larger initial impact depths. Under these conditions, the a-C:H with high H/E was too brittle, with extensive lateral fracture from 750 mN. WC/C and other carbon-based coatings with moderate hardness and relatively low H/E on hardened tool steel were, however, resistant to radial cracking and lateral fracture (Beake et al., 2020; McMaster et al., 2020). It appears that the greater ductility of the hardened tool steel substrate than cemented carbide aids damage tolerance of the coating systems at higher load, since on higher plasticity coatings (moderate hardness, relatively low H/E), lateral fracture was completely suppressed under the test conditions. The combination of a coating with high plasticity index and a tough, damage tolerant substrate is beneficial for impact resistance. Several other investigations into the behavior of DLC coatings on hardened steel substrates in nano- (Beake & Smith, 2004), micro- (Beake et al., 2020; McMaster et al., 2020; 2020), or macro-impact tests (Ramírez et al., 2015) have reported similar results where there was no benefit to having high H^3/E^2, and coatings with lower hardness, H/E and H^3/E^2 were consistently more impact-resistant and damage-tolerant. A clear correlation between coating performance in nano- and micro-impact tests and sand erosion resistance was reported for DLC coatings on hardened steel (McMaster et al., 2020). Ramírez et al. reported enhanced impact resistance for soft W-doped carbon films in comparison to TiN when both were deposited on cold-work steel (Ramírez et al., 2015). Under reciprocating sliding conditions, however, several groups have reported that Si and W-doped DLC coatings show higher wear rates than undoped DLC although the latter was susceptible to occasional spalling within the wear tracks (Bai et al., 2021; Beake, Isern, et al., 2021; Beake, McMaster, et al., 2021; Beake, Vishnyakov, et al., 2021).

Weikert et al. performed high cycle indentation fatigue tests on single-layer tungsten doped (a-C:H:W) and multilayer silicon oxide containing (a-C:H:Si:0/a-C:H)$_{25}$ amorphous carbon coatings on hardened high-speed steel (Weikert et al., 2021). Under these conditions—cyclic loading with conical indenter with 0.83 μm radius—relatively less severe fatigue with no permanent substrate deformation—the multilayer coating with higher H^3/E^2 performed much better. This is likely to be due to (1) much lower bending stresses during the test, (2) multilayer structure with softer layers to relieve stress and more columnar structure in the a-C:H:W. He et al. (2019) altered the modulation period in WC/a-C coatings on stainless steel, finding an optimum modulation period of 10 nm where coating hardness and H^3/E^2 increased and the coating showed less volume loss in impact but a more brittle response. Wang et al. (2023) reported that small increases in the H^3/E^2 in soft carbon films (graphite-like carbon) deposited on stainless steel altered the failure mechanism in a dynamic sliding test, from interfacial failure at lower H^3/E^2 to cohesive failure at higher H^3/E^2, though all the films studied had hardness ~6 GPa.

Nano-impact tests on PVD coatings with a sharp cube corner diamond indenter (i.e., under high t/R conditions) have been shown to be very sensitive to small changes in coating brittleness and excellent correlation to coating performance in applications has been reported (Bouzakis et al., 2011, 2012, 2017; Fox-Rabinovich et al., 2006, 2009; Skordaris et al., 2014, 2017). Under these high localized stresses, some coating plasticity can occur—which is not typically the case in macro-scale impact tests where the substrate is assumed to deform elastoplastically and the coating to simply deform as a thin elastic plate. This may be one reason why small-scale impact tests show correlation with high-speed machining where finite element modeling of the cutting process has shown that coatings are overstressed. In tests on (Ti,Al)N-based coatings on cemented carbide, a clear correlation between reduced wear of the coated tools in high-speed machining and better resistance to fracture in the nanoimpact test has been reported in several studies (Bouzakis et al., 2011, 2012, 2017; Fox-Rabinovich et al., 2006, 2009; Skordaris et al., 2014, 2017). In comparison to microimpact tests that produce chipping and fracture accompanied by delamination and substrate exposure, the smaller forces and sharper probe geometry (higher t/R) in the nanoimpact test produced chipping/fracture but without interfacial failure as the smaller contact size concentrated stresses within the coatings.

Fracture resistance and cutting performance are also influenced by several other factors including coating nano/microstructure and its compressive stress. Intrinsic thin film stress and stress measurement have been reviewed by Windischmann (1992) and Abadias et al. (2018). A pronounced columnar structure may not be desirable as the column boundaries can act as lines of weakness for the through-thickness cracks that can result in extensive chipping. Coating design by multilayering and grain refinement to minimize weak columnar boundaries are promising strategies. Multilayered (including nanomultilayered) PVD nitride coatings have consistently displayed improved performance in tribological tests and machining applications than monolayered coatings (Chang et al., 2013; Chen et al., 2011; Mendibide et al., 2004, 2006; Ou et al., 2015; Roa et al., 2014; Zha et al. 2018; Zhang et al., 2018, Vereschaka & Grigoriev, 2017, Vereschaka et al., 2020). Increasing the number of layers significantly enhanced resistance to repetitive nanoimpact and increased cutting life in multilayered 8 μm $Al_{0.54}Ti_{0.46}N$ coatings (Bouzakis et al., 2004, 2017; Skordaris et al., 2015). Multilayered TiAlCrSiYN/TiAlCrN with higher H^3/E^2 showed better impact resistance than monolayered TiAlCrSiYN coatings with lower H^3/E^2 (Fox-Rabinovich, Beake, et al., 2010; Fox-Rabinovich, Endrino, et al., 2012; Fox-Rabinovich, Yamamoto, et al., 2010; Fox-Rabinovich, Yamamoto, et al., 2012). The influence of compressive stresses developed during microblasting on the brittleness of $Al_{0.6}Ti_{0.4}N$ on WC-Co has been assessed by nanoimpact testing (Bouzakis et al., 2011, 2012). High impact resistance in the nanoscale impact tests correlated with longer cutting tool life when milling hardened steel. The nano-impact test data showed the same relative ranking of cutting performance after microblasting with ZrO_2 and Al_2O_3 and also reproduced a switch in relative performance between 0.2 and 0.4 MPa microblasting pressure (Bouzakis et al., 2011, 2012). The influence of residual stress on the cutting performance of $Al_{0.55}Ti_{0.45}N$ on cemented carbide was studied by Skordaris et al. (2017). Coatings with differing residual stress were produced by altering bias voltage during deposition, with the coating deposited with lowest compressive stress being subsequently annealed to introduce some tensile stress. The best cutting performance was shown for a coating with optimum (rather than too high) level of compressive residual stress, consistent with too much compressive stress being detrimental to performance as has been reported elsewhere (Wang & Zhang, 2014). There was a direct correspondence between the final nanoimpact depth and the cutting tool life as illustrated by Fig. 3.8.

Increasing the impact force (and hence energy through impact energy = applied force x accelerating distance) in the nanoimpact test results in more severe test conditions with greater coating bending stresses and increasingly crack-dominated behavior. Coatings with higher H^3/E^2 often perform better than coatings with lower H^3/E^2 at low force (due to greater load carrying capacity) but less well as the impact force is increased. At high force, coatings with higher H^3/E^2 may display greater cracking compared with coatings with lower H^3/E^2 if they do not have an associated microstructural advantage (such as multilayering) to dissipate the high strain.

Yang and McKellar (2015) have shown how the number of layers (and interfaces) in PVD nitride coatings can not only change the rate of solid particle erosion but also whether the maximum erosion rate occurred at 90 or 30 degrees impact of the angular alumina erodent. A maximum in erosion rate at 90 degrees is commonly observed for a brittle material and a maximum at 30 degrees for a ductile material. The erosion rate was greater at 90 degrees for 12-layer CrAlTiN-AlTiN. Increasing the number of layers to 38 improved erosion resistance at both angles but with higher erosion resistance now at the lower angle. Nanomultilayered CrAlTiN coatings showed further slight improvements in erosion resistance at 90 degrees but were more susceptible to damage at 30 degrees.

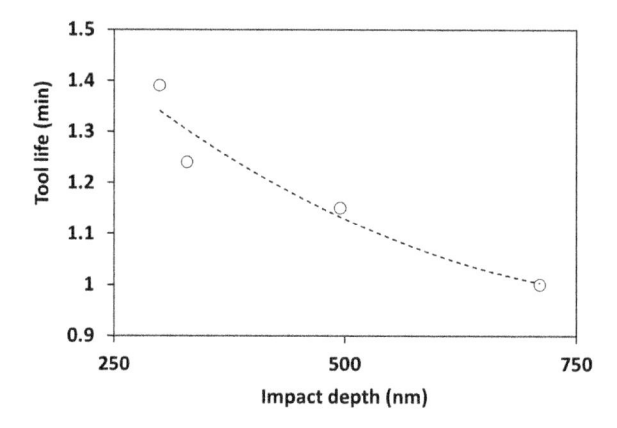

Figure 3.8 Relationship between cutting tool life in turning AISI 1045 steel and nanoimpact depth in $Al_{0.55}Ti_{0.45}N$ coatings with differing levels of residual stress. Cemented carbide substrate.

3.2.6 Experimental simulation of contact conditions – high-temperature testing

Hard coatings operate at high temperatures in many applications (Beake & Fox-Rabinovich, 2014; Jantschner et al., 2013; Schalk et al., 2022; Tkadletz et al., 2016). To improve productivity in high-performance machining, there is a continual drive to develop more wear-resistant coatings to allow faster cutting, and machining of difficult-to-cut materials, under more tribologically extreme conditions. Frictional heating results in high cutting temperatures, which can reach 800°C–1000°C or more at high cutting speeds. To extend tool life, coatings can display adaptive behavior by tribo-film formation. The mechanical properties of the coating at the operating temperature can be critically important as they affect the stability of the environment where the mixed (e.g., AlO_x/TiO_x on (Ti,Al)N coatings) protective and lubricious tribofilms form, which enables the coating system to display adaptive behavior and consequently lower tool wear. The temperature dependence of the mechanical properties of the coating and substrate affects the location and the magnitude of developed stresses in highly loaded contact, which can influence the predominant deformation mechanism and control wear rate. In general, coating hardness reduces with increasing temperature. The decrease in hardness can be due to relaxation of high compressive residual stresses, grain boundary sliding, and thermally activated deformation mechanisms, such as dislocation glide (Giuliani et al., 2019; Hultman & Mitterer, 2006).

For experimental convenience, it is common to measure coating mechanical properties at room temperature even though the in-use temperature may be very high. However, for this tribologically extreme elevated temperature application, the coating mechanical properties determined at room temperature may be less relevant than those measured directly in high-temperature tests. Using room-temperature data to reliably predict relative coating performance under high-temperature conditions requires that the coating ranking at room temperature remains the same at the application temperature. However, there is no guarantee of this as coatings with the highest hardness at room temperature may soften more rapidly with temperature than other coatings. This behavior has been reported for PVD and CVD coatings in hot microhardness tests (Inspektor & Salvador, 2014; Jinhal et al., 1999) and for PVD coatings in elevated temperature nanoindentation tests (Beake & Fox-Rabinovich, 2014). Inspektor and Salvador investigated the influence of cutting speed (and indirectly cutting temperature) on the tool life of CVD-coated tools when turning 1045 steel. They found that at low cutting speed, TiC and TiN coatings had longer life but at speed >300 m min^{-1}, an alumina coating with higher hot hardness outperformed them (Inspektor & Salvador, 2014). Highly stressed coatings with high strain hardening coming from high defect densities where the defects act as obstacles for dislocation movement and create high compressive residual stresses are less suitable for high-temperature applications since there is a gradual reduction in the strain hardening contribution resulting in a decrease in hardness at temperatures above the deposition temperature (Mayrhofer et al., 2006).

Critical issues to consider and strategies to obtain reliable data from nanomechanical measurements at, or near to, application temperatures are discussed in detail in references (Beake & Harris, 2019; Beake, 2021; Wheeler & Michler, 2013; Wheeler et al., 2015). The mechanical properties of the PVD coatings typically decrease with increasing temperature. To illustrate this, Fig. 3.9 shows the temperature dependence of H/E for 2–3 μm thick PVD monolayer and multilayer nitride coatings on cemented carbide determined from nanomechanical measurements to 600°C. Since the elastic modulus is less strongly influenced by temperature, the temperature dependence of H/E broadly follows that of the hardness, but there was a larger decrease in H^3/E^2.

The coating high-temperature mechanical properties have a strong influence on the performance of coated tools in high-performance machining applications. The influence of annealing for 2 h in vacuum at 700°C–900°C on the lifetime of $Al_{0.67}Ti_{0.33}N$-coated cemented carbide tools in continuous and interrupted cutting tests on steels has been studied by Fox-Rabinovich et al. (2006, 2008). Annealing increased the hot hardness of the $Al_{0.67}Ti_{0.33}N$ coating, particularly after 2 h in vacuum at 900°C. Annealing at 700°C increased tool life in continuous and interrupted contact cutting tests on steels. Annealing at 900°C was only favorable for continuous turning and led to lower tool life in interrupted turning. Different coatings have different balance of properties, and their relative importance in different types of cutting tests determines their actual performance in these operations.

The cutting temperatures are very high in continuous high-speed turning operations (Naskar & Chattopadhyay, 2018). Maintaining a relatively high coating hardness and resistance to plastic deformation at elevated temperature is important in protecting against substrate softening. In high-speed dry turning of low and high carbon steel plastic deformation induced necking and ridge formation has been observed on the worn flank faces of PVD TiAlN-coated cemented carbide tools

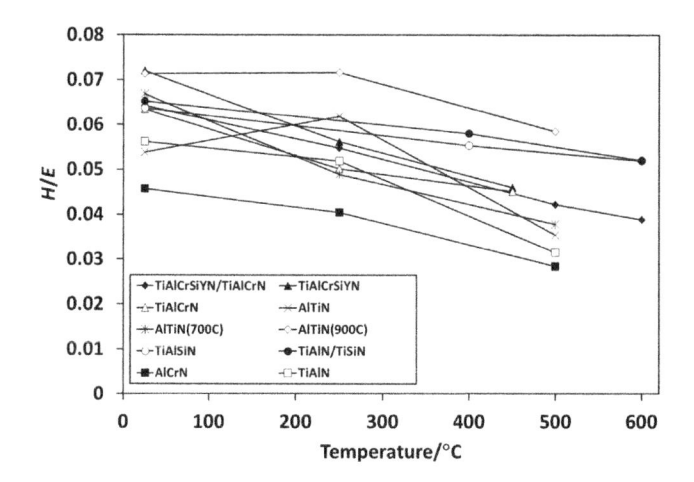

Figure 3.9 Temperature dependence of H/E in monolayer and multilayer nitride coatings on cemented carbide.

(Naskar & Chattopadhyay, 2018). Coatings with high hot hardness (e.g., AlTiN after annealing at 900°C, multilayer TiAlCrSiYN/TiAlCrN) have shown longer tool life in high-speed turning of hardened steel and hard-to-cut aerospace alloys such as ME 16 superalloy. In the high-speed machining of hardened steels, high-temperature coating properties are critical as they influence the environment where the tribo-films form, providing a stable environment for the system to display adaptive behavior when there is reduced wear and fracture. Under these conditions TiAlCrN, a coating with metal-covalent character, high hot hardness, and lower plasticity index (higher H/E) exhibits longer cutting tool life AlTiN with its more metallic bonding, low hot hardness, and higher plasticity index (lower H/E) (Beake & Fox-Rabinovich, 2014).

In interrupted cutting operations, temperatures are typically lower. A suitable combination of crack resistance and hot hardness is needed. Cracking is clear in images of worn tools by interrupted cutting, for example, milling of 4140 steel (Santhanam et al., 1996) or hardened H13 steel (Fox-Rabinovich et al., 2014; Fox-Rabinovich, Beake, et al., 2010; Fox-Rabinovich, Yamamoto, et al., 2010). In comparison to continuous cutting, interrupted cutting conditions alter the relative importance of high temperature hardness (or high temperature yield stress) and plasticity (or high temperature fracture resistance) in determining the ultimate tool life.

It was suggested that high hot hardness in ternary and quaternary nitride coatings should be combined with slightly improved plasticity when end milling of hardened steels to extend tool life (Beake et al., 2009). High Al-fraction monolayered PVD nitride coatings having higher plasticity than $Ti_{0.50}Al_{0.50}N$—such as TiAlCrN, AlTiN and AlCrN—outperform it under interrupted cutting conditions. For monolayer nitride coatings, optimum values of hardness and plasticity measured at 450°C–500°C to achieve longer tool life in interrupted cutting conditions such as end milling were found to be $H \sim 20$ GPa, $H/E_r \sim 0.06$, and plasticity index ~ 0.6 (plasticity index = plastic work/total work), that is, a combination of sufficiently high hardness together with some capability for plastic deformation to relieve stress. Subsequently, it was realized that the idea of an optimum coating hardness combined with optimum H/E is actually not dissimilar to the concept of an optimized H^3/E^2 since the latter can be considered as simply $H \times (H/E)^2$ (Beake, 2021). Finding optimum values for longer tool life at moderate-to-high hardness and plasticity index—rather than ultrahigh hot hardness and low plasticity index—is likely to be a result of the requirement to combine high hot hardness and fracture resistance in this application.

The quaternary Si-containing nitrides and the Si-containing nanomultilayer coatings shown in Fig. 3.9 have better high-temperature mechanical properties than the ternary nitrides, supporting their enhanced performance in high-temperature tribological tests and machining applications. In TiAlSiN, the addition of Si induces grain refinement resulting in a coating with a dense microstructure (nc-TiAlSiN structure with c-TiAlN crystals in a-Si_3N_4 matrix) that is resistant to oxidation and contact fatigue (Fuentes et al., 2014). In microcompression tests at 500°C, another

Si-containing nitride coating, TiAlCrSiN, performed better than ternary nitride coatings such as AlCrN (Wheeler et al., 2014).

The nanomultilayer coatings typically show better tool life than the monolayered coatings, particularly at higher cutting speed. This is because they have improved high-temperature mechanical properties (higher H^3/E^2 providing enhanced load-carrying capability and resistance to crack initiation) and nano/microstructural advantages (dense microstructure, eliminating/minimizing weak columnar boundaries and nanolaminate structure providing resistance to through-thickness cracking with crack deflection along interfaces providing an additional route to relieving strain accumulation in heavy loaded cutting). In contrast, the cutting speed is more limited in monolayer coatings with columnar microstructure so they cannot as readily combine high H^3/E^2 with a mechanism that avoids fast crack propagation leading to large scale fracture under increasingly severe conditions.

3.2.7 Elevated temperature microtribology and impact testing

The temperature dependence of coating and substrate mechanical properties affects the magnitude and position of the induced stresses in mechanical contact, and the relative importance of fracture and plastic deformation on the overall deformation, which can change the dominant deformation mechanism. Elevated temperature tribological tests confirm that the test temperature has an important role on the wear rate and dominant deformation mechanism (Allsopp & Hutchings, 2001; Blau, 2010; Jantschner et al., 2013; Pauschitz et al., 2008; Polcar et al., 2006; Tkadletz et al., 2016; Veverkova & Hainsworth, 2008; Voevodin et al., 2014). Low friction is important to reduce contact temperatures during dry sliding. Designing for solid lubrication by low friction protective layers is increasingly important industrially with self-adaptive mechanisms of low friction generation by formation of tribo-oxides at high temperature (e.g., mullites, Magneli phases) being effective in reducing the thermal load on the coating and substrate and hence wear (Mayrhofer et al., 2006). Coated components in machining applications operate at high stresses under a wide spectrum of normal and tangential loads (Gsellmann et al., 2020). High-temperature nano- and micro-scale scratch tests can replicate the deformation in highly loaded sliding/abrasive contacts and high-temperature nano- and microimpact replicate the interrupted highly loaded mechanical contacts in milling operations. Due to limitations with the durability of diamond indenters, which graphitize in air at high temperature, the maximum scratch and impact test temperatures have typically been limited to 500°C−600°C. While performing tests at these temperatures does not fully replicate all of the potential coating adaptive mechanisms that may be operative at higher temperatures (such as formation of protective tribofilms), it been able to reveal features of the coating deformation at high temperature that could not be predicted from tests at room temperature. The behavior at high temperature has been supported by mechanistic understanding through simulation of the stresses generated in contact. Chavoshi and Xu have reviewed high-temperature nanoscratch and nanotribological tests (Chavoshi & Xu, 2018).

Reduced coating and substrate mechanical properties at high temperatures lower the stresses that develop in sliding contact. In micro-scratch tests of 2 μm PVD monolayer TiAlSiN and nanomultilayer TiAlN/TiSiN coatings on a cemented carbide with 10 wt.% Co, the critical load for coating failure was found to be relatively insensitive to temperature. The large Co fraction in the cemented carbide was responsible for a more rapid decrease in hardness with temperature than is found for cemented carbides with lower percentage Co (Milman et al., 1999). The apparent lack of sensitivity of the critical load to temperature presumably reflects the increased coating bending due to lower load support from substrate softness at high temperature being offset by reduced brittleness (Beake, Bird, et al., 2019, Beake, Isern, et al., 2019). In micro-scratch tests on ternary nitrides on a cemented carbide with 6 wt.% Co (whose hardness decreases more slowly with increasing temperature due to the lower cobalt fraction in the composite), the softening of the coating was more significant than substrate softening. At 500°C, the critical load for film failure was higher than at room temperature for $Al_{0.7}Cr_{0.3}N$ and $Al_{0.67}Ti_{0.33}N$ AlTiN coatings, but lower on $Ti_{0.5}Al_{0.5}N$. Coating wear rates in 1 N repetitive micro-scratch tests at 500°C with the same 25 mm radius probe followed the same trends, being lower than at room temperature on $Al_{0.7}Cr_{0.3}N$ and $Al_{0.67}Ti_{0.33}N$ (Beake, Endrino, et al., 2017, Beake, Vishnyakov, et al., 2017).

Table 3.3 illustrates the importance of temperature on the deformation behavior. The maximum stresses in the simulations of micro-scratch tests on AlTiN were determined with inputs from micro-scratch and nanoindentation tests at 500°C. At this temperature, there was plastic flow in the AlTiN coating. The maximum von Mises stress generated in the substrate was lower than its yield stress. Increasing the load to L_{c2} causes more extensive plastic flow extending to, and hence weakening, the interface. Table 3.3 shows that the contact stresses are lower at elevated temperature. The calculated maximum shear stress at the L_{c2} failure was also calculated at 25°C and 500°C. A lower shear stress was found at 500°C implying a reduction in the coating − substrate bonding strength at this temperature, consistent with the expected weakening in the strength of the atomic bonds as temperature increases (Beake, Endrino, et al., 2017; Beake, Vishnyakov, et al., 2017). The increase in the experimentally measured critical load at 500°C is not due to any improvement in the intrinsic adhesion strength of the coating but primarily a result of the different stress distribution at high temperature. The analytical modeling of

Table 3.3 Calculated maximum stresses on the scratch center-axis in microscratch tests on PVD $Al_{0.67}Ti_{0.33}N$ on WC-Co at 25 and 500°C (Probe radius = 25 μm).

Temp. (°C)	Applied load (N)	von Mises stress in coating (GPa)	von Mises stress in substrate (GPa)	Normal stress at surface (GPa)
25	1	13.09	9.97	11.75
	4.5 (L_{c2})	11.14	9.79	9.18
500	1	5.95	5.02	5.20
	5.8 (L_{c2})	5.92	5.53	4.80

the main stresses acting at the interface reveals how differences in coating high temperature mechanical properties influence the initial interfacial weakening (i.e., by initial substrate or coating yielding or both, depending on how high the coating high temperature hardness is) and thus play a key role in the deformation failure mechanism at elevated temperature. Simulations show that at 500°C the interface between $Ti_{0.5}Al_{0.5}N$ and WC-Co is weakened by plastic flow of coating and substrate at lower load, while the higher high-temperature hardness of AlCrN resulted in interfacial weakening at higher load, from the substrate side. Coating behavior in the elevated-temperature microscratch tests was found to be consistent with cutting tests where the higher Al-fraction coatings, $Al_{0.7}Cr_{0.3}N$ and $Al_{0.67}Ti_{0.33}N$ outperform $Ti_{0.5}Al_{0.5}N$ in cutting tests (Beake et al., 2007; Endrino et al., 2006; Kalss et al., 2006). $Al_{0.67}Ti_{0.33}N$ coatings have generally been effective when the cutting conditions require a relatively high coating plasticity index (i.e., lower H/E than, e.g., $Ti_{0.5}Al_{0.5}N$ or TiAlCrN) to minimize intensive adhesive − fatigue interaction with workpiece materials. Its lower brittleness appears to be a factor in providing a low wear environment for tribofilm formation protecting the cutting tool surface against chipping. As-deposited $Al_{0.67}Ti_{0.33}N$ coatings have relatively poor high-temperature hardness at >250°C but retain sufficient load support below this temperature to be effective in interrupted cutting of Ti alloys where temperatures are lower. Depending on cutting speed, the temperatures in interrupted cutting of Ti6Al4V can be well under 400°C (Michailidis, 2016). In cutting applications that need higher hot hardness, such as continuous turning or end milling of hardened steel, as-deposited $Al_{0.67}Ti_{0.33}N$ (without post-deposition high-temperature annealing to improve its hot hardness) shows lower tool life than tools with coatings that retain higher hardness at elevated temperature. Examples of these include $Al_{0.7}Cr_{0.3}N$ (Kalss et al., 2006), $Al_{0.67}Ti_{0.33}N$ after annealing at 700°C or 900°C for 2 h in vacuum (Fox-Rabinovich et al., 2008; Fox-Rabinovich, Beake, et al., 2010; Fox-Rabinovich, Yamamoto, et al., 2010), TiAlCrN (Fox-Rabinovich et al., 2009), TiAlCrSiYN (Fox-Rabinovich et al., 2008), and TiAlCrSiYN/TiAlCrN (Fox-Rabinovich, Beake, et al., 2010; Fox-Rabinovich, Endrino, et al., 2012; Fox-Rabinovich, Yamamoto, et al., 2010; Fox-Rabinovich, Yamamoto, et al., 2012), all of which have longer tool life than $Al_{0.67}Ti_{0.33}N$ in these tests. Combining high hot hardness and fracture resistance, the nanomultilayered TiAlCrSiYN/TiAlCrN performed best out of coatings studied in turning the hard-to-cut aerospace materials ME 16 Ni-base superalloy and Inconel DA 718 (Fox-Rabinovich, Beake, et al., 2010; Fox-Rabinovich, Yamamoto, et al., 2010).

High-temperature nano- and micro-impact tests have been performed on hard nitride coatings (Beake et al., 2007; Beake, Bird, et al., 2019; Beake, Endrino, et al., 2017; Beake, Iscrn, et al., 2019; Beakc, Vishnyakov, et al., 2017). Nano-impact tests with a sharp cube corner indenter were performed at 500°C on PVD $Ti_{0.5}Al_{0.50}N$ and $Al_{0.67}Ti_{0.33}N$ coatings deposited on H10A cemented carbide (a grade with low Co fraction and hence relatively small decrease in hardness with temperature). As $Ti_{0.5}Al_{0.5}N$ and $Al_{0.67}Ti_{0.33}N$ both soften considerably at 500°C, the impact stresses in the high temperature tests are lower, which resulted in significantly reduced fracture than was found at room temperature. The reduced fracture of $Al_{0.67}Ti_{0.33}N$ in elevated-temperature nano-impact tests correlated with it having a higher cracking threshold (L_{c1} critical load) in the micro-scratch test at elevated temperature than at room temperature (Beake, Endrino, et al., 2017; Beake, Vishnyakov, et al., 2017). The $Al_{0.67}Ti_{0.33}N$ coating exhibited better resistance to fracture in the elevated temperature nano-impact test and improved cutting performance than the $Ti_{0.5}Al_{0.50}N$ (Beake et al., 2007). Micro-impact tests with a 20 μm radius probe have been performed on 2 μm PVD monolayer TiAlSiN and nanomultilayer TiAlN/TiSiN coatings on a cemented carbide substrate containing 10 wt.% Co (Beake, Bird, et al., 2019; Beake, Isern, et al., 2019). Both coatings show a brittle response at room temperature with fatigue and a transition to more rapid wear. At higher temperatures, the dominant fatigue mechanism switched from fracture-dominated to plasticity-dominated deformation, by 600°C for the monolayer TiAlSiN and by 500°C for the nanomultilayer TiAlN/TiSiN. This is due to significant substrate (rather than coating) softening, since (1) high temperature nanoindentation to 600°C showed relatively small decrease in coating hardness with temperature; (2) micro-scratch with the same 20 μm radius probe showed considerable plasticity at temperature, consistent with the expected greater decrease in hardness with temperature for the substrate with higher Co fraction. In these elevated-temperature impact test studies (Beake et al., 2007; Beake, Bird, et al., 2019; Beake, Isern, et al., 2019), increasing temperature lowered the mechanical properties of the coating systems, which resulted in reduced fracture in the impact tests. Interestingly, for $Ti_{0.5}Al_{0.50}N$ and $Al_{0.67}Ti_{0.33}N$ on H10A cemented carbide, this was primarily due to coating softening, and for the TiAlSiN coatings on the cemented carbide with higher cobalt fraction, it was primarily due to substrate softening. This difference was partly due to changing mechanical properties and partly due to the sharpness of the probes used.

3.3 Conclusions

The effectiveness of different strategies in coating design to achieve high wear resistance by combining hardness and toughness is dependent on the severity and dominant type of mechanical contact (e.g., impact/erosion or sliding/abrasion) in the application. Coatings that are designed to perform well in one situation may perform poorly in another. H/E and H^3/E^2 often are at least as useful as hardness as screening parameters in coating optimization. Although increasing H^3/E^2 can be beneficial in enhancing resistance to crack initiation, H^3/E^2 should generally not be considered as a proxy for fracture toughness.

Nano-/microscale (single asperity) contact tests have been performed to understand why these H/E and H^3/E^2 coating ratios can be effective predictors of wear resistance in some complex wear situations and less effective in others. Insights from these experimental tests combined with analytical modeling of the stresses developed in sliding contact have indicated that a relatively low coating stiffness can be an important factor. The coating mechanical properties control how the coating − substrate interface is weakened (by initial substrate or coating yielding or both) at high load and resultant failure mechanism. Low stiffness is beneficial by (1) load spreading over a wider area (through higher H^3/E^2), (2) avoiding high tensile stresses that develop in highly loaded sliding contact when E_c/E_s is $\gg 1$. Lower coating stiffness, which reduces the modulus mismatch with the substrate, is an important factor in improving crack resistance. For this reason optimizing coating H/E through nanocomposite (rather than "ceramic" where higher H is accompanied by higher E) coating design is expected to be more effective on lower modulus substrates (e.g., steel, aluminum, titanium alloys, silicon) rather than on stiff cemented carbide substrate, as was originally proposed by Leyland and Matthews.

When developing structure − properties − performance relationships, it is useful to consider that microstructural changes often are responsible for the changing coating mechanical properties. Microstructural control for higher density, with more complex coating architectures and graded mechanical properties can produce advanced multifunctional coating systems where higher hardness is combined with enhanced toughness and microstructural advantages (e.g., multilayering) to exploit their improved mechanical properties. If PVD coatings can be designed to be denser with less pronounced columnar structure, then this removes the through-thickness crack propagation along columnar boundaries as the weakest link and provides a microstructural advantage for higher coating H^3/E^2 to be effective. Designing coatings with in-built dissipative structures and mechanisms to combat stress—including low-friction tribofilm formation as an example of self-adaptive behavior—can be a more successful approach to increase their wear resistance than aiming to minimize plastic deformation by simply increasing hardness.

It is important to consider the coating system as a whole and not only optimize the properties of the top layer. The role of the substrate—its brittleness and load support—also can be critical since accommodating substrate deformation without cracking and failure is key to performance of coated components. For example, DLC coatings on hardened tool steel experience will greater bending strains than coatings on deposited on the harder and stiffer cemented carbide. Under these conditions (on steel), there appears to be less benefit in designing coatings to be as hard as possible with very high H^3/E^2. It is important to consider the severity of the test. High coating H^3/E^2 can be detrimental under highly loaded contact when deposited on a low stiffness substrate as it will experience high bending strains, if there is no compensating microstructural/dissipative advantage to compensate for the high brittleness.

Contact size effects can be critical in coated contacts as the severity of loading in the test influences the development of the stresses including the depth of von Mises stresses, which affect substrate plasticity and coating bending. Closer simulation of the single asperity contact to the actual deformation by, for example, increasing strain rate and temperature can increase the usefulness of the data obtained. Temperature is important as it can alter the balance between deformation and fracture. Simulated stress distributions, for example, of sliding contact, show reduced stresses are present in high temperature contact than at room temperature, particularly when the coating or substrate properties are highly temperature-dependent.

The extent of the decrease in hardness, elastic modulus, H/E, and H^3/E^2 at elevated temperature has a significant influence on the performance of PVD nitride coatings on cemented carbide in high-temperature mechanical contact applications—such as in tribologically extreme conditions of high-speed machining—which have different relative requirements in terms of hot hardness and fracture resistance to ensure high resistance to wear. In nano/microscale tribotests, a switch from brittle fracture-dominated deformation to more plasticity-dominated deformation was found as the temperature increased. In the coating systems studied experiments showed that this was primarily a result of either coating or substrate softening in these cases.

A coating design strategy for high toughness could combine a high load threshold for the initiation of cracking with mechanisms to minimize and retard crack propagation. The enhanced performance shown in tribological tests and machining applications by nanomultilayered nitride coatings compared with monolayered nitride coatings is considered to be due to a combination of microstructural advantages with a dense microstructure to eliminate/minimize weak columnar boundaries and nanolaminate structure providing resistance to through-thickness cracking by crack deflection between layers, and excellent high temperature mechanical properties, with high H^3/E^2 giving greater load-carrying capability and resistance to crack initiation.

References

Abadias, G., Chason, E., Keckes, J., Sebastiani, M., Thompson, G. B., Barthel, E., Doll, G. L., Murray, C. E., Stoessel, C. H., & Martinu, L. (2018). Stress in thin films and coatings: Current status, challenges, and prospects. *Journal of Vacuum Science & Technology. A, Vacuum, Surfaces, and Films: An Official Journal of the American Vacuum Society, 36*, 020801, 49pp.

Allsopp, D. N., & Hutchings, I. M. (2001). Micro-scale abrasion and scratch response of PVD coatings at elevated temperatures. *Wear, 251*(1–12), 1308–1314. Available from https://doi.org/10.1016/s0043-1648(01)00755-4.

Archard, J. F. (1953). Contact and rubbing of flat surfaces. *Journal of Applied Physics, 24*(8), 981–988. Available from https://doi.org/10.1063/1.1721448.

Bai, M., Yang, L., Li, J., Luo, L., Sun, S., & Inkson, B. (2021). Mechanical and tribological properties of Si and W doped diamond like carbon (DLC) under dry reciprocating sliding conditions. *Wear., 484–485*. Available from https://doi.org/10.1016/j.wear.2021.204046.

Baker, M. A., Klose, S., Rebholz, C., Leyland, A., & Matthews, A. (2002). Evaluating the microstructure and performance of nanocomposite PVD TiAlBN coatings. *Surface and Coatings Technology, 151–152*, 338–343. Available from https://doi.org/10.1016/s0257-8972(01)01657-7.

Bantle, R., & Matthews, A. (1995). Investigation into the impact wear behaviour of ceramic coatings. *Surface and Coatings Technology, 74–75*(2), 857–868. Available from https://doi.org/10.1016/0257-8972(95)08314-6.

Bartosik, M., Rumeau, C., Hahn, R., Zhang, Z. L., & Mayrhofer, P. H. (2017). Fracture toughness and structural evolution in the TiAlN system upon annealing. *Scientific Reports*, *7*(1). Available from https://doi.org/10.1038/s41598-017-16751-1.

Beake, B. D. (2021). Elevated temperature nanomechanics of coatings for high-temperature applications: A review. *Emergent Materials.*, *4*(6), 1531−1545. Available from https://doi.org/10.1007/s42247-021-00255-w, Springer.com/journal/42247.

Beake, B. D., Endrino, J. L., Kimpton, C., Fox-Rabinovich, G. S., & Veldhuis, S. C. (2017). Elevated temperature repetitive micro-scratch testing of AlCrN, TiAlN and AlTiN PVD coatings. *International Journal of Refractory Metals and Hard Materials*, *69*, 215−226. Available from https://doi.org/10.1016/j.ijrmhm.2017.08.017.

Beake, B. D., & Fox-Rabinovich, G. S. (2014). Progress in high temperature nanomechanical testing of coatings for optimising their performance in high speed machining. *Surface and Coatings Technology*, *255*, 102−111. Available from https://doi.org/10.1016/j.surfcoat.2014.02.062.

Beake, B. D., Fox-Rabinovich, G. S., Veldhuis, S. C., & Goodes, S. R. (2009). Coating optimisation for high speed machining with advanced nanomechanical test methods. *Surface and Coatings Technology*, *203*(13), 1919−1925. Available from https://doi.org/10.1016/j.surfcoat.2009.01.025.

Beake, B. D., & Harris, A. J. (2019). Nanomechanics to 1000°C for high temperature mechanical properties of bulk materials and hard coatings. *Vacuum*, *159*, 17−28. Available from https://doi.org/10.1016/j.vacuum.2018.10.011.

Beake, B. D., Liskiewicz, T. W., Bird, A., & Shi, X. (2020). Micro-scale impact testing − A new approach to studying fatigue resistance in hard carbon coatings. *Tribology International*, *149*. Available from https://doi.org/10.1016/j.triboint.2019.04.016, http://www.elsevier.com/inca/publications/store/3/0/4/7/4.

Beake, B. D., & Liskiewicz, T. W. (2013). Comparison of nano-fretting and nano-scratch tests on biomedical materials. *Tribology International*, *63*, 123−131. Available from https://doi.org/10.1016/j.triboint.2012.08.007.

Beake, B. D., Liskiewicz, T. W., Vishnyakov, V. M., & Davies, M. I. (2015). Development of DLC coating architectures for demanding functional surface applications through nano- and micro-mechanical testing. *Surface and Coatings Technology*, *284*, 334−343. Available from https://doi.org/10.1016/j.surfcoat.2015.05.050.

Beake, B. D., Smith, J. F., Gray, A., Fox-Rabinovich, G. S., Veldhuis, S. C., & Endrino, J. L. (2007). Investigating the correlation between nano-impact fracture resistance and hardness/modulus ratio from nanoindentation at 25−500°C and the fracture resistance and lifetime of cutting tools with Ti1 − xAlxN (x = 0.5 and 0.67) PVD coatings in milling operations. *Surface and Coatings Technology*, *201*(8), 4585−4593. Available from https://doi.org/10.1016/j.surfcoat.2006.09.118.

Beake, B. D., & Smith, J. F. (2004). Nano-impact testing − An effective tool for assessing the resistance of advanced wear-resistant coatings to fatigue failure and delamination. *Surface and Coatings Technology*, *188−189C*, 594.

Beake, B. D., Vishnyakov, V. M., Valizadeh, R., & Colligon, J. S. (2006). Influence of mechanical properties on the nanoscratch behaviour of hard nanocomposite TiN/Si_3N_4 coatings on Si. *Journal of Physics D: Applied Physics*, *39*(7), 1392−1397. Available from https://doi.org/10.1088/0022-3727/39/7/009.

Beake, B. D. (2022). The influence of the H/E ratio on wear resistance of coating systems−Insights from small-scale testing. *Surface and Coatings Technology*, *442*. Available from https://doi.org/10.1016/j.surfcoat.2022.128272.

Beake, B. D., Bird, A., Isern, L., Endrino, J. L., & Jiang, F. (2019). Elevated temperature micro-impact testing of TiAlSiN coatings produced by physical vapour deposition. *Thin Solid Films*, *688*, 137358, 9pp.

Beake, B. D., Isern, L., Endrino, J. L., & Fox-Rabinovich, G. S. (2019). Micro-impact testing of AlTiN and TiAlCrN coatings. *Wear*, *418–419*, 102–110.

Beake, B. D., Isern, L., Endrino, J. L., Liskiewicz, T. W., & Shi, X. (2021). Micro-scale impact resistance of coatings on hardened tool steel and cemented carbide. *Materials Letters*, *284*, 129009.

Beake, B. D., McMaster, S. J., Liskiewicz, T. W., & Neville, A. (2021). Influence of Si- and W- doping on micro-scale reciprocating wear and impact performance of DLC coatings on hardened steel. *Tribology International*, *160*, 107063.

Beake, B. D., Vishnyakov, V. M., & Harris, A. J. (2011). Relationship between mechanical properties of thin nitride-based films and their behaviour in nano-scratch tests. *Tribology International*, *44*, 468.

Beake, B. D., Vishnyakov, V. M., & Harris, A. J. (2017). Nano-scratch testing of (Ti, Fe)Nx thin films on silicon. *Surface and Coatings Technology*, *309*, 671–679.

Beake, B. D., Vishnyakov, V. M., & Liskiewicz, T. W. (2021). Integrated nanomechanical characterisation of hard coatings. In S. Zhang, J.-M. Ting, & W.-Y. Wu (Eds.), *Protective thin coatings technology* (pp. 95–140). CRC Press.

Bernoulli, D., Wyss, A., Raghavan, R., Thorwarth, K., Hauert, R., & Spolenak, R. (2015). Contact damage of hard and brittle thin films on ductile metallic substrates: an analysis of diamond-like carbon on titanium substrates. *Journal of Materials Science*, *50*(7), 2779–2787. Available from https://doi.org/10.1007/s10853-015-8833-3.

Blau, P. J. (2010). Elevated-temperature tribology of metallic materials. *Tribology International*, *43*(7), 1203–1208. Available from https://doi.org/10.1016/j.triboint.2010.01.003.

Bousser, E., Benkahoul, M., Martinu, L., & Klemberg-Sapieha, J. E. (2008). Effect of microstructure on the erosion resistance of Cr−Si−N coatings. *Surface and Coatings Technology*, *203*(5–7), 776–780. Available from https://doi.org/10.1016/j.surfcoat.2008.08.012.

Bousser, E., Martinu, L., & Klemberg-Sapieha, J. E. (2014). Solid particle erosion mechanisms of protective coatings for aerospace applications. *Surface and Coatings Technology*, *257*, 165–181.

Bouzakis, K.-D., Asimakopoulos, A., Skordaris, G., Pavlidou, E., & Erkens, G. (2007). The inclined impact test: A novel method for the quantification of the adhesion properties of PVD films. *Wear*, *262*(11–12), 1471–1478. Available from https://doi.org/10.1016/j.wear.2007.01.027.

Bouzakis, K.-D., Klocke, F., Skordaris, G., Bouzakis, E., Gerardis, S., Katirtzoglou, G., & Makrimallakis, S. (2011). Influence of dry micro-blasting grain quality on wear behaviour of TiAlN coated tools. *Wear*, *271*(5–6), 783–791. Available from https://doi.org/10.1016/j.wear.2011.03.010.

Bouzakis, K. D., Michailidis, N., Skordaris, G., Bouzakis, E., Biermann, D., & M'Saoubi, R. (2012). Cutting with coated tools: Coating technologies, characterization methods and performance optimization. *CIRP Annals - Manufacturing Technology*, *61*(2), 703–723. Available from https://doi.org/10.1016/j.cirp.2012.05.006.

Bouzakis, K.-D., Siganos, A., Leyendecker, T., & Erkens, G. (2004). Thin hard coatings fracture propagation during the impact test. *Thin Solid Films*, *460*(1–2), 181–189. Available from https://doi.org/10.1016/j.tsf.2004.02.009.

Bouzakis, K. D., Skordaris, G., Bouzakis, E., Kotsanis, T., & Charalampous, P. (2017). A critical review of characteristic techniques for improving the cutting performance of coated tools. *Journal of Machine Engineering*, *17*, 25−44.

Brazhkin, V. V., & Solozhenko, V. L. (2019). Myths about new ultrahard phases: Why materials that are significantly superior to diamond in elastic moduli and hardness are impossible. *Federation Journal of Applied Physics*, *125*(13). Available from https://doi.org/10.1063/1.5082739, http://scitation.aip.org/content/aip/journal/jap.

Bromark, M., Hedenqvist, P., & Hogmark, S. (1995). The influence of substrate material on the erosion resistance of TiN coated steels. *Wear*, *186−187*, 189−194.

Bull, S. J. (2005). Nanoindentation of coatings. *Journal of Physics D: Applied Physics*, *38*(24), R393−R413. Available from https://doi.org/10.1088/0022-3727/38/24/r01.

Bull, S. J. (2006). Using work of indentation to predict erosion behavior in bulk materials and coatings. *Journal of Physics D: Applied Physics*, *39*(8), 1626−1634. Available from https://doi.org/10.1088/0022-3727/39/8/023.

Bull, S. J. (1999). Can scratch testing be used as a model for the abrasive wear of hard coatings? *Wear*, *233−235*, 412−423. Available from https://doi.org/10.1016/s0043-1648(99)00207-0.

Bull, S. J. (2015). A simple method for the assessment of the contact modulus for coated systems. *Philosophical Magazine*, *95*(16−18), 1907−1927. Available from https://doi.org/10.1080/14786435.2014.909612.

Bull, S. J. (2019). Microstructure and indentation response of TiN coatings: The effect of measurement method. *Thin Solid Films*, *688*. Available from https://doi.org/10.1016/j.tsf.2019.137452.

Chang, Y.-Y., & Wu, C.-J. (2013). Mechanical properties and impact resistance of multilayered TiAlN/ZrN coatings. *Surface and Coatings Technology*, *231*, 62−66.

Chavoshi, S. Z., & Xu, S. (2018). A review on micro- and nanoscratching/tribology at high temperatures: Instrumentation and experimentation. *Journal of Materials Engineering and Performance*, *27*(8), 3844−3858. Available from https://doi.org/10.1007/s11665-018-3493-5, http://www.springerlink.com/content/1059-9495/.

Chen, J., Ji, R., Khan, R. H. U., Li, X., Beake, B. D., & Dong, H. (2011). Effects of mechanical properties and layer structure on the cyclic dynamic loading of TiN-based coatings. *Surface and Coatings Technology*, *206*(2−3), 522−529. Available from https://doi.org/10.1016/j.surfcoat.2011.07.079.

Chen, X., Du, Y., & Chung, Y.-W. (2019). Commentary on using H/E and H^3/E^2 as proxies for fracture toughness of hard coatings. *Thin Solid Films*, *688*, 137265.

Cheng, Y. T., & Cheng, C. M. (1998). Relationships between hardness, elastic modulus, and the work of indentation. *Applied Physics Letters*, *73*(5), 614−616. Available from https://doi.org/10.1063/1.121873.

Cheng, Y.-T., & Cheng, C.-M. (2004). Scaling, dimensional analysis, and indentation measurements. *Materials Science and Engineering R*, *44*, 91−149.

Cheng, Y.-T., Li, Z., & Cheng, C.-M. (2002). Scaling relationships for indentation measurements, 2004 *Philosophical Magazine*, *82*, 1821−1830.

Chudoba, T. (2006). Measurement of hardness and Young's modulus of coatings by nanoindentation. In A. Cavaleiro, & J. T. M. De Hosson (Eds.), *Nanostructuredcoatings* (pp. 216−260). New York: Springer.

Chudoba, T., Schwarzer, N., & Richter, F. (2002). Steps towards a mechanical modelling of layered systems. *Surface and Coatings Technology*, *154*, 140.

Chychko, A., García, J.,Ciprés, V. C., Holmström, E., & Blomqvist, A. (2022). HV-KIC property charts of cemented carbides: A comprehensive data collection. *International*

Journal of Refractory Metals and Hard Materials, 103. Available from https://doi.org/10.1016/j.ijrmhm.2021.105763.

Diao, D. F., Kato, K., & Hayashi, K. (1994). The maximum tensile stress on a hard coating under sliding friction. *Tribology International, 27*, 267–272.

Endrino, J. L., Fox-Rabinovich, G. S., & Gey, C. (2006). Hard AlTiN, AlCrN PVD coatings for machining of austenitic stainless steel. *Surface and Coatings Technology, 200*, 6840–6845.

Fischer-Cripps, A. C. (2006). Critical review of analysis and interpretation of nanoindentation test data. *Surface and Coatings Technology, 200*(14–15), 4153–4165. Available from https://doi.org/10.1016/j.surfcoat.2005.03.018.

Fischer-Cripps, A. C. (2014). Measurement of hardness of very hard materials. *Solid Mechanics and its Applications, 203*, 53–62. Available from https://doi.org/10.1007/978-94-007-6919-9_3, http://www.springer.com/series/6557.

Fischer-Cripps, A. C., Bull, S. J., & Schwarzer, N. (2012). Critical review of claims for ultrahardness in nanocomposite coatings. *Philosophical Magazine, 92*, 1601–1630.

Fox-Rabinovich, G. S., Endrino, J. L., Beake, B. D., Aguirre, M. H., Veldhuis, S. C., Quinto, D. T., Bauer, C. E., Kovalev, A. I., & Gray, A. (2008). Effect of annealing below 900°C on structure, properties and tool life of an AlTiN coating under various cutting conditions. *Surface and Coatings Technology, 202*.

Fox-Rabinovich, G. S., Endrino, J. L., Beake, B. D., Kovalev, A. I., Veldhuis, S. C., Ning, L., Fontaine, F., & Gray, A. (2006). Impact of annealing on microstructure, properties and cutting performance of an AlTiN coating. *Surface and Coatings Technology, 201* (6), 3524–3529. Available from https://doi.org/10.1016/j.surfcoat.2006.08.075.

Fox-Rabinovich, G. S., Kovalev, A. I., Aguirre, M. H., Beake, B. D., Yamamoto, K., Veldhuis, S. C., Endrino, J. L., Wainstein, D. L., & Rashkovskiy, A. Y. (2009). Design and performance of AlTiN and TiAlCrN PVD coatings for machining of hard to cut materials. *Surface and Coatings Technology, 204*(4), 489–496. Available from https://doi.org/10.1016/j.surfcoat.2009.08.021.

Fox-Rabinovich, G. S., Beake, B. D., Yamamoto, K., Aguirre, M. H., Veldhuis, S. C., Dosbaeva, G., Elfizy, A., Biksa, A., Shuster, L. S., & Rashkovskiy, A. Y. (2010). Structure, properties and wear performance of nano-multilayered TiAlCrSiYN/TiAlCrN coatings during machining of Ni-based aerospace superalloys. *Surface and Coatings Technology, 204*, 3698–3706, 2010.

Fox-Rabinovich, G. S., Endrino, J. L., Agguire, M. H., Beake, B. D., Veldhuis, S. C., Kovalev, A. I., Gershman, I. S., Yamamoto, K., Losset, Y., Wainstein, D. L., & Rashkovskiy, A. Y. (2012). Mechanism of adaptability for the nano-structured TiAlCrSiYN-based hard physical vapor deposition coatings under extreme frictional conditions. *Journal of Applied Physics, 111*, 064306.

Fox-Rabinovich, G. S., Kovalev, A., Aguirre, M. H., Yamamoto, K., Veldhuis, S., Gershman, I., Rashkovskiy, A., Endrino, J. L., Beake, B. D., Dosbaeva, G., Wainstein, D., Yuan, J., & Bunting, J. W. (2014). Evolution of self-organization in nano-structured PVD coatings under extreme tribological conditions. *Applied Surface Science, 297*, 22–32.

Fox-Rabinovich, G. S., Yamamoto, K., Beake, B. D., Kovalev, A. I., Aguirre, M. I., Veldhuis, S. C., Dosbaeva, G. K., Wainstein, D. L., Biksa, A., & Rashkovskiy, A. Y. (2010). Emergent behavior of nano-multilayered coatings during dry high speed machining of hardened tool steels. *Surface and Coatings Technology, 204*, 3425–3435.

Fox-Rabinovich, G. S., Yamamoto, K., Beake, B. S. D., Gershman, I. S., Kovalev, A. I., Veldhuis, S. C., Aguirre, M. H., Dosbaeva, G., & Endrino, J. L. (2012). Hierarchical adaptive nanostructured PVD coatings for extreme tribological applications: The quest for nonequilibrium states and emergent behaviour. *Science and Technology of Advanced Materials*, *13*, 043001, 26pp.

Fuentes, G. G., Almandoz, E., Rodríquez, R. J., Dong, H., Qin, Y., Mato, S., et al. (2014). Vapour deposition technologies for the fabrication of hot-forming tools. *Manufacturing Review*, *1*, 20.

Giuliani, F., Ciurea, C., Bhakhri, V., Werchota, M., Vandeperre, L. J., & Mayrhofer, P. H. (2019). Deformation behaviour of TiN and Ti−Al−N coatings at 295 to 573 K. *Thin Solid Films*, *688*. Available from https://doi.org/10.1016/j.tsf.2019.06.013, http://www.journals.elsevier.com/journal-of-the-energy-institute.

Goltsberg, R., & Etsion, I. (2013). A model for the weakening effect of very thin hard coatings. *Wear*, *308*, 10−16.

Goltsberg, R., Etsion, I., & Davidi, G. (2011). The onset of plastic yielding in a coated sphere compressed by a rigid flat. *Wear*, *271*, 2968−2977.

Greenwood, J. A., & Williamson, J. P. (1966). Contact of nominally flat surfaces. *Proceedings of the Royal Society A*, *295*, 300−319.

Gsellmann, M., Klünsner, T., Mitterer, C., Marsoner, S., Skordaris, G., Bouzakis, K., Leitner, H., & Ressel, G. (2020). Near-interface cracking in a TiN coated high speed steel due to combined shear and compression under cyclic impact loading. *Surface and Coatings Technology*, *394*. Available from https://doi.org/10.1016/j.surfcoat.2020.125854.

Hahn, R., Bartosik, M., Soler, R., Kirchlechner, C., Dehm, G., & Mayrhofer, P. H. (2016). Superlattice effect for enhanced fracture toughness of hard coatings. *Scripta Materialia*, *124*, 67−70. Available from https://doi.org/10.1016/j.scriptamat.2016.06.030.

Hamilton, G. M., & Goodman, L. E. (1966). The stress field created by a circular sliding contact. *Journal of Applied Mechanics*, *33*(2), 371−376. Available from https://doi.org/10.1115/1.3625051.

Hassani, S., Bielawski, M., Beres, W., Martinu, L., Balazinski, M., & Klemberg-Sapieha, J. E. (2008). Predictive tools for the design of erosion resistant coatings. *Surface and Coatings Technology*, *203*(3−4), 204−210. Available from https://doi.org/10.1016/j.surfcoat.2008.08.050.

He, D., Li, W., Wang, L., Lu, Z., Zhang, G., & Cai, Z. (2019). Impact wear behaviour of WC/a-C nanomultilayers. *Materials Research Express*, *6*.

Holmberg, K., Ronkainen, H., Laukkanen, A., & Wallin, K. (2007). Friction and wear of coated surfaces — Scales, modelling and simulation of tribomechanisms. *Surface and Coatings Technology*, *202*(4−7), 1034−1049. Available from https://doi.org/10.1016/j.surfcoat.2007.07.105.

Holmberg, K., Ronkainen, H., Laukkanen, A., Wallin, K., Erdemir, A., & Eryilmaz, O. (2008). Tribological analysis of TiN and DLC coated contacts by 3D FEM modelling and stress simulation. *Wear*, *264*, 877−884.

Huang, X., Kasem, H., Shang, H. F., Shao, T. M., & Etsion, I. (2012). Experimental study of a potential weakening effects in spheres with thin hard coatings. *Wear*, *296*, 590−597.

Huang, H., Lawn, B. R., Cook, R. F., & Marshall, D. B. (2020). Critique of materials-based models of ductile machining in brittle solids. *Journal of the American Ceramic Society*, *103*(11), 6096−6100. Available from https://doi.org/10.1111/jace.17344, http://onlinelibrary.wiley.com/journal/10.1111/(ISSN)1551-2916.

Huang, X., Etsion, I., & Shao, T. (2015a). Effects of elastic modulus mismatch between coating and substrate on the friction and wear properties of TiN and TiAlN coating systems. *Wear*, *338−339*, 54−61.

Huang, X., Etsion, I., & Shao, T. (2015b). Indentation pop-in as a potential characterization of weakening effect in coating/substrate systems. *Wear*, *338−339*, 325−331.

Hultman, L., & Mitterer, C. (2006). Nanostructured coatings. In A. Cavaleiro, & J. T. M. De Hosson (Eds.), *Chapter 11, "Thermal stability of advanced nanostructured wear-resistant coatings"* (pp. 464−510). New York: Springer.

Hutchings, I. M. (1992). Ductile-brittle transitions and wear maps for the erosion and abrasion of brittle materials. *Journal of Physics D: Applied Physics, 25*(1A), A212−A221. Available from https://doi.org/10.1088/0022-3727/25/1a/033.

INDICOAT. (2001). *Determination of Hardness and Modulus of Thin Films and Coatings by Nanoindentation (INDICOAT)*, European project, Contract No. SMT4-CT98−2249, NPL Report MATC(A) 24, May 2001.

Inspektor, A., & Salvador, P. A. (2014). Architecture of PVD coatings for metalcutting applications: A review. *Surface and Coatings Technology, 257*, 138−153. Available from https://doi.org/10.1016/j.surfcoat.2014.08.068, http://www.journals.elsevier.com/surface-and-coatings-technology/.

ISO 14577. (2015). *Metallic materials—Instrumented indentation test for hardness and materials parameters*, Part 4—Test method for metallic and non-metallic coatings.

Jantschner, O., Walter, C., Muratore, C., Voevodin, A. A., & Mitterer, C. (2013). V-alloyed ZrO$_2$ coatings with temperature homogenization function for high-temperature sliding contacts. *Surface and Coatings Technology, 228*, 76−83. Available from https://doi.org/10.1016/j.surfcoat.2013.04.009.

Jiang, J., & Arnell, R. D. (1998). On the running-in behaviour of diamond-like carbon coatings under the ball-on-disk contact geometry. *Wear, 217*, 190−199.

Jinhal, P. C., Santhanam, A. T., Schleinkofer, U., & Shuster, A. F. (1999). Performance of PVD TiN, TiCN, and TiAlN coated cemented carbide tools in turning. *International Journal of Refractory Metals and Hard Materials, 17*, 163−170.

Johnson K.L. (1985). *Contact mechanics*. 9780521255769.

Kalss, W., Reiter, A., Derflinger, V., Gey, C., & Endrino, J. L. (2006). Modern coatings in high performance cutting applications. *International Journal of Refractory Metals and Hard Materials, 24*(5), 399−404. Available from https://doi.org/10.1016/j.ijrmhm.2005.11.005.

Kato, K. (1995). Microwear mechanisms of coatings. *Surface and Coatings Technology, 76−77*(2), 469−474. Available from https://doi.org/10.1016/0257-8972(95)02570-7.

Kim, D. K., Jung, Y.-G., Peterson, I. M., & Lawn, B. R. (1999). Cyclic fatigue of intrinsically brittle ceramics in contact with spheres. *Acta Materialia, 47*, 4711−4725.

Knotek, O., Bosserhoff, B., Schrey, A., Leyendecker, T., Lemmer, O., & Esser, S. (1992). A new technique for testing the impact load of thin films: The coating impact test. *Surface and Coatings Technology, 54−55*, 102−107. Available from https://doi.org/10.1016/S0257-8972(09)90035-4.

Komvopoulos, K. (1989). Elastic-plastic finite element analysis of indented layered media. *Journal of Tribology, 111*(3), 430−439. Available from https://doi.org/10.1115/1.3261943.

Krella, A. (2020). Resistance of PVD coatings to erosive and wear processes: A review. *Coatings, 10*(10). Available from https://doi.org/10.3390/coatings10100921.

Lanigan, J. L., Wang, C., Morina, A., & Neville, A. (2016). Repressing oxidative wear within Si doped DLCs. *Tribology International, 93*, 651−659. Available from https://doi.org/10.1016/j.triboint.2014.11.004.

Lavigne, O., Cinca, N., Ther, O., & Tarrés, E. (2022). Effect of binder nature and content on the cavitation erosion resistance of cemented carbides. *International Journal of Refractory Metals and Hard Materials, 109*. Available from https://doi.org/10.1016/j.ijrmhm.2022.105978.

Lawes, S. D. A., Hainsworth, S. V., & Fitzpatrick, M. E. (2010). Impact wear testing of diamond-like carbon films for engine valve-tappet surfaces. *Wear*, *268*, 1303−1308.

Leyland, A., & Matthews, A. (2000). On the significance of the H/E ratio in wear control: a nanocomposite coating approach to optimised tribological behaviour. *Wear*, *246*(1−2), 1−11. Available from https://doi.org/10.1016/s0043-1648(00)00488-9.

Leyland, A., & Matthews, A. (2004). Design criteria for wear-resistant nanostructured glassy-metal coatings. *Surface and Coatings Technology*, *177−178*, 317−324.

Lo, W.-L., Hsu, S.-Y., Lin, Y.-C., Tsai, S.-Y., Lai, Y.-T., & Duh, J.-G. (2020). Improvement of high entropy alloy nitride coatings (AlCrNbSiTiMo)N on mechanical and high temperature tribological properties by tuning substrate bias. *Surface and Coatings Technology*, *401*, 126247.

Lorenz, L., Chudoba, T., Makowski, S., Zawischa, M., Schaller, F., & Weihnacht, V. (2021). Indentation modulus extrapolation and thickness estimation of ta-C coatings from nanoindentation. *Journal of Materials Science*, *56*(33), 18740−18748. Available from https://doi.org/10.1007/s10853-021-06448-2.

Malzbender, J. (2005). Comment on the determination of mechanical properties from the energy dissipated during indentation. *Journal of Materials Research*, *20*(5), 1090−1092. Available from https://doi.org/10.1557/jmr.2005.0162.

Malzbender, J., & de With, G. (2002). Indentation load−displacement curve, plastic deformation, and energy. *Journal of Materials Research*, *17*(2), 502−511. Available from https://doi.org/10.1557/jmr.2002.0070.

Mayrhofer, P. H., Mitterer, C., Hultman, L., & Clemens, H. (2006). Microstructural design of hard coatings. *Progress in Materials Science*, *51*(8), 1032−1114. Available from https://doi.org/10.1016/j.pmatsci.2006.02.002.

McMaster, S. J., Liskiewicz, T. W., Neville, A., & Beake, B. D. (2020). Probing fatigue resistance in multi-layer DLC coatings by micro- and nano-impact: Correlation to erosion tests. *Surface and Coatings Technology*, *402*. Available from https://doi.org/10.1016/j.surfcoat.2020.126319, http://www.journals.elsevier.com/surface-and-coatings-technology/.

Meindlhumer, M., Brandt, L. R., Zalesak, J., Rosenthal, M., Hruby, H., Kopecek, J., Salvati, E., Mitterer, C., Daniel, R., Todt, J., Keckes, J., & Korsunsky, A. M. (2021). Evolution of stress fields during crack growth and arrest in a brittle-ductile CrN-Cr clamped-cantilever analysed by X-ray nanodiffraction and modelling. *Materials & Design*, *198*. Available from https://doi.org/10.1016/j.matdes.2020.109365.

Mendibide, C., Fontaine, J., Steyer, P., & Esnouf, C. (2004). Dry sliding wear model of nanometer scale multilayered TiN/CrN PVD hard coatings. *Tribology Letters*, *17*(4), 779−789. Available from https://doi.org/10.1007/s11249-004-8086-9.

Mendibide, C., Steyer, P., Fontaine, J., & Goudeau, P. (2006). Improvement of the tribological behavior of PVD nanostratified TiN/CrN coatings − An explanation. *Surface and Coatings Technology*, *201*, 4119−4124.

Michailidis, N. (2016). Variations in the cutting performance of PVD-coated tools in milling Ti6Al4V, explained through temperature-dependent coating properties. *Surface and Coatings Technology*, *304*, 325−329.

Michler, J., & Blank, E. (2001). Analysis of coating fracture and substrate plasticity induced by spherical indentors: Diamond and diamond-like carbon layers on steel substrates. *Thin Solid Films*, *381*(1), 119−134. Available from https://doi.org/10.1016/s0040-6090(00)01340-7.

Milman, Y. V., Luyckx, S., & Northrop, I. T. (1999). Influence of temperature, grain size and cobalt content on the hardness of WC-Co alloys. *International Journal of Refractory Metals and Hard Materials, 17,* 39–44.

Miranda, P., Pajares, A., Guiberteau, F., Deng, Y., & Lawn, B. R. (2003). Designing damage-resistant brittle-coating structures: I. Bilayers. *Acta Materialia, 51*(14), 4347–4356. Available from https://doi.org/10.1016/S1359-6454(03)00290-8.

Mishra, A. K., Gopalan, H., Hans, M., Kirchlechner, C., Schneider, J. M., Dehm, G., & Jaya, B. N. (2022). Strategies for damage tolerance enhancement in metal/ceramic thin films: Lessons learned from Ti/TiN. *Acta Materialia, 228.* Available from https://doi.org/10.1016/j.actamat.2022.117777, http://www.journals.elsevier.com/acta-materialia/.

Mo, J. L., Zhu, M. H., Leyland, A., & Matthews, A. (2013). Impact wear and abrasion resistance of CrN, AlCrN and AlTiN PVD coatings. *Surface and Coatings Technology, 215,* 170–177. Available from https://doi.org/10.1016/j.surfcoat.2012.08.077.

Musil, J., & Jirout, M. (2007). Toughness of hard nanostructured ceramic thin films. *Surface and Coatings Technology, 201*(9–11), 5148–5152. Available from https://doi.org/10.1016/j.surfcoat.2006.07.020.

Musil, J. (2006). *Physical and mechanical properties of hard nanocomposite films prepared by reactive magnetron sputtering* (pp. 407–463). Springer Nature. Available from 10.1007/978-0-387-48756-4_10.

Musil, J. (2012). Hard nanocomposite coatings: Thermal stability, oxidation resistance and toughness. *Surface and Coatings Technology, 207,* 50–65.

Musil, J. (2015). Advanced hard coatings with enhanced toughness and resistance to cracking. In S. Zhang (Ed.), *Thin films and coatings: Toughening and toughness characterisation* (pp. 378–463). CRC Press, p383.

Mutafov, P., Lanigan, J., Neville, A., Cavaleiro, A., & Polcar, T. (2014). DLC-W coatings tested in combustion engine — Frictional and wear analysis. *Surface and Coatings Technology, 260,* 284–289, 2014.

Naskar, A., & Chattopadhyay, A. K. (2018). Investigation on flank wear mechanism of CVD and PVD hard coatings in high speed dry turning of low and high carbon steel. *Wear, 396–397,* 98–106.

Ou, Y. X., Lin, J., Che, H. L., Sproul, W. D., Moore, J. J., & Lei, M. K. (2015). Mechanical and tribological properties of CrN/TiN multilayer coatings deposited by pulsed dc magnetron sputtering. *Surface and Coatings Technology, 276,* 152–159. Available from https://doi.org/10.1016/j.surfcoat.2015.06.064.

Palacio, J. F., Bull, S. J., Neidhardt, J., & Hultman, L. (2006). Nanoindentation response of high performance fullerene-like CNx. *Thin Solid Films, 494*(1–2), 63–68. Available from https://doi.org/10.1016/j.tsf.2005.08.212.

Pauschitz, A., Roy, M., & Franek, F. (2008). Mechanisms of sliding wear of metals and alloys at elevated temperatures. *Tribology International, 41,* 584–602.

Pei, Y. T., Galvan, D., De Hosson, J., & Th, M. (2005). Nanostructure and properties of TiC/a-C:H composite coatings. *Acta Materialia, 53,* 4505–4521.

Polcar, T., Novak, R., & Siroky, P. (2006). The tribological characteristics of TiCN coating at elevated temperatures. *Wear, 260,* 40.

Ramírez, G., Jiménez-Piqué, E., Mestra, A., Vilaseca, M., Casellas, D., & Llanes, L. (2015). A comparative study of the contact fatigue behaviour and associated damage micromechanisms of TiN- and WC:H- coated cold-work tool steel. *Tribology International, 88,* 263–270.

Ranade, A. N., Rama Krishna, L., Li, Z., Wang, J., Korach, C. S., & Chung, Y. W. (2012). Relationship between hardness and fracture toughness in Ti-TiB$_2$ nanocomposite coatings. *Surface and Coatings Technology.*, *213*, 26−32. Available from https://doi.org/10.1016/j.surfcoat.2012.10.007.

Rebholz, C., Ziegele, H., Leyland, A., & Matthews, A. (1999). Structure, mechanical and tribological properties of nitrogen-containing chromium coatings prepared by reactive magnetron sputtering. *Surface and Coatings Technology*, *115*(2−3), 222−229. Available from https://doi.org/10.1016/s0257-8972(99)00240-6.

Rhee, Y. W., Kim, H. W., Deng, Y., & Lawn, B. R. (2001). Brittle fracture versus quasi plasticity in ceramics: A simple predictive index. *Journal of the American Ceramic Society*, *84*(3), 561−565. Available from https://doi.org/10.1111/j.1151-2916.2001.tb00698.x, http://www.blackwellpublishing.com/aims.asp?ref = 0002-7820.

Roa, J. J., Jiménez-Piqué, E., Martínez, R., Ramírez, G., Tarragó, J. M., Rodríguez, R., & Llanes, L. (2014). Contact damage and fracture micromechanisms of multilayered TiN/CrN coatings at micro- and nano-length scales. *Thin Solid Films*, *571*(2), 308−315. Available from https://doi.org/10.1016/j.tsf.2014.04.018.

Saha, R., & Nix, W. D. (2002). Effects of the substrate on the determination of thin film mechanical properties by nanoindentation. *Acta Materialia*, *50*(1), 23−38. Available from https://doi.org/10.1016/S1359-6454(01)00328-7.

Santhanam, A. T., Quinto, D. T., & Grab, G. P. (1996). Comparison of the steel-milling performance of carbide inserts with MTCVD and PVD TiCN coatings. *International Journal of Refractory Metals and Hard Materials*, *14*, 31−40.

Schalk, N., Tkadletz, M., & Mitterer, C. (2022). Hard coatings for cutting applications: Physical vs. chemical vapor deposition and future challenges for the coatings community. *Surface and Coatings Technology*, *429*, 127949.

Schwarzer, N. (2006). Analysing nanoindentation unloading curves using Pharr's concept of the effective indenter shape. *Thin Solid Films*, *494*(1-2), 168−172. Available from https://doi.org/10.1016/j.tsf.2005.08.253.

Schwarzer, N., Duong, Q.-H., Bierwisch, N., Favaro, G., Fuchs, M., Kempe, P., Widrig, B., & Ramm, J. (2011). Optimization of the scratch test for specific coating designs. *Surface and Coatings Technology*, *206*(6), 1327−1335. Available from https://doi.org/10.1016/j.surfcoat.2011.08.051.

Shi, K., Wang, C., Gross, C., & Chung, Y. W. (2015). Reversing the inverse hardness-toughness trend using W/VC multilayer coatings. *Surface and Coatings Technology*, *284*, 80−84. Available from https://doi.org/10.1016/j.surfcoat.2015.06.086, http://www.journals.elsevier.com/surface-and-coatings-technology/.

Skordaris, G., Bouzakis, K. D., & Charalampous, P. (2015). A dynamic FEM simulation of the nano-impact test on mono- or multi-layered PVD coatings considering their graded strength properties determined by experimental-analytical procedures. *Surface and Coatings Technology*, *265*, 53−61.

Skordaris, G., Bouzakis, K. D., Charalampous, P., Bouzakis, E., Paraskevopoulou, R., Lemmer, O., & Bolz, S. (2014). Brittleness and fatigue effect of mono- and multi-layer PVD films on the cutting performance of coated cemented carbide inserts. *CIRP Annals − Manufacturing Technology*, *63*(1), 93−96. Available from https://doi.org/10.1016/j.cirp.2014.03.081, http://www.elsevier.com/wps/find/journaldescription.cws_home/709764/description#description.

Skordaris, G., Bouzakis, K. D., Kotsanis, T., Charalampous, P., Bouzakis, E., Breidenstein, B., Bergmann, B., & Denkena, B. (2017). Effect of PVD film's residual stress on their mechanical properties, brittleness, adhesion and cutting performance of coated tools. *CIRP Journal of Manufacturing Science and Technology*, *18*, 145−151.

Strauss, H. W., Chromik, R. R., Hassani, S., & Klemberg-Sapieha, J. E. (2011). In situ tribology of nanocomposite Ti-Si-C-H coatings prepared by PE-CVD. *Wear*, *272*(1), 133–148. Available from https://doi.org/10.1016/j.wear.2011.08.001.

Sun, Y., Bloyce, A., & Bell, T. (1995). Finite element analysis of plastic deformation of various TiN coating/substrate systems under normal contact with a rigid sphere. *Thin Solid Films*, *271*, 122–131.

Tkadletz, M., Schalk, N., Daniel, R., Keckes, J., Czettl, C., & Mitterer, C. (2016). Advanced characterization methods for wear resistant hard coatings: A review on recent progress. *Surface and Coatings Technology*, *285*, 31–46. Available from https://doi.org/10.1016/j.surfcoat.2015.11.016.

Tomastik, J., Ctvrtlik, R., Ingr, T., Manak, J., & Opletalova, A. (2018). Effect of nitrogen doping and temperature on mechanical durability of silicon carbide thin films. *Scientific Reports*, *8*(1). Available from https://doi.org/10.1038/s41598-018-28704-3.

Tsui, T. Y., Pharr, G. M., Oliver, W. C., Bhatia, C. S., White, R. L., Anders, S., Anders, A., & Brown, I. G. (1995). Nanoindentation and nanoscratching of hard carbon coatings for magnetic disks. *MRS Proceedings*, *383*. Available from https://doi.org/10.1557/proc-383-447.

Veprek, S., & Veprek-Heijman, M. J. G. (2008). Industrial applications of superhard nanocomposites. *Surface and Coatings Technology*, *202*, 5063–5073.

Vereschaka, A., Tabakov, V., Grigoriev, S., Sitnikov, N., Milovich, F., Andreev, N., Sotova, C., & Kutina, N. (2020). Investigation of the influence of the thickness of nanolayers in wear-resistant layers of Ti-TiN-(Ti, Cr, Al)N coating on destruction in the cutting and wear of carbide cutting tools. *Surface and Coatings Technology*, *385*, 125402.

Vereschaka, A. A., & Grigoriev, S. N. (2017). Study of cracking mechanisms in multi-layered composite nano-structured coatings. *Wear*, *378-379*, 43–57.

Veverkova, J., & Hainsworth, S. V. (2008). Effect of temperature and counterface on the tribological performance of W-DLC on a steel substrate. *Wear*, *264*(7-8), 518–525. Available from https://doi.org/10.1016/j.wear.2007.04.003.

Voevodin, A. A., Muratore, C., & Aouadi, S. M. (2014). Hard coatings with high temperature adaptive lubrication and contact thermal management: Review. *Surface and Coating Technol.*, *257*, 247–265.

Wang, Y. X., & Zhang, S. (2014). Toward hard yet tough ceramic coatings. *Surface and Coatings Technology*, *258*, 1–16. Available from https://doi.org/10.1016/j.surfcoat.2014.07.007, http://www.journals.elsevier.com/surface-and-coatings-technology/.

Wang, Y. X., & Zhang, S. (2015). Present status of hard-yet-tough ceramic coatings. In S. Zhang (Ed.), *Thin films and coatings: Toughening and toughness characterisation* (pp. 1–45). CRC Press, p15.

Wang, Y. X., Zhang, S., Lee, J.-W., Lew, W. S., & Li, B. (2013). Toughening effect of Ni on nc-CrAlN/a-SiNx hard nanocomposite. *Applied Surface Science*, *265*, 418–423.

Wang, Z., Ma, D., Wang, H., Jiang, X., Wang, Y., & Li, Z. (2023). Effects of intrinsic mechanical properties on the dynamic bonding strength of GLC films undergoing ball bearing contact fatigues. *Diamond and Related Materials*, *136*. Available from https://doi.org/10.1016/j.diamond.2023.109977.

Wänstrand, O., Larsson, M., & Hedenqvist, P. (1999). Mechanical and tribological evaluation of PVD WC/C coatings. *Surface and Coatings Technology*, *111*, 247–254.

Weikert, T., Wartzack, S., Baloglu, M. V., Willner, K., Gabel, S., Merle, B., Pineda, F., Walczak, M., Marian, M., Rosenkranz, A., & Tremmel, S. (2021). Evaluation of the surface fatigue behavior of amorphous carbon coatings through cyclic nanoindentation.

Surface and Coatings Technology, 407. Available from https://doi.org/10.1016/j.surfcoat.2020.126769.

Wheeler, J. M., & Michler, J. (2013). Indenter materials for high temperature nanoindentation. *Review of Scientific Instruments, 84*, 101301.

Wheeler, J. M., Raghavan, R., Chawla, V., Morstein, M., & Michler, J. (2014). Deformation of hard coatings at elevated temperatures. *Surface and Coatings Technology, 254*, 382−387. Available from https://doi.org/10.1016/j.surfcoat.2014.06.048.

Wheeler, J. M., Armstrong, D. E. J., Heinz, W., & Schwaiger, R. (2015). High temperature nanoindentation: The state of the art and future challenges. *Current Opinion in Solid State and Materials Science, 19*, 354−366.

Whitehouse, D., & Archard, J. (1970). The properties of random surfaces of significance in their contact. *Proceedings of the Royal Society A, 316*, 97−121.

Windischmann, H. (1992). Intrinsic stress in sputter-deposited thin films. *Critical Reviews in Solid State and Materials Sciences, 17*(6), 547−596. Available from https://doi.org/10.1080/10408439208244586.

Xu, Y. X., Riedl, H., Holec, D., Chen, L., Du, Y., & Mayrhofer, P. H. (2017). Thermal stability and oxidation resistance of sputtered Ti Al Cr N hard coatings. *Surface and Coatings Technology, 324*, 48−56. Available from https://doi.org/10.1016/j.surfcoat.2017.05.053.

Yang, J., Botero, C. A., Cornu, N., Ramírez, G., Mestra, A., & Llanes, L. (2013). Mechanical response under contact loads of AlCrN-coated tool materials. *IOP Conference Series: Materials Science and Engineering, 48*(1). Available from https://doi.org/10.1088/1757-899x/48/1/012003.

Yang, Q., & McKellar, R. (2015). Nanolayered CrAlTiN and multilayered CrAlTiN−AlTiN coatings for solid particle erosion protection. *Tribology International., 83*, 12−20. Available from https://doi.org/10.1016/j.triboint.2014.11.002.

Yang, Q., Seo, D. Y., Zhao, L. R., & Zeng, X. T. (2004). Erosion resistance performance of magnetron sputtering deposited TiAlN coatings. *Surface and Coatings Technology, 188-189*(1-3), 168−173. Available from https://doi.org/10.1016/j.surfcoat.2004.08.012.

Yoon, S. Y., Yoon, S. Y., Chung, W. S., & Kim, K. H. (2004). Impact-wear behaviors of TiN and Ti-Al-N coatings on AISI D2 steel and WC-Co substrates. *Surface and Coatings Technology, 177-178*, 645−650. Available from https://doi.org/10.1016/j.surfcoat.2003.08.067, http://www.journals.elsevier.com/surface-and-coatings-technology/.

Zak, S., Trost, C. O. W., Kreiml, P., & Cordill, M. J. (2022). Accurate measurement of thin film mechanical properties using nanoindentation. *Journal of Materials Research, 37*(7), 1373−1389. Available from https://doi.org/10.1557/s43578-022-00541-1.

Zawischa, M., Weihnacht, V., Kaspar, J. J., & Zimmermann, M. (2022). Effect of doping elements to hydrogen-free amorphous carbon coatings on structure and mechanical properties with special focus on crack resistance. *Materials Science and Engineering: A, 857*. Available from https://doi.org/10.1016/j.msea.2022.144086.

Zha, X., Jiang, F., & Xu, X. (2018). Investigating the high frequency fatigue failure mechanisms of mono and multilayer PVD coatings by the cyclic impact tests. *Surface and Coatings Technology, 344*, 689−701.

Zhang, Q., Xu, Y., Zhang, T., Wu, Z., & Wang, Q. (2018). Tribological properties, oxidation resistance and turning performance of AlTiN/AlCrSiN multilayer coatings by arc ion plating. *Surface and Coatings Technology, 356*, 1−10. Available from https://doi.org/10.1016/j.surfcoat.2018.09.027, http://www.journals.elsevier.com/surface-and-coatings-technology/.

Zhang, S., Bui, X. L., Fu, Y., Butler, D. L., & Du, H. (2004). Bias-graded deposition of diamond-like carbon for tribological applications. *Diamond and Related Materials, 13*, 867−871.

Zhang, S., Bui, X. L., Fu, Y., & Du, H. (2004). Development of carbon-based coating on extremely high toughness with good hardness. *International Journal of Nanoscience, 3*, 571−578.

Zhang, S., Sun, D., Fu, Y., & Du, H. (2004). Effect of sputtering target power on microstructure and mechanical properties of nanocomposite nc-TiN/a-SiNx thin films. *Thin Solid Films, 447−8*, 462−467.

Zhang, X., Beake, B. D., & Zhang, S. (2015). Toughness evaluation of thin hard coatings and films. In S. Zhang (Ed.), *Thin films and coatings: Toughening and toughness characterisation* (pp. 48−121). CRC Press, July 2015, p 53.

Zimowski, S., Kot, M., Wiazania, G., & Moskalewicz, T. (2019). The ability of nanocomposite carbon coatings to withstand friction under severe conditions. *Tribologia, 5*, 115−124.

Section II

Experimental methods and techniques

Nanoindentation—strategies for reliable measurements on coated systems

Thomas Chudoba
ASMEC GmbH, Dresden, Germany

4.1 Preconditions for correct and reliable measurements

The aim of every material testing method is to get correct, reliable, and reproducible testing results. However, with decreasing size of the material, this becomes more and more challenging. Especially in the nano range, that is typically specified by dimensions below 100 nm (in nanoindentation below 200 nm), effects have to be considered, which normally don't play a role in the macro world. The standard for instrumented indentation testing (IIT) ISO 14577 (2015) describes the necessary preconditions for correct and reliable measurements for all dimensions from nano to macro, but the importance of the necessary corrections varies with the applied force and the realized displacement. Part 4 of the standard is dedicated to the measurement of metallic and nonmetallic coatings while part 2 describes the verification and calibration of testing machines.

The most important points of part 2 will be explained in the following. This will be made with the background knowledge that by far not all instruments in use are verified according to the requirements of the standard. The remarks shall help to identify possible discrepancies and to improve the verification process.

According to the standard, a force and displacement calibration of the instrument should be done at least every 3 years or in a shorter period when an indirect verification using reference materials is not satisfactory. A minimum of 16 evenly distributed points in the test force or displacement range of the instrument shall be calibrated, that is, 16 points during application and 16 during removal of the test force or during increase and decrease of the travel range. This shall be repeated three times. If the maximum force of an instrument is, for instance, 100 mN, this would mean that the first calibration point is at 6.25 mN, and every 6.25 mN follows another point. Since calibrated weights are normally not available in such increments, it is recommended to use an electronic balance with the necessary resolution that can be verified by calibrated weights, which can be traced back to the national mass standard. When using weights for the instrument calibration, more points could be used that would better fit to the available weight increments. However, in practice, this is rarely done since it is time-consuming, and mounting the weight at the indenter shaft is a difficult procedure.

For the displacement calibration, a laser interferometer is recommended that has nowadays a resolution of a few picometre.

The deviation from the result of the calibration device must not be larger than 1% over the complete measuring range or 2.5% for force and 2 nm for displacement in the nano range. However, 1% tolerance is also highly recommended in the nano range as well. For the 100 mN instrument, this means that the deviation at the point with smallest force shall not be larger than 60 µN. An example for a force calibration result is given in Fig. 4.1. It shows that a deviation from the reference far beyond 1% can be reached with modern instruments for a 100 mN force range. Every single point of the measurement above 6.25 mN should fulfill the 1% requirement. Forces below are not mandatory.

An overall precision of the calibration can be calculated when the force data from the calibration device are shown on one axis of a graph and the force data from the nanoindenter on the other axis. A linear fit through the data should then ideally have a slope of 1. Typically, the slope deviates less than 1% from 1 for such a calibration shown in Fig. 4.1.

The calibration of force and displacement is a prerequisite of other necessary calibrations or verifications. This comprises the calibration of instrument compliance and indenter area function. A change of the force or displacement calibration of the instrument (e.g., by an adjustment of the calibration factor that relates a voltage to a force or depth signal) automatically requires a recalibration of instrument compliance and indenter area function. A verification of the instrument compliance calibration should also be done after every change of the indenter. Every new indenter requires first a calibration of both functions since there is some

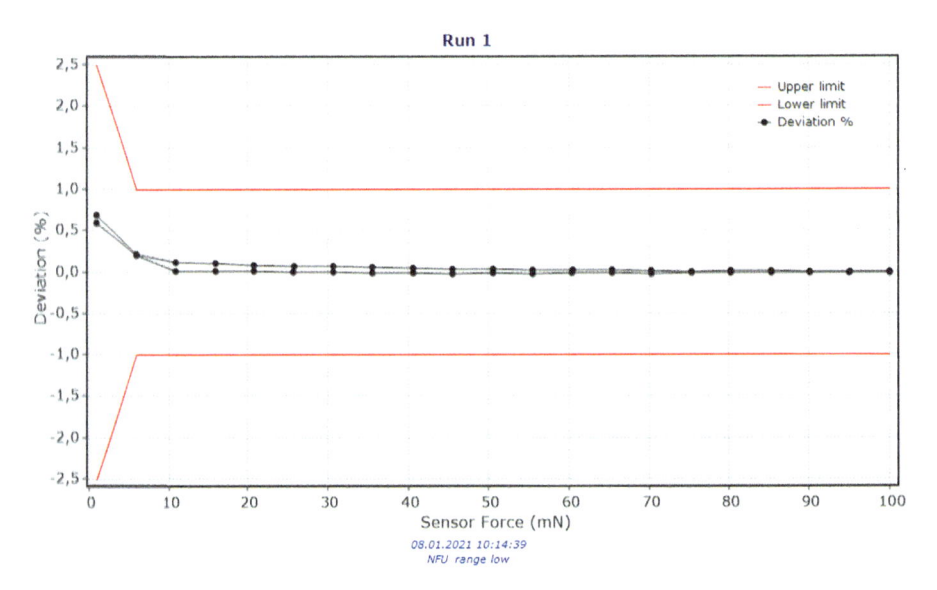

Figure 4.1 Example for a force calibration result as deviation from the reference device over a force range of 100 mN with 21 points.

contribution of the indenter to the total compliance of the instrument, and this contribution can be different from one indenter to the other.

The importance of a correct instrument compliance calibration is increasing with increasing force while the importance of a correct area function is increasing with decreasing depth. For indents with a depth above 6 µm, no calibration is required according to the standard; however, a calibration independent on the maximum depth is highly recommended. An additional result of such an area function calibration is the opening angle of the indenter. For a Berkovich indenter, the conical equivalent for the mean angle between indenter axis and face is 70.3 degrees. The correct face angle (no average) for a modified Berkovich (the mostly used Berkovich version), which is normally certified by the manufacturer, is 65.27 degrees ± 0.3 degrees. It is a clear sign for an inaccurate instrument compliance calibration when the indenter angle result of an area function calibration deviates more than 0.3 degrees for larger depth from the mean angle or from the back-calculated certified value of the manufacturer. The area function calibration gives the information whether a new instrument compliance calibration or even a new force or displacement calibration is necessary.

The state of an instrument or tip may change with time. Especially abrasion of the tip takes place when many measurements on hard materials are done or when the tip accidentally scratches along the surface. A contamination of the tip can also influence measurements. A time period of 1 year or more between verifications is therefore much too long to ensure the precision of measurements. For this purpose, an indirect verification should be done at least monthly or weakly, especially when measurements with high accuracy and low forces shall be done. In an indirect verification, measurements on reference materials with certified material parameters are done. In the nano and micro range (up to 2 N), these are Young's modulus and Poisson's ratio while in the macro range also hardness of reference blocks can be used. According to the standard, two different reference materials with a property ratio of 2 or more should be used at two different forces. The reason for this is that with one force and one material accidentally a correct number can be obtained because two error contributions compensate each other. Even with two forces and one material, the measurement results can be in the tolerance limits of the reference material, especially when it was also used for area function calibration. Only with two materials it can be confirmed that both calibrations: area function and instrument compliance, are correct. It is clear that the outcome of an area function calibration should not depend on the material. However, in practice, it is often observed that only one material is used for the indirect verification (typically fused silica), and then inaccurate results for a material with much higher modulus are published because the instrument compliance has a larger uncertainty.

Typically, the acceptable deviation from the reference value in an indirect verification according to the standard in the version from 2015 is 5% for the average of five measurements. However, in a new revision of the standard 10 measurements are recommended, and every single measurement should be in the tolerance limit. This method makes the probability of a false acceptance of an instrument much lower. For fused silica with a reference modulus of 72 GPa as an example, this would mean that all modulus measurement results should be in the range between 68.4 and 75.6 GPa. For modern instrument, this is normally not difficult to achieve.

4.2 Application of a radial displacement correction

Since 2015, a radial displacement correction is normative for the calculation of the contact area. This correction caused some confusion since it was not well explained in the standard. It shall therefore be explained in more detail in the following. The maximum change of the contact area by this correction is about 10%, which can result in up to 10% deviation in hardness and 5% deviation in modulus results, especially for soft materials. Only with radial displacement correction, it is possible to guarantee a material independency of the area function calibration results.

When Warren Oliver and George Pharr developed the method for the analysis of indentation measurement in their pioneering paper of 1992 (Oliver & Pharr, 1992), they used the contact mechanical solution of Sneddon for the elastic deformation of an infinite half-space by a conical indenter (Sneddon, 1948). This solution was incomplete because it only considered the elastic deformation perpendicular to the surface. A complete solution for the cone was later given by Hanson (1992a). It comprises the elastic deformation in all three spatial directions as well as all stress components. He also gave a complete solution for a spherical contact (Hanson, 1992b) and could even consider friction, which results in transverse isotropy. During loading, a surface element is not only pushed downward but also dragged inward.

It should be mentioned here that all complete analytical contact mechanical solutions do not really consider a physical indenter of a certain shape. They first derive a contact pressure distribution that such an indenter would produce on an infinite and homogeneous half-space. In a second step, they derive the stresses and deformations that such a pressure distribution on the surface would produce. The shape of the absolute surface deformation is therefore not mirroring the shape of a rigid cone or sphere and deviates also from that of a deformable indenter with different elastic properties. This is only important for the surface area below the contact area and can have a slight influence on the contact stiffness derived from the analytical model in relation to the realized contact stiffness from a stiffer physical indenter. However, the elastic inward deformation (radial displacement) at the contact edge can be correctly derived. The amount of radial displacement depends on the elastic properties and the contact pressure, which is equal to hardness in a hardness test. The formula was already given by Johnson (1985):

$$\frac{u}{a} = \frac{-(1-2\nu)(1+\nu)}{2}\frac{H}{E} \tag{4.1}$$

with u—radial displacement, a—contact radius, ν—Poisson's ratio, E—Young's modulus, and H—hardness. Instead of hardness the contact pressure p should be used for spherical indenters. It can be recognized that the relative radial displacement u/a is large for materials with high H/E ratio and small Poisson's ratio like fused silica. This formula is valid for any rotationally symmetric indenter, and it makes no difference whether a cone or a sphere is used if contact radius and pressure are equal. This is important since the pressure distribution changes drastically after plastic deformation with a pyramid in relation to the pressure distribution for an elastic deformation with a cone. The infinite stress in the center of a sharp tip is reduced by the plastic deformation, and the final pressure distribution is more similar to that of an elastic deformation with a sphere. In reality, it is more complex and a correlation can be found between the unloading exponent of the load − displacement curve and the pressure distribution. This correlation is also represented by a variable epsilon factor, which is included in the standard as normative correction since 2015. Details of that can be found in Chudoba and Jennett (2008).

Two methods are possible for the determination of the area function. In the direct measurement method, the tip is scanned with an atomic force microscope or another tool and the projected contact area as function of the contact depth is derived from the height−surface area data pairs. The other, indirect measurement method, uses a reference material with known elastic properties and calculates the contact area from the measured depth, force, and contact stiffness. A correct instrument compliance calibration is presupposed. Both methods should give the same results. This is only possible when in the indirect method the area function is derived for the unloaded state, that means, when the amount of radial displacement is added to the contact area under load. Only this gives the true dimension of the pressure distribution on the surface for which all derived parameters (depth, contact stiffness) are valid. Contrary, from the contact mechanical point of view, the measurable contact stiffness is related to the projected contact area under load that is smaller than the contact area of the sample in the unloaded state.

That has several consequences. First, it is impossible to get the same area function from reference materials with a clear difference in properties, when the radial displacement correction (and variable epsilon factor) is not considered. This can only be mentioned when a comparison is made with a directly scanned area or when the calibration is done with two reference materials. Since most of the calibration methods of the different nanoindenter types use only one material, it was hitherto not seen as a problem. When the 10% larger area for fused silica is not used in a comparison with the direct method (atomic force microscope (AFM) scan), an agreement over a certain range of the area function can only be reached by artificially adapting the instrument compliance. However, if such an incorrect determined compliance function is applied to a second material, no agreement with the AFM area function can be realized.

Second, the area of an imprint after unloading, measured by optical or other means like for Vickers hardness is bigger than the calculated contact area under load, especially for very elastic materials. A conversion of the indentation hardness, which is defined as a hardness under load, to Vickers hardness would require an additional correction of the radial displacement (originally it was assumed that the area under load is equal to the area after unloading and then hardness under load and hardness after unloading are the same). The converted Vickers hardness from the indentation hardness of fused silica becomes than about 10% smaller, which correlates well with experimental findings, while there is nearly no change for soft metals.

Third, the measured contact stiffness is valid for the area under load and the area function that is saved for the unloaded state, must be reduced by the radial displacement correction for the calculation of indentation modulus (and indentation hardness, since it is defined under load). When the area function calibration is done with fused silica and later fused silica is measured, the remaining correction is zero. It is therefore no difference between a use and a neglection of the radial displacement correction. A clear difference can only be seen when a material with very different radial displacement is measured such as aluminum. For instance, when the area during the calibration with fused silica becomes 10% larger and it is later applied to a material where the shrinking of the area under load is only 0.2%, then the effective correction is 9.8%. This is mainly the case for soft metals. Due to the application of the radial displacement correction according to the standard and a calibration with fused silica, the hardness of many metals becomes 7%–9% and the modulus 3.5%–4.5% smaller. In contrast, the effect of the correction on the results of hard and elastic coatings such as diamond like carbon (DLC) is negligible.

4.3 Combined calibration of area function and instrument compliance

The calibration of indenter area function and instrument compliance under consideration of radial displacement correction shall be outlined in the following. The total measured compliance, C_T, is the sum of contact compliance C_S of the sample and frame compliance C_F, thus: $C_T = C_S + C_F$.

The contact depth h_c is obtained from maximum dept h_{max} and force F according to

$$h_c = h_{max} - F \cdot C_F - \varepsilon(m) \times F \times (C_T - C_F) \qquad (4.2)$$

The epsilon factor ε depends on the unloading exponent m. Using the reduced modulus E_r with subscript i for indenter and s for sample

$$\frac{1}{E_r} = \frac{1-\nu_i^2}{E_i} + \frac{1-\nu_s^2}{E_s} \qquad (4.3)$$

the contact area A_C in the unloaded state can be calculated via

$$\sqrt{A_C} = \sqrt{\pi} \cdot a = \frac{\sqrt{\pi}}{2} \cdot \frac{1}{E_r} \cdot \frac{1}{C_T - C_F} \cdot (1 + u_r) \qquad (4.4)$$

Where u_r is the elastic radial displacement [see Chudoba and Jennett (2008)]

$$u_r = \frac{(1-2\nu)(1+\nu)}{2} \cdot \frac{F}{E \cdot a^2} \cos\left(\arctan\left(\frac{h_0}{a}\right)\right) \qquad (4.5)$$

with a as contact radius and h_0 as depth after unloading.

A plot of C_T versus $A^{-1/2}$ is linear for a given material, and the intercept with the compliance axis is a measure of the instrument compliance when the correct area function is known. The best estimation of C_F is obtained when $A^{-1/2}$ is small, namely, when the contact depth is large and the influence of a tip rounding is small. An iterative procedure will finally result in both: a correct area function and instrument compliance. This is the method, developed in Oliver and Pharr, (1992), that was included in the standard (ISO 14577, 2015) in Annex D of ISO 14577 section D.3.3 or D.3.5. The difference between D.3.3 and D.3.5 is that in the first method, only one reference material is necessary while in the second, two materials are used.

These methods have two weaknesses:

1. It is assumed that the instrument compliance is constant and not force-dependent.
2. It is not confirmed that the area function is independent of the reference material.

A method that is based on the comparison of the area functions from two different reference materials with a large difference in modulus can eliminate both weaknesses. Using the measurement results for h, F, and C_T, contact depth h_C and square root of contact area are calculated with Eqs. (4.2–4.5), starting with zero compliance. The result is demonstrated in Fig. 4.2 for an instrument with larger frame compliance where the difference between the two area functions is significant. Fused silica and sapphire single crystal with (0001) surface have been used as reference materials. The value for the frame compliance in Eqs. (4.2) and (4.4) is now increased for every point i of the stiffer material until $\sqrt{A_i(h_{C,i})}$ is intersecting the

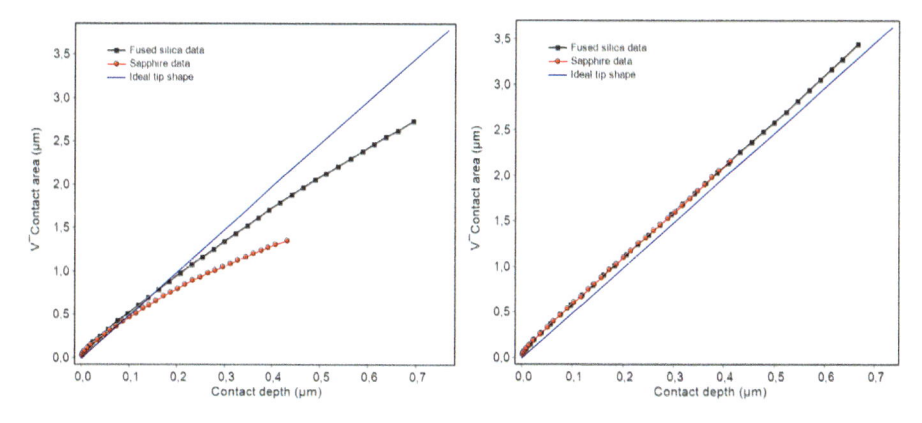

Figure 4.2 Area functions for a low and high modulus material calculated without correction of instrument compliance (left) and after correct compliance calculation (right).

curve of the less stiff fused silica, starting with the point with largest h_C. When this is done for all points down to that with the smallest depth, a set of results $C_{F,i}$ (F_i) is available. The new force-dependent frame compliance results are now applied to fused silica. Typically, both materials are measured with the same forces. If not, a linear interpolation between two neighboring points using the C_F (F) function is used. Since the modulus of material 1 is lower, the reaction of the compliance change is weaker, and the shifting of the points to the upper left side in the graph is slower. After several iterations (typically less than 10), both curves agree, and the final frame compliance function is obtained. This method can also be used to check the correct face angle of the indenter. For larger depths, the slope of the final curves should be approximately in parallel to the curve for an ideal tip. If not, this would be an indication that incorrect reference values are used or that the force or displacement calibration of the instrument is wrong or that the tip is defect. In Chudoba et al. (2022), it was also shown that such an area function agrees with the area function from a direct method using an AFM scan. Modulus and hardness results after calibration of many Berkovich indenters are compared in this paper for low forces and depths below 250 nm.

4.4 The influence of a tip rounding

The geometry of the pyramidal indenters needs a more detailed investigation because it determines if comparable hardness values can be measured. For simplification, in the following a conical tip with the same depth to area ratio like a Berkovich or Vickers indenter is considered. The face angle is then 70.3 degrees. In a first approximation, the outermost end can be described by a spherical cap. The transition between spherical end with tip radius R and the conical part is located at a depth h_S for a face angle α:

$$h_s = R(1 - \sin\alpha) \tag{4.6}$$

For $\alpha = 70.3$ degrees, the transition depth is located at 5.85% of the radius. For a radius of 250 nm, this would give a depth of 14.6 nm. The tip can also be characterized by a tip blunting height Δh (see Fig. 4.3) that can be calculated according to

$$\Delta h = R\left(\frac{\cos\alpha}{\tan\alpha} + \sin\alpha - 1\right) \tag{4.7}$$

Δh is 15.5 nm for a tip radius of 250 nm. After the transition point, h_S is the contact radius offset Δa to an infinite sharp conical tip constant and independent of depth.

$$\Delta a = \sqrt{2Rh_s - h_s^2} - h_s \cdot \tan\alpha \tag{4.8}$$

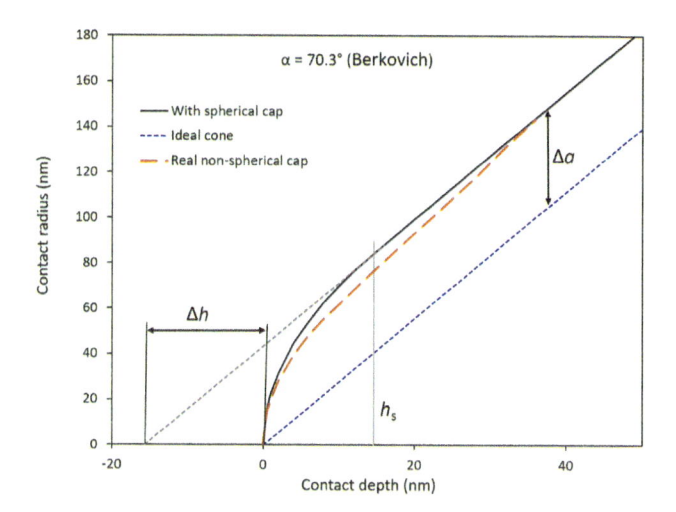

Nanoindentation—strategies for reliable measurements on coated systems

Figure 4.3 Tip geometry for a spheroconical indenter of 70.3 degrees opening half-angle and 250 nm tip radius. Δa is the contact radius offset to an ideal conical tip, h_S is the transition point from spherical to conical shape, and Δh is the tip blunting height. Additionally, an example for a real tip with nonspherical cap is shown.

For a tip radius of 250 nm is $\Delta a = 43.4$ nm. However, for real tips it is rarely the case that the tip geometry can be described by one radius and one face angle, especially because the three corners cannot merge to a sphere in the same way like the faces. A possible tip shape for a nonspherical end is shown in Fig. 4.3. After the transition point, an effective tip radius R can be assigned to every contact radius offset by the equation

$$R = \Delta a / \cos\alpha \qquad (4.9)$$

For a tip with an ideal spherical cap, the radius would be constant, but in reality, there is a larger transition range far beyond h_s and Δa varies over a longer distance. It is therefore insufficient to describe the quality of a Berkovich indenter only by one tip radius because the transition range can be very different from tip to tip and at low depth the effective radius is smaller than at larger depth. A better measure for the sharpness of a tip is the contact radius offset Δa at a certain depth, for instance, at 200 nm. Even when the tip radius is only 50 nm, the offset Δa can be large at a depth of 200 nm because the face angle below 200 nm deviates from the certified face angle. This transition range with different face angle determines from which depth onward a correct hardness can be measured because below this depth the self-similarity of the tip is violated.

Fig. 4.4 shows the depth-dependent effective tip radius according to Eq. (4.9) as function of contact depth for three different tips. The data are converted from calibrated area functions. More details for the area function calibration of these tips can be found in Chudoba et al. (2022). The tip radii derived for a depth where Δa is no

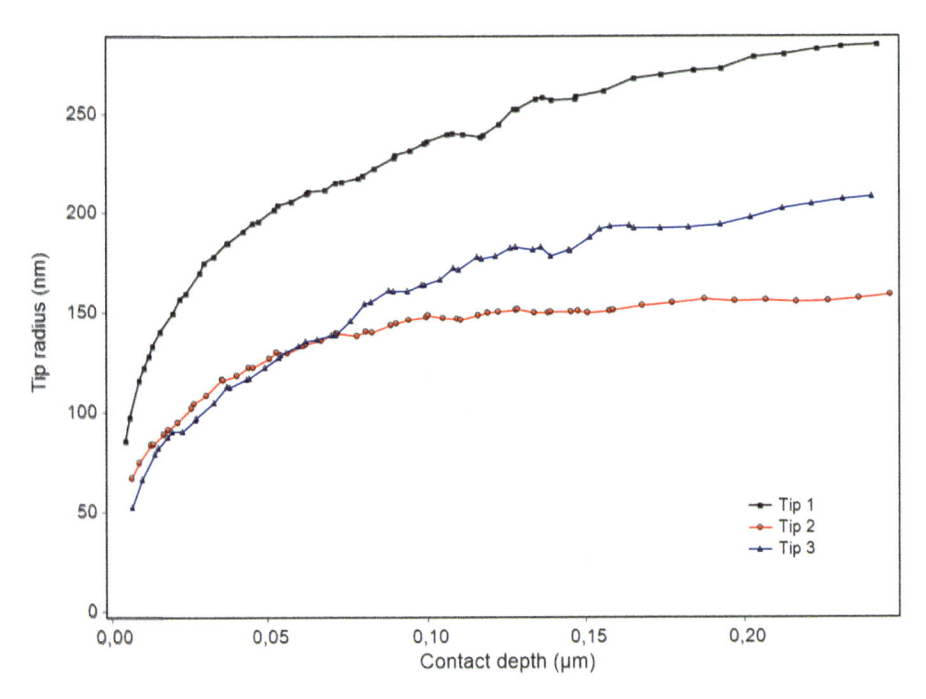

Figure 4.4 Effective tip radius according to Eq. (4.9) as function of contact depth for three different tips. The data are converted from calibrated area functions.

longer increasing, lay between 160 and 300 nm and the transition depth h_s is between 9.4 and 17.6 nm. Points below would be incorrect since Eq. (4.9) is only valid above the transition point. It can be seen that besides tip 2 even at a depth of 250 nm Δa and therefore the effective radius is not constant. Tip 3 exhibits a radius step at about 70 nm, which indicates an abrupt shape modification at this depth that can be caused by a flaw or roughness peak.

A tip rounding can drastically increase the apparent hardness when no area function calibration is done. A precise calculation was given by Olaf and Ritter for a Vickers indenter where the tip rounding is not described by a spherical cap but was modeled by a cross-shaped vault (Olaf & Ritter, 1992). The result for different tip radii is shown in Fig. 4.5. The error at a depth equal to the tip radius would be 16%. The use of a correct calibrated area function reduces the error for the hardness. However, a depth-independent hardness (assuming a material with depth-independent properties) can only be obtained when the self-similar region of the tip shape is reached, and this is normally much deeper than the transition point from sphere to cone. Practice has shown that at least an indentation depth of 20% of the tip radius is necessary to reach a constant hardness range but, in some cases, even 40%–50% is necessary.

For the modulus calculation, no self-similarity of the indenter is necessary because the model is correct for any approximately rotational symmetric indenter

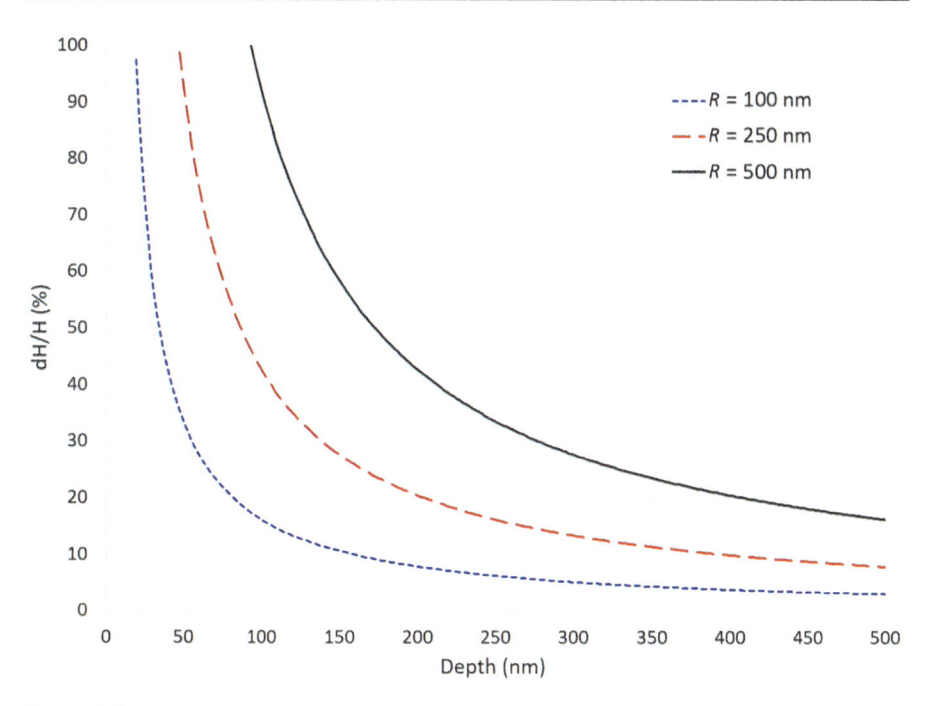

Figure 4.5 Apparent hardness increase caused by the tip rounding of a Vickers indenter when no area function is used.

and the modulus error is only determined by the precision of the area function. However, a nonsymmetric shape, for instance, caused by a line of conjunction between opposite faces for a Vickers indenter violates the model assumptions and a modulus error occurs.

4.5 Theoretical considerations for the measurement of coatings

The measurement of coatings is specified in part 4 of the standard ISO 14577 (ISO 14577, 2015). Test method for metallic and nonmetallic coatings. It describes the requirements for a correct measurement of hardness and modulus of a single coating on a substrate. Multilayers or graded coatings can also be measured, but then it is only possible to give a result for the topmost layer or the upper range of the graded coating. It is impossible to measure the pure properties of a second or third layer in a layer stack since the test method is an integral method, and the results are influenced by all the material above the indention depth and also from a certain region below.

The aim for the selection of the test parameter is to prevent any substrate influence. Further, only the plastic properties of the coating shall be measured by the hardness number and not the fracture behavior without any plastic deformation. To realize that, some preliminary considerations are necessary.

It is known from the analysis of macroscopic Vickers indents that the plastic zone in steel is about 10 times deeper than the indent itself (Weiler, 1986). When the plastic zone in a coating reaches the interface to the substrate, a beginning substrate influence can be expected. This is the source of the 1/10th rule or Bückle rule. It says that the indentation depth into a coating should only be 1/10th of the film thickness to prevent substrate influence. It is only a rule of thumb and especially for soft materials on a harder substrate, it can be weakened while for very hard materials on soft substrates 1/10th can already be too much. Therefore it is necessary to check if substrate influence exists. For this reason, the standard requires the measurement of at least three different forces but recommends five forces or more. At every force at least five indents should be made and averaged but 10 indents or more are recommended. The contact depth h_c for all forces should be smaller than half of the film thickness t_c: $0 < h_c/t_c < 0.5$ and approximately evenly distributed. This requires the knowledge of the film thickness and preliminary tests to determine roughly the contact depth. When the film thickness is not known, even more different forces would be necessary to cover a span where some points are within the 1/10th rule. This can, for instance, be realized by using a dynamic test method (see further).

The theoretical hardness profile of a homogeneous hard coating on a softer substrate should look like that in Fig. 4.6. For very small depth below 20% of the tip radius, the contact pressure increases until a plateau is reached where a depth-independent coating hardness can be measured and after that only a compound hardness is obtained. Considering the 1/10th rule, the minimum film thickness where the correct hardness can be measured is about two times the tip radius. For a

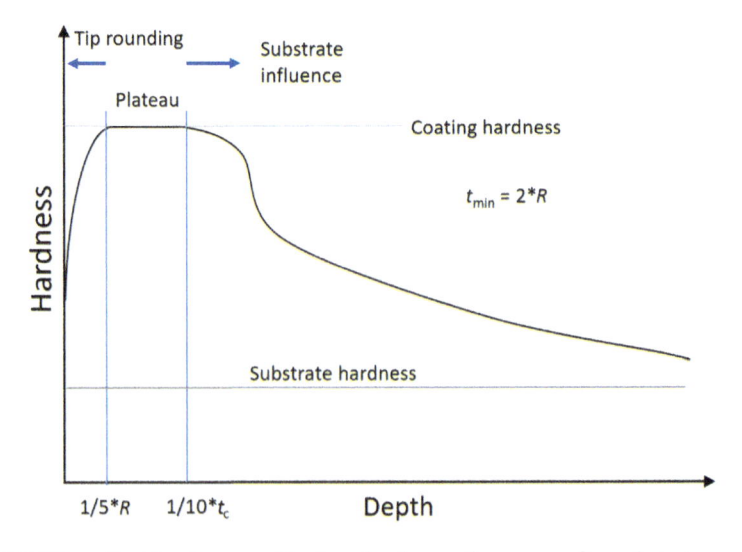

Figure 4.6 Theoretical hardness profile for a harder coating on a softer substrate.

tip with 250 nm radius, this would be 0.5 μm. Any thinner coating requires sharper tips. When the coating is too thin for the tip radius, both limits, tip rounding and substrate influence, will overlap and the plateau is not reached. An example is shown in fig. 6 of part 4 of the standard ISO 14577 for a 460 nm thin DLC coating on steel.

For the modulus things are different. The measurement does not require a self-similar tip shape, and the tip radius is not important for the calculation of the modulus. The decision tree in the standard recommends even to choose a larger radius. This, however, would require a tip change for every coating were hardness and modulus shall be characterized and is therefore not much used in practice. The reason for the recommendation is that a larger radius would allow purely elastic indents that cannot be influenced by some negative effects that may occur during plastic deformation such as pile-up, sink-in, or cracking. A larger radius, however, is connected with a larger information depth and less influence of the coating on the result but more importance of surface roughness. The selection of the tip radius is therefore a compromise between surface sensitivity and the exclusion of negative effects.

There exists no 1/10th rule for the modulus because, contrary to the plastic zone, the range of the elastic field is theoretically infinite and there are contributions far away from the contact zone. Substrate influence has therefore always to be considered. Its effect depends on the modulus difference to the coating and the coating thickness. Like for hardness, the standard requires measurements with at least three but better five different forces. For soft coatings, the contact radius a should be smaller than 1.5 times the film thickness t_c, which corresponds for the Berkovich indenter to $h_c/t_c < 0.54$. For hard coatings on a softer substrate, the ratio should be $a/t_c < 2$ and $h_c/t_c < 0.72$ for a Berkovich, respectively.

A theoretical modulus profile is shown in Fig. 4.7. The steepness of the modulus curve is largest at the beginning and causes an underestimation of the real coating modulus already at low depths, typically smaller then 10% of the film thickness. For thin coatings and a large modulus difference to the substrate a depth of only 2% of the film thickness may already be too much. The standard requires therefore an extrapolation of the modulus − depth data from the measurements with different forces to zero depth by using a linear fit. The fit can be more accurate when more than three datasets are available. Dynamic test methods provide typically 30 datasets or more, and this would allow to apply more sophisticated nonlinear fit methods. Lorenz et al. (2021) have shown that a sigmoid fit model is best suited to describe the course of the modulus − depth curve for ta-C coatings on 100Cr6 steel. This fit even could be used to estimate the thickness of the coatings.

It should be mentioned here that the reduced modulus E_r (see Eq. 4.3) or the plain strain modulus is analyzed. It can be converted to the indentation modulus or Young's modulus when the Poisson's ratio of the coating is known or when a reasonable assumption can be made.

Another effect has to be considered for the measurement of coatings that is often named as egg shell effect. It is of increasing importance with decreasing film thickness and increasing hardness difference between coating and substrate where the coating is the harder part. It is further more important for rounded than for sharp

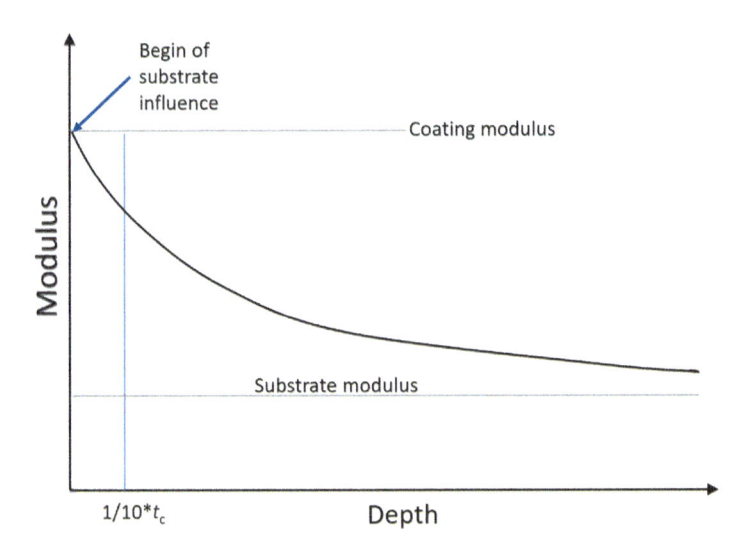

Figure 4.7 Theoretical modulus profile for a stiffer coating on a more compliant substrate.

tips. The condition for a correct hardness measurement of a coating is that plastic deformation takes place in the coating and not in the substrate. When plastic deformation first takes place in the substrate, a harder and brittle coating may fracture and the pieces are pressed into the substrate. This can hardly be identified in the load − displacement curve, but it will lead to wrong hardness results. The meaning of the effect is explained in Fig. 4.8 for 300 nm thin coatings on a steel substrate. The von Mises stress as yield criteria is calculated along the depth axis Z because there is the position of the maximum stress when no lateral forces occur. The first calculation is done for a (theoretical) TiN coating with a modulus of 450 GPa and a hardness of 35 GPa. The yield strength of a ceramic coating material can be estimated using the expanding cavity model of Marsh (1964) with the modification of Bushby and Swain (1995). A value of 21 GPa was used for the simulation. The force is chosen in a way that the maximum von Mises stress is equal to the yield strength of the coating material. It can now be derived if plastic deformation first takes place in coating or substrate when the yield strength of the substrate is known. The maximum stress in the substrate is located just at the interface and reaches more than 6 GPa for a 1 μm tip radius. Assuming a yield strength of steel of 1.5 GPa, even the 0.5 μm radius tip would first cause plastic deformation in the substrate while the 0.2 μm radius tip is sharp enough to measure the hardness of the coating because plastic deformation in the substrate is prevented.

The requirements on the substrate strength can be reduced when a coating of the same depth has a lower modulus than the substrate. For example, a yield strength of 6 GPa is estimated for a modulus of 70 GPa und a hardness of 8.7 GPa for a glassy coating. The more compliant coating can absorb more of the force, and the stress in the substrate is reduced in comparison to a stiffer coating and the same force. Such a coating could be measured even with a tip of 0.5 μm radius.

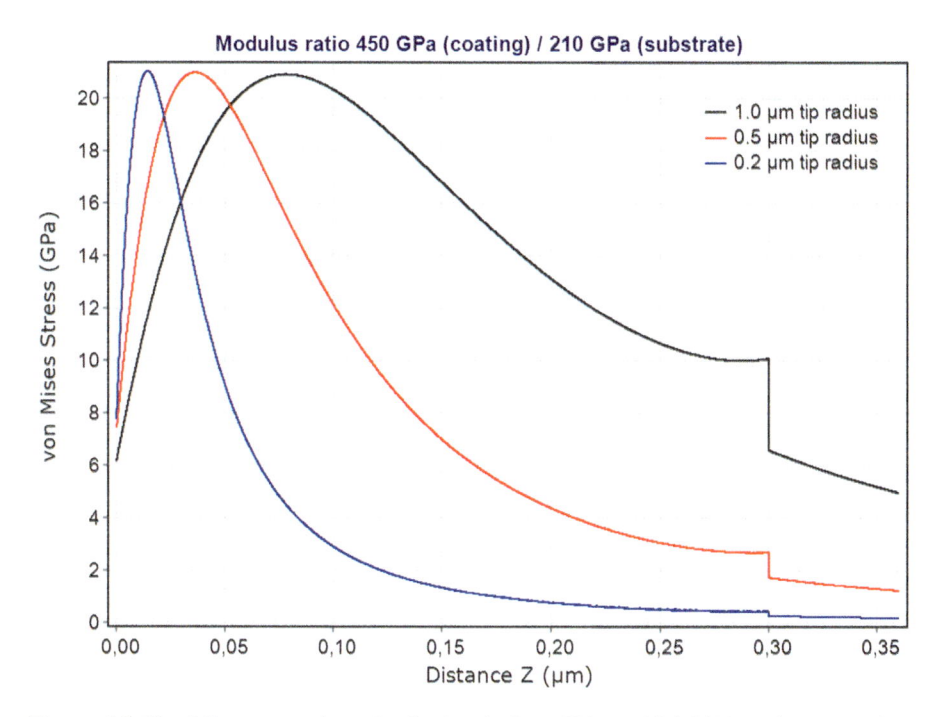

Figure 4.8 Von Mises stress along the depth axis for a 300 nm thick TiN coating on steel and for contact with indenters of different radius and a force at which the yield strength of the coating is reached.

These calculations are done for the beginning of the indentation process when plastic deformation starts. In the course of the expansion of the plastic zone, the effective coating thickness becomes smaller and plastic deformation starts easier in the substrate. This may happen before a depth of 20% of the radius is reached. Hardness measurements on a 300 nm thin coating require therefore a tip radius smaller than 150 nm.

4.6 Modulus measurements by fully elastic indentations

Fully elastic indentations with spherical indenter have the advantage that no disturbing effects take place such as pile-up, sink-in, densification, phase transitions, or fracture. The load − displacement curve can be calculated by an analytical model and compared with the measurement data. For soft materials, the elastic − plastic transition takes place already at very low forces in the µN range when typical indenter radii <50 µm are used. This could be compensated by larger radii; however, the influence of surface roughness is increasing with increasing radius, and it is very hard to realize an agreement between loading and unloading curve better than 5 nm. The method is therefore preferably applied to hard and smooth materials such as optical coatings on glass or ceramic or DLC coatings on silicon wafers.

The Hertzian contact model can no longer be used for coated materials because the assumption of a homogeneous and isotropic half-space is violated. An extended Hertzian model for coated materials was developed by Schwarzer and coworkers (Schwarzer, 2000; Schwarzer et al., 1999) that is based on a Hertzian pressure distribution. Later he extended the model to other pressure distributions (Schwarzer, 2006) as well. Here we focus on the Hertzian pressure distribution since it is valid for fully elastic indentations.

It is not well known that a modulus mismatch between coating and substrate changes the unloading exponent m of the fully elastic load – displacement curve and therefore the epsilon factor that is used for the calculation of the contact depth:

$$h_c = h_{\max} - \varepsilon(m) \cdot \frac{F}{S} \qquad (4.10)$$

For a homogeneous substrate and a spherical indenter, the exponent m is 1.5, the ratio between contact depth and maximum depth h_c/h is always 0.5 and ε is 0.75. The change of the exponent caused by coatings was first published in 1999 (Chudoba et al., 1999). The result for the dependence of m on the contact radius a to film thickness t ratio is shown in Fig. 4.9 for modulus ratios between coating and substrate of 4:1 and 0.25:1. A Poisson's ratio of 0.3 was used for both materials. The largest deviation from 1.5 of about 15% is reached close to an a/t ratio of 1. The deviation will reduce for smaller modulus mismatch. The modulus difference is not only influencing the epsilon factor, but also the h_c/h ratio (see Fig. 4.10). The calculation was done with the software package Elastica 5 of ASMEC GmbH

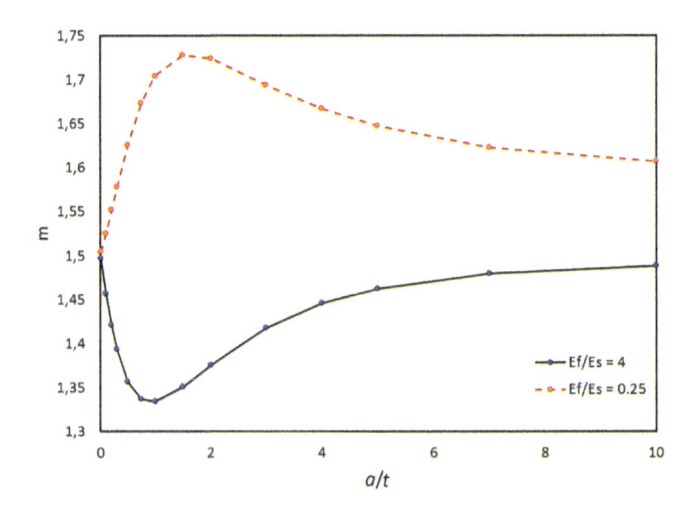

Figure 4.9 Exponent of the elastic load – displacement curve in dependence on contact radius to film thickness ratio for a stiff coating with four times higher modulus than the substrate and vice versa.

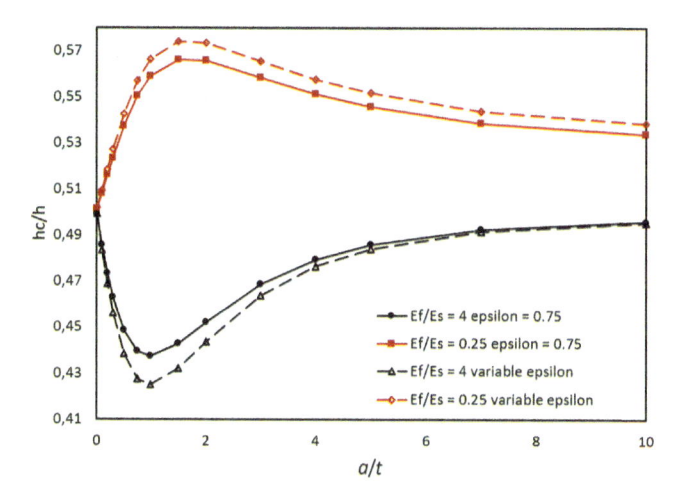

Figure 4.10 Ratio between contact depth and absolute depth of an elastic spherical contact in dependence on contact radius to film thickness ratio for a stiff coating with four times higher modulus than the substrate and vice versa. The calculation was done for a constant epsilon factor of 0.75 (bold lines) and for a variable epsilon (dashed lines).

(ELASTICA, 2024), once for a constant epsilon factor and once for a variable one. The variable epsilon has a minor influence and changes the result in maximum by 3%. However, the absolute deviation to 0.5 reaches 15% close to $a/t = 1$ for the stiffer coating and close to $a/t = 1.5$ for the softer one. The contact depth was calculated using Eq. (4.10), and the contact stiffness was calculated from the slope of the loading curve (which is identical to the unloading curve) at maximum force. A fit of an elastic load − displacement curve of a coating with the conventional Hertzian contact model would deliver wrong results if the contact radius is not much smaller than the film thickness.

The influence of the substrate is also reflected in the result for the reduced modulus when it is calculated with the well-known formula $E_r = S/(2a)$. To differentiate it from the reduced modulus of a pure substrate, it is named reduced composite modulus E_{cr}. The results for the same modulus ratios like in the previous figures are given as function of a/t in Fig. 4.11. The reduced modulus for a 400 GPa material and a diamond sphere is 317.7 GPa and for a 100 GPa material, it is 100.3 GPa. This are the start values for $a/t = 0$ in the graph. Interestingly, the composite modulus can reach values for the softer coating, which are higher than the reduced substrate modulus. In general, the course of the curves reflects the theoretical modulus profile for plastic deformations in Fig. 4.7. It also demonstrates that a nearly linear behavior can be assumed for $a/t < 1$ and that a linear fit in this range is reasonable.

The extended Hertzian contact module can now be used to analyze the Young's modulus of ultrathin coatings down to a few nanometers where the conventional method has no chance since the substrate influence is always too high, even for the smallest measurable forces (Chudoba et al., 2000, 2004). This method requires first an accurate calibration of the area function and instrument compliance, for instance,

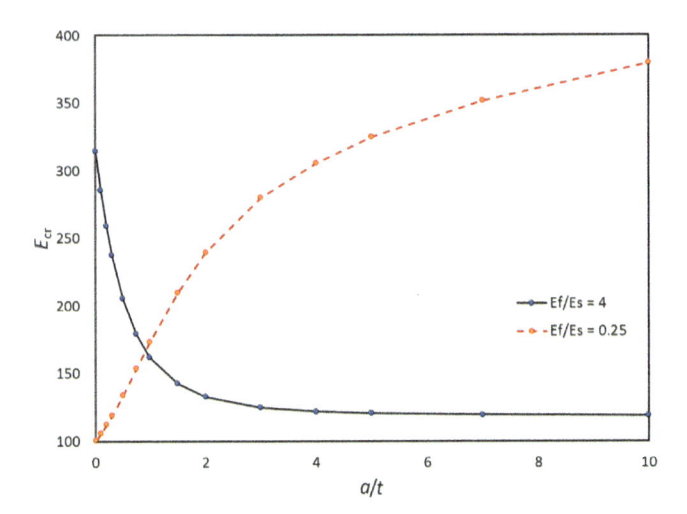

Figure 4.11 Reduced composite modulus of a coated substrate with a modulus ratio of film modulus Ef to substrate modulus Es of 4 and 0.25 in dependence on contact radius to film thickness ratio for fully elastic indents with a spherical tip.

with the method described in Section 4.2. Then a radius function is calculated that is replacing the area function for further analysis. There exists no diamond tip with ideal spherical shape and therefore a different radius has to be used for every depth, which is calculated with the extended Hertzian contact model. An example for a relatively good sphere is given in Fig. 4.12. The radius is decreasing from 7.4 to 6.2 μm at 400 nm depth. The effective radius is defined as the radius of an ideal sphere that is giving the same depth like the nonideal sphere. The effective depth is the absolute depth minus the indenter deformation and therefore only the deformation of the sample itself. The effective radius was calculated for the sample materials fused silica and sapphire to demonstrate that the result is material-independent. To apply the method to coatings with a thickness of 50 nm or less, high precision is necessary for the measurement of the elastic load – displacement curves. This can be realized by doing many indents and averaging the curves. Further, very precise correction of zero point and thermal drift are necessary. In the given example, a 20 nm tick Al_2O_3 layer on a (100) silicon single crystal substrate was used, produced by atomic layer deposition (ALD). First, the pure substrate must be measured to get the substrate modulus for the subsequent calculations with coating. Forty-five measurements have been averaged. The statistical depth error was 0.35 nm, and the modulus result was 163.6 ± 0.4 GPa. This is very close to the Hill average of 162.1 GPa for a silicon single crystal when the single crystal constants from Landolt-Börnstein (1992) are used. The statistical depth error from the average of 43 indents on the coated samples was 0.15 nm, and the maximum depth difference between coated and uncoated sample is 2.6 nm at a force of 50 mN. This is shown in Fig. 4.13 together with the fit curves. The mean difference of all depth positions between measurement and fit is 0.5 nm for the pure substrate and 0.4 nm for the coated sample. The Young's modulus result for the coating was 333.4 ± 7.8 GPa.

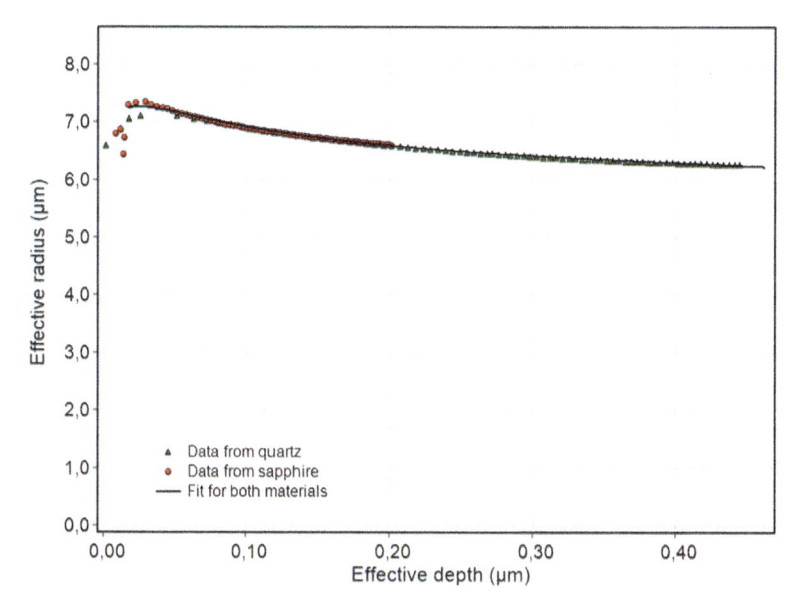

Nanoindentation—strategies for reliable measurements on coated systems 131

Figure 4.12 Radius function for the outer end of a spherical indenter with a nominal radius of 10 μm, obtained from elastic measurements on fused silica (red points) and sapphire (dark green). Radius function of a spherical indenter.

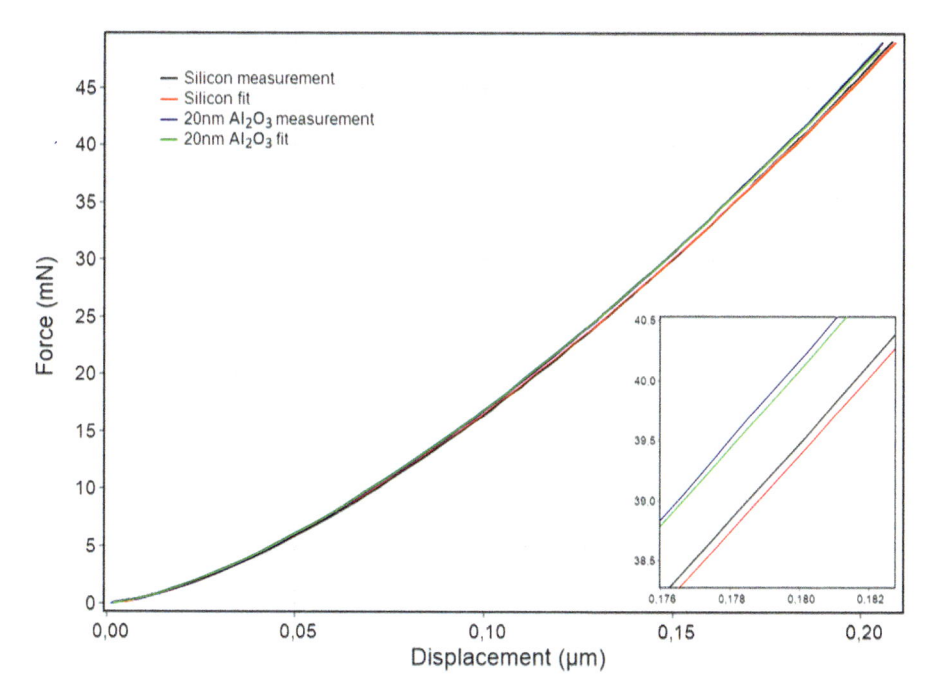

Figure 4.13 Elastic load – displacement curves from uncoated Si (100) and a 20 nm thick Al$_2$O$_3$ layer on Si together with the fit curves with the extended Hertzian contact model. The spherical indenter from Fig. 4.12 was used.

4.7 The use of dynamic test methods for the measurement of hardness and modulus

The measurement of five different forces with 10 indents per force, as recommended by the standard, requires relatively much measurement time and may nevertheless give not the necessary information to exclude substrate influence. An alternative strategy to get enough results along the depth axis is a cyclic measurement with many load − unload cycles (10 or more). An even better alternative, however, is a dynamic test method where more data points along the depth can be obtained in less time with higher accuracy than in a cyclic measurement. The method is called continuous stiffness measurement (CSM) and the patent for it dates back to 1989 (United States Patent 4848141 from 18.7.1989, Method for continuous determination of the elastic stiffness of contact between two bodies, 1989). Despite the long application time, a standard for the test method did not exist until 2022. In this year, part 5 of the standard ISO 14577 was published for linear elastic dynamic instrumented indentation testing (DIIT) (ISO 14577-5, 2022). In this test method, a small harmonic oscillatory force or displacement is applied to the indenter while the indenter is continuously loaded to a prescribed target load or target depth. The amplitude of the oscillation is analyzed with a lock-in amplifier and the ratio between force and displacement amplitude is converted to the contact stiffness that is normally obtained from the fit of the unloading curve in a quasistatic experiment. The conversion requires knowledge of the dynamic stiffness of the indenter shaft supporting springs, the instrumented damping coefficient as functions of the frequency and the oscillating mass of the actuator (indenter plus shaft). In a careful investigation of the test method (Merle et al., 2017), it was shown that the continuous quasistatic loading results in periodic discontinuities in the slope of the raw displacement signal and a deviation from the sinusoidal shape, especially in the displacement signal. This leads to an error in the lock-in analysis and an apparent phase shift even for nonviscous materials. The plasticity error during nanoindentation with continuous stiffness measurement was also modeled by Phani et al. (2020). The error becomes all the larger, the faster the static loading rate $\dot{F}$ is in relation to the oscillation frequency f. Further, the error of the lock-in amplifier becomes smaller for a larger amplitude ΔF since the plastic part of every oscillation cycle decreases in relation to the elastic part. This can be summarized in the ratio

$$\frac{\dot{F}}{\Delta F \cdot f} \ll 1 \tag{4.11}$$

The ratio is more important for soft materials with a large E/H ratio while for elastic materials like fused silica even a ratio of 2 may be acceptable. For soft materials, the deviation of the dynamic stiffness result from the unloading stiffness may reach more than 80% [see Merle et al. (2017)] if this ratio is not kept small.

The phase shift between force and displacement signal is used in the standard as criterion, if the ratio (Eq. 4.11) is small enough. It should be below 5% for metals or other hard materials (not for viscoelastic materials). It is recommended to increase the frequency when the angle is larger than 5%. However, too large frequencies have also a disadvantage. The stiffness correction that is necessary for a moving mass is proportional to the frequency squared, and this requires larger corrections and may result in larger error contributions for high frequencies. Further, the frequency should always be lower than the resonance frequency of the measuring head.

For good measurement conditions in dynamic tests, it is useful to increase the force amplitude proportional to the static force. For pyramidal indenters, a linear force amplitude increase proportional to the square root of the static force will result in an approximately constant, depth-independent displacement amplitude. The displacement amplitude should not be greater than $1-5$ nm to prevent a loss of contact during unloading at shallow depths.

An alternative to the CSM method is the quasicontinuous stiffness measurement (QCSM), developed by ASMEC GmbH [see, for instance, Refs. Chudoba et al. (2022) and Lorenz et al. (2021)]. The QCSM method uses a stepwise loading, and the oscillation is turned off between the static load steps and starts only when the defined static force is reached. A dwell time per point between 0.5 and 5 s (in dependence on frequency and chosen accuracy) is used for the periods with oscillation. In this mode, the force of every single point is exactly controlled, and therefore all points in the loading segment have the same force in repeated tests within some micronewton accuracy so that an averaging of several measurements can easily be done. Doing the oscillation, a constant static force has the advantage that the amplitudes of several oscillations can be averaged. The lock-in time constant is always smaller than $1/f$ so that every single amplitude of the oscillation can be analyzed without limitations. In contrast to the CSM method, an additional plastic deformation of the sample takes only place during the first few oscillations when the static load step was just reached, followed by purely elastic oscillations. Therefore the first 20% of the oscillations are not used for analysis. The measurement uncertainty of the averaged displacement amplitudes (obtained from the standard deviation of the amplitudes) is typically between 0.05 and 0.08 nm and can be reduced further by longer dwell times. This allows very precise dynamic stiffness measurements even at shallow depths of a few nanometers. A phase shift caused by additional plastic deformation in every cycle due to the static load increase does not occur. This does not mean that the phase shift for nonviscous materials is always zero because also stronger creep effects during an oscillation cycle can deform the sinusoidal shape of an oscillation which result in a phase shift after lock-in analysis. However, Eq. (4.11) must not be used since $\dot{F}$ is zero.

In the QCSM method, the oscillation is also used during a creep segment, and the transition between loading and creep segment is not accompanied by a step in the dynamic stiffness result. This allows a direct comparison between the dynamic contact stiffness from the last point with oscillations and the quasistatic stiffness from the unloading curve, analyzed according to the standard. The difference between the two stiffness results can be used for an additional correction of the dynamic data. Typically, the difference is below 5%.

In the application of dynamic test methods, it must be considered that the dynamic indentation hardness is obtained before creep while the measurement of the indentation hardness according to the standard always requires a creep segment that is long enough that creep effects cannot influence the slope of the unloading curve. The hardness drop can be seen in the depth-dependent hardness curves when an oscillation is also used in the creep segments as will be shown further. In contrast to hardness, the modulus results during the creep segment should not change. This can be used as criterion for the validity of the dynamic modulus measurements.

4.8 Examples for the measurement of coatings

The following results have all be obtained with a ZwickRoell nanoindenter ZHN and a measuring head with 2 N maximum force (ZHN nanoindenter, 2019) and a Berkovich indenter. The calibration of indenter area function and instrument compliance was done with the method described in Section 4.2 using fused silica and sapphire single crystal as reference materials. A zero-point correction was carried out and a thermal drift correction was done using a hold period at 10% of the maximum force. The dynamic test method QCSM was used to get enough points along the depth axis. The oscillation frequency was 40 Hz and the displacement amplitudes were always smaller than 5 nm but mostly 2−3 nm. The amplitude for the first three points was smaller and started between 0.7 and 1.5 nm. Typically, 30 points have been used during loading and 8 points in the creep time at maximum force. Quadratic load steps have been chosen to get nearly equidistant depth results and for nickel a loading with constant strain rate was used. Between 7 and 15 measurements per force have been done and averaged. For every signal, a standard deviation is calculated that is used to derive the statistical uncertainty per point for hardness and modulus. Only the averaged curves have been used for further analysis.

The unloading curve was analyzed according to the standard using a fit to extract contact stiffness and unloading exponent for the calculation of a variable epsilon factor. The difference between contact stiffness and dynamic stiffness at the last point of the creep segment was used to correct the dynamic stiffness. The difference was mostly below 2%. A radial displacement correction was also applied. The loading was done in 50 s, and the creep time was 15 s. The maximum force was adapted to the film thickness and is given below the figures.

In a first step, it was checked that the dynamic test method is giving depth-independent modulus results for homogeneous materials without coating (Fig. 4.14). For these measurements, a Berkovich indenter was used that had a tip radius of 145 at 25 nm depth and a radius of 260 nm when it is calculated from the maximum contact radius offset Δa with Eq. (4.9). The modulus of fused silica and sapphire is depth-independent, and the results are 72.0 ± 0.8 GPa and 411.3 ± 8.6 GPa, respectively. This can be expected because both materials have

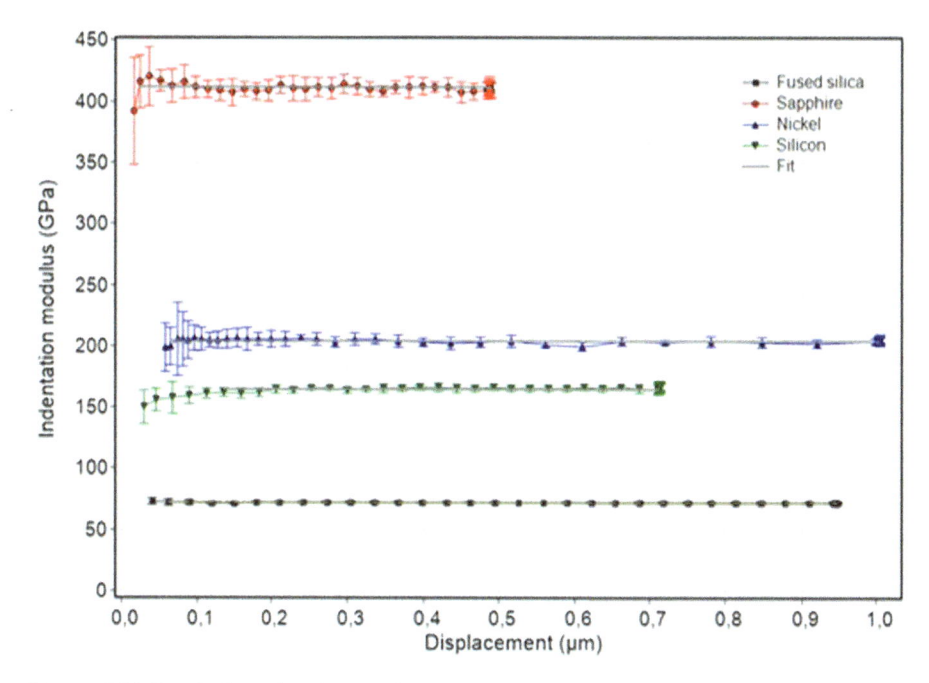

Figure 4.14 Depth-dependent indentation modulus results for four different homogeneous materials measured with the dynamic QCSM method and a maximum force of 100 mN. The mean modulus result is represented by a gray horizontal line (depicted as fit). The error bars represent the statistical uncertainty from the average of seven measurements.

been used for the calibration. The result for nickel is also depth-independent with a mean value of 204.6 ± 6.7 GPa and that of silicon single crystal with (100) surface is 165.6 ± 3.0 GPa for a depth larger than 200 nm. Below 200 nm, the modulus of silicon is slightly decreasing. This can be related to the phase transition during the expansion of the plastic zone [see for instance Refs. Bradby et al. (2003) and Weppelmann et al. (1993)]. For all materials, the modulus is also constant during the creep period and not different from the results during loading.

The indentation hardness of all four materials is not constant (Fig. 4.15). For fused silica and silicon, it is decreasing with decreasing depth, probably because the contact radius offset Δa of the Berkovich tip is still increasing until about 300 nm depth, and the opening angle of the faces is not constant in this region. For fused silica, also densification plays a role. The hardness of sapphire and nickel is increasing. The reason for nickel is the indentation size effect that occurs especially at metals (Durst et al., 2005; Nix & Gao, 1998; Pharr et al., 2010). For sapphire, it is mainly the elastic − plastic transition that takes place at low depths. A single crystal without dislocations in the indentation area can reach the theoretical strength of about $E/10$. For a 250 nm radius tip, the transition would be in a depth range between 20 and 25 nm, and this is that what can be observed [see also Chudoba et al. (2022)].

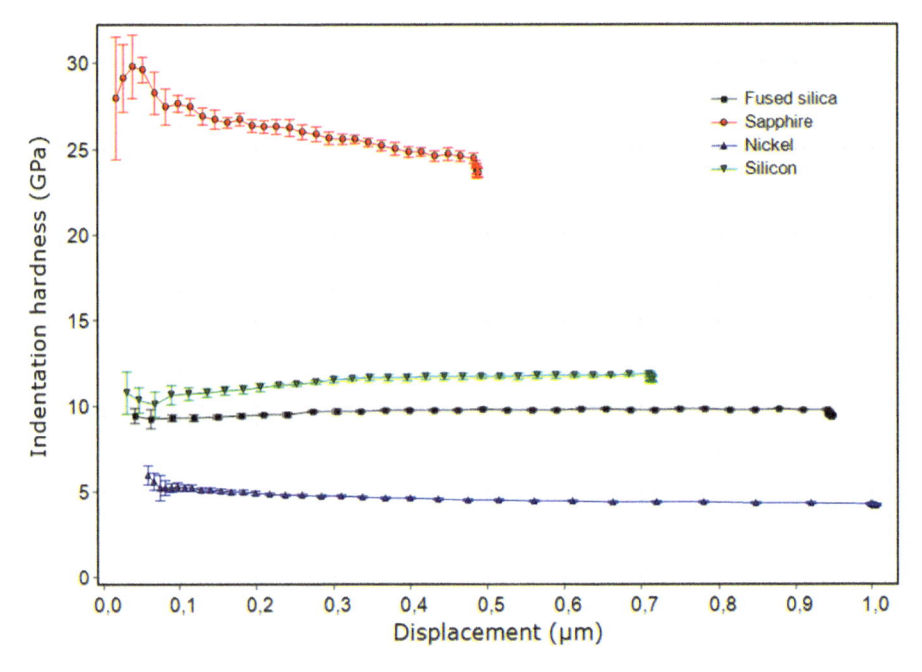

Figure 4.15 Depth-dependent indentation hardness results for four different homogeneous materials measured with the dynamic QCSM method and a maximum force of 100 mN. The error bars represent the statistical uncertainty from the average of seven measurements.

For all materials, the hardness is dropping during the creep time at the right end of the curves. For fused silica, for instance, the hardness is reduced from 9.75 to 9.38 GPa and for nickel from 4.27 to 4.19 GPa.

The same test method is now applied to different coating systems. The Berkovich indenters and the calibrations were not always the same because the data have been collected from different time periods. However, calibration and analysis have been done in the same manner.

The first example is from a 10 μm thick chromium and a nonstochiometric CrNO coating (depicted with CrX) on a steel substrate. The maximum force of the measurement was 500 mN. Twelve measurements have been averaged, and the scatter of the results is represented by the error bars. The measurement of such a thick coating is normally not a problem, but the example shows that the correct modulus of the coating material can only be measured at very low depths or when the extrapolation method is used. The first measurement point is obtained at a force of 0.8 mN. The correct chromium modulus of 250 GPa is measured for the pure chromium coating (Fig. 4.16). At 1/10th of the film thickness, the modulus result is only that of steel. For larger depth than 1/10th, it is even smaller than that of steel, probably because the model assumptions are violated. For the stiffer coating, a linear modulus increase toward the surface can be observed, and a film modulus of 322.3 ± 10.5 GPa is obtained by extrapolation.

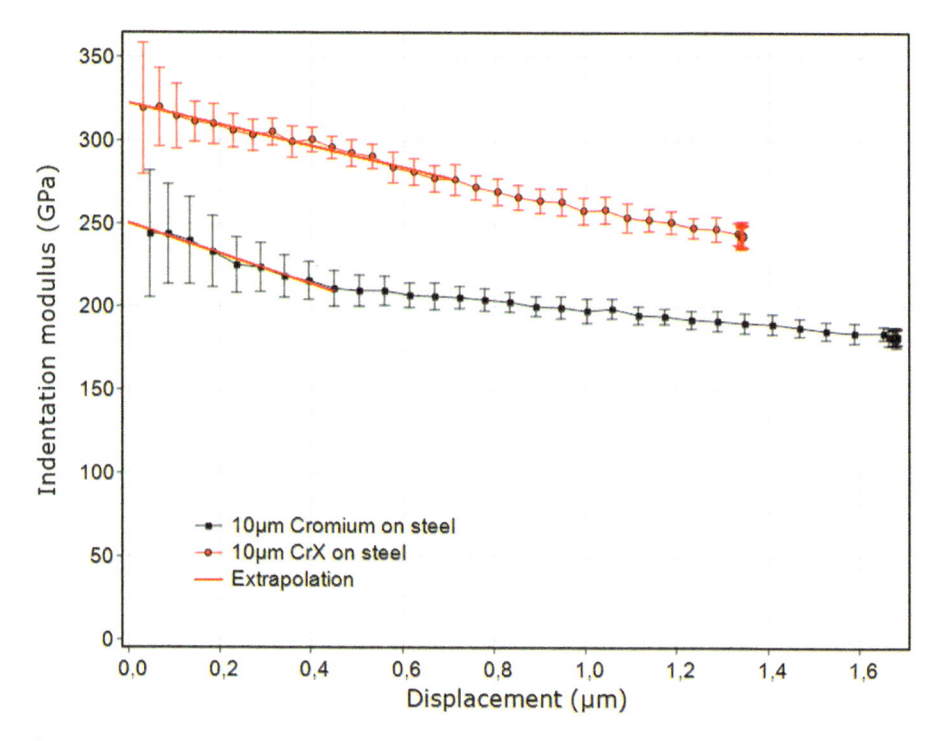

Figure 4.16 Indentation modulus as function of depth for a 10 μm thick chromium and a CrN$_x$O$_y$ coating on steel with extrapolation to zero depth.

The hardness in the 1/10th range is nearly constant (Fig. 4.17). Only a slight indentation size effect can be observed. For the harder coating, a beginning substrate influence can be seen between 0.9 and 1.2 μm.

The second example is given for an only 350 nm thick Si$_3$N$_4$ coating on glass. The maximum force was 20 mN, and the result is an average of 12 single measurements. The surface roughness was very low so that good measurement conditions have been available. The depth-dependent indentation modulus and hardness functions agree well with the predictions from Figs 4.6 and 4.7. The modulus extrapolation to zero depth (neglecting point 1, which is a measurement artifact) gives a modulus of 184.1 ± 5.0 GPa for the coating while the glass modulus is 71.9 ± 1.0 GPa (Fig. 4.18).

The indentation hardness profile (Fig. 4.19) shows first a pressure increase due to tip rounding before a fully developed plastic zone is realized at a depth of about 40 nm and a plateau is reached with a hardness of 16.73 ± 0.60 GPa. The calibrated tip radius was 173 nm. This corresponds to a measurable hardness for depths larger than 35 nm. A hardness decrease can only be observed after 100 nm. In this case, the 1/10th rule can be weakened and the correct hardness is measured up to 28% of the film thickness. The hardness of the glass is 8.0 GPa before creep and 7.7 GPa

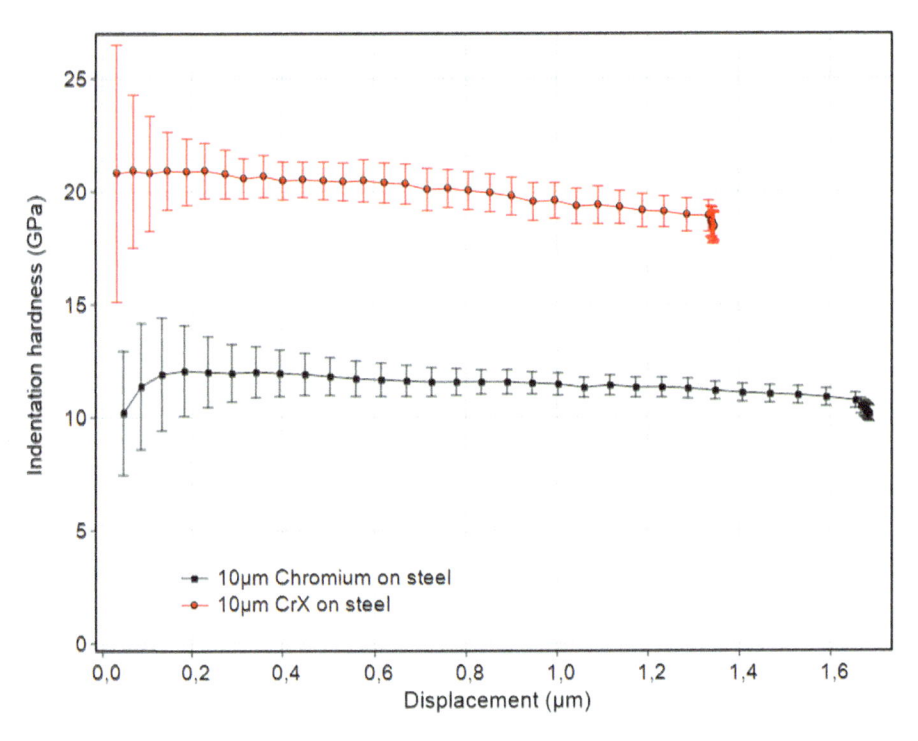

Figure 4.17 Indentation hardness as function of depth for a 10 μm thick chromium and a CrN$_x$O$_y$ coating on steel.

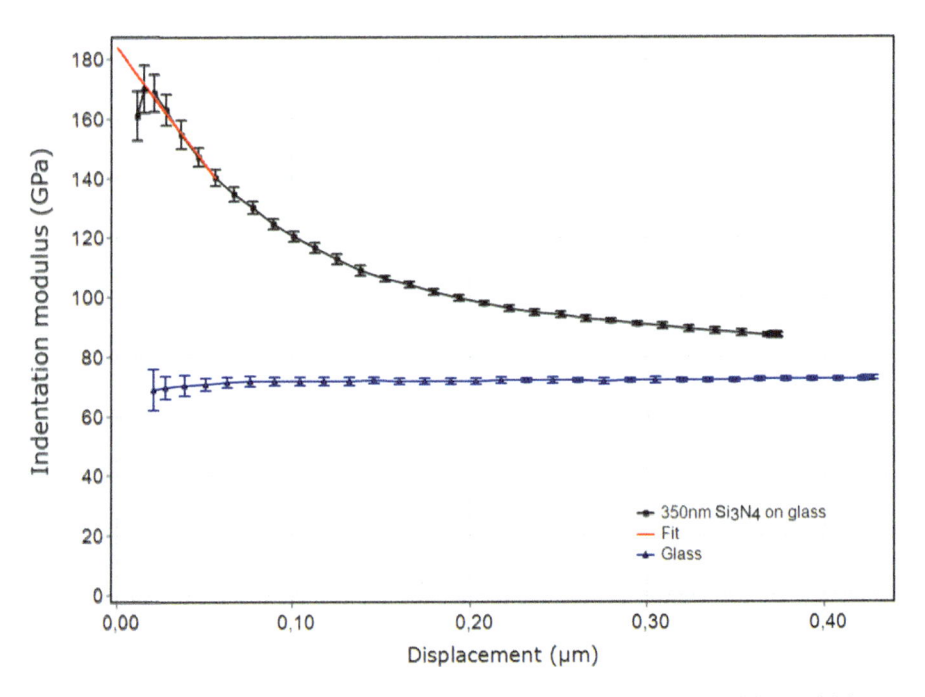

Figure 4.18 Indentation modulus as function of depth for pure glass and a 350 nm thick Si$_3$N$_4$ coating on glass with extrapolation to zero depth. The coating modulus is 184 GPa.

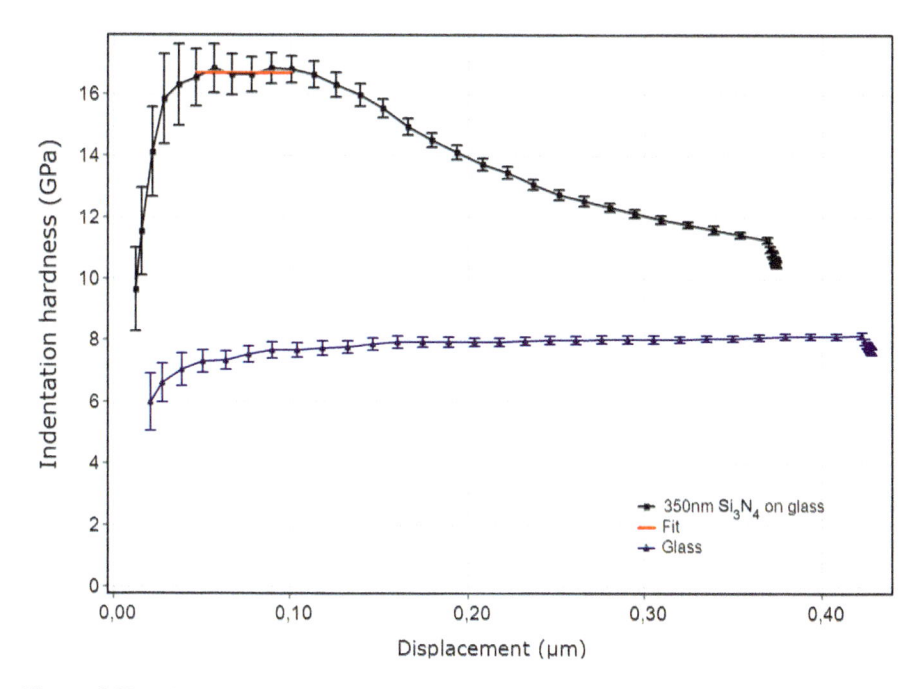

Figure 4.19 Indentation hardness as function of depth for pure glass and a 350 nm thick Si_3N_4 coating on glass. The coating hardness is represented by the plateau (marked by red line).

after the creep period. For small depths, the hardness is slightly decreasing, probably due to tip geometry influence and densification effects.

The examples demonstrate that more than only three data points along the depth axis are necessary to characterize coatings mechanically and to exclude any substrate influence on the results. The higher number of data points can be generated by many measurements with different forces. Dynamic measurements, especially using the QCSM method, deliver the necessary number of results along the depth with high precision and the course of the hardness and modulus profile corresponds to theoretical predictions. For hardness measurements, it has to be considered that dynamic tests give the hardness before creep and only when tip oscillations are also used during a creep segment the last data point can be compared with the conventional indentation hardness obtained from the analysis of the unloading curve.

References

Bradby, J. E., Williams, J. S., & Swain, M. V. (2003). In situ electrical characterization of phase transformations in Si during indentation. *Physical Review B, 67*(8). Available from https://doi.org/10.1103/physrevb.67.085205.

Bushby, A. J., & Swain, M. V. (1995). Spherical indentation as a means for investigating the plastic deformation of ceramics. In R. C. Bradt, C. A. Brookes, & J. L. Routbort (Eds.),

Plastic Deformation of Ceramics (pp. 161−172). Boston: Springer. Available from https://doi.org/10.1007/978-1-4899-1441-5_14.

Chudoba, T., Griepentrog, M., Dück, A., Schneider, D., & Richter, F. (2004). Young's modulus measurements on ultra-thin coatings. *Journal of Materials Research, 19*(1), 301−314. Available from https://doi.org/10.1557/jmr.2004.19.1.301.

Chudoba, T., & Jennett, N. M. (2008). Higher accuracy analysis of instrumented indentation data obtained with pointed indenters. *Journal of Physics D: Applied Physics, 41*(21) 215407. Available from https://doi.org/10.1088/0022-3727/41/21/215407.

Chudoba, T., Schwarzer, N., & Richter, F. (1999). New possibilities of mechanical surface characterization with spherical indenters by comparison of experimental and theoretical results. *Thin Solid Films*, 284−289. Available from https://doi.org/10.1016/s0040-6090 (99)00445-9.

Chudoba, T., Schwarzer, N., & Richter, F. (2000). Determination of elastic properties of thin films by indentation measurements with a spherical indenter. *Surface and Coatings Technology, 127*(1), 9−17. Available from https://doi.org/10.1016/S0257-8972(00) 00552-1.

Chudoba, T., Schwenk, D., Reinstädt, P., & Griepentrog, M. (2022). High-precision calibration of indenter area function and instrument compliance. *JOM: the journal of the Minerals, Metals & Materials Society, 74*(6), 2179−2194. Available from https://ui. adsabs.harvard.edu/link_gateway/2022JOM....74.2179C/doi:10.1007/s11837-022-05291-3.

Durst, K., Backes, B., & Göken öGöken, M. (2005). Indentation size effect in metallic materials: Correcting for the size of the plastic zone. *Scripta Materialia, 52*(11), 1093−1097. Available from https://doi.org/10.1016/j.scriptamat.2005.02.009.

ELASTICA, (2024). Software, Version 5. explained at www.asmec.de.

Hanson, M. T. (1992a). The elastic field for spherical hertzian contact including sliding friction for transverse isotropy. *Journal of Tribology, 114*(3), 606−611. Available from https://doi.org/10.1115/1.2920924.

Hanson, M. T. (1992b). The elastic field for conical indentation including sliding friction for transverse isotropy. *Journal of Applied Mechanics, Transactions ASME, 59*(2), S123−S130. Available from https://doi.org/10.1115/1.2899476.

ISO 14577. (2015). Metallic materials - Instrumented indentation test for hardness and materials parameters. ISO Central Secretariat, Geneva, Switzerland, 2015.

ISO 14577-5. (2022). Metallic materials - Instrumented indentation test for hardness and materials parameters, Part 5: Linear elastic dynamic instrumented indentation testing (DIIT). ISO Central Secretariat Geneva, Switzerland, 2022.

Johnson, K. L. (1985). *Contact Mechanics* (p. 62ff) Cambridge University Press.

Landolt-Börnstein. (1992). In Nelson (Ed.), *New Series, Group III: 29(a)*. Berlin: Verlag.

Lorenz, L., Chudoba, T., Makowski, S., Zawischa, M., Schaller, F., & Weihnacht, V. (2021). Indentation modulus extrapolation and thickness estimation of ta-C coatings from nanoindentation. *Journal of Materials Science, 56*(33), 18740−18748. Available from https://doi.org/10.1007/s10853-021-06448-2, http://www.springer.com/journal/10853.

Marsh, D. M. (1964). Plastic flow in glass. *Proceedings of the Royal Society, 279*.

Merle, B., Maier-Kiener, V., & Pharr, G. M. (2017). Influence of modulus-to-hardness ratio and harmonic parameters on continuous stiffness measurement during nanoindentation. *Acta Materialia Inc, Germany Acta Materialia, 134*, 167−176. Available from https:// doi.org/10.1016/j.actamat.2017.05.036, http://www.journals.elsevier.com/acta-materialia/.

Nix, W. D., & Gao, H. (1998). Indentation size effects in crystalline materials: A law for strain gradient plasticity. *Journal of the Mechanics and Physics of Solids, 46*(3), 411−425. Available from https://doi.org/10.1016/S0022-5096(97)00086-0.

Olaf, J., & Ritter, B. (1992). Fehler beim registrierenden Härtemessen. *Materials Testing*, *34*(5), 143−146. Available from https://doi.org/10.1515/mt-1992-340504.

Oliver, W. C., & Pharr, G. M. (1992). An improved technique for determining hardness and elastic modulus using load and displacement sensing indentation experiments. *Journal of Materials Research*, *7*(6), 1564−1583. Available from https://doi.org/10.1557/jmr.1992.1564.

Phani, P. S., Oliver, W. C., & Pharr, G. M. (2020). Understanding and modelling the plasticity error during nanoindentation with continuous stiffness measurement. *Materials*, *194*, 108923. Available from https://doi.org/10.1016/j.matdes.2020.108923.

Pharr, G. M., Herbert, E. G., & Gao, Y. (2010). The indentation size effect: A critical examination of experimental observations and mechanistic interpretations. *Annual Review of Materials Research*, *40*, 271−292. Available from https://doi.org/10.1146/annurev-matsci-070909-104456.

Schwarzer, N. (2000). Arbitrary load distribution on a layered half space. *Journal of Tribology*, *122*(4), 672−681. Available from https://doi.org/10.1115/1.1310330.

Schwarzer, N. (2006). The extended Hertzian theory and its uses in analyzing indentation experiments. *Philosophical Magazine*, *86*(33−35), 5179−5197. Available from https://doi.org/10.1080/14786430600690507.

Schwarzer, N., Richter, F., & Hecht, G. (1999). The elastic field in a coated half-space under Hertzian pressure distribution. *Surface and Coatings Technology*, *114*(2−3), 292−303. Available from https://doi.org/10.1016/s0257-8972(99)00057-2.

Sneddon, I. N. (1948). Boussinesq's problem for a rigid cone. *Mathematical Proceedings of the Cambridge Philosophical Society*, *44*(4), 492−507. Available from https://doi.org/10.1017/S0305004100024518.

United States Patent 4848141 from 18.7.1989, Method for continuous determination of the elastic stiffness of contact between two bodies.

Weiler, W. (1986). On the definition of a new hardness scale based on hardness values determined under test force. *Materialprüfung*, *28*, 217−220. Available from https://doi.org/10.1515/mt-1986-287-811.

Weppelmann, E. R., Field, J. S., & Swain, M. V. (1993). Observation, analysis, and simulation of the hysteresis of silicon using ultra-micro-indentation with spherical indenters. *Journal of Materials Research*, *8*(4), 830−840. Available from https://doi.org/10.1557/JMR.1993.0830.

ZHN nanoindenter. (2019). https://www.zwickroell.com/products/hardness-testing-machines/instrumented-indentation-test/zhn-nanoindenter/.

Indentation energy-based analysis methods 5

Steve J. Bull
School of Engineering, Newcastle University, Newcastle upon Tyne, United Kingdom

5.1 Introduction

Continuously recorded indentation tests are now used widely for the assessment of the mechanical properties of bulk materials and coatings at high spatial resolution. The instruments generally produce a record of load against indenter displacement, and this information is used to calculate desired mechanical properties such as hardness and Young's Modulus. There is general agreement on the methodology for analysis due to Oliver and Pharr (1992), but this method has some problems associated with material behavior and calibration, and there has been interest in methods that address some of these issues and allow the assessment of other important properties such as fracture toughness and adhesion using the same input data.

For instance, CRIT experiments also allow measurement of the plastic work of indentation (Wp) (Page & Hainsworth, 1993), which is given by the area enclosed by the loading and unloading curves. In addition, techniques such as atomic force microscopy (AFM) may allow the residual plastically deformed volumes of indentations to be estimated (e.g., Hainsworth et al., 1997). Thus, it may become possible to meaningfully define hardnesses by a work of indentation per unit volume approach (e.g., Hainsworth et al., 1997; Nowak & Sakai, 1993; Twigg et al., 1995). New insights into understanding the energy dissipation mechanisms involved in the deformation of coated systems are produced and are a reasonable basis for modeling the indentation response of the coating/substrate system. It may be useful for determining fracture and adhesion energy (see later).

5.2 Hardness and Young's modulus

CRIT data for hardness were adjusted to take into consideration the conventional hardness measurement method of load divided by area of the impression. It is assumed that the area of the impression can be derived from the area of the impression calculated from the measured dimensions of the indent under load (e.g., the indenter diagonals for a Vickers indenter (Fig. 5.1)) and these do not change on

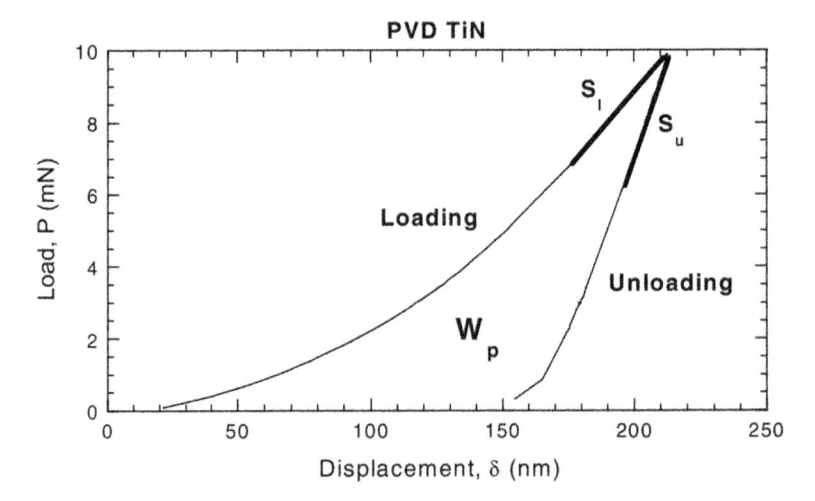

Figure 5.1 Load–displacement curve showing the unloading (S_u) and loading (S_l) slopes used in the calculation of hardness and Young's modulus. Also indicated is the plastic work of indentation W_p, which is the area bounded by the loading and unloading curves and the displacement axis.
Source: From Bull, S. J. (2005). Nanoindentation of coatings. *Journal of Physics D: Applied Physics*, 38, R393–R413. http://iopscience.iop.org/0022-3727/38/24/R01.

unloading of the indenter (i.e., no elastic recovery of the impression), which is a reasonable assumption for relatively large indents in soft, plastic materials such as metals.

5.2.1 Conventional nanoindentation assessment

The most commonly used analysis method for obtaining hardness and Young's modulus was developed by Oliver and Pharr (1992). In this approach, the total penetration depth is given by the sum of the plastic depth (contact depth), δ_c, and the elastic depth, δ_e, which represents the elastic flexure of the surface during loading. Thus, the total penetration depth, δ, is given by

$$\delta = \delta_c + \delta_e \tag{5.1}$$

and

$$\delta_e = \varepsilon\, P\, S_u, \tag{5.2}$$

where S_u is the slope of the unloading curve at maximum load (see Fig. 5.1), and ε is a constant, which depends on indenter geometry ($\varepsilon = 0.75$ for Berkovich indenters).

δ_c can be calculated from the equations of Nowak and Sakai (1993) and the contact area, A_c can be calculated from this if the relationship is known. For an ideal Berkovich indenter, $A_c = 24.5\, \delta^2_c$, but most indenters are blunted at their end and a more complex tip area function is then required (see later). The hardness, H, is then given by:

$$H = P/A_c. \tag{5.3}$$

Young's modulus can be determined from the slope of the unloading curve using a modified form of Sneddon's flat punch equation (Sneddon & Welch, 1963)

$$S_u = \gamma\beta \frac{2}{\sqrt{\pi}} E_r \sqrt{A_c}, \tag{5.4}$$

where E_r is the contact modulus, which can be derived from Young's modulus and Poisson's ratio of the indenter and the test material through

$$\frac{1}{E_r} = \frac{(1-\nu_m^2)}{E_m} + \frac{(1-\nu_i^2)}{E_i}, \tag{5.5}$$

where the subscripts m and i refer to the test material and indenter, respectively.

The constant γ was introduced by Joslin and Oliver (1990) to account for deviations from the ideal Sneddon behavior predicted by finite element investigations. The constant β was introduced to correct for the fact that the Berkovich indenters generally used for this sort of experiment are not axially symmetric; in this study, $\beta = 1.03$ (King, 1987). Using Eqs. (5.1–5.5), it is possible to determine the Young's modulus of the test material if the Poisson's ratio of the test material is known, which is not always the case, but for most metals and ceramics, an estimated value of $\nu_m = 0.25$ will give only a small error as Eq. (5.5) shows that E_r is not a sensitive function of ν_m. A much more important concern is to accurately determine the relationship between A_c and δ_c since this is rarely ideal. Oliver and Pharr (1992) provide a method for achieving this calibration using indentation tests made in fused silica, which is elastically isotropic and can be assumed to have a constant Young's modulus with depth. This produces a tip area function of the form

$$A_c = c_0 \delta_c^2 + c_1 \delta_c + c_2 \delta_c^{0.5} + c_3 \delta_c^{0.25} + c_4 \delta_c^{0.125} + \cdots. \tag{5.6}$$

There are some problems with this approach because fused silica can have a thin water-softened layer on its surface with a lower Young's modulus than the bulk, but for most practical indenters, the area function is reasonably accurate above a contact depth of 20 nm. It is critical that regular calibrations are performed as the detailed geometry of the indenter can be changed by a single indentation cycle. Careful tip calibration is essential for accurate work where residual indentation depths are less than 100 nm.

The need for this sort of detailed and time-consuming calibration suggests that an ideal analysis method would not be reliant on knowing the precise $A_c - \delta_c$

relationship. If the area function is known, the Oliver and Pharr method produces reasonable hardness and Young's modulus values for hard materials such as TiN (Fig. 5.2A) although nickel (Fig. 5.2B) has some problems due to pileup.

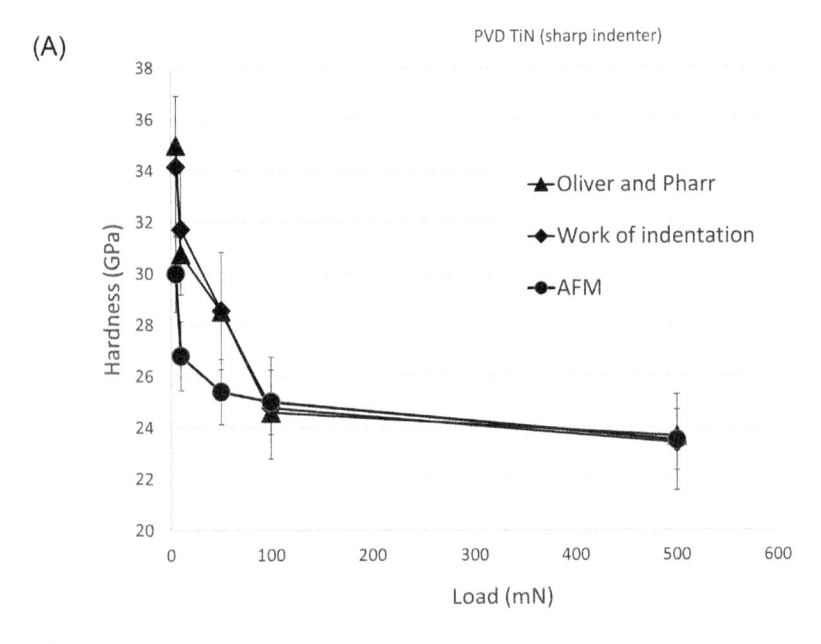

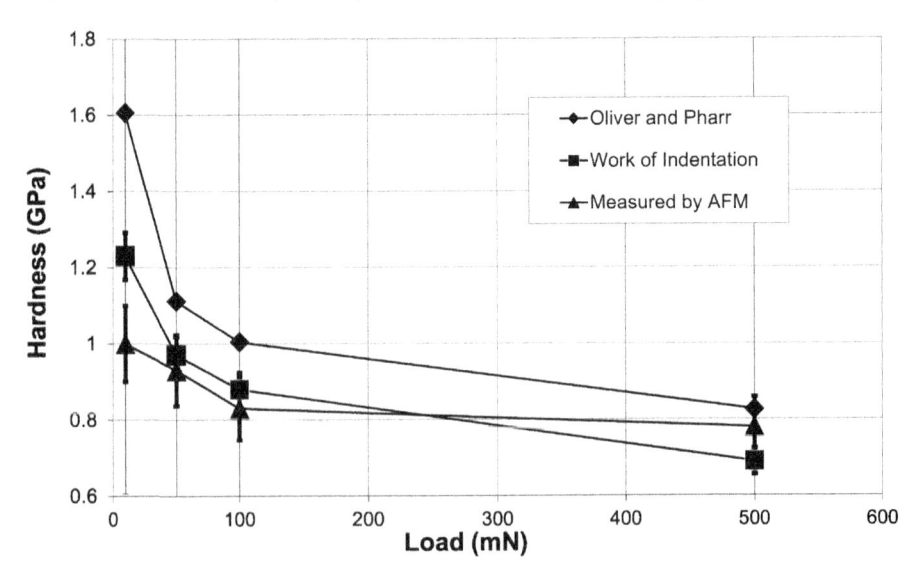

Figure 5.2 Comparison of hardness of coatings determined by different techniques. (A) PVD TiN and (B) electroplated nickel.

A potential problem with the Oliver and Pharr analysis method is that it cannot account for pileup around the indenter, which effectively means that the true contact depth is measured from a position above the original surface of the material (Fig. 5.3). This is a particular problem for indentation testing of very soft coatings such as nickel where the Oliver and Pharr method greatly overestimates the measured hardness compared with the traditional direct measurement.

5.2.2 Work of indentation approach

According to Stillwell and Tabor (1961), the standard definition of hardness in Eq. (5.3) can be rewritten as

$$H = W/V, \tag{5.7}$$

where W is the plastic work of indentation, and V is the total plastic deforming volume. W is very easy to measure from the area between the loading and unloading curves in continuously recording indentation tests, and its value is very accurate and less influenced by the position of the surface. However, a precise determination of V is less easy.

The work of indentation is just the integral of the force applied to the indenter with respect to displacement. One approach to determining hardness from the work of indentation is to rewrite Eq. (5.3) and integrate such that

$$W = H \int_0^{\delta_c} A_c d\delta, \tag{5.8}$$

where the integral in Eq. (5.8) is equivalent to V and can be evaluated explicitly from the tip calibration function. Since the powers of all terms are increased by this approach, the first few terms, which are more accurately known, dominate the volume calculation. However, the quality of the tip function is still critical in

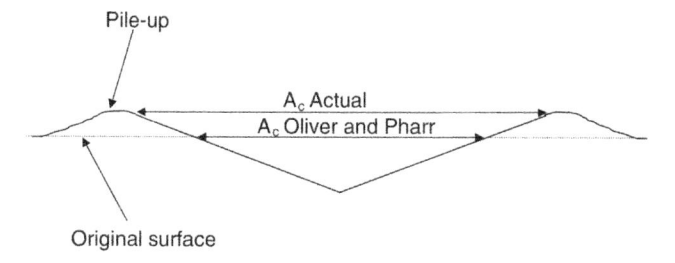

Figure 5.3 Cross section through an indentation in a soft material showing the effect of pileup. The Oliver and Pharr method underestimates the contact area and thus overestimates hardness and Young's modulus.
Source: From Bull, S. J. (2005). Nanoindentation of coatings. *Journal of Physics D: Applied Physics*, *38*, R393−R413. http://iopscience.iop.org/0022-3727/38/24/R01.

determining the hardness, and Eq. (5.8) is only valid if the hardness is independent of depth. This is most likely to occur when the hardness is determined at large indentation depths for soft materials.

Previous work has shown that Eq. (5.8) can be simplified if it is assumed that the indenter shape is ideal (Tuck et al., 2001). In this case, the hardness is given by

$$H = \frac{1}{9c_0}\frac{P^3}{W^2}. \tag{5.9}$$

Hardnesses determined from Eq. (5.9) show good agreement to values obtained by the Oliver and Pharr method for TiN coatings at high loads where indenter geometry effects are not important (Fig. 5.2A). The agreement with hardness determined from Eq. (5.8) is good over a wider range of loads but only if the tip area function is accurately known. The advantage of the work of indentation approach is that measured work is much less sensitive to displacement offsets caused by soft surface layers or elasticity at low loads. However, the volume calculation in Eq. (5.8) is still very sensitive to tip shape and the precise plastic depth.

Contact depth has been eliminated in Eq. (5.9), and this makes the formulation very useful for analyzing the indentation data of soft materials where pileup has significant effects on the validity of values obtained by the Oliver and Pharr method (Oliver & Pharr, 1992) (see Fig. 5.2).

5.3 Fracture toughness and fracture energy

Indentation fracture testing has been reviewed many times (e.g., Chen & Bull, 2007). There is a considerable difference between the assessment of thick coatings and bulk materials and the techniques, which can be used for thin coatings where the conventional cracking associated with indentation is affected by the presence of the coating; in many thin coatings (<2 μm), the required crack radius is much greater than the coating thickness. In such cases, two sorts of coating fracture are observed:

1. Through-thickness fracture, which generally runs normal to the coating/substrate interface for thin coatings but may run at a lower angle to it as the coating thickness increases. Through-thickness fracture is exacerbated by the bending stresses, which arise once plastic deformation of the substrate occurs, and the coating is bent into the impression or over the material piled up around it.
2. Interfacial fracture, which occurs at or near the coating/substrate interface or the weakest interface in a multilayer stack. Interfacial detachment may occur around the impression during loading and in the contact zone during unloading.

In fact, in nanoindentation, there are two distinct types of through-thickness fracture: radial cracks which follow the indenter edge, and picture frame cracks, which follow the edge of the impression (Fig. 5.4). For sharp indenters, the bending of the coatings around the indenter edges means that these are more likely to occur at

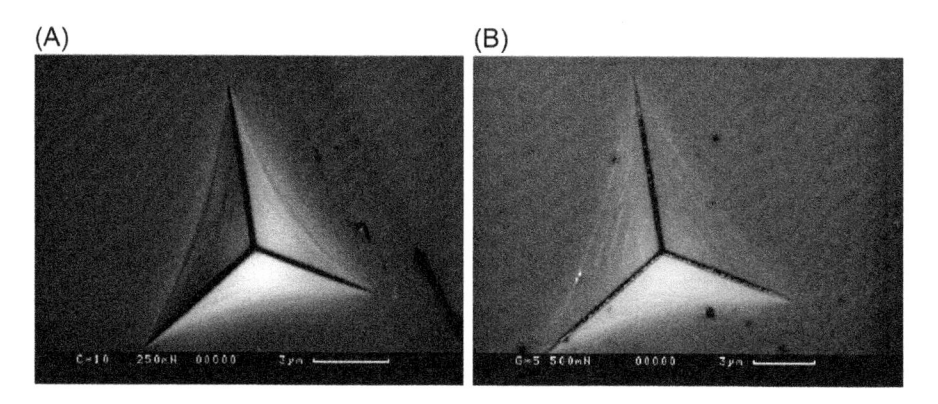

Figure 5.4 Cracks around nanoindentations in thin oxide coatings on glass. (A) radial cracks and (B) picture-frame cracks.

lower loads, whereas the bending of the coating into the plastic impression in the substrate is less severe, and these are more apparent at higher loads.

5.3.1 Fracture toughness

There are a number of different indentation fracture events, which might occur in a bulk material or coating − substrate system (Table 5.1), the occurrence of which will depend on the relative toughness of the material and any interfaces present. These failure modes will be altered by plastic deformation in the material, which can lead to the superposition of bending stresses on to those generated by the indentation. Indentation methods have been developed for assessing the fracture toughness of brittle materials over many years.

Conventional indentation toughness methods were initially developed for monolithic bulk ceramic materials tested by microindentation where well-developed radial cracks form (e.g., Anstis et al., 1981; Chantikul et al., 1979). The toughness K_{Ic} is related to the applied load P (Chantikul et al., 1979) and the crack dimension c such that

$$K_{Ic} = \chi \left(\frac{E}{H}\right)^{1/2} \frac{P}{c^{3/2}}, \tag{5.10}$$

where E and H are Young's modulus and hardness of the material. For Berkovich and Vickers indenters, $\chi = 0.016$. This method has been extended to coated systems where radial cracks are well developed by some authors (e.g., Malzbender et al., 2000; den Toonder et al., 2002; Volinsky et al., 2003), generally for a coating much thicker than 1 μm. The values of toughness obtained by this method will depend on the residual stress in the coating since Eq. (5.10) is strictly valid only in the absence of internal stresses. For hard coatings on harder and stiffer substrates, it may be reasonable to assume that the residual stress only modifies the crack shape.

Table 5.1 Fracture events associated with coating/substrate system.

Substrate	Coating	Fracture type
Ductile	Brittle	Through-thickness fracture. brittle fracture in coating parallel to interface, ductile interfacial fracture, microfracture in coating.
Brittle	Brittle	Bulk chipping, through-thickness fracture, brittle interfacial fracture, microfracture in coating, microfracture in substrate.
Ductile	Ductile	None.

Source: Nanoindentation of coatings. *Journal of Physics D: Applied Physics, 38*, R393–R413. http://iopscience.iop.org/0022-3727/38/24/R01.

However, it is necessary to point out that this traditional method will be invalid when the cracks are confined to the indentation impression, and coated systems consisting of a harder coating on softer substrate are usually an example of this. Chen and Bull (2006) have shown that the radial cracks that run along the indenter edges in hard coatings on softer substrates are generally confined within the indentation impression provided that the substrate is not sufficiently brittle that coating cracks can propagate into it. Therefore, the conventional indentation method is not valid as the cracks produced are not sufficiently well developed.

An alternative approach to assess coating toughness was proposed by Li and coworkers (Li et al., 1997; Li & Bhushan, 1998) and used if a step in the nanoindentation load–displacement curve is obtained under load control, which is associated with chipping. In this model, the load–displacement curve is extrapolated from the step start point (assumed to be the onset of fracture) to its end point. This is illustrated in Fig. 5.5, where the curve OABD is the measured loading curve, and OAC is the extrapolated loading curve from the initial loading part OA where no cracks occur. The difference between the extrapolated curve and the actual curve (i.e., area ABC in Fig. 5.5) is assumed to be the fracture dissipated energy.

Then, the coating toughness is given by

$$K_{\mathrm{Ic}} = \left[\frac{E_{\mathrm{c}} U_{\mathrm{fr}}}{(1-\nu_{\mathrm{c}}^2) A_{\mathrm{crack}}} \right]^{1/2}, \tag{5.11}$$

where U_{fr}, A_{crack} are the fracture dissipated energy and the fracture area; E_{c} and ν_{c} are Young's modulus and Poisson's ratio of the coating. A measurement of fracture area may be obtained by the microscopical analysis of the fracture around the indentation.

There is an assumption that any pop-in event that produces a measurable change in indenter displacement is due to fracture, but this is only usually achieved at high indenter displacement and in fact such an event is often due to plasticity at indenter loads less than 1 mN. In fact, for a typical used Berkovich indenter (with tip end-radius >250 nm), the initiation load for fracture in brittle materials such as sapphire is greater than 3 mN, and just the existence of a pop-in cannot be sufficient to

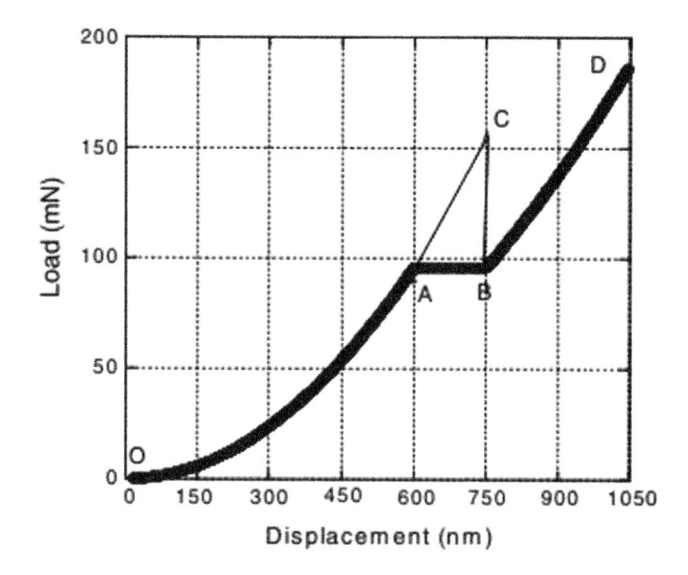

Figure 5.5 Schematic of the extrapolation of a load versus displacement curve to determine the energy dissipated in fracture ABC (the ld–dp method developed by Li and co-workers (Li et al., 1997; Li & Bhushan, 1998)).
Source: From Bull, S. J. (2005). Nanoindentation of coatings. *Journal of Physics D: Applied Physics*, 38, R393–R413. http://iopscience.iop.org/0022-3727/38/24/R01.

measure toughness. It is for this reason that the use of sharper indenters, such as the cube corner indenter, is preferred for toughness assessment.

It is often argued that a step in the load–displacement curve obtained under load control indicates the loss of contact caused by a transient event such as fracture in brittle materials, but it cannot be assumed that the energy dissipated during the displacement excursion is due to the transient event only because of the possible additional permanent deformation associated with it (Warren & Wyrobek, 2005). In contrast, the load drop in a displacement-controlled experiment was argued to be unambiguously related to loss of contact attributed to fracture. Comparisons show that displacement control tests eliminate the additional deformation inherent with fracture. Also, it has been argued that the displacement control tests will be more sensitive to the fracture initiation for this reason.

The method also completely ignores the change in the elastic–plastic behavior of the coated system when fracture occurs. den Toonder et al. (2002) also argued that the area ABC was not the actual energy dissipated by fracture. Furthermore, this method cannot be applied to tests carried out under displacement control. To eliminate the elastic–plastic deformation influence from substrate, it is preferred to perform small indentations in thin coatings. For this reason, the cracks may be not well developed compared with the indentation size. Since existing methods cannot work well in this aspect, a new method has been developed to assess this kind of ultrasmall cracks in a very thin coating (Chen & Bull, 2006). Fracture behavior may be assessed from a plot of total work of indentation versus displacement (Wt—dp) curve.

The total work of indentation is determined by the area under the load—displacement curve up to a given displacement. The method to determine fracture events is explained in Fig. 5.6. First, we extrapolate the initial Wt—dp curve from the crack start point A to the crack end point C, we get the work difference CD after fracture; then we extrapolate the Wt—dp curve after trracking backward to the crack start point, and thus, we get the work difference AB at the onset of fracture. AB represents the difference between the work of elastic—plastic deformation of the material before and after fracture, whereas CD represents the work of elastic—plastic deformation plus the work of fracture not including any contribution from relaxation of residual stress in the coating at the end of the crack event.

The difference between the two (i.e., CD minus AB) is the total work dissipated in the fracture event. It may be assumed that to calculate the strain energy release rate and fracture toughness, we may substitute the work dissipated in fracture for U_{fr} in Eq. (5.11). To do this, an accurate measurement of crack area is necessary. The approach has been used to assess a range of oxide coatings on glass (Table 5.2) with thicknesses in the range 200—400 nm using a cube corner indenter, which produces fracture in the coatings at loads of around 1 mN (Li & Bhushan, 1998). Although the conventional indentation method (Anstis et al., 1981; Malzbender et al., 2000) cannot work well for these very thin layers, it is also used for

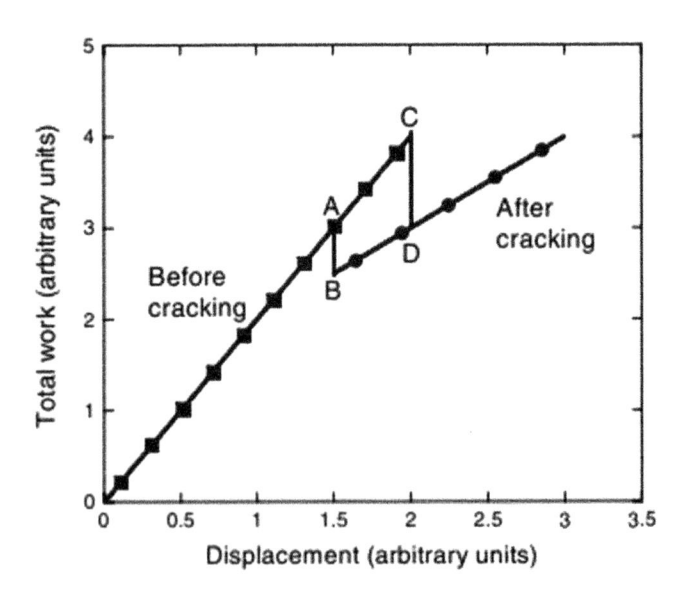

Figure 5.6 Schematic of extrapolating the total work versus displacement curve before and after cracking to determine the fracture dissipated energy CD—AB.
Source: From Bull, S. J. (2005). Nanoindentation of coatings. *Journal of Physics D: Applied Physics*, *38*, R393—R413. http://iopscience.iop.org/0022-3727/38/24/R01.

Table 5.2 The energy release rates and toughness calculated for the solar control coating components based on radial through-thickness fracture.

	Energy release rate for coating, G_R (Jm^{-2}), calculated by the Wt–dp method	Toughness of coating, K_{Ic} (MPa m$^{1/2}$)	
		Calculated by Wt–dp model	Estimated by the Lawn method
400 nm TiO$_x$N$_y$/glass	24.4 ± 1.4	1.8 ± 0.2	0.9 ± 0.1
240 nm ITO/glass	36.3 ± 8.2	2.2 ± 0.3	0.9 ± 0.1
400 nm ITO/glass	32.7 ± 4.4	2.1 ± 0.2	0.7 ± 0.1
400 nm SnO$_2$/glass	29.3 ± 9.8	1.9 ± 0.3	1.3 ± 0.1

Source: Nanoindentation of coatings. *Journal of Physics D: Applied Physics*, *38*, R393–R413. http://iopscience.iop.org/0022-3727/38/24/R01.

comparison here. The coefficient χ for the cube corner indenter is 0.0319 (Harding et al., 1995). These results are summarized in Table 5.2.

The data from the Wt–dp model is reasonable for the brittle coatings under investigation, whereas the traditional indentation method just returns the toughness of the glass substrate in most cases.

5.3.2 Interfacial fracture energy and adhesion

Coating detachment from the substrate during indentation tests can occur during loading or during unloading of the indenter when the adhesion is relatively poor; otherwise, mainly through-thickness fracture is observed (Fig. 5.7). When the indenter is loaded on to the coated sample, the coating around the indenter is under compressive stress. A thin coating may buckle to relieve the compressive stress and become detached from the substrate. The buckles that form on loading occur adjacent to the impression where the coating is not pressed into the surface by the indenter and the applied load. A small pop-in may be observed associated with this. On unloading these buckles propagate further and can detach the region under the indenter—the buckled material may push back on the indenter giving the linear unloading often observed in the $P-\delta$ curve. Alternatively, if the coating is relatively thick and has a high stiffness, a shear crack originates in it and propagates down to the interface causing the coating to detach from the substrate around the indenter (see SEM micrographs in Figs. 5.7B and 5.8C). This is apparent in the load–displacement curve, where a long pop-in can be observed in the loading curve as the indenter pushes the detached material out from underneath it. To quantify the adhesion in such cases, the expression proposed by den Toonder et al. (2002) may be used, in which the curved geometry of the chipped segment of the coating and

(A)

Compressive stress

Compressive stress

Detached region

(B)

(C)

Fracture during loading

Fracture during unloading

Linear unloading

Load, mN

Displacement, nm

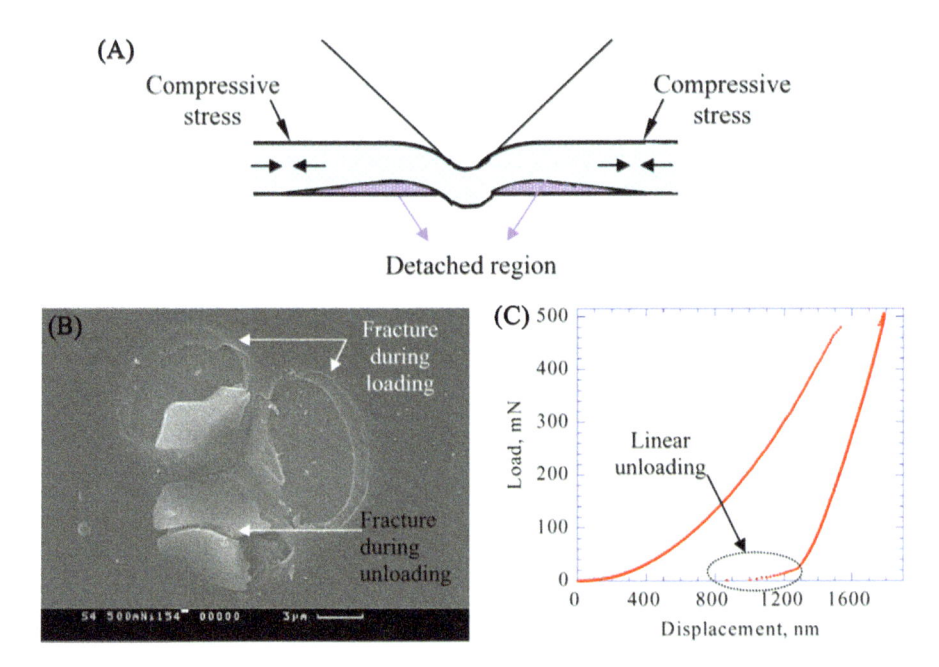

Figure 5.7 Detachment of a thin coating due to buckling in response to compressive stress around the indenter. (A) Schematic diagram of the detachment of a thin coating due to buckling in response to compressive stress around the indenter. (B) SEM micrograph for a 1 μm thick CNx coating on Si(001) showing coating fracture during loading and unloading of the indenter at 500 mN applied load. (C) Corresponding P–δ curve showing linear unloading at the bottom of the unloading curve associated with buckling of the coating.

the residual stress in the coating are used to calculate the adhesion energy or interfacial fracture energy, γ_i. The interfacial fracture energy is calculated using

$$\gamma_i = 1.42 \frac{Et^5}{L^4} \left(\frac{(a/L) + (\beta_c \pi/2)}{(a/L) + \beta_c \pi} \right)^2 + \frac{t(1-\nu)\sigma_r^2}{E} + 3.36 \frac{(1-\nu)t^3\sigma_r}{L^2} \left(\frac{(a/L) + (\beta_c \pi/2)}{(a/L) + \beta_c \pi} \right),$$

(5.12)

where E is the Young's modulus of the coating; t is the thickness of the coating; ν is the Poisson's ratio of the coating; σ_r is the residual stress in the coating; and a, L and β_c define the geometry of the chipped piece (Fig. 5.9). The higher the value of γ_i, the better the adhesion between coating and substrate.

This equation is very sensitive to the precise values of thickness and crack length measured, and care must be taken to determine these to a high level of accuracy. This expression is based on a model suggested by Thouless (1998), which assumes that (1) the in-plane load on the delaminated sector due to indentation causes the growth of the delamination area and (2) the coating chips at the moment of buckling of the sector are due to the same in-plane load. The results for the interfacial fracture energy are presented

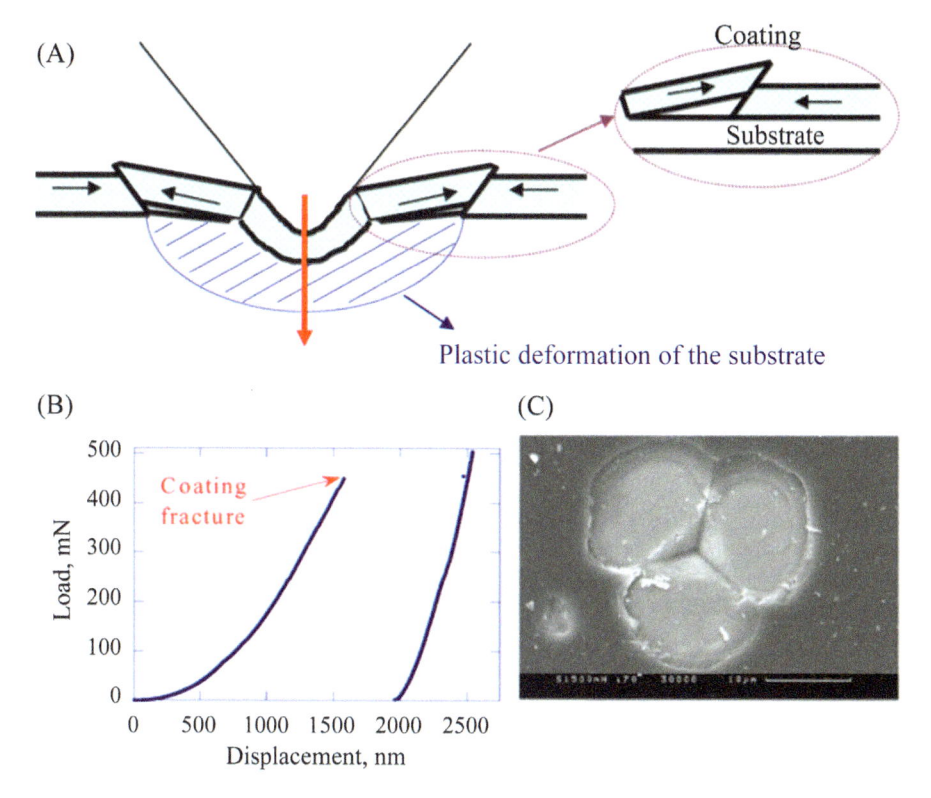

Figure 5.8 Coating detachment during loading of the indenter. (A) Schematic diagram showing coating detachment during loading of the indenter. (B) Step in the loading curve due to through-thickness cracking of the coating and spalling. (C) SEM micrograph showing coating detachment during loading of the indenter for a 1 μm thick CN_x coating on 3C SiC (001) substrate.

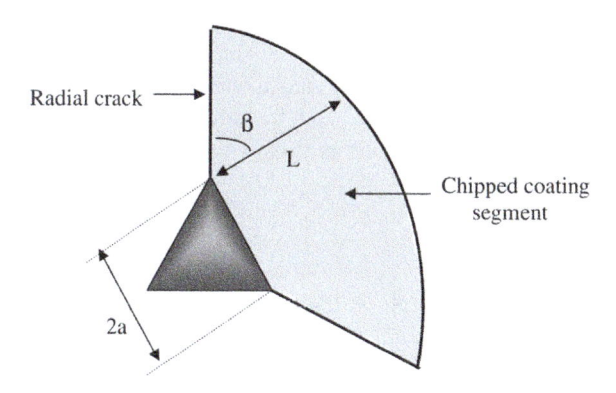

Figure 5.9 Schematic of the geometry of a chipped segment showing the dimensions used in calculating interfacial toughness.
Source: From Bull, S. J. (2005). Nanoindentation of coatings. *Journal of Physics D: Applied Physics*, *38*, R393–R413. http://iopscience.iop.org/0022-3727/38/24/R01.

Table 5.3 Interfacial fracture energy of CN_x coatings (1 μm thick) deposited on different substrates.

Substrate	γ_i (J m^{-2})
Si (001)	27 ± 5
Al$_2$O$_3$	21 ± 4
1 μm 3C SiC on (001) silicon	5.1 ± 0.9
1 μm 3C SiC on (111) silicon	5.3 ± 0.9

Source: Nanoindentation of coatings. *Journal of Physics D: Applied Physics, 38*, R393–R413. http://iopscience.iop.org/0022-3727/38/24/R01.

in Table 5.3. In the case of the coatings of SiC on silicon, the detachment occurs at or near the CN_x/SiC interface as determined by XPS analysis of the fracture surfaces. The values obtained are somewhat higher than those obtained for pure metals on the same substrates (typically $1-10$ J m^{-2}) but are typical of harder metal and ceramic coatings where considerable plastic deformation occurs at the crack tip during delamination.

5.4 Conclusions

The accuracy of the hardness and Young's modulus determined from nanoindentation by the method of Oliver and Pharr (1992) is critically dependent on the accuracy of some of the parameters used on their calculation, which are not necessarily known reliably. The alternative energy-based methods are much less susceptible to the same inaccuracies but are still reliant on a measurement of the deforming volume around the indentation, which has its own inaccuracies. For this reason, the Oliver and Pharr approach is still the methodology of choice for all but very soft materials.

The use of the work of indentation to determine fracture properties of materials shows much promise, but the changes in energy associated with many indentation fracture events are quite small, and the area of fracture is not well-known in many cases. However, in cases where well-defined indentation fracture occurs, very reasonable values for crack driving force and fracture toughness are observed. Even very thin coating toughness can be successfully assessed by indentation methods.

References

Anstis, G. R., Chantikul, P., Lawn, B., & Marshall, D. B. (1981). A critical evaluation of indentation techniques for measuring fracture toughness: I, direct crack measurements. *Journal of the American Ceramic Society, 64*, 533–538.

Chantikul, P., Marshall, D. B., Lawn, B. R., & Drexhage, M. G. (1979). Strength of partially leached glass rods after indentation. *American Ceramic Society Bulletin, 58*, 378–378.

Chen, J., & Bull, S. J. (2006). Assessment of the toughness of thin coatings using nanoindentation under displacement control. *Thin Solid Films, 494*(1–2), 1–7. Available from https://doi.org/10.1016/j.tsf.2005.08.176.

Chen, J., & Bull, S. J. (2007). Indentation facture and toughness assessment for thin optical coatings of glass. *Journal of Physics D: Applied Physics, 40*, 5401−5417.
den Toonder, J., Malzbender, J., de With, G., & Balkenende, R. (2002). Fracture toughness and adhesion energy of sol-gel coatings on glass. *Journal of Materials Research, 17*, 224−233.
Hainsworth, S. V., Sjostrom, H, Page, T., & Sundgren, J.-E. (1997). Hardness and deformation mechanisms of highly elastic carbon nitride thin films as studied by nanoindentation. *Materials Research Society Symposium − Proceedings, 436*, 275−280.
Harding, D. S., Oliver, W. C., & Pharr, G. M. (1995). Cracking during nanoindentation and its use in the measurement of fracture toughness. *Materials Research Society Symposium − Proceedings, 356*, 663−668.
Joslin, D., & Oliver, W. C. (1990). A new method for analyzing data from continuous depth-sensing microindentation tests. *Journal of Materials Research, 5*, 123−126.
King, R. B. (1987). Elastic analysis of some punch problems for a layered medium. *International Journal of Solids and Structures, 23*, 1657−1664.
Li, X. D., & Bhushan, B. (1998). Measurement of fracture toughness of ultra-thin amorphous carbon films. *Thin Solid Films, 315*, 214−221.
Li, X. D., Diao, D. F., & Bhushan, B. (1997). Fracture mechanisms of thin amorphous carbon films in nanoindentation. *Acta Materialia, 44*, 4453−4461.
Malzbender, J., & den Toonder, J. M. J. (2000). The P-h^2 relationship in indentation. *Journal of Materials Research, 15*, 1209−1212.
Nowak, R., & Sakai, M. (1993). Energy principle of indentation contact - the application to sapphire. *Journal of Materials Research, 8*, 1068−1078.
Oliver, W. C., & Pharr, G. M. (1992). An improved technique for determining hardness and elastic-modulus using load and displacement sensing indentation experiments. *Journal of Materials Research, 7*, 1564−1583.
Page, T. F., & Hainsworth, S. V. (1993). Using nanoindentation techniques for the characterization of coated systems - a critique. *Surface and Coatings Technology, 61*, 201−208.
Sneddon, I. N., & Welch, J. T. (1963). A note on the distribution of stress in a cylinder containing a penny-shaped crack. *International Journal of Engineering Science, 1*, 411−419.
Stillwell, N. A., & Tabor, D. (1961). Elastic recovery of conical indentations. *Proceedings of the Physical Society of London, 78*, 169.
Thouless, M. D. (1998). An analysis of spalling in the microscratch test. *Engineering Fracture Mechanics, 61*(1), 75−81. Available from https://doi.org/10.1016/s0013-7944(98)00049-6.
Tuck, J. R., Korsunsky, A. M., Bull, S. J., & Davidson, R. I. (2001). On the application of the work-of-indentation approach to depth-sensing indentation experiments in coated systems. *Surface and Coatings Technology, 137*, 217−224.
Twigg, P. C., McGurk, M. R., Hainsworth, S. V., & Page, T. F. (1995). Apparent Indentation Plasticity in Ceramic Coated Systems. In R. C. Bradt, C. A. Brookes, & J. L. Routbort (Eds.), *Plastic Deformation of Ceramics* (pp. 219−229). Boston, MA: Springer. Available from https://doi.org/10.1007/978-1-4899-1441-5_19.
Volinsky, A. A., Vella, J. B., & Gerberich, W. W. (2003). Fracture toughness, adhesion and mechanical properties of low-K dielectric thin films measured by nanoindentation. *Thin Solid Films, 429*, 201−210.
Warren, O. L., & Wyrobek, T. J. (2005). Nanomechanical property screening of combinatorial thin-film libraries by nanoindentation. *Measurement Science and Technology, 16*(1), 100−110. Available from https://doi.org/10.1088/0957-0233/16/1/014, http://www.iop.org/EJ/journal/0957-0233.

Nano-scratch testing

Ben Beake[1] and Tomasz Liskiewicz[2]
[1]Micro Materials Ltd., Wrexham, United Kingdom, [2]Department of Engineering, Manchester Metropolitan University, Manchester, United Kingdom

6.1 Introduction

Sensitivity to coating and interfacial properties can be improved by performing tests at smaller contact size, which reduces the influence of substrate deformation on the overall coating system response. This can be achieved by reducing the test probe radius and using instrumentation with greater sensitivity at lower load to perform nano- or microscale scratch tests, which has been termed "dimensioning." Nano-scratch testing (e.g., with 5 μm or sharper probes) is more suitable for thinner coatings, and microscale scratch testing (e.g., with 25 μm probes) is more suitable for thicker coatings (e.g., >2 μm). The most common types of scratch test involve either (1) ramped (or progressive) load scratch tests where critical loads are recorded corresponding to yield or failure events and (2) repetitive constant load tests where the number of cycles to failure at a load lower than that causing failure in the ramped load test is determined. More complex tests can be performed by the addition of low load topographical scans (pre- and post- scans), which enable the nonelastic deformation generated by the ramped or constant load scratches to be quantified. A typical example of a ramped scratch with pre- and post- scratch topographic (low load) scans is shown in Fig. 6.1A. A typical example of a repetitive constant load scratch test with interspersed topographic (low load) scans is shown in Fig. 6.1B.

Aided by experience gained in round-robin interlaboratory intercomparisons in the EU NANOINDENT and NANOINDENT-PLUS projects involving all the major nanomechanical test instruments in common use the project partners developed best practice guidelines, which are summarized in the CEN ISO standard (CEN/TS, 2021; NANOINDENT-PLUS, 2013). The standard focuses on (1) progressive load nano-scratch tests performed as "three-pass" procedures with pre- and post- topographic scans and (2) repetitive constant load tests, with examples on bulk materials and coating systems. Particular issues highlighted include the influence of experimental conditions (scan speed and loading rate), the effect of probe radius, estimation of contact pressure, and the influence of plowing on friction. These are discussed in detail below.

In all of these tests, the scratches are performed in the same track. An alternative approach for multiple-scratch test experiments to be designed involving statistically

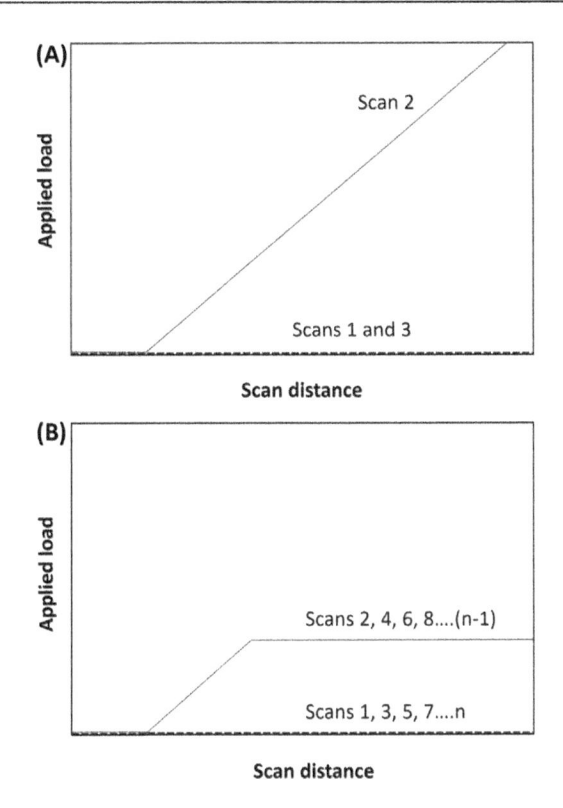

Figure 6.1 Examples of scratch tests. (A) Typical example of a ramped scratch with pre- and post- scratch topographic (low load) scans. (B) Typical example of a repetitive constant load scratch test with interspersed topographic (low load) scans.

distributed arrays of parallel scratches. In this type of repetitive scratch test, the damage from previous scratches interacts with subsequent scratches so that the accumulated damage builds up and the cycle-by-cycle development of an abraded region of the surface can be studied. The statistical (or randomly) distributed nano-scratch test appears to more closely simulate abrasion where material removal can be influenced by interaction between damage produced by different contacts close to each other than is possible by performing multiple scratches in the same track (Beake et al., 2024).

6.2 Experimental considerations

The choice of scanning load in the topographic pre- and post- scans depends slightly on the type of test/sample and the environment, but in general, it should be as low as practically possible to minimize the amount of plastic and elastic deformation occurring so that a true surface profile is obtained. For very long scratches with highly sloped samples and/or in laboratories with high vibration levels, it may be desirable to increase this to ensure that the probe remains on the surface throughout the test.

In "three-scan" tests, it is most common for the direct in each pass to be the same, that is, at the end of the scan, the probe is removed and replaced at the start before starting the next pass. Alternatively it is possible, and quicker, to keep the probe in contact with the surface throughout by performing the scratch in the opposite direction. However, this is less desirable on rougher surfaces and/or for probe asymmetry reasons since no spherical probe is perfectly spherical, and with Berkovich, the scratches would be, for example, edge-on and the topographic scans face-on.

High surface roughness can present a challenge for nano-scratch tests as it does for nanoindentation. Higher stresses are generated on the peaks of the asperities, which, as shown later, can result in earlier failure lowering the critical load. In ductile materials, the effect of roughness is to produce changing (including higher) attack angles so that cutting behavior might be observed when it would not under the same conditions with a flat sample—that is, the roughness changes the deformation mode. Thin films are often deposited on highly polished substrates (e.g., silicon wafers) to avoid these problems, but results can be more meaningful when the coating is deposited on the actual substrate in the application. Strategies to mitigate its effect include (1) lengthening the levelling distance before the scratch ramp and/or combining with a second levelling period at the end of the scratch, (2) instrumentation design with higher/optimized lateral rigidity, and (3) in three-scan methods, it may be preferable to correct for slope and instrument compliance but not correct for the initial topography. The metrology requirements for performing nano- and microscale scratch tests have been reviewed, with high lateral rigidity of the loading head being identified as important to minimize artifacts on rough surfaces (Beake et al., 2013a, 2013b; CEN/TS, 2021). Insufficient instrument lateral stiffness can result in curved scratch tracks (Brinckmann et al., 2015).

Sample behavior in the nano-scratch test is a strong function of the test probe geometry. The choice of probe radius can be a compromise between having sufficient force range available to produce the desired deformation (e.g., film failure) and minimizing/eliminating tip wear. Larger radius probes require more load to create the same high stresses but are less susceptible to tip wear. Tip wear is potentially a problem when scratching hard and rough surfaces. It is also important not to rely on a manufacturer-specified tip radius but to calibrate the tip geometry by spherical nanoindentation (e.g., into fused silica and single crystal aluminum reference samples) over the depth range of interest. For manufacturing reasons, spherical tips can often be slightly sharper near the apex, and it is useful therefore to determine the effective radius as a function of depth.

Instrumentation limitations include (1) its lateral stiffness and its influence on smooth and rough surfaces, (2) available force range, (3) scan distance (4) method restrictions—for example, load then move or 3-scan but in different directions. Scratches with short distances may not be able to capture the influence of crystallographic orientation as easily as scratching over longer distances. Instruments differ in terms of how they perform nano-scratch tests. In one type, the load is applied statically, and then the sample motion commences, and in another type, the sample is moving and the load is gradually applied. The former approach enables stiction forces to be studied but can lose much other useful information regarding the onset of plasticity.

Although macroscale scratch tests have been shown to be somewhat sensitive to the experimental conditions, nano-scratch tests appear to be less so, at least for thin hard coatings. Critical loads and friction coefficients at failure have been shown to be relatively insensitive to changing loading rate and scan speed for diamond-like carbon (DLC) coatings (Beake, Ogwu, et al., 2006) on glass or Si (Beake et al., 2022). This is a useful result as it enables results of studies with different loading conditions to be compared, for example, so the influence of probe geometry can be assessed quantitatively.

6.3 General features

In the following sections, the 3-scan scratch test is introduced before discussing (1) the influence of probe radius; (2) contact pressure calculation; (3) how much of the load is supported on the back half of the probe; (4) friction, scratch recovery; and plowing (5) models for determining adhesion energy and interfacial toughness; (6) constant load scratch tests: indentation to sliding transition; (7) scratch hardness; (8) surface roughness and scratch orientation relative to grinding marks; (9) scratch size effects including the influence of probe geometry on size effects, size effects in yield; and (10) size effects in coatings including film thickness effects. The chapter finishes with (1) a discussion of multipass scratch testing and (2) the influence of test temperature.

6.3.1 Topography – progressive load scratch – topography – the 3-scan test

An improvement on the basic progressive load nano-scratch test is the 3-scan procedure (topography scan – progressive scratch – topography scan) that enables identification of failure mechanisms, the role of stress in particular, in more detail. The first reported multipass test of this type was described in detail by Wu and coworkers from IBM in 1989 (Wu, 1991; Wu et al., 1988). It was later shown that by removal of the instrument compliance contribution to the measured deformation, the true nano-scratch and nanowear depth data could be displayed after levelling (Beake, Vishnyakov, et al., 2006; Beake, Ogwu, et al., 2006). By performing three-scan progressive load nano-scratch tests, it is possible to determine the critical load for the onset of nonelastic deformation since this is the load at which the residual scratch depth is no longer zero.

When the contributions of frame compliance and sample slope are removed from the depth data, the loading curve in the scratch test can be compared directly to that in an indentation contact with the same probe. When friction and surface

[Figure: Loading curves from indentation and scratch tests, plotting Applied Load (mN) vs Depth (nm)]

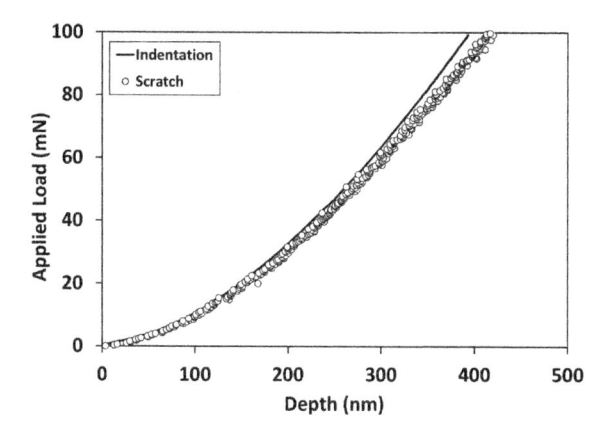

Figure 6.2 Loading curves from indentation and scratch tests. Comparison of loading curves from indentation and scratch tests on DLC film with $R = 6.5$ μm radius probe.

roughness are sufficiently low and the sample is mechanically homogeneous across the surface, the curves can be surprisingly similar. As an example, Fig. 6.2 shows loading curves from indentation and scratch tests performed with a 6.5 μm radius probe on a 462 nm ECR-CVD DLC film deposited on Si(100) (Beake & Liskiewicz, 2017). Alternatively, comparison between indentation and scratch loading curves can be used to investigate the influence of tangential loading on yield.

6.3.2 Probe radius dependence

Contact mechanics analysis shows that for a flat surface in elastic/plastic contact with a rigid ball of radius R, the yield pressure (P_y) is given by Eq. (6.1) (Johnson, 1985; Tsui et al., 1995).

$$P_y = 0.78 R^2 \left(H^3 / E^2 \right) \tag{6.1}$$

If yield (or cracking) of a bulk material occurs when a critical pressure is reached, the critical load will scale with R^2. For coating systems with different mechanical properties, thickness, etc., the situation is more complex since (1) the location of yield varies with probe sharpness and (2) failure can be controlled by a critical bending strain in the coating, which also is radius-dependent. For example, in nano-scratch tests of thin ta-C films on Si, the dependence of the critical load for total film failure was found to follow a similar power law dependence but with a slightly smaller exponent.

$$L_c = x R^m \tag{6.2}$$

where m is a best fitting parameter. For example, best fitting values of $x = 8.4$ and $m = 1.76$ were reported for a 80 nm ta-C film (Beake et al., 2009). In this case, because the film was relatively thin, the yield point was in the substrate for all the probes used. However, with thicker coatings altering the probe radius can change the location of yield relative to the coating − substrate interface. This behavior has been clearly shown for 3−4 μm thick DLC coatings on hardened steel in scratch tests with 5 and 25 μm probes (Beake et al., 2022).

6.3.3 Contact pressure calculation

It is possible to estimate the contact pressure during the scratch test from indentation contact mechanics. The contact depth (h_p) in a spherical indentation contact is given by:

$$h_p = \frac{(h_t + h_r)}{2} \tag{6.3}$$

where h_p is the contact depth, h_t is the on-load scratch depth, and h_r is the residual depth from the final scan. The contact radius (a) is determined by:

$$a = \sqrt{\left(2Rh_p - h_p^2\right)} \tag{6.4}$$

where R is the indenter radius. The contact pressure, P_m, can be calculated at any point along the scratch track by:

$$P_m = \frac{L}{\pi a^2} \tag{6.5}$$

where L is the applied load. At the onset of nonelastic deformation, $P_m = 1.1\ Y$, so that the yield stress can be determined.

To apply this approach to the nano-scratch data, it is necessary to assume that certain conditions are met (Beake et al., 2009). These are: (1) the presence of a tangential load does not influence the pressure distribution too greatly so that the measured friction coefficient is well below 0.3; (2) the radius of the indenter is constant; (3) the sliding speed is sufficiently slow and contact sufficiently close to elastic that the load is supported on the rear of the indenter; and (4) the indenter can reach the bottom of the scratch track in the final topographic scan.

When using spherical probes, it appears that these conditions can be met in the nano-scratch test, although the approximation obviously becomes less accurate as the contact geometry moves away from Hertzian conditions due to increasing friction or plasticity, and the load is supported progressively more on the contact over front half of the sliding probe and less on the back half. The methodology has been validated, for example, for thin films on silicon wafers, with good agreement found between (1) the scratch hardness independently determined from optical measurements of scratch widths and (2) contact pressures for film and substrate yield events.

For example, with very thin films on Si (e.g., 80 nm ta-C deposited by FCVA), the onset of yield is in the silicon substrate in nano-scratch tests performed with diamond indenters with tip radii R in the range 1.1–9.0 μm (Beake & Lau, 2005; Beake et al., 2009; Beake et al., 2013a, 2013b). The corresponding depth at yield is $\sim 0.47a$, where a is the contact radius. The calculated mean pressure (P_y) at yield was typically $\sim 12-13$ GPa. The yield stress of silicon is 11.3 GPa, so 1.1 $Y = 12.4$ GPa, which is almost the same as the pressure required for phase transformation (12 GPa) and hardness (12.5 GPa) emphasizing the importance of yield by phase transformation in the Si substrate.

The approach described earlier can also provide an estimate of the contact pressure during the progressive load nano-scratch test, at least at low contact forces where the friction coefficient is sufficiently low. The Hertzian treatment is well suited to the nano-scratch testing of DLC films with spherical probes of ~ 5 μm end radius due to their intrinsic low friction and high so that contact remains elastic or close to it over a wide load range. Once failure and delamination occur, it is not possible to apply this treatment.

6.3.4 How much load is supported on the back half of the probe in the scratch test?

Many models of the macroscale scratch test have assumed that post-yield that the load is only supported in the front face. Although when scratching ductile metals with sharp indenters (highly plastic contacts with little elastic recovery; low H/E), much more of the load is supported on the front, with spherical probes and highly elastic materials (high H/E materials such as DLC and Si), a significant fraction of the load is supported on the rear half of the probe—that is, there is significant scratch recovery as in indentation contact. Clear evidence for this load support from the rear half comes from several sources including (1) comparison of loading curves, (2) friction measurements, (3) simulations, (4) calculations of mean pressures, and (5) scratch recovery. After correcting the displacement data for sample slope and frame compliance, the scratch recovery can be determined from the on-load depth (h_t) and residual depth (h_r) data according to Eq. (6.6):

$$\text{Scratch recovery} = 100\% \times (h_t - h_r)/h_r \tag{6.6}$$

6.3.5 Friction, scratch recovery, and plowing

Friction measurements on coated systems can differ dramatically when measured at different length scales. However, the agreement is much better when comparing tests with "comparable deformation." For example, at coating failure, the friction coefficients of TiN coatings have been reported to be ~ 0.2 in nano- and macroscale scratch tests (Beake et al., 2013a, 2013b). In the absence of significant cracking, the friction coefficient depends on the probe geometry and extent of plowing in addition to the interfacial friction component. In the elastic regime, Bowden and Tabor (Bowden & Tabor, 1986) have shown that the friction is directly related to the junction shear strength and the contact area. In this regime, the friction coefficient should vary with $L^{-1/3}$ provided that there is no plowing component. At the start of the fully elastic regime, decreasing friction coefficient with increasing load has been observed experimentally in tests on Si with 5 and 100 μm probes (Zou, 2012), tests with 0.6–6.5 μm probes on DLC (Bandorf et al., 2004), and tests with 5 μm probes on high-entropy alloy thin films (Beake et al., 2024).

Brinckmann et al. identified three regimes in nano-scratching of austenite with a 5 μm indenter (Brinckmann et al., 2015). In the low load ($\leq$2 mN), regime deformation was elastic. Between 2 and 20 mN, plastic plowing was the dominant deformation mechanism and above 20 mN, plastic wear occurs. In their study, friction was scattered in the low load regime, increasing steadily in the plowing regime and at a faster rate with load in the plastic wear regime.

Martínez-Nogués and coworkers investigated the nano-scratch behavior of different CoCrMo alloys (Martínez-Nogués, 2016; Martínez-Nogués, Nesbitt, et al., 2016). They reported that in nano-scratch tests on CoCrMo alloys with 5 and 200 μm end radii diamond probes, the friction coefficient was independent of load for the blunter probe (Martínez-Nogués, 2016). It was strongly load dependent for the sharper probe, with higher friction coefficients as the load increased. They showed that the increased friction was correlated with higher degree of penetration (on-load depth/contact radius).

The Bowden and Tabor model (Eq. 6.7) enables the interfacial (shear) and plowing components to the total friction to be deconvoluted (Kamminga & Janssen, 2007; Zhang et al., 2016).

$$\mu_t = \mu_p + \mu_a \tag{6.7}$$

The plowing term has been determined from the geometrical comparison of normal (A_N) and lateral (A_L) contact areas (Eq. 6.8).

$$\mu_p = A_L/A_N \tag{6.8}$$

Equations have been developed depending on whether the scratch width or scratch depth is being recorded. In nano-scratch testing, the scratch depth is more convenient. A generalized equation for plowing friction that can account for variable elastic recovery is Eq. (6.9).

$$\mu_p \approx q \frac{\left(R^2 \cos^{-1}\left(\frac{(R-h)}{R}\right) - (R-h)\sqrt{h(2R-h)}\right)}{(\pi(2Rh - h^2))} \tag{6.9}$$

The prefactor q is 2 when there is no elastic recovery, and the load is supported only on the front ½ of the sliding probe. In the fully elastic regime, the load is equally supported on front and back ½ of the sliding probe so that $q = 1$. Kamminga and Janssen showed that the load carrying area was approximately circular in scratch tests of CrN-coated nitrided steel with a 200 μm diamond under relatively mild scratch conditions (Kamminga & Janssen, 2007). The interfacial friction component can be determined from low load measurements where the plowing component is minimized. At the onset of nonelastic deformation, the load starts to become less supported on the back ½ of the sliding contact and the measured friction increases more rapidly with increasing scratch depth and applied load. Experimental nano- and microscratch test data, for example, on uncoated Si(100) with a 6.5 μm probe or on DLCs with 25 μm probe, clearly show increased friction at the onset of nonelastic behavior marked in the residual topographic scan (Beake et al., 2022). Hence, when experiments are performed without a residual topographic scan, an inflexion in the friction coefficient can provide a robust estimate of the critical load for the onset of plasticity. The importance of elastic recovery on friction has been used to explain bilinear friction vs load behavior in AFM nanoscratch tests on Cu, Ti6Al4V, and Al (Lafaye & Troyon, 2006). Improved accuracy in fitting the experimental AFM data was obtained simulating the degree of elastic recovery through a varying rear angle. An alternative approach is determine elastic recovery more directly from the scratch recovery.

In modeling the nano and microscale friction of DLC, Si-doped DLC and WC/C coatings, the friction model (Eq. 6.9) was used accounting for differences in elastic recovery by allowing q to be a fitting parameter that varied with depth (Beake, McMaster, et al., 2021). Since the on-load and residual depths were measured, it was possible to determine the depth dependence of the scratch recovery and use this to improve the closeness of fit to the experimental nano- and microscale scratch friction data. For simplicity, a constant probe radius was assumed and surface topography neglected.

Fig. 6.3 shows how frictional data on DLC, Si-DLC, and WC/C coatings with the 25 μm probe can be simulated and the more rapid increase in friction from the onset of nonelastic deformation reproduced. The two fitting parameters were (1) the interfacial friction, determined from best fit in the elastic regime; and (2) the μ shows how frictional data on DLC, Si-DLC, and WC/C coatings with the 25 μm probe can be simulated, and the more rapid increase in friction from the onset of nonelastic deformation reproduced. The two fitting parameters were (1) the interfacial friction, determined from best fit in the elastic regime and (2) the q parameter, which gradually increased in line with the experimental scratch recovery. The scratch recovery was 100% at ∼300 nm ($q = 1$) decreasing to ∼65% at 1500 nm ($q = 1.57$). The fitting was restricted to on-load depths that were high enough to minimize topographic influence, but below those where cracking occurred. At lower penetration depth friction was more variable due to the higher geometrical influence of the surface topography as discussed by Achanta, Drees, and Celis (Achanta et al., 2005, 2008; Drees et al., 2004). Finite element modeling of the contact boundary in scratch tests with 50−200 μm diamond probes on a similar

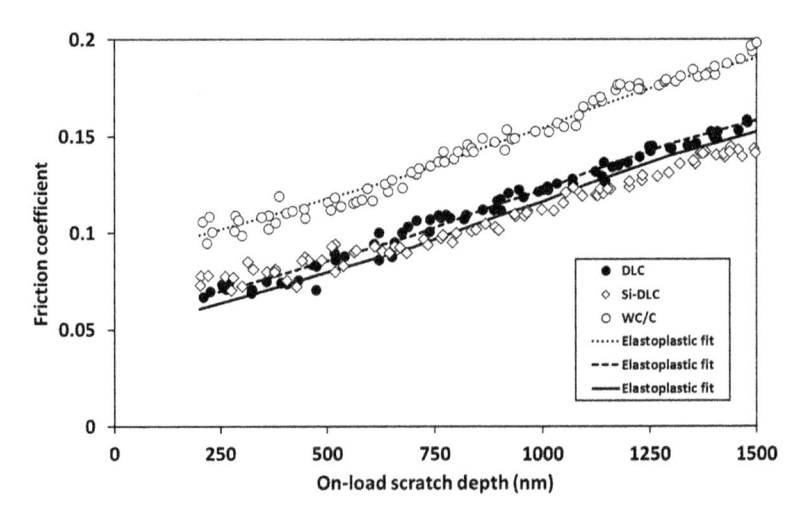

Figure 6.3 Friction coefficient in scratch experiments. Variation in friction coefficient with on-load scratch depth for DLC, Si-DLC, and WC/C coatings in microscratch tests with a 25 μm probe.

multilayered Si-DLC coating also determined that contact is a little more than half supported on the rear of the probe (Pagnoux et al., 2014), in good agreement with our experimental scratch recovery and q values used in the fitting.

As a further test for the robustness of the analysis, the interfacial friction coefficients were determined, and these values combined with the experimental scratch recovery data using a sharper $R = 5$ μm probe to fit the friction vs. depth data with this probe. The interfacial friction coefficients vs. diamond were estimated as 0.072 for WC/C, 0.040 for Si-DLC, and 0.034 for DLC. With the sharper probe, there were differences in scratch recovery between the coatings due to their different mechanical properties, which were not found with the blunter probe since in the latter case, the elastic substrate deformation dominated. Over the depth range simulated, the deformation with the $R = 5$ μm probe was elastic on the harder and higher H^3/E^2 DLC, with nonelastic behavior for the softer coatings from above 400 nm for Si-DLC and 300 nm for WC/C. The fit was not perfect but nevertheless encouraging, with maximum deviation in the friction coefficient being within 0.03 across the entire depth range. The method was also applied to fitting the friction vs depth data with a 6.5 μm probe on (1) a 1 μm DLC on Si(100) and (2) Si(100). It was possible to fit data on the coating to 400 nm with $q = 1$ as deformation was very close to elastic over this depth range. On Si(100), a more rapid rise in friction coefficient starting from ∼200 nm depth required an increase in q, which on-load and residual depth data showed marked the onset of plasticity/fracture. Carreon and Funkenbusch (2018) adapted the friction coefficient model of Lafaye and Troyon to study the scratch behavior of brittle materials with a Berkovich indenter in the edge-forward orientation to incorporate the effect of additional contact area through material-dependent pile-up at the sides of the scratch track (which presumably included the formation of brittle debris).

When the scratch hardness is equal to the normal hardness, the friction coefficient is given by

$$\mu = (2/\pi)\cot\theta \qquad (6.10)$$

where θ is the semi-apical angle. Ben Tkaya et al. used analytical and numerical simulations to show how with increasing attack angle, the friction coefficient increased, and there was a gradual transition from plowing to cutting in tests in an Al alloy (Ben Tkaya et al., 2007).

When there is particulate debris in the scratch track, the "third body" will influence the friction and scratch resistance. Bhushan and Nosonovsky (Bhushan, 2005; Nosonovsky & Bhushan, 2007) have noted that the plowing contribution in the third-body friction mechanism is composed of the sum of the asperity summit deformation and particles deformation component. In comparison to the two-body situation, there is an increase in plowing component due to the presence of trapped particles (Eq. 6.11), where μ_{sd} = summits deformation, μ_{pd} = particles deformation.

$$\mu_{total} = \mu_{sd} + \mu_{pd} + \mu_{interfacial} \qquad (6.11)$$

Some of these transiently trapped particles act as miniature rolling bearings in reducing friction, but it is considered that in most engineering situations, only 10% of particles roll so that the third-body mechanism leads to an increase in friction (Nosonovsky & Bhushan, 2007).

Rojacz et al. investigated the influence of mechanical properties and microstructure on the behavior of the fcc-metals austenite, aluminum, and copper in nano-scratch tests (Rojacz et al., 2023). These three materials have different hardness and H^3/E^2 but almost identical H/E. These authors showed that scratch resistance correlated with H^3/E^2 and load-dependent friction correlation. The transition to cutting was dependent on their mechanical properties. The calculated contact stress variation with load was calculated assuming no recovery and found to reach higher values than the nanoindentation hardness but potentially extent of scratch recovery will also have affected these since the values were closest for aluminum, which exhibits little elastic recovery.

6.3.6 Models for determining adhesion energy and interfacial toughness

Botero et al. (2023) performed nano-scratch tests on $3Al_2O_3 \cdot 2SiO_2$ coatings on SiC. Since they observed cracking before delamination and the post-test profile after chipping corresponded to the coating thickness, they felt justified in using the critical load as a measure of adhesion and implemented five different models to determine the adhesion energy and interface fracture toughness. They found some variability between them and values of both adhesion and interface toughness that were generally lower than comparative values determined by nanoindentation.

6.3.7 Constant load scratch tests: indentation to sliding transition

Most commonly, the load in a nano-scratch test is applied at a constant sliding speed, enabling for example, critical loads to be determined. An alternative approach for performing constant load scratch tests is to pause the scanning, then apply the constant load, and then start sliding. In this load-then-move approach, several authors have reported a transition region where the depth initially increases before decreasing to a constant value in the steady-state region (Tsybenko et al., 2020). Brazil et al. have studied the effects of H/E and elastic recovery on friction during this transition from sink-in in static friction to plowing in sliding (Brazil et al., 2021).

6.3.8 Scratch hardness

It is possible to define various different measures of scratch hardness (Williams, 1996). The standard approach is to divide the normal load by the loading-bearing area projected in the surface-normal direction. Often the load bearing area is estimated from the scratch track width assuming that no elastic recovery occurs. Other alternatives include defining a plowing hardness as tangential force/projected area. Tsybenko et al. recently investigated the scratch hardness determined with 5 and 20 µm radii probes using four different definitions of contact area for materials with a range of properties (Tsybenko et al., 2021). Scratch hardness determined from contact depth showed different dependencies on load for ductile materials, where a size effect similar to indentation was found, and for harder materials an opposite effect was observed. These differences were interpreted as being due to densification and reverse phase transformation, but it is also possible that differences in load-dependent elastic recovery play a part since they assumed no elastic recovery occurred. This neglect of the scratch recovery may have also explained the significant differences in the ratio of scratch to indentation contact areas they found between the materials.

Tayebi et al. have noted that after correcting for interfacial shear stress and elastic recovery, the nano-scratch method gives the same hardness as nanoindentation (Tayebi et al., 2003). They developed a model for thin films, which suggested that the effect of substrate influence on the measured hardness in the nano-scratch test was delayed compared with that in nanoindentation tests at comparable contact depth to thickness ratios (Tayebi et al., 2004a, 2004b).

6.3.9 Surface roughness and scratch orientation relative to grinding marks

Directional surface roughness can influence scratch resistance. Beake et al. investigated the influence of the direction of scratching relative to grinding marks on the behavior of 1500 nm TiN coatings on hardened M42 steel in nano-scratch tests with a spheroconical diamond indenter with a 4.4 µm radius (Beake et al., 2011).

The critical load for delamination failure was dependent on scratch orientation relative to polishing marks on the surface made prior to coating deposition. When scratching perpendicular to the grinding marks, the critical load was 22% lower than when scratching parallel to them. Coating failure was more gradual when the scratch direction was either parallel to or aligned at an intermediate angle to the grinding grooves than when it was perpendicular. Orientation dependence has also been reported in scratch testing of thicker TiN coated high-speed steel with 200 μm probes. The decrease in critical load of $\sim 20\%-25\%$ between scratches made across or parallel to grinding grooves for a TiN film with $R_a = 0.12$ μm in the nano-scratch test is consistent with the macroscale result reported by Larsson et al. (2000), who found that below $R_a = 0.1$ μm, there was little influence of scan direction, but the critical load decreased by over 50% when R_a reaches 0.5 μm.

6.3.10 Scratch size effects

In sliding contacts on metallic materials, there are similar size effects to those observed in indentation. Kareer et al. noted that size effects were found during nano-scratch testing of single crystal copper scratched by a diamond Berkovich (pyramidal) indenter, whether the indenter was sliding face-forward or edge-forward (Kareer et al., 2016a, 2016b). The higher scratch hardness was ascribed as a lateral size effect. Subsequent crystal plasticity finite element simulations confirmed the deformation state in sliding was different from that in indentation (Kareer et al., 2020). The lateral size effect was smaller for small-grained materials than for larger grained or single crystal materials. The authors note that this provides evidence that the length scales combine rather than superimpose as they do in indentation, and it is this combined length scale that is responsible for the observed lateral size effect.

Adams et al studied the influence of contact strain and strain rate in constant load nanos-cratching the viscoelastic polymer PMMA with a Berkovich indenter edge-on (a 13.0 degrees attack angle) and face-on (24.7 degrees attack angle) and comparing the differences in pileup between the different tip orientations and also to that in a static indentation at the same load (Adams et al., 2001). The strain rate in the scratch test has been defined by sliding velocity/scratch width. Strain rates thus defined were of the order of 10 s^{-1} for both tip orientations, that is, much higher than in a normal nanoindentation contact. The imposed strain is given by 0.2 tan θ, where θ is the attack angle, corresponding to 0.05 and 0.09 for the edge and face-on scratches, respectively. The scratch depths under load were similar, but scratches performed edge-on showed greater scratch recovery and smaller pile-up. The recovery in the nano-scratch test was 87% and 75% for edge- and face-on scratches compared with only 40% in indentation. These differences relate to the greater strain rate in the scratch tests, with the larger recovery for the edge-on due to the smaller applied strain.

Microstructural variations, for example, in Ti6Al4V (Budinski, 1991), have been implicated in variability in abrasion resistance at larger scale. One advantage of nano-/microscale testing is that tests can be performed in single grains or across several grains and, with the help of EDX, a direct correlation between the tribological performance and the grain orientation determined. For example, Shugurov et al. studied the influence of crystallographic orientation of polycrystalline titanium on plowing in nano-scratch testing, finding a strong influence on the residual depth and the pileup (Shugurov et al., 2018). They noted that the most abrasion resistant grains were those that favored strain hardening of the material within the scratch groove, and that H/E correlated better with residual scratch depth than either H or H^3/E^2. Pöhl et al. schematically illustrated the deformation after a single scratch in terms of induced plastic strain (Pöhl et al., 2019). Transition between micro plowing (no wear) and a combined microplowing/microcutting mode (where wear can occur) occurred at a critical induced level of plastic strain.

6.3.11 Probe geometry and size effects

In practice, "spherical" diamond indenters are usually spheroconical. The transition between the spherical end cap and the cone depends on the cone angle. For example, for a 5 μm end radius probe with a 90 degrees cone angle, the transition occurs at 1.46 μm (0.293 R). Changing the angle to 60 degrees results in more of the spherical end cap, with the transition occurring at 2.5 μm (0.5 R). When the scratch depth under load remains in the spherical regime, increases in scratch depth change the contact strain and can therefore influence the dominant deformation mechanism. With increasing scratch depth, the dominant deformation mechanism can change from microplowing to microcutting. In nano-scratch tests of fcc-metals, Rojacz et al. reported that the transition to cutting was dependent on their mechanical properties (Rojacz et al., 2023). In a study of the influence of carbon content and hardness on scratch behavior of tempering steels, Pöhl et al. reported that the microcutting transition was pushed to greater scratch depths when steel hardness was increased (Pöhl et al., 2019). Williams and Xie noted that rougher surfaces produce higher attack angles than smoother surfaces, which can result in more severe deformation than would otherwise be expected (Williams & Xie, 1992).

Coatings may experience greater bending stresses in scratch tests with sharper probes. In sliding contacts on coated systems, Diao and coworkers proposed the ratio of coating thickness (t) divided by contact radius (a), t/a as an alternative severity index (Diao et al., 1994; Kato, 1995). Finite element (FE) modeling showed that the magnitude of the tensile stresses generated at the rear of the contact was dependent on E_c/E_s and t/a (Diao et al., 1994; Kato, 1995).

6.3.11.1 Size effects in yield

Nano-scratch tests have been performed on highly polished biomedical grade Ti6Al4V, 316L stainless steel, and wrought high-carbon CoCrMo alloy samples using a 3.7 μm end radius sphero-conical diamond indenter (Beake & Liskiewicz, 2013). The relationship between the mechanical properties of the biomaterials determined from nanoindentation tests at 100 mN with a Berkovich indenter and their friction and nano-/microscale scratch resistance was investigated. The CoCrMo alloy had the highest hardness and stiffness ($H = 6.0 \pm 0.4$ GPa, $E = 277 \pm 8$ GPa) among them. Ti6Al4V showed intermediate hardness and low elastic modulus ($H = 4.4 \pm 0.4$ GPa, $E = 146 \pm 10$ GPa), and 316L showed the lowest hardness and intermediate stiffness ($H = 3.1 \pm 0.2$ GPa, $E = 222 \pm 10$ GPa). Although it was not as hard as the CoCrMo alloy, the Ti alloy had the highest H/E and H^3/E^2 due to its relatively low elastic modulus.

In ramped load scratch tests to 500 mN, there was a predominantly ductile response to scratching with plowing and pile-up and debris at the sides of the scratch track, with the onset of chipping events at increasing applied load for all the alloys. CoCrMo, the hardest alloy, showed the best scratch resistance, and the softest alloy, 316L stainless steel the lowest scratch resistance. By performing the nano-scratch tests as 3-pass (topography − scratch − topography) experiments, it was possible to estimate of the contact pressure at low contact force where the friction coefficient is sufficiently low for the Hertzian analysis to be applied. The mean pressures at yield (determined as the first point in the final topographic scan where the depth was >0) in the nano-scratch were 2.3 GPa for Ti6Al4V, 2 GPa for CoCrMo, and 1 GPa for the steel. The mean contact pressure required to produce plastic deformation is slightly less than 1.1 times the yield stress due to shear stress. There is a strong lateral size effect on plastic yielding. Due to the size effect, the yield stresses are higher at small scale. The macroscopic yield stresses on these alloys ranged from 0.3 GPa on the steel to 1 GPa on the Ti6Al4V (Goldberg & Gilbert, 1997; Mirghany & Jin, 2004; Raimondi et al., 2001).

6.3.11.2 Scratch size effects in coated systems

The film thickness is a key parameter influencing the critical load in the nano-scratch test. In principle, film thickness can have two opposing effects: (1) thicker films that are harder than the underlying substrate provide more load support and so delay the onset of the substrate deformation that is often the precursor of film failure (higher critical load); (2) thicker films can be more highly stressed and more easily through-thickness crack and delaminate when deformed (lower critical load) since the driving force for spallation to reduce stored elastic energy is greater. Changes to coating thickness can also alter microstructure.

Nano-scratch tests of 500 and 1500 nm TiN coatings on hardened M42 tool steel have been performed using a spheroconical indenter with a 4.4 μm radius. Yield stresses determined from nanoindentation confirmed that these films were mechanically almost identical. In general, high surface roughness will decrease the critical load (by creating stress raisers), but the load-carrying capability of a higher thickness coating can have a much greater effect. Despite being much smoother, the critical load on 500 nm TiN was only ∼50% of the critical load on the 1500 nm TiN.

The critical load for total film failure in the nano-scratch test can be strongly correlated with film thickness reflecting enhanced load support and substrate protection with thicker coatings. In nano-scratch tests on 200−1000 nm a-C films deposited on Si (Shi et al., 2008) and 150−600 nm Si:a-C:H films deposited on glass (Beake, Vishnyakov, et al., 2006; Beake, Ogwu, et al., 2006), the ratio L_c/t_f

was approximately constant when the films were deposited with relatively low stress. However, this is not a general result since, thin films can be highly stressed, and these high internal stresses when coupled with shear stresses generated in the scratch test can lead to poor adhesion and extensive coating spallation. For example, in nano-scratch tests with either 3.7 or 6.5 μm radius probes on 450 and 962 nm ECR-CVD DLC films deposited on Si without any adhesion promoting interlayers, it was found that the thicker film failed at around 60% of the critical load of the thinner film to due high stress.

6.4 Multi-pass scratch testing

The test methodology for assessing repetitive subcritical constant load scratch tests has been developed enabling S−N type relationships to be determined for coating systems undergoing clear failure. This approach is outlined in the recent CEN standard (CEN/TS, 2021). In tests on coated systems where the load was close to L_{c1}, the entire scratch track typically failed within 1−2 cycles from the initiation of failure at a specific location, but at lower load failure was more localized. Using the cycle-by-cycle evolution in the mean values (of h_t, h_t, scratch recovery, surface roughness, friction coefficient) from the entire track length under the constant repetitive load provided a convenient graphical method to follow the development of the failure process. A limitation of this approach is that the averaging of the properties over the entire track length means that the precise details of localized failure and how it evolves with each scratch can be hidden so it can also be useful to graphically display the properties over the entire scratch track.

When testing coatings, the repetitive constant load nano-scratch test method has the benefit that the applied load can be controlled to locate the maximum von Mises stress to be close to the coating − substrate interface and minimize the substrate deformation as a precursor of coating debonding which typically occurs in ramped load tests. Repetitive scratch tests can be more sensitive to adhesion differences than single-pass scratch tests. After a number of passes, the coating failure over the entire constant load track can occur during a single scratch or over a few scratches (Beake et al., 2011).

Beake et al. investigated the behavior of DLC films on Si in repetitive scratch tests at different subcritical loads (Beake et al., 2022). Although the coatings studied were very soft and compliant in comparison to more standard DLC films, they were nevertheless an ideal model system for studying parameter sensitivity as the delaminated consistently with very low standard deviation in the critical loads. The films were (1) 450 nm DLC with $H = 4.4$ GPa, $E = 31.6$ GPa and (2) 962 nm DLC with $H = 4.1$ GPa, $E = 28.5$ GPa. They were deposited on Si without any adhesion-promoting interlayer by electron cyclotron resonance chemical vapor deposition at BAM. Nano-scratch tests were performed with a 6.5 μm radius probe. The critical loads to failure were nearly double for the thinner film. The increase in critical load required for total coating failure after the initial cracking ($L_{c2} - L_{c1}$) was low for the thicker film and almost zero for the thinner film.

Fig. 6.4A shows the variation in number of cycles to failure in repetitive scratch tests with constant scratch load. The differences in $L_{c2} - L_{c1}$ for the two soft DLC films discussed earlier influence their behavior in the repetitive scratch test. Fig. 6.4B compares the number of cycles to failure vs the fractional load (L/L_c) for these coatings. Under the subcritical loads, the thicker coating shows relatively improved behavior. This improved damage tolerance is consistent with the higher $L_{c2} - L_{c1}$ found for this coating. In the repetitive scratch tests, the on-load depth at failure was lower than in ramped scratch tests. The difference was dependent on L/L_c and the film thickness. For the 450 nm DLC, the depth at failure was much lower than for ramped tests. The scratch recovery also decreased with each cycle.

Korres et al. developed a geometric model correlating wear track widening and friction, which assumed no elastic recovery (Korres et al., 2013). Interestingly,

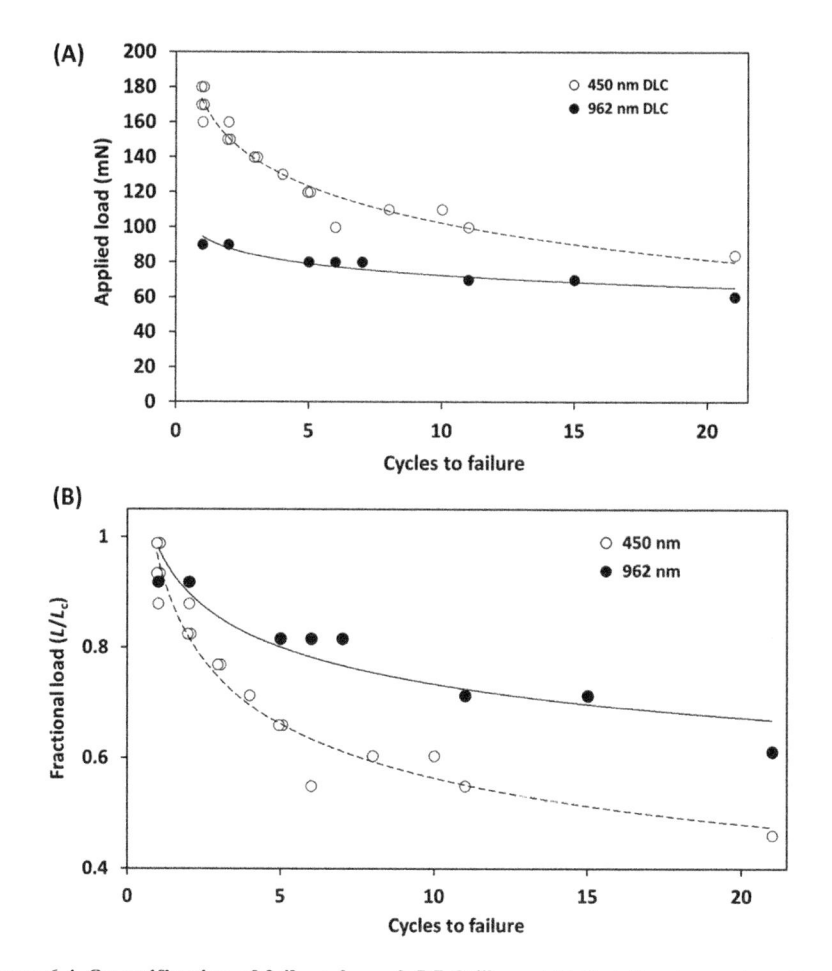

Figure 6.4 Quantification of failure for soft DLC films. (A) Variation in number of cycles to failure for soft DLC films (B) equivalent fractional load vs. cycles.

although the contact approaches conformal with continued cycles, it was still possible to correlate the friction with changing plowing through changes to the track width. In the nano-scratch tests, we also have direct access to the scratch recovery data. In the tests on the DLC films, it can be seen that the scratch recovery is gradually decreasing with number of passes, indicating some microscale cracking and/or densification, so the Korres model potentially might be a slight oversimplification, at least for the first few cycles of a repetitive test. Nevertheless, when there is little change in depth with cycles, as occurred in tests at <70 mN on the 962 nm DLC, the friction coefficient did not change significantly, remaining near to 0.1 throughout the 20-cycle test.

The relationship between evolving friction and cycling is more complex when failure occurs and/or wear becomes important. The variation in friction coefficient and on-load probe depth during a 20-cycle test at 70 mN on the thicker film shown in Fig. 6.5 is a typical example when coating failure occurs. After initial reduction in plowing component over a few cycles, subsequent frictional changes are due to a combination of factors including (1) changing contact area, (2) formation of transfer layers, (3) third body effects, (4) changing probe geometry (e.g., moving from spherical end to conical part of the probe as penetration depth increases). At coating failure, there is an increase in mean friction and subsequent decrease as damaged material is plowed out of the scratch track. Similar behavior was observed in multipass nano-scratch tests of 1.5 μm TiN coatings on hardened M42 tool steel (Beake et al., 2011).

In the absence of significant fracture, the magnitude of the friction force in a repetitive scratch test and its variation over the scratch track tends to gradually decrease with each cycle due to a progressive reduction in plowing component. This has been observed in several materials besides coatings, including bulk metallic alloys such as 316L stainless steel, CoCrMo (Beake & Liskiewicz, 2013) copper

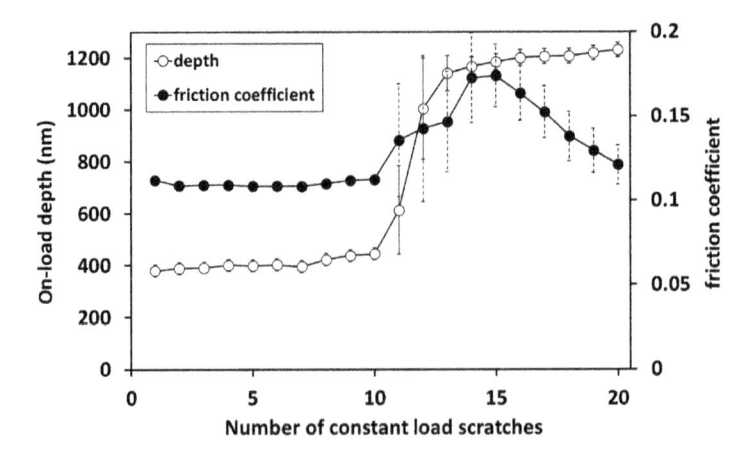

Figure 6.5 Friction coefficient and on-load probe depth during scratch test. Variation in friction coefficient and on-load probe depth during a 20 cycle test at 70 mN on 962 nm DLC film on Si(100).

(Singh et al., 2011), and Ag alloy (Beake, Harris, et al., 2021). An exception is when the repetitive scratching breaks up a passive protective oxide resulting in severe wear and continuing high friction since a hardened tribo-layer (TTS) cannot effectively form, for example, on Ti6Al4V.

Beake and Liskiewicz showed that significant differences in friction and wear occur for 316L stainless steel, CoCrMo, and Ti6Al4V in 10-pass repetitive nano-scratch tests with a 3.7 μm probe at 30 mN (Beake & Liskiewicz, 2013). Fig. 6.6A shows how friction and Fig. 6.6B shows how on-load depth varied with number of scratch passes. The gradual reduction in friction CoCrMo and 316L stainless steel during the initial wear cycles of this low-cycle test is explained by changes to the plowing component of friction and smoothing of asperities, both of which alter the contact area and therefore can influence the friction (Achanta et al., 2009). For CoCrMo and 316L stainless steel, the mechanical behavior on repetitive sliding is

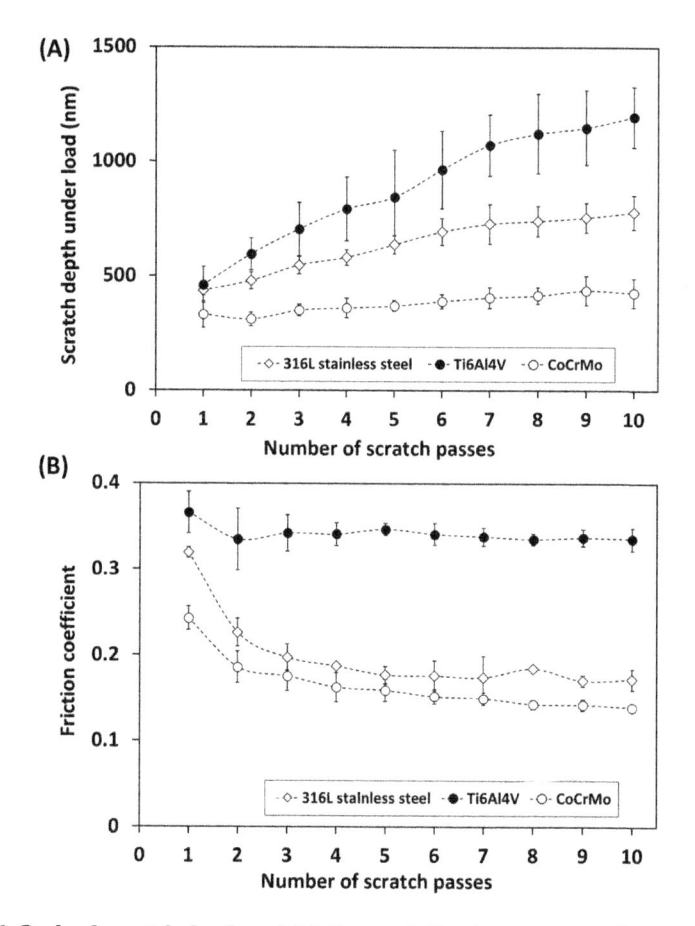

Figure 6.6 On-load scratch depth and friction evolution in nano-scratch tests. Evolution of (A) on-load scratch depth and (B) friction, in 10-cycle constant load nano-scratch tests of 316L stainless steel, Ti6Al4V and CoCrMo at 30 mN using a 3.7 μm radius probe.

consistent with surface hardening within the scratch track as has been observed in microscale sliding on 304 stainless steel in a pin-on-disk test (Kitsunai et al., 1990). Grain refinement resulting in an increase in flow stress and hardness within the scratch track has also been reported on Cu (Hanlon et al., 2005; Singh et al., 2011). The relative importance of yield stress (hardness) and microstructure (grain size) on the evolution of friction and wear of metallic materials such as Ti, Cu, and Ni during microscale repetitive sliding has been investigated by several authors (Hanlon et al., 2005; Singh et al., 2011; Wang et al., 2012). The conclusion from these studies was that yield stress plays the dominant role on the evolution of friction as the friction was almost independent of the grain size but decreased with increasing hardness (Hanlon et al., 2005; Singh et al., 2011; Wang et al., 2012).

In repetitive microscratch testing of copper at 1N with a 100 μm probe, Bellemare et al. observed that eight cycles of repetitive sliding were sufficient to minimize the plowing component and found that the total friction could be further reduced by lubricating the contact with isostearic acid (Bellemare et al., 2008). This reduced $\mu_{(interfacial)}$ to 0.14 when dry and 0.11 when lubricated. Twenty-cycle nano-scratch experiments have been performed at 50 and 200 mN on an Ag-10 wt.% Cu alloy disk with a diamond indenter of 18 μm end radius sliding at 5 μm/s over a 1 mm track. The alloy disk had a periodic surface profile due to the manufacturing process; to mitigate this high surface roughness, a relatively long scratch track was chosen. The multipass scratch tests were designed with slow loading rate to combine sufficient regions of ramped and constant loading in the scratch track. This approach enabled the friction coefficient to be determined from single point measurements during the ramp rather than be averaged over the entire track at constant load. The similarity in the friction coefficients obtained from the ramped and constant load sections of the scratches on this alloy suggests a potentially more efficient testing protocol than using only mean values from constant load segments in multicycle tests. Over 20 cycles there was a clear gradual reduction in friction, with some load dependence in that (1) friction coefficients increased with load (2) the reduction with cycling was greater at higher load. $\mu_{(1st\ cycle)} = 0.13$ and $\mu_{(20\ cycles)} = 0.11$; at 50 mN, $\mu_{(1st\ cycle)} = 0.17$ and $\mu_{(20\ cycles)} = 0.13$; at 200 mN, $\mu_{(1st\ cycle)} = 0.19$ and $\mu_{(20\ cycles)} = 0.17$.

A single-term adhesion model for friction has previously been used to estimate the friction coefficient in clean, smooth, unlubricated noble metal contacts (Aukland et al., 2000). When plowing is minimal, according to this model, the friction force is the product of shear strength and the area of welded junctions between contacting asperities (i.e., the force necessary to shear these junctions). The coefficient of friction is then expressed as the ratio of shear stress to normal stress. Assuming that the shear strength is approximately half of the yield strength, Y, and the contact pressure is $\sim 3\ Y$ when asperities are in plastic deformation (Tabor constraint factor = 2.8 for metallic materials with low elastic recovery), results in an estimate of 0.17 for the friction coefficient. Notwithstanding the simplicity of this approach, when plowing is minimized (through repetitive scratches), the measured friction coefficient is quite close to this on the silver alloy and in the tests with the sharper probes on CoCrMo and 316L stainless steel.

It can be seen in Fig. 6.6 that friction is higher on the Ti6Al4V and the decrease in friction with increasing scratch cycles was much smaller than for either CoCr or 316L stainless steel. On titanium, high friction ($\mu \sim 0.5$) has been observed in microscale tests. In repetitive scratch testing with larger probes, titanium (98 cycles at 1.25 N, 25 μm/s, 500 μm track length, $R = 800$ μm sapphire indenter, unidirectional sliding) showed little (Hanlon et al., 2005) decrease in friction coefficient. In microtribological tests, Wang et al. reported (Wang et al., 2012) no decrease in friction coefficient. In all these tests on pure titanium metal and Ti6Al4V, the absence of a marked reduction in friction can be explained by a lower scan-on-scan decrease in the plowing contribution. The differences in how friction and wear change with continued scratches on stainless steel, CoCr, and TiAl4V shown in Fig. 6.2 are consistent with differences in their stacking fault energies. These are much lower for stainless steel and CoCrMo (promoting rapid work hardening) than titanium, which does not work harden to the same degree, resulting in little reduction in plowing and friction on repetitive nano-scratching.

In a study of a wide range of hard metals and ceramics, Gee has shown that scratch resistance in low-load multiple-pass scratch tests was not correlated with hardness, as it was in single scratches, due to differences in the contribution of fracture to the damage development (Gee, 2001). In the repetitive nano-scratch tests on the 316L stainless steel, Ti6Al4V, CoCrMo, and 316L stainless steel, the hardest material showed the best performance, but 316L stainless steel showed better performance than the Ti alloy despite being significantly softer.

In studying wear mechanisms during repetitive scratching of steels with a 30 μm spheroconical tungsten carbide indenter, Kato and coworkers (Hokkirigawa et al., 1988; Kitsunai et al., 1990) observed that increasing load and contact strain resulted in a transition in wear mode from plowing to wedge formation and then cutting at higher cone angles. To explain these changes in wear behavior, they introduced a degree of penetration parameter, D_p, defined as the on-load contact depth/contact radius, to characterize the severity of the abrasive contact. They investigated wear mechanisms during low-pass (1–10 cycles) scratching of steels finding that the wear process could shift to milder behavior after scanning for a few cycles, presumably due to the sharp decrease in contact pressure. However, with sharper probes at sufficiently high load, the increase in depth through the test is also accompanied by an increase in contact strain, and this increase in attack angle can promote more severe deformation modes. Nimishakavi and Gee have observed large decreases in contact pressure in nano-scratch testing of WC/Co hardmetals, for example, from 8 GPa to under 0.5 GPa in tests at 40 mN with a 1 μm radius probe (Nimishakavi & Gee, 2011).

6.5 Influence of test temperature

Although nano-scratch tests are usually performed at room temperature, they have been performed at temperatures well above (to 750°C) or below room temperature (e.g., to −30°C). Smith et al. reported nanoscale friction measurements of TiN-based coatings and a Cr_2AlC MAX-phase coating sliding against a WC ball of

350 μm radius at 25°C, 400°C, and 750°C (Smith et al., 2013). Maximum friction was found at 400°C with the decrease in friction at 750°C being ascribed to the formation of lubricious oxides and oxidation-induced surface roughening, which lowered the contact area between the probe and the sample. Chavoshi and Xu have reviewed high temperature nano-scratch and nanotribological tests (Chavoshi & Xu, 2018). Goel et al. have suggested that in view of the limited high-temperature thermal stability of diamond, and reactivity against ferrous alloys, that cBN might be a better alternative probe material (Goel et al., 2012). Chavoshi et al. performed high-temperature nano-scratch tests of Si(110) to 500°C with a Berkovich indenter under reduced oxygen environment (Chavoshi et al., 2017). Chen et al. studied the nano-scratch and nanomechanical behavior of a functionally graded Ti-doped DLC coating on M2 tool steel over the temperature range from 25°C down to −30°C (Chen et al., 2011). The functionally graded coating was deposited with the multilayered structure a-C:H(Ti)/TiCN/TiN/Ti to improve its interfacial shear strength. In the room-temperature nano-scratch tests to 400 mN with a 8 μm radius probe, the multilayered DLC coating failed by cracking and spallation. Analysis of fracture sections showed these failures originated from, or close to, the interface between the top a-C:H(Ti) layer and the TiCN layer. Decreasing the test temperature increased the mechanical properties of the DLC coating and significantly improved its resistance to cracking in the nano-scratch tests.

6.6 Conclusions

The nano-scratch test has become sufficiently well established to be standardized with the 3-scan ((1) pre-scratch topographic scan (2) scratch (3) post-scratch topographic scan) scratch test being an improvement over a single ramped scratch. Friction measurements provide additional information about deformation and failure mechanisms. When using spherical probes incorporating the effect of scratch recovery and partial load support on the rear half of the probe improves the accuracy of friction models. In using nano-scratch tests to inform coating design, an integrated approach combining experiments and simulations has proved effective in understanding how the contact size can influence the yield and failure behavior of coated systems.

Multipass scratch tests involving repetitive scratches at constant load(s) lower than that causing failure in the ramped load test reduce the influence of substrate deformation on the overall coating system response and can increase sensitivity to interfacial failure. The number of cycles to failure as a function of applied load can be determined enabling S−N type relationships to be determined for coating systems undergoing clear failure.

References

Achanta, S., Drees, D., & Celis, J.-P. (2005). Friction and nanowear of hard coatings in reciprocating sliding at milli-Newton loads. *Wear, 259*, 719−729.

Achanta, S., Drees, D., & Celis, J.-P. (2008). Friction from nano to macroforce scales analysed by single and multiple-asperity contact approaches. *Surface and Coatings Technology*, *202*, 6127–6135.

Achanta, S., Liskiewicz, T., Drees, D., & Celis, J.-P. (2009). Friction mechanisms at the micro-scale. *Tribology International*, *42*(11–12), 1792–1799. Available from https://doi.org/10.1016/j.triboint.2009.04.018.

Adams, M. J., Allan, A., Briscoe, B. J., Doyle, P. J., Gorman, D. M., & Johnson, S. A. (2001). An experimental study of the nano-scratch behaviour of poly (methyl methacrylate). *Wear*, *250–251*, 1579–1583. Available from https://doi.org/10.1016/S0043-1648(01)00798-0.

Aukland, N., Hardee, H., & Lees, P. (2000). Sliding wear experiments on clad gold-nickel material systems lubricated with a 6-ring polyphenyl ether. *Proceedings of the 46th IEEE Holm Conference on Electrical Contacts*, 27–35.

Bandorf, R., Lüthje, H., & Staedler, T. (2004). Influencing factors on microtribology of DLC films for MEMS and microactuators. *Diamond and Related Materials*, *13*, 1491–1493.

Beake, B. D., Davies, M. I., Liskiewicz, T. W., Vishnyakov, V. M., & Goodes, S. R. (2013a). Nano-scratch, nanoindentation and fretting tests of 5–80 nm ta-C films on Si(100). *Wear*, *301*(1–2), 575–582. Available from https://doi.org/10.1016/j.wear.2013.01.073.

Beake, B. D., Harris, A. J., & Liskiewicz, T. W. (2013b). Review of recent progress in nano-scratch testing. *Tribology – Materials Surfaces & Interfaces*, *7*, 87–96.

Beake, B. D., Goodes, S. R., & Shi, B. (2009). Nanomechanical and nanotribological testing of ultra-thin carbon-based and MoST films for increased MEMS durability. *Journal of Physics D: Applied Physics*, *42*, 065301.

Beake, B. D., Harris, A. J., Liskiewicz, T. W., Wagner, J., McMaster, S. J., Goodes, S. R., Neville, A., & Zhang, L. (2021). Friction and electrical contact resistance in reciprocating nano-scale wear testing of metallic materials. *Wear*, *474–475*, 203886.

Beake, B. D., & Lau, S. P. (2005). Nanotribological and nanomechanical properties of 5–80 nm tetrahedral amorphous carbon films on silicon. *Diamond and Related Materials*, *14*(9), 1535–1542. Available from https://doi.org/10.1016/j.diamond.2005.04.002.

Beake, B. D., & Liskiewicz, T. W. (2013). Comparison of nano-fretting and nano-scratch tests on biomedical materials. *Tribology International*, *63*, 123–131. Available from https://doi.org/10.1016/j.triboint.2012.08.007.

Beake, B. D., & Liskiewicz, T. W. (2017). Nanomechanical characterization of carbon films. In A. Tiwari, & S. Natarajan (Eds.), *Applied nanoindentation inadvanced materials* (pp. 19–68). Wiley.

Beake, B. D., McMaster, S. J., & Liskiewicz, T. W. (2022). Contact size effects on the friction and wear of amorphous carbon films. *Applied Surface Science Advances*, *9*(100248), 19p.

Beake, B. D., McMaster, S. J., Liskiewicz, T. W., & Neville, A. (2021). Influence of Si- and W- doping on micro-scale reciprocating wear and impact performance of DLC coatings on hardened steel. *Tribology International*, *160*, 107063.

Beake, B. D., Ogwu, A. A., & Wagner, T. (2006). Influence of experimental factors and film thickness on the measured critical load in the nanoscratch test. *Materials Science and Engineering: A*, *423*, 70–73.

Beake, B. D., Shi, B., & Sullivan, J. L. (2011). Nanoscratch and nanowear testing of TiN coatings on M42 steel. *Tribology – Materials Surfaces & Interfaces*, *5*, 141.

Beake, B. D., Vishnyakov, V. M., Goodes, S. R., & Rahmati, A. T. (2024). Statistically distributed nano-scratch testing of AlFeMnNb, AlFeMnNi and TiN/Si$_3$N$_4$ thin films on silicon. *Journal of Vacuum Science & Technology. A, Vacuum, Surfaces, and Films: An Official Journal of the American Vacuum Society*, *42*, 013104.

Beake, B. D., Vishnyakov, V. M., Valizadeh, R., & Colligon, J. S. (2006). Influence of mechanical properties on the nanoscratch behaviour of hard nanocomposite TiN/Si$_3$N$_4$ coatings on Si. *Journal of Physics D: Applied Physics, 39*(7), 1392–1397. Available from https://doi.org/10.1088/0022-3727/39/7/009.

Bellemare, S. C., Dao, M., & Suresh, S. (2008). Effects of mechanical properties and surface friction on elasto-plastic sliding contact. *Mechanics of Materials, 40*, 206–219.

Bhushan, B. (2005). Nanotribology and nanomechanics. *Wear, 259*, 1507–1531.

Botero, C. A., Cabezas, L., Sarin, V. K., Llanes, L., & Jiménez-Piqué, E. (2023). Nanoscratch testing of 3Al$_2$O$_3 \cdot$ 2SiO$_2$ EBCs: Assessment of induced damage and estimation of adhesion strength. *Ceramics, 6*(1), 664–677. Available from https://doi.org/10.3390/ceramics6010040, https://www.mdpi.com/journal/ceramics.

Bowden, F. P., & Tabor, D. (1986). *The friction and lubrication of solids*. Oxford Clarendon Press.

Brazil, O., Pethica, J. B., & Pharr, G. M. (2021). The contribution of plastic sink-in to the static friction of single asperity microscopic contacts. *Proceedings of the Royal Society A, 477*, 20210502.

Brinckmann, S., Fink, C. A. C., & Dehm, G. (2015). Nanotribology in austenite: Normal force dependence. *Wear, 338–339*, 430–435.

Budinski, K. G. (1991). Tribological properties of titanium alloys. *Wear, 151*(2), 203–217. Available from https://doi.org/10.1016/0043-1648(91)90249-t.

Carreon, A. H., & Funkenbusch, P. D. (2018). Material specific nanoscratch ploughing friction coefficient. *Tribology International, 126*, 363–375. Available from https://doi.org/10.1016/j.triboint.2018.05.027, http://www.elsevier.com/inca/publications/store/3/0/4/7/4.

CEN/TS. (2021). 17629:2021 Nanotechnologies – Nano- and micro- scale scratch testing.

Chavoshi, S. Z., Gallo, S. C., Dong, H., & Luo, X. (2017). High temperature nanoscratching of single crystal silicon under reduced oxygen condition. *Materials Science and Engineering: A, 684*, 385–393. Available from https://doi.org/10.1016/j.msea.2016.11.097, http://www.elsevier.com.

Chavoshi, S. Z., & Xu, S. (2018). A review on micro- and nanoscratching/tribology at high temperatures: Instrumentation and experimentation. *Journal of Materials Engineering and Performance, 27*(8), 3844–3858. Available from https://doi.org/10.1007/s11665-018-3493-5, http://www.springerlink.com/content/1059-9495/.

Chen, J., Bell, G. A., Beake, B. D., & Dong, H. (2011). Low temperature nano-tribological study on a functionally graded tribological coating using nanoscratch tests. *Tribology Letters, 43*(3), 351–360. Available from https://doi.org/10.1007/s11249-011-9813-7, http://www.springerlink.com/(snpxut45gxflnr45vb2gia45)/app/home/journal.asp?referrer = parent&backto = searchpublicationsresults,1,2.

Diao, D. F., Kato, K., & Hayashi, K. (1994). The maximum tensile stress on a hard coating under sliding friction. *Tribology International, 27*, 267–272.

Drees, D., Celis, J.-P., & Achanta, S. (2004). Friction of thin coatings on three length scales under reciprocating sliding. *Surface and Coatings Technology, 188–189*, 511–518.

Gee, M. G. (2001). Low load multiple scratch tests of ceramics and hard metals. *Wear, 250* (1–12), 264–281. Available from https://doi.org/10.1016/s0043-1648(01)00591-9.

Goel, S., Stukowski, A., Goel, G., Luo, X., & Reuben, R. L. (2012). Nanotribology at high temperatures. *Beilstein Journal of Nanotechnology, 3*(1), 586–588. Available from https://doi.org/10.3762/bjnano.3.68.

Goldberg, J. R., & Gilbert, J. L. (1997). Electrochemical response of CoCrMo to high-speed fracture of its metal oxide using an electrochemical scratch test method. *Journal of Biomedical Materials Research, 37*(3), 421–431. Available from https://doi.org/10.1002/(SICI)1097-4636.

Hanlon, T., Chokshi, A. H., Manoharan, M., & Suresh, S. (2005). Effects of grain refinement and strength on friction and damage evolution under repeated sliding contact in nanostructured metals. *International Journal of Fatigue, 27*(10−12), 1159−1163. Available from https://doi.org/10.1016/j.ijfatigue.2005.06.036.

Hokkirigawa, K., Kato, K., & Li, Z. Z. (1988). The effect of hardness on the transition of the abrasive wear mechanism of steels. *Wear, 123*, 241−251.

Johnson K. L. (1985). Contact mechanics. 9780521255769.

Kamminga, J.-D., & Janssen, G. C. A. M. (2007). Experimental discrimination of plowing friction and shear friction. *Tribology Letters, 25*(2), 149−152. Available from https://doi.org/10.1007/s11249-006-9135-3.

Kareer, A., Hou, X. D., Jennett, N. M., & Hainsworth, S. V. (2016a). The existence of a lateral size effect and the relationship between indentation and scratch hardness in copper. *Philosophical Magazine, 96*, 3396−3413.

Kareer, A., Hou, X. D., Jennett, N. M., & Hainsworth, S. V. (2016b). The interaction between Lateral size effect and grain size when scratching polycrystalline copper using a Berkovich indenter. *Philosophical Magazine, 96*, 3414−3429.

Kareer, A., Tarleton, E., Hardie, C., Hainsworth, S. V., & Wilkinson, A. J. (2020). Scratching the surface: Elastic rotations beneath nanoscratch and nanoindentation tests. *Acta Materialia, 200*, 116−126. Available from https://doi.org/10.1016/j.actamat.2020.08.051, http://www.journals.elsevier.com/acta-materialia/.

Kato, K. (1995). Microwear mechanisms of coatings. *Surface and Coatings Technology, 76−77*, 469−474.

Kitsunai, H., Kato, K., Hokkirigawa, K., & Inoue, H. (1990). The transitions between microscopic wear during repeated sliding friction observed by a scanning electron microscope tribosystem. *Wear, 135*, 237−249.

Korres, S., Feser, T., & Dienwiebel, M. (2013). A new approach to link the friction coefficient with topography measurements during plowing. *Wear, 303*, 202−210.

Lafaye, S., & Troyon, M. (2006). On the friction behaviour in nanoscratch testing. *Wear, 261* (7−8), 905−913. Available from https://doi.org/10.1016/j.wear.2006.01.036.

Larsson, M., Olsson, M., Hedenqvist, P., & Hogmark, S. (2000). Mechanisms of coating failure as demonstrated by scratch and indentation testing of TiN coated HSS − On the influence of coating thickness, substrate hardness and surface topography. *Surface Engineering, 16*, 436−444.

Martínez-Nogués, V. (2016). *Nanoscale tribocorrosion of CoCrMo alloys* (PhD Thesis). nCATS, University of Southampton, UK.

Martínez-Nogués, V., Nesbitt, J. M., Wood, R. J. K., & Cook, R. B. (2016). Nano-scale wear characterization of CoCrMo biomedical alloys. *Tribology International, 93*, 563−572.

Mirghany, M., & Jin, Z. M. (2004). Prediction of scratch resistance of cobalt chromium alloy bearing surface, articulating against ultra-high molecular weight polyethylene, due to third body wear particles. *Proceedings of the Institution of Mechanical Engineers, Part H, 218*, 41−50.

NANOINDENT-PLUS. (2013). NANOINDENT-PLUS project standardising the nano-scratch test, FP7/2007−2013, grant agreement NMP-2012-CSA-6−319208.

Nimishakavi, L., & Gee, M. G. (2011). Model single point abrasion experiments on WC/Co hardmetals. *International Journal of Refractory Metals and Hard Materials, 29*, 1−9.

Nosonovsky, M., & Bhushan, B. (2007). Multiscale friction mechanisms and hierarchical surfaces in nano- and bio-tribology. *Materials Science and Engineering A R, 58*, 162−193.

Pagnoux, G., Fouvry, S., Peigney, M., Delattre, B., & Mermza-Rollet, G. (2014). Mechanical behaviour of DLC coatings under various scratch conditions. *Proceedings of the 3rd International Conference on Fracture, Fatigue and Wear, 2*, 308−313.

Pöhl, F., Hardes, C., & Theisen, W. (2019). Deformation behavior and dominant abrasion micro mechanisms of tempering steel with varying carbon content under controlled scratch testing. *Wear, 422–423*, 212–222.

Raimondi, M. T., Vena, P., & Pietrabissa, R. (2001). Quantitative evaluation of the prosthetic head damage induced by microscopic third-body particles in total hip replacement. *Journal of Biomedical Materials Research – Part B: Applied Biomaterials, 58*, 436–448.

Rojacz, H., Nevosad, A., & Varga, M. (2023). On wear mechanisms and microstructural changes in nano-scratches of fcc metals. *Wear, 526–527*, 204928.

Shi, B., Sullivan, J. L., & Beake, B. D. (2008). An investigation into which factors control the nanotribological behaviour of thin sputtered carbon films. *Journal of Physics D: Applied Physics, 41*, 045303.

Shugurov, A., Panin, A., Dmitriev, A., & Nikonov, A. (2018). The effect of crystallographic grain orientation of polycrystalline Ti on ploughing under scratch testing. *Wear, 408–409*, 214–221. Available from https://doi.org/10.1016/j.wear.2018.05.013.

Singh, A., Dao, M., Lu, L., & Suresh, S. (2011). Deformation, structural changes and damage evolution in nanotwinned copper under repeated frictional contact sliding. *Acta Materialia, 59*, 7311–7324.

Smith, J. F., Vishnyakov, V. M., Davies, M. I., & Beake, B. D. (2013). Nanoscale friction measurements up to 750°C. *Tribology Letters, 49*(3), 455–463. Available from https://doi.org/10.1007/s11249-013-0102-5.

Tayebi, N., Conry, T. F., & Polycarpou, A. A. (2003). Determination of hardness from nano-scratch experiments: Corrections for interfacial shear stress and elastic recovery. *Journal of Materials Research, 18*, 2150–2162.

Tayebi, N., Conry, T. F., & Polycarpou, A. A. (2004a). Reconciliation of nanoscratch hardness with nanoindentation hardness including the effects of interface shear stress. *Journal of Materials Research, 19*, 3316–3323.

Tayebi, N., Polycarpou, A. A., & Conry, T. F. (2004b). Effects of the substrate on the determination of hardness of thin films by the nano-scratch and nanoindentation techniques. *Journal of Materials Research, 19*, 1791.

Ben Tkaya, M., Zahouani, H., Mezlini, S., Kapsa, P. H., Zidi, M., & Dogui, A. (2007). The effect of damage in the numerical simulation of a scratch test. *Wear, 263*(7–12), 1533–1539. Available from https://doi.org/10.1016/j.wear.2007.01.083.

Tsui, T. Y., Pharr, G. M., Oliver, W. C., Bhatia, C. S., White, R. L., Anders, S., Anders, A., & Brown, I. G. (1995). Nanoindentation and nanoscratching of hard carbon coatings for magnetic disks. *MRS Proceedings, 383*. Available from https://doi.org/10.1557/proc-383-447.

Tsybenko, H., Farzam, F., Dehm, G., & Brinckmann, S. (2021). Scratch hardness at a small scale: Experimental methods and correlation to nanoindentation hardness. *Tribology International, 163*. Available from https://doi.org/10.1016/j.triboint.2021.107168.

Tsybenko, H., Xia, W., Dehm, G., & Brinckmann, S. (2020). On the commensuration of plastic ploughing at the microscale. *Tribology International, 151*, 10647.

Wang, C. T., Gao, N., Gee, M. G., Wood, R. J. K., & Langdon, T. G. (2012). Effect of grain size on the micro-tribological behavior of pure titanium processed by high-pressure torsion. *Wear, 280–281*, 28–35. Available from https://doi.org/10.1016/j.wear.2012.01.012.

Williams, J. A. (1996). Analytical models of scratch hardness. *Tribology International, 29*, 675–694.

Williams, J. A., & Xie, Y. (1992). The generation of wear surfaces by the interaction of parallel grooves. *Wear*, *155*(2), 363–379. Available from https://doi.org/10.1016/0043-1648(92)90095-p.

Wu, T. W. (1991). Micro-scratch and load relaxation tests for ultra-thin films. *Journal of Materials Research*, *6*, 407–426.

Wu, T. W., Burn, R. A., Chen, M. M., & Alexopoulos, P. S. (1988). Micro-indentation and micro-scratch tests on sub-micron carbon films. *Materials Research Society Symposium Proceedings*, *130*, 117–122.

Zhang, F., Meng, B., Geng, Y., Zhang, Y., & Li, Z. (2016). Friction behavior in nanoscratching of reaction bonded silicon carbide ceramic with Berkovich and sphere indenters. *Tribology International*, *97*, 21–30. Available from https://doi.org/10.1016/j.triboint.2016.01.013.

Zou, M. (2012). Effects of micro- and nanoscale texturing on surface adhesion and friction. In Y.-W. Chung (Ed.), *Micro- and nanoscale phenomena in tribology* (pp. 103–152). ISBN 978-1-4398-3922-5.

Reciprocating nano- and micro-scale wear testing

Ben Beake[1] and Tomasz Liskiewicz[2]
[1]Micro Materials Ltd., Wrexham, United Kingdom, [2]Department of Engineering, Manchester Metropolitan University, Manchester, Manchester, United Kingdom

7.1 Introduction

Reciprocating contacts occur in a wide variety of practical wear situations including hip joints and electrical contacts. In developing tribological tests for candidate materials with improved durability in these contacts, it is desirable that the contact conditions (e.g., contact size, number of reciprocating cycles, sliding speed/oscillation frequency) can be reproduced. Nano- and microscale tests can be more appropriate than macroscale tests as the individual contacts are small, and protective coating systems involve ultrathin layers. At this scale, the topography of the contacting surfaces should also be considered. With the advent of methods for precision topographic control, for example, to produce textured surfaces for friction control, there has been an increasing interest in studying the influence of surface topography on friction and wear. This can be conveniently studied by nanomechanical tribology (using atomic force microscopes or nanoindenters), but this is typically limited to relatively small number of cycles (<100) and small track lengths of the order of 100 μm. Accordingly, the capabilities of commercial instruments have been developed and expanded to perform nano- and microscale reciprocating tests over a wider range of forces, contact sizes, track lengths, speeds, etc. Typical test conditions with one commercial instrument (NanoTest system, Micro Materials Ltd.) are shown in Table 7.1.

Tests have been described as nanofretting when performed using an additional oscillating stage unit that employs a multilayer piezo stack to generate sample motion with the piezo voltage produced by signal generator−amplifier−transformer combination. The nanofretting terminology was introduced since the wear depths are nano- but as the contact size can vary as tip radii vary from ~ 50 nm (sharp Berkovich) to hundreds of microns, microscale wear tests are also possible, and the tests have also variously been described as "accelerated nano-wear," "small-scale fretting," "nano-tribology," "nano-wear," and "micro-wear." In this chapter, they will be described as nanofretting for simplicity.

More recently, the piezo stack has been replaced with a greater level of control, and an increase in maximum track length (from ~ 20 to ~ 90 μm), by using an integrated SPM-nanopositioning stage to generate the oscillation. An additional stage (NanoTriboTest) is used for tests with the larger sliding speeds and larger sliding

Table 7.1 Typical test conditions in nano- and microtribological tests.

	AFM wear	Nano-scratch	Nanofretting	NanoTriboTest
Motion	Reciprocating	Unidirectional	Reciprocating	Reciprocating
Sliding speed (mm/s)	0.001−0.25	0.001−0.1	0.01	1−10
Track length (mm)	0.001−0.1	0.01−1	0.02	1−10
Number of cycles	1−50	1−20	1000−200,000	100−30,000
Total sliding distance (m)	0.000001−0.001	0.00001−0.01	0.01−0.1	1−300
Probe radius (μm)	0.02−1	0.05−25	0.05−200	25−2500

distances, enabling, for example, longer duration, high contact pressure testing. The higher sliding speeds enable direct replication of those in MEMS contacts (Williams & Le, 2006).

7.1.1 Nanofretting

In the nanofretting test, the large number of cycles that can be conveniently run allows durability testing at lower stresses (through larger probe radii and/or smaller contact loads) than are needed to induce failure in nano-scratch tests with much lower wear cycles. The small track length means that total sliding distance is relatively low, but the transition from fretting/partial slip to gross slip can be studied (Wilson, 2008; Wilson, Smith, et al., 2008).

In the nano-scratch test, the combination of relatively high load and smaller probe radii resulted in high contact pressure (typically >1 GPa) during the test. The measured probe depth at film failure depends on the probe radius and coating bending and is typically larger than the film thickness when the probe radius is over ~1 μm. In contrast, when failure occurs at lower contact pressure due to fatigue in the microwear test, the measured probe depth is below that where failure occurs due to overload.

Wilson and coworkers performed nanofretting tests on thin (0.01−2 μm) carbon films on Si using a ruby test probe with 150 μm radius and the residual wear depths determined by post-test AFM imaging of wear scars (Wilson et al., 2008, 2009; Wilson & Sullivan, 2009; Wilson, 2008), and contact pressures were estimated from the radius of a clean region surrounding the wear scars. On the thinner films, the maximum von Mises stress was far into the Si substrate, and the AFM imaging revealed an abrasive wear mechanism with a gradually increasing wear scar.

When the probe depth during the test is monitored, it may be possible to determine the number of cycles to failure, but the observance of discrete fracture-based failure or more gradual wear processes in nanofretting tests can be highly dependent on the probe sharpness. For thin coatings, for example, under 200 nm, failure is typically gradual when probes with, for example, 10–25 μm end radii are used and abrupt when, for example, using Berkovich indenters. Extended tests can be performed with blunter probes. In a nanofretting test of a 1 μm a-C film on Si (50 mN; 53.1 Hz; 22 μm track length; 10 μm probe), complete coating failure was observed after ∼480,000 cycles (Liskiewicz et al., 2010).

Nanofretting tests have typically been designed with a short load ramp followed by a much longer period of oscillation at constant load, but it is also possible to perform ramped load fretting tests where there is oscillation during loading. For example, to investigate the fatigue behavior of a multilayered coating (nominal structure 20 nm Si_3N_4/20 nm NiCr/80 nm Si_3N_4) deposited on float glass, nanofretting tests were performed with a Berkovich indenter under a slowly increasing load (Liskiewicz et al., 2010). Under these conditions, there were changes in wear rate as the applied load increased when the probe penetrated through the different layers in the multilayer stack coating. Distinct plateau regions were observed consistent with failures, or wearing through, of the individual layers.

7.1.2 Reciprocating microscale sliding

To bridge the gap between the length scales usually probed in nano- or macroscale tribological tests and enable the study of running-in processes in greater detail, the NanoTriboTest was developed to provide a fully instrumented capability for rapid high-cycle, high-sliding distance linear reciprocating nano-/microscale sliding wear tests. Typical test conditions are given in Table 7.1. The design retains the high level of lateral rigidity of the loading head and has sufficient stability to perform nano- or microscale wear tests for extended duration (e.g., several hours, up to 300 m sliding). As single-asperity contacts, the microtribological tests provide unique opportunities to study the interplay between the severity of the test, surface topography, and mechanical properties on the friction and wear that develop. The probe displacement is monitored continuously and recorded over the entire wear track enabling in-situ wear measurements. Several authors have reiterated the potential usefulness in measuring wear in situ rather than relying on posttest profilometry measurements (Kamps et al., 2017; Korres et al., 2013). In wear prediction, it is often assumed that the wear rate is linear, but this is generally not the case, particularly when there is a change in the predominant wear mechanism during the test or a tribologically transformed structure (TTS) layer forms (Kamps et al., 2017; Liu et al., 2019).

Zhu et al. studied the friction and wear of $Ag/MoS_2/WS_2$ composite under reciprocating microscale sliding with $R = 25$ μm, 3 mm track, 200 μm s^{-1},

50−400 mN (Zhu et al., 2023). The measured friction coefficient was deconvoluted into interfacial and plowing components, which reduced with continued cycling. At higher load, there was formation of a thick lubricating MoS_2 and WS_2-based transfer film.

Incorporating electrical contact resistance (ECR) measurement during the tests increases the multisensing capability to detect changes in deformation behavior. There are obvious advantages in combining ECR and friction measurements since they may be able to provide complementary information about the wear process. Although the benefits of ECR measurement have been criticized (Tian et al., 1991), it has generally proved extremely useful, as in fretting tests (Fouvry et al., 2017; Laporte et al., 2015; Liskiewicz & Fouvry, 2005; Liskiewicz et al., 2005; Rybiak et al., 2008). Echeverrigaray et al. reported that ECR monitoring provided a valuable tool for detecting premature wear of an electrically nonconductive DLC coating sliding against steel in a pin-on-disk test by dielectric breakdown (Echeverrigaray et al., 2020).

Beake et al. used ECR measurement in reciprocating microwear tests of electrical connectors and novel alloys being investigated for this application (Beake et al., 2021). Tribological behavior of electrical connectors is key to the reliability of electronic systems. Understanding sliding failure due to frequent mating-unmating is important in developing improved connector systems. To achieve low contact resistance, the real area of contact must be very high, so soft metals are pushed together at high force, and conditions for electrical power contacts are more severe than most tribological contacts resulting in very short lifetimes (low sliding distance to failure) (Kassman & Jacobson, 1993). Tian et al. reported a multistage deformation in sphere-on-flat reciprocating sliding tests between electroplated gold contacts (Au−Ni vs Au−Ni) on brass (Tian et al., 1991). Typically, the failure was rapid with the Au layer removed within 100 cycles and the substrate exposed within ~ 200 cycles.

Electroplated multilayer sliding interconnectors with layer architecture ~ 50 nm Au/~ 2.5 μm Ag/nickel underplate/Cu alloy substrate have shown similarly rapid failure in sphere-on-flat microscale reciprocating tests (10 mN vs 1 mm steel ball, sliding at 2 mm s^{-1} over 5 mm track; fixed current 0.1 A). Fig. 7.1A shows the raw ECR data from a typical test. Mean ECR and friction averaged over each cycle indicate a multistage tribological failure process (Fig. 7.1B): (1) in stage 1, the friction coefficient was initially ~ 0.25, and the resistance gradually increased to $\sim 1 \, \Omega$, which marks a transition; (2) in stage 2, the friction and resistance increase rapidly, reaching a high friction coefficient of ~ 0.7; (3) in stage 3, friction and resistance are high and variable with periodic increasing/decreasing resistance; and (4) in stage 4, increases in resistance and slight reduction in friction. Microscopy showed probe wear/debris and fracture-dominated wear tracks.

Extended microscale reciprocating tests ($\sim 2-46$ h duration) have been performed on bulk gold and silver alloys. Reciprocating contacts between bulk noble metal alloys (non-axisymmetric Au alloy pin vs Au alloy disk and non-axisymmetric Ag alloy pins vs Ag alloy disk) have shown improved performance with number of sliding cycles required for failure of the bulk alloy disk being much larger than for the electroplated thin film interconnections.

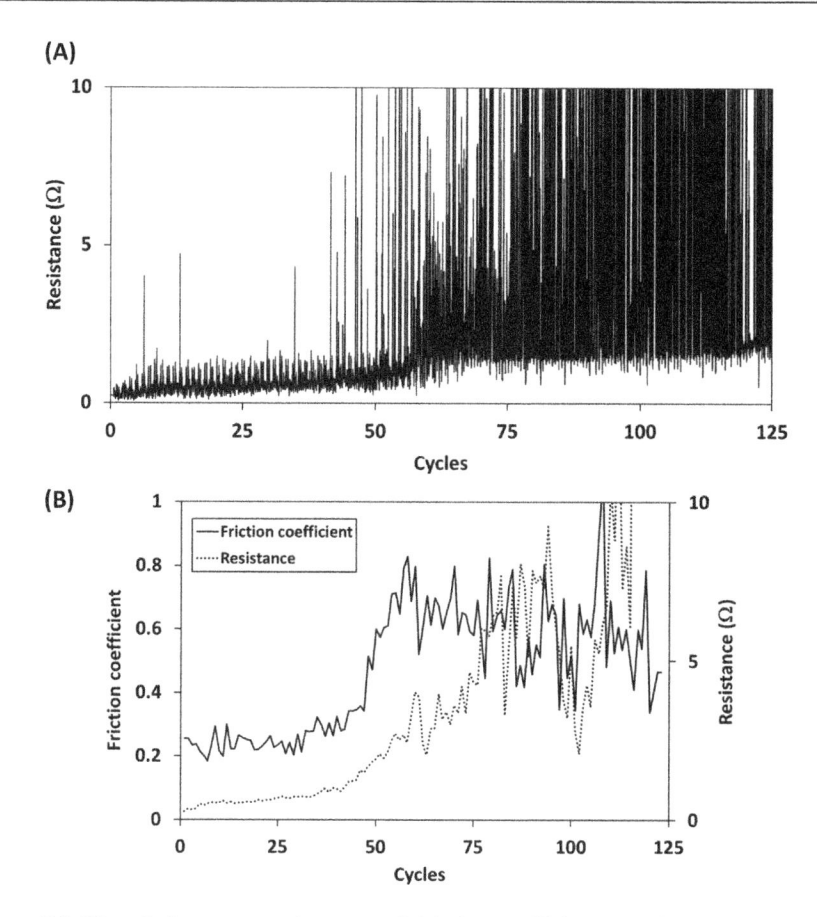

Figure 7.1 Electrical contact resistance and friction coefficient data from a typical test. Reciprocating test between steel ball and electroplated gold multilayer sliding interconnector. (A) Raw (non-averaged) electrical contact resistance, (B) mean friction coefficient and electrical contact resistance (averaged over entire track length under full velocity) versus cycles.

As an example, a 35,000 cycle test between an Ag alloy probe, Ag-20 wt.% Cu-2 wt.% Ni, and Ag alloy (Ag-10 wt.% Cu) disk (Load = 200 mN, 5 mm track, Set I = 0.1 A) is shown in Fig. 7.2. In this test, periodic dramatic increases in resistance were accompanied by simultaneous rises in friction. More detailed analysis reveals that a more complex picture emerged. Aside from these events, however, the friction and resistance were strongly inversely correlated, apart from at the start of the test. Over the first few reciprocating cycles, the friction and contact resistance generally reduce. The decrease in contact resistance is consistent with the asperities on the probe and disk being smoothed out, which increases the contact area. The decrease in friction occurs due to a reduction in plowing, with a possible contribution from grain refinement (Cai & Bellon, 2013). Correlation between the

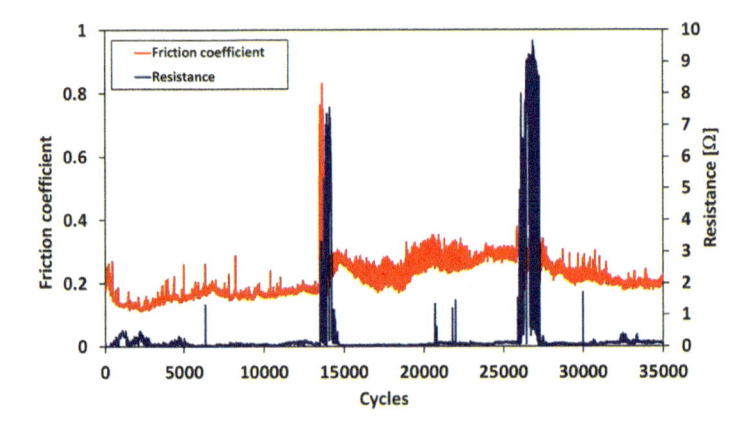

Figure 7.2 Friction and resistance for Ag/Ag tribo-couple. Mean friction and mean resistance per cycle in a 35,000-cycle test between Ag probe and Ag disk.

friction and ECR, when both increase dramatically and subsequently recover, is likely to be largely topographic in nature, with nonconducting non-noble metal oxide debris trapped in the contact increasing the contact area. The two spikes in friction and resistance after 13,500 and 26,000 cycles may relate to oxide debris being trapped in contact that eventually is removed from the sliding contact. The inverse correlation observed otherwise may be related to changing metallic/nonmetallic composition of the contact, for example, where the metallic contact area is decreasing, this could be due to more (previously buried) oxide fragments in the contact, which was indicated in the reciprocating wear of copper sliding contacts (Prasad et al., 2011). The general behavior on the Au—Au was similar with a decrease in resistance and increase in friction preceding the first isolated failure of the Au—Au couple suggestive of an increase in metallic contact area.

7.1.3 Friction loops—fretting and gross slip

The appearance of the friction loop in a fretting test is dependent on lateral contact size through the sliding ratio. Fig. 7.3 illustrates this schematically. Fouvry et al. (1996), defined a sliding ratio $e = \delta_g/a$, where δ_g = sliding amplitude, a = contact radius. The tribosystem remains in the true fretting regime when the unexposed surface is maintained at the center of the fretted surface ($e < 1$). The system moves into the reciprocating sliding regime when the center of the contact area becomes exposed to the atmosphere ($e > 1$). Note that the sliding amplitude can be different from the displacement amplitude due to the contact and testing rig compliance. In studies of small-scale fretting, a marked reduction in track length has been reported at higher load (Martínez-Nogués, 2016; Raeymaekers et al., 2010), which may be connected to changing contact stiffness of the tribosystem. The slope of the near-vertical section at the beginning and end of the friction loops is the total contact stiffness of the tribosystem. In microscale reciprocating tests on DLC coatings at 500 mN, this was ~ 2000 N m^{-1} (Beake et al., 2021).

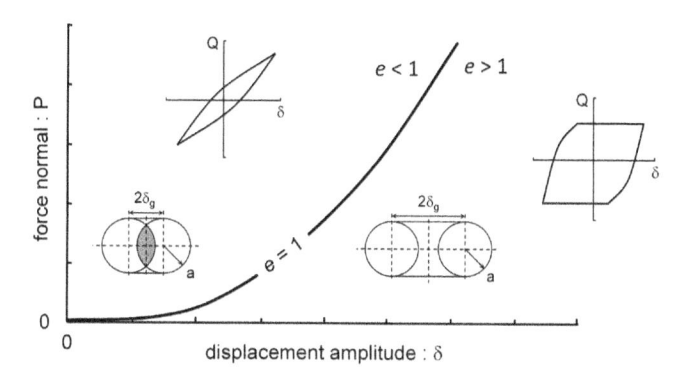

Figure 7.3 Sliding ratio in fretting.

Friction coefficients during the nanofretting tests have been determined from friction loops by the method of Burris and Sawyer, that is, $\mu = (\mu_{forward} + \mu_{reverse})/2$ to eliminate any potential transducer misalignment issues (Burris & Sawyer, 2009). To date, many nanofretting tests have been performed under gross slip conditions. Potentially one of the reasons is that in many cases, the wear rates are higher under gross slip conditions, as mechanisms where lower wear due to stabilized third body or restricting tribo-oxidation as occur in partial slip are less common (Fouvry et al., 2009).

Fouvry and Liskiewicz showed that the friction coefficient in macroscale fretting and full sliding reciprocating tests can be determined from the frictional energy dissipation (Fouvry et al., 2017; Laporte et al., 2015; Liskiewicz & Fouvry, 2005; Liskiewicz et al., 2005; Rybiak et al., 2008) by Eq. (7.1) where E_d is frictional energy dissipation, L is the applied load and l is track length.

$$\mu = \frac{E_d}{2Ll} \tag{7.1}$$

To determine the dynamic and static friction coefficient in the microscale tests separately, it is possible to combine the Fouvry and Liskiewicz and Burris and Sawyer approaches. The (dynamic) friction coefficient can be calculated from the energy dissipation during full sliding (Eq. 7.2) where l_{fs} is the track length under full sliding velocity

$$\mu = \frac{E_d}{2Ll_{fs}} \tag{7.2}$$

The total track length is replaced by the total track at the full velocity (e.g., 90% of the track). The static friction coefficient is obtained from the maximum friction at the turnaround points (where velocity drops to zero) using a similar approach to Burris and Sawyer, which accounts for any transducer misalignment (Burris & Sawyer, 2009). The apparent static friction coefficient may not solely be a function of the tribocontact and be affected by the topography and instrumentation used. Although friction loops revealed the presence of occasional track-position sensitive changes in friction, determining an average friction coefficient from the energy dissipation over the entire track sliding at full velocity proved a reliable approach enabling differences between samples as a function of load and cycles to be investigated. Case studies on industrially relevant material systems from the biomedical and automotive sectors are presented further. In each of these studies, the influence of length scale (contact size) on the observed behavior is investigated.

7.2 Case studies on biomedical materials—Ti6Al4V, CoCrMo, and 316L stainless steel—influence of passivating oxide layer in reciprocating tests over different length scales

In hip implants, wear occurs mainly on bearing surfaces leading to adverse metallic ions and particulate debris generation. The rate at which an implant experiences wear depends on the bearing surfaces selected for the ball-and-socket components, which typically include metal-on-metal, metal-on-polyethylene, ceramic-on-polyethylene, and ceramic-on-ceramic implant designs, each combination having unique advantages and distinct drawbacks (Buford & Goswami, 2004). Although artificial joints are designed as fully lubricated systems, asperity-to-asperity contact occurs leading to partially dry contact at microscopic level (i.e., potential areas with boundary lubrication). To understand the failure mechanism of artificial joints, it is necessary to understand the mechanism that governs roughening of metallic surface and subsequent damage nucleation and wear at the bearing interface. Scratching of the metallic surface by entrapped wear debris leading to increased UHMWPE wear rates has been recognized as one of the main causes of early failure of TJRs (Najjer et al., 2000). Surface scratches on retrieved (i.e., failed) protheses are typically only a few microns wide and <1 µm deep and show significant pile-up (Raimondi et al., 2001). The hard particles responsible for the scratches can be zirconia or complex carbides present in the as-cast femoral head. In reciprocating sliding tests, Co- and Mo-carbides are fractured and torn off the surfaces resulting in additional surface fatigue and abrasion. From the presence of these scratches, it has been suggested that current biomedical materials do not provide adequate load support (Li et al., 2004).

To develop full mechanistic understanding for reliable artificial joint design, it is necessary to investigate the mechanical and tribological properties of biomedical materials *at the relevant contact scale* as these properties are size-dependent. Accordingly, reciprocating wear tests have been performed on CoCrMo, Ti6Al4V, and 316L stainless steel at different length scales (from nano- to macro-) and in combination with nanoindentation, nano-scratch, and EDX analysis of wear scars.

7.2.1 Nanofretting

Nano-scratch and nanofretting tests have been performed on highly polished biomedical-grade Ti6Al4V, 316L stainless steel, and wrought high-carbon CoCrMo alloy samples using a 3.7 μm end radius spheroconical diamond indenter in a commercial nanomechanical test system (NanoTest system) (Beake & Liskiewicz, 2013). By performing repetitive constant load unidirectional and bidirectional scratches in addition to the single scratches, it was attempted to simulate the repetitive and complex contact that can occur in the body when abrasion occurs by third body wear. The relationship between the mechanical properties of the biomaterials and their friction and nano-/microscale wear behavior was investigated. Nanoindentation tests were performed with a Berkovich indenter at 100 mN, that is, a sufficiently high load that the native surface oxide (1.9 nm for 316L stainless steel (Tardio et al., 2015), 1–4 nm CoCrMo and 4–5 nm for Ti6Al4V (Callen et al., 1995; Sittig et al., 1999) does not affect the results. Tests at lower load (10 mN) showed a typical indentation size effect in hardness but not elastic modulus. At 100 mN, CoCrMo had the highest hardness and stiffness ($H = 6.0 \pm 0.4$ GPa, $E = 277 \pm 8$ GPa). Ti6Al4V showed intermediate hardness and low elastic modulus ($H = 4.4 \pm 0.4$ GPa, $E = 146 \pm 10$ GPa), and 316L showed the lowest hardness and intermediate stiffness ($H = 3.1 \pm 0.2$ GPa, $E = 222 \pm 10$ GPa). Although it was not as hard as the CoCrMo alloy, the Ti alloy had the highest H/E and H^3/E^2 due to its relatively low elastic modulus. Macroscopic yield stresses on these alloys ranged from 0.3 GPa on the steel to 1 GPa on the Ti6Al4V (Goldberg & Gilbert, 1997; Mirghany & Jin, 2004; Raimondi et al., 2001). Due to the size effect, the yield stresses are higher at small scale. Mean pressures at yield in the nano-scratch were 2.3 GPa for Ti6Al4V, 2 GPa for CoCrMo, and 1 GPa for the steel.

The 3000-cycle nanofretting experiments were performed with a 10 μm track length and 10 Hz oscillation frequency at 1–30 mN. The load was linearly increased to the test load in 10 s and held for 290 s before unloading (i.e., total number of fretting cycles during the loading = 100 and hold segments = 2900). The same probe was used for fretting, scratch, and spherical nanoindentation tests. This enabled the on-load depth in each test (after correction for sample slope and compliance) to be used as a direct comparison of the extent of deformation in each test. The approach clearly shows that tangential loading promotes deformation. For example, the on-load depth in tests on CoCrMo at 30 mN ranged from 0.25 μm (indentation), 0.29 μm (scratch) to 0.92 and 1.8 μm at the start and end of the constant load segment of a nanofretting test, respectively.

Wear resistance of the 316L steel and Ti6Al4V markedly deteriorates as the fretting load increases. Scanning electron microscopy (SEM) images of fretting scars after 2900 cycles at 1 mN applied load revealed an appreciable plastic fretting track on 316L stainless steel though the fretting tracks were barely visible on CoCrMo and Ti6Al4V. At 2 mN, the fretting scars are still not distinct on CoCr and Ti6Al4V, but there was a transition to a more severe fretting wear on the 316L stainless steel. At 3 mN, the fretting scars are more developed on CoCrMo and on Ti6Al4V, and the 316L stainless steel exhibits more pronounced delamination wear. The probe depth under load was used to track the fretting wear. At the beginning of the test, the stainless steel performed better than the Ti6Al4V alloy. A marked increase in probe depth occurred after typically ~200 repeat wear cycles on the stainless steel. Abrupt increases in probe depth were commonly observed on both Ti6Al4V and the 316L stainless steel but were absent on the CoCr alloy.

CoCrMo exhibited the highest wear resistance and smallest—and more gradual—increase in depth during the constant load period. To investigate the role of alloy processing on the gradual wear on CoCrMo in more detail, Martinez-Nogues et al. performed 3000 cycle nanowear tests of four CoCrMo alloys (forged, as-cast, and two as-cast alloys subject to different thermal treatments) (Martinez-Nogues et al., 2015). Tests were performed with similar conditions (1—30 mN load, 10 μm track length, 7.5 Hz, 5 μm diamond probe). Martinez-Nogues et al. chose 10 μm displacement amplitude on the basis of clinical studies that demonstrated micromotions between the femoral stem and the cement are typically <40 μm (Ebramzadeh et al., 2005; Maloney et al., 1989). To shorten test duration, tests were run at 7.5 Hz rather than the physiological 1 Hz. The 5 μm diamond probe was used to be representative of a single-asperity contact between hard inert ZrO_2 radiopacifier agglomerates (mean diameter 5—10 μm) and the CoCrMo stem surface. In principle, any low-depth long-duration test data can be affected by thermal drift or an inaccurate thermal drift correction. In this study, the final depth data from the instrument (after correction for any thermal drift) was cross-validated against AFM and white light interferometric measurements with excellent agreement between them. This is a useful result because it implies that instrument depth measurements during the test also should be quantitatively accurate, so, for example, changing wear rate could be studied. Below 15 mN, there was no influence of the manufacturing process or thermal history, but above 15 mN there were differences in wear depth and wear debris production. The as-cast alloy showed higher wear depth above 15 mN despite being the only alloy to show clear wear scars at the lowest loads. After thermal treatment, there was more wear debris, which the authors suggested could accelerate the nanowear process by acting as a third body between the contacting surfaces. Running-in and steady state friction decreased with load, with the influence of surface topography greatest at 5 mN and almost absent at 20 mN. Martínez-Nogués et al. reported that an increase in applied load from 50 to 500 mN in nanofretting of as-cast CoCrMo alloy resulted in less sliding with a reduction in track length from 10 to 2 μm and accompanied changes to the shape of the friction loops (Martinez-Nogues et al., 2016). Sun et al. studied the correlation between the deformation around nano-scratch tracks in simulated body fluid and CoCrMo from retrieved MoM prostheses, concluding that the nano-scratch test could simulate the in vivo deformation as similar features were found in both (Sun et al., 2009).

A range of wear mechanisms were observed in the nanofretting tests as the load was increased from 1 to 30 mN, from ironing at very low load to plowing and then more extreme cutting and delamination wear as the load increased. This transition was observed during the tests in the probe depth versus time plots and in the post-test SEM imaging (Fig. 7.6). Between 3 and 4 mN, there was an abrupt transition to a more severe wear mode. The variability above this is another contact size effect, that is, the complex surface/near-surface microstructure in Ti6Al4V is variable at the scale of the tests. Microstructural variations in Ti6Al4V have been considered responsible for differences in abrasion resistance at larger scale (Budinski, 1991). Hardness mapping (by a grid of 400 indentations to 50 mN with 20 μm spacing) on this sample revealed regions of varying mechanical properties; it seems likely that they could be a factor in the variability in wear depth seen in the wear and scratch tests at the nano-/microscale.

Similar load-dependent transitions were observed on 316L stainless steel. As the on-load probe depth increases as wear progresses in the higher load tests, there is a transition from the spherical part of the probe to its conical part. This sphere − cone transition increases the attack angle promoting more severe wear modes. Kato and coworkers investigated wear mechanisms during repetitive scratching of steels with a 30 μm WC-Co indenter at 70 mN to 1.5 N (Hokkirigawa et al., 1988; Kitsunai et al., 1990). They observed that increasing load and contact strain resulted in a transition in wear mode from plowing to wedge formation and then cutting. Increased steel hardness promoted the transition to cutting wear, which then occurred at lower attack angle. The nanofretting wear scars from tests at high load show some similarities to the wedge and cutting wear in the low-cycle repetitive scratch tests on steels reported by Kato and coworkers. These authors noted that the wear process could shift to milder behavior after scanning for a few cycles, presumably due to the sharp decrease in contact pressure. However, in the nanofretting tests with the sharper probe and much greater number of wear cycles, at sufficiently high load, the increase in depth through the test is also accompanied by an increase in contact strain, and it is this increase in attack angle that results in the observed transition to the wedge and cutting/delamination wear.

Debris on CoCr appeared generally smaller in size, and Ti6Al4V and 316L stainless steel exhibited what appear to be sheets of material that are progressively being removed from the contact zone during the nanofretting test, consistent with their lower shear stress. The contact pressure required for the onset of this delamination wear is of the order of 1−2 GPa on the Ti6Al4V and 316L stainless steel materials. Differences in stacking fault energies may also influence friction and wear. CoCr has a very low stacking fault energy resulting in little cross-slip and rapid work hardening; the imposed strain cannot be accommodated solely by dislocation flow, and deformation occurs by mechanical twinning and shear banding, which is thought to be the origin of a nanocrystalline layer observed in high-wear regions of worn samples (Rainforth et al., 2012). Stainless steel also has low stacking fault energy (Rainforth et al., 2012). Shear banding has also been observed in the worn surface of stainless steel and has been implicated in the production of wear debris (Rainforth et al., 1992). In contrast, the titanium alloy has a much higher stacking fault energy of $\sim 110\ \mathrm{J\ m^{-2}}$ (Lee et al., 2012) and does not work harden to the same degree, resulting in little reduction in plowing and friction on repetitive nanoscratching. The thickness and structure of the passive oxide films formed on the alloys play an important role in their nanoscale tribological behavior. Oxide integrity is important in influencing how the tribologically transformed structure (TTS) layer develops. Ti6Al4V has a harder and thicker surface oxide (Moharrami et al., 2013), which effectively protects the underlying alloy at low load. With fretting cycles at higher load, the hard oxide layers break exposing the Ti6Al4V surface causing an abrupt increase in wear rate. Rainforth et al. noted that FIB-TEM imaging through wear tracks formed by low cycle repetitive nano-scratch tests on CoCr has been used to show that the contact-induced formation of a subsurface nanocrystalline (TTS) layer did not occur straightaway but within a short number of wear cycles (Rainforth et al., 2012). In addition to its high hardness and yield stress, the ability of the CoCr to show a type of adaptive behavior with the formation of a subsurface nanocrystalline layer on repetitive scratching (Pourzal et al., 2009; Zeng et al., 2015) appears beneficial to its wear resistance.

7.2.2 Reciprocating microscale tests on 316L stainless steel and Ti6Al4V

Reciprocating nanowear experiments were performed with a NanoTest Vantage fitted with a NanoTriboTest module at 10−500 mN on biomedical grade 316L stainless steel and Ti6Al4V alloy samples (Beake et al., 2021). In total, 500 cycle tests were performed with a diamond indenter of 25 μm end radius over a 1 mm track at a maximum sliding velocity of 0.5 mm s^{-1}. The velocity was at its maximum over the central 90% of the track and linearly reduced to zero at the turn-around points.

There was ductile behavior on the stainless steel with increasing pile-up at the sides of the wear track at higher load. Pile-up is more significant in sliding contact than indentation and is also influenced by H/E and a/R (where a is the contact radius and R is the indenter radius) (Hanlon et al., 2005; Singh et al., 2011). There is more pile-up on 316L stainless steel due to its lower H/E. At lower load, and after a higher number of wear cycles, there was some variability with track position over a single cycle due to surface topography. Friction gradually decreased with cycling to $\mu \sim 0.05-0.07$ after 500 cycles. At higher load, the friction was higher around the turn-around points where the velocity was lower, indicating a transition where static friction(μ_{st}) > dynamic friction(μ_d). The static friction increased with load and cycles reaching $\mu_{st} \sim 0.45$ at 500 mN. The friction and wear behavior was completely different on Ti6Al4V. At low load, the friction was $\mu \sim 0.04-0.05$. At ≥ 50 mN, the friction was also initially low ($\mu \sim 0.1$) but after $\sim 10-20$ cycles over a couple of cycles, it increased rapidly to reach $\mu \sim 0.5$. An example at 100 mN where the transition from $\mu \sim 0.1$ started over 10 cycles reaching $\mu \sim 0.5$ four cycles later is shown in Fig. 7.4A. Friction loops before the transition (dotted line), immediately after (dashed line), and after 250 cycles (full line) are shown in Fig. 7.4B. There was a direct correlation between the transition to higher friction and the onset of roughening in the wear track and significantly higher final wear depth.

Friction (Fig. 7.5) and SEM images (Fig. 7.6) of wear tracks at 30 and 50 mN (above and below the transition on Ti6Al4V) show a marked correlation between friction coefficient and the observed deformation, with low friction for ductile plowing and a transition to very high friction at 50 mN on Ti6Al4V.

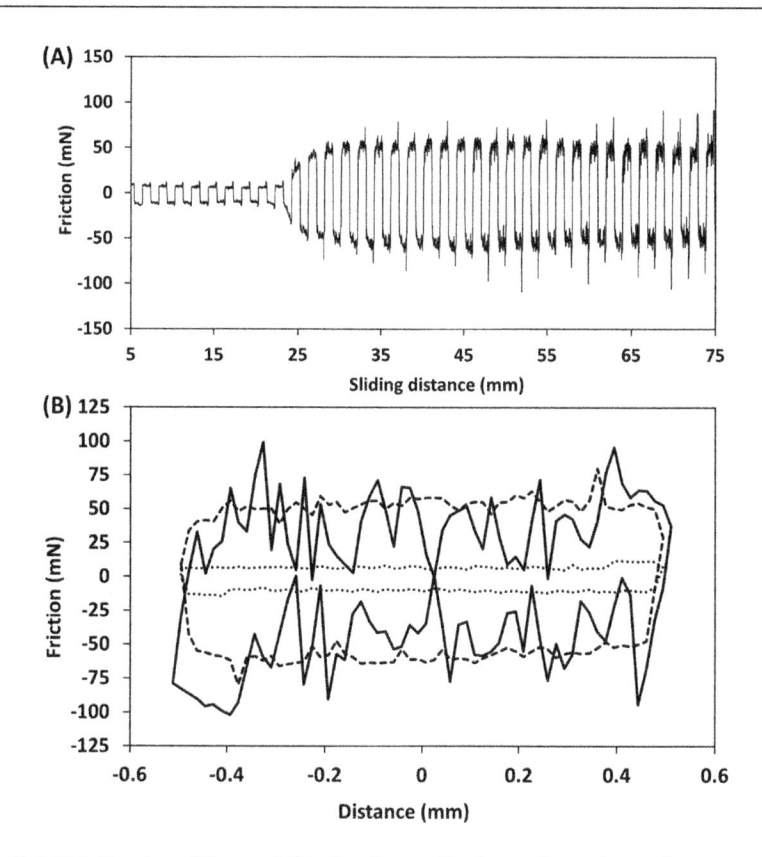

Reciprocating nano- and micro-scale wear testing 199

Figure 7.4 Friction transition and friction loops. Reciprocating microscale test on Ti6Al4V. (A) Friction versus sliding distance showing rapid transition to high friction and (B) friction loops before the transition (dotted line), immediately after (dashed line), and after 250 cycles (full line).

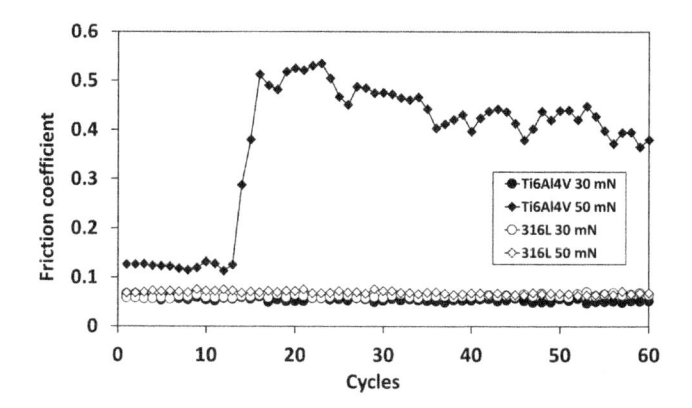

Figure 7.5 Friction in reciprocating microscale tests. Friction versus cycles in reciprocating microscale tests.

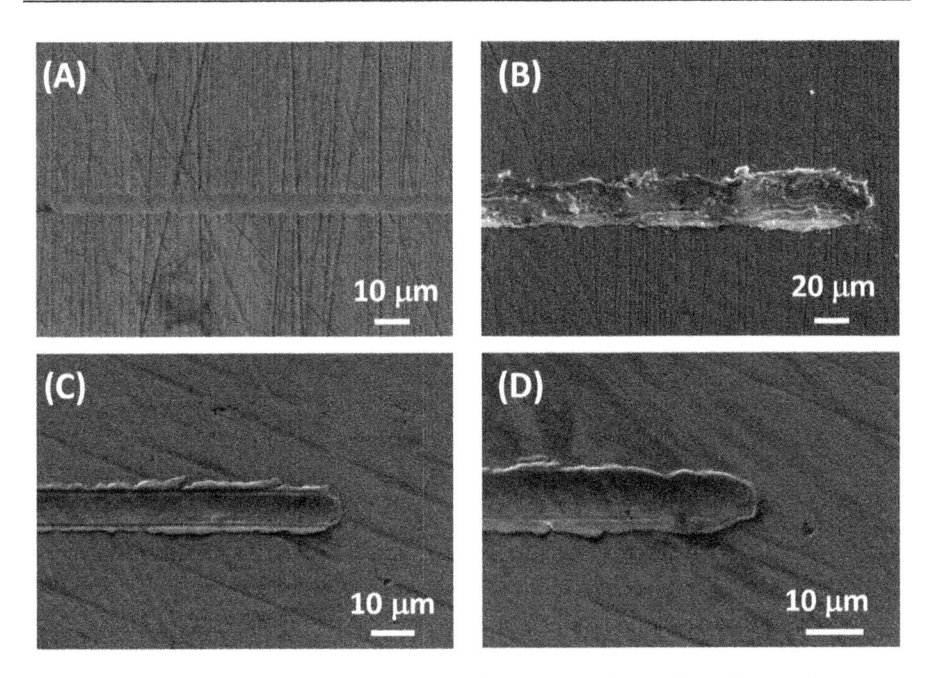

Figure 7.6 SEM images of wear tracks after 500-cycle reciprocating microscale tests. (A) 30 mN on Ti6Al4V, (B) 50 mN on Ti6Al4V, (C) 30 mN on 316L stainless steel, (D) 50 mN on 316L stainless steel.

Ti6Al4V is somewhat harder but lower in elastic modulus than the 316L stainless steel. This combination of mechanical properties results in higher elastic-strain-to-break (H/E) and resistance to plastic deformation, H^3/E^2. At lower load, wear appears to be controlled by its mechanical properties, resulting in more elastic deformation and lower wear depth than 316L stainless steel.

EDX oxygen mapping across the wear tracks provides information on the nature of the tribologically transformed structure (TTS) formed. The oxygen enrichment was load dependent on both alloys. On the 316L stainless steel, there was no oxygen enrichment at ≤ 100 mN and only very slight enhancement at the edges of the track at ≥ 200 mN. On Ti6Al4V, oxygen enrichment within the wear track was strongly load-dependent with no enrichment in the wear track at ≤ 40 mN, slight enrichment at 50 mN and more significant oxygen enrichment at ≥ 70 mN. With continued cycling marked roughening in the wear track accompanied these changes in friction on Ti6Al4V.

The TTS that forms below the surface after fretting wear of 316L stainless steel and Ti6Al4V have been investigated using FIB imaging, EDX (Liu et al., 2019), and nanomechanical measurements (Liskiewicz et al., 2017). Liu et al. (2019) used a custom-built electrodynamic shaker powered fretting rig to perform 10–100,000 cycle fretting tests (10 N, 8 mm radius alumina ball, 1 Hz, ± 100 μm displacement amplitude resulting in gross slip). On Ti6Al4V, the TTS is much thicker, and EDX shows that it is accompanied by significant surface oxidation, it is much thinner on

316L stainless steel, and the extent of tribo-oxidation revealed by EDX is much reduced. These structures form rapidly and initially increase hardness on both materials. After further wear cycles on Ti6Al4V, there was a transition to a softer, lower modulus, more porous TTS (Liskiewicz et al., 2017; Liu et al., 2019). Liu et al. introduced a novel method to improve the accuracy of the Archard wear model (Liu et al., 2019). In the modified treatment, the initial hardness is replaced by the hardness developing in the TTS through wear. This resulted in a dramatic improvement in the predictive ability on 316L steel and pure copper where hardening was observed in the fretting track. The model was less successful in predicting wear on Ti6Al4V due to the added complexity from changing microstructure, oxidation, porosity, and cracking (Liu et al., 2019).

The EDX measurements in the reciprocating microscale tests are in good agreement with the macroscale data (Liu et al., 2019) and provide further support for the influence of the TTS on the measured friction. When the native oxide is still intact in the wear track, it protects the underlying alloy enabling a hard, low friction TTS to form. With continued wear at higher load, the depth increases on Ti6Al4V; hence, there is a transition from the spherical end cap to the conical part of the tip, which increases the attack angle and contact strain, which can alter the predominant wear mechanism. As the attack angle increases, there is a transition from plowing wear to cutting. This more severe contact results in breakup of the protective oxide on the Ti6Al4V leading to much higher friction, with an increase in oxygen and generation of a rougher and more porous surface with reduction in mechanical properties.

Friction and wear in the microscale tests on Ti6Al4V have some parallels with what has been observed in nanofretting (gross slip) and repetitive nano-scratch (unidirectional) tests on this alloy (Beake & Liskiewicz, 2013). Transitions to a more severe wear mechanism with cycling were strongly load-sensitive in nanofretting tests with a sharper $R = 3.7$ μm probe on 316L and Ti6Al4V. In contrast, when reciprocating tests are performed with larger probes, the attack angle is initially much lower, and abrupt oxide-film breakdown was only seen for Ti6Al4V. In repetitive nano-scratch tests at 30 mN with the same 3.7 μm probe, the friction and wear rate on Ti6Al4V remained high for 10 cycles at 30 mN. This was accompanied by a large decrease in scratch recovery, consistent with less effective formation of a protective, load-supporting TTS. In comparison, on 316L stainless steel and CoCrMo, the wear rate and friction rapidly reduce and the reduction in scratch recovery is much less, indicating formation of a protective TTS. SEM images of these nano-scratch tracks on Ti6Al4V reveal similarities with the higher load reciprocating tests. There was non-smooth profile within the track (i.e., indicative of cold-welding/adhesive stick-slip) consistent with the highly variable scratch depth along with it (Beake & Liskiewicz, 2013).

The surface/near surface microstructure of the alloys may also evolve during sliding and influence the friction. With continued sliding, grain refinement or formation of a TTS has been reported for many alloys (Cai & Bellon, 2013; Laporte et al., 2015; Liskiewicz et al., 2017; Liu et al., 2019 Sauger, Fouvry, et al., 2000; Sauger, Ponsonnet, et al., 2000; Singh et al., 2011). Suresh and coworkers have reported grain refinement resulting in an increase in hardness within the scratch track on Cu (Hanlon et al., 2005; Singh et al., 2011). Cai and Bellon have shown that pin-on-disk wear tests of eutectic Ag−Cu (Ag-28.1 wt.% Cu) resulted in grain refinement and work hardening within the wear track (Cai & Bellon, 2013). Laporte et al. found that a mechanically mixed layer (TTS) forms in multilayer sliding contacts with Ag top layers, with significant tribo-oxidation within the wear track (Laporte et al., 2015).

Complex microstructure and susceptibility to brittle deformation make wear difficult to predict in Ti6Al4V tribosystems (Dong & Bell, 1999; Molinari et al., 1997; Li et al., 2016; Zivic et al., 2011). Titanium has been shown to be susceptible to delamination in macroscale fretting wear (Waterhouse & Taylor, 1974). On Ti6Al4V, several factors influence its tribological behavior with marked transition to more severe wear as the load increases. It has a harder surface oxide (Moharrami et al., 2013), which effectively protects the underlying alloy at low load. With continuing wear cycles at higher load, the hard oxide layer is removed exposing the Ti alloy surface. In fretting-wear studies, the wear rate of Ti6Al4V is typically an order of magnitude higher than 316L stainless steel, even though the latter has significantly lower hardness (Liskiewicz et al., 2017; Liu et al., 2019). Molinari et al. observed that the TTS, which forms in dry sliding of Ti6Al4V, can be very thick, but oxygen tends to embrittle the matrix reducing its mechanical strength (Molinari et al., 1997). Li et al. found that the tribolayers on Ti6Al4V were relatively porous containing a high number of cracks (Li et al., 2016). In pin-on-disk tests with an alumina ball sliding against Ti6Al4V, Dong and Bell reported alloy wear that was three orders of magnitude higher than when sliding against hardened steel (Dong & Bell, 1999), considering this due to the breakup of the weak surface oxide on the titanium surface. Li et al. (2004) reported that in reciprocating tests with 4 mm WC probes, Ti6Al4V showed the most debris, and both of which and a 316L stainless steel showed large chip-like debris while CoCr showed smaller debris. In contrast, in microscale reciprocating tests on 316L stainless steel, the oxide remains intact through the load range, resulting in plowing and low friction (Beake et al., 2021). Under these conditions, a hard, thin TTS can develop with continued cycling.

7.3 Case study on diamond-like carbon coatings on steel

Diamond-like carbon (DLC) coatings are applied to components in the power train of an automotive engine including tappets, pistons, piston rings, and fuel injectors (Erdemir & Donnet, 2006; Gåhlin et al., 2001; Lawes et al., 2010). Under severe mechanical loading, the ability of DLC coatings to protect these components is limited by their resistance to contact damage (Bernoulli et al., 2015). Despite being hard and elastic (with higher H/E than ceramic hard coatings), they generally show poor performance at higher load. In pin-on-disk tests, DLC films that exhibited very low rates of wear under low contact pressure sliding were susceptible to abrupt increases in wear rate when the pressure increased above a critical threshold at higher load (Jiang & Arnell, 1998). Low-friction Si-doped DLC coatings have been explored for a range of applications (Lanigan et al., 2016), and WC/C (a-C:H:W) coatings have been shown to perform well under several different tribological conditions (Mutafov et al., 2014; Ramírez et al., 2015; Wänstrand et al., 1999).

Beake, Liskiewicz, and coworkers have investigated the mechanical and nano- and microtribological behavior of Si-doped DLC and WC/C coatings on hardened tool steel and compared them with the behavior of a typical hard hydrogenated DLC (Beake et al., 2015, 2021; Liskiewicz et al., 2013). For these studies, the multilayer carbon coatings with a-C:H (undoped DLC) and Si-a-C:H (Si-doped DLC) top layers were deposited on hardened M2 tool steel with a PECVD Flexicoat 850 system (Hauzer Techno Coating, the Netherlands). WC/C (W-doped DLC) was a commercial coating, Balinit C Star, deposited on hardened M2 tool steel by Oerliken Balzers with a CrN sublayer for adhesion promotion (Oerliken Balzers Technical Note, 2010).

Coating-only values were determined from nanoindentation following the ISO14577 approach. These were (1) DLC $H = 23.4$ GPa, $E = 234$ GPa, $H/E = 0.1$, $H^3/E^2 = 0.23$ GPa (2) Si-DLC $H = 16.8$ GPa, $E = 151$ GPa, $H/E = 0.11$, $H^3/E^2 = 0.21$ GPa (3) WC/C $H = 13.0$ GPa, $E = 149$ GPa, $H/E = 0.09$, $H^3/E^2 = 0.1$ GPa. The length scale in the tribological tests was varied by changing the probe radius and applied load. The nanofretting and nano-scratch tests were performed with 5 μm probes, and microscratch and reciprocating tests with 25 μm probes (Beake et al., 2015, 2021; Liskiewicz et al., 2013). The nanofretting and reciprocating tests were performed at loads well below those causing cracking in ramped scratch tests. The WC/C coating showed higher critical loads for cracking and total film failure in microscratch tests. The 4500-cycle nanofretting tests on the coatings were performed with a 5 μm probe, 100 mN load, 13.4 μm track length, and 5 Hz frequency (Liskiewicz et al., 2013). The tests were designed with 10 s hold periods so that the initial on-load (before fretting), final on-load (after fretting), and residual depths could be determined. The steps were (1) load ramp to 100 mN load at 5 mN s^{-1}, (2) 10 s at 100 mN without oscillation, (3) 4500 cycle fretting at 100 mN, (4) 10 s without oscillation, (5) unloading at 5 mN s^{-1} to 10 mN, (6) 300 s for thermal drift correction, (7) full unloading at 5 mN s^{-1}, and (8) 10 s at full unload.

At the start of the nanofretting tests, contact was either elastic or close to elastic, supported by nano-scratch and simulation data. Thereafter the on-load depth increased, but the rate of fretting wear was not constant during the test, decreasing as the contact pressure reduced with increasing wear depth. The increase in depth during the test (determined from the final on-load depth minus the initial on-load depth) matched the residual depth data for the coatings indicating that the increase in depth was plastic and that the initial contact was elastic or close to it. The relative performance of the coatings correlated with hardness. The residual wear depths were (288 ± 10) nm on the DLC, (608 ± 113) nm on Si-DLC, and (752 ± 154) nm for WC/C. For all three coatings, the friction reduced with cycling. The increase in friction through the test was much greater on WC/C consistent with its larger wear. In general, a higher wear rate was associated with increasing friction and a transition to lower wear rate more steady friction.

A NanoTest Vantage (Micro Materials Ltd., Wrexham, UK) fitted with a NanoTriboTest module including a reciprocating stage (PI, Germany) controlled within the NanoTest software was used for the reciprocating tests (Beake et al., 2021). The 500-cycle reciprocating wear tests were performed at 10–500 mN with a diamond indenter of 25 μm end radius as the test probe, 1 mm track length, and 0.5 mm s^{-1} maximum sliding velocity. The sliding velocity was at its maximum value over the central 90% of the track and linearly reduced to zero at the turn-around points. The friction force and the raw (i.e., unlevelled, uncorrected for any instrumental drift or frame compliance contribution) probe displacements were monitored continuously and recorded over the entire wear track. The instrumentation has the sensitivity to detect local changes in friction along the scratch track, but for convenient comparison between the coatings and load- and cycle- dependent changes in friction, the mean friction coefficient averaged over the 1 mm track.

The initial friction coefficient was 0.08–0.10 in every test but marked differences emerged with continued reciprocating cycles. At 10 mN, the mean friction coefficient averaged over the wear track showed more fluctuations than at higher load for all three coatings. Friction and wear differences between the coatings became clearer >10 mN. On DLC, there was a slight decrease in friction with continued cycling at 50–200 mN, but there was a transition to higher friction during the tests at 500 mN. For DLC, the wear was relatively low in two of the three tests at 500 mN, but in the other test, there was coating failure over a large part of the track. A typical transition region between fractured and nonfractured wear track is shown in Fig. 7.7. Energy dispersive X-ray (EDX) analysis revealed Cr and W exposure in the fractured regions of the track. On moving between the greater and lesser worn regions, the friction changed abruptly. Measurements of the wear depth during the test showed that the coating fractured over progressively more of the track with more cycles. On Si-DLC, there was little change in friction with continued cycling at 50–200 mN, but at ≥ 300 mN, there was a transition to a more periodic friction, although the mean coefficient of friction was unchanged. On Si-DLC abrasive marks in the wear track and debris at the sides of the track were more

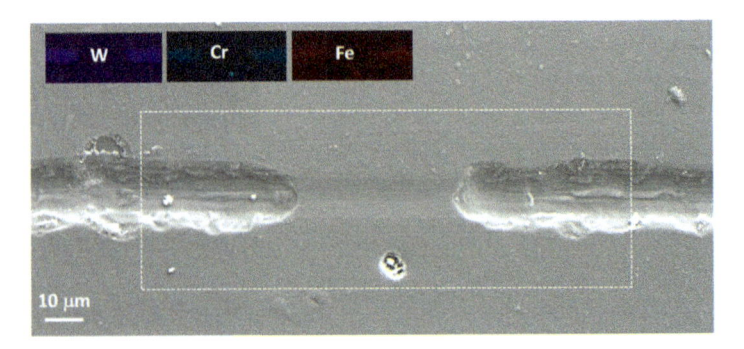

Figure 7.7 SEM image of wear track at 500 mN on DLC showing fractured and nonfractured regions. The EDX maps were from the marked central region.

pronounced. On WC/C, there was an initial increase in friction coefficient at 100 mN to 0.1 before a gradual decrease to $\mu = 0.09$. At 500 mN, there was a gradual decrease to a final friction coefficient of ~ 0.075. On WC/C, the track was smoother with no debris at the sides. SEM images and confocal microscopy showed clear differences in track width in tests at ≥ 100 mN, with WC/C being the widest and DLC being narrowest, although the Si-DLC coating performed relatively better at lower load.

Fig. 7.8 shows that although the friction at 500 mN is initially similar, there are clear differences thereafter with (1) a transition to high friction on DLC, (2) cyclic and variable friction on Si-DLC, and (3) gradual reduction on WC/C. The relative wear resistance of the three coatings was dependent on the type and severity of the test and the contact length scale as controlled by the applied load and test probe radius as the optimum coating mechanical properties (balance of load support and fracture resistance) vary with the test. The WC/C performed best in terms of higher L_{c1} and L_{c2} in the scratch tests, but in the nanofretting and reciprocating tests, the coating ranking changed, with the hardest coating showing the highest wear resistance and WC/C the least (Beake et al., 2015, 2021; Liskiewicz et al., 2013). In the reciprocating and nanofretting tests, the initial contact pressures were $\sim 11-14$ GPa, gradually decreasing with reciprocating cycles through wear.

The nanofretting and reciprocating tests were performed at loads well below those causing cracking in ramped scratch tests. At nano/microscale plastic deformation is a major contributor to apparent wear. In the nanofretting and reciprocating tests at low L/L_c and small scratch depths, the tensile stresses generated in the surface behind the moving probe were much lower than at higher loads. Under these conditions, the coating wear resistance was largely controlled by their resistance to plastic deformation (i.e., hardness). As the test severity increases, tensile stresses become important, and wear is more influenced by microcracking. SEM images of wear tracks show grooving wear and debris to the sides of the track on Si-DLC, which had the lowest L_{c1} in scratch tests with same $R = 25$ μm geometry probe.

Analytical modeling of nano- and microscratch data can provide an indication of the initial stress state developed in the reciprocating sliding tests. Differences in

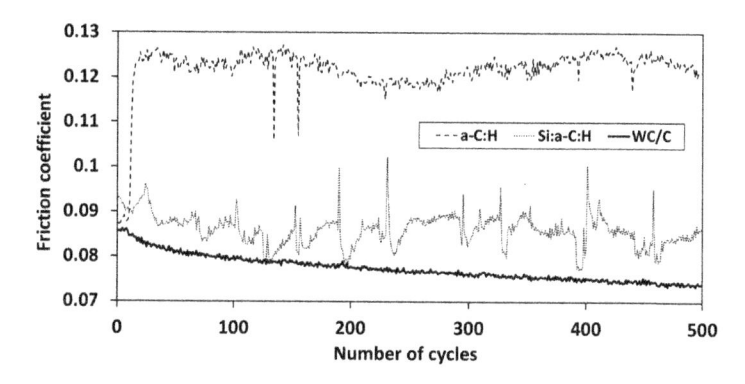

Figure 7.8 Friction evolution for three coating systems. Friction versus cycles at 500 mN.

reciprocating wear behavior between the three coatings can be understood with reference to the stress distribution during initial sliding contact, which is determined from the probe radii, applied load, friction coefficient, and mechanical properties of the indenter, coating, and substrate. The deformation in nano/microscale sliding tests on the DLC coatings has been modeled as single-asperity contact with plowing friction and no third body wear as the diamond indenters are precision polished to low surface roughness, and the DLC coating surfaces are also very smooth with R_a ~11–12 nm. This modeling treatment assumes that the coatings are monolayered, single-asperity sphere-on-flat contact and that the surfaces were perfectly smooth. Nevertheless, it has proved able to explain how contact size and coating properties combine to control the yield location in the various nano/microscale scratch tests, nanofretting tests, and reciprocating microscale tests. The results of the simulations show the regions where the developed von Mises stresses in the contact are greater than the yield stresses in these regions. The system is overloaded in these regions, and plastic flow is expected.

Analytical modeling of the initial stress distribution was able to explain differences in coating wear in the nanofretting experiments at 100 mN with a 5 μm end radius probe (Beake et al., 2021). At 100 mN, the contact was elastic on the DLC, but on Si-DLC within the coating, there was a small yielded region and a more extensive yielded region in the WC/C coating. The analytical results are consistent with differences in their critical load for yielding and in their on-load and residual depths in nano-scratch and in nanofretting tests. Despite the lower wear rate on the DLC coating, the contact was not fully elastic in the nanofretting test, presumably due to higher contact pressure at asperities on the coating surface. During sliding contact with a $R = 25$ μm diamond probe, the deformation was initially elastic or very close to it. The 10–500 mN applied loads used in the reciprocating tests ranged from well below those resulting in plastic deformation in a single microscratch with a probe of the same 25 mm nominal geometry (L_y ~ 400 mN), to slightly above, although at 500 mN, the residual depths were only <50 nm (Beake et al., 2015). At 500 mN, maximum von Mises stresses were located within the coatings, but they remain high within the substrate and due to its lower yield stress they result in yield. On the lower yield stress WC/C coating, the modeling indicated appreciable yielding occurring within the coating as well (9.6 GPa peak von Mises stress in coating at 400 mN) as the substrate. The modeling shows that despite the high stresses developed in the DLC and Si-DLC coatings, they remain below the coating yield stress. In microscratch tests with same nominal probe geometry, there was no noticeable differences in depth between the three coatings to ≥ 1500 mN, confirming the importance of the substrate properties on the deformation. Plastic deformation and wear reduced the initially very high (~14 GPa) pressures in the sliding contact. SEM track width measurements showed the largest decrease to ~4 GPa on the WC/C, and the smallest reduction to reach ~7 GPa on the DLC.

The shear stress distribution in sliding contact is asymmetric, with the maximum compressive stress in front of the probe and the maximum tensile stress located behind the probe. Cracking occurs when the maximum tensile stress exceeds the fracture strength of the coating (Chudoba et al., 2002, 2008; Hamilton & Goodman,

1966; Holmberg et al., 2007). In a scratch test with the same probe geometry ($R = 25$ μm), a load ≥ 1800 mN was required for cracking (L_{c1}) at the rear of the contact due to high tensile stress of ~ 6 GPa. The 10–500 mN loads used in the reciprocating tests were much lower than this. However, below the critical load crack formation is still possible with continued wear cycles due to low cycle fatigue. Schiffmann has observed that periodic loading in reciprocating sliding leads to an accumulation of plastic deformation and/or densification of the material that gradually increases the subsurface stress in the coating (Schiffmann, 2004; Schiffmann & Hieke, 2003). After a certain number of wear cycles, microcracks may be formed even though the critical load is not exceeded. Apparent wear at the microscale is therefore typically a combination of plastic deformation and real material removal (Schiffmann & Hieke, 2004; Stoyanov et al., 2010; Stoyanov & Chromik, 2017). Microscopy and friction measurements indicate that wear of the coatings was dominated by plastic deformation with wear being only a small contributor, particularly on the WC/C. The extent of wear is largely controlled by the coating resistance to plastic deformation (i.e., its hardness) rather than its resistance to fracture. The differences in relative wear of the coatings in the reciprocating microscale tests match those found in the nanofretting tests with a sharper probe, where wear resistance also correlated with hardness (Liskiewicz et al., 2013). Gee et al. also reported low wear resistance for WC/C and for other relatively soft coatings (MoST and Graphit-IC from Teer Coatings) in comparison with harder DLC coatings and ceramic coatings in reciprocating tests with a sharper $R = 1$ μm diamond probe (Gee et al., 2011).

At the start of the reciprocating tests, the friction coefficient was $\sim 0.08-0.10$. At 10 mN, there was greater variability in the friction coefficient between cycles than was found in tests at higher loads. Achanta and coworkers also reported higher variability in friction in smaller scale contact (Achanta et al., 2005, 2008; Drees et al., 2004). These authors considered that the local surface topography influences the frictional variability through two effects: (1) geometrical—local slope/ratchet mechanism and (2) adhesive—contact area. The small rise in friction during the first ~ 50 cycles for all three coatings before reaching an approximately constant value is likely to be topographical in origin, with increasing contact area and the difficulty in forming a friction reducing transfer layer in light multiasperity contact on rough surfaces (Holmberg et al., 2015). Santner and coworkers have described geometric changes in friction where sliding probes encounter topographic features (Meine et al., 2002a, 2002b; Santner et al., 2006a, 2006b). These authors observed sharp increases in friction sliding up steps and similar reduction sliding down steps.

At >10 mN, running-in effects influencing friction during the first few cycles are related to increasing contact area as asperities break down. Adhesion dominates friction when there is high surface conformity and the average gap between the surfaces is very low, so friction scales with contact area (Hsia et al., 2020). Achanta et al. explained the increasing friction during the fretting testing of a hard DLC coating against an Si_3N_4 ball under gross slip conditions (100 mN, $R = 2.5$ mm, displacement amplitude $= 300$ μm, 0.2 Hz) was due to changing roughness and breakdown of asperities on both surfaces. The limiting value was associated with reduced average asperity inclination in the multiasperity contact (Achanta et al., 2005). Subsequent constant friction was explained as a result of third body wear with microparticles rolling in the contact. This mechanism could explain the appearance of the wear track on Si-DLC at 500 mN, with grooving and little plowing and smoothening of the wear track that would reduce friction.

McMaster et al. showed similar trends in performance for DLC, Si-DLC, and W-doped DLC coatings on hardened M2 tool steel in 15,000 cycle fretting tests with 5 mm AISI 52100 steel balls at 20 N, 250 mm s^{-1}, 5 Hz (McMaster et al., 2023). In reciprocating wear tests (108,000 cycles, 4 N, 4 mm diameter Al$_2$O$_3$ ball, 1−36 Hz frequency), Bai et al. also reported similar behavior (Bai et al., 2021). Zhou et al. investigated the influence of substrate hardness (stainless steel or cemented carbide) and interlayer (presence or absence of AlTiN) on the behavior of DLC coatings in long duration reciprocating wear (Zhou et al., 2021). They found that high substrate hardness improved wear resistance during the early stage of abrasion, but the lower hardness steel, and the AlTiN interlayer, reduced the contact stress and lowered the tendency of sp^3 to sp^2 transition in the wear track.

In the absence of clear fracture, there was less influence of the local surface topography of the wear track on the mean friction coefficient at higher load. In the tests at 100−500 mN on WC/C and 50−200 mN on DLC, the friction continued to decrease with further wear cycles after the running-in stage. On WC/C, the wear track at 500 mN was very smooth with no debris at the edges, consistent with a reduction in the plowing component to the measured friction. When there is minimal third body wear, there is a correlation between track widening and a reduction in the plowing component of friction (Korres et al., 2013). The WC/C coating is reported (Oerliken Balzers product note, 2010) to have a microlamellar structure with alternating C- and WC-rich layers produced by rotation during deposition, which may shear easily, providing low friction, consistent with the smooth appearance of the wear track.

However, there was more influence of wear and fracture in tests at higher load on DLC and Si-DLC. Friction was higher and more variable at 500 mN on DLC, even in tests without dramatic fracturing regions. This is consistent with microcracking that may be the start of the low cycle fatigue process that is sensitive to areas of coating inhomogeneity. On Si-DLC, the mean friction was unchanged with load from 50−500 mN, although at ≥ 300 mN periodic changes indicating some microcracking were clear, consistent with SEM images showing abrasive/grooving wear.

In low-cycle repetitive nano-scratch tests on amorphous carbon films, the friction coefficient initially decreases due to the reduction in the plowing contribution, before, provided there is no adhesive or cohesive failure, stabilizing within a few cycles at around ~0.08 (Beake et al., 2021). On DLC, the greater reduction at 50−200 mN to reach values of 0.04−0.08 may be due to formation of a low-friction, predominantly graphitic tribolayer. In reciprocating sliding on DLC-DLC contacts, Holmberg et al. also found decreasing friction to reach $\mu \sim 0.04-0.06$, which was a greater decrease than in pin-on-disk tests (Holmberg et al., 2015). Tribolayer formation may be responsible for the friction reduction. In sliding on

DLC, a partially graphitized low-friction tribolayer can form (Singer et al., 2013). In a study of unrepeated and repeated reciprocating sliding, where each cycle was on a virgin region of the surface in the unrepeated sliding tests, Hsia et al. found that repeated sliding promotes the formation of a third body composed of compressed wear particles that stabilizes the friction (Hsia et al., 2020). This mechanism for friction reduction does not appear to be as effective at 500 mN where the abrupt increases soon after the start of the test may be a consequence of microcracking that results in more topographical influence on the friction. Initial increasing friction before levelling out has also been observed in nanofretting tests on DLC (Liskiewicz et al., 2013). Bai et al. contrasted the friction versus cycles behavior of DLC, Si-DLC, and W-doped DLC in reciprocating tests at 1–36 Hz (Bai et al., 2021) using the Raman I_D/I_G ratio (a measure of sp^2/sp^3 ratio) to determine the extent of graphitisation in the wear track. Decreasing friction with increasing velocity on DLC and W-doped DLC was associated with a higher I_D/I_G ratio and graphitization. In contrast, a periodic and higher friction was found for Si-DLC at 36 Hz consistent with a lack of change in I_D/I_G and an absence of graphitization for this coating.

On the assumption that the majority of the frictional losses occur through plastic deformation and cracking, the dissipated energy during reciprocating sliding contacts has been used as a measure of wear rate (Huq & Celis, 2002; Liskiewicz & Fouvry, 2005). This approach has mixed success with microscale reciprocating contacts. In nanofretting tests, the friction increases more rapidly with cycles on WC/C, which exhibited much greater wear than the other two coatings. Friction forces in nanofretting experiments were higher than in the reciprocating microscale tests, particularly for WC/C. The higher values found in the smaller scale tests reflect the dominant contribution of changing attack angle when wearing to greater depth for the sharper probe (Lafaye & Troyon, 2006). The correlation between friction and wear was less clear in the microscale single-asperity contacts at 500 mN where the highest wear was found for WC/C, which has the lowest friction. In reciprocating tests on a range of soft and hard coatings sliding against a $R = 1$ μm diamond probe, Gee et al. also reported that the microscale friction coefficient was 0.04−0.08, but with no particular correlation to either the macroscale friction or microscale wear (Gee et al., 2011). Overall, the similarities in the friction evolution in all these contacts with spheroconical diamond probes with 1−25 μm end radii can be understood in terms of the varying contributions of plowing, changing contact area and microwear as the various tests progress.

7.4 Conclusions

The thickness and structure of the passive oxide films formed on the biomaterials alloys play an important role in their nano- to microscale tribological behavior. Maintaining the oxide integrity is important in enabling an effective hard TTS layer to form and protect the underlying alloy. At smaller scale, there is greater microstructural variability and a more pronounced influence of surface roughness on the measured friction forces. With sphero-conical indenters, the probe sharpness and applied load influence not only the contact pressure but also the attack angle, which can result in dramatic transitions to more severe wear modes.

The tests on DLC coatings on hardened steel showed the importance of mechanical properties on the wear resistance. Analytical modeling of nano- and microscratch data provided an indication of the initial stress state developed in the reciprocating sliding tests, which could explain differences in reciprocating wear resistance. Under the test conditions, the deformation in the wear track was largely controlled by plastic deformation and hence hardness since microscale fatigue wear was only a small contributor. Si- and W-doped DLC coatings, which are softer than the undoped DLC, therefore showed lower wear resistance in reciprocating sliding.

References

Achanta, S., Drees, D., & Celis, J.-P. (2005). Friction and nanowear of hard coatings in reciprocating sliding at milli-Newton loads. *Wear, 259*, 719−729.

Achanta, S., Drees, D., & Celis, J.-P. (2008). Friction from nano to macroforce scales analysed by single and multiple-asperity contact approaches. *Surface and Coatings Technology, 202*, 6127−6135.

Bai, M., Yang, L., Li, J., Luo, L., Sun, S., & Inkson, B. (2021). Mechanical and tribological properties of Si and W doped diamond like carbon (DLC) under dry reciprocating sliding conditions. *Wear, 484−485*. Available from https://doi.org/10.1016/j.wear.2021.204046.

Beake, B. D., Harris, A. J., Liskiewicz, T. W., Wagner, J., McMaster, S. J., Goodes, S. R., Neville, A., & Zhang, L. (2021). Friction and electrical contact resistance in reciprocating nano-scale wear testing of metallic materials. *Wear, 474−475*203886.

Beake, B. D., & Liskiewicz, T. W. (2013). Comparison of nano-fretting and nano-scratch tests on biomedical materials. *Tribology International, 63*, 123−131. Available from https://doi.org/10.1016/j.triboint.2012.08.007.

Beake, B. D., Liskiewicz, T. W., Vishnyakov, V. M., & Davies, M. I. (2015). Development of DLC coating architectures for demanding functional surface applications through nano- and micro-mechanical testing. *Surface and Coatings Technology, 284*, 334−343. Available from https://doi.org/10.1016/j.surfcoat.2015.05.050.

Beake, B. D., McMaster, S. J., Liskiewicz, T. W., & Neville, A. (2021). Influence of Si- and W- doping on micro-scale reciprocating wear and impact performance of DLC coatings on hardened steel. *Tribology International, 160*. Available from https://doi.org/10.1016/j.triboint.2021.107063, http://www.elsevier.com/inca/publications/store/3/0/4/7/4.

Bernoulli, D., Wyss, A., Raghavan, R., Thorwarth, K., Hauert, R., & Spolenak, R. (2015). Contact damage of hard and brittle thin films on ductile metallic substrates: an analysis of diamond-like carbon on titanium substrates. *Journal of Materials Science, 50*(7), 2779−2787. Available from https://doi.org/10.1007/s10853-015-8833-3.

Budinski, J. G. (1991). Tribological properties of titanium alloys. *Wear, 151*, 203−217.

Buford, A., & Goswami, T. (2004). Review of wear mechanisms in hip implants: Paper I— General. *Materials & Design, 25*, 385−393.

Burris, D. L., & Sawyer, W. G. (2009). Addressing practical challenges of low friction coefficient measurements. *Tribology Letters, 35*(1), 17–23. Available from https://doi.org/10.1007/s11249-009-9438-2.

Cai, W., & Bellon, P. (2013). Subsurface microstructure evolution and deformation mechanism of Ag–Cu eutectic alloy after dry sliding wear. *Wear, 303*(1–2), 602–610. Available from https://doi.org/10.1016/j.wear.2013.04.006.

Callen, B. W., Lowenberg, B. F., Lugowski, S., Sodhi, R. N. S., & Davies, J. E. (1995). Nitric acid passivation of Ti6A14V reduces thickness of surface oxide layer and increases trace element release. *Journal of Biomedical Materials Research, 29*(3), 279–290. Available from https://doi.org/10.1002/jbm.820290302.

Chudoba, T., Schwarzer, N., & Richter, F. (2002). Steps towards a mechanical modelling of layered systems. *Surface and Coatings Technology, 154*, 140.

Dong, H., & Bell, T. (1999). Tribological behaviour of alumina sliding against Ti6Al4V in unlubricated contact. *Wear, 225–229*(II), 874–884. Available from https://doi.org/10.1016/s0043-1648(98)00407-4.

Drees, D., Celis, J.-P., & Achanta, S. (2004). Friction of thin coatings on three length scales under reciprocating sliding. *Surface and Coatings Technology, 188–189*, 511–518.

Ebramzadeh, E., Billi, F., Sangiorgio, S. N., Mattes, S., Schmoelz, W., & Dorr, L. (2005). Simulation of fretting wear at orthopaedic implant interfaces. *Journal of Biomechanical Engineering, 127*(3), 357–363. Available from https://doi.org/10.1115/1.1894121.

Echeverrigaray, F. G., de Mello, S. R. S., Leidens, L. M., Boeira, C. D., Michels, A. F., Braceras, I., & Figueroa, C. A. (2020). Electrical contact resistance and tribological behaviors of self-lubricated dielectric coating under different conditions. *Tribology International, 143*. Available from https://doi.org/10.1016/j.triboint.2019.106086.

Erdemir, A., & Donnet, C. (2006). Tribology of diamond-like carbon films: Recent progress and future prospects. *Journal of Physics D: Applied Physics, 39*(18), R311–R327. Available from https://doi.org/10.1088/0022-3727/39/18/r01.

Fouvry, S., Kapsa, P., & Vincent, L. (1996). Quantification of fretting damage. *Wear, 200*, 186–205.

Fouvry, S., Laporte, J., Perrinet, O., Jedrzejczyk, P., Graton, O., Alquier, O., & Sautel, J. (2017). France Fretting wear of low current electrical contacts: Quantification of electrical endurance. *Electrical Contacts, Proceedings of the Annual Holm Conference on Electrical Contacts, Institute of Electrical and Electronics Engineers Inc.* Available from https://doi.org/10.1109/HOLM.2017.8088056, 9781538610916, 1–11.

Fouvry, S., Paulin, C., & Deyber, S. (2009). Impact of contact size and complex gross–partial slip conditions on Ti–6Al–4V/Ti–6Al–4V fretting wear. *Tribology International, 42*, 461–474.

Gåhlin, R., Larsson, M., & Hedenqvist, P. (2001). Me-C:H coatings in motor vehicles. *Wear, 249*, 302–309.

Gee, M. G., Nunn, J. W., Muniz-Piniella, A., & Orkney, L. P. (2011). Micro-tribology experiments on engineering coatings. *Wear, 271*(9–10), 2673–2680. Available from https://doi.org/10.1016/j.wear.2011.02.031.

Goldberg, J. R., & Gilbert, J. L. (1997). Electrochemical response of CoCrMo to high-speed fracture of its metal oxide using an electrochemical scratch test method. *Journal of Biomedical Materials Research, 37*(3), 421–431. Available from https://doi.org/10.1002/(SICI)1097-4636(19971205)37:3%3C421::AID-JBM13%3E3.0.CO;2-E.

Hamilton, G. M., & Goodman, L. E. (1966). The stress field created by a circular sliding contact. *Journal of Applied Mechanics, 33*(2), 371–376. Available from https://doi.org/10.1115/1.3625051.

Hanlon, T., Chokshi, A. H., Manoharan, M., & Suresh, S. (2005). Effects of grain refinement and strength on friction and damage evolution under repeated sliding contact in nanostructured metals. *International Journal of Fatigue*, *27*(10−12), 1159−1163. Available from https://doi.org/10.1016/j.ijfatigue.2005.06.036.

Hokkirigawa, K., Kato, K., & Li, Z. Z. (1988). The effect of hardness on the transition of the abrasive wear mechanism of steels. *Wear*, *123*, 241−251.

Holmberg, K., Laukkanen, A., Ronkainen, H., Waudby, R., Stachowiak, G., Wolski, M., Podsiadlo, P., Gee, M., Nunn, J., Gachot, C., & Li, L. (2015). Topographical orientation effects on friction and wear in sliding DLC and steel contacts, part 1: Experimental. *Wear*, *330−331*, 3−22. Available from https://doi.org/10.1016/j.wear.2015.02.014.

Holmberg, K., Ronkainen, H., Laukkanen, A., & Wallin, K. (2007). Friction and wear of coated surfaces — scales, modelling and simulation of tribomechanisms. *Surface and Coatings Technology*, *202*(4−7), 1034−1049. Available from https://doi.org/10.1016/j.surfcoat.2007.07.105.

Hsia, F. C., Elam, F. M., Bonn, D., Weber, B., & Franklin, S. E. (2020). Wear particle dynamics drive the difference between repeated and non-repeated reciprocated sliding. *Tribology International*, *142*. Available from https://doi.org/10.1016/j.triboint.2019.105983, http://www.elsevier.com/inca/publications/store/3/0/4/7/4.

Huq, M. Z., & Celis, J.-P. (2002). Expressing wear rate in sliding contacts based on dissipated energy. *Wear*, *252*(5−6), 375−383. Available from https://doi.org/10.1016/s0043-1648(01)00867-5.

Jiang, J., & Arnell, R. D. (1998). On the running-in behaviour of diamond-like carbon coatings under the ball-on-disk contact geometry. *Wear*, *217*(2), 190−199. Available from https://doi.org/10.1016/s0043-1648(98)00178-1.

Kamps, T. J., Walker, J. C., & Plint, A. G. (2017). In-situ stylus profilometer for high frequency reciprocating tribometer. *Surface Topography: Metrology and Properties*, 5034004.

Kassman, Å., & Jacobson, S. (1993). Surface damage, adhesion and contact resistance of silver plated copper contacts subjected to fretting motion. *Wear*, *165*(2), 227−230. Available from https://doi.org/10.1016/0043-1648(93)90339-n.

Kitsunai, H., Kato, K., Hokkirigawa, K., & Inoue, H. (1990). The transitions between microscopic wear during repeated sliding friction observed by a scanning electron microscope tribosystem. *Wear*, *135*, 237−249.

Korres, S., Feser, T., & Dienwiebel, M. (2013). A new approach to link the friction coefficient with topography measurements during plowing. *Wear*, *303*, 202−210.

Lafaye, S., & Troyon, M. (2006). On the friction behaviour in nanoscratch testing. *Wear*, *261*(7−8), 905−913. Available from https://doi.org/10.1016/j.wear.2006.01.036.

Lanigan, J. L., Wang, C., Morina, A., & Neville, A. (2016). Repressing oxidative wear within Si doped DLCs. *Tribology International*, *93*, 651−659.

Laporte, J., Perrinet, O., & Fouvry, S. (2015). Prediction of the electrical contact endurance of silver-plated coatings subject to fretting wear, using a friction energy density approach. *Wear*, *330−331*, 170−181.

Lawes, S. D. A., Hainsworth, S. V., & Fitzpatrick, M. E. (2010). Impact wear testing of diamond-like carbon films for engine valve-tappet surfaces. *Wear*, *268*, 1303−1308.

Lee, W.-S., Chen, T.-S., & Huang, S.-C. (2012). Impact deformation behaviour of Ti-6Al-4V alloy in the low temperature regime. *Journal of Nuclear Materials*, *402*, 1−7.

Li, X., Wang, X., Bondokov, R., Morris, J., An, Y. H., & Sudarshan, T. S. (2004). Micro/nanoscale mechanical and tribological characterisation of SiC for orthopaedic applications. *Journal of Biomedical Materials Research Part B*, *72*, 353−361.

Li, X. X., Zhang, Q. Y., Zhou, Y., Liu, J. Q., Chen, K. M., & Wang, S. Q. (2016). Mild and severe wear of titanium alloys. *Tribology Letters, 61*(2). Available from https://doi.org/10.1007/s11249-015-0637-8.

Liskiewicz, T. W, Beake, B. D., & Smith, J. F. (2010). In situ accelerated micro-wear - A new technique to fill the measurement gap. *Surface and Coatings Technology, 205,* 1455−1459.

Liskiewicz, T. W., Beake, B. D., Schwarzer, N., & Davies, M. I. (2013). Short note on improved integration of mechanical testing in predictive wear models. *Surface and Coatings Technology, 237,* 212−218. Available from https://doi.org/10.1016/j.surfcoat.2013.07.044.

Liskiewicz, T. W., & Fouvry, S. (2005). Development of a friction energy capacity approach to predict the surface coating endurance under complex oscillating sliding conditions. *Tribology International, 38,* 69−79.

Liskiewicz, T. W., Fouvry, S., & Wendler, B. (2005). Development of a Wöhler-like approach to quantify the Ti(CxNy) coatings durability under oscillating sliding conditions. *Wear, 259,* 835−841.

Liskiewicz, T. W., Kubiak, K., & Comyn, T. (2017). Nano-indentation mapping of fretting-induced surface layers. *Tribology International, 108,* 186−193.

Liu, Y., Liskiewicz, T. W., & Beake, B. D. (2019). Dynamic changes of mechanical properties induced by friction in the Archard wear model. *Wear, 428−429,* 366−375.

Maloney, W. J., Jasty, M., Burke, D. W., O'Connor, D. O., Zalenski, E. B., Bragdon, C., & Harris, W. H. (1989). Biomechanical and histologic investigation of cemented total hip arthroplaties. A study of autopsy-retrieved femurs after in vivo cycling. *Clinical Orthopaedics and Related Research, 249,* 129−140.

Martínez-Nogués V. (2016). *Nanoscale tribocorrosion of CoCrMo alloys.* PhD Thesis, nCATS.

Martinez-Nogues, V., Nesbitt, J. M., Wood, R. J. K., & Cook, R. B. (2016). Nano-scale wear characterization of CoCrMo biomedical alloys. *Tribology International, 93,* 563−572. Available from https://doi.org/10.1016/j.triboint.2015.03.037.

McMaster, S. J., Kosarieh, S., Liskiewicz, T. W., Neville, A., & Beake, B. D. (2023). Utilising H/E to predict fretting wear performance of DLC coating systems. *Tribology International, 185.* Available from https://doi.org/10.1016/j.triboint.2023.108524, http://www.elsevier.com/inca/publications/store/3/0/4/7/4.

Meine, K., Schneider, T., Spaltmann, D., & Santner, E. (2002a). The influence of roughness on friction Part II. The influence of multiple steps. *Wear, 253,* 733−738.

Meine, K., Schneider, T., Spaltmann, D., & Santner, E. (2002b). The influence of roughness on friction Part I: The influence of a single step. *Wear, 253*(7−8), 725−732. Available from https://doi.org/10.1016/S0043-1648(02)00159-X.

Mirghany, M., & Jin, Z. M. (2004). Prediction of scratch resistance of cobalt chromium alloy bearing surface, articulating against ultra-high molecular weight polyethylene, due to third body wear particles. *Proceedings of the Institution of Mechanical Engineers, 218,* 41−50.

Moharrami, N., Langton, D. J., Sayginer, O., & Bull, S. J. (2013). Why does titanium alloy wear cobalt chrome alloy despite lower bulk hardness: A nanoindentation study? *Thin Solid Films, 549,* 79−86. Available from https://doi.org/10.1016/j.tsf.2013.06.020.

Molinari, A., Straffelini, G., Tesi, B., & Bacci, T. (1997). Dry sliding wear mechanisms of the Ti6Al4V alloy. *Wear, 208*(1−2), 105−112. Available from https://doi.org/10.1016/s0043-1648(96)07454-6.

Mutafov, P., Lanigan, J., Neville, A., Cavaleiro, A., & Polcar, T. (2014). DLC-W coatings tested in combustion engine — Frictional and wear analysis. *Surface and Coatings Technology, 260,* 284−289, 2014.

Najjer, D., Behnamghader, A., Iost, A., & Migaud, H. (2000). Influence of a foreign body on the wear of metallic femoral heads and polyethylene acetabular cups of total hip prostheses. *Journal of Materials Science, 35*, 4583−4588.

Oerliken Balzers Technical Note. (2010). *Coated components: greater performance and reliability* (3rd Edition).

Pourzal, R., Theissmann, R., Williams, S., Gleising, B., Fisher, J., & Fischer, A. (2009). Subsurface changes of a MoM hip implant below different contact zones. *Journal of the Mechanical Behavior of Biomedical Materials, 2*(2), 186−191. Available from https://doi.org/10.1016/j.jmbbm.2008.08.002.

Prasad, V.S., Misra, P., & Nagaraju, J. (2011). An experimental study to show the behaviour of electrical contact resistance and coefficient of friction at low current sliding electrical interfaces, *57th Holm Conference on Electrical Contacts (Holm)*, IEEE 2011. Available from https://doi.org/10.1109/HOLM.2011.6034813.

Raeymaekers, B., Helm, S., Brunner, R., Fanslau, E. B., & Talke, F. E. (2010). Investigation of fretting wear at the dimple/gimbal interface in a hard disk drive suspension. *Wear, 268*(11−12), 1347−1353. Available from https://doi.org/10.1016/j.wear.2010.02.003.

Raimondi, M. T., Vena, P., & Pietrabissa, R. (2001). Quantitative evaluation of the prosthetic head damage induced by microscopic third-body particles in total hip replacement. *Journal of Biomedical Materials Research Part B: Applied Biomaterials, 58*, 436−448.

Rainforth, W. M., Stevens, R., & Nutting, J. (1992). Deformation structures induced by sliding contact. *Philosophical Magazine A, 66*, 621−641.

Rainforth, W. M., Zeng, P., Ma, L., Valdez, A. N., & Stewart, T. (2012). Dynamic surface microstructural changes during tribological contact that determine the wear behaviour of hip prostheses: Metals and ceramics. *Faraday Discussions, 156*, 41−57. Available from https://doi.org/10.1039/c2fd00002d.

Ramírez, G., Jiménez-Piqué, E., Mestra, A., Vilaseca, M., Casellas, D., & Llanes, L. (2015). A comparative study of the contact fatigue behaviour and associated damage micromechanisms of TiN- and WC:H- coated cold-work tool steel. *Tribology International, 88*, 263−270.

Rybiak, R., Fouvry, S., Liskiewicz, T., & Wendler, B. (2008). Fretting wear of a TiN PVD coating under variable relative humidity conditions—Development of a 'composite' wear law. *Surface and Coatings Technology, 202*(9), 1753−1763. Available from https://doi.org/10.1016/j.surfcoat.2007.07.103.

Santner, E., Klaffke, D., Meine, K., Polaczyk, C., & Spaltmann, D. (2006a). Effects of friction on topography and vice versa. *Wear, 261*, 101−106.

Santner, E., Klaffke, D., Meine, K., Polaczyk, C., & Spaltmann, D. (2006b). Demonstration of topography modification by friction processes and vice versa. *Tribology International, 39*(5), 450−455. Available from https://doi.org/10.1016/j.triboint.2005.04.029.

Sauger, E., Fouvry, S., Ponsonnet, L., Kapsa, P., Martin, J. M., & Vincent, L. (2000). Tribologically transformed structure in fretting. *Wear, 245*, 39−52.

Sauger, E., Ponsonnet, L., Martin, J. M., & Vincent, L. (2000). Study of the tribologically transformed structure created during fretting tests. *Tribology International, 33*, 743−750.

Schiffmann, K. I. (2004). Phenomena in microwear experiments on metal-free and metal-containing diamond-like carbon coatings: Friction, wear, fatigue and plastic deformation. *Surface and Coatings Technology, 177−178*, 453−458. Available from https://doi.org/10.1016/j.surfcoat.2003.08.064, http://www.journals.elsevier.com/surface-and-coatings-technology/.

Schiffmann, K. I., & Hieke, A. (2003). Analysis of microwear experiments on thin DLC coatings: Friction, wear and plastic deformation. *Wear, 254*(5−6), 565−572. Available from https://doi.org/10.1016/S0043-1648(03)00188-1.

Singer, I. L., Dvorak, S. D., Wahl, K. J., & Scharf, T. W. (2013). Role of third bodies in friction and wear of protective coatings. *Journal of Vacuum Science & Technology A, Vacuum, Surfaces, and Films: An Official Journal of the American Vacuum Society, 21*, S232−S240.

Singh, A., Dao, M., Lu, L., & Suresh, S. (2011). Deformation, structural changes and damage evolution in nanotwinned copper under repeated frictional contact sliding. *Acta Materialia, 59*(19), 7311−7324. Available from https://doi.org/10.1016/j.actamat.2011.08.014.

Sittig, C., Textor, M., Spencer, N. D., Wieland, M., & Vallotton, P. H. (1999). Surface characterization of implant materials c. p. Ti, Ti-6Al-7Nb and Ti-6Al-4V with different pretreatments. *Journal of Materials Science: Materials in Medicine, 10*(1), 35−46. Available from https://doi.org/10.1023/A:1008840026907.

Stoyanov, P., & Chromik, R. R. (2017). Scaling effects on materials tribology: From macro to micro scale. *Materials, 10*(5). Available from https://doi.org/10.3390/ma10050550, http://www.mdpi.com/1996-1944/10/5/550/pdf.

Stoyanov, P., Chromik, R. R., Gupta, S., & Lince, J. R. (2010). Micro-scale sliding contacts on Au and Au-MoS$_2$ coatings. *Surface and Coatings Technology, 205*(5), 1449−1454. Available from https://doi.org/10.1016/j.surfcoat.2010.07.026.

Sun, D., Wharton, J. A., & Wood, R. J. K. (2009). Micro-abrasion mechanisms of cast CoCrMo in simulated body fluids. *Wear, 267*, 1845−1855.

Tardio, S., Abel, M. L., Carr, R. H., Castle, J. E., & Watts, J. F. (2015). Comparative study of the native oxide on 316L stainless steel by XPS and ToF-SIMS. *Journal of Vacuum Science and Technology A: Vacuum, Surfaces, and Films, 33*(5). Available from https://doi.org/10.1116/1.4927319, http://scitation.aip.org/content/avs/journal/jvsta.

Tian, H., Saka, N., & Rabinowicz, E. (1991). Friction and failure of electroplated sliding contacts. *Wear, 142*, 57−85.

Wänstrand, O., Larsson, M., & Hedenqvist, P. (1999). Mechanical and tribological evaluation of PVD WC/C coatings. *Surface and Coatings Technology, 111*, 247−254, 1999.

Waterhouse, R. B., & Taylor, D. E. (1974). Fretting debris and the delamination theory of wear. *Wear, 29*(3), 337−344. Available from https://doi.org/10.1016/0043-1648(74)90019-2.

Williams, J. A., & Le, H. R. (2006). Tribology and MEMS. *Journal of Physics D: Applied Physics, 39*(12), R201−R214. Available from https://doi.org/10.1088/0022-3727/39/12/R01.

Wilson, G.M. (2008). An investigation of thin amorphous carbon-based sputtered coatings for MEMS and Micro-engineering applications, *PhD Thesis*, Aston University.

Wilson, G. M., Smith, J. F., & Sullivan, J. L. (2008). A nanotribological study of thin amorphous C and Cr doped amorphous C coatings. *Wear, 265*, 1633−1641, 2008.

Wilson, G. M., Smith, J. F., & Sullivan, J. L. (2009). A DOE nano-tribological study of thin amorphous carbon-based films. *Tribology International, 42*, 220−228.

Wilson, G. M., & Sullivan, J. L. (2009). An investigation into the effect of film thickness on nanowear with amorphous carbon-based coatings. *Wear, 266*, 1039−1043.

Zeng, P., Rana, A., Thompson, R., & Rainforth, W. M. (2015). Subsurface characterisation of wear on mechanically polished and electro-polished biomedical grade CoCrMo. *Wear, 332−333*, 650−661.

Zhou, Y., Ma, W., Geng, J., Wang, Q., Rao, L., Qian, Z., Xing, X., & Yang, Q. (2021). Exploring long-run reciprocating Wear of diamond-like carbon coatings: Microstructural, morphological and tribological evolution. *Surface and Coatings Technology, 405*. Available from https://doi.org/10.1016/j.surfcoat.2020.126581.

Zhu, X., Zhang, S., Zhang, L., He, Y., Zhang, X., & Kang, X. (2023). Frictional behaviour and wear mechanism of Ag/MoS2/WS2 composite under reciprocating microscale sliding. *Tribology International, 185*.

Zivic, F., Babic, M., Mitrovic, S., & Vencl, A. (2011). Continuous control as alternative route for wear monitoring by measuring penetration depth during linear reciprocating sliding of Ti6Al4V alloy. *Journal of Alloys and Compounds, 509*(19), 5748−5754. Available from https://doi.org/10.1016/j.jallcom.2011.02.158.

Nanowear by atomic force microscopy

8

Jiaxin Yu[1], Linmao Qian[2] and Hongtu He[1]
[1]Key Laboratory of Testing Technology for Manufacturing Process, Ministry of Education, Southwest University of Science and Technology, Mianyang, P.R. China, [2]Tribology Research Institute, The State Key Laboratory of Rail Vehicle System, School of Mechanical Engineering, Southwest Jiaotong University, Chengdu, P.R. China

8.1 Single-asperity nanotribology by atomic force microscopy

Surface forces and the corresponding surface phenomena at nanoscale are usually considered as the dominated role to enable or hinder the functionality of a nanodevice. As one example, silicon-based nanoelectromechanical systems (NEMS) usually exhibited undesirable high adhesion, friction, and wear at nanoscale during its operation process, which can cause the degradation of the stability and lifetime of those devices (Kim et al., 2007; Maboudian & Howe, 1997). Therefore it is of great importance to understand the nanoscale adhesion, friction, and wear of Si materials for NEMS application and thus finding possible ways to improve the stability and lifetime of NEMS devices. Generally, most tribology tests are conducted at macroscale where multiple-asperity contacts are made due to the surface roughness of two contact surfaces, while the nanotribology is aiming to reveal the friction, adhesion, lubrication, and wear at contacts of nanometer sizes where the single-asperity contact is usually made. Single-asperity contact measurements have been a very useful tool in the nanotribology studies, because there is a well-defined contact pressure and contact area, which can avoid the uncertainty of multiple-asperities contact at macroscale. This can also facilitate the execution of meaningful comparisons between experiment, theory, and simulation in nanotribology tests. Furthermore, based on the nanotribology behaviors of individual asperities, the tribological behavior of more complex, multiasperity interfaces at nano/macroscale can be well understood or predicted (Szlufarska et al., 2008).

Atomic force microscopy (AFM) is the most widely used tool for single-asperity tribology studies at nanoscale, which can be used to reveal tribology at a fundamental level since it can provide a controllable and single-asperity contact between the tip and the sample. In addition, during the single-asperity nanotribology measurement, the forces and displacements can be measured with atomic-level precision and accuracy, and the environmental conditions can be precisely controlled over a wide range. The general design of the AFM has been reviewed in many literature publications (Alsteens et al., 2017; Krieg et al., 2019; Tseng, 2011), so we will focus on those aspects that are critical for single-asperity nanotribology studies Fig. 8.1.

In the AFM, a sharp tip, with a radius typically between 10 and 100 nm, is integrated with an AFM cantilever near its free end (Fig. 8.1). After the AFM tip is brought into close enough to the sample's surface, the forces between the tip and sample can result in the deflections of the cantilever. The cantilever can bend vertically (i.e., toward and/or away from the sample) in response to attractive and/or repulsive forces acting on the tip. This vertical deflection of the cantilever from its equilibrium position is proportional to the normal load applied to the tip by the AFM cantilever. In contrast, during the nanoscratch and/or nanowear tests, the lateral or friction forces will cause the twisting of the AFM cantilever from its equilibrium position, where this tilting can ensure the tip still make contact with the sample during the nanoscratch and/or nanowear tests (Britz-Grell et al., 2021; Hauser et al., 2017). In additions, single-asperity nanotribology by AFM can be performed in a variety of environments such as ambient air, controlled atmosphere (Wu et al., 2021), liquids (Cai et al., 2022), or ultrahigh vacuum (UHV) (Franceschi et al., 2023).

Regardless of numerous AFM-related studies focused on the adhesion and topography measurements, nanofriction and nanoscratch (or nanowear) have also been studied by AFM in single-asperity nanotribological research studies. Generally, both the single-pass and reciprocating multipass line scanning can be performed by AFM. The single-pass scanning mode is more like a nanoscratch test, which can simulate the ultraprecision cutting process; in contrast, the reciprocating multipass scanning mode can cause the nanowear of substrate materials, which can be used to reveal the material removal mechanism by a single abrasive nanoparticle in precision polishing. To compare the nanoscale friction and wear in single-pass and multipass sliding, Yu et al. (2015) investigated the nanoscale friction and wear of phosphate glass with the variation of normal load and number of friction cycles

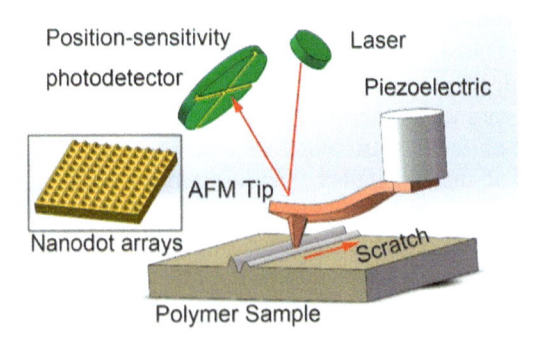

Figure 8.1 Schematic of a typical nanoscratch (wear) tests on a polymer sample by atomic force microscopy.
Source: From Yan et al. (2019). Scratch on polymer materials using AFM tip-based approach: A Review. *Polymers, 11*(10), 1590. https://doi.org/10.3390/polym11101590.

upon a spherical diamond AFM tip, and it was found that in the single-pass nanoscratch of phosphate glass, the friction coefficient and wear depth showed strongly load dependence, compared with those in the multipass reciprocating sliding, and the critical transition of friction mechanism from interfacial friction dominated to plowing dominated friction occurred at a lower normal load in the single-pass nanoscratch, compared with that in the multipass reciprocating sliding (Fig. 8.2). In the multipass reciprocating sliding, the interfacial friction dominated in all friction

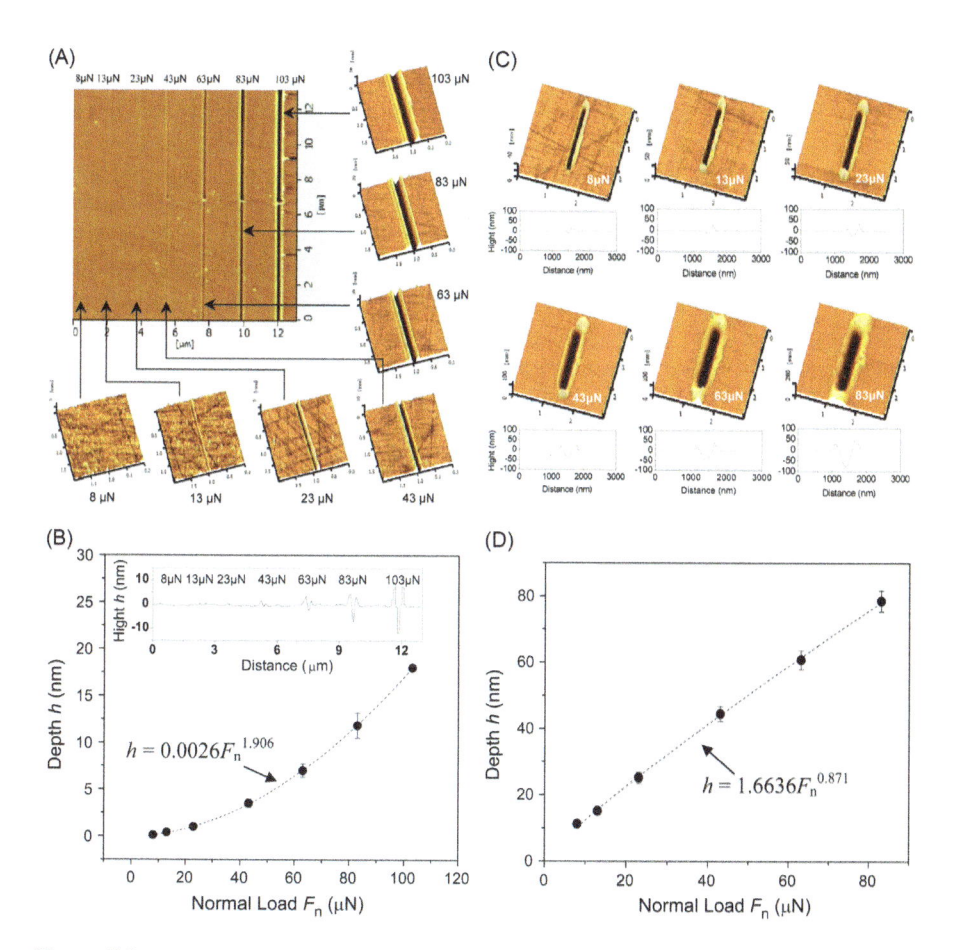

Figure 8.2 Atomic force microscopy topography images and average depth of wear mark on phosphate glass surface under various normal loads, (A and B) single-pass nanoscratch scars, and (C and D) multipass reciprocating wear scars. The dotted lines in (B) and (D) are the fitting curves. The average wear depths are measured from the cross-sectional profile curves of scars in (B) and (C).
Source: From Yu, J., Hu, H., Jia, F., Yuan, W., Zang, H., Cai, Y., & Ji, F. (2015). Quantitative investigation on single-asperity friction and wear of phosphate laser glass against a spherical AFM diamond tip. *Tribology International*, *81*, 43–52. https://doi.org/10.1016/j.triboint.2014.07.020.

cycles under low load conditions, and with the increase in the number of sliding cycles, the role of interfacial friction mechanism became more significant. However, under the high load with increasing cycles, the friction mechanism transformed from the plowing dominated friction to the combination of interfacial friction and plowing friction. Due to the plowing contribution becoming more under the high load with increasing cycles, wear depth of phosphate glass showed strongly cycle dependence under the high load conditions, compared with that under the low load conditions.

To understand the single-asperity contact behavior in Si-based MEMS/NEMS, Bhushan and coworkers (Bhushan et al., 1995; Zhao & Bhushan, 1998) studied the nanowear behavior of single-crystalline silicon with an AFM and a diamond tip, and they found that the nanowear of silicon was mainly due to the abrasive wear with plastic deformation. Later, Yu et al. (2009a) investigated the tangential nanofretting behaviors of monocrystalline silicon (100) with an AFM and spherical SiO_2 tips in atmosphere and vacuum conditions, and it was found that since the adhesion force and the applied normal load in nanofretting were at the same scale, the adhesion force can induce the increase in the maximum static friction force and thus prevent the contact pair from slipping. In addition, in contrast to fretting wear, the generation of hillocks on silicon surface was observed in nanofretting under the given conditions, and the height of the hillocks on silicon surfaces first increased and then attained a constant value with the increase in the displacement amplitude in slip regime. Further analyses revealed that compared with chemical reactions at sliding interfaces, the amorphization of crystalline silicon during nanoscratching played the dominated role in the formation of hillocks on silicon surface during the nanofretting in vacuum (Yu et al., 2009b; Yu, Li, et al., 2012). Even though both the formation of hillocks and groove on silicon surface depended on the applied load, the formation conditions were completely different: the generation of hillocks usually occurred at low load, and the formation of grooves occurred at high load. As the nominal Hertzian contact pressure between the diamond tip and silicon surface increased to a value closed to the hardness of silicon, the grooves would be observed in nanofretting wear of silicon instead of the hillocks.

The nanowear of solid materials upon AFM tip rubbing or sliding depends on not only intrinsic properties of solid materials such as hardness and bond energy but also many extrinsic factors of the sliding interfaces involving counter-surface and adsorbent. Take the silicon as one example, the silicon surface can resist mechanical wear as long as the contact pressure was lower than the hardness of silicon regardless of the counter-surface chemistry (diamond or SiO_2) and ambient gas type (vacuum, N_2, O_2, air) (Yu, Kim, et al., 2012). In these conditions, a hillock can be formed on the sliding contact region at silicon surface. However, with the increase in relative humidity (RH), the tribochemical wear of the silicon surface took place even at a contact pressure that is much lower than the hardness, which is due to the "solid-like" water layer that can facilitate the Si−O bond dissociation at silicon interfaces. The increase in nanowear volume of silicon was found when the RH increased from 0% to ∼50% RH, but the nanowear volume of silicon was found to decrease below the detection limit of AFM when the RH increased further to above 85% or in water (Fig. 8.3) (Wang et al., 2015). This can be explained by that at high RHs or liquid water conditions, the role of the formation of "Si−O−Si" chemical bonds (bridges) at sliding interface was different, where such tribochemical wear of silicon surface was suppressed due to the thickness of the adsorbed interfacial water layer at high RHs or in liquid water was too thick so that the formation of "Si−O−Si" chemical bonds (bridges) at sliding interface was prevented. These findings imply that the tribochemical wear of silicon surface at nanoscale depends on the formation and dissociation of "Si−O−Si" chemical bonds (bridges) at sliding interface, and such mechanism has been used to explain the nanoscale tribochemical wear of silicon surface under various conditions such as the dependence of the sliding speed (Chen, He, et al., 2015), the presence of surface oxide layer (Chen et al., 2016), crystal plane orientation (Xiao et al., 2017), and coadsorption of water and alcohol vapor (Chen, Yang, et al., 2015).

Even though the tribochemical reactions induced nanoscale wear will not cause the direct subsurface damage in crystalline materials since the nominal contact pressure is relatively small (Fig. 8.3), but some subsurface damage can exist in the case of tribochemical wear of noncrystalline materials. For instance, after the mechanochemical wear of oxide glass materials, the subsurface densification underneath the mechanochemical wear mark of oxide glass materials was found by the fact that the subsurface densification can be fully recovered by sub-T_g annealing treatment (Fig. 8.4); this is because the subsurface densification is generally caused by the interfacial force transferring into the subsurface region, and it can be fully relaxed and recovered at sub-T_g annealing conditions (He et al., 2020). Since the nanoscratch-induced deformation of oxide glass in the completely ductile range could be classified into three independent aspects, including plastic damage, densification, and chemistry-enhanced material removal (Fu et al., 2018), therefore it is

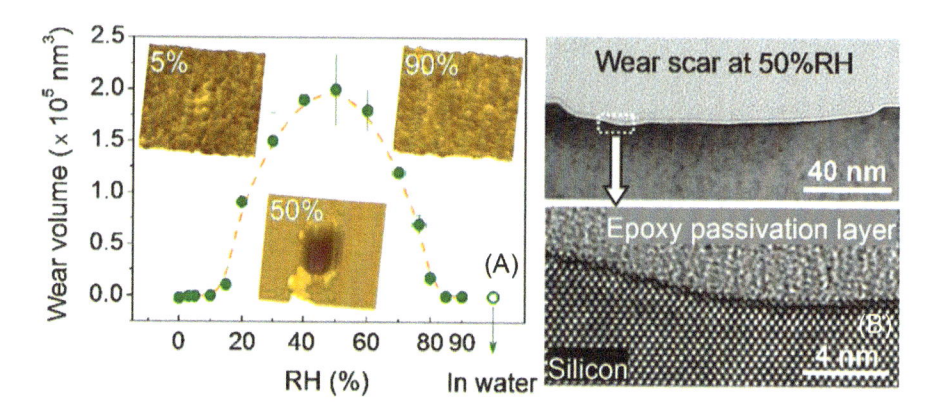

Figure 8.3 (A) Nanowear volume of silicon as a function of RH. (B) TEM image showing the cross section of the nanowear mark on silicon surfaces. *RH*, Relative humidity.
Source: From Wang, X., Kim, S. H., Chen, C., Chen, L., He, H., & Qian, L. (2015). Humidity dependence of tribochemical wear of monocrystalline silicon. *ACS Applied Materials and Interfaces*, 7(27), 14785−14792. https://doi.org/10.1021/acsami.5b03043.

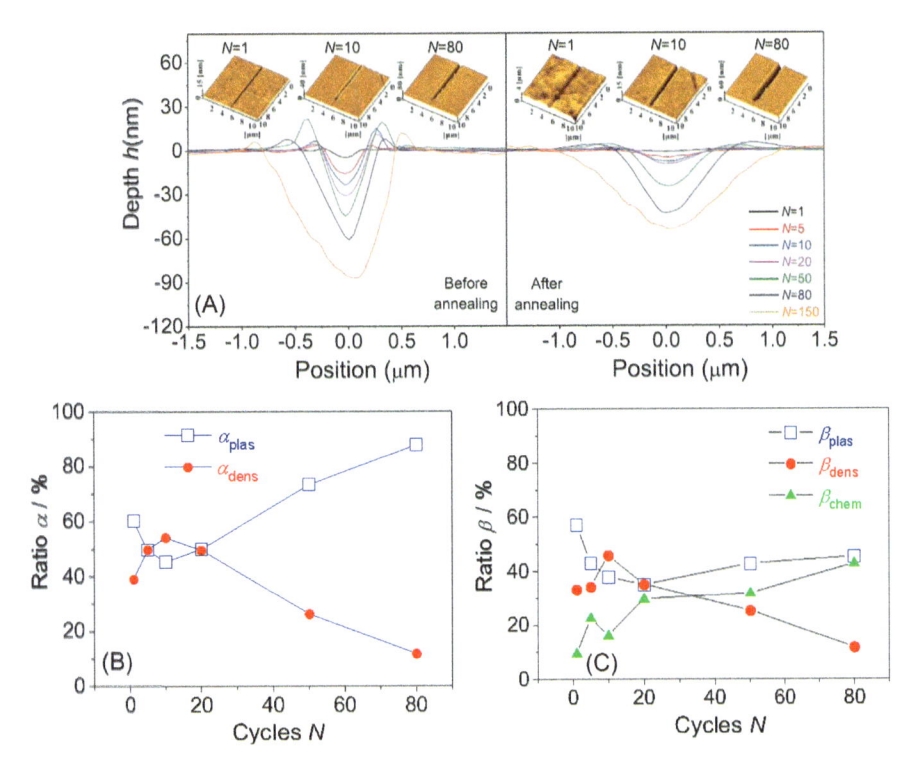

Figure 8.4 (A) Cross-sectional profile lines of nanoscratch tracks in phosphate laser (PL) glass at various scratching cycles, N, before and after annealing in dry air, and the insets selectively show the atomic force microscopy images of the nanoscratch tracks in PL glass for various scratching cycles ($N = 1$, 10, and 80). (B) The contribution ratio of plastic damage and densification to the total deformation of PL glass, as a function of scratching cycles in dry air. (C) The contribution ratio of plastic damage, densification, and chemistry-enhanced material-removal to the total deformation of PL glass, as a function of scratching cycles in humid air.
Source: From Fu, J., He, H., Yuan, W., Zhang, Y., & Yu, J. (2018). Towards a deeper understanding of nanoscratch-induced deformation in an optical glass. *Applied Physics Letters*, *113*(3). https://doi.org/10.1063/1.5030848.

possible to calculate the contributions of the three independent factors to the total deformation of glass by a diamond tip through comparing the deformation volume of a phosphate laser (PL) glass subject to different conditions, namely, in dry or humid air and before or after annealing (Fig. 8.4A). It was found that in dry air, the densification process can dominate the PL glass deformation in fewer-pass scratch, while plastic damage dominated the glass deformation in multipass scratching (Fig. 8.4B). In humid air, the combined action of plastic damage and densification dominated the glass deformation in fewer-pass scratching, while plastic damage and chemistry-enhanced material-removal together determine the final deformation volume for multipass scratching in humid air (Fig. 8.4C).

8.2 Limitations and strengths of atomic force microscopy

In an AFM, an extremely small tip located at the end of a microfabricated cantilever is brought into contact or close proximity to a specimen, and the forces acting on the tip due to the interactions with the substrate, such as van der Waals, mechanical, magnetic, or electrical interactions, can cause the deflections of the cantilever. Therefore based on such interactions between the tip and the substrate, AFM has been widely used as a tool not only to observe surface topography with high spatial resolution but also to characterize the properties such as mechanical, magnetic, ferroelectric, and piezoelectric at nanoscale (Chung, 2014).

In addition to the conventional nanoindentation tests, AFM-based nanoindentation tests have also been used to investigate the atomic-scale yield, dislocation nucleation, and single-slide events in solid materials, which is benefited from the high force resolution ($\sim$nN), displacement resolution ($\sim$subangstrom), and sharp indenting tips (a few nanometers in tip radius) by AFM (Filleter & Bennewitz, 2007; Paul et al., 2013). As one example, Yu et al. (2021) investigated the yield stress, onset of yielding, and atomic-scale plastic flow of a platinum-based bulk metallic glass (BMG) with "atomically flat" surfaces, and they found that the yield stresses of platinum-based BMG by AFM-based nanoindentation are higher than in conventional nanoindentation testing, and the subsequent flow was then established to be homogeneous without exhibiting collective shear localization or loading rate dependence. These findings imply that the AFM-based nanoindentation techniques can be used to study the deformation behaviors of solid materials at atomic scale without a need to decrease the sample's dimensions (Yu et al., 2021).

Recently, advances in microscopy and superresolution techniques in AFM enable the collection of even more than two measurands. For instances, the conductive atomic force microscopy (CAFM) can evaluate the surface current changes including defect generating, and this can be helpful to reveal the nanoscratch or nanowear mechanism of solid materials. Using CAFM, Wu et al. (2020) found that higher current was detected at the edge of scratches on GaAs surface and the current was increased with the applied normal load in nanoscratching tests. Moreover, the increasing current signals were detected from the scratch as the etching progressed, and the long-time etching can result in local current breakover of unscratched GaAs surface. This is because the lattice defects of scratched area are responsible for abrupt change of the current by CAFM, and the insertion of conductive ions is expected to make the current increase during the selective etching. Photothermal atomic force microscopy coupled with infrared spectroscopy (AFM-IR) is a breakthrough in infrared (IR) spectroscopic imaging with nanoscale image resolution, enabling us to detect the chemical structure at nanoscale through the thermal expansion of substrate material at nanoscale by IR (Dazzi & Prater, 2017). The application of AFM-IR to silica and silicate glass surfaces has recently been demonstrated to characterize the subsurface structural changes and surface heterogeneity due to mechanical stresses from physical contacts, as well as chemical alterations manifested in surface layers through aqueous corrosion (Lin et al., 2022). Nanoscale IR spectroscopy analysis coupled with AFM can also be analyzed with nea-SNOM microscopy where the tip-scattered IR signal can be used to characterize the near-surface chemical structure change of solid materials at nanoscale (Huber et al., 2009). The nanoscale infrared spectroscopy combined with scattering-type scanning near-field optical microscopy as well as reactive molecules dynamic simulations has demonstrated that the Si−O bond length of silica glass can be elongated with the physical contact in the elastic regime at nanoscale (He et al., 2021). Moreover, the hybrid methods based on the AFM and complementary techniques can be found in other applications such as scanning near-field ellipsometry microscopy (SNEM), tip-enhanced Raman spectroscopy (TERS), scanning thermal microscopy (SThM), AFM combined with quartz crystal microbalance (QCM), Kelvin probe force microscopy (KPFM), and more details about these techniques and applications can be found in Handschuh-Wang et al. (2017).

Although the AFM and its coupled techniques have been widely used in surface engineering and tribology applications, the wear of the tip is a critical concern that is limiting its further applications. For example, the blunted tip due to the wear of tip can lead to the significant degradation in the AFM images (Strahlendorff et al., 2019), nanofriction (Zavedeev et al., 2018), and nanowear of substrates (Colaço & Serro, 2020). Thus it is of great importance to understood the wear characteristics of the AFM tip, which is usable not only for the reliable measurement of material properties using an AFM but also for the realization of AFM-based applications. Wear of tip has been reported for the silicon-based tip (such as silica tip, silicon tip, silicon nitride tip), carbon-based tip (such as diamond tip, CNT tip, Si-DLC tip), and metal-coated tip (such as Ti, Pt, and Au coated tip, Au-Ni, Pt-Ir, and Pt-Ni coated tips). The most accepted wear mechanisms of AFM tip are adhesive wear, abrasive wear, fatigue, and oxidation, while atom-by-atom attrition has recently been proposed as the main wear mechanism at single-asperity contact such as for the case of the AFM tip (Jacobs & Carpick, 2013). It should be noted that the key factor that affects the wear characteristics of the AFM tip is the contact stress that is depended on instinct properties of the substrate material, contact geometry, and applied load. In addition, the environmental condition such as relative humidity can affect the wear characteristics for the case of the silicon-based tips due to the tribochemical wear of silica with water molecules impinging from the gas phase. In the case of metal-coated tip, the effect of the electrical stress should also be taken into account (Chung, 2014).

Significant progress has been made in the fabrication of wear-resistant tips. In addition to the diamond-coated tip, the SiC, Si-DLC, and ultrananocrystalline diamond (UNCD) tips with radius comparable with those of typical Si-based tips have been recently demonstrated (Chung, 2014). It has been showed that the wear rates of these tips were much lower than that of the typical silicon-based tips, and therefore it may be appropriate for AFM applications such as nanolithography and nanotribological studies. The remaining challenge for those tips would be the efficient fabrication with reduced time and cost, such as for the case of the carbon nanotube (CNT) tip.

8.3 Friction and wear of ultrathin films

Ordered molecular films can cover solid surfaces in a dense and orderly manner with a thickness of only a few to tens of nanometers, and their properties can be designed according to various applications. Ordered molecular membrane technology can provide an important way for boundary lubrication, which is very important for solving lubrication problems in MEMS and computer disk storage systems. Experimental studies have found that the nanoscale friction coefficient decreases with molecular chain length of the self-assembled membrane, and when the molecular chain length increases to a certain extent, the friction coefficient no longer decreases. Xiao et al. (1996) used silicon nitride tips and AFM to study the nanotribological properties of silane self-assembled monolayer films on mica surfaces, and it was found that the friction coefficient of the self-assembled membrane of short-chain silane molecules (C3 and C6) was much higher than that of long-chain silane molecules (C18). Since the tail groups of the aforementioned film-forming molecules are the same, they believed that the reason for the significant difference in friction was that the van der Waals interaction between the molecules of the short-chain self-assembled membrane was smaller, and the molecular membrane order and bulk density were lower, so it had more multiple energy dissipation modes (chain bending, tilting, torsion, twisting, etc.), consequently, it showed larger friction. In contrast, the van der Waals interaction between the molecules of the long-chain self-assembly membrane was stronger, and the molecular membrane arrangement was denser, the strength was higher; as a result, the lubrication effect was better.

The adhesion and friction properties of self-assembled membranes are closely related to the chemical characteristics of the tail groups of the molecular membrane. Frisbie et al. (1994) found that there was a great difference in adhesion and friction between self-assembled membrane-modified probes and the substrate surface of different end groups ($-COOH$ and $-CH_3$). The order of adhesion was COOH/COOH > CH_3/CH_3 > $COOH/CH_3$, and the nanofriction was in the same order as the adhesion. This was due to the strong hydrogen bonding interactions between COOH/COOH, which hinders sliding contact. Kim et al. (1999) found that self-assembled membranes with CH_3 as the dangling group had lower friction than self-assembled films with CF_3 as the dangling group. Although the CH_3 group was larger, it will generate more energy consumption in friction. However, the CF_3 group had a large polarity, resulting in the increases in adhesion force and thus greater friction.

In addition to the chain length and dangling group, the environment conditions can play important roles in the nanoscale friction. Tsukruk and Bliznyuk (1998) used SPM technology to study the nanotribological properties of organosilane self-assembled membranes with dangling groups of CH_3, NH_2, and SO_3H in aqueous solutions at different pH values (pH = 2 ~ 10). It was found that when pH = 4 ~ 8, the adhesion and friction between the surfaces were larger, while when pH > 9 and pH < 3, the adhesion and friction were relatively small. This finding could be explained with the combined effect of the electric double layer on the surface and the van der Waals action. When the pH of the solution pH was far from neutral point, the surface will adsorb ions to form an electric double layer, resulting in the electrostatic repulsion and thus reducing the adhesion and friction. When the solution pH was close to neutral point, the ion concentration was greatly reduced, the surface electrostatic repulsion was weakened, and the van der Waals force played a major role, so it had a high friction force. Qian et al. (2003) and Xiao and Qian (2000) used AFM to study the adhesion and friction of hydrophilic silica surfaces and hydrophobic silane self-assembly membrane surfaces at different RH conditions, as shown in Fig. 8.5. Since there was no water adsorption on the hydrophobic self-assembly membrane surface, the adhesion was basically unchanged with RH. For hydrophilic silica surfaces, under low humidity conditions, the adhesion gradually increased with RH, and when the humidity is 70%, the adhesion reached the maximum, and the humidity continued to increase, and the adhesion gradually decreased. In nanotribology, adhesion and applied load were usually on the same order of magnitude, so adhesion can affect the friction significantly. The experimental results showed that the change of friction with RH had a similar law to the adhesion force. For hydrophilic surfaces, as RH increased, the area of water film adsorbed between the AFM tip and the sample increased, and adhesion and friction increased. When the thickness of the water film reached certain level, it can play

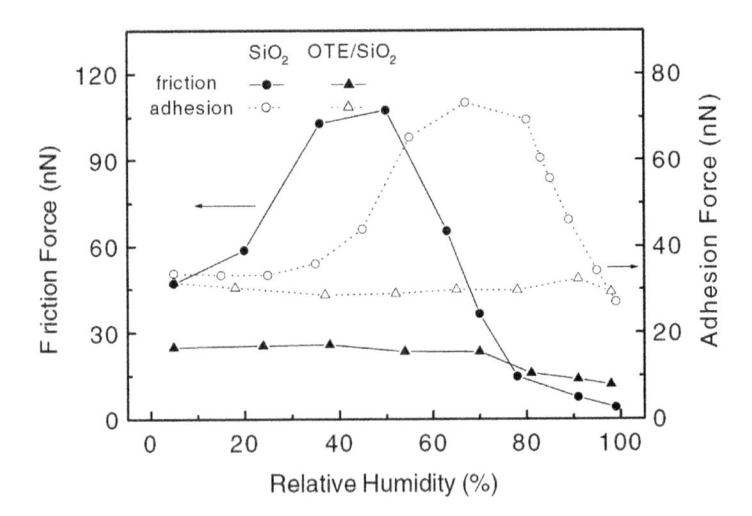

Figure 8.5 The friction and adhesion at an external load of 100 nN versus humidity for SiO_2 (solid and open circles, respectively) and for OTE/SiO_2 (solid and open triangles, respectively).
Source: From Xiao. X., Hu. J., Charych. D. H., & Salmeron, M. (1996). Chain length dependence of the frictional properties of alkylsilane molecules self-assembled on mica studied by atomic force microscopy. *Langmuir*, *12*(2), 235−237. https://doi.org/10.1021/la950771u.

a lubricating role. Therefore even if the adhesion continued to increase, the friction had begun to decrease with RH.

Owing to its excellent mechanical properties and stable chemical behaviors, the diamond-like carbon (DLC) coating has been extensively used as a low-frictional and wear-resistant protective coating in MEMS. To investigate the efficacy on resisting nanofretting damage of silicon substrate, Chen et al. (2011) studied the nanofretting behaviors of ultrathin DLC coating and its Si(100) substrate against SiO$_2$ microsphere was investigated under vacuum and humid air environments with AFM (Fig. 8.6). It was found that due to the absence of oxygen and water molecule in vacuum, both the adhesion and friction forces on DLC coating were similar as those on Si(100) substrate. However, while the nanofretting was tested in air, the DLC coating could significantly reduce the adhesion and friction forces of Si(100) substrate. Even though the coating was only 2 nm in thickness, it could effectively protect the silicon substrate from nanofretting damage. In vacuum, the ultrathin DLC coating can prevent the formation of hillock on Si(100) substrate. In air, the wear depth on DLC coating was <5% of that on Si(100) substrate under the same nanofretting conditions.

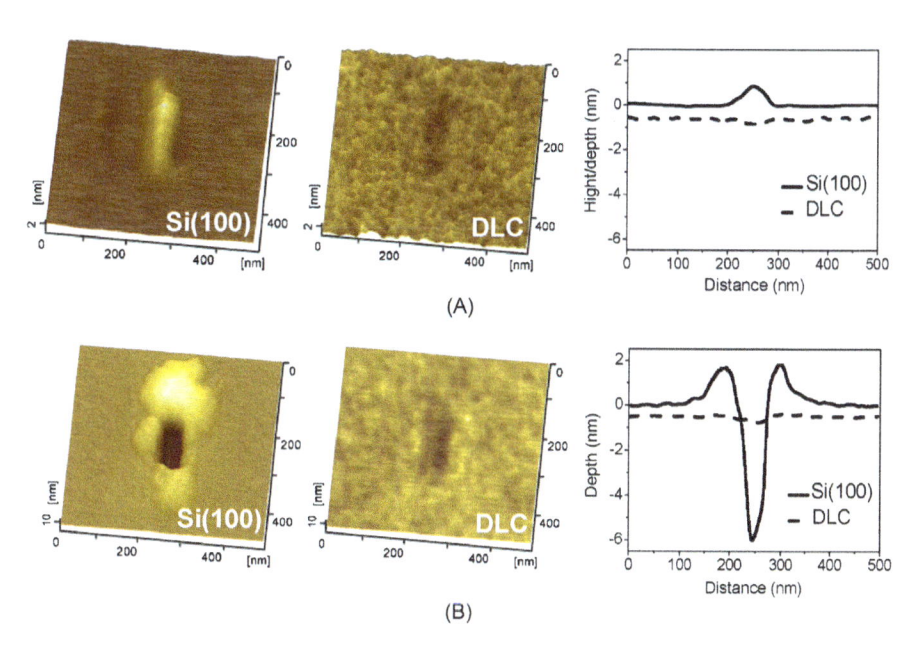

Figure 8.6 Atomic force microscopy images and cross-section profiles of nanofretting scars on Si(100) surface and DLC coating after nanofretting under $F_n = 2$ μN and $\mu = 2$ μN and $D = 100$ nm: (A) in vacuum and (B) in air. *DLC*, Diamond-like carbon.
Source: From Chen, L., Yang, M., Yu, J., Qian, L., & Zhou, Z. (2011). Nanofretting behaviours of ultrathin DLC coating on Si(100) substrate. *Wear*, *271*(9–10), 1980–1986. https://doi.org/10.1016/j.wear.2010.11.016.

In addition to the nanofretting, the DLC film can also affect the running-in behavior during nanowear tests. Shi et al. (2021) investigated the running-in behaviors of an H-DLC/Al$_2$O$_3$ pair through a controllable single-asperity contact study using an AFM (Fig. 8.7). It was found that after 200 sliding cycles, a thin transfer layer was formed on the Al$_2$O$_3$ tip. Compared with a clean tip, this modified tip showed a significantly lower adhesion force and friction force on the original H-DLC film, which confirmed the contribution of the transfer layer formation in the friction reduction during running-in process. It was also found that the friction coefficient of the H-DLC/Al$_2$O$_3$ pair decreased linearly as the oxygen concentration of the H-DLC substrate surface decreased. This phenomenon can be explained by a change in the contact surface from an oxygen termination with strong hydrogen bonding interactions to a hydrogen termination with weak van der Waals interactions.

Graphene is atomically thin and flexible conductor, with very large stiffness and strength, and with good lubricating properties, which is a promising material for future NEMS devices such as nanoelectromechanical resonators, piezoresistive sensors, and nanoelectromechanical switches (Bae et al., 2013; Chen et al., 2009; Kim et al., 2009). This is because the graphene is an excellent choice for the mechanical protection of underlying substrates and nanoobjects, including the wear protection and friction reduction (Berman et al., 2013; Klemenz et al., 2014). Wear of graphene as well as wear protection by graphene was mainly studied on micro- and macroscale, on large graphene sheets obtained by the solution processing (Berman et al., 2014), the epitaxial growth on SiC (Wählisch et al., 2013), or the chemical vapor deposition (CVD) on copper (Won et al., 2013). In order to explore the wear of graphene at nanoscale, and the efficiency of graphene for the wear protection of an underlying substrate, the nanoscale wear of graphene has also been studied by

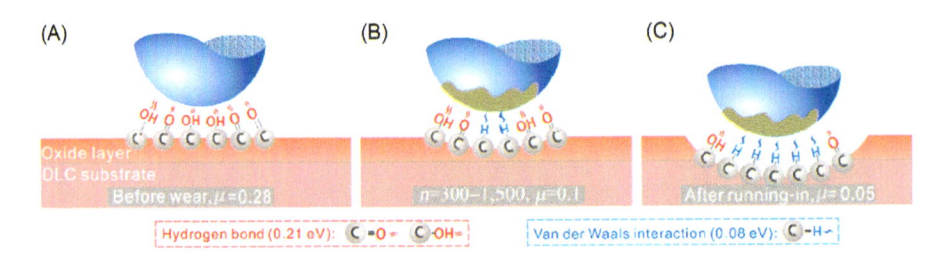

Figure 8.7 Schematics illustrating the running-in process of an H-DLC film against an Al$_2$O$_3$ tip under vacuum. (A) Initial stage ($n = 0$, before wear): high friction due to the strong hydrogen bonds between a new Al$_2$O$_3$ tip and the oxide-covered H-DLC surface. (B) Intermediary stage ($n = 300-1500$ cycles): low friction due to the transfer layer formation and partial oxide layer removal. (C) The final stage ($n > 1500$ cycles, after running-in): lowest friction due to the weak van der Waals interactions between the modified tip and the exposed H-DLC substrate. *DLC*, Diamond-like carbon.
Source: From Shi, P., Sun, J., Liu, Y., Zhang, B., Zhang, J., Chen, L., & Qian, L. (2021). Running-in behavior of a H-DLC/Al$_2$O$_3$ pair at the nanoscale. *Friction*, 9(6), 1464–1473. https://doi.org/10.1007/s40544-020-0429-5.

AFM-based scratching (Vasić et al., 2017). It was found that the plastic deformation of graphene increased with applied normal force and some defects in graphene lattice could be found. For high enough normal loads, the breaking strength of graphene dropped down, leading to a sudden fracture, an uncontrolled tearing and subsequent peeling off, approximately within the scratching area. For wear protection of the underlying substrate, at least 5 nm thick graphene flakes were necessary, while thin graphene flakes, around 1 nm thick (single and bilayer), can only enhance the mechanical capacity of the underlying substrate.

To further reveal the relationship between the thickness and nanowear resistance of graphene, Liu et al. (2022) performed nanowear tests on fluorinated graphene (or named fluorographene, FG) covered on a Si substrate with a Si_3N_4 AFM probe, and the nanowear of FG was compared as the thickness increases from ∼0.8 to ∼4.2 nm, which corresponds to the range from monolayer (1 L) to seven layers (7 L). It was found that before the onset of local wear, the friction signal remained low. From the load dependence of the friction force measured before wear occurs, FG showed a slight decrease in COF with the number of layers, which could be explained by the decrease in puckering around the contact region as the layer became stiffer, or the decrease in the pinning force due to atomic-scale reconfigurations inside the contact area (Lee et al., 2010; Li et al., 2016). It is quite surprising to see that as the number of FG layers decreased, the wear resistance (i.e., durability) increased. This negative thickness dependence of wear resistance of FG was completely opposite to the wear behavior of mechanically exfoliated graphene layers, which showed enhanced durability with increasing layer thickness (Cui et al., 2020). Further, DFT-D calculations suggested that such correlations could be explained with the degree of charge transfer between FG and substrate, which affected the adhesion energy. The DFT-D calculations also showed that the degree of charge transfer was the largest for 1 L FG as compared with 2 L and 3 L FG, which could explain the negative dependence of wear resistance on the FG layer thickness (Liu et al., 2022).

To further reveal the thickness dependence of other 2D films, Tang et al. (2023) using AFM and Si_3N_4 tip to study the thickness dependence of the nanowear resistance of graphene oxide (GO) when deposited on silicon and H-DLC substrate. It was found that when the substrate was silicon, an inverse thickness dependence of the nanowear resistance occurred for few-layer GO (<4 layers), whereas a direct correlation between thickness and nanowear resistance was observed for thicker nanosheets (>4 layers). In contrast, for an H-DLC substrate with a low surface energy, a direct correlation between thickness and nanowear was found for all thicknesses considered. Combined with density functional theory calculations, the wear resistance of few-layer GO was found to correlate with the substrate's surface energy, which can be traced back to substrate-dependent adhesive strengths of GO that was correlated with the GO thickness originating from differences in the interfacial charge transfer. These findings can propose a strategy of optimization of the thickness of layered materials and tuning the nanoscale chemical interactions of the sample/substrate interfaces to improve the wear resistance and prolong the lifetime of 2D materials.

8.4 Line and area scanning comparison

The reciprocating line scanning is a highly common method for the nanowear test by AFM (Fig. 8.8A) (Zhang et al., 2018). However, when the reciprocating sliding or line scanning was conducted for many times, the experimental result can be affected by the change in contact profile caused by severe wear and an irregular wear profile induced by thermal drift due to long-time contact history (Stevens et al., 2006). On the other hand, the area scanning by AFM was introduced to generate a square wear scar to reveal the nanotribological properties of substrate surface. This kind of wear method can avoid the significant change of contact profile and reduce the influence of thermal drift on experimental data by tuning the scan parameters (Fig. 8.8B) (Lapshin, 2007).

To investigate the effect of wear methods (line scanning vs area scanning) on the nanowear behavior, both wear methods were performed on copper and silicon surface. These two materials were selected because the nanowear mechanism of copper with a diamond tip in vacuum was usually considered as the mechanical wear, while the nanowear mechanism of oxide-free silicon against SiO_2 tip in pure

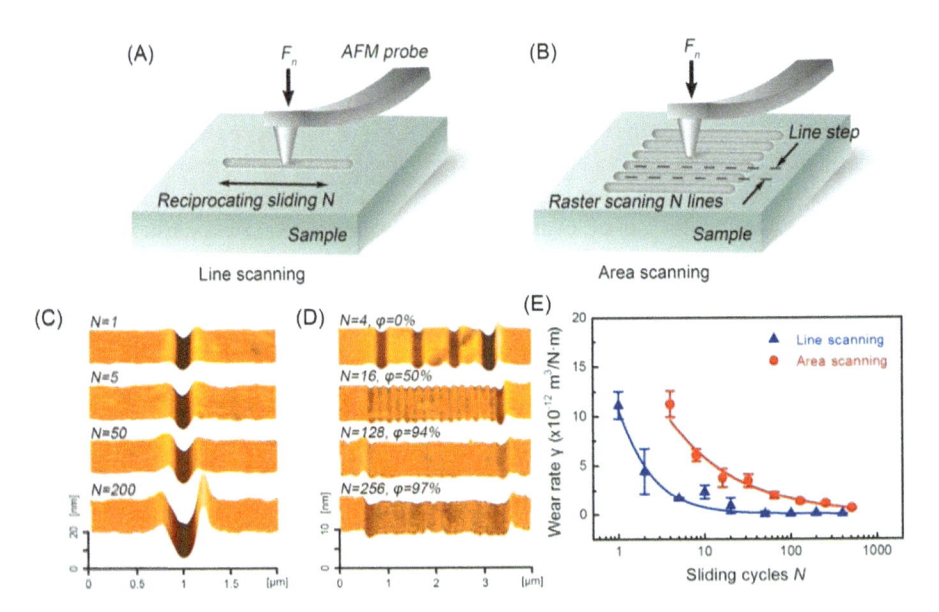

Figure 8.8 Illustration of scanning methods based on AFM: (A) line scanning and (B) area scanning. AFM images of wear scar on copper surface against diamond tip in vacuum based on (C) line scanning and (D) area scanning. (E) wear rate on copper surface with sliding cycles under different scanning methods. Error bars represent data standard deviations. The nanowear tests were performed at a normal load of 10 μN and scan frequency of 1 Hz. *AFM*, Atomic force microscopy.
Source: From Zhang, P., Chen, C., Xiao, C., Chen, L., & Qian, L. (2018). Comparison of wear methods at nanoscale: Line scanning and area scanning. *Wear, 400–401*, 137–143. https://doi.org/10.1016/j.wear.2018.01.004.

water was usually considered as the tribochemical wear (Wang et al., 2011; Zhang et al., 2018).

As shown in Fig. 8.8C–E, regardless of line scanning or area scanning, the wear rate of copper was not proportional to sliding cycles. The wear rate of line scanning of copper decreased by ~98% during the initial 10 sliding cycles and then maintains a stable value, while the wear rate of area scanning gradually decreased with the increase of sliding cycles and scan overlap rate (Fig. 8.8E). When the line step became smaller and the successive scan lines almost overlapped, the wear rate finally decreased by ~95% (Fig. 8.8E). These results confirmed that two scanning methods can cause clearly differences in nanowear behaviors.

For pure mechanical wear, the material removal is mainly dependent on the contact pressure and indentation volume of the diamond tip on copper surface (Liu et al., 2002). If the different scanning methods were used for this nanowear test, the scan path of tip on substrate surface would show significant difference. Further analyses indicated that in the line scanning, the average contact pressure will inevitably decrease owing to the increase in contact area between the tip and the substrate (Yu & Blanchard, 1996). The repeated plastic indent sliding at the same location can lead to the local "work hardening" of copper surface (Chen et al., 2017), which may further restrain the substrate material removal. As a result, the wear rate of the copper surface gradually decreases and finally reaches a stable value. In the case of area scanning, the line step will decrease with scanning lines, and no overlap occurred between the successive scan lines when scan lines $N = 4$, and the average wear rate of copper surface was nearly same as that of 1 sliding cycle in line scan case owing to the high contact pressure at initial sliding contact. When scan line $N = 8$, the line step was very close to the contact diameter. Even the theoretical overlap rate equals zero, the real influencing zone may exceed the separation distance between the lines for the plastic indent sliding process, and the wear depth of the successive scan lines can be influenced by the scanning history. The scanned area will gradually generate the local "work hardening." The residual wear depth of copper surface was always less than 3 nm even if the overlap rate over 95%. Thus the wear rate of copper surface will eventually reach a low value after undergoing a decline trend.

In the case of nanoscale tribochemical wear of oxide-free silicon surface in pure water, the wear rate of line scanning decreased by ~ 90% during the initial five sliding cycles and then maintains a stable value, while the wear rate of area scanning decreases gradually with the sliding cycles (Fig. 8.9C). There was no plastic deformation or work hardening occurs on the silicon surface because of the direct atomic removal mechanism at silicon surface that involves counter-surface chemistry and adsorbed water molecules (Takahagi et al., 1988; Wen et al., 2016). The contact area during the tribochemical wear of silicon under elastic contact condition was the key factor affecting wear rate, since the wear rate was not proportional to the contact area in both wear methods (Fig. 8.9C). In addition, the line scanning needed only five sliding cycles to complete a "running-in" process of Si/SiO_2 interface, whereas the area scanning can cause a larger wear area. Nevertheless, the line scanning was a simple and efficient nanowear test method, continuous reciprocating

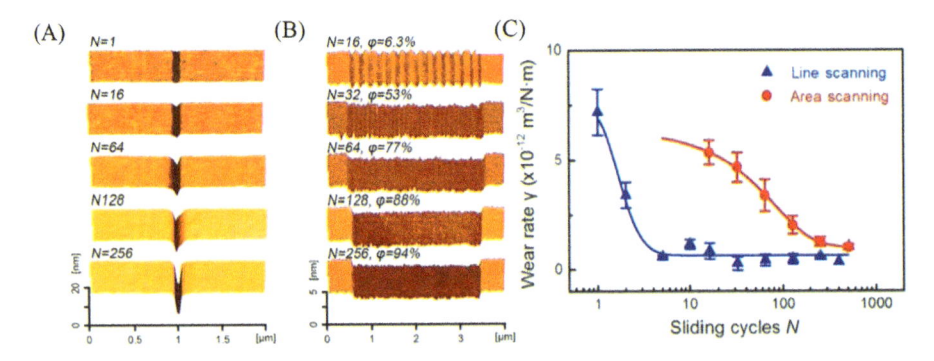

Figure 8.9 Atomic force microscopy images of wear scar on silicon surface against SiO$_2$ tip in water based on (A) line scanning and (B) area scanning. (C) Variation of wear rate on silicon surface with sliding cycles based on different scanning methods. Error bars represent data standard deviations. The nanowear tests was performed at a normal load of 3 µN and scan frequency of 1 Hz.
Source: From Zhang, P., Chen, C., Xiao, C., Chen, L., & Qian, L. (2018). Comparison of wear methods at nanoscale: line scanning and area scanning. *Wear*, *400−401*, 137−143. https://doi.org/10.1016/j.wear.2018.01.004.

sliding can promptly damage the specimen surface and produce a deep groove, inducing a significant change of wear profile, which is not conducive to the analysis of interface contact mechanics due to the conforming contact (nonsphere-flat contact) between tip and substrate (Mao et al., 2013). In contrast, the area scanning method can provide the adjustable scan paths. By altering the scan size and the number of scan lines, the scan overlap rate was determined as a crucial parameter. If the scan path had no any overlapping, the wear behaviors presented by experimental results should be consistent with the result of a single sliding. However, once the reciprocation or overlap occurred between the successive scan lines, the interfacial contact state and chemical properties will change. By setting a low overlap rate in area scan case, the "running-in" wear process of single-asperity contact can be extended. It is reasonable to analyze the change in interfacial chemistry and its effect on mechanochemical wear behavior at the nanoscale. In addition, the influence of device thermal drift or local material defects on the experimental results can be reduced in area scan case by tuning the overlap rate (Lapshin, 2007). Therefore the selection of suitable wear methods and scan parameters is essential to obtain reliable experimental data to achieve the aim of nanotribological research.

References

Alsteens, D., Gaub, H. E., Newton, R., Pfreundschuh, M., Gerber, C., & Müller, D. J. (2017). Atomic force microscopy-based characterization and design of biointerfaces. *Nature*

Reviews Materials, 2(5). Available from https://doi.org/10.1038/natrevmats.2017.8, http://www.nature.com/natrevmats/.

Bae, S. H., Lee, Y., Sharma, B. K., Lee, H. J., Kim, J. H., & Ahn, J. H. (2013). Graphene-based transparent strain sensor. *Carbon*, 51(1), 236−242. Available from https://doi.org/10.1016/j.carbon.2012.08.048.

Berman, D., Deshmukh, S. A., Sankaranarayanan, S. K. R. S., Erdemir, A., & Sumant, A. V. (2014). Extraordinary macroscale wear resistance of one atom thick graphene layer. *Advanced Functional Materials*, 24(42), 6640−6646. Available from https://doi.org/10.1002/adfm.201401755, http://onlinelibrary.wiley.com/journal/10.1002/(ISSN)1616-3028.

Berman, D., Erdemir, A., & Sumant, A. V. (2013). Few layer graphene to reduce wear and friction on sliding steel surfaces. *Carbon*, 54, 454−459. Available from https://doi.org/10.1016/j.carbon.2012.11.061.

Bhushan, B., Israelachvili, J. N., & Landman, U. (1995). Nanotribology: Friction, wear and lubrication at the atomic scale. *Nature*, 374(6523), 607−616. Available from https://doi.org/10.1038/374607a0.

Britz-Grell, A. B., Saumer, M., & Tarasov, A. (2021). Challenges and opportunities of tip-enhanced Raman spectroscopy in liquids. *Journal of Physical Chemistry C*, 125(39), 21321−21340. Available from https://doi.org/10.1021/acs.jpcc.1c05353, http://pubs.acs.org/journal/jpccck.

Cai, W., Xu, D., Zhang, F., Wei, J., Lu, S., Qian, L., Lu, Z., & Cui, S. (2022). Intramolecular hydrogen bonds in a single macromolecule: Strength in high vacuum versus liquid environments. *Nano Research*, 15(2), 1517−1523. Available from https://doi.org/10.1007/s12274-021-3696-1, http://www.springer.com/materials/nanotechnology/journal/12274.

Chen, C., Rosenblatt, S., Bolotin, K. I., Kalb, W., Kim, P., Kymissis, I., Stormer, H. L., Heinz, T. F., & Hone, J. (2009). Performance of monolayer graphene nanomechanical resonators with electrical readout. *Nature Nanotechnology*, 4(12), 861−867. Available from https://doi.org/10.1038/nnano.2009.267, http://www.nature.com/nnano/index.html.

Chen, C., Xiao, C., Wang, X., Zhang, P., Chen, L., Qi, Y., & Qian, L. (2016). Role of water in the tribochemical removal of bare silicon. *Applied Surface Science*, 390, 696−702. Available from https://doi.org/10.1016/j.apsusc.2016.08.175, http://www.journals.elsevier.com/applied-surface-science/.

Chen, H., Zhao, D., Wang, Q., Qiang, Y., & Qi, J. (2017). Effects of impact energy on the wear resistance and work hardening mechanism of medium manganese austenitic steel. *China Friction*, 5(4), 447−454. Available from https://doi.org/10.1007/s40544-017-0158-6, http://www.springer.com/engineering/mechanical + engineering/journal/40544.

Chen, L., He, H., Wang, X., Kim, S. H., & Qian, L. (2015). Tribology of Si/SiO$_2$ in humid air: Transition from severe chemical wear to wearless behavior at nanoscale. *Langmuir: The ACS Journal of Surfaces and Colloids*, 31(1), 149−156. Available from https://doi.org/10.1021/la504333j, http://pubs.acs.org/journal/langd5.

Chen, L., Yang, M., Yu, J., Qian, L., & Zhou, Z. (2011). Nanofretting behaviours of ultrathin DLC coating on Si(100) substrate. *Wear*, 271(9−10), 1980−1986. Available from https://doi.org/10.1016/j.wear.2010.11.016.

Chen, L., Yang, Y. J., He, H. T., Kim, S. H., & Qian, L. M. (2015). Effect of coadsorption of water and alcohol vapor on the nanowear of silicon. *Wear*, 332−333, 879−884. Available from https://doi.org/10.1016/j.wear.2015.02.052.

Chung, K. H. (2014). Wear characteristics of atomic force microscopy tips: A reivew. *International Journal of Precision Engineering and Manufacturing*, 15(10), 2219−2230. Available from https://doi.org/10.1007/s12541-014-0584-6, http://www.springerlink.com/content/2234-7593/.

Colaço, R., & Serro, A. P. (2020). Nanoscale wear of hard materials: An overview. *Current Opinion in Colloid and Interface Science*, *47*, 118−125. Available from https://doi.org/10.1016/j.cocis.2020.01.001, http://www.elsevier.com/inca/publications/store/6/2/0/0/5/3.

Cui, T., Mukherjee, S., Sudeep, P. M., Colas, G., Najafi, F., Tam, J., Ajayan, P. M., Singh, C. V., Sun, Y., & Filleter, T. (2020). Fatigue of graphene. *Nature Materials*, *19*(4), 405−411. Available from https://doi.org/10.1038/s41563-019-0586-y, http://www.nature.com/nmat/.

Dazzi, A., & Prater, C. B. (2017). AFM-IR: Technology and applications in nanoscale infrared spectroscopy and chemical imaging. *Chemical Reviews*, *117*(7), 5146−5173. Available from https://doi.org/10.1021/acs.chemrev.6b00448, http://pubs.acs.org/journal/chreay.

Filleter, T., & Bennewitz, R. (2007). Nanometre-scale plasticity of Cu. *Nanotechnology*, *18*. Available from https://doi.org/10.1088/0957-4484/18/4/044004.

Franceschi, G., Kocán Kocán, P., Conti, A., Brandstetter, S., Balajka, J., Sokolović, I., Valtiner, M., Mittendorfer, F., Schmid, M., Setvín, M., & Diebold, U. (2023). Resolving the intrinsic short-range ordering of K+ ions on cleaved muscovite mica. *Nature Communications*, *14*(1). Available from https://doi.org/10.1038/s41467-023-35872-y, https://www.nature.com/ncomms/.

Frisbie, C. D., Rozsnyai, L. F., Noy, A., Wrighton, M. S., & Lieber, C. M. (1994). Functional group imaging by chemical force microscopy. *Science (New York, N.Y.)*, *265* (5181), 2071−2074. Available from https://doi.org/10.1126/science.265.5181.2071.

Fu, J., He, H., Yuan, W., Zhang, Y., & Yu, J. (2018). Towards a deeper understanding of nanoscratch-induced deformation in an optical glass. *Applied Physics Letters*, *113*(3). Available from https://doi.org/10.1063/1.5030848, http://scitation.aip.org/content/aip/journal/apl.

Handschuh-Wang, S., Wang, T., & Zhou, X. (2017). Recent advances in hybrid measurement methods based on atomic force microscopy and surface sensitive measurement techniques. *RSC Advances*, *7*(75), 47464−47499. Available from https://doi.org/10.1039/c7ra08515j.

Hauser, M., Wojcik, M., Kim, D., Mahmoudi, M., Li, W., & Xu, K. (2017). Correlative super-resolution microscopy: New dimensions and new opportunities. *Chemical Reviews*, *117*(11), 7428−7456. Available from https://doi.org/10.1021/acs.chemrev.6b00604, http://pubs.acs.org/journal/chreay.

He, H., Chen, Z., Lin, Y. T., Hahn, S. H., Yu, J., van Duin, A. C. T., Gokus, T. D., Rotkin, S. V., & Kim, S. H. (2021). Subsurface structural change of silica upon nanoscale physical contact: Chemical plasticity beyond topographic elasticity. *Acta Materialia*, *208*. Available from https://doi.org/10.1016/j.actamat.2021.116694, http://www.journals.elsevier.com/acta-materialia/.

He, H., Hahn, S. H., Yu, J., Qiao, Q., van Duin, A. C. T., & Kim, S. H. (2020). Friction-induced subsurface densification of glass at contact stress far below indentation damage threshold. *Acta Materialia*, *189*, 166−173. Available from https://doi.org/10.1016/j.actamat.2020.03.005, http://www.journals.elsevier.com/acta-materialia/.

Huber, A. J., Ziegler, A., Köck, T., & Hillenbrand, R. (2009). Infrared nanoscopy of strained semiconductors. *Nature Nanotechnology*, *4*(3), 153−157. Available from https://doi.org/10.1038/nnano.2008.399, http://www.nature.com/nnano/index.html.

Jacobs, T. D. B., & Carpick, R. W. (2013). Nanoscale wear as a stress-assisted chemical reaction. *Nature Nanotechnology*, *8*(2), 108−112. Available from https://doi.org/10.1038/nnano.2012.255, http://www.nature.com/nnano/index.html.

Kim, H. I., Graupe, M., Oloba, O., Koini, T., Imaduddin, S., Lee, T. R., & Perry, S. S. (1999). Molecularly specific studies of the frictional properties of monolayer films: A systematic comparison of CF3-, (CH3)2CH-, and CH3-terminated films. *Langmuir: The ACS Journal of Surfaces and Colloids*, *15*(9), 3179−3185. Available from https://doi.org/10.1021/la981497h.

Kim, K. S., Zhao, Y., Jang, H., Lee, S. Y., Kim, J. M., Kim, K. S., Ahn, J. H., Kim, P., Choi, J. Y., & Hong, B. H. (2009). Large-scale pattern growth of graphene films for stretchable transparent electrodes. *Nature*, *457*(7230), 706−710. Available from https://doi.org/10.1038/nature07719.

Kim, S., Asay, D., & Dugger, M. (2007). Nanotribology and MEMS. *Nanotoday*, *2*(5), 22−29. Available from https://www.sciencedirect.com/science/article/pii/S1748013207701408.

Klemenz, A., Pastewka, L., Balakrishna, S. G., Caron, A., Bennewitz, R., & Moseler, M. (2014). Atomic scale mechanisms of friction reduction and wear protection by graphene. *Nano Letters*, *14*(12), 7145−7152. Available from https://doi.org/10.1021/nl5037403, http://pubs.acs.org/journal/nalefd.

Krieg, M., Fläschner, G., Alsteens, D., Gaub, B. M., Roos, W. H., Wuite, G. J. L., Gaub, H. E., Gerber, C., Dufrêne, Y. F., & Müller, D. J. (2019). Atomic force microscopy-based mechanobiology. *Nature Reviews Physics*, *1*(1), 41−57. Available from https://doi.org/10.1038/s42254-018-0001-7, http://nature.com/natrevphys/.

Lapshin, R. V. (2007). Automatic drift elimination in probe microscope images based on techniques of counter-scanning and topography feature recognition. *Federation Measurement Science and Technology*, *18*(3), 907−927. Available from https://doi.org/10.1088/0957-0233/18/3/046, http://www.iop.org/EJ/journal/0957-0233.

Lee, C., Li, Q., Kalb, W., Liu, X. Z., Berger, H., Carpick, R. W., & Hone, J. (2010). Frictional characteristics of atomically thin sheets. *Science (New York, N.Y.)*, *328*(5974), 76−80. Available from https://doi.org/10.1126/science.1184167.

Li, S., Li, Q., Carpick, R. W., Gumbsch, P., Liu, X. Z., Ding, X., Sun, J., & Li, J. (2016). The evolving quality of frictional contact with graphene. *Nature*, *539*(7630), 541−545. Available from https://doi.org/10.1038/nature20135, http://www.nature.com/nature/index.html.

Lin, Y. T., He, H., Kaya, H., Liu, H., Ngo, D., Smith, N. J., Banerjee, J., Borhan, A., & Kim, S. H. (2022). Photothermal atomic force microscopy coupled with infrared spectroscopy (AFM-IR) analysis of high extinction coefficient materials: A case study with silica and silicate glasses. *Analytical Chemistry*, *94*(13), 5231−5239. Available from https://doi.org/10.1021/acs.analchem.1c04398, http://pubs.acs.org/journal/ancham.

Liu, Y., Jiang, Y., Sun, J., Wang, Y., Qian, L., Kim, S. H., & Chen, L. (2022). Inverse relationship between thickness and wear of fluorinated graphene: Thinner is better. *Nano Letters*, *22*(14), 6018−6025. Available from https://doi.org/10.1021/acs.nanolett.2c01043, http://pubs.acs.org/journal/nalefd.

Liu, Z., Sun, J., & Shen, W. (2002). Study of plowing and friction at the surfaces of plastic deformed metals. *Tribology International*, *35*(8), 511−522. Available from https://doi.org/10.1016/S0301-679X(02)00046-4.

Maboudian, R., & Howe, R. T. (1997). Critical review: Adhesion in surface micromechanical structures. *Journal of Vacuum Science and Technology B: Microelectronics and Nanometer Structures*, *15*(1), 1−20. Available from https://doi.org/10.1116/1.589247, http://scitation.aip.org/content/avs/journal/jvstb.

Mao, K., Sun, Y., & Bell, T. (2013). Contact mechanics of engineering surfaces: State of the art. *Surface Engineering*, *10*(4), 297−306. Available from https://doi.org/10.1179/sur.1994.10.4.297.

Paul, W., Oliver, D., Miyahara, Y., & Grütter, P. H. (2013). Minimum threshold for incipient plasticity in the atomic-scale nanoindentation of Au(111). *Physical Review Letters*, *110* (13). Available from https://doi.org/10.1103/PhysRevLett.110.135506, http://oai.aps.org/filefetch?identifier = 10.1103/PhysRevLett.110.135506&component = fulltext&description = markup&format = xml, Canada.

Qian, L., Tian, F., & Xiao, X. (2003). Tribological properties of self-assembled monolayers and their substrates under various humid environments. *Tribology Letters*, *15*(3), 169−176. Available from https://doi.org/10.1023/A:1024868532575.

Shi, P., Sun, J., Liu, Y., Zhang, B., Zhang, J., Chen, L., & Qian, L. (2021). Running-in behavior of a H-DLC/Al$_2$O$_3$ pair at the nanoscale. *Friction*, *9*(6), 1464−1473. Available from https://doi.org/10.1007/s40544-020-0429-5, http://www.springer.com/engineering/mechanical + engineering/journal/40544.

Stevens, F., Langford, S. C., & Dickinson, J. T. (2006). Tribochemical wear of sodium trisilicate glass at the nanometer size scale. *Journal of Applied Physics*, *99*(2). Available from https://doi.org/10.1063/1.2166646.

Strahlendorff, T., Dai, G., Bergmann, D., & Tutsch, R. (2019). Tip wear and tip breakage in high-speed atomic force microscopes. *Ultramicroscopy*, *201*, 28−37. Available from https://doi.org/10.1016/j.ultramic.2019.03.013, http://www.elsevier.com/locate/ultramic.

Szlufarska, I., Chandross, M., & Carpick, R. W. (2008). Recent advances in single-asperity nanotribology. *Journal of Physics D: Applied Physics*, *41*(12), 123001. Available from https://doi.org/10.1088/0022-3727/41/12/123001.

Takahagi, T., Nagai, I., Ishitani, A., Kuroda, H., & Nagasawa, Y. (1988). The formation of hydrogen passivated silicon single-crystal surfaces using ultraviolet cleaning and HF etching. *Journal of Applied Physics*, *64*(7), 3516−3521. Available from https://doi.org/10.1063/1.341489.

Tang, C., Jiang, Y., Chen, L., Sun, J., Liu, Y., Shi, P., Aguilar-Hurtado, J. Y., Rosenkranz, A., & Qian, L. (2023). Layer-dependent nanowear of graphene oxide. *ACS Nano*, *17*(3), 2497−2505. Available from https://doi.org/10.1021/acsnano.2c10084, http://pubs.acs.org/journal/ancac3.

Tseng, A. A. (2011). Advancements and challenges in development of atomic force microscopy for nanofabrication. *Nano Today*, *6*(5), 493−509. Available from https://doi.org/10.1016/j.nantod.2011.08.003.

Tsukruk, V. V., & Bliznyuk, V. N. (1998). Adhesive and friction forces between chemically modified silicon and silicon nitride surfaces. *Langmuir: The ACS Journal of Surfaces and Colloids*, *14*(2), 446−455. Available from https://doi.org/10.1021/la970367q, http://pubs.acs.org/journal/langd5.

Vasić, B., Matković, A., Ralević, U., Belić, M., & Gajić, R. (2017). Nanoscale wear of graphene and wear protection by graphene. *Carbon*, *120*, 137−144. Available from https://doi.org/10.1016/j.carbon.2017.05.036, http://www.journals.elsevier.com/carbon/.

Wählisch, F., Hoth, J., Held, C., Seyller, T., & Bennewitz, R. (2013). Friction and atomic-layer-scale wear of graphitic lubricants on SiC(0001) in dry sliding. *Wear.*, *300*(1−2), 78−81. Available from https://doi.org/10.1016/j.wear.2013.01.108.

Wang, X., Kim, S. H., Chen, C., Chen, L., He, H., & Qian, L. (2015). Humidity dependence of tribochemical wear of monocrystalline silicon. *ACS Applied Materials and Interfaces*, *7*(27), 14785−14792. Available from https://doi.org/10.1021/acsami.5b03043, http://pubs.acs.org/journal/aamick.

Wang, Y. G., Zhang, L. C., & Biddut, A. (2011). Chemical effect on the material removal rate in the CMP of silicon wafers. *Wear*, *270*(3−4), 312−316. Available from https://doi.org/10.1016/j.wear.2010.11.006.

Wen, J., Ma, T., Zhang, W., Psofogiannakis, G., van Duin, A. C. T., Chen, L., Qian, L., Hu, Y., & Lu, X. (2016). Atomic insight into tribochemical wear mechanism of silicon at the Si/SiO$_2$ interface in aqueous environment: Molecular dynamics simulations using ReaxFF reactive force field. *Applied Surface Science*, *390*, 216–223. Available from https://doi.org/10.1016/j.apsusc.2016.08.082, http://www.journals.elsevier.com/applied-surface-science/.

Won, M. S., Penkov, O. V., & Kim, D. E. (2013). Durability and degradation mechanism of graphene coatings deposited on Cu substrates under dry contact sliding. *Carbon*, *54*, 472–481. Available from https://doi.org/10.1016/j.carbon.2012.12.007.

Wu, L., Yu, B., Fan, Z., Zhang, P., Feng, C., Chen, P., Ji, J., & Qian, L. (2020). Effects of normal load and etching time on current evolution of scratched GaAs surface during selective etching. *Materials Science in Semiconductor Processing*, *105*, 104744. Available from https://doi.org/10.1016/j.mssp.2019.104744.

Wu, S., Zhang, Z., Watanabe, K., Taniguchi, T., & Andrei, E. Y. (2021). Chern insulators, van Hove singularities and topological flat bands in magic-angle twisted bilayer graphene. *Nature Materials*, *20*(4), 488–494. Available from https://doi.org/10.1038/s41563-020-00911-2.

Xiao, C., Guo, J., Zhang, P., Chen, C., Chen, L., & Qian, L. (2017). Effect of crystal plane orientation on tribochemical removal of monocrystalline silicon. *Scientific Reports*, *7*. Available from https://doi.org/10.1038/srep40750, http://www.nature.com/srep/index.html.

Xiao, X., Hu, J., Charych, D. H., & Salmeron, M. (1996). Chain length dependence of the frictional properties of alkylsilane molecules self-assembled on mica studied by atomic force microscopy. *Langmuir: The ACS Journal of Surfaces and Colloids*, *12*(2), 235–237. Available from https://doi.org/10.1021/la950771u, http://pubs.acs.org/journal/langd5.

Xiao, X., & Qian, L. (2000). Investigation of humidity-dependent capillary force. *Langmuir: The ACS Journal of Surfaces and Colloids*, *16*(21), 8153–8158. Available from https://doi.org/10.1021/la000770o.

Yu, B., Li, X., Dong, H., Chen, Y., Qian, L., & Zhou, Z. (2012). Towards a deeper understanding of the formation of friction-induced hillocks on monocrystalline silicon. *Journal of Physics D: Applied Physics*, *45*(14), 145301. Available from https://doi.org/10.1088/0022-3727/45/14/145301.

Yu, J., Datye, A., Chen, Z., Zhou, C., Dagdeviren, O. E., Schroers, J., & Schwarz, U. D. (2021). Atomic-scale homogeneous plastic flow beyond near-theoretical yield stress in a metallic glass. *Communications Materials*, *2*(1). Available from https://doi.org/10.1038/s43246-021-00124-3, https://www.nature.com/commsmat/.

Yu, J., Hu, H., Jia, F., Yuan, W., Zang, H., Cai, Y., & Ji, F. (2015). Quantitative investigation on single-asperity friction and wear of phosphate laser glass against a spherical AFM diamond tip. *Tribology International*, *81*, 43–52. Available from https://doi.org/10.1016/j.triboint.2014.07.020, http://www.elsevier.com/inca/publications/store/3/0/4/7/4.

Yu, J., Kim, S. H., Yu, B., Qian, L., & Zhou, Z. (2012). Role of tribochemistry in nanowear of single-crystalline silicon. *ACS Applied Materials and Interfaces*, *4*(3), 1585–1593. Available from https://doi.org/10.1021/am201763z.

Yu, J., Qian, L., Yu, B., & Zhou, Z. (2009a). Nanofretting behavior of monocrystalline silicon (100) against SiO$_2$ microsphere in vacuum. *Tribology Letters.*, *34*(1), 31–40. Available from https://doi.org/10.1007/s11249-008-9385-3.

Yu, J. X., Qian, L. M., Yu, B. J., & Zhou, Z. R. (2009b). Nanofretting behaviors of monocrystalline silicon (100) against diamond tips in atmosphere and vacuum. *Wear*, *267* (1–4), 322–329. Available from https://doi.org/10.1016/j.wear.2008.11.008.

Yu, W., & Blanchard, J. P. (1996). An elastic-plastic indentation model and its solutions. *Journal of Materials Research*, *11*(9), 2358−2367. Available from https://doi.org/10.1557/JMR.1996.0299, http://journals.cambridge.org/action/displayJournal?jid = JMR.

Zavedeev, E. V., Jaeggi, B., Zuercher, J., Neuenschwander, B., Zilova, O. S., Shupegin, M. L., Presniakov, M. Y., & Pimenov, S. M. (2018). Effects of AFM tip wear on frictional images of laser-patterned diamond-like nanocomposite films. *Wear*, *416−417*, 1−5. Available from https://doi.org/10.1016/j.wear.2018.09.008.

Zhang, P., Chen, C., Xiao, C., Chen, L., & Qian, L. (2018). Comparison of wear methods at nanoscale: Line scanning and area scanning. *Wear*, *400−401*, 137−143. Available from https://doi.org/10.1016/j.wear.2018.01.004.

Zhao, X., & Bhushan, B. (1998). Material removal mechanisms of single-crystal silicon on nanoscale and at ultralow loads. *Wear*, *223*(1−2), 66−78. Available from https://doi.org/10.1016/S0043-1648(98)00302-0.

Nano-impact testing

Xiangru Shi[1] and Jian Chen[2]
[1]School of Materials Science and Engineering, Hohai University, Nanjing, Jiangsu Province, P.R. China, [2]School of Materials Science and Engineering, Southeast University, Nanjing, Jiangsu Province, P.R. China

9.1 Introduction

Coating technology is widely used in industrial manufacturing, forming a special thin film on the surface of components to provide better performance and longer service life. In the field of automotive manufacturing, hard coatings provide long-term corrosion resistance, scratch resistance, and superior friction properties (Deng et al., 2020; Schmauder et al., 2006; Veprek & Veprek-Heijman, 2008). In the aerospace field, the coated aircraft engine blades can increase the thrust of aircraft engines, extend engine life, and improve fuel efficiency (Wellman & Nicholls, 2007). The development of modern industry and advanced manufacturing puts forward higher requirements and challenges for the coatings technology. In addition to the superior mechanical properties, desired friction characteristics, and long-term corrosion resistance, the impact fatigue resistance is a critical factor in many applications of coating systems involving highly loaded mechanical contact (Beake et al., 2004; Biswas et al., 2010; Yang et al., 2016). For instance, various physical vapor deposition (PVD) coatings have been deposited on the surface of cutting tools, where they can be subjected to the fatigue impact of the cyclic load when the coated tools cut into the workpieces (Sun et al., 2016; Yang et al., 2010; Zhong-Lei et al., 2004). In a diesel engine system, the repetitive impacts in the powertrain, including fuel injectors, tappets, pistons, and piston rings can cause the diamond-like carbon (DLC) coatings peeling, deteriorate coatings performance, and increase fracture failure of the components (Lawes et al., 2010).

Current research has demonstrated that the cyclic impact tests are a convenient and effective testing method to characterize and evaluate the impact resistance of coating/substrate system under dynamic loading (Beake, 2022; Beake et al., 2001, 2007, 2011; Beake & Smith, 2004). In this regard, different impact testing apparatuses mainly based on piezoelectric ceramics and ultrasonic devices have been established to simulate the dynamic loading of the coated system under actual service environment (Wang et al., 2021). From the perspective of the failure mechanism, it is found that the impact energy and the geometry of the test probe have an important effect on the impact deformation and damage of the coating/substrate system (Bernoulli et al., 2015; Michler & Blank, 2001). The failure mechanism of the indenter on the coated system largely depends on the ratio of coating thickness t to the indenter radius R (t/R), which can be dominated by the coatings, the overall performance of the coating/substrate system, or the substrate over different contact sizes. Therefore, impact tests at different scales including tests at the millimeter, micrometer, and nanometer have been developed. At the macroscale impact testing, the hardened steel or cemented carbide spherical indenters with radii of 1−10 mm were commonly used to evaluate the impact resistance of thin PVD coatings (Bantle & Matthews, 1995; Bouzakis et al., 2004; Lamri et al., 2013). Although simple in principle and operation, there are some drawbacks that limit the further usefulness of these macroscale impact tests. First, they are not depth sensing and can only determine whether the coating has failed or survived after the experiment is completed, rather than real-time monitoring of coating failure. Then, the impact fatigue response of the coatings is averaged over a large area of the coating surface due to the large size of the test probe, making it difficult to obtain the impact resistance sensitivity at different positions of the coatings. Most important of all, the values of t/R are very low (≈ 0.001) and the impact loads are very high (> 100 N) for the macroscale impact tests. Under these conditions, the peak von Mises stresses are located deep into the substrate. The failure of the coating is mainly caused by plastic deformation of the substrate material, which is not closely related to the performance of the coating itself.

In recent years, the development of nano-impact testing utilizing the depth-sensing device of nanoindentation systems can be used to improve understanding of coating damage evolution at higher t/R (≈ 10) with much sharper probes (Bouzakis et al., 2013; Chen, Shi, et al., 2016; Chen, Li, et al., 2016; Qin et al., 2019; Wheeler & Gunner, 2013; Wheeler et al., 2019). These nanoscale impact devices can conduct repetitive impact tests on coatings at strain rates several orders of magnitude higher than quasistatic nanoindentation. The high strain rate contact in nanoscale impact testing can record the curves of impact depth with impact cycles in situ at the same position of the coating, thus simulating the performance and damage evolution of the coating system more closely under high load intermittent contact conditions. Compared with traditional macroscale impact testing, nano-impact has several advantages in evaluating the fatigue resistance of coatings. First, the evolution of impact damage in coatings can be explored in a relatively short period of time, as a cube corner diamond indenter with high contact strain caused by geometric shapes that are typically selected as test probes. Second, multiple rapid experimental evaluations can be conducted on thin and small volume coatings due to the smaller size (≈ 50 nm) of the indenter, so it is suitable to investigate the coating surface sensitivity influenced by their nano/microstructure under dynamic load. Besides, a high-precision displacement-sensing device can record the cycles to failure in real time, thereby accurately obtaining information on the fatigue failure mechanism of the coating. The most crucial thing is that in both nano- and macroscale tests, the peak position of von Mises stress relative to the coating − substrate interface is completely different. In general, the impact load of nano-impact is relatively small, and the impact resistances are mainly dominated by coating performance in application, so that the influence of substrate can be ignored during the impact process.

In addition to the independent influence of the substrate and coating on impact failure, the interface performance of the coating/substrate is also an essential factor affecting the overall performance of the system. By changing the applied load and probe geometry of nano-impact, a new microimpact test has recently been developed to move the peak position of von Mises stress concentrated near the coating − substrate interface (Beake et al., 2020, 2021). The microscale impact retains the original high-precision depth-sensing capability of nano-impact enabling the progression of the wear process to be monitored throughout the test, combined with the allowable impact load range of increased to 0.5−5 N. The maximum impact energy provided by each impact is about two orders of magnitude higher than that of nano-impact testing. The development of microimpact test bridges the gap in t/R between the nano and macroranges and provides closer simulation of the performance of coating systems under highly loaded intermittent contact and the evolution of wear under these conditions than tests at a lower strain rate.

Regarding practical applications, the micro/nano-impact tests can also be divided into multiple impact and single impact mode. Originating from the evaluation of the degradation performance of coatings under local repeated stresses, multiple impact tests have been applied in many situations as a reliable simulation tool, including (1) the evaluation of erosion resistance of hard coating (Chen et al., 2012; McMaster et al., 2020), (2) construction of impact fatigue failure map (Wheeler & Gunner, 2013), (3) calculation of fracture toughness values of structural ceramics (Frutos et al., 2016), and (4) comparison of energy absorption density of the surface of natural tough materials (Huang et al., 2020). However, due to the lack of high-precision time − displacement resolution, most studies mainly qualitatively evaluate the fatigue failure of materials based on the depth changes of multiple impact curves. The understanding of the single impact process is not deep enough, especially the lack of quantitative analysis of its internal physical processes. The introduction of a high-speed data acquisition system can observe the complete mechanical response during a single impact process (Chen, Shi, et al., 2016; Shi et al., 2020). By fitting and analyzing the high-precision single-impact curves, the mechanical properties of materials, such as energy absorption, dynamic hardness, dynamic toughness, and strain rate sensitivity, can be quantitatively evaluated. Currently, different physical models and data analysis methods for single-impact tests have been developed (Arreguin-Zavala et al., 2013; Constantinides et al., 2008; Somekawa & Schuh, 2012, 2013). Although it does not measure the accurate dynamic mechanical properties due to some extra energy dissipation, it can provide a quantitative assessment of resistance to fatigue fracture or plastic deformation under repetitive loading.

The goal of this chapter is to review the principle and application of micro/nanoscale impact testing in various field. The experimental setup and operating principles of the micro/nano test technique are presented in the first section of the chapter. The application of multiple micro/nano-impact testing including the evaluation of impact resistance of hard DLC coatings, simulation of erosion failure of thermal barrier coatings (TBC), and elevated temperature impact testing is discussed in the second section. Then, the application of high-precision single nano-impact testing to obtain the dynamic hardness and fracture toughness of materials is analyzed in detail.

9.2 Experimental setup of micro/nano-impact testing

9.2.1 Experimental setup

Similar to the traditional pendulum dynamic indentation testing process, the micro/nano-impact experimental setup (NanoTest, Micro Materials Ltd., Wrexham, UK) is also a pendulum-based depth-sensing system, with the samples mounted vertically and the load applied electromagnetically (Fig. 9.1A). Current in the coil

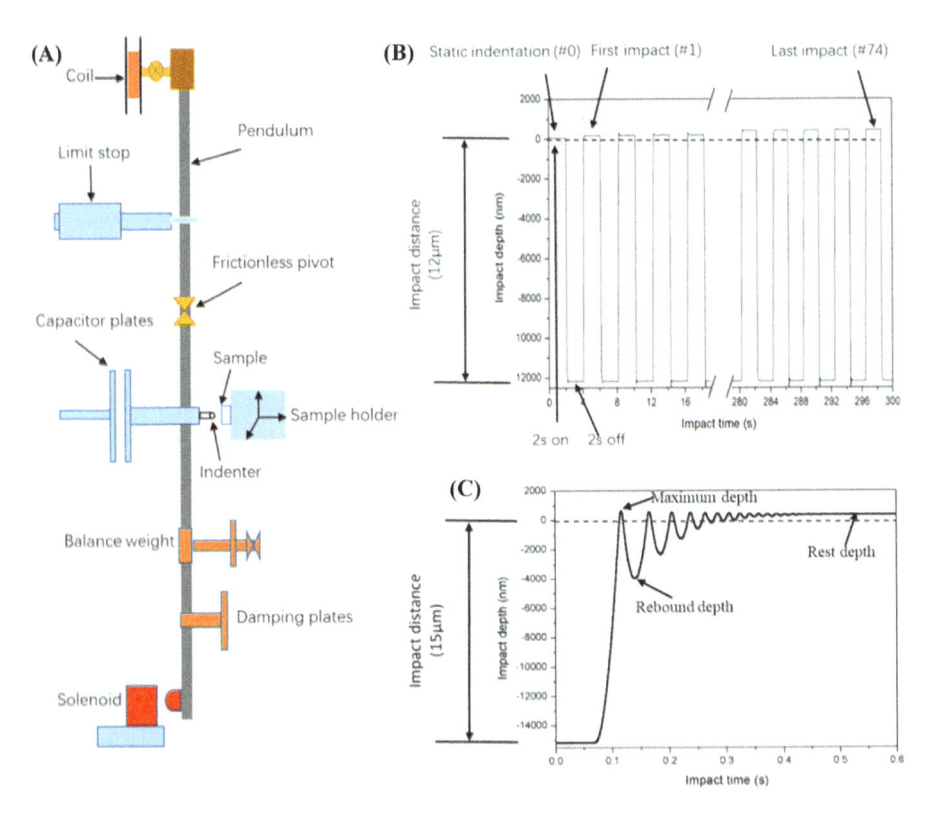

Figure 9.1 (A) Schematic illustration of the nanotest system showing the pendulum configuration for impact test, (B) typical impact depth − time curves of "multiple impulse" mode, and (C) typical impact depth − time curves of "dynamic hardness" mode
Source: From Shi, X., Chen, J., Beake, B. D., Liskiewicz, T. W., & Wang, Z. (2021). Dynamic contact behavior of graphite-like carbon films on ductile substrate under nano/micro-scale impact. *Surface and Coatings Technology*, 422. https://doi.org/10.1016/j.surfcoat.2021.127515.

causes the pendulum to rotate on its frictionless pivot so that the diamond indenter is withdrawn to a set distance from the sample surface and then rapidly accelerated to produce repetitive impacts at the same position with precisely controlled force. The depth-sensing capability with sub-nm resolution is used to monitor the test probe displacement in real time under repeated high strain rate loading conditions.

At the beginning of the multiple impact tests, an initial surface contact is produced under a minimum contact load (0.03 mN) to determine the position of depth zero. The static coil force is then applied (5–50 mN) producing elastoplastic deformation by indentation. The corresponding initial static indentation depth, h_0, including elastic and plastic deformation, is used to confirm that the depth zero is measured correctly and the test did not impact in an anomalous region of the surface. By activating the solenoid, the impact probe is rapidly removed to a certain distance from the sample surface and then accelerated over this distance to impact the surface causing additional deformation due to the higher inertial force. Automation enables repetitive impacts at the same position on the sample surface with a set frequency, typically at 4 s intervals, including 2 s on and 2 s off (see Fig. 9.1B). The impact test records the evolution of impact-induced damage with time (i.e., elastic and plastic deformation, fatigue wear, and fracture) on repetitive impact. In general, the abrupt increases in probe depth during the tests correspond to the occurrence of coating fracture. Repeat tests are performed for each sample at each load to assess any variability.

The "Dynamic Hardness" option of the impact module of the NanoTest system is performed in a single nano-impact test, and an instantaneous change of probe displacement as a function of impact time can be recorded by using a high data acquisition rate system. As with the multiple impact tests, the nano-impact configuration utilizes a specially designed solenoid fixed on the worktable and a ferromagnetic bead on the bottom of the pendulum. The indenter can be positioned at a set acceleration distance away from the sample surface, held by the solenoid connected to a timed relay. Unlike multiple impact test, there is no initial quasistatic load indentation depth for a single impact. Once the solenoid is activated, the indenter is released and accelerated toward the sample surface driven by the applied load. In this stage, the velocity of indenter will increase gradually, resulting in increasing kinetic energy. Subsequently, the tip will penetrate into the sample until the velocity decreases to zero, corresponding to the indentation stage. During this process, the kinetic energy is transferred into the reversible elastic work, irreversible plastic work, and other dissipation processes such as heat loss and frictional work. Subsequently, the stored elastic energy is released, leading to the rebound of the tip. The rebound height will depend on the elastoplastic properties of the sample. Then a new contact cycle with much reduced impact distance will begin. Due to the influence of material resistance, the indenter undergoes a series of impact oscillations, and the impact energy is completely dissipated, ultimately stopping at the surface of the sample. A typical single impact curve is shown in Fig. 9.1C. By extracting the parameters such as impact depth and velocity before and after the first contact cycle, the dynamic hardness and fracture toughness of material can be evaluated quantitatively based on the energy analysis method.

9.2.2 Controllable parameters

In a micro/nano-impact test, the controllable parameters mainly include the applied load, acceleration distance, probe geometry, and the total number of impact cycles. The impact load and acceleration distance are mainly used to determine the impact energy delivered to the samples. Generally, the applied load of nano-impacts is in the range of 1–200 mN, while the applied load of microimpact is between 100 and 5000 mN. The acceleration distance of microimpact is also greater than that of nano-impact. The former is 10–15 μm, while the latter can be adjusted within the range of 10–60 μm. The higher impulse forces and acceleration distances result in microimpact with 2–3 orders of magnitude higher impact energy than nano-impact test, which can accelerate the occurrence of fracture in thicker coatings.

The sharpness of impact indenter has a crucial influence on the deformation and failure mechanism of the coating. The most commonly used indenter in nano-impact is a cube corner diamond indenter with a small tip radius of about 50 nm. The shape indenter helps to generate high contact stress under small load, thereby driving the coating to fracture in a relatively short test time. Due to higher impact energy in microimpact, the spheroconical diamond indenter with tip radius of 8–100 μm has been commonly used to generate sufficiently high stress to cause coating failure. The advantage of using a large spheroconical diamond indenter is that it can significantly reduce the wear of indenters and save experimental costs.

Besides, the impact time is also an important effecting parameter, which monitored the damage evolution of impact samples. For the NanoTest system, the impact cycle for both nano-impact and microimpact is adjustable, with a maximum impact capacity of nearly a thousand times during a multiple-impact testing. A typical nano-impact test duration is 300 s with 75 impact cycles. In addition to these conventional experimental parameters, the impact angle, impact frequency, and test temperature can also be changed in the system, which requires more experimental exploration.

9.3 Multiple nano/microimpact testing

9.3.1 Research progress in multiple-impact testing

Multiple-impact testing can simulate the actual service environment of components by repeatedly loading at high strain rates at designated positions of the samples. In 2001, Beake et al. (2001) first proposed a nanoscale impact technology with small-sized probes to evaluate the toughness, bonding performance, erosion wear resistance, and dynamic hardness of thin film materials under dynamic loading environments. After more than 20 years of development, this technology has been further developed and applied to evaluate the fatigue fracture performance of different materials. In 2004, Beake et al. (2007) and Beake et al. (2004) studied the fracture and fatigue wear properties of ultrathin ta-C films deposited on Si substrates using multiple-impact testing. By analyzing the impact depth – time curves, they divided the nano-impact response of brittle thin film materials into four processes including (1) the occurrence of plastic deformation and nucleation of subsurface cracks caused by initial impact; (2) the further nucleation and growth of microcracks during the fatigue stage, during which the impact depth remains basically unchanged or changes slightly; (3) the combination of cracks leads to brittle fracture of the material, manifested as a sudden and significant change in impact depth; (4) the next fatigue fracture cycle process of the material. By analyzing the depth mutation in the impact curves, the impact fatigue resistance of materials can be compared. In 2007, Beake et al. (2007) used a heating platform to study the relationship between the nano-impact fracture resistance of hard PVD coatings and the ratio of hardness and elastic modulus (H/E_r) at different temperature. The results showed that TiAlN films with higher H and H/E_r were more prone to fracture under impact load than AlCrN films. In 2011, Beake et al. (2011) proposed four key parameters on multiple impact curves to evaluate the impact fatigue performance of materials, namely: the initial contact depth h_s between the indenter and the sample surface, the first impact depth h_1, the final residual depth h_f, and the depth change $h_f - h_1$ caused by continuous impact. It was found that under low impact loads, thin films with high plastic deformation resistance are more resistant to the formation of impact cracks, while the thin film cannot provide protection against the substrate under high impact loads as the impact continues.

In 2011, Chen et al. (2011) collaborated with Beake to study the dynamic mechanical behavior of TiN-based hard coatings under repeated nano-impact loads. The results indicate that the failure mechanism of thin films under dynamic conditions is significantly different from that under quasistatic conditions. The generation of cracks tends to be transverse cracks, and the failure of thin films is cohesive failure. Multilayer TiAlSiN films exhibit higher crack formation resistance compared with single-layer TiN films due to their higher H/E_r and H (Schmauder et al., 2006)/E_r (Veprek & Veprek-Heijman, 2008) values. In 2012, Chen et al. (2012) used repeated nano-impact testing to simulate the erosion effect of zirconia thermal barrier coatings under actual operating conditions. The results showed that there was a clear relationship between the impact resistance and their erosion volume, which can be used as a rapid test method for evaluating the erosion resistance of thermal barrier coatings. In 2016, Chen, Li, et al. (2016) proposed using a new dimensionless parameter $(h_1 - h_s)/(h_f - h_1)$ to evaluate the impact resistance of materials when studying the effect of applied load on the impact performance of TiN and AlTiN hard coatings. It was found the mechanical properties and microstructure of the coating play a positive role in improving its impact resistance under low impact loads. Under high impact loads, the noncylindrical structure of the coating also improves its impact resistance during the initial impact process, while its limited ductility affects the damage tolerance of the coating under subsequent sustained impact conditions.

Faisal et al. (2012) studied the effect of applied load on the nano-impact characteristics of amorphous TiNi films in 2012. The results indicate that when the impact load is less than 1 mN, the impact behavior is mainly determined by the properties of the thin film material, and the impact depth increases with the increase of impact cycles. When the impact load increases to 1 and 10 mN, the performance of interface between the film and substrate determines the nano-impact response. In the early stage of impact, the plastic deformation and crack formation in the film/substrate interface led to an increase in impact depth. As the impact cycles increase, the strain mismatch at the film/substrate interface causes the thin film to peel off and bubble from the substrate, resulting in a downward trend in impact depth. When the impact load increases to 100 mN, the impact response of the film/substrate system is dominated by the mechanical properties of the substrate, and the impact depth increases with the increase of the impact cycles. In 2014, Faisal et al. (2012) further studied the effect of indenter shape on the nano-impact behavior of DLC thin films, and the results showed that the impact depth of the film only showed a decreasing trend with impact cycling under the use of spherical indenters and low load conditions. In other cases, it increased with the increase of impact cycles. They divided the failure process of DLC film under impact load into five stages: (1) plastic deformation of film/substrate system under initial impact; (2) formation of surface cracks in films and strain mismatch at the film/substrate interface; (3) vertical propagation of surface cracks to the film/substrate interface; (4) further separation of film/substrate interface; (5) obvious film failure and separation between the film and substrate.

Besides, Bouzakis et al. (2011, 2013) studied the fracture mechanism of TiAlN thin films using multiple nano-impact testings and compared it with finite element simulation to further demonstrate the correlation between the fatigue failure mechanism of the thin film and impact load. Wheeler and Gunner (2013) studied the nano-impact properties of sol − gel films in 2013. By analyzing various impact parameters, it is proposed to use impact fatigue damage maps to characterize the impact fatigue performance of thin films. Ma et al. (2018) proposed a method to estimate the energy estimation of crack initiation as a damage criterion in Sn-Ag-Cu(Ni) solder bump based on nano-impact testing. Ravi et al. (2017) used nano-impact technology to characterize the fracture resistance of TiN, TiN/NC, and nc-TiAlN/a-Si_3N_4 nanocomposite films during cyclic impact. The results indicate that the columnar structure of TiN thin films exhibits the best nano-impact performance.

In summary, multiple nano-impact techniques can efficiently evaluate the fatigue fracture, erosion wear, and interface bonding properties of materials. However, the research on the nano-impact performance of materials is still in its early stages, and there is no unified criterion for evaluating impact damage. The following will provide a brief introduction to the analysis methods and applications of multiple nano-impact testings using specific examples.

9.3.2 Analysis method of multiple impact depth − time curves

The impact test records the evolution of impact-induced damage (i.e., elastic and plastic deformation, fatigue wear, and fracture) caused by repeated impacts on the test probe over time. The depth change of the probe is recorded over the impact cycle in real time by a subnanometer precision displacement sensor. Two typical multiple impact test depth − time curves are shown in Fig. 9.2. The initial impact

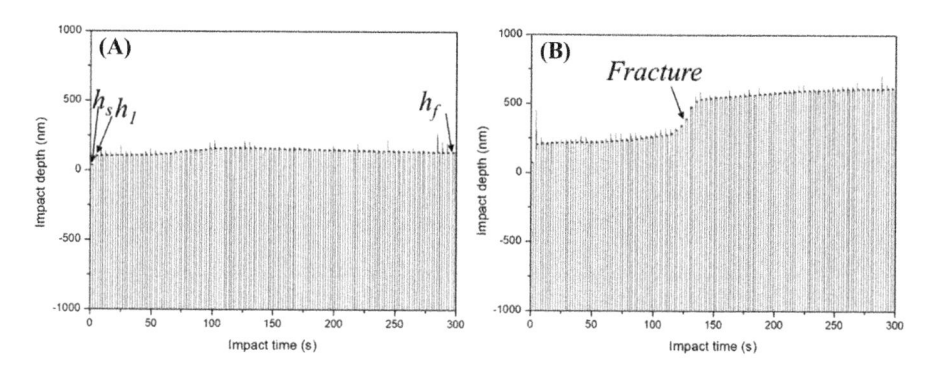

Figure 9.2 Two typical multiple impact test depth − time curves on (A) ductile and (B) brittle materials.

response is very similar on all the samples tested, in which the probe penetration depth increases gradually over the first several impact cycles. However, there are often abrupt increases in probe depth during the test for a brittle material indicative of material removal by fracture or interfacial failure, and there was little change in probe depth throughout the rest of the test (Fig. 9.2B). In contrast, the impact depth always changes slowly with impact time for a ductile material (Fig. 9.2A).

To evaluate the impact resistance of brittle materials, the variation of fracture probability with impact cycles under different loads was plotted by Beake et al. (2004). In their method, multiple repeated tests under the same conditions are first applied to different positions of the sample to improve the statistical significance of the results. Subsequently, the number of impact cycles corresponding to the abrupt steps in impact depth is counted in each test. Fracture probability is estimated by ranking the time-to-failure events in order of increasing fatigue resistance and then assigning a probability of failure $P(f) = n/(N + 1)$ to the nth ranked failure event in a total sample size of N. Plotting the curves of the impact fracture probability of different materials over impact cycles can significantly compare their impact resistance. As an example, Fig. 9.3 indicates that the ta-C thin film systems display some damage tolerance by delaying Si substrate failure and slightly reducing its severity when impacted by a spherical indenter with applied load of 15 mN (Shi et al., 2019).

Although the number of impacts required for failure to occur in 50% of the tests has been very used in studies of the distribution of the critical load for fracture of brittle materials at impact load, the impact resistance of ductile materials cannot be compared through the fracture probability due to the slowly increasing impact curve without obvious fracture step. In this case, the static indentation depth h_s, the first impact depth h_1, and the final impact depth h_f were proposed, as shown in Fig. 9.2A. The static indentation depth h_s is the penetration depth recorded (under load) on the initial contact between the indenter and the surface under quasistatic load before the first impact, which reflects the load support capability of samples. The first impact depth h_1 enables the role of material property on initial resistance

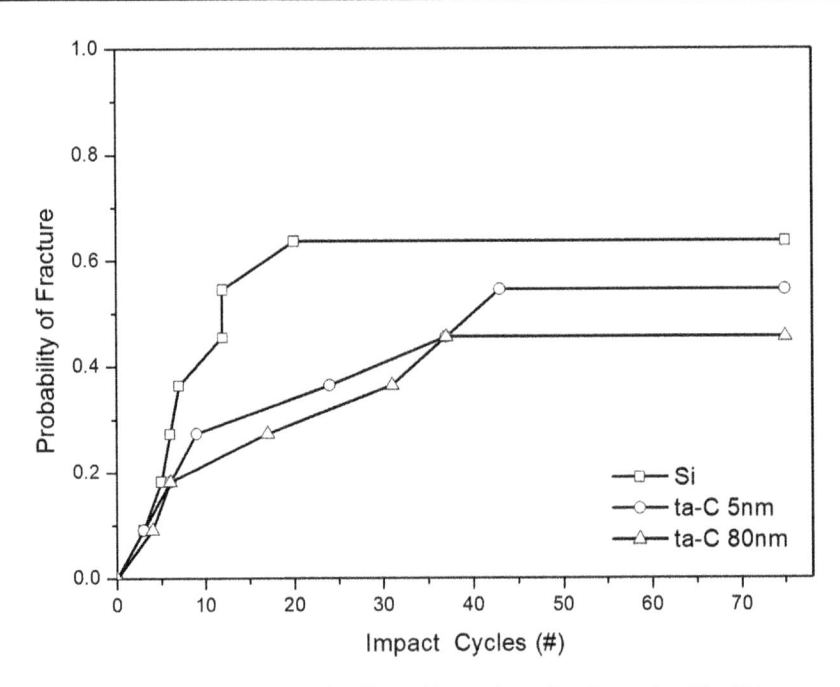

Figure 9.3 Variation in fracture probability with number of cycles under 15 mN impact loads on Si substrate, 5 and 80 nm ta-C films.
Source: Modified from Shi, X., Beake, B. D., Liskiewicz, T. W., Chen, J., & Sun, Z. (2019). Failure mechanism and protective role of ultrathin ta-C films on Si (100) during cyclic nano-impact. *Surface and Coatings Technology, 364*, 32–42. https://doi.org/10.1016/j.surfcoat.2019.02.082.

to crack nucleation, while the value of $h_f - h_1$ presents the resistance of subsequent crack propagation in the impact samples. As an example, Fig. 9.4 shows the different impact parameters of TiN and AlTiN coatings at different impact load (Chen, Li, et al., 2016). Obviously, the AlTiN coating shows enhanced impact resistance in comparison with the TiN coating throughout the impact load range studied, consistent with previous studies showing better impact behavior for more complex nitrides than TiN (Chen et al., 2011; Shi et al., 2019).

When nano-impact tests are done with self-similar indenters, a dimensionless index $(h_1 - h_s)/(h_f - h_1)$ can also provide a measure of the partitioning between initial impact resistance and subsequent damage tolerance and give a useful assessment of differences in load sensitivity between coatings. At low load, if the coating remains crack-resistant within the duration of the test, then the parameter is high (and depending on indenter sharpness it can be >1). This is due to effective impact force in the first impact being greater than the static indentation force and there being little change in depth with subsequent impacts, so that $(h_f - h_1)$ is small. When coatings have high surface roughness and/or they show a combination of crack-free or cracking within the repeat tests, then this results in high variability. At

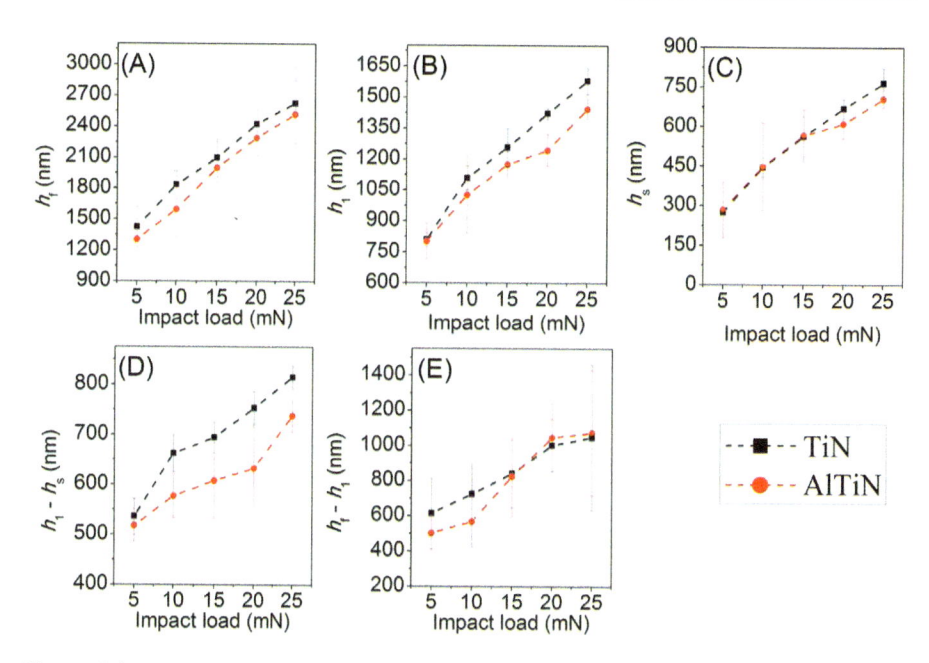

Figure 9.4 Impact parameters (A) final impact depth after 44 impacts h_f (B) initial impact depth h_1, (C) static depth h_s, (D) $h_1 - h_s$, (E) $h_f - h_1$ for TiN and AlTiN coatings.
Source: From Chen, J., Li, H., & Beake, B. D. (2016). Load sensitivity in repetitive nano-impact testing of TiN and AlTiN coatings. *Surface and Coatings Technology*, *308*, 289–297. https://doi.org/10.1016/j.surfcoat.2016.05.094.

higher load, there appears to be two distinct regimes—the first being where the initial impact is not enough to fail coating but with continued impact clear failure occurs, and at yet higher load, a transition to a situation where more significant fracture occurs on the first impact event (Chen, Li, et al., 2016).

9.3.3 Applications of multiple nano/microimpact testings

9.3.3.1 Evaluation of impact resistance of hard diamond-like carbon coatings

DLC coatings have been widely used as protective films in different industrial fields due to their high hardness, low friction coefficient, high wear resistance, and chemical stability (Bewilogua & Hofmann, 2014; Erdemir & Martin, 2018; Vetter, 2014). For example, in microelectromechanical systems (MEMS), complex mechanical behaviors such as high brittleness and phase structure transition under high contact stress limit the reliable performance of silicon-based systems. However, the application of high-hardness ultrathin DLC films (2–100 nm) can significantly improve the fracture resistance and service life of MEMS systems without affecting the substrate performance (Luo et al., 2007). Besides, thick DLC coating can be deposited on the surfaces of mechanical components such as cutting tools, stamping abrasives, and automotive engines, effectively reducing the friction coefficient, improving wear resistance and impact resistance (Erdemir & Donnet, 2006). Therefore, evaluating the service life and failure mechanism of DLC coatings with different thicknesses under repeated high strain rate impact loads has become a crucial aspect.

In this case, the multiple nano-impact testings are used to evaluate the failure mechanism and protective role of ultrathin ta-C films (5—80 nm) deposited on Si substrate, while the repetitive microimpact testing is used to investigate the impact fatigue resistance of thick graphite-like carbon (GLC) films (0.5—1.5 μm) deposited on 316 L stainless steel. In the nano-impact testing, a spheroconical diamond indenter with end radius ~4.6 μm was accelerated from a distance of 12 μm above the sample surface to generate 75 repetitive impacts. The applied impact load was set in the range of 5—40 mN and 10 repeat impact tests were performed at different positions for each load. For the multiple microimpact testings, a spherical diamond indenter with end radius of 40 μm was chosen, and the acceleration distance was increased to 40 μm with the applied force in the range of 0.1—2 N.

The failure behavior and damage mechanism of ultrathin ta-C thin films are highly correlated with impact loads. At low impact load, the impact energy is insufficient to cause fracture failure of Si substrate, but the abrupt depth change close to the thickness of the film (80 nm) indicates the delamination of the ta-C film. The corresponding SEM proves the fatigue failure and material removal of the 80 nm ta-C films at 5 mN impact load. In nano-impact tests on these films with a blunt Berkovich indenter at very low impact forces (100—300 μN), film-dominated behavior was also observed (Beake et al., 2004). There was gradual film failure after a period of fatigue, but subsequent substrate failure was not observed. The impact depth after the test was greater for the taC-80 than for taC-5, by an amount close to the difference in their thickness. Low load (100—1000 μN) nano-impact tests have also been performed on a 100 nm sputtered DLC on silicon with Berkovich and 10 μm conical indenters by Ahmed et al. (2012) and Faisal et al. (2012). They summarized the failure mechanisms at these very low loads as: (1) some plastic deformation occurs in the film and substrate, and microcracks nucleate in the films after initial impact; (2) the cracks continue to nucleate, coalesce, and propagate to the film/substrate interface with little or no probe depth change during the first few impact cycles; (3) debonding occurs at the interface with a more rapid depth increase due to the interfacial mismatch when the plastic deformation occurred and the relaxation of preexisting residual stress in the film. In all these studies, the loads are sufficiently low that the stress fields were more concentrated in the film and permanent deformation of the Si was low (Fig. 9.5).

At high-impact load, the films exhibited a similar type of impact response to the uncoated substrate, with the rapid increases in depth during the test far exceeding the film thickness. However, lower mean impact depth and the larger number of impact cycles required before the rapid depth change indicate that the thin film systems display some damage tolerance by delaying substrate failure and slightly reducing its severity, particularly for the 80 nm ta-C coating. Substrate fracture

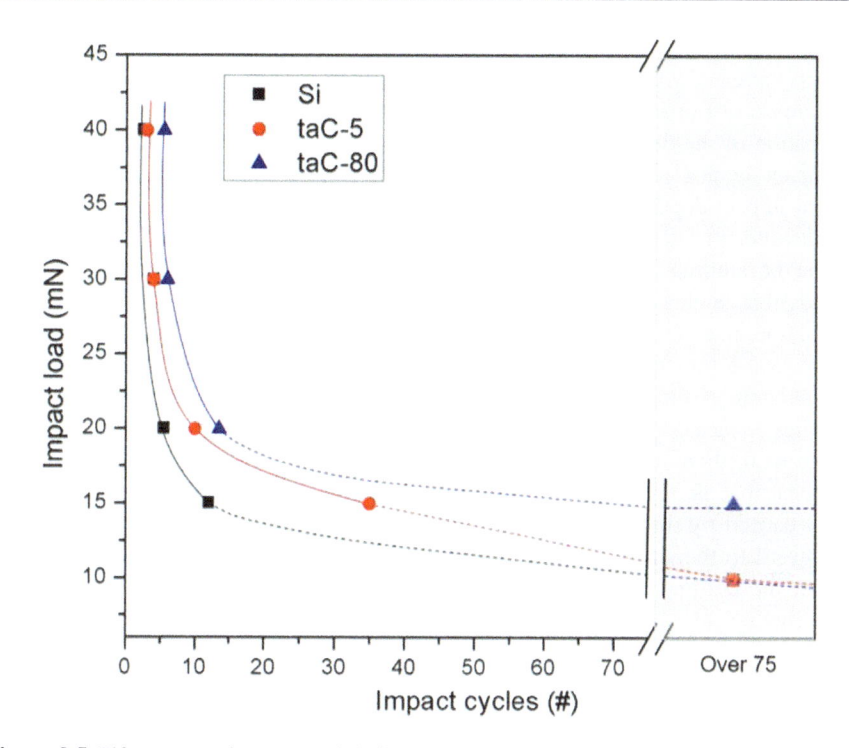

Figure 9.5 Fifty percent fracture probability under different impact loads as a function of impact cycles for 5 and 80 nm ta-C film on Si and Si.
Source: From Shi, X., Beake, B. D., Liskiewicz, T. W., Chen, J., & Sun, Z. (2019). Failure mechanism and protective role of ultrathin ta-C films on Si (100) during cyclic nano-impact. *Surface and Coatings Technology, 364*, 32−42. https://doi.org/10.1016/j.surfcoat.2019.02.082.

occurred at higher loads through a failure mechanism involving initially plastic deformation/phase transformation during the first few impact cycles, with subsequent brittle fracture after the completed plastic deformation. The improvement of ta-C films could be attributed to the high hardness of the ta-C films, which restricts the occurrence and development of phase transformation in the silicon substrate by providing load support, lowering the contact-induced stresses in the substrate. Besides, fracture and delamination of the films will absorb part of the impact energy, which may also reduce the stresses reaching the substrate, retarding the initiation and propagation of the cracks in the substrate (Shi et al., 2019). Based on statistical analysis of impact curves, the 50% fracture probability under different impact loads as a function of impact cycles (Woehler-like curves) was plotted in Fig. 9.5.

The results of microimpact testing on the thick GLC films deposited on ductile substrate indicate that all the GLC samples with different thickness show similar behavior, where the impact depth increases rapidly at the first few impact cycles and then increases more slowly to reach an approximately constant plateau depth

(Shi et al., 2021). The plateau depth and the required impact number to reach this platform increase with the increasing of acceleration forces. Different from the sudden increase of impact depth in the ultrathin ta-C coated on Si substrate, the plastic deformation of the ductile substrate affects the total kinetic energy minimally when the impact cycles reach a certain number, corresponding to the onset of near-constant impact depth until the end of the test. For a brittle thin film coated on ductile substrate, the general deformation-damage evolution to repetitive impact contact can be summarized by three stages: (1) plastic deformation of the substrate with possible nucleation of microcracks in the thin films, (2) suppression of plastic deformation and the further growth of subsurface cracks, (3) crack coalescence and fracture in the thin films. This type of contact damage has been reported on DLC films, ceramic coating, and amorphous TiNi film by using nano-impact and macro-impact tests (Faisal et al., 2012; Wheeler & Gunner, 2013).

In general, there are three types of cracks within the hard thin film on compliant substrates, that is, radial cracks, ring cracks, and lateral cracks. Radial cracks are usually caused by the radial tensile stress at the film/substrate interface and propagate toward to the surface of films, while the ring crack is dependent on the radial stress at the periphery of the indenter and grows downward to the interior of the films (Singh et al., 2008; Xie et al., 2007). In this case, the thinnest film has higher sensitivity to radial cracks in the inner zone of the indent, despite its higher intrinsic hardness. In contrast, the thickest GLC film provided more load capability, but more pronounced ring cracks at the edge of the impact crater caused by more defects and looser microstructure. The medium thickness film seems to display the best impact fatigue resistance due to the combined effects of stress distribution, microstructure, and mechanical properties.

9.3.3.2 Simulation of erosion failure of thermal barrier coating

As the "protective clothing" for aircraft engines, TBCs operating in extreme environments may be affected by various damage modes and microstructure changes, including oxidation, erosion, foreign object damage (FOD), compaction, calcium−magnesium−aluminum−silicate erosion (CMAS), and sintering. With the oxidation of the bonding layer, the erosion of TBCs has been recognized as one of the key factors affecting their service life (Vecchione et al., 2009; Wellman et al., 2005, 2009). The resistance to erosion attack of TBCs produced by electron-beam physical vapor deposition (EB-PVD) and air plasma spray (APS) has been investigated by several researchers (Janos et al., 1999; Wellman et al., 2005). However, erosion resistance is commonly investigated using test apparatus and approaches originally designed for bulk materials. Testing is costly and time-consuming (Hassani et al., 2008), and difficulties may arise due to the interactions of many factors. There is a clear need to develop a rapid laboratory test method that can be used for the prediction of erosion resistance of TBCs, thus aiding in their development.

In this case, a series of nano-impact tests have been performed to investigate whether there was a direct relationship between the results of the dynamic nano-impact testing and the erosion performance of the TBCs (Chen et al., 2012). The samples are as-received, heat-treated YSZ EB-PVD TBCs. The nano-impact tests were carried out with the multiple impulse impact module of the NanoTest system. A well-worn (blunt) cube corner diamond probe with a radius of 11.5 μm was accelerated from a set distance (~12 μm) away from the surface to produce repetitive impacts at 1, 10, or 20 mN. The total impact time for each test is 300 s with 75 impact cycles, and each impact test was repeated at different positions on the TBC surface at least five times.

Typical nano-impact curves with the as-received and heat treated YSZ EB-PVD TBCs are shown in Fig. 9.6B. In general, the impact probe depth increases with the number of impacts and the applied load. The as-received TBC sample shows the lowest impact depth and the depth vs. number of impacts curve is smooth indicating a gradual damage mode. Impact resistance became significantly worse for the HT1100°C100h (Al_2O_3). The impact resistance of the HT1500°C24h (ZrO_2) was slightly worse than that of the HT1100°C100h (Al_2O_3). Abrupt increases in depth were commonly observed for all the thermally aged samples suggesting a more

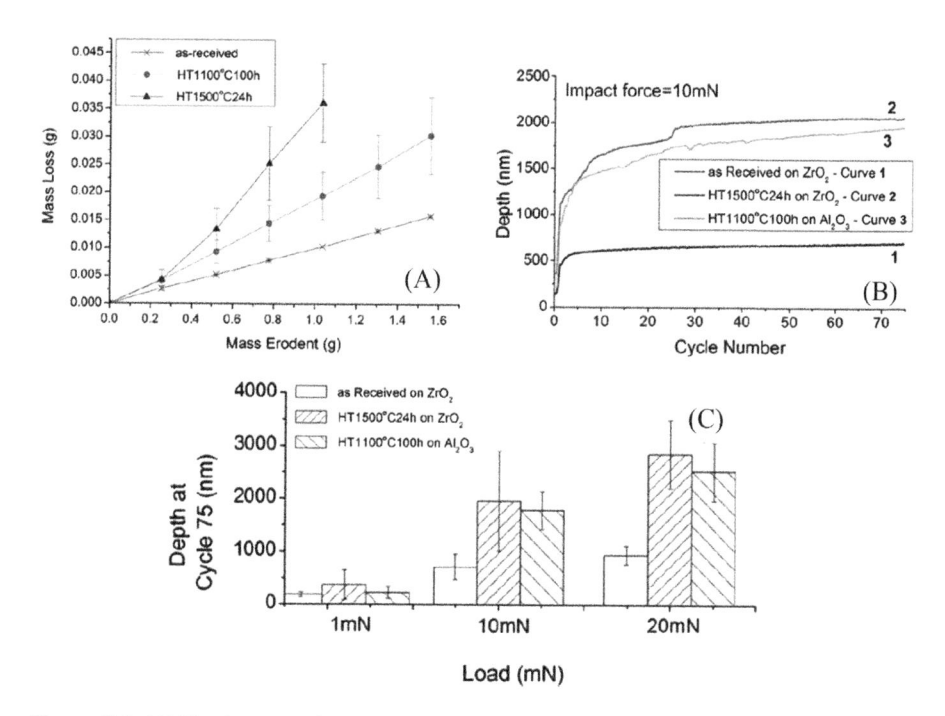

Figure 9.6 (A) Erosion mass loss data of as-received and heat treated TBCs, (B) typical nano-impact curves of as-received and heat-treated TBCs at 10 mN, (C) depth at 75th impact cycles.
Source: From Chen, J., Beake, B. D., Wellman, R. G., Nicholls, J. R., & Dong, H. (2012). An investigation into the correlation between nano-impact resistance and erosion performance of EB-PVD thermal barrier coatings on thermal ageing. *Surface and Coatings Technology*, 206(23), 4992–4998. https://doi.org/10.1016/j.surfcoat.2012.06.011.

cracking-dominated impact response. From the impact depth plots in Fig. 9.5, it is possible to rank (from high to low) the impact resistance as as-received $\gg$ HT1100°C100h (Al$_2$O$_3$) $\geq$ HT1500°C24h (ZrO $\geq$) $\geq$ HT1500°C24h (ZrO$_2$). To capture the statistics of the evolution of the impact probe depth with progressive impacts, the depths at cycle 75 were determined from the data and are summarized in Fig. 9.6C. This confirms that the HT1500°24h (ZrO$_2$) sample generally showed slightly worse resistance than the HT1100°C100h (Al$_2$O$_3$) sample, and the as-received sample showed the lowest impact depth and scattering at all impact forces. These results are consistent with the erosion data of the same samples shown in Fig. 9.6A.

The good agreement between the nano-impact results and the erosion data can be attributed to the matching of the contact deformation size in the two types of tests. The ratio between the contact diameter of the impacting particle (D) and the column diameter of the coating (d) is a critical parameter to determine the erosion failure mechanism. When D/d is less than 2, an erosion mechanism (lateral crack/sublayer crack) plays a dominant role. Based on the geometry of the blunt diamond probe used for the nano-impact tests, the maximum ratio of contact radius against the column size was around 2, and for most of the measurements it was considerably lower, and the coating failed though the formation of lateral cracks. The blunt probe used in the nano-impact enabled closer matching of contact geometry in the impact test than the sharp probe geometry used in the indentation. The static loading depth during nano-impact, produced by quasistatic loading with the blunt probe, generally showed a higher penetration resistance of the as-received TBC compared with the aged samples. This difference might be attributed to the larger contact diameter by the blunt probe and different stress distribution under the probe than in the nanoindentation with the sharp Berkovich probe. The nano-impact test, as repetitive dynamic test producing damage on the correct length scale, has been shown to be effective in simulating the erosion mechanism and could be used as a rapid screening test to evaluate the erosion resistance.

9.4 High-precision single nano-impact testing

9.4.1 Research progress in single nano-impact testing

In the 1940s, Tabor (1951) dropped a hard ball from a high altitude and hit the metal surface. They found that the impact energy of the ball was proportional to the volume of metal surface deformation, indicating that the metal surface provided a constant resistance to the ball. Therefore, the dynamic hardness H$_D$ of the material surface can be obtained from the following Eq. (9.1):

$$H_D = \frac{AE}{V_r} \tag{9.1}$$

AE represents the energy difference between the falling and rebounding processes of the small ball, which is the energy absorbed by the metal surface due to plastic deformation; V_r represents the volume of residual indentation on the metal surface. Compared with traditional Hopkinson compression bars, this method not only obtains dynamic hardness values that match it but also greatly simplifies the dynamic mechanical testing process.

Although the measurement of dynamic hardness H_D was initially based on methods of high load (kN) and large depth changes (mm) at the macroscale, the emergence of nano-impact has enabled its application at the nano or microscale. Based on Andrews' one-dimensional model (Andrews et al., 2002), Constantinides et al. (2008) evaluated the dynamic hardness H_{imp} of Al (1100) using a high data acquisition rate single nano-impact technique, as described in Eq. (9.2):

$$H_{imp} = \frac{W_p}{V_r} = \frac{1/2 m(v_{in}^2 - v_{out}^2)}{8.2 x_r^3} \tag{9.2}$$

W_p is the energy absorbed by the plastic deformation of the material, which is equal to $1/2\, mv_{in}^2$ of the total kinetic energy when the indenter contacts the material surface minus $1/2\, mv_{out}^2$ of the energy stored by the elastic deformation of the material. V_r is the residual indentation volume on the surface of the material, and its relationship with the residual depth x_r is approximately $V_r = 8.2 x_r^3$. The dynamic hardness of Al (1100) obtained through this formula is about 0.29 GPa, slightly higher than its quasistatic hardness of 0.27 GPa.

Trelewicz and Schuh (2008) used a similar method to evaluate the dynamic hardness H_D of Ni-W alloys with different grain sizes at high strain rates. The results indicate that the dynamic hardness evolution of Ni-W alloy is consistent with the quasistatic results under large nano grain sizes, following the classical Hall − Petch relationship. When the grain size is around 20 nm, the H_D of the alloy reaches a peak, which clearly violates the Hall − Petch relationship. Somekawa and Schuh (2012) and (2013) analyzed the dynamic hardness of fine-grained magnesium alloys at high strain rates using a single nano-impact technique. The results showed that there was a good linear relationship between the hardness values of H_D and quasistatic with the change of strain rate, and different alloy elements had varying degrees of influence on their dynamic hardness values. Arreguin-Zavala et al. (2013) found that the dynamic hardness H_D of Al-Si coatings tends to a stable value of 1.57 GPa with the increase of residual depth, and this value is not closely related to the grain size of Al-Si coatings. By comparing with the quasistatic hardness H, it was found that there is a better correlation between the value of H_D/E and the dry sliding wear of the coating. Ghosh et al. (2017) also studied the indentation size effect of nano-impact dynamic hardness of materials. Slightly different from the results of Zavala et al., Ghosh et al. found that the dynamic hardness of three different materials (AA7075-T6, A7075-Annealed, and Ni-200) showed a decreasing trend with increasing residual depth of indentation, similar to the indentation size effect of materials under quasistatic conditions.

In summary, the high data acquisition single nano-impact technology has been successfully applied to evaluate the dynamic hardness of materials. However, currently the main focus is on the study of bulk materials, and there are few reports on the dynamic hardness evaluation of coatings. In addition, there is a lack of in-depth research on the quantitative evaluation of dynamic fracture toughness of materials using nano-impact. Therefore, using nano-impact technology to study the dynamic mechanical properties (dynamic hardness, dynamic toughness) of coatings not only enables the development of dynamic testing techniques but also has positive significance for deepening the understanding of the performance evolution of coatings under dynamic conditions and optimizing coatings design. The following will elaborate on the physical model of high-precision single nano-impact and the evaluation of dynamic hardness and dynamic fracture toughness of coating materials using specific cases.

9.4.2 The physical mode of single nano-impact

By using a high-speed data acquisition nano-impact device, the depth − time curves of the material under instantaneous (800 Ms) impact response can be accurately recorded, as shown in Fig. 9.7. It can be observed that under the set pendulum impact distance, the indenter initiates a rapid impact on the surface of the sample under accelerated load. Due to the influence of material resistance, the indenter undergoes a series of impact oscillations, and the impact energy is completely dissipated, ultimately stopping at the surface of the sample. Comparing the depth peaks of each impact oscillation, it can be found that the impact depth of the material reaches its maximum value during the first impact contact process. Therefore, it

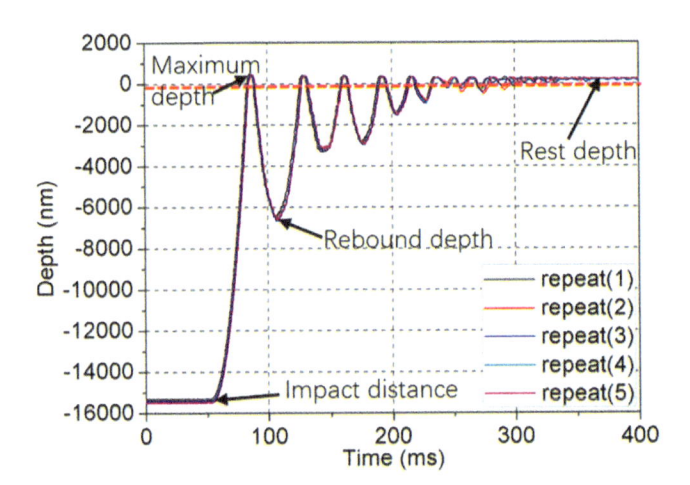

Figure 9.7 Typical dynamic impact depth − time curve of single nanoimpact.
Source: From Shi, X., Li, H., Beake, B. D., Bao, M., Liskiewicz, T. W., Sun, Z., & Chen, J. (2020). Dynamic fracture of CrN coating by highly-resolved nano-impact. *Surface and Coatings Technology, 383*. https://doi.org/10.1016/j.surfcoat.2019.125288.

can be assumed that the plastic deformation of the material has been completed during the first impact contact, and the subsequent impact contact mainly manifests as an elastic deformation. In fact, Wheeler et al. (2019) recently confirmed this hypothesis by preventing subsequent rebound impacts on Cu samples.

Fig. 9.8 magnifies the displacement − time curve of the material during the first impact contact and the velocity − time curve obtained by fitting the derivative. By analyzing the impact contact process, the entire impact can be divided into four stages, and each stage can be limited by critical velocity and displacement.

1. The indenter moves toward the surface of the sample through the set impact distance under a given load. At this stage, the velocity of the indenter is continuously increased by the acceleration load. When the indenter contacts the surface of the sample, the velocity reaches its maximum value $v_0 = v_{in}$, and the impact depth $h_0 = 0$. This stage is called the acceleration stage.
2. In the second stage, the indenter continues to press into the material under the action of accelerated kinetic energy. At this time, some of the kinetic energy is converted into reversible stored elastic work, while the other part is absorbed by irreversible plastic deformation of the material. The velocity of the indenter gradually decreases. When the depth of penetration reaches the maximum value h_m, the velocity v_1 of the indenter decreases to 0. This stage is called the push-in stage.
3. In the third stage, due to the release of energy stored in the elastic deformation, the indenter rebounds from inside the material. When the elastic work is released, the rebound speed v_2 of the indenter reaches its maximum value v_{out}. At this point, the indenter begins to separate from the material and leaves a residual depth h_r on the surface of the sample. This stage is called the rebound stage.

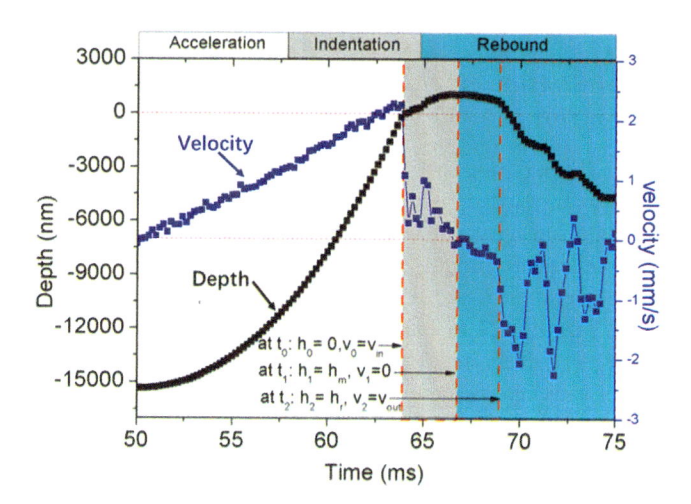

Figure 9.8 A typical curve of depth and velocity over time during the first impact contact. *Source*: From Shi, X., Li, H., Beake, B. D., Bao, M., Liskiewicz, T. W., Sun, Z. & Chen, J. (2020). Dynamic fracture of CrN coating by highly-resolved nano-impact. *Surface and Coatings Technology, 383*. https://doi.org/10.1016/j.surfcoat.2019.125288.

4. Finally, the rebound speed of the indenter decreases continuously under the impact load and air resistance until it reaches 0, and the indenter also reaches its maximum distance from the sample surface. This stage is called the deceleration stage, and the entire impact cycle is completed. Then, under the accelerated load, the indenter initiates a new round of impact oscillation. As the impact distance continues to decrease, the kinetic energy at the initial contact also gradually decreases, resulting in the indenter finally stopping at the surface of the sample after undergoing a series of contact cycles.

9.4.3 Applications of single nano-impact testing

The high strain rate pendulum-based instrumented nano-impact technique has been used to record the dynamic mechanical response of various materials (Chen, Li, et al., 2016; Ravi et al., 2017) It was found that the performance in real applications where high strain rate intermittent contact occurs such as erosion, wear, fatigue, and cutting can be more correlated to the evolution of the depth during repeated impacting than to quasistatic mechanical properties from nanoindentation due to closer simulation of the dynamic conditions in the impact test (Chen et al., 2012; Saber-Samandari & Gross, 2013; Wheeler & Gunner, 2013). However, most of studies were based upon the statistical analysis of depth changes or the indentation compliance rather than extracting the dynamic mechanical properties (hardness and toughness) (Chen, Shi, et al., 2016). As a clearer metric of strain rate sensitivity in resistance to plastic deformation, H_D (dynamic hardness) and K_{IC} (dynamic toughness) is more attractive to the scientific community as it not only provides a quantitative measure of the resistance to plastic deformation, which compared with quasistatic properties can also provide indirect information for deformation and failure mechanism through the strain rate sensitivity.

9.4.3.1 Determination of dynamic hardness of graphite-like carbon films

In this case, the GLC coating with thickness of 1.5 μm was deposited on 316 L stainless steel by using closed-field unbalanced magnetron sputtering system (Shi et al., 2021). Single nano-impact tests with a speroconical diamond indenter of 5 μm end radius were performed to evaluate the dynamic hardness of GLC film. In this study, the acceleration distance was set to 15 μm, and the acceleration force was varied from 2 to 40 mN to obtain a range of impact energies. Five repeated tests were conducted at different positions for each load (Fig. 9.9)

Fig. 9.9 summarizes the average maximum impact depth h_m, residual depth h_r, impact contact velocity v_{in}, separation velocity v_{out} between the indenter and the sample, as well as their ratios h_r/h_m and v_{out}/v_{in}, obtained by fitting and deriving the impact curves of the stainless steel substrate (ss) and GLC coating under different impact loads. It can be observed that compared with uncoated SS substrate, GLC film exhibits lower h_m and h_r under corresponding loads, indicating that the deposition of GLC film provides more load support for the substrate and reduces the plastic deformation effect on the substrate material. The h_r/h_m value to some extent also

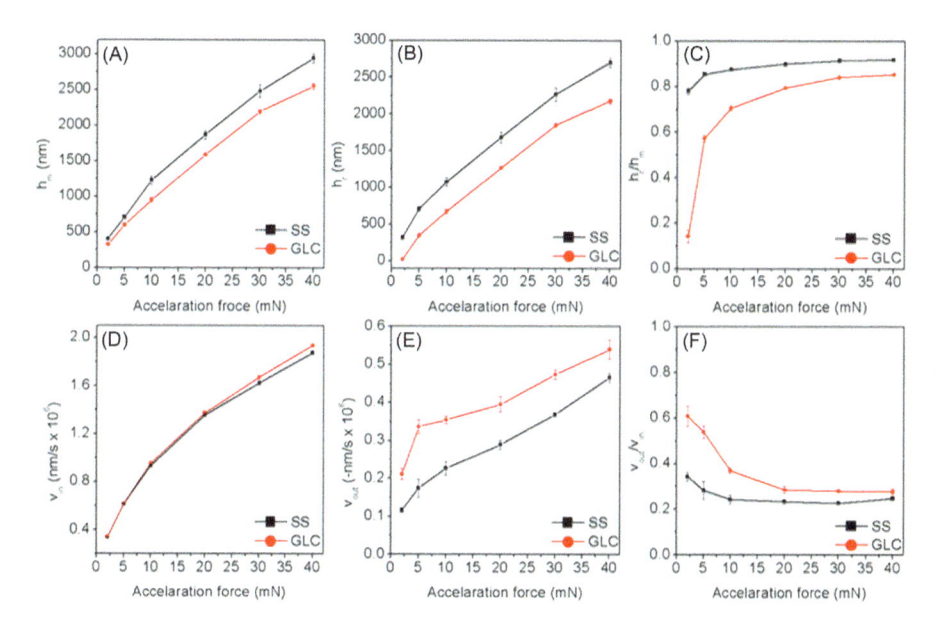

Figure 9.9 The average maximum impact depth h_m, residual depth h_r, impact contact velocity v_{in}, separation velocity v_{out}, as well as their ratios h_r/h_m and v_{out}/v_{in}, of the stainless steel substrate (ss) and GLC coating under different impact loads.
Source: Modified from Shi, X., Chen, J., Beake, B. D., Liskiewicz, T. W., & Wang, Z. (2021). Dynamic contact behavior of graphite-like carbon films on ductile substrate under nano/micro-scale impact. *Surface and Coatings Technology*, 422. https://doi.org/10.1016/j.surfcoat.2021.127515.

reflects the material's ability to resist plastic deformation. In Fig. 9.9(E), the GLC film exhibits a higher separation rate v_{out} than the uncoated substrate, indicating that the plastic deformation of the substrate absorbs more indenter kinetic energy during the impact. The value of v_{out}/v_{in} is commonly used to represent the energy loss of a material during the impact process. Fig. 9.9(F) shows that under low acceleration load, the high hardness of GLC film can provide more load support for the substrate, resulting in lower energy loss during the impact. As the acceleration load increases, the energy loss of GLC film becomes closer to the energy loss of the uncoated substrate.

Based on Tabor's energy calculation method, the dynamic hardness H_d is expressed as the plastic work absorbed per unit deformation volume, as shown in Eq. (9.3):

$$H_d = \frac{W_p}{Vr} \tag{9.3}$$

Due to the impact indentation of GLC coating under spherical indenters exhibiting a spherical crown morphology, the deformation volume V_r can be calculated using Eq. (9.4):

$$V_r = \frac{\pi}{3} \times (3R - h_r) \times h_r^2 \tag{9.4}$$

Where R is the radius of the spherical indenter, h_r is the depth of the spherical crown, which is equal to the residual depth of the indentation in this experiment.

Through the analysis of the first cycle in the impact process, the total energy KE_0 of the impact contact can be divided into three parts:

$$KE_0 = W_e + W_p + W_0 \tag{9.5}$$

Where W_e represents the energy stored during the elastic deformation. When the indenter separates from the material, the stored elastic energy completely released and converted into the kinetic energy of indenter, which is equal to 1/2 mv_{out}^2. W_0 represents other energy losses, such as various heat and friction losses, as the relative proportion of this energy is small and can be ignored. Therefore, the W_p absorbed by plastic deformation of the material can be derived:

$$W_P = \frac{1}{2}mv_{in}^2 - \frac{1}{2}mv_{out}^2 \tag{9.6}$$

By combining the aforementioned formula, the dynamic hardness H_d of GLC film and SS substrate can be calculated as shown in Fig. 9.10A. It can be observed that the dynamic hardness of the GLC film is much greater than the dynamic hardness value of the substrate at low residual depths, and as the residual depth increases, it becomes closer to the H_d value of the substrate material. It should be noted the calculated dynamic hardness is not only the value of GLC film, but the

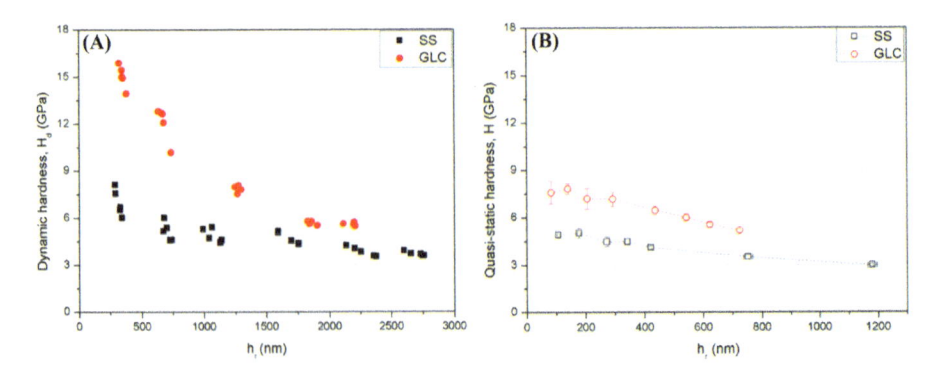

Figure 9.10 (A) The calculated dynamic hardness H_d and (B) quasistatic hardness H of the GLC coating and SS substrate.

comprehensive result of the coating/substrate system. At low acceleration load, the GLC film can provide more load support for the substrate, and the film hardness plays a dominant role in the coated system. With the increase of acceleration load, the plastic deformation of the substrate becomes more and more obvious, and the hardness of the substrate material occupies a dominant role. This phenomenon is consistent with the quasistatic hardness results in Fig. 9.10B.

Besides, by comparing Fig. 9.10A and B, it can be found that for both the GLC coating and the uncoated substrate, the calculated dynamic hardness H_d is a little higher than the quasistatic hardness H at the same residual depth, which is consistent with the previous results obtained on single crystal Al (Qin et al., 2019). This is first attributed to overestimating the energy absorbed by material plastic deformation when calculating dynamic hardness. The energy loss caused by friction and heat loss during the contact are ignored in the calculation, especially under low impact loads where this part of the energy accounts for a large proportion, resulting in the calculated plastic absorption work W_p being greater than the actual absorbed energy. Second, the formation of microcracks at GLC films may also absorb partial energy, and the proportion increases with the increasing of acceleration force. Further, the high dynamic hardness may also be caused by high strain rates. Several metals exhibit an increase in yield strength at high strain rates, known as work hardening. In a nano-impact test, the instantaneous strain rate of the indenter is very high, reaching 10 s^{-1} (Chen, Shi, et al., 2016). Therefore, the strain rate sensitivity of plastic materials leads to an increase in dynamic hardness. The dynamic hardness of gold obtained by Constantinides et al. by using nano-impact was higher than the quasistatic hardness due to the rate-dependent energy absorption (Constantinides et al., 2008). Wheeler et al. (2019) also found the dynamic hardness of Cu sample is higher than that at quasistatic condition, particularly clear at shallower impact depth. By comparing the strain rate sensitivity exponent with the value reported in literature, it was suggested that the dynamic hardness obtained from nano-impact indentation is reasonable.

From the aforementioned, the dynamic hardness of coating/substrate system or coatings with high thickness at high strain rate can be obtained reasonably by using single nano-impact technology, but the impact energy should be controlled in a suitable range to avoid the high calculation error at low impact energy and the fracture of coatings at high impact energy. To calculate the exact value of dynamic hardness, further work should be done to extract the dissipated energy W_0.

9.4.3.2 Evaluation of fracture toughness of hard CrN coating

In this case, the CrN coating with thickness of 2 μm was deposited using cathodic arc evaporation on 42 $CrMo_4$ high-speed steel (Shi et al., 2020). The single nano-impact with a blunt cube corner indenter of 5 μm end radius was carried out to extract the energy dissipation caused by the fracture of CrN coating. The acceleration forces were set in the range of 10−30 mN with accelerating distance (AD) of 15 μm from the sample. Five repeated tests were conducted at different positions for each load (Fig. 9.11).

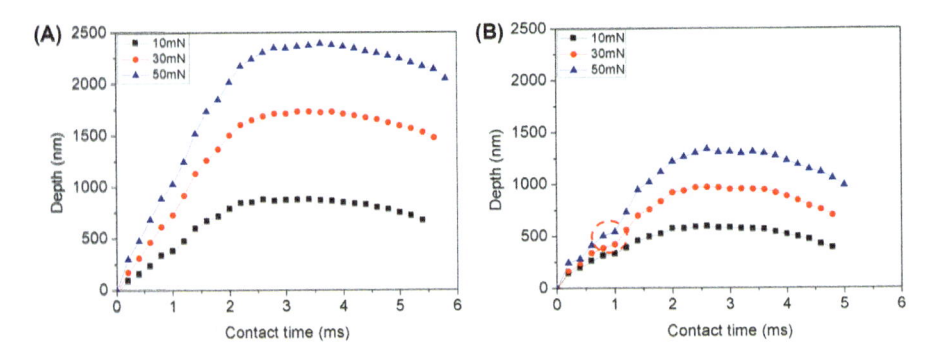

Figure 9.11 The $t-h$ curves during first contact with different acceleration forces (A) HSS, (B) CrN/HSS (the circle highlights the sudden changes during indentation stage). *Source*: From Shi, X., Li, H., Beake, B. D., Bao, M., Liskiewicz, T. W., Sun, Z., & Chen, J. (2020). Dynamic fracture of CrN coating by highly-resolved nano-impact. *Surface and Coatings Technology, 383*. https://doi.org/10.1016/j.surfcoat.2019.125288.

Fig. 9.11 shows the $t-h$ curves of the first impact contact of HSS substrate and CrN/HSS sample under different loads. For the HSS substrate, it is interesting to notice that all the $t-h$ curves for the HSS and that for CrN/HSS at 10 mN show continuous smooth curves, while those for the CrN/HSS at higher AF (30 and 50 mN) exhibit a short irregular "plateau" period during the indentation stage as indicated by the circle in Fig. 9.11B. A similar phenomenon has been reported widely in the multiple nano-impact testings corresponding to fracture and removal of the coating (Wheeler & Gunner, 2013; Wheeler et al., 2019). Surface morphological observation indicates the ring-like cracking induced by nano-impact on CrN/HSS at high impact energy.

Based on the "pop in" phenomenon of the load − displacement curve caused by sudden displacement of the indenter during the nanoindentation process, Li and Bhushan (1998, 1999) and Li et al. (1997) proposed a calculation method for evaluating the fracture toughness K_{IC} of hard coatings, as shown in Eq. (9.7):

$$K_{IC} = \sqrt{\left[\frac{U_{\text{fra}} \times E}{(1-v^2) \times S}\right]} \quad (9.7)$$

Where U_{fra} represents the fracture absorption energy of the coating, E and v are the elastic modulus and Poisson's ratio of the coating, respectively. $S = 2\pi C_R t$ represents the fracture area of the coating, while C_R and t represent the radius of the circumferential crack and the effective thickness of the coating, respectively.

Based on the analysis of the first contact cycle during the impact, the total impact energy W_t can be divided into the following four parts:

$$W_t = W_e + W_p + U_{\text{fra}} + W_o \quad (9.8)$$

where W_e is elastic storage work, W_p is the work of plastic deformation, U_{fra} is the fracture dissipated energy, and W_0 represents other energy loss such as the dissipated heat and friction work. The W_t and W_e can be measured by the kinetic energy at the contact and detachment respectively. Thus, the fracture energy U_{fra} can be simplified as:

$$U_{\text{fra}} = \frac{1}{2}mv_0^2 - \frac{1}{2}mv_1^2 - W_p - W_0 \tag{9.9}$$

Clearly, the calculation of fracture dissipation work U_{fra} depends on the acquisition of irreversible work W_p and W_0. Because the deformation of CrN/HSS mainly manifests as plastic deformation of the substrate, the energy loss W_0 of CrN/HSS is not significantly different from the loss caused by pure substrate deformation. Therefore, it can be estimated through the energy conversion of HSS substrate:

$$W_0 = W_t - W_p - W_e \tag{9.10}$$

Chen and Bull (2007, 2008) and Chen (2012) proposed that the influence of cracking for a coated system can be averaged over the whole loading curve as the E/H remains almost constant and the ring-like cracking occurs during loading. Despites the presence of crack, the total indentation work will not be strongly affected because fracture only plays a role in converting some stored elastic energy into irreversible work. In light of this, the W_p can be roughly estimated by Eq. (9.11):

$$\frac{W_p}{W_t} = 1 - \gamma \frac{H}{E_r} \tag{9.11}$$

Combined with Eqs. (9.7)−(9.11), the fracture dissipated work and the fracture toughness of the coating can be approximately estimated. The detailed data for the calculated energy at acceleration force of 50 mN were concluded in Table 9.1 To avoid the influence of indenter shape, the indenter geometry constant γ for CrN/HSS sample was taken as 5.7−6.4 in the calculation process, and the value of K_{IC} was estimated in the range of 2.75−7.74 MPa m$^{1/2}$, which is comparable with the reported data (K_{IC} = 3.13 MPa m$^{1/2}$) of Ms-PVD CrN coating determined by single-cantilever bending and pillar splitting experiments (Sebastiani et al., 2015).

Table 9.1 The calculated fracture dissipated energy and fracture toughness for CrN coating.

Sample	W_t (nJ)	W_e (nJ)	γ	W_p (nJ)	W_0 (nJ)	W_f (nJ)	K_{IC} (MPa m$^{1/2}$)
HSS	484 ± 5.0	24 ± 0.7	7	407 ± 4.8	54 ± 0.6	/	/
CrN	543 ± 4.8	51 ± 2.3	5.7∼6.4	422 ± 3.6 ∼435 ± 3.8	54 ± 0.6	2 ± 0.5 ∼16 ± 0.7	2.75 ± 0.3 ∼7.74 ± 0.5

Source: From Shi, X., Li, H., Beake, B. D., Bao, M., Liskiewicz, T. W., Sun, Z., & Chen, J. (2020). Dynamic fracture of CrN coating by highly-resolved nano-impact. *Surface and Coatings Technology*, 383. https://doi.org/10.1016/j.surfcoat.2019.125288.

In summary, we provide a new insight to evaluate the dynamic fracture toughness of hard coatings based on the energy method by using nano-impact testing. However, the accurate extraction of irreversible work W_p during the impact process is still a great challenge, which is also the focal point of much debate in indentation fracture method. Besides, there are also some errors in the calculation of other dissipated energy W_0 due to the different impact depth of CrN/HSS and uncoated HSS substrate at the same acceleration load. Thus, further investigation will be done on the acquisition of indenter geometry constant γ and irreversible plastic work W_p during nano-impact for more precise quantification of the dynamic fracture toughness.

9.5 Conclusions

Multiple micro/nano-impact testings can qualitatively evaluate the fatigue fracture, erosion wear, and interface bonding properties of materials. In this test mode, the extracting of impact depth at different impact cycles, especially the confirming of the fracture depth, is crucial to evaluate the impact resistance. Two case studies on ultrathin ta-C deposited on Si substrate and thick GLC on ductile 316 L stainless steel substrate have been used to show the effect of substrate material on the impact response of hard coatings. For ta-C deposited on Si substrate, the brittle fracture of the substrate causes the sudden change of the impact depth, while the plastic deformation of the ductile stainless steel substrate affects the total kinetic energy minimally when the impact cycles reach a certain number, corresponding to the onset of near-constant impact depth until the end of the test. The case of TBCs coating shows that the micro/nano-impact is effective in simulating the erosion mechanism and could be used as a rapid screening test to evaluate the erosion resistance.

High data acquisition single nano-impact technology can be applied to quantitatively evaluate the dynamic mechanical properties of materials. The fitting of key parameters such as depth and velocity of the indenter at each stage of the impact process is the crucial to analyze the impact energy. Two cases of single nano-impact on DLC and CrN have elaborated on the evaluation methods for dynamic hardness and dynamic toughness, respectively. Although there are still some shortcomings in the analysis method, it provides us with a new perspective to evaluate material properties under high strain rate dynamic loading conditions. Besides, the combination of these two different impact modes can effectively and accurately simulate the high strain rate intermittent loading of coating materials under actual working conditions in the laboratory.

References

Ahmed, R., Fu, Y. Q., & Faisal, N. H. (2012). Fatigue at nanoscale: An integrated stiffness and depth sensing approach to investigate the mechanisms of failure in diamondlike carbon film. *Journal of Tribology, 134*(1). Available from https://doi.org/10.1115/1.4005774.

Andrews, E. W., Giannakopoulos, A. E., Plisson, E., & Suresh, S. (2002). Analysis of the impact of a sharp indenter. *International Journal of Solids and Structures*, *39*(2), 281−295. Available from https://doi.org/10.1016/s0020-7683(01)00215-3.

Arreguin-Zavala, J., Milligan, J., Davies, M. I., Goodes, S. R., & Brochu, M. (2013). Characterization of nanostructured and ultrafine-grain aluminum−silicon claddings using the nanoimpact indentation technique. *JOM*, *65*(6), 763−768. Available from https://doi.org/10.1007/s11837-013-0593-4.

Bantle, R., & Matthews, A. (1995). Investigation into the impact wear behaviour of ceramic coatings. *Surface and Coatings Technology.*, *74−75*(2), 857−868. Available from https://doi.org/10.1016/0257-8972(95)08314-6.

Beake, B. D. (2022). Micro, scale impact testing of hard coatings: A review. *Coatings*, *12*.

Beake, B. D., Goodes, S. R., Smith, J. F., Zhang, A., & Jisheng, E. (2001). Micro-impact testing: A novel nano-/micro-tribological tool for assessing the performance of wear-resistant coatings under impact/fatigue conditions. *Science in China Series A*, 418−422.

Beake, B. D., & Smith, J. F. (2004). Nano-impact testing − An effective tool for assessing the resistance of advanced wear-resistant coatings to fatigue failure and delamination. *Surface and Coatings Technology*, *188−189*(1−3), 594−598. Available from https://doi.org/10.1016/j.surfcoat.2004.07.016.

Beake, B. D., Lau, S. P., & Smith, J. F. (2004). Evaluating the fracture properties and fatigue wear of tetrahedral amorphous carbon films on silicon by nano-impact testing. *Surface and Coatings Technology*, *177−178*, 611−615. Available from https://doi.org/10.1016/S0257-8972(03)00934-4, http://www.journals.elsevier.com/surface-and-coatings-technology/.

Beake, B. D., Smith, J. F., Gray, A., Fox-Rabinovich, G. S., Veldhuis, S. C., & Endrino, J. L. (2007). Investigating the correlation between nano-impact fracture resistance and hardness/modulus ratio from nanoindentation at 25−500°C and the fracture resistance and lifetime of cutting tools with Ti1 − xAlxN (x = 0.5 and 0.67) PVD coatings in milling operations. *Surface and Coatings Technology*, *201*(8), 4585−4593. Available from https://doi.org/10.1016/j.surfcoat.2006.09.118.

Beake, B. D., Vishnyakov, V. M., & Colligon, J. S. (2011). Nano-impact testing of TiFeN and TiFeMoN films for dynamic toughness evaluation. *Journal of Physics D: Applied Physics*, *44*(8). Available from https://doi.org/10.1088/0022-3727/44/8/085301.

Beake, B. D., Liskiewicz, T. W., Bird, A., & Shi, X. (2020). Micro-scale impact testing − A new approach to studying fatigue resistance in hard carbon coatings. *Tribology International*, *149*. Available from https://doi.org/10.1016/j.triboint.2019.04.016, http://www.elsevier.com/inca/publications/store/3/0/4/7/4.

Beake, B. D., Isern, L., Endrino, J. L., Liskiewicz, T. W., & Shi, X. (2021). Micro-scale impact resistance of coatings on hardened tool steel and cemented carbide. *Materials Letters*, *284*. Available from https://doi.org/10.1016/j.matlet.2020.129009.

Bernoulli, D., Wyss, A., Raghavan, R., Thorwarth, K., Hauert, R., & Spolenak, R. (2015). Contact damage of hard and brittle thin films on ductile metallic substrates: an analysis of diamond-like carbon on titanium substrates. *Journal of Materials Science*, *50*(7), 2779−2787. Available from https://doi.org/10.1007/s10853-015-8833-3.

Bewilogua, K., & Hofmann, D. (2014). History of diamond-like carbon films − From first experiments to worldwide applications. *Surface and Coatings Technology*, *242*, 214−225. Available from https://doi.org/10.1016/j.surfcoat.2014.01.031.

Biswas, S., Satapathy, A., & Patnaik, A. (2010). Erosion wear behavior of polymer composites: A review. *Journal of Reinforced Plastics and Composites*, *29*(19), 2898−2924. Available from https://doi.org/10.1177/0731684408097786.

Bouzakis, K.-D., Siganos, A., Leyendecker, T., & Erkens, G. (2004). Thin hard coatings fracture propagation during the impact test. *Thin Solid Films, 460*(1−2), 181−189. Available from https://doi.org/10.1016/j.tsf.2004.02.009.

Bouzakis, K.-D., Gerardis, S., Skordaris, G., & Bouzakis, E. (2011). Nano-impact test on a TiAlN PVD coating and correlation between experimental and FEM results. *Surface and Coatings Technology, 206*(7), 1936−1940. Available from https://doi.org/10.1016/j.surfcoat.2011.08.015.

Bouzakis, K.-D., Skordaris, G., Gerardis, S., & Bouzakis, E. (2013). Nano-impact test on PVD-coatings with graded mechanical properties for assessing their brittleness. *Materialwissenschaft und Werkstofftechnik, 44*(8), 684−690. Available from https://doi.org/10.1002/mawe.201300176.

Chen, J., & Bull, S. J. (2007). Indentation fracture and toughness assessment for thin optical coatings on glass. *Journal of Physics D: Applied Physics, 40*(18), 5401−5417. Available from https://doi.org/10.1088/0022-3727/40/18/s01.

Chen, J., & Bull, S. J. (2008). Investigation of the relationship between work done during indentation and the hardness and Young's modulus obtained by indentation testing. *International Journal of Materials Research, 99*(8), 852−857. Available from https://doi.org/10.3139/146.101709.

Chen, J., Ji, R., Khan, R. H. U., Li, X., Beake, B. D., & Dong, H. (2011). Effects of mechanical properties and layer structure on the cyclic dynamic loading of TiN-based coatings. *Surface and Coatings Technology, 206*(2−3), 522−529. Available from https://doi.org/10.1016/j.surfcoat.2011.07.079.

Chen, J., Beake, B. D., Wellman, R. G., Nicholls, J. R., & Dong, H. (2012). An investigation into the correlation between nano-impact resistance and erosion performance of EB-PVD thermal barrier coatings on thermal ageing. *Surface and Coatings Technology, 206*(23), 4992−4998. Available from https://doi.org/10.1016/j.surfcoat.2012.06.011.

Chen, J., Shi, X., Beake, B. D., Guo, X., Wang, Z., Zhang, Y., Zhang, X., & Goodes, S. R. (2016). An investigation into the dynamic indentation response of metallic materials. *Journal of Materials Science, 51*(18), 8310−8322. Available from https://doi.org/10.1007/s10853-016-0031-4.

Chen, J., Li, H., & Beake, B. D. (2016). Load sensitivity in repetitive nano-impact testing of TiN and AlTiN coatings. *Surface and Coatings Technology, 308*, 289−297. Available from https://doi.org/10.1016/j.surfcoat.2016.05.094.

Chen, J. (2012). Indentation-based methods to assess fracture toughness for thin coatings. *Journal of Physics D: Applied Physics, 45*(20). Available from https://doi.org/10.1088/0022-3727/45/20/203001.

Constantinides, G., Tweedie, C. A., Holbrook, D. M., Barragan, P., Smith, J. F., & Van Vliet, K. J. (2008). Quantifying deformation and energy dissipation of polymeric surfaces under localized impact. *Materials Science and Engineering: A., 489*(1−2), 403−412. Available from https://doi.org/10.1016/j.msea.2007.12.044, http://www.elsevier.com.

Deng, Y., Chen, W., Li, B., Wang, C., Kuang, T., & Li, Y. (2020). Physical vapor deposition technology for coated cutting tools: A review. *Ceramics International, 46*(11), 18373−18390. Available from https://doi.org/10.1016/j.ceramint.2020.04.168, https://www.journals.elsevier.com/ceramics-international.

Erdemir, A., & Donnet, C. (2006). Tribology of diamond-like carbon films: Recent progress and future prospects. *Journal of Physics D: Applied Physics, 39*(18), R311−R327. Available from https://doi.org/10.1088/0022-3727/39/18/R01.

Erdemir, A., & Martin, J. M. (2018). Superior wear resistance of diamond and DLC coatings. *Current Opinion in Solid State and Materials Science*, *22*(6), 243−254. Available from https://doi.org/10.1016/j.cossms.2018.11.003.

Faisal, N. H., Ahmed, R., & Fu, R. Y. (2012). Nano-impact (fatigue) characterization of as-deposited amorphous nitinol thin film. *Coatings*, *2*(3), 195−209. Available from https://doi.org/10.3390/coatings2030195, https://res.mdpi.com/coatings/coatings-02-00195/article_deploy/coatings-02-00195.pdf?filename = &attachment = 1.

Frutos, E., González-Carrasco, J. L., & Polcar, T. (2016). Repetitive nano-impact tests as a new tool to measure fracture toughness in brittle materials. *Journal of the European Ceramic Society*, *36*(13), 3235−3243. Available from https://doi.org/10.1016/j.jeurceramsoc.2016.04.026.

Ghosh, A., Jin, S., Arreguin-Zavala, J., & Brochu, M. (2017). Characterization and investigation of size effect in nano-impact indentations performed using cube-corner indenter tip. *Journal of Materials Research*, *32*(12), 2241−2248. Available from https://doi.org/10.1557/jmr.2017.170, http://journals.cambridge.org/action/displayJournal?jid = JMR.

Hassani, S., Bielawski, M., Beres, W., Martinu, L., Balazinski, M., & Klemberg-Sapieha, J. E. (2008). Predictive tools for the design of erosion resistant coatings. *Surface and Coatings Technology*, *203*(3−4), 204−210. Available from https://doi.org/10.1016/j.surfcoat.2008.08.050.

Huang, W., Shishehbor, M., Guarín-Zapata, N., Guarín-Zapata, N. D., Kirchhofer, J., Li, L., Cruz, T., Wang, S., Bhowmick, D., Stauffer, P., & Manimunda, P. (2020). A natural impact-resistant bicontinuous composite nanoparticle coating. *Nature Materials*, *19*, 1236−1243. Available from https://doi.org/10.1038/s41563-020-0768-7.

Janos, B. Z., Lugscheider, E., & Remer, P. (1999). Effect of thermal aging on the erosion resistance of air plasma sprayed zirconia thermal barrier coating. *Surface and Coatings Technology*, *113*(3), 278−285. Available from https://doi.org/10.1016/s0257-8972(99)00002-x.

Lamri, S., Langlade, C., & Kermouche, G. (2013). Damage phenomena of thin hard coatings submitted to repeated impacts: Influence of the substrate and film properties. *Materials Science and Engineering: A*, *560*, 296−305. Available from https://doi.org/10.1016/j.msea.2012.09.070.

Lawes, S. D. A., Hainsworth, S. V., & Fitzpatrick, M. E. (2010). Impact wear testing of diamond-like carbon films for engine valve-tappet surfaces. *Wear.*, *268*(11−12), 1303−1308. Available from https://doi.org/10.1016/j.wear.2010.02.011.

Li, X., & Bhushan, B. (1998). Measurement of fracture toughness of ultra-thin amorphous carbon films. *Thin Solid Films*, *315*(1−2), 214−221. Available from https://doi.org/10.1016/S0040-6090(97)00788-8, http://www.journals.elsevier.com/journal-of-the-energy-institute.

Li, X., Diao, D., & Bhushan, B. (1997). Fracture mechanisms of thin amorphous carbon films in nanoindentation. *Acta Materialia*, *45*(11), 4453−4461. Available from https://doi.org/10.1016/S1359-6454(97)00143-2, http://www.journals.elsevier.com/acta-materialia/.

Li, X., & Bhushan, B. (1999). Evaluation of fracture toughness of ultra-thin amorphous carbon coatings deposited by different deposition techniques. *Thin Solid Films*, *355−356*, 330−336. Available from https://doi.org/10.1016/s0040-6090(99)00446-0.

Luo, J. K., Fu, Y. Q., Le, H. R., Williams, J. A., Spearing, S. M., & Milne, W. I. (2007). Diamond and diamond-like carbon MEMS. *Journal of Micromechanics and Microengineering*, *17*(7), S147−S163. Available from https://doi.org/10.1088/0960-1317/17/7/s12.

Ma, Z., Belhenini, S., Joly, D., Chalon, F. (2018). Energy density estimation of crack initiation in Sn-Ag-Cu(Ni) solder by nano-impact. International Conference on Electronic Packaging Technology & High Density Packaging.

McMaster, S. J., Liskiewicz, T. W., Neville, A., & Beake, B. D. (2020). Probing fatigue resistance in multi-layer DLC coatings by micro- and nano-impact: Correlation to erosion tests. *Surface and Coatings Technology, 402*. Available from https://doi.org/10.1016/j.surfcoat.2020.126319, http://www.journals.elsevier.com/surface-and-coatings-technology/.

Michler, J., & Blank, E. (2001). Analysis of coating fracture and substrate plasticity induced by spherical indentors: Diamond and diamond-like carbon layers on steel substrates. *Thin Solid Films, 381*(1), 119−134. Available from https://doi.org/10.1016/s0040-6090(00)01340-7.

Qin, L., Li, H., Shi, X., Beake, B. D., Xiao, L., Smith, J. F., Sun, Z., & Chen, J. (2019). Investigation on dynamic hardness and high strain rate indentation size effects in aluminium (110) using nano-impact. *Mechanics of Materials, 133*, 55−62. Available from https://doi.org/10.1016/j.mechmat.2019.03.008.

Ravi, N., Markandeya, R., & Joshi, S. V. (2017). Fracture behaviour of nc-TiAlN/a-Si$_3$N$_4$ nanocomposite coating during nanoimpact test. *Surface Engineering, 33*(4), 282−291. Available from https://doi.org/10.1080/02670844.2016.1239853, http://www.tandfonline.com/loi/ysue20#.VwHcBE1f1Qs.

Saber-Samandari, S., & Gross, K. A. (2013). Contact nanofatigue shows crack growth in amorphous calcium phosphate on Ti, Co-Cr and Stainless steel. *Acta Biomaterialia, 9*(3), 5788−5794. Available from https://doi.org/10.1016/j.actbio.2012.11.011, http://www.journals.elsevier.com/acta-biomaterialia.

Schmauder, T., Nauenburg, K.-D., Kruse, K., & Ickes, G. (2006). Hard coatings by plasma CVD on polycarbonate for automotive and optical applications. *Thin Solid Films, 502*(1−2), 270−274. Available from https://doi.org/10.1016/j.tsf.2005.07.296.

Sebastiani, M., Johanns, K. E., Herbert, E. G., & Pharr, G. M. (2015). Measurement of fracture toughness by nanoindentation methods: Recent advances and future challenges. *Current Opinion in Solid State and Materials Science, 19*(6), 324−333. Available from https://doi.org/10.1016/j.cossms.2015.04.003.

Shi, X., Beake, B.D., Liskiewicz, T.W., Chen, J., & Sun, Z.. Failure mechanism and protective role of ultrathin ta-C films on Si (100) during cyclic nano-impact. Available from http://www.journals.elsevier.com/surface-and-coatings-technology/

Shi, X., Li, H., Beake, B. D., Bao, M., Liskiewicz, T. W., Sun, Z., & Chen, J. (2020). Dynamic fracture of CrN coating by highly-resolved nano-impact. *Surface and Coatings Technology, 383*. Available from https://doi.org/10.1016/j.surfcoat.2019.125288, http://www.journals.elsevier.com/surface-and-coatings-technology/.

Shi, X., Chen, J., Beake, B. D., Liskiewicz, T. W., & Wang, Z. (2021). Dynamic contact behavior of graphite-like carbon films on ductile substrate under nano/micro-scale impact. *Surface and Coatings Technology, 422*. Available from https://doi.org/10.1016/j.surfcoat.2021.127515, http://www.journals.elsevier.com/surface-and-coatings-technology/.

Singh, R. K., Tilbrook, M. T., Xie, Z. H., Bendavid, A., Martin, P. J., Munroe, P., & Hoffman, M. (2008). Contact damage evolution in diamondlike carbon coatings on ductile substrates. *Journal of Materials Research, 23*(1), 27−36. Available from https://doi.org/10.1557/jmr.2008.0030.

Somekawa, H., & Schuh, C. A. (2012). High-strain-rate nanoindentation behavior of fine-grained magnesium alloys. *Journal of Materials Research*, *27*(9), 1295–1302. Available from https://doi.org/10.1557/jmr.2012.52.

Somekawa, H., & Schuh, C. A. (2013). Nanoindentation behavior and deformed microstructures in coarse-grained magnesium alloys. *Scripta Materialia*, *68*(6), 416–419. Available from https://doi.org/10.1016/j.scriptamat.2012.11.010.

Sun, H., Xiao, H., & Li, L. (2016). Experimental Study on Cutting Force and Cutting Power in High Feed Milling of Ti5Al5Mo5VCrFe. *Materials Science Forum.*, *836–837*, 88–93. Available from https://doi.org/10.4028/http://www.scientific.net/msf.836-837.88.

Tabor, D. (1951). The Hardness of Metals. *Measurement Technique*.

Trelewicz, J. R., & Schuh, C. A. (2008). The Hall-Petch breakdown at high strain rates: Optimizing nanocrystalline grain size for impact applications. *Applied Physics Letters*, *93*(17). Available from https://doi.org/10.1063/1.3000655.

Vecchione, N., Wasmer, K., Balint, D. S., & Nikbin, K. (2009). Characterization of EB-PVD yttrium-stabilised zirconia by nanoindentation. *Surface and Coatings Technology*, *203*(13), 1743–1747. Available from https://doi.org/10.1016/j.surfcoat.2008.11.016.

Veprek, S., & Veprek-Heijman, M. J. G. (2008). Industrial applications of superhard nanocomposite coatings. *Surface and Coatings Technology*, *202*(21), 5063–5073. Available from https://doi.org/10.1016/j.surfcoat.2008.05.038.

Vetter, J. (2014). 60 years of DLC coatings: Historical highlights and technical review of cathodic arc processes to synthesize various DLC types, and their evolution for industrial applications. *Surface and Coatings Technology*, *257*, 213–240. Available from https://doi.org/10.1016/j.surfcoat.2014.08.017.

Wang, T., Zha, X., Chen, F., Wang, J., Li, Y., & Jiang, F. (2021). Mechanical impact test methods for hard coatings of cutting tools: A review. *The International Journal of Advanced Manufacturing Technology.*, *115*(5-6), 1367–1385. Available from https://doi.org/10.1007/s00170-021-07219-8.

Wellman, R. G., & Nicholls, J. R. (2007). A review of the erosion of thermal barrier coatings. *Journal of Physics D: Applied Physics*, *40*(16), R293–R305. Available from https://doi.org/10.1088/0022-3727/40/16/R01.

Wellman, R. G., Deakin, M. J., & Nicholls, J. R. (2005). The effect of TBC morphology and aging on the erosion rate of EB-PVD TBCs. *Tribology International*, *38*(9), 798–804. Available from https://doi.org/10.1016/j.triboint.2005.02.008.

Wellman, R. G., Nicholls, J. R., & Murphy, K. (2009). Effect of microstructure and temperature on the erosion rates and mechanisms of modified EB PVD TBCs. *Wear*, *267*(11), 1927–1934. Available from https://doi.org/10.1016/j.wear.2009.04.002.

Wheeler, J. M., & Gunner, A. G. (2013). Analysis of failure modes under nano-impact fatigue of coatings via high-speed sampling. *Surface and Coatings Technology*, *232*, 264–268. Available from https://doi.org/10.1016/j.surfcoat.2013.05.028.

Wheeler, J. M., Dean, J., & Clyne, T. W. (2019). Nano-impact indentation for high strain rate testing: The influence of rebound impacts. *Extreme Mechanics Letters*, *26*, 35–39. Available from https://doi.org/10.1016/j.eml.2018.11.005.

Xie, Z.-H., Singh, R., Bendavid, A., Martin, P. J., Munroe, P. R., & Hoffman, M. (2007). Contact damage evolution in a diamond-like carbon (DLC) coating on a stainless steel substrate. *Thin Solid Films*, *515*(6), 3196–3201. Available from https://doi.org/10.1016/j.tsf.2006.01.035.

Yang, K., Rong, J., Liu, C., Zhao, H., Tao, S., & Ding, C. (2016). Study on erosion-wear behavior and mechanism of plasma-sprayed alumina-based coatings by a novel slurry injection method. *Tribology International*, *93*, 29−35. Available from https://doi.org/10.1016/j.triboint.2015.09.007, http://www.elsevier.com/inca/publications/store/3/0/4/7/4.

Yang, Z., Zhang, D., Huang, X., Yao, C., & Ren, J. (2010). The simulation of cutting force and temperature in high-speed milling of Ti-6Al-4V. *Advanced Materials Research*, *139−141*, 768−771. Available from https://doi.org/10.4028/www.scientific.net/AMR.139-141.768.

Zhong-Lei, M., Ning, H.E., Kai, W.U., Liang, L.I., Cheng-Yu, J. (2004). Study on the force of high speed milling of Ti alloy under different cutting media.

Nanomechanical testing methods to understand the effects of residual stress on coating's performance

Marco Sebastiani[1,2], Edoardo Rossi[1,2] and Saqib Rashid[1]
[1]Università degli Studi Roma Tre, Department of Civil, Computer Science and Aeronautical Technologies Engineering, Rome, Italy, [2]Consorzio Interuniversitario Nazionale per la Scienza e Tecnologia dei Materiali (INSTM), Florence, Italy

Abbreviations

CMAS	Calcium-Magnesium-Alumino-Silicate
CZ	Cohesive Zone
DIC	Digital Image Correlation
DLC	Diamond Like Carbon
EBSD	Electron Backscatter Diffraction
FEM	Finite Element Modelling
FIB	Focused Ion Beam
FWHM	Full Width Half Maximum
LEFM	Linear Elastic Fracture Mechanics
MAXS	Maximum Stress
PVD	Physical Vapor Deposition
SEM	Scanning Electron Microscopy
SF	Sensitivity Function
TBC	Thermal Barrier Coatings
TGO	Thermally Grown Oxide
TRISO	TRIstructural-ISOtropic
XRD	X-ray Diffraction
YSZ	Yttria-Stabilized Zirconia

10.1 Introduction

The significance of residual stresses within coatings engineering and materials science is pivotal, underpinning the performance and durability of coated systems. Residual stresses, emanating as remnants of the intricate interplay between thermal, chemical, and mechanical forces during the coating deposition process (Withers & Bhadeshia, 2001a), profoundly influence structural, mechanical, and functional attributes. Indeed, the overall stress state in a system is the sum of all the applied external loading stresses and residual stresses.

The residual stresses can be defined as macroscopic, microscopic, and submicroscopic stresses. Macro- residual stresses, which are often referred to as *Type I*, are in equilibrium across any cross section that leads to macroscopic dimensional alterations. Micro- residual stresses, referred to as *Type II*, are localized over small areas such as a few grains in polycrystalline materials. Submicroscopic residual stresses, which are classified as *Type III*, affecting atomic scale areas, these stresses are nonuniform and confined to minute volumes. Changes in *Type III* stresses do not result in macroscopic size changes (Withers & Bhadeshia, 2001b).

The total residual stress within a structure is the sum of these three contributions (Eq. 10.1):

$$\sigma_{tot} = \sigma_I + \sigma_{II} + \sigma_{III} \tag{10.1}$$

A comprehensive elucidation of the residual stress profile within coatings emerges as an essential prerogative, directing the optimization of coating fabrication, the augmentation of coating longevity, and the facilitation of technological progress reliant upon coated materials.

The imperative of comprehending the residual stress landscape within coatings is multifaceted and profound, underscored by the ensuing rationales:

- **Mechanical performance modulation:** The tensor of residual stresses engenders discernible ramifications upon the mechanical characteristics of coatings, including hardness, resilience, and resistance to cyclic loading. The existence of unmitigated or excessive residual stresses can induce untimely debilitation and failure of operational life.
- **Adhesion and delamination propensity:** residual stresses influence the adhesive interactions of the coating – substrate interface. High tensile stresses furnish a means for delamination and crack formation, compromising the adhesive efficacy and overall structural robustness.
- **Thermal endurance dynamics:** residual stresses potentiate exacerbating disparities in thermal expansion coefficients, potentially culminating in thermal fatigue and fracture proclivities. Moreover, during thermal exposure, residual stresses may relax through mechanisms such as creep or plastic deformation, altering the coating's thermal response and potentially leading to dimensional changes.
- **Surface characteristics and wear behaviour**: the textural attributes, wear resistance, and frictional comportment of coatings are profoundly modulated by residual stresses, bearing notable implications in applications encompassing tribological domains and wear safeguarding.
- **Functional traits perturbation**: in coatings engineered to embody functional attributes encompassing corrosion resistance, electrical conductivity, or thermal insulation, residual stresses wield a transformative influence over the manifestation of desired functional characteristics.

Eigenstrains represent an instrumental concept within the ambit of residual stress analysis, characterizing the net effect of all mechanisms for inelastic deformation in a generic body after its processing: structure modification, for example, amorphization; defect population, for example, dislocations and vacancies, phase transformation, and thermal strains (Jun & Korsunsky, 2010). In other words, the eigenstrain (ε^*) is the permanent (and invariant) inelastic strain that generates the residual stress. It represents the material "memory" of the deformation and processing history that developed the residual stress state (Korsunsky, 2017).

In a generic solid, the total strain can be additively decomposed in the sum of the elastic and inelastic parts:

$$\varepsilon_{tot} = \varepsilon^* + \varepsilon_{el} \tag{10.2}$$

The eigenstrain (ε^*) represents the internally "frozen" component, while the elastic part depends on the external load. A residual stress state is established to maintain compatibility and equilibrium in the body, according to the scheme reported in Fig. 10.1. In the most common situation, the residual stress state in a body is the known (or measured) quantity. The eigenstrain distribution is the unknown quantity that needs to be determined. This is the so-called "inverse problem" of residual stress analysis.

The measurement of residual stresses in thin coatings represents a relevant challenge due to the limited thickness of the coating, in most advanced cases, and the need for high spatial resolution. Over the years, various techniques have been developed to measure residual stresses in thin coatings, each with advantages, limitations, and state-of-the-art advancements. Some of the prominent methods utilized in the current state-of-the-art are listed below:

- X-ray diffraction (XRD) (Perry et al., 1996): X-ray diffraction is a well-established method for residual stress measurement. It involves analyzing the lattice spacing of crystalline planes in the coating material using X-ray beams. Changes in lattice spacing caused by residual stresses lead to shifts in diffraction angles ($\sin^2\psi$ method). XRD is

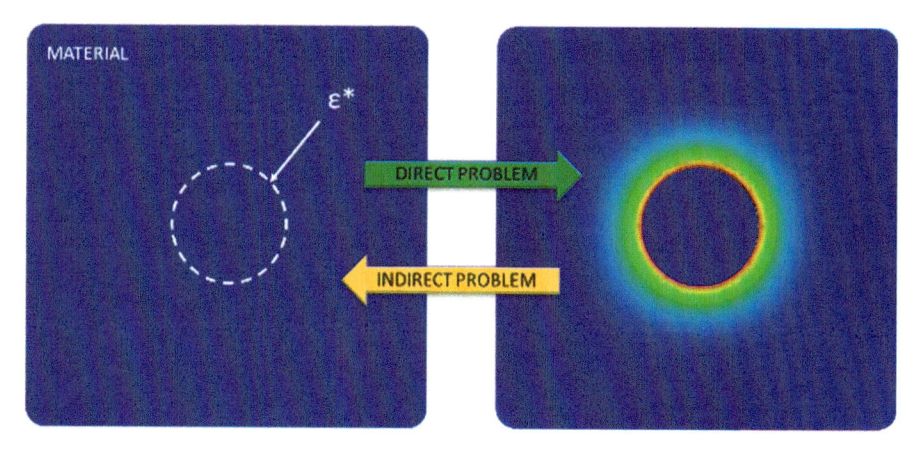

Figure 10.1 The concept of eigenstrain as the invariant source of residual stresses (ε^* can be any form of inelastic strain field that is generated during the material processing (Jun & Korsunsky, 2010)).

nondestructive and can provide depth-resolved stress profiles using grazing incidence techniques or multiple diffraction orders. However, it requires specialized equipment and is not suitable for amorphous coatings.
- Synchrotron radiation: this method offers deep penetration into materials, allowing for both surface and bulk stress analysis with high resolution and speed. The technique involves energy dispersive or angle dispersive diffraction methods to obtain detailed stress profiles. Advantages include superior spatial and strain resolution (Fitzpatrick & Lodini, 2003).
- Neutron diffraction (Albertini et al., 1999): Neutron diffraction is like XRD but employs neutrons instead of X-rays. Neutrons are particularly useful for coatings containing light elements and depth-resolved measurements in thicker coatings. However, neutron sources are limited, and the technique requires specialized facilities.
- Micro-Raman spectroscopy (Mohrbacher et al., 1996): Raman spectroscopy measures shifts in vibrational modes due to residual stresses. It can be used for both crystalline and amorphous coatings for small residual stresses. Micro-Raman setups offer a high spatial resolution, making them suitable for thin coatings. However, it requires careful calibration and may be influenced by coating composition and other factors.
- Instrumented indentation (Bhushan & Li, 2003): Depth-sensing indentation involves indenting the coating surface and analyzing the resulting elastic deformation. Residual stresses can be calculated using indentation models by measuring the indentation depth and load. This technique is relatively simple but may be affected by substrate effects and surface defects. Moreover, a stress relieved sample may be required as reference.
- Wafer curvature (Abadias & Daniel, 2021): This method measures the curvature of a substrate before and after coating deposition. The change in curvature is related to the residual stress in the coating through the Stoney equation. It offers a direct measurement but requires accurate curvature measurements and is limited to coatings with a thermal expansion coefficient like the substrate.
- FIB-DIC: this method, recently introduced, allows for high-resolution residual stress measurement by controlled focused ion beam (FIB) material removal, coupled with digital image correlation (DIC) for mapping of the consequent relaxation strain.

In this chapter, we will review the FIB-DIC method for residual stress mapping at the micron scale, with specific focus on recent advances in understanding the stress calculation procedures and the developments in the automation of the method for industrial exploitation.

Additionally, the focus will also be given on microscale fracture toughness assessment, since the presence of residual stress in a coating predominantly affects crack propagation resistance and fracture toughness. For such reason, and especially in case of coatings, it is of paramount importance to understand the net effect of residual stress state on crack propagation. Indeed, the resistance to crack propagation is intimately connected to the presence of stresses and can significantly influence a coating's fracture behavior, directly affecting its apparent fracture toughness. In this context, the importance of accurately measuring fracture toughness on nearly stress-free microspecimens becomes apparent. FIB-based techniques, such as microcantilever bending, micropillar splitting, and nanoindentation, are commonly employed (Ast et al., 2019). These methods precisely determine fracture toughness in specific coating regions, revealing how residual stresses impact the material's overall mechanical properties.

Following this concept, this chapter will focus on the use of microsize pillars to understand the effects of residual stress in coatings. In particular, the pillar splitting method will be reviewed as a robust approach for obtaining spatially resolved fracture toughness measurements within coatings. By inducing fracture in a micropillar by sharp nanoindentation, the method determines fracture toughness from the critical load needed for crack unstable propagation, using linear elastic fracture mechanics (LEFM) to relate this to the coating's resistance to cracking directly.

Understanding the variability and heterogeneity of coating properties is crucial, especially in critical applications where failure can have significant consequences. The pillar-splitting method offers a reliable and efficient way to gain these insights, thanks to:

1. **Small-scale testing**: It allows microscale testing, reducing material consumption and enabling localized measurements.
2. **Consistent loading**: The method ensures uniform loading conditions for controlled crack propagation, producing reproducible and reliable results.
3. **Statistical significance**: The method provides statistically significant data by testing multiple pillars, in a fast and spatially localized fancy, yielding a comprehensive understanding of the coating's behavior.
4. **Spatial resolution**: It enables fracture toughness measurements at specific locations within a coating, facilitating the identification of variations in fracture behaviour properties of different locations.

This stated, the method stands out for its unique combination of efficiency, reliability, and convenient positioning during fabrication in fracture toughness measurement, making it indispensable for spatially resolved fracture toughness assessments in heterogeneous materials and within different layers of thick coatings (Beirau et al., 2021; Mughal et al., 2016).

10.2 Recent advances in residual stress measurement at micron scale: the focused ion beam ring-core method

The focused ion beam (FIB) localized residual stress measurement technique is a reliable, accurate, and robust way to evaluate residual stresses among others at the micron and submicron scale, which has emerged as a semi-destructive analysis for localized, high-throughput means particularly suited for thin films (Sebastiani et al., 2011; Winiarski & Withers, 2012; Winiarski & Gholinia, Tian, et al., 2012).

In the early stages of its development, the milling geometries were inspired by macroscopic residual stress evaluation methods, such as single-slot and hole drilling. However, these approaches often led to strain gradients near the milled areas, adversely affecting the DIC algorithm's accuracy and necessitating an assessment of ion implantation effects on stress values. The single-slot geometry, much convenient in terms of simplicity, was limited to analyzing only one stress component perpendicular to the slot. Advancements in the technique led to the adoption of the ring-core geometry (Taking the stress out of nanomaterials design et al., 2024), involving incremental milling of annular trenches. This approach offers two advantages: (i) the pillar's surface remains largely undisturbed, ideal for DIC analysis, and (ii) the strain relief is more uniform across the pillar's diameter, beneficial for analyzing non-equibiaxial residual stress distributions. Further refinements in the method included FEM modeling to optimize the trench's depth and size.

Regarding residual stress inspection objectives, two primary goals are identified:

1. Measuring the mean residual stress at a specific depth.
2. Determining the residual stress depth profile.

Therefore, the method exploits micromachining and strain analysis principles to discern residual stresses at the microscale, with a spatial resolution down to ∼50 nm and an achievable lateral resolution for modern FIB-SEM microscopes down to 1 μm. This section summarizes the theoretical foundation, underlying assumptions, practical implementation, and comparative limitations of the FIB ring-core method with respect to the other methodologies.

The FIB ring-core method builds upon the principles of mechanical deformation and strain analysis, integrating the concept of eigenstrains (Korsunsky, 2018). It employs focused ion milling to meticulously eliminate material in a concentric circular configuration encircling a central core within the material. This targeted removal induces localized relaxations, prompting the liberated strain to reequilibrate. By quantifying the resultant deformation and correlating it with the elastic properties, the original residual stresses can be extrapolated.

The methodology encompasses several pivotal stages (Fig. 10.2), combining focused ion beam (FIB) milling, SEM imaging, digital image correlation (DIC) analysis, and finite element modeling:

- Ion milling: A focused ion beam is directed onto the surface, eradicating concentric rings of material encircling a central core zone in a step-based procedure. The ion beam generates localized material removal, thereby inducing strain relaxation.

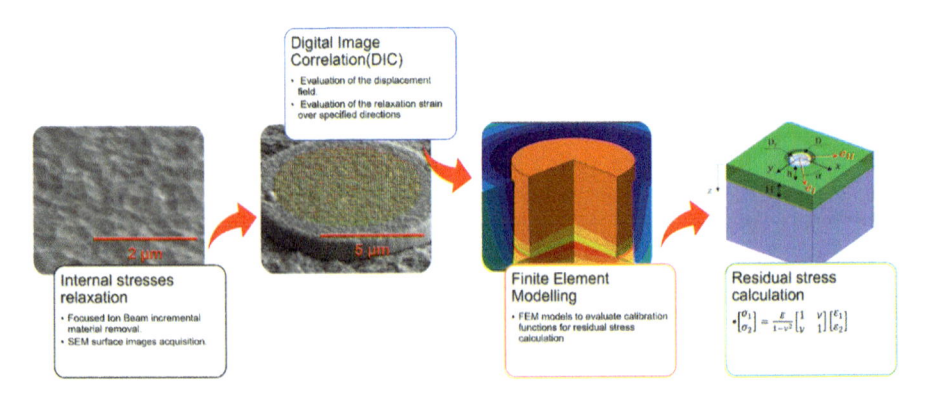

Figure 10.2 Focused ion beam-digital image correlation ring-core measurement workflow.

- Strain measurement: High-resolution electron microscopy (SEM) is harnessed to gauge deformation within the core region of the isolated material surface after each removal step.
- Stress calculation: The strain data, material property knowledge, and geometrical parameters facilitate residual stress calculation utilizing pertinent stress − strain relationships (Hooke's law).
- Finite element modelling: used to calibrate the material relaxation to the pre-existing eigenstrain field, ultimately elucidating coefficients or integral forms to reconstruct the residual stress field. FEM simulates the material's response to the incremental milling process, predicting how stress redistributes and how the material deforms as layers are removed in the presence of unitary eigenstrains.

Some assumptions underpin the FIB ring-core technique, the two main of which are: (1) pure elastic deformation: it presumes the coating and substrate manifest elastic behavior during ion milling, negating significant plastic deformation or inelastic behaviours; (2) plane stress condition: this considers a two-dimensional plane stress state where stresses fluctuate solely within the coating plane, while it neglects stress components along the direction of the coating's growth;

The technique offers distinct advantages over conventional methodologies such as X-ray diffraction and incremental hole drilling, including its capability to gauge microscale stresses and locally measure in defined volumes of interest with precise spatially resolved positioning and semi-destructive character. Nevertheless, certain limitations warrant consideration:

- Sample size constraints: the method predominantly suits diminutive samples, constraining its applicability to thick coatings or heterogeneous materials.
- Complex data analysis: the approach mandates sophisticated strain analysis and numerical modeling to interpret gauged deformation accurately and extract dependable stress values.
- Surface topology: surface irregularities and topography can influence strain measurements, possibly culminating in erroneous stress calculations.
- Subsurface stress sensitivity and general in-depth sensitivity limitations: the FIB ring-core method principally captures near-surface residual stresses, potentially neglecting substantial subsurface stress gradients.

As anticipated, the concept of ***eigenstrains*** is one of the cornerstones of FIB-DIC stress measurements to back-calculate residual stress depth profiles from FIB-DIC micro-ring-core experiments and to understand the effects of FIB damage on FIB-DIC experiments (Fig. 10.3) (Salvati et al., 2016).

10.2.1 FEM analysis for average stress and stress gradient analysis

At the beginning of its development, the FIB-DIC ring-core method was meant to understand the average stresses within its local probing volume. Sebastiani et al. (2017) developed a procedure to convert the relieved measured deformations to the original residual stress in the material, in the case of incremental material removal, applying the most used approach proposed by (Schajer, 1981; 2013). Based on it,

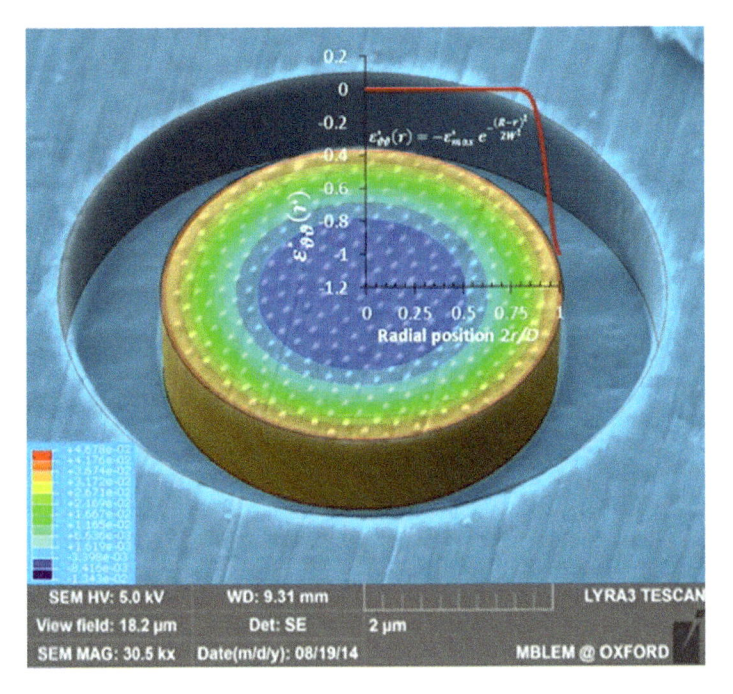

Figure 10.3 The use of the eigenstrain approach to quantify focused ion beam-induced residual stress after milling of a micropillar (Salvati et al., 2016).

the relaxation strain $\varepsilon(h)$ induced on the surface of the *pillar*, due to an annular trench of depth h, is given by the integral of the infinitesimal strain relief components deriving from material removal at all depths in the range $0 \leq z \leq h$:

$$\varepsilon(h) = \int_0^h \{A(z,h)*[\sigma_1(z) + \sigma_2(z)] + B(z,h)*[\sigma_1(z) - \sigma_2(z)]*\cos 2\alpha_k(h)\}dz \tag{10.3}$$

Where, $\sigma_1(z)$ and $\sigma_2(z)$ are the principal residual stresses acting at depth z, α is the angular coordinate measured anticlockwise from the maximum principal stress direction to the measuring direction, and $A(z,h)$, $B(z,h)$ are the influence functions, which are usually determined by finite element modelling. In the case of incremental milling, the milling process is divided into n finite steps and the integral method can be written as follows:

$$\varepsilon_n = \sum_{i=1}^n A_{ni}(\sigma_{1i} + \sigma_{2i}) + \sum_{i=1}^n B_{ni}(\sigma_{1i} - \sigma_{2i})\cos 2\alpha_i \tag{10.4}$$

Where, σ_{1i} and σ_{2i} are the principal residual stresses in the ith layer, and α_i is the angle from the stress σ_{1i} to the measuring direction. In the discrete step case, $A(z,h)$, $B(z,h)$ become vectors of coefficients that relate the strain measured on the surface for n milling steps to the principal stresses acting in the i layer. Those coefficients consider the material properties, the annular trench aspect ratio, the lateral slope of the isolated material stub (*pillar*), and other practical aspects of the measuring process.

The evaluation of the calibration coefficients A_{ni} and B_{ni} is done by FEM analysis of the milling process, considering an axisymmetric model subjected to a hydrostatic calibration stress field ($\sigma_1 = \sigma_2 = -1$) to evaluate the A_{ni} coefficients and pure shear stress field ($\sigma_2 = -\sigma_1$) to evaluate the B_{ni} coefficients.

During FEM modelling, a critical observation, valid for all material property combinations, made in the case of equal-biaxial stress states constant through the thickness of the film and applicable for an arbitrary in-plane residual stress distribution, has been made. Indeed, what was found is that the normalized milling stress relief (normalised to the biaxial modulus ($E/(1-\nu^2)$)) approaches unity for $h/D > 1$ (where h is the total milling depth and D is the pillar diameter), meaning that full stress relief (i.e., complete saturation) is achieved when the trench depth is at least equal to its diameter.

All the conclusions above are valid only if the *pillar* diameter (D) is equal or is smaller than the coating thickness (t). Otherwise, a dependence of the relief strain profile on the elastic properties of the substrate may be observed.

10.2.2 Measurement of the mean residual stress value at a certain depth

In the first development stages of the technique, for biaxial residual stress, based on the FEM modeling results, the shape of the strain relief curve was found to be universal (i.e., independent from the properties of the sample being studied). This is an important observation that leads to the use of a simple approximating function, "master function" that fits the data obtained from the DIC analysis, to have a robust evaluation of the saturation strain ($\Delta\varepsilon_\infty$):

$$f(\Delta\varepsilon_\infty, z) = 1.12 \Delta\varepsilon_\infty * \frac{z}{1+z}\left[1 + \frac{2}{1+z^2}\right] \tag{10.5}$$

The "master function" is expressed in terms of the normalized depth parameter $z = \left(h/(0.42 * D)\right)$.

A nonequi-biaxial residual stress state needs the evaluation of the relaxation stain over three different directions when the ones of the principal stresses are not known a priori. The Mhor's circle principles are followed to evaluate the in-plane principal stress components. If data are collected for three directions forming an angle of 45 degrees (rectangular rosetta), the principal maximum strain reliefs $\overline{\Delta\varepsilon_1}$ and $\overline{\Delta\varepsilon_2}$ are calculated using the following relations:

$$\overline{\Delta\varepsilon_1}, \overline{\Delta\varepsilon_2} = \frac{1}{2}\left[(\varepsilon_\alpha + \varepsilon_\beta) \pm \sqrt{2(\varepsilon_\alpha - \varepsilon_\beta)^2 + 2(\varepsilon_\gamma - \varepsilon_\beta)^2}\right] \tag{10.6}$$

The principal residual stresses, known the principal maximum strain reliefs, are evaluated as follows:

$$\overline{\sigma_1} = -\frac{E}{(1-\nu^2)}\left[\overline{\Delta\varepsilon_1} + \nu\overline{\Delta\varepsilon_2}\right] \tag{10.7}$$

$$\overline{\sigma_2} = -\frac{E}{(1-\nu^2)}\left[\overline{\Delta\varepsilon_2} + \nu\overline{\Delta\varepsilon_1}\right] \tag{10.8}$$

10.2.3 Measurement of the residual stress depth profile

Theoretically, the residual stress depth profile can be investigated by directly applying the integral method described in Section 10.2.1. This procedure has two main weaknesses:

- It is mathematically "ill-conditioned" due to the numerical inversion of the calibration coefficient matrices. Indeed, the measured relief strain vector (containing values for the ith milling depth) relates to the residual stresses vector through the matrices of the calibration coefficients. For an equi-biaxial stress state:

$$\overline{\sigma} = A^{-1} \cdot \overline{\varepsilon} \tag{10.9}$$

Unstable results are observed for $h/D > 1$ (i.e., considerable stress jumps even for a small strain noise).
- Time-consuming and sample-specific finite element modelling (FEM) calculations are needed.

Therefore, a novel theoretical approach (Korsunsky, 2018), instead of focusing on the residual stress depth profile, concentrates on the reconstruction of the *eigenstrain*.

As talked about in the introductory part of this theoretical section, unlike residual stresses that are extrinsic, *eigenstrains* aren't sensitive to the changes in the sample geometry, so the *eigenstrain* analysis of each incremental milling step can be conducted separately, leading to complete decoupling between increments: the relationship between the intrinsic *eigenstrain* and the measured relief strain at the ith milling step reduces to an algebraical equation, while the overall profile consists of the integral of each evaluation at the ith step:

$$\varepsilon\left(\frac{h}{D}\right) = \int_0^{\frac{h}{D}} F\left(\frac{z}{D}\right) \epsilon^*(z) d\left(\frac{z}{D}\right) \tag{10.10}$$

Where h/D is the normalised milling depth, z/D is the normalised depth at which the *eigenstrain* is being determined, and $\epsilon^*(z)$ is the inspected *eigenstrain* distribution. $F(z/D)$ is the single variable (due to the decoupling effect) master influence function that describes the incremental contribution to the surface strain relief from the *eigenstrain* acting at the normalised depth value z ($\epsilon^*(z)$). Behind this novel technique's mathematical basis is the assumption that the ion beam damage is negligible. Studies of the milling process effect have shown that the ion implantation effect is limited to a superficial layer of approximatively ~30 nm, which can be neglected for ring-core sizes that exceed ~1 μm.

The method needs to be calibrated by performing numerical calculations on a simplified reference case to obtain the values of the master influence function. For this purpose, a uniform negative *eigenstrain* of unity strength is considered to exist within the *pillar* volume, leading to the so-called cumulative strain relief function:

$$\varepsilon\left(\frac{h}{D}\right)_{\epsilon^*=-1} = f\left(\frac{h}{D}\right) = \int_0^{\frac{h}{D}} F\left(\frac{z}{D}\right) d\left(\frac{z}{D}\right) \tag{10.11}$$

Some observations for the master influence function must be done.

- The latter is generic and material-independent, removing the strong dependence on the material elastic properties.
- It overcomes the need for matrices-based calculations.

The FIB-DIC averages the strain relief over the radial position. Still, typically, the averaging isn't done over the entire area to prevent the insertion of data from being compromised by ion milling artifacts. For this reason, a reduction factor (φ) is introduced to define an effective area during FEM calculation. The values of φ influence the cumulative master influence function parameters.

The incremental master **influence function** can be found by analytical differentiation of the relief curve to the normalized depth variable, as follows:

$$F\left(\frac{h}{D}\right) = \frac{df\left(\frac{h}{D}\right)}{d\left(\frac{h}{D}\right)} = \exp\left(-\alpha\left(\frac{h}{D}\right)\right) * \left[\alpha - \beta + \alpha\beta\left(\frac{h}{D}\right) + 2\gamma\left(\frac{h}{D}\right) - \alpha\gamma\left(\frac{h}{D}\right)^2\right] \tag{10.12}$$

Once the values of the master influence function are extracted for the system under analysis, employing interpolation of the data points obtained from the DIC analysis, using the cumulative strain relief function proposed, the conversion between the *eigenstrain* field and the residual elastic strain field, that corresponds to the residual stress state, amounts only to changing sign:

$$\varepsilon = -\epsilon^* \tag{10.13}$$

This last step can be done under specific conditions: (i) the substrate is sufficiently thick; (ii) the ratio between the characteristic length (ring-core diameter D) and the local curvature ρ is small.

The sensitivity (S) of the stress profiling technique depends on the magnitude of the master influence function $F(h/D)$:

$$S = \frac{\text{output}}{\text{input}} = \frac{d\varepsilon\left(\frac{h}{D}\right)/d\left(\frac{h}{D}\right)}{\epsilon^*(h)} = F\left(\frac{h}{D}\right) \tag{10.14}$$

With reference to the values assumed by the master influence function with respect to the normalized milling depth, it's clear that S has optimal values for h/D over a restricted range between 0.1 and 0.2. Therefore, the residual stress profile within a thin coating system could be, in principle, inspected using multiple ring-cores of different diameters to maintain a high and constant sensitivity of the technique. Indeed, ring-cores with smaller diameter contribute to the residual strain data near the sample surface. In contrast, ring-cores of larger diameter allow residual elastic strain evaluation in the deeper subsurface layers.

Obviously, also in the case of the above described *eigenstrain* reconstruction approach, measurements of the relaxation strain from three different directions in a rosette configuration is mandatory to reconstruct principal residual stress components and retrive their direction with respect to a reference system coordinates.

The described *eiganstrain* procedure must, therefore, be adapted to be applicable over the hydrostatic and deviatoric decompositions of the residual stress tensor field. A novel procedure has been developed by Salvati et al., which represents a further improvement of the ring-core residual stress depth profiling method in the case of nonequi-biaxial stresses (Salvati et al., 2019).

As evidenced in previous sections, the incremental strain relief tensor can be decomposed into hydrostatic and deviatoric parts:

$$\overline{\overline{\Delta \varepsilon_i}} = \overline{\overline{\Delta \varepsilon_{H,i}}} + \overline{\overline{\Delta \varepsilon_{D,i}}} \qquad (10.15)$$

The residual elastic strain measured at the normalized depth h_i^*/D for the equi-biaxial case can be calculated as the incremental strain relief divided by the corresponding value of the influence function $F_{H,i}$ (which, in this case, is referred to as the hydrostatic influence function), changed in sign to pass from eigenstrain to residual elastic strain. For this case, therefore, this incremental strain relief corresponds to the hydrostatic (or equi-biaxial) part of the strain state. For the deviatoric part of the strain state the appropriate influence function needs to be evaluated, denoted as $F_{D,i}$.

The principal elastic residual strains can then be obtained as:

$$\begin{pmatrix} e_{1,i} \\ e_{2,i} \end{pmatrix} = -\frac{1}{F_{H,i}} \begin{pmatrix} \Delta \varepsilon_{H,i} \\ \Delta \varepsilon_{H,i} \end{pmatrix} - \frac{1}{F_{D,i}} \begin{pmatrix} \Delta \varepsilon_{D_1,i} \\ -\Delta \varepsilon_{D_1,i} \end{pmatrix} \qquad (10.16)$$

where F_H and F_D are the influence functions, respectively, for the hydrostatic and deviatoric parts.

Only the hydrostatic term is relevant when the strain state is purely equi-biaxial. In the other opposite case, the deviatoric term is the only relevant one when a pure in-plane shear strain state is present. In the general case, both the equi-biaxial (hydrostatic) and pure shear (deviatoric) influence functions must be applied.

A 3D FEM model was used to extract relief strain profiles, in the two cases of equibiaxial residual stress state and pure share stress state, as a function of the area coverages.

It was found that a scale factor (SF) can be applied to the F_H functions that allows a quick determination of the approximate form of the deviatoric influence function (F_D):

$$F_H\left(\frac{h}{D}\right) \cong \text{SF} * F_D\left(\frac{\frac{h}{D}}{\text{SF}}\right) \quad F_D\left(\frac{h}{D}\right) \cong F_H\left(\text{SF} \times \frac{h}{D}\right)/\text{SF} \quad (10.17)$$

10.2.4 Recent advances, applications and opportunities

10.2.4.1 Ultrathin coatings with nonequibiaxial stress states

One of the major challenges for the FIB-DIC ring-core technique at present is the achievement of high-resolution measurement for in-depth stresses while limiting the number of required tests ensuring reliability and high-throughput. The optimised approach described in the previous sections presents a major challenge indeed: a careful analysis of the strain distribution over the pillar surface during relaxation shows that the calculated influence function can be significantly different depending on the fraction of central pillar area that is selected for DIC analysis. In more detail, the area% for DIC is determined by selecting a certain percentage of the pillar's radius—from here on called *radius%* (see Fig. 10.4A and B). This corresponds to different influence functions *F(z/D)* that are reported in Fig. 10.4C and D, which show that a maximum exists for each of the influence functions at the different values of the *radius%* selected for calculation. To avoid inaccuracies during stress calculation, a value of 80% was suggested to guarantee a sufficiently high strain sensitivity in the depth range of $0.015 < h/D < 0.2$ (as elucidated in the previous sections of this book) for both the hydrostatic and deviatoric stress components. Such functions show that the method's sensitivity to an eigenstrain depth variation decreases remarkably for *h/D* values higher than 0.2 for all considered values of the *radius%*. Therefore, one again remarked, the $0.015 < h/D < 0.2$ range for effective depth profiling has been suggested for optimal residual stress depth profiling. This way, a depth resolution of 50 nm was demonstrated for pillar diameters in the $1-15$ μm (Korsunsky, 2018).

However, for very thin multilayer films (<300 nm) with individual layers about 10–40 nm thick, the method encounters limitations, as pillars smaller than 1 μm could lead to inaccuracies (Korsunsky, 2016).

To overcome these challenges, Sebastiani proposed an improved micro ring-core method (Sebastiani, 2020). This method develops a unique influence function by varying the pillar's radius percentage for DIC analysis continuously during processing, retaining the maximum strain sensitivity for in-depth calculations. This enabled nanoscale non-equibiaxial residual stress profiling in ultrathin films. Finally, the method is applied to an Si_3N_4/Ag/ZnO multilayer

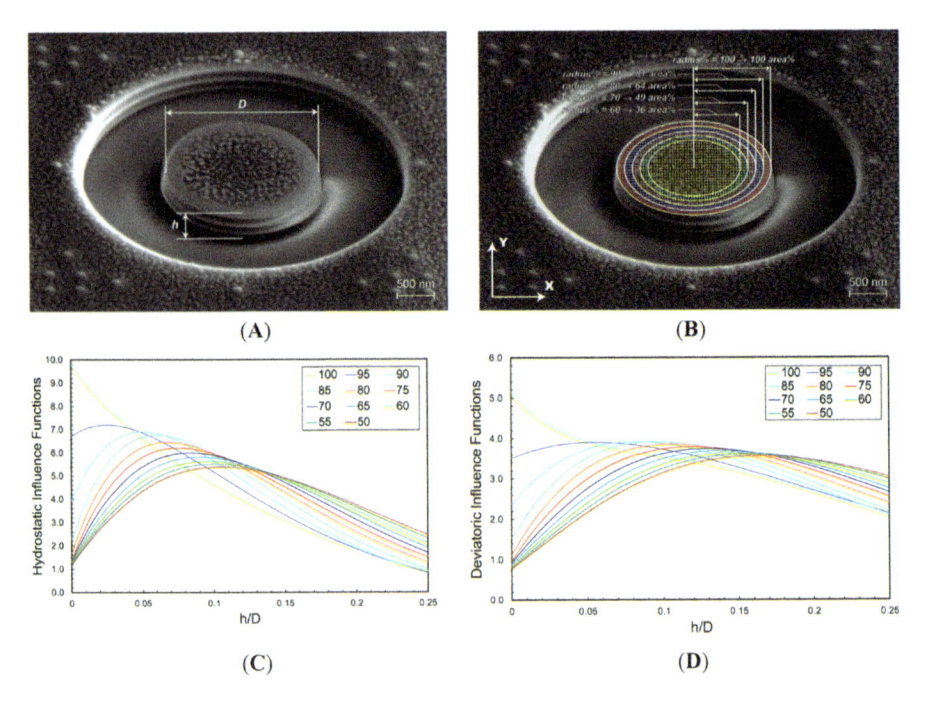

Figure 10.4 The novel approach for multi-area analysis: (A) representative ring-core focused ion beam (FIB) milling experiment, (B) multiple *area%* analysis ROIs evidenced for digital image correlation (DIC), (C) FEM calculated hydrostatic influence function F(z/D) for different values of the *radius%*, (D) FEM calculated deviatoric influence functions for different values of the *radius%*.

stack on a glass substrate, where the influence of residual stress depth gradient on film adhesion (scratch critical load) is clearly identified (for the same average stress and hardness of the film). The results showed that the nanoscale residual stress depth profile can be the main design parameter to be controlled for the optimisation of adhesion in multilayer low-emissivity thin films on glass substrates.

The principles of in-depth resolved measurements of residual stresses in thin films have been further applied since the aforementioned improvement. A recent work from (Dang, 2022) focused on the properties of TiN/Ti duplex coatings created using high-power impulse magnetron sputtering (HiPIMS). The coatings analysed in the study consist of TiN thin films with a 600 nm thickness and Ti adhesion films with a thickness of 50 nm. These coatings were deposited on titanium (Ti) substrates prepared by cold spraying. In these systems the accurate trough-thickness measurement of residual stresses in each precisely measured layer is vital for evaluating the efficacy of the coating process, and the FIB-DIC method was successfully applied to understand the interplay between col sprayed substrates and film deposition parameters.

10.2.4.2 Localized, position-resolved residual stress measurements

The ultimate need for spatially resolved and localized residual stress measurement at the micron and submicron scale is given by those applications where geometrical and microstructural variables play a role in determining the functional performances of coatings. The simple example among the applications in this sense is given by components experiencing complex 3D shapes (where curvature effects influence the stress profile). In the realm of microscale components, coated micro- and nano-particles are the maxima representatives.

A valuable recent application is presented, within this context, by Leide (2023), for TRIstructural-ISOtropic (TRISO) particles, consisting of a fuel kernel, typically made of uranium, surrounded by multiple layers of carbon and ceramic materials. Spatially resolving stresses in coated nuclear fuel particles, like TRISO particles, is crucial due to their multilayered structure, the strong geometrical curvature, and the differing mechanical properties of each layer. These particles are designed to contain fission products and maintain structural integrity under extreme conditions. By carefully positioning ring-core measurements along the different layers of the coating cross sections (Fig. 10.5B), FIB-DIC allowed the authors to extract spatially resolved critical insights into the stress distributions within these particles, finding tensile residual hoop stresses in pyrolytic carbon layers and compressive stress in the silicon carbide layer. These findings are key to understanding the behavior of

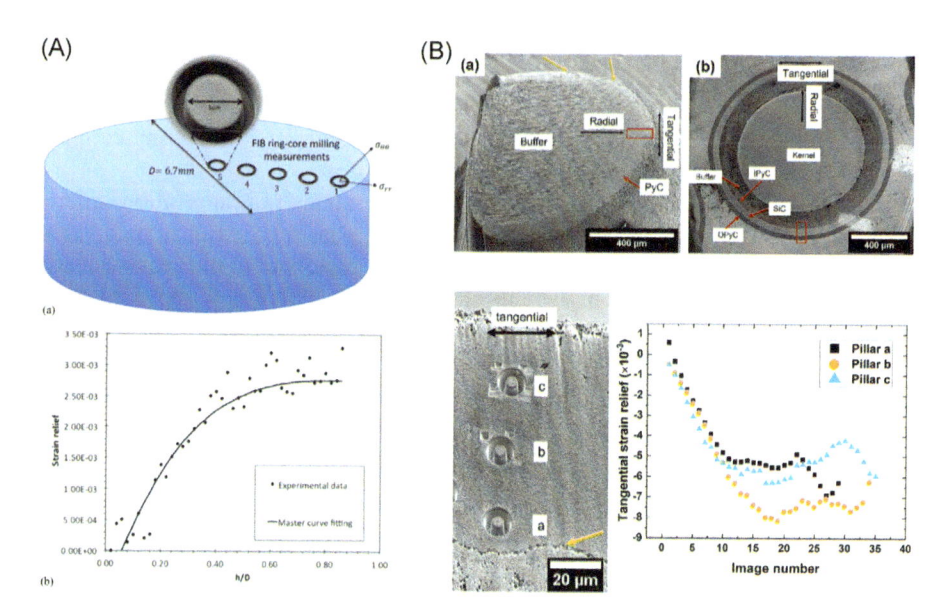

Figure 10.5 (A) Illustration of the five milling positions along the radial direction, with the inset showing a representative image of the FIB-DIC marker for spatially resolved measurements in bulk metallic glasses (Korsunsky et al., 2016). (B) TRISO particle structure with three pillars in the pyrolytic carbon coating (Leide, 2023).

nuclear fuel particles under operational conditions and are instrumental in advancing manufacturing and operational strategies for these materials, considering tensile stresses as major drivers for crack propagation and, therefore, fission products emission.

10.2.4.3 High-speed nanoindentation mapping coupled residual stress measurements

One of the most important advantage of the FIB-DIC ring-core method lies in its localised nature, enabling high lateral resolution in in-plane stresses measurement over specific portions of a surface. In their seminal work aimed at demonstrating the capabilities of the FIB-DIC ring-core technique, (Korsunsky et al., 2016) proved that the measurement technique was able to track the residual stress profiles along the radial direction of rapidly quenched bulk metallic glasses (Fig. 10.5A). A further step, fundamental for heterogeneous material has recently been given by advanced nanoindentation techniques such as the high-speed mapping capabilities of novel nanoindentation systems (Rossi et al., 2023).

High-speed nanoindentation mapping is an advanced technique that combines the principles of traditional nanoindentation with optimised testing parameters and automation to achieve rapid mechanical property measurements across a sample's surface. The method's efficiency lies in its ability to conduct multiple indentation tests quickly, as fast as one indent per second, while retaining accuracy and reliability. Integrating high-speed nanoindentation mapping with the focused ion beam (FIB) ring-core method presents a potent approach to perform residual stress mapping. This section elucidates the significance of this coupling, explores the theoretical framework, and demonstrates how high-speed nanoindentation mapping of the elastic modulus can enhance spatial resolution by correlating with FIB ring-core pillar positioning.

Indeed, heterogeneous coatings often exhibit variations in composition, microstructure, and mechanical properties across different phases or regions. Such variations can significantly influence the residual stress distribution and, consequently hinder a correct design for optimised mechanical performance and improved durability. Coupling high-speed nanoindentation mapping with the FIB ring-core method enables a comprehensive understanding of how mechanical properties correlate with residual stresses within distinct phases, revealing stress concentrations, stress gradients, and potential failure mechanisms. These coupling allows for phase-specific stress assessment. A more comprehensive and accurate stress map can be generated by analyzing how the elastic modulus correlates with residual stresses in different phases.

The contribution of high-speed nanoindentation mapping to the problem of enhancing spatial resolution in residual stress measurement is multifaceted:

1. **Multi-parameter correlation:** residual stress mapping needs a corresponding correlation with the evolution of elastic properties and their distribution over the region of interest. High-speed nanoindentation mapping provides a statistically significant dataset that enhances the confidence and reliability of the spatially resolved residual stress

measurements when coupled with FIB ring-core analysis. A multiparameter analysis must be performed by correlating the statistical distribution of the elastic modulus obtained from nanoindentation with the positions of FIB ring-core pillars.
2. **Microstructural insights:** The correlation between elastic modulus and FIB ring-core pillar positioning enables the identification of stress concentrations or variations related to specific microstructural features. This information aids in pinpointing regions prone to stress-induced failure.

10.3 Micron-scale fracture toughness analysis: the pillar splitting method

Pillar-splitting represents a fairly new technique introduced by Sebastiani et al. (Sebastiani, Johanns, Herbert & Carassiti, & Pharr, 2015; Sebastiani, Johanns, Herbert, & Pharr, 2015) within the spectra of nanoindentation-based methodologies for assessing fracture toughness at micrometer and submicrometer levels. The method is based on the sharp indentation of focused ion beam machined pillars fabricated within the target material until unstable crack propagation is reached. The technique overcomes most of the issues affecting traditional indentation cracking-based methodologies (Cook & Pharr, 1990) such as (1) substrate influences and residual stress state(indeed, a complete relaxation of the residual stress field is achieved, in the case of the pillar splitting, for aspect ratios $\left(\frac{h}{D}\right)$ equal to or greater than 1 as the method is strictly connected to the geometry elucidated for the FIB-DIC residual stress measurement under the same conditions, (2) post-test determination of crack system and size, and (3) low time-consuming fabrication of many reproducible samples.

Indeed, a crack nucleates beneath the indenter tip while the pillar is progressively loaded at its center in a precisely controlled way, as shown in Fig. 10.6. Crack nucleation is primarily driven by the elastic and plastic stress state developed beneath the sharp tip, while the crack propagation geometry is governed by the stress concentrations induced by the sides of the indenter tip (generally, a three-side diamond indenter). It becomes unstable due to the additional increment of stress intensity factor attributed to the presence of the pillar edges (at a length value that in most cases corresponds to 60% of the pillar radius), propagating in a Mode I (while Mode II and Mode III are negligible) way due to the opening action exerted by the tip edges penetrating through the material.

The splitting phenomenon (so called as the pillar splits, upon unstable cracking, into three axi-symmetric segments if perfectly loaded at its center) is measurable through direct identification of a cusp point within the load − displacement curve (where the right and left derivatives of the experimental data do not match anymore) corresponding to a pop-in event called "critical splitting load" (P_c), as elucidated in Fig. 10.7.

The theoretical modelling background of the problem starts from the simplest model adopted for the determination of the stress intensification factor K_I in the

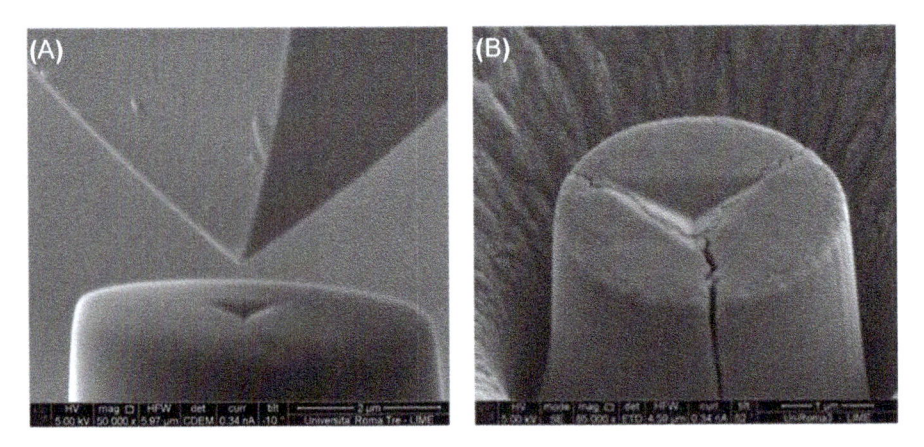

Figure 10.6 (A) Cube-corner diamond tip indenting the micropillar top surface. (B) Postmortem image of the micropillar after unstable propagation of crack and splitting process.

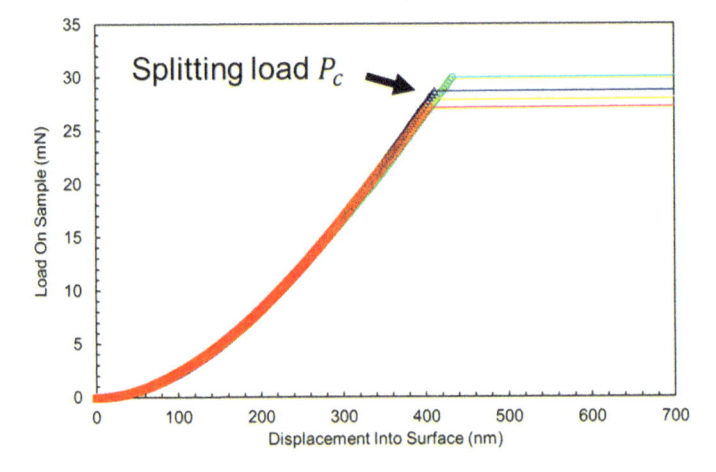

Figure 10.7 Representative load – displacement behavior for 4 μm pillars with highlighted discontinuity and pop-in event.

case of a semielliptical crack placed on the surface of an isotropic and homogeneous material in a semiinfinite space is given by:

$$K_I = A \frac{P}{c^{3/2}} \tag{10.18}$$

Where A is a parameter for which dependence on the sample surface and crack in-depth position is found. In the case of a semifinite space having a pillar geometry, to take into account the influence of the edges of the pillar in the ideal model of Eq. (10.18), Shah and Kobayashi have developed a model (Eq. 10.19) that allows the calculation of K_I introducing the amplification factor M_f Shah and Kobayashi (1973) equal to:

$$M_f = \left(1 - \frac{c}{R}\right)^{-1} \tag{10.19}$$

Which, substituted in Eq. (10.18), gives:

$$K_{IC} = \frac{A}{1 - \frac{c}{R}} \frac{P}{c^{3/2}} \tag{10.20}$$

A dimensional argument on Eq. (10.20) leads to an expression for determination of the fracture toughness from a pillar-splitting experiment:

$$K_c = \gamma \frac{P_c}{R^{\frac{3}{2}}} \tag{10.21}$$

γ represents a calibration factor computed through cohesive zone elements finite element modelling (CZ-FEM) analysis on the model geometry (Fig. 10.8).

Specifically, bilinear traction-separation-based cohesive zone (CZ) modeling is used at the edges of the indenter tip contact (crack planes) with an elastic-perfectly plastic material constitutive low (for the modelled material the yield strength is chosen such as its value matches the measured hardness). The MAXS criterion (where damage is considered to initiate at the maximum value of the normal and shear directions to the crack plane) is generally used to simulate the onset of debonding of cohesive elements. At the same time, fracture energy determines crack nucleation (Johanns et al., 2013). Additionally, the CZs' behavior is determined by their strength, stiffness, and viscosity parameters, which must be carefully selected to exceed the elastic modulus (in the case of the stiffness) to avoid elements

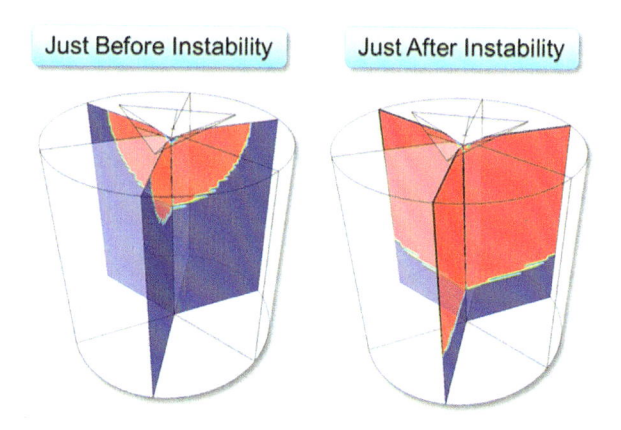

Figure 10.8 CZ-FEM simulation of the crack propagation during a pillar-splitting.

compliance artifacts and assure convergence during softening/debonding (viscosity), allowing for stresses to be outside the limits set by the traction-separation law for a short amount of time.

The value of the mode I energy release rate, the area beneath the traction-separation low curve, is related to the fracture toughness through LEFM by Eq. (10.22):

$$K_c = \frac{\sqrt{EG_{Ic}}}{(1-v^2)} \tag{10.22}$$

Calculation of the γ parameter is performed by determining the instability load in the simulation, with all the other variables being independent parameters based on the specific material properties, from Eq. (10.21).

Therefore, the γ parameter is dependent upon the material properties such as the elastic modulus E, yield strength (hardness equivalence) σ_y, Poisson's ratio v, and indenter geometry (half cone equivalent angle ψ). Determination of the dependence of γ from indenter geometry and Poisson's ratio values have been extensively performed by Ghidelli et al. (2017).

10.3.1 Experimental considerations for pillar splitting

Besides its simplicity compared with other techniques for fracture toughness determination, pillar-splitting has some critical experimental issues that must be discussed to clarify the steps performed in its optimization in this work.

As investigated by Lauener et al. (2018), two major issues are found:

1. FIB damage due to ion implantation:
    Reproduced from the work of (Salvati et al., 2016) a schematic visualization of the FIB-machining induced damage zone for the FIB pillar milling considered geometry is reported in Fig. 10.9. The damage is limited to the lateral walls of the pillar geometry for an extension dependent upon the acceleration voltage and beam current used for production (which finally determines the beam's FWHM). Thermal effects can further reconfigure atoms at the end of the collision cascade (Korsunsky, 2016; Salvati, 2018).
2. Positioning accuracy:
    Positioning accuracy dramatically influences the reliability of the results. Indeed, the average positioning accuracy for most piezo-stage actuated indenters is about 1 μm. Deviances from the pillar centre higher than 10% of its diameter will drop the fracture toughness values' precision, as demonstrated by Best et al. (2016). This work has tackled specific testing strategies and hardware setups for precise decupling of experimental error artefacts from other relevant investigation parameters and evidenced how positioning, specimen geometry, and ion source effects could influence the reliability of the methodology to an extent.

Additionally, a recent work by Durst et al. evidenced a third influencing factor. While most pillar-splitting experiments are ex-situ, positioning into specific complex geometries or the analysis of the crack patch and failure modes may require direct observation of the process within an SEM system. The resulting electron

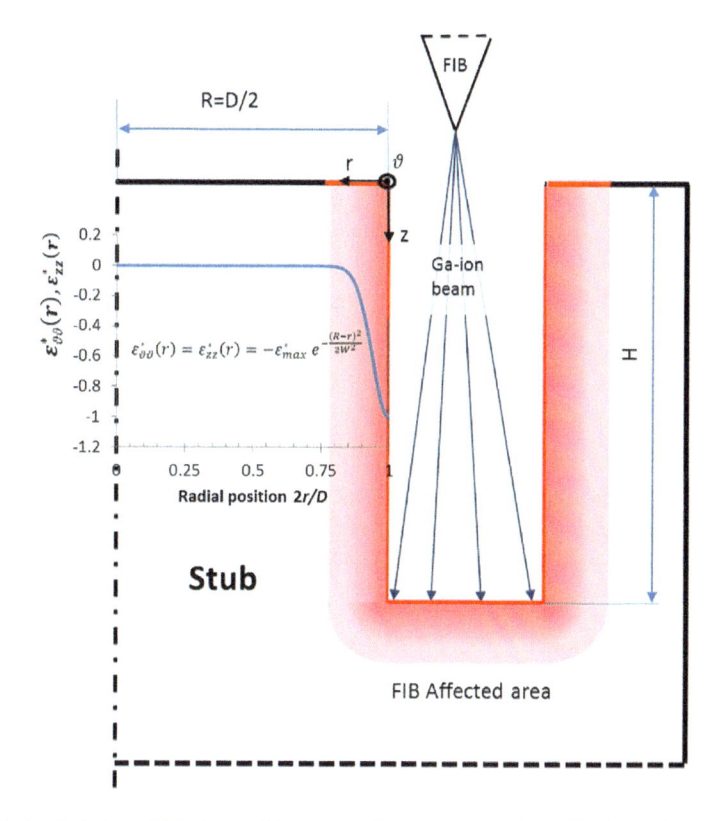

Figure 10.9 Modeling of Ga focused ion beam damage in annular milled trenches (side view).

bombardment may lead, for specific systems (e.g. glassy systems), to a rearrangement of atoms at the crack tip, ultimately influencing the fracture toughness (Bruns et al., 2020). While this phenomenon was known to an extent within cantilever bending experiments, where direct observation of the crack tip opening during bending would likely lead to effects in its propagation, Bruns et al. also extended the finding to pillar-splitting experiments.

10.3.2 Recent advances and applications

Pillar-splitting has been established through the years since its early development (2015) as one of the pivotal techniques for heterogeneous coatings and fast yet reliable (high-statistics) testing of complex ceramic material systems for spatially resolved analysis. Indeed, it recently played a crucial role in advancing the understanding of material properties in various scientific fields. By enabling the measurement of fracture toughness at micro- and nanoscales, this technique provides invaluable insights into the mechanical behaviour and reliability of advanced materials, facilitating the development of more robust and efficient applications.

In this section, the early critical areas of application will be reviewed. These include:

1. **Nanocomposite coatings**: pillar-splitting in this field can enhance the understanding of how the reinforcement can modify the fracture mechanics of films, elucidating the statistics behind the interwinted role of reinforcement distribution, toughening mechanisms (reinforcement cracking, bridging, crack deflection, change in crack geometry and system, etc.) and statistical reliability.
2. **Ceramic thin films**: first developed to quickly prompt the fracture toughness of ceramic thin films, pillar-splitting remains one of the most reliable techniques for fast screening of these materials. Recent developments focused on the assessment of high-temperature fracture toughness, an area in which the pillar approach solves most of the issues encountered by other geometries.
3. **Thermal barrier coatings (TBCs)**: spatially resolved and statistically reliable measurements of TBC's heterogeneous and porous layers culminated recently in the assessment of doping compositional gradients through the thickness of these coatings, especially under thermal cycling, understanding how the local heterogeneities contribute to the overall fracture performances by tackling properties of each TBC constituent.
4. **Multilayered systems**: in this category, pillar-splitting helps in understandinghow crack could propagate under complex tri-axial stress states through the varius layers that forms the systems, decoupling the analysis from the residual stress states and providing averaged values over the layers in a design oriented convinient way. To this scope, the technique could provide insights on how the fracture properties could be statistically correlated to other design characteristics (e.g. optical performances).

10.3.2.1 Nanocomposite coatings

In this sub-section, the reader's attention is brought to a representative example of the multifaceted problem of toughening mechanisms in nanocomposite coatings. The recent work by Jiang et al. (2018), in which the authors innovatively use pillar-splitting tests to evaluate the fracture toughness of TiO_2 nanoparticle films infiltrated with polystyrene, is representative of the capabilities of pillar splitting in understanding how compositional and microstructural heterogeneity (reinforcement type and distribution) needs highly-efficient techniques from the statistical point of view. The article analyses how polymer volume fractions, polymer type, and nanoparticle interactions can be analyzed via pillar splitting and fracture mechanisms understood. The authors demonstrate a significant increase in toughness, nearly by an order of magnitude, especially at low polymer volume fractions. This enhancement is attributed to increased interparticle contact area, crack deflection, and polymer bridging in nanoparticle films. The methodology provides a nuanced understanding of toughness in nanoparticle packings, which is relevant for lightweight composites and coatings.

10.3.2.2 Ceramic thin films

Beyond the classical use of pillar-splitting as an established standard for ceramic thin films fracture toughness assessment, as mentioned several times in this section, in this subsection we focus on a remarkable aspect of the technique for this class of

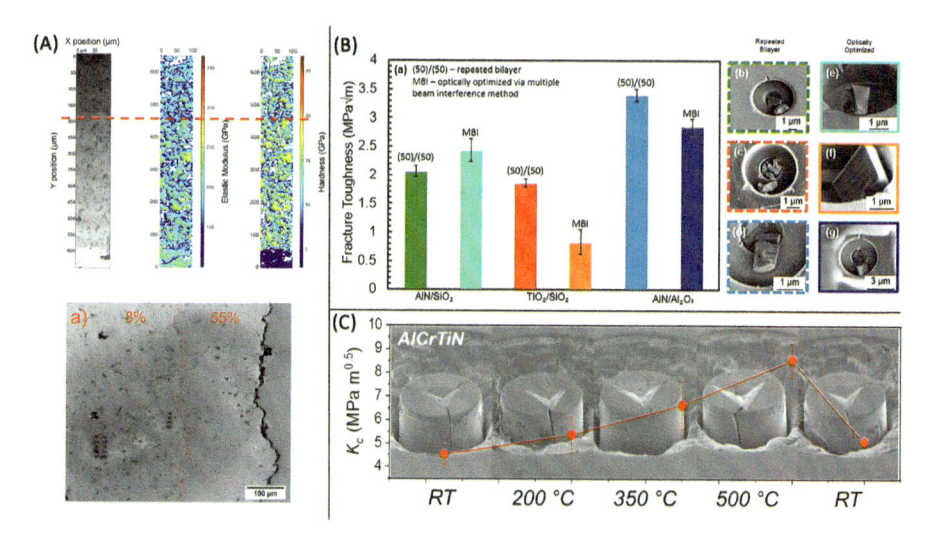

Figure 10.10 (A) Spatially resolved fracture toughness measurement of YSZ TBC coating via pillar splitting coupled with high-speed nanoindentation mapping. (B) Apparent fracture toughness measurement of complex multilayered coatings for optical applications, with crack propagation path assessment. (C) High-temperature pillar-splitting of AlCrTiN PVD coatings.

materials. Being increased the search for elevated temperature materials (aerospace, aeronautics, energy production), their high-temperature failure characterization became pivotal. The pillar-splitting methodology excels in high-temperature fracture toughness measurement compared with traditional beam-based testing geometries due to its reduced ion impregnation risk, which is vital for maintaining the material's intrinsic properties at elevated temperatures. It enables highly localised stress application, crucial for materials with significant property variations under thermal stress, where larger-scale testing might be less effective. To this scope, notable advancements have been made in the field of high-temperature testing of ceramic films. A significant contribution has been made by Best et al. works (Best et al., 2016; 2019) focused on the fracture toughness of thin physical vapour-deposited ceramic-nitride coatings, specifically at high temperatures up to 500°C, using micropillar splitting (Fig. 10.10C). It highlights the innovative approach to reduce ion impregnation during fracture initiation, as afore mentioned crucial for accurate toughness assessment. The studies are notable for revealing the temperature-dependent toughness increase in nanostructured ceramic coatings: AlCrTiN shows a particularly high toughness at room temperature and a significant increase with temperature.

10.3.2.3 Thermal barrier coatings

As previously introduced, pillar-splitting is particularly proficient in material systems requiring high-statistical, spatially resolved information. This is the case of innovative TBC's hetero systems for which gradient properties from the bond coat to the top layer are realized and in which mechanical and environmental wear effects must be studied locally and separately for each phase. To this field, two notable contributions are reported. The paper by Bolelli et al. (2019) investigates the impact of thermal cycling on the nanomechanical properties of thermal barrier coatings (TBCs). They focus on plasma-sprayed yttria-stabilized zirconia (YSZ) top layers and the thermally grown oxide (TGO) layer formed on a vacuum plasma-sprayed NiCoCrAlY bond coat. High-speed nanoindentation and micropillar splitting tests assess the mechanical properties, including fracture toughness. The study finds that TGO's fracture toughness initially increases but decreases significantly after a critical thickness is reached. This leads to damage in the TBC system, including delamination and transverse cracking in the YSZ topcoat. The paper by Bursich et al. (2024) investigates the performance of yttria-stabilized zirconia (YSZ) coatings with different powder morphologies for thermal barrier coatings (TBCs). The study evaluates the impact of powder morphology and chemical composition on the resistance of ceramic monolayer and bilayer coatings to CMAS corrosion and thermal cycling. Various powder morphologies such as agglomerated and sintered, hollow spherical, and fused and crushed are explored. The coatings' mechanical properties, including fracture toughness, are assessed using nanoindentation and pillar-splitting tests (Fig. 10.10A).

10.3.2.4 Multilayered systems

Pillar-splitting is particularly important for assessing fracture toughness in multilayered systems because it provides a homogenized value across specifically controlled volumes within the pillar itself. This is crucial in multilayered systems where layers might contribute differently to the overall toughness. Moreover, pillar-splitting can measure fracture toughness without being affected by the residual stress distributions that often arise from the layering process. Two innovative applications are reported to this scope.

In the study by Xia et al. (Xia et al., 2021), the fracture toughness and microstructure of microscaled multilayer diamond-like carbon (DLC)/silicon systems are explored. The authors investigate the effects of different numbers of DLC coating layers on the fracture behaviour and toughness enhancement. The study finds that the fracture toughness of coated samples significantly increases compared with bare silicon, with a notable increase in protective effect with more DLC layers. Finite element cohesive zone modelling simulations align closely with experimental results, showing the effectiveness of multilayer coatings in improving fracture toughness.

A multivariate approach is represented by the work of White et al. which explores the mechanical performance of optically tuned ceramic nanomultilayers (White et al., 2023), using techniques such as microtensile testing, nanoindentation, and pillar-splitting (Fig. 10.10B). The study focuses on different configurations of AlN/SiO$_2$, TiO$_2$/SiO$_2$, and AlN/Al$_2$O$_3$ nanomultilayers, examining how optical optimisation influences mechanical performance. It reveals that layer thickness, volume fraction, and interfacial crystallinity significantly impact optical and mechanical properties, suggesting the possibility of tuning film characteristics for joint optimisation of optomechanical nano-multilayered coatings. Moreover, a qualitative assessment of the fracture morphology is performed over post-mortem images of the split pillars. These insights are ultimately correlated with failure modes under micro-tensile testing. Remarkably, in the latter work, several stress conditions are compared, and the fracture mechanics of the analysed coatings are in agreement.

10.4 Discussion and conclusions

Recent residual stress measurement techniques developments, particularly in thin coatings, have shown remarkable progress. The focused ion beam ring-core (FIB ring-core) method is a standout example, representing a significant advancement over traditional methods. This innovative approach leverages the precision of ion milling combined with 3D strain analysis, allowing for a deeper, microscale understanding of stress distribution within materials. This is especially critical in fields such as microelectronics and flexible electronics, where the behavior of thin coatings under stress can significantly impact performance and longevity.

The FIB ring-core method addresses a longstanding gap—the need for spatially and in-depth resolved stress measurements. Traditional techniques often provided a more generalized view of stress distribution, which, while useful, could overlook critical microscale stress variations that significantly impact material behavior. By enabling a more granular view, the FIB ring-core method has proven a tool to understand and optimize stress distribution in coatings, accounting for localized geometrical features and heterogeneous structures. One of the main points within the innovation lies within the improvement of its in-depth measurement capabilities, leveraging stresses in ultrathin coatings (as demonstrated by the latest work by Sebastiani et al. (2020)), down to outstanding 20 nm. The main challenge sees both hardware improvements and mathematical modeling of the strain relaxation phenomenon: data interpolation, sensitivity analysis, and error propagation. To the latter, (Lunt & Korsunsky, 2015) provided a calculation framework that accounts for errors within the whole measurement chain of an FIB-DIC measurement: this implies understanding effects of mechanical properties measurements on the accuracy of a stress measurement, effects of DIC tracking errors, and, ultimately, the mathematical modeling of the inverse problem for eigenstrain.

Moreover, the latest advancements in nanoindentation hardware enabled fast mapping of wide areas (high-speed mapping protocols) the feature of residual stresses techniques foresees the coupling of localized FIB-DIC automated mapping experiments with the mapping of mechanical properties through high-speed nanoindentation to understand cross-correlations between phase distributions and residual stresses. Early signs of the latter trend (correlation between multiple techniques with FIB-DIC measurements) are reported in the work by (Archie et al., 2018), where EBSD has been coupled with localized stress measurement trough ring-core to understand the relationship between mechanical performances and lath martensite crystallographic habits in high-strength steels.

Similarly, the pillar-splitting technique's unique foundation in fracture mechanics and its spatially resolved measurements showcases its pioneering role in assessing fracture toughness at the microscale. Its application extends beyond conventional coatings to materials engineered with intricate microstructures, such as additive manufacturing components or composite materials. This method's capability to unravel fracture behavior in highly localized regions brings new insights into material failure mechanisms, contributing to the design and optimization of structures with enhanced damage tolerance.

Sebastiani, Johanns, Herbert, Carassiti, and Pharr (2015) pioneered the application of the pillar splitting technique to measure the fracture toughness of thin ceramic coatings. Their method addressed the challenges of substrate effects and residual stresses, common in the characterization of thin films (Sebastiani et al., 2015). This approach is particularly suited to ceramics, which are prone to brittle fracture, allowing for a direct measurement of fracture toughness without the need for posttest crack measurements.

Despite its advantages, pillar-splitting is not without limitations. Its efficacy is primarily seen in brittle materials, and its application to ductile materials or composites can be challenging. Ductile materials may undergo plastic deformation rather than crack propagation, complicating the measurement of fracture toughness (Sebastiani, Johanns, Herbert, & Pharr, 2015). Moreover, the interaction between different material phases in composites presents a new set of challenges for the pillar-splitting method.

Recent advancements have aimed to address some of the method's drawbacks. For instance, integrating simulation, machine learning, and experimental approaches has been proposed to characterize fracture instability in indentation pillar-splitting, as seen in Athanasiou's work (Athanasiou et al., 2023).

Their work combines experimental observations with high-fidelity simulations and machine learning to create a comprehensive picture of the fracture process. They focus on the unique transition from stable to unstable crack propagation in this method, a phenomenon not fully understood before. Their integration of simulation results and machine learning, specifically Gaussian process regression, leads to a predictive model for critical indentation load, enhancing the practical application of the pillar-splitting technique in material characterization.

In conclusion, while the pillar-splitting technique has provided valuable insights into the mechanical properties of brittle materials, its application to other materials and environmental conditions presents challenges that need to be addressed. Future research may focus on refining the pillar-splitting method to extend its applicability and to mitigate the influence of external variables on the results.

References

Abadias, G., & Daniel, R. (2021). *Stress in physical vapor deposited thin films: Measurement methods and selected examples. Handbook of modern coating technologies: Advanced characterization methods* (pp. 359–436). Available from http://doi.org/10.1016/B978-0-444-63239-5.00008-1.

Albertini, G., Bruno, G., Carradò, A., Fiori, F., Rogante, M., & Rustichelli, F. (1999). Determination of residual stresses in materials and industrial components by neutron diffraction. *Measurement Science & Technology*, *10*(3), R56. Available from https://doi.org/10.1088/0957-0233/10/3/006.

Archie, F., Mughal, M. Z., Sebastiani, M., Bemporad, E., & Zaefferer, S. (2018). Anisotropic distribution of the micro residual stresses in lath martensite revealed by FIB ring-core milling technique. *Acta Materialia*, *150*, 327–338. Available from https://doi.org/10.1016/j.actamat.2018.03.030.

Ast, J., Ghidelli, M., Durst, K., Göken, M., Sebastiani, M., & Korsunsky, A. M. M. (2019). A review of experimental approaches to fracture toughness evaluation at the micro-scale. *Materials and Design*, *173*, 107762. Available from https://doi.org/10.1016/j.matdes.2019.107762.

Athanasiou, C. E., Liu, X., Zhang, B., Cai, T., Ramirez, C., Padture, N. P., Lou, J., Sheldon, B. W., & Gao, H. (2023). Integrated simulation, machine learning, and experimental approach to characterizing fracture instability in indentation pillar-splitting of materials. *Journal of the Mechanics and Physics of Solids*, *170*, 105092. Available from https://doi.org/10.1016/J.JMPS.2022.105092.

Beirau, T., Rossi, E., Sebastiani, M., Oliver, W. C., Pöllmann, H., & Ewing, R. C. (2021). Fracture toughness of radiation-damaged zircon studied by nanoindentation pillar-splitting. *Applied Physics Letters*, *119*(23), 231903. Available from https://doi.org/10.1063/5.0070597.

Best, J. P., Wehrs, J., Polyakov, M., Morstein, M., & Michler, J. (2019). High temperature fracture toughness of ceramic coatings evaluated using micro-pillar splitting. *Scripta Materialia*, *162*, 190–194. Available from https://doi.org/10.1016/j.scriptamat.2018.11.013.

Best, J. P., Zechner, J., Wheeler, J. M., Schoeppner, R., Morstein, M., & Michler, J. (2016). Small-scale fracture toughness of ceramic thin films: The effects of specimen geometry, ion beam notching and high temperature on chromium nitride toughness evaluation. *Philosophical Magazine*, *96*(32–34), 3552–3569. Available from https://doi.org/10.1080/14786435.2016.1223891.

Bhushan, B., & Li, X. (2003). Nanomechanical characterisation of solid surfaces and thin films. *International Materials Reviews*, *48*(3), 125–164. Available from https://doi.org/10.1179/095066003225010227.

Bolelli, G., et al. (2019). Damage progression in thermal barrier coating systems during thermal cycling: A nano-mechanical assessment. *Materials and Design*, *166*, 107615. Available from https://doi.org/10.1016/j.matdes.2019.107615.

Bruns, S., Petho, L., Minnert, C., Michler, J., & Durst, K. (2020). Fracture toughness determination of fused silica by cube corner indentation cracking and pillar splitting. *Materials and Design*, *186*, 108311. Available from https://doi.org/10.1016/J.MATDES.2019.108311.

Bursich, S., et al. (2024). The effect of ceramic YSZ powder morphology on coating performance for industrial TBCs. *Surface and Coatings Technology*, *476*, 130270. Available from https://doi.org/10.1016/j.surfcoat.2023.130270.

Cook, R. F., & Pharr, G. M. (1990). Direct observation and analysis of indentation cracking in glasses and ceramics. *Journal of the American Ceramic Society*, *73*(4), 787–817. Available from https://doi.org/10.1111/j.1151-2916.1990.tb05119.x.

Dang, N. M., et al. (2022). Mechanical properties and residual stress measurement of TiN/Ti duplex coating using HiPIMS TiN on cold spray Ti. *Coatings*, *12*(6). Available from https://doi.org/10.3390/coatings12060759.

Fitzpatrick, M. E., & Lodini, A. (2003). *Analysis of residual stress by diffraction using neutron and synchrotron radiation*. CRC Press.

Ghidelli, M., Sebastiani, M., Johanns, K. E., & Pharr, G. M. (2017). Effects of indenter angle on micro-scale fracture toughness measurement by pillar splitting. *Journal of the American Ceramic Society*, *100*(12), 5731–5738. Available from https://doi.org/10.1111/jace.15093.

Jiang, Y., Hor, J. L., Lee, D., & Turner, K. T. (2018). Toughening nanoparticle films via polymer infiltration and confinement. *ACS Applied Materials & Interfaces*, *10*(50), 44011–44017. Available from https://doi.org/10.1021/acsami.8b15027.

Johanns, K. E., Lee, J. H., Gao, Y. F., & Pharr, G. M. (2013). An evaluation of the advantages and limitations in simulating indentation cracking with cohesive zone finite elements. *Modelling and Simulation in Materials Science and Engineering*, *22*(1), 015011. Available from https://doi.org/10.1088/0965-0393/22/1/015011.

Jun, T. S., & Korsunsky, A. M. (2010). Evaluation of residual stresses and strains using the Eigenstrain reconstruction method. *International Journal of Solids and Structures*, *47*(13), 1678–1686. Available from https://doi.org/10.1016/j.ijsolstr.2010.03.002.

Korsunsky, A. M., et al. (2016). Quantifying eigenstrain distributions induced by focused ion beam damage in silicon. *Materials Letters*, *185*, 47–49. Available from https://doi.org/10.1016/j.matlet.2016.08.111.

Korsunsky, A. M. A teaching essay on residual stresses and Eigenstrains. Matthew Deans, 2017.

Korsunsky, A. M., et al. (2018). Nanoscale residual stress depth profiling by focused ion beam milling and eigenstrain analysis. *Materials and Design*, *145*, 55–64. Available from https://doi.org/10.1016/j.matdes.2018.02.044.

Korsunsky, A. M., Sui, T., Salvati, E., George, E. P., & Sebastiani, M. (2016). Experimental and modelling characterisation of residual stresses in cylindrical samples of rapidly cooled bulk metallic glass. *Materials and Design*, *104*, 235–241. Available from https://doi.org/10.1016/j.matdes.2016.05.017.

Lauener, C. M., Petho, L., Chen, M., Xiao, Y., Michler, J., & Wheeler, J. M. (2018). Fracture of silicon: Influence of rate, positioning accuracy, FIB machining, and elevated temperatures on toughness measured by pillar indentation splitting. *Materials and Design*, *142*, 340–349. Available from https://doi.org/10.1016/j.matdes.2018.01.015.

Leide, A. J., et al. (2023). Measurement of residual stresses in surrogate coated nuclear fuel particles using ring-core focussed ion beam digital image correlation. *Nuclear Materials and Energy*, *36*. Available from https://doi.org/10.1016/j.nme.2023.101470.

Lunt, A. J. G., & Korsunsky, A. M. (2015). A review of micro-scale focused ion beam milling and digital image correlation analysis for residual stress evaluation and error estimation. *Surface and Coatings Technology*, *283*, 373–388. Available from https://doi.org/10.1016/j.surfcoat.2015.10.049.

Mohrbacher, H., Van Acker, K., Blanpain, B., Van Houtte, P., & Celis, J. P. (1996). Comparative measurement of residual stress in diamond coatings by low-incident-beam-angle-diffraction and micro-Raman spectroscopy. *Journal of Materials Research*, *11*(7), 1776–1782. Available from https://doi.org/10.1557/JMR.1996.0222/METRICS.

Mughal, M. Z., Moscatelli, R., Amanieu, H. Y., & Sebastiani, M. (2016). Effect of lithiation on micro-scale fracture toughness of $Li_xMn_2O_4$ cathode. *Scripta Materialia*, *116*, 62–66. Available from https://doi.org/10.1016/j.scriptamat.2016.01.023.

Perry, A. J., Sue, J. A., & Martin, P. J. (1996). Practical measurement of the residual stress in coatings. *Surface and Coatings Technology*, *81*(1), 17–28. Available from https://doi.org/10.1016/0257-8972(95)02531-6.

Rossi, E., Wheeler, J. M., & Sebastiani, M. (2023). High-speed nanoindentation mapping: A review of recent advances and applications. *Current Opinion in Solid State and Materials Science*, *27*(5). Available from https://doi.org/10.1016/j.cossms.2023.101107.

Salvati, E., et al. (2018). Nanoscale structural damage due to focused ion beam milling of silicon with Ga ions. *Materials Letters*, *213*, 346−349. Available from https://doi.org/10.1016/j.matlet.2017.11.043.

Salvati, E., Romano-Brandt, L., Mughal, M. Z., Sebastiani, M., & Korsunsky, A. M. (2019). Generalised residual stress depth profiling at the nanoscale using focused ion beam milling. *Journal of the Mechanics and Physics of Solids*, *125*, 488−501. Available from https://doi.org/10.1016/j.jmps.2019.01.007.

Salvati, E., Sui, T., Lunt, A. J. G., & Korsunsky, A. M. (2016). The effect of eigenstrain induced by ion beam damage on the apparent strain relief in FIB-DIC residual stress evaluation. *Materials and Design*, *92*, 649−658. Available from https://doi.org/10.1016/j.matdes.2015.12.015.

Schajer, G. S. (1981). Application of finite element calculations to residual stress measurements. *Journal of Engineering Materials and Technology, Transactions of the ASME*, *103*(2), 157−163. Available from https://doi.org/10.1115/1.3224988.

Schajer, G. S. (2013). Hole Drilling and Ring Coring. In *Practical residual stress measurement methods*. Hoboken, New Jersey, U.S.: John Wiley & Sons. Available from http://doi.org/10.1002/9781118402832.

Sebastiani, M., et al. (2020). Nano-scale residual stress profiling in thin multilayer films with non-equibiaxial stress state. *Nanomaterials*, *10*(5). Available from https://doi.org/10.3390/nano10050853.

Sebastiani, M., Eberl, C., Bemporad, E., & Pharr, G. M. (2011). Depth-resolved residual stress analysis of thin coatings by a new FIB−DIC method. *Materials Science and Engineering: A*, *528*(27), 7901−7908. Available from https://doi.org/10.1016/j.msea.2011.07.001.

Sebastiani, M., Johanns, K. E., Herbert, E. G., Carassiti, F., & Pharr, G. M. (2015). A novel pillar indentation splitting test for measuring fracture toughness of thin ceramic coatings. *Philosophical Magazine*, *95*(16−18), 1928−1944. Available from https://doi.org/10.1080/14786435.2014.913110.

Sebastiani, M., Johanns, K. E., Herbert, E. G., & Pharr, G. M. (2015). Measurement of fracture toughness by nanoindentation methods: Recent advances and future challenges. *Current Opinion in Solid State and Materials Science*, *19*(6), 324−333. Available from https://doi.org/10.1016/j.cossms.2015.04.003.

Sebastiani, M., Sui, T., & Korsunsky, A. M. (2017). Residual stress evaluation at the micro- and nano-scale: Recent advancements of measurement techniques, validation through modelling, and future challenges. *Materials and Design*, *118*, 204−206. Available from https://doi.org/10.1016/j.matdes.2017.01.025.

Shah, R. C., & Kobayashi, A. S. (1973). Stress intensity factors for an elliptical crack approaching the surface of a semi-infinite solid. *International Journal of Fracture*, *9*(2), 133−146. Available from https://doi.org/10.1007/BF00041855.

Taking the stress out of nanomaterials design | ISTRESS Project | Results in brief | FP7 | CORDIS | European Commission. (2024). Available: https://cordis.europa.eu/article/id/202149-taking-the-stress-out-of-nanomaterials-design, (Accessed January 01, 2024).

White, D. E., Appleget, C. D., Rossi, E., Sebastiani, M., & Hodge, A. M. (2023). The mechanical performance of optically tuned ceramic nanomultilayers. *Materials and Design*, *231*. Available from https://doi.org/10.1016/j.matdes.2023.112014.

Winiarski, B., Gholinia, A., Tian, J., Yokoyama, Y., Liaw, P. K., & Withers, P. J. (2012). Submicron-scale depth profiling of residual stress in amorphous materials by incremental focused ion beam slotting. *Acta Materialia, 60*(5), 2337–2349. Available from https://doi.org/10.1016/j.actamat.2011.12.035.

Winiarski, B., & Withers, P. J. (2012). Micron-scale residual stress measurement by microhole drilling and digital image correlation. *Experimental Mechanics, 52*(4), 417–428. Available from https://doi.org/10.1007/s11340-011-9502-3.

Withers, P. J., & Bhadeshia, H. K. D. H. (2001a). Residual stress. Part 2 – Nature and origins. *Materials Science and Technology, 17*(4), 366–375. Available from https://doi.org/10.1179/026708301101510087.

Withers, P. J., & Bhadeshia, H. K. D. H. (2001b). Residual stress. Part 1 – Measurement techniques. *Materials Science and Technology, 17*(4), 355–365. Available from https://doi.org/10.1179/026708301101509980.

Xia, Y., Hirai, Y., & Tsuchiya, T. (2021). Effect of alternating a-C:H multilayer full coating on fracture behavior of single-crystal silicon-based microstructure in tensile and toughness tests. *Materials Science and Engineering: A, 827*. Available from https://doi.org/10.1016/j.msea.2021.142054.

Multi-sensing approaches

Jan Tomastik[1,2], Adrian Harris[3], Vladimir Vishnyakov[4] and Ben Beake[3]
[1]Joint Laboratory of Optics of Palacký University and Institute of Physics AS CR, Palacký University in Olomouc, Olomouc, Czech Republic, [2]Institute of Physics of the Czech Academy of Sciences, Joint Laboratory of Optics of Palacký University and Institute of Physics AS CR, Olomouc, Czech Republic, [3]Micro Materials Ltd., Wrexham, United Kingdom, [4]Inomorph Ltd., Huddersfield, United Kingdom

11.1 Introduction

Multi-sensing approaches where monitoring of multiple signals during the test (including indentation or wear depth, friction, acoustic emission, and electrical contact resistance) together with post-test microscopic analysis of the indentation or wear tracks can provide further information about the onset and progression of deformation/wear processes at a highly localized scale. Allied to multi-sensing techniques are approaches where environmental control is used. Nanomechanical properties are usually determined by measurements under standard laboratory conditions with controlled temperature and humidity. The mechanical properties of materials should be characterized under environmentally relevant conditions.

This chapter focusses on acoustic emission, electrical contact resistance, and friction measurements. The chapter concludes with a short overview of various experimental approaches where nanomechanical test instruments have been used to perform more complex experiments with additional spectroscopic, electrochemical measurements and pulsing electrical current, in some cases involving measurements at elevated temperature.

11.2 Multi-sensing approach using simultaneous record of acoustic emission

Acoustic emissions (AEs) are the phenomenon of elastic acoustic waves that are emitted during the irreversible changes in materials such as the deformation and/or degradation. Techniques for testing materials and products based on sound sensations have been known since antiquity (pottery, weapons, etc.). In modern industrial times, they were successfully employed as a detection technique for the structural diagnosis at the "macro-level" in large scale objects, such as pressure vessels safety measurement, damage detection in the primary circuit of nuclear power plants

(Svoboda, 1992), or as a part of geomechanical investigation in mines (Becker et al., 2010; Hesser et al., 2015) and many others. However, microlevel applications for similar purposes are also expanding as they are desirable due to the continuous progress of miniaturization in industry. On microscale, they can provide valuable information about the processes in internal structure of materials, such as microcrack initiation and spreading, failures of adhesion/cohesion in coating − substrate systems but also more subtle events such as dislocation movement or force-induced phase transformation (Ctvrtlik et al., 2019; Nazarchuk et al., 2017; Scruby, 1987; Tomastik et al., 2015).

Acoustic emissions can be frequently initiated during nanoindentation and scratch testing of surfaces and coatings. Stronger AEs are emitted during testing of harder and more brittle materials, where deformation usually leads to more severe cracks (Tomastik, Ctvrtlik, Ingr, et al., 2018). On the contrary, AE may not be detectable when testing ductile samples of lower hardness, either due to a missing initiation effect or because of an insufficient signal strength. Research efforts are underway to increase the detection threshold, which may be close to the level of signal noise. At the same time, the operation of the nanoindentation or scratch test instrument itself generates a lot of parasitic vibrations and sound wave emissions. This was particularly a problem with the early accelerometer-based sensors introduced around 1980s. The development thus progressed toward AE resonant transducers operating in the frequency range of 30−1000 kHz with most parasitic signals tending to be low-frequency signals below 50 kHz (Julia-Schmutz & Hintermann, 1991). Nowadays, the main component of modern AE sensors is usually a broadband piezoelectric detector capable of sensing high-frequency signals up to 1 MHz, which arises from very localized deformation processes such as those produced in nanoindentation-based experiments (Choudhary & Mishra, 2016; Proctor, 1982).

The location of the sensor relative to the AE initiation area is crucial. In the case of macroscopic measurements where intense AEs are generated, the sensors can be placed along the body of the sample (pipes, etc.), whereas for more localized techniques such as nanoindentation, the positioning of the AE sensor needs to be placed more closely to the initiation point. For nanoindentation methods where the tip interacts with the sample, this situation can be practically approached in three ways —(1) miniaturized sensors can be placed directly on the indenter tip or very close to its shaft (Choudhary & Mishra, 2016; Hiroyuki, 2012; von Stebut et al., 1999), (2) glued to the surface of sample (Gallego et al., 2005, 2007), or (3) use of specific sample holder with embedded AE sensor (Beake et al., 2020; Ctvrtlik et al., 2019, 2022; Němeček et al., 2023; Tomastik et al., 2015; Tomastik, Ctvrtlik, Ingr, et al., 2018). All solutions are used in practice and possess some advantages and disadvantages. A sensor mounted directly on the indenter body has the advantage that it is always at the same distance from the tip itself and, therefore, from the point of initiation of acoustic emissions. On the other hand, there are several drawbacks. Even though initiated AEs are omnidirectional, they are transferred through the small contact area between the tip and the sample surface. It is also quite complicated to mount the sensor on the tip, which in turn can affect handling, and the weight of the sensor (and attached cables) can affect its movement. A frequently

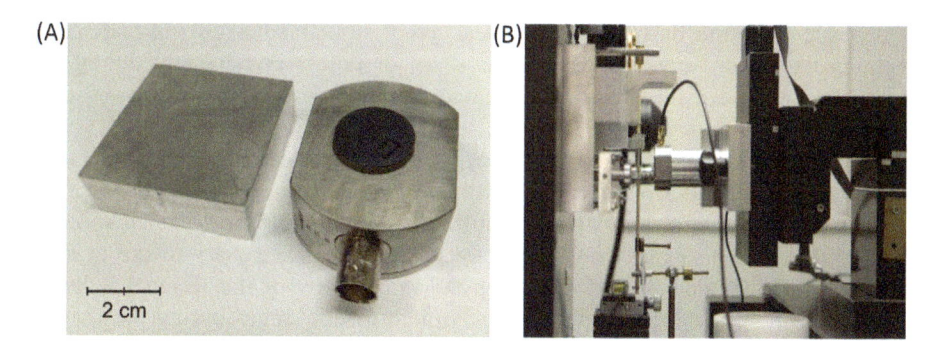

Figure 11.1 AE detection with embedded piezoelectric acoustic emission detector in sample holder. (A) Comparison of standard 5 × 5 cm specimen holder (left) with special holder with embedded piezoelectric acoustic emission detector. (B) Attachment of acoustic emission holder into the NanoTest instrument.
Source: From Tomastik, J., Ctvrtlik, R., Drab, M., Manak, J. (2018). On the importance of combined scratch/acoustic emission test evaluation: SiC and SiCN thin films case study. *Coatings*, *8*(5). https://doi.org/10.3390/coatings8050196.

used option is to attach the AE sensor on the sample using pressure, magnetic force, or by gluing. This approach usually brings higher sensitivity due to the larger contact area with AE sensor. The mounting method must be chosen appropriately as contact between sensor and sample must be as tight as possible without voids or excessively thick layers of adhesive that could interfere with AE transmission. The disadvantage of this solution is the varying distance of the sensor from the AE initiation sources during measurement in different areas of the sample. Sensor positioning is also important as contact with the indenter must be avoided. The latter method, with the built-in sensor in a special sample holder, allows for the easiest handling during experiments and brings especially the possibility to use the already fixed sample for other types of experiments without the need to reattach the sensor. Its disadvantages may be the longer distance of the sensor from the AE initiation point and multiple interfaces between the source and the sensor; however, with a suitable design and highly sensitive piezo sensor, these problems can be overcome. This solution (shown in Fig. 11.1) will be further discussed in this chapter and presented on several practical cases.

11.2.1 Acoustic emission in scratch testing

The scratch test evaluation is standardly performed using depth change record and postprocess microscopic evaluation. The use of one evaluation method may be sufficient when a relatively well-known coating − substrate system is being tested or when the objective is to investigate a distinctive phenomenon such as coating cracking or large delamination of coating from the substrate. In such cases, the acoustic emission detection that accompanies these significant destructive events is a suitable standalone alternative method for such evaluation. Simultaneous detection of acoustic emission during a scratch test has a similar advantage as indenter's depth-change records, as it identifies the onset of failure immediately during the test without the need for subsequent time-consuming microscopic evaluation (e.g., see Fig. 11.2).

When performing the basic and applied research on the unknown coating − substrate systems or systems differing in composition, structure, and properties, the need for more thorough evaluation (albeit time-consuming) arises. The demands for an exact evaluation of the scratch test increase, and it is necessary to use at least both standard methods—depth recording and microscopy of the residual indentation. However, given the complexity of the deformation zones during the scratch test, even two standardized techniques may not give accurate results. For example, in situations where microcracks are formed at the interface between substrate and opaque coating, both microscopic evaluation and depth change record may not successfully detect the very first onset of failure.

This can be demonstrated in Tomastik, Ctvrtlik, Drab, et al. (2018) dealing with hard a-SiC and a-SiCN thin films deposited on Si substrates that were studied after annealing using scratch test and subsequently evaluated by depth change record, microscopic evaluation, and simultaneous acoustic emissions detection, while the latter was utilized using the embedded sensor in the body of the sample holder. Fig. 11.3A shows the progressive load scratch test with maximum load 500 mN leading to a large-scale delamination seemingly emerging from point "A." It is the subsequent methods that reveal the true dynamics of the failure process in the film − substrate system. The depth change record (see Fig. 11.3C) with on-load curve (blue) and the final topography curve (red) indicates to film's gross spallation originating from "L_{C2}" point expanding backward to the point "A." This analysis is then strongly supported by subsequent AE record (see Fig. 11.3B). AE data show an isolated peak at c. 17th second, which can be explained by interaction and breaking of the surface asperity, which is actually confirmed from the first topographic

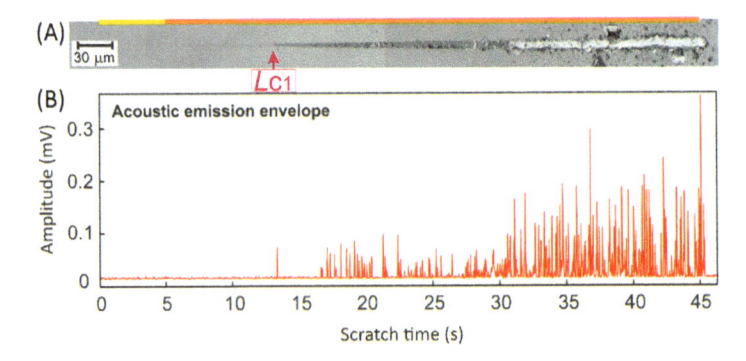

Figure 11.2 Evaluation of ramped scratch test on nanodiamond coating using (A) microscopic evaluation of residual scratch and (B) a simultaneous acoustic emission record. Clear correlation between the onset of coating failure and first burst of acoustic emission. Acoustic emission is higher for subsequent substrate fracture.

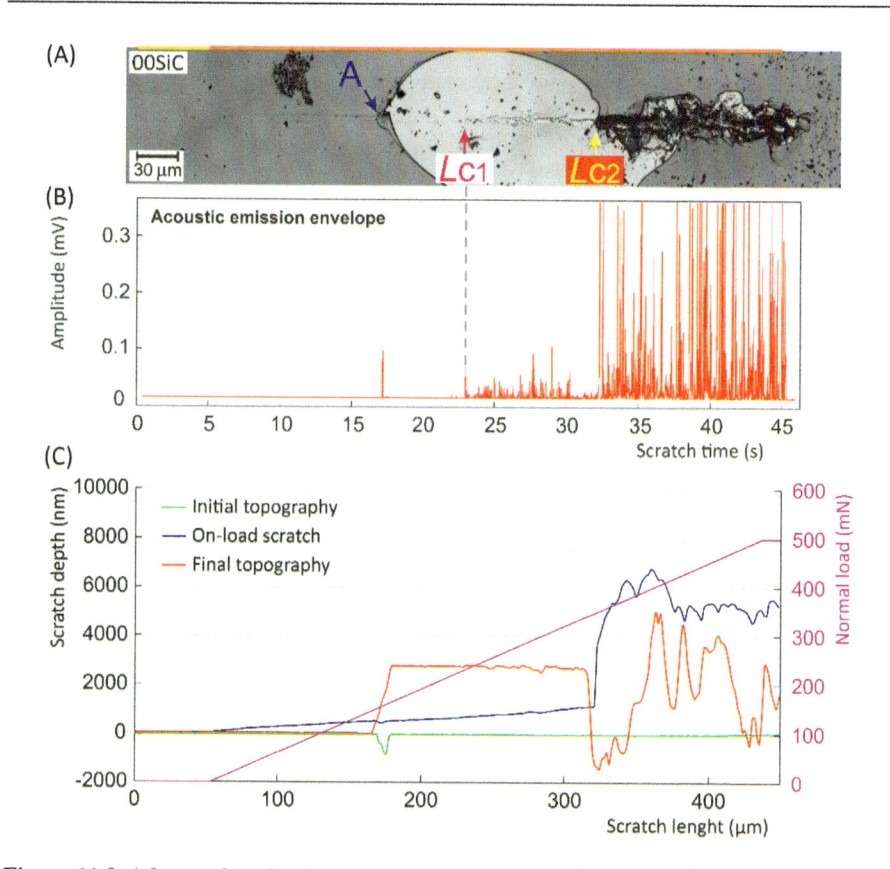

Figure 11.3 Advanced evaluation of ramped nano-scratch test on a-SiC thin film on silicon. (A) Microscopic observation of scratch track, (B) acoustic emissions record, and (C) depth-change record with marked critical loads.
Source: From Tomastik, J., Ctvrtlik, R., Ingr, T., Manak, J., Opletalova, A. (2018). Effect of nitrogen doping and temperature on mechanical durability of silicon carbide thin films. *Scientific Reports*, 8(1), (2018). https://doi.org/10.1038/s41598-018-28704-3.

curve in the depth record (see green curve in Fig. 11.3C). Thus, point A is in fact not the origin of any type of failure, it is a random impurity limiting the backward delamination of the film. Actual critical failure occurs only at the "L_{C2}" referring to critical load causing a subsequent pronounced cleaving of both the film and the substrate (see Fig. 11.3A), which is clearly identifiable also from the increase in acoustic emission intensity that occurred at the 33rd second of the AE record (see Fig. 11.3B). The capabilities of the simultaneous acoustic emission recording during scratch test become even more evident on the detection of the initial failure (marked as "L_{C1}"). AE signal of lower amplitude (than "L_{C2}") arising between 23rd and 33rd second (see Fig. 11.3B) cannot be easily identified from the depth record (see Fig. 11.3C). It is slightly apparent from the microscopic image as a substrate cracking inside the scratch track; however, if there was no backward delamination of the opaque film, this type of failure would remain hidden in microscopic image. The acoustic emission recording thus demonstrates the ability to detect failures at the film − substrate interface or in the substrate itself, even though they may not be apparent from other evaluation methods, thus more precisely quantifying critical loads in a scratch test.

Another demonstration of AE sensitivity can be shown in case of subsurface cracking of more ductile SiCN thin film sample. While microscopic evaluation of the residual scratch cannot distinguish the onset of subsurface cracking (see Fig. 11.4A), the resulting acoustic emissions provide information about some critical event in the coating substrate system (see Fig. 11.4B). The origin of the acoustic emissions was confirmed by subsequent analysis of the residual indentation by transverse sectioning using the focused ion beam (FIB) method. In the detail marked "A" of Fig. 11.4C, subsurface cracks can be seen propagating from the substrate into the coating, but not reaching its surface. The cracks only reached the surface in the later part between areas A and B, but acoustic emission revealed cracking earlier.

11.2.2 Acoustic emission in bulk analysis

Acoustic emission can be also employed for study of more complex heterogeneous material phases of different mechanical properties as shown in Němeček et al. (2023) focused on fracture toughness evaluation of hydration products of ordinary Portland cement paste at the microscale.

Nanoindentation test coupled with AE record has shown that AE signal emerged only in "portlandite phase" of this complex system. Given the structure of the phase, it is likely to be caused by slipping of the planes of the portlandite crystal (see Figs. 11.5 and 11.6). This was supported by a scratch test with simultaneous AE recording, which detected an AE signal only when the tip was crossed the portlandite phase. This suggests that the deformation mechanisms for the remaining cement paste phases are different (see Fig. 11.7).

11.2.3 Acoustic emission in impact testing

Beake et al. used AE monitoring during repetitive nanoimpact tests on the bulk technical ceramics alumina and MgO-partially stabilized zirconia (PSZ) to more deeply investigate how their mechanical properties and microstructure influence the crack evolution in impact tests with different probe geometries and applied loads (Beake et al., 2020). The mechanical properties of the ceramics were determined by nanoindentation with a Berkovich indenter. For alumina, these were $H = (33.0 \pm 2.4)$ GPa; $E = (465 \pm 32)$ GPa; $H/E = 0.071$, $H^3/E^2 = 0.166$ GPa. PSZ was softer and less stiff: $H = (14.8 \pm 0.9)$ GPa; $E = (240 \pm 10)$ GPa; $H/E = 0.061$, $H^3/E^2 = 0.056$ GPa. Impact depth data and corresponding AE are shown in Table 11.1 for tests with cube corner indenter at 5 mN and a 5 μm spherical indenter at 5 and 50 mN. The AE monitoring clearly showed that in the tests with the cube corner probe, the large displacement bursts were accompanied by single

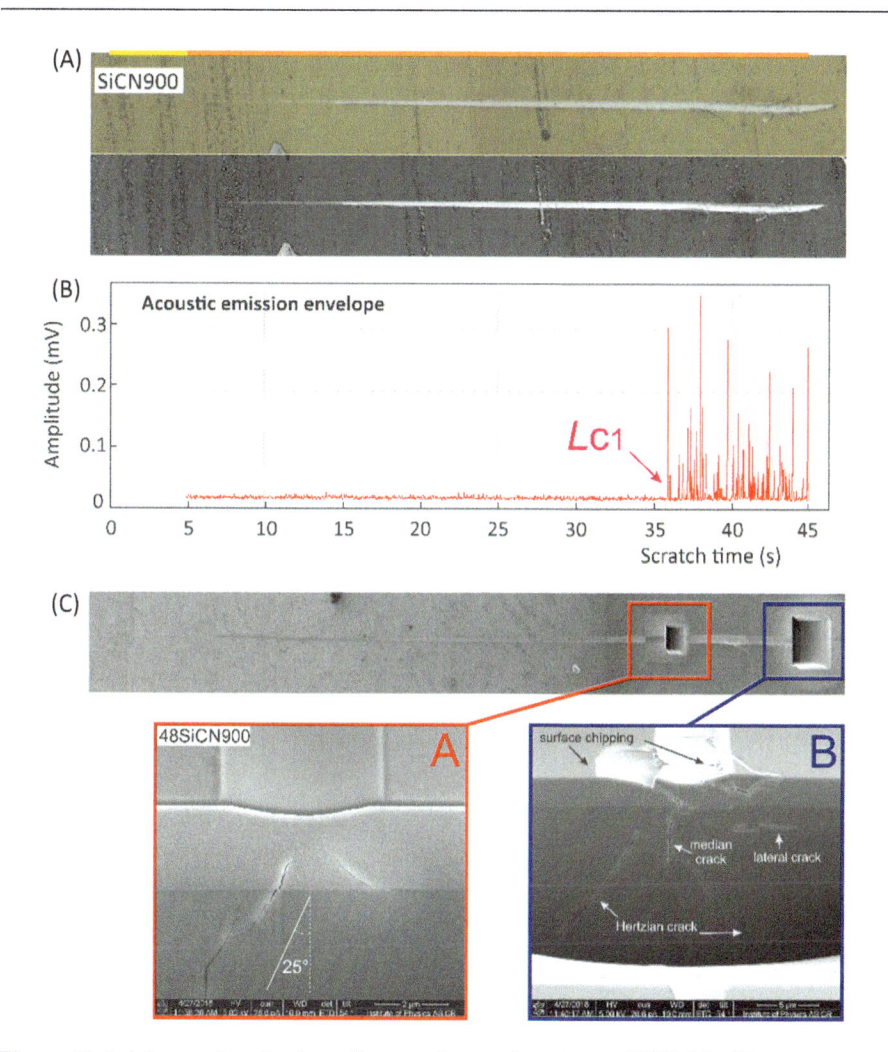

Figure 11.4 Advanced evaluation of ramped scratch test on a-SiCN thin film on silicon. (A) microscopic observation (in color an confocal mode), (B) acoustic emissions record, and (C) cross section of two residual scratch areas using focused ion beam, revealing subsurface cracking at the burst of acoustic emission.
Source: Modified from Tomastik, J., Ctvrtlik, R., Ingr, T., Manak, J., Opletalova, A. (2018). Effect of nitrogen doping and temperature on mechanical durability of silicon carbide thin films. *Scientific Reports*, 8(1), (2018). https://doi.org/10.1038/s41598-018-28704-3.

high-energy bursts of AE, confirming that these relate to cracking and material removal.

In the tests with a 5 μm radius spherical indenter at 50 mN, the damage progression was more complex and differed between the ceramics with more AE events near the start of the test on alumina and more later in the test for PSZ. Typical

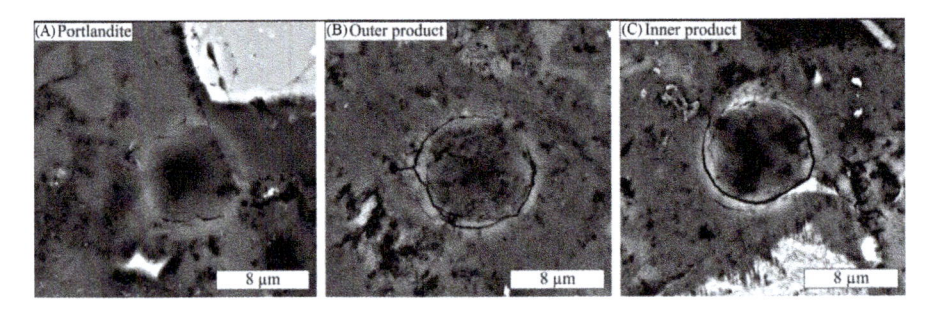

Figure 11.5 SEM-BSE images of nanoindentation into different phases of Portland cement paste. (A) Portlandite, (B) Outer product, (C) Inner product.
Source: From Němeček, J., Čtvrtlík, R., Václavek, L., Němeček, J. (2023). Fracture toughness of cement paste assessed with micro-scratch and acoustic emission. *Powder Metallurgy Progress*, 22(1), 7−13, https://doi.org/10.2478/pmp-2022-0002, http://www.degruyter.com/view/j/pmp.

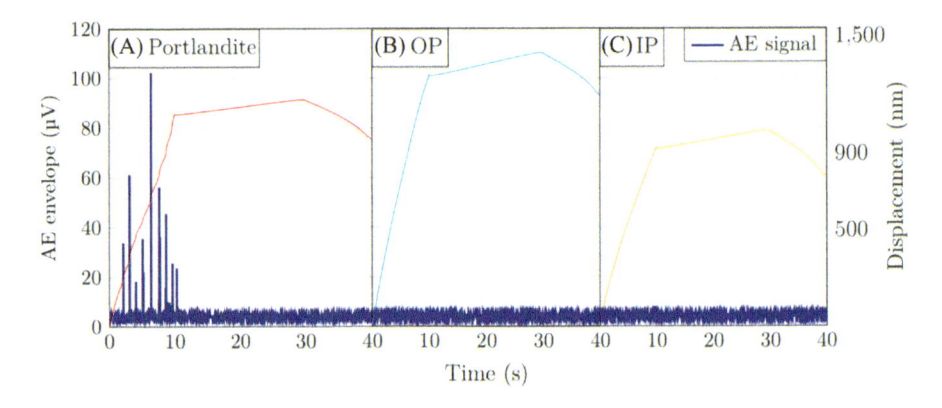

Figure 11.6 Displacement − time curves with recorded acoustic emission signals. (A) Portlandite, (B) Outer product, (C) Inner product.
Source: From Němeček, J., Čtvrtlík, R., Václavek, L., Němeček, J. (2023). Fracture toughness of cement paste assessed with micro-scratch and acoustic emission. *Powder Metallurgy Progress*, 22(1), 7−13, https://doi.org/10.2478/pmp-2022-0002, http://www.degruyter.com/view/j/pmp.

examples are shown in Figs. 11.8 and 11.9. The resulting impact craters at 50 mN both samples can be seen in angled views in Fig. 11.10. The initial depth is lower on alumina due to its higher hardness and H^3/E^2. However, a burst of AE on alumina was observed, which was absent on PSZ sample. After the initial impact, the depth changed little for the next seven or eight impacts before increasing on both ceramics, more rapidly for alumina. A focused ion beam was used to mill sections through the impact craters. This revealed subsurface intergranular cracking on alumina and subsurface transgranular cracking on PSZ. On alumina, the network of

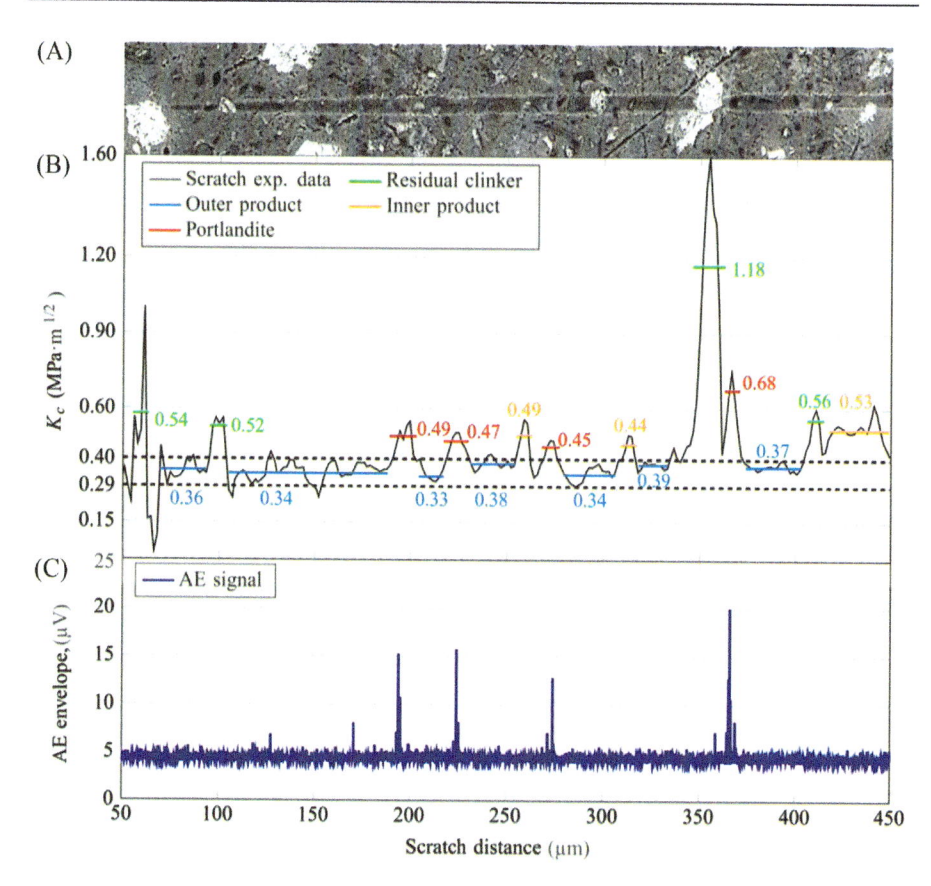

Figure 11.7 Nanoscratch test of cement paste. (A) SEM-BSE image of the scratch line; (B) fracture toughness calculated from the scratch test with the highlighted mean values of phases; (C) record of acoustic emission signal.
Source: From Němeček, J., Čtvrtlík, R., Václavek, L., Němeček, J. (2023). Fracture toughness of cement paste assessed with micro-scratch and acoustic emission. *Powder Metallurgy Progress*, 22(1), 7–13, https://doi.org/10.2478/pmp-2022-0002, http://www.degruyter.com/view/j/pmp.

subsurface cracking revealed by FIB is consistent with an initiation process where cracks form subsurface, but they were not joined up sufficiently to cause material removal under the indenter, and several impacts are required before material removal. Once established, the driving force for propagation of intergranular cracks on alumina is reduced, so that further impacts produce little change in depth or AE activity. The transformation toughening mechanism available in PSZ initially relieves the stresses responsible for cracking but with continued reloading plasticity is exhausted within a few cycles and transgranular cracking ensues. The transgranular cracks appear to propagate efficiently with a gradual increase in the impact depth (so that the depth plateau seen on alumina is absent), extensive lateral cracking, and bursts of AE seen for nearly all the impacts.

Table 11.1 Impact depth and acoustic emission data in nano-impact tests on alumina and partially stabilized zirconia.

Sample	Probe	Applied load (mN)	h_1/μm	h_f/μm	Total AE hits
Alumina	Cube corner	5	0.7 ± 0.0	2.0 ± 0.5	18 ± 6
PSZ	Cube corner	5	0.9 ± 0.0	2.1 ± 0.3	24 ± 14
Alumina	$R = 5$ μm spherical	5	0.1 ± 0.0	0.1 ± 0.0	5 ± 6
Alumina	$R = 5$ μm spherical	50	0.4 ± 0.0	2.5 ± 0.2	125 ± 12
PSZ	$R = 5$ μm spherical	5	0.2 ± 0.0	0.2 ± 0.1	0.4 ± 0.5
PSZ	$R = 5$ μm spherical	50	0.8 ± 0.1	2.9 ± 0.3	215 ± 37

Notes: h_1, on-load impact depth after the first impact. h_f, on-load impact depth at the end of the test.

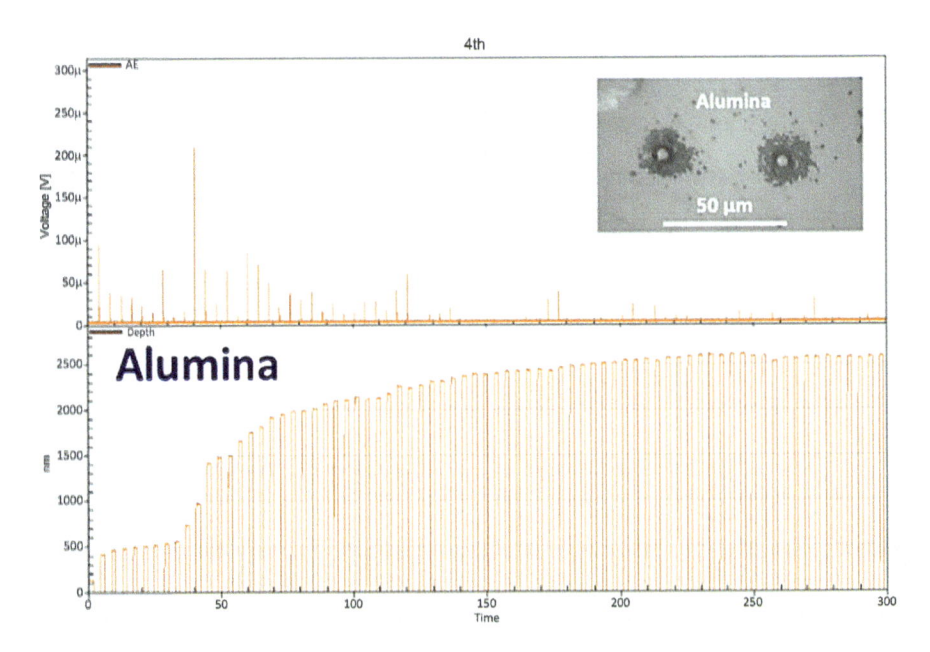

Figure 11.8 Depth and acoustic emission from nano-impact tests on alumina at 50 mN with a 5 μm spherical indenter. On alumina bursts of acoustic emission occur before the increase in impact depth. Typical impact craters are shown in the inset.

11.2.4 Advanced evaluation of acoustic emission during scratch test

The detection and analysis of acoustic emissions presented in the previous cases utilize in practice only the compressed time record or the so-called "**envelope**."

Multi-sensing approaches 311

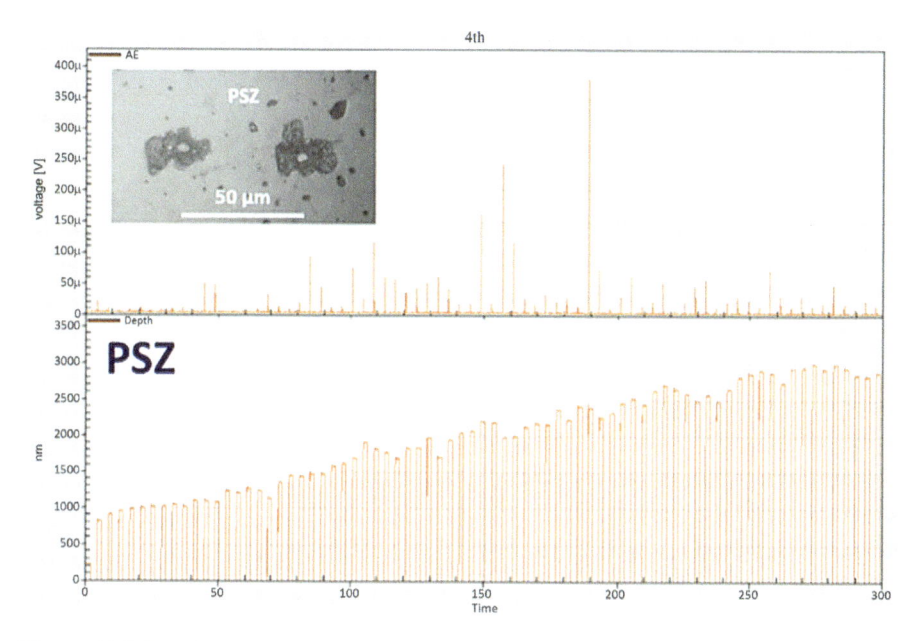

Figure 11.9 Depth and acoustic emission from nano-impact tests on partially stabilized zirconia at 50 mN with a 5 μm spherical indenter. On partially stabilized zirconia bursts of AE occur later in the test. Typical impact craters are shown in the inset.

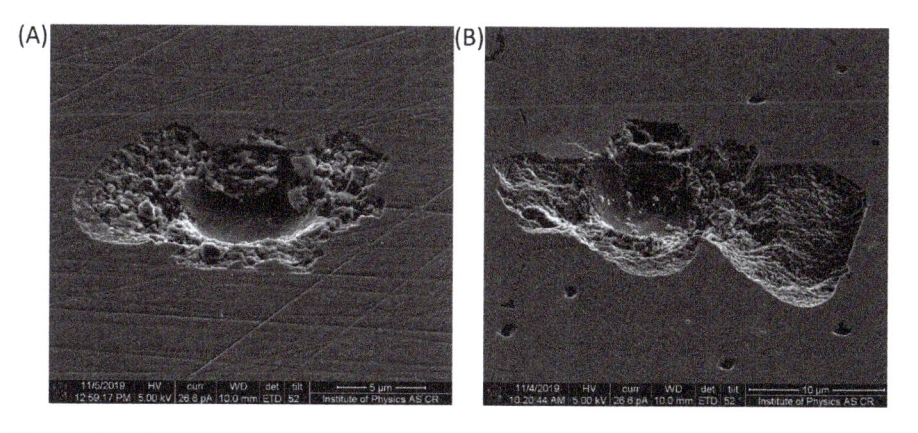

Figure 11.10 Impact craters on (A) alumina and (B) partially stabilized zirconia. 50 mN with a 5 μm spherical indenter SEM images of impact craters on alumina and partially stabilized zirconia.

It represents suitable visualization that can be compared with other scratch test outputs such as microscopic image or depth change record thanks to the synchronised time or length coordinates. However, the acoustic emission process can be analyzed more thoroughly, and further information can be obtained, as shown in Fig. 11.11.

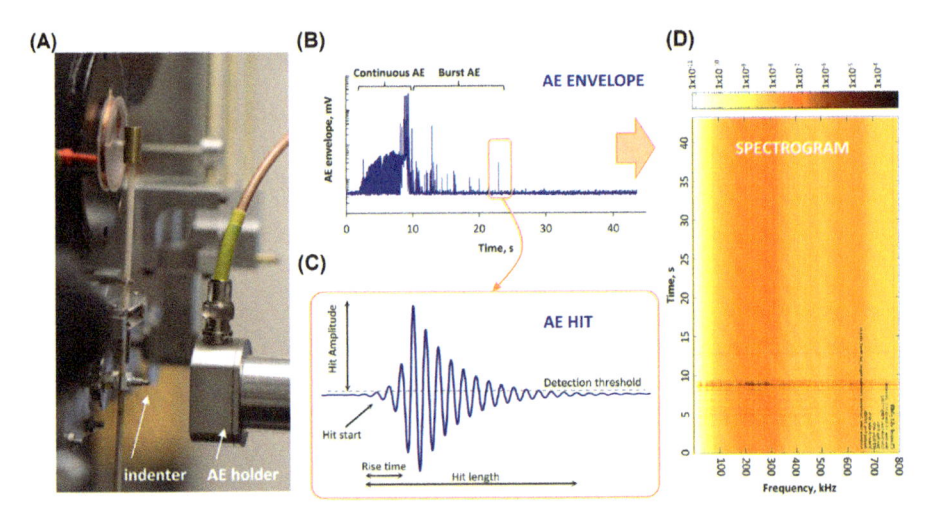

Figure 11.11 Acoustic emission measurement on the NanoTest system. (A) Acoustic emission holder implemented in the NanoTest system, (B) acoustic emission envelope (compressed signal), (C) acoustic emission hit waveform (full resolution signal), and (D) spectrogram corresponding to the crushing of the silicalite-1 zeolite crystal by a cube-corner indenter.
Source: From Ctvrtlik, R., Tomastik, J., Vaclavek, L., Beake, B. D., Harris, A. J., Martin, A. S., Hanak, M., Abrham, P. (2019). High-resolution acoustic emission monitoring in nanomechanics, *JOM*, *71*(10), 3358−3367, https://doi.org/10.1007/s11837-019-03700-8.

Depending on the incidence of the signal in the envelopes, one can distinguish between burst emission characterized by discrete peaks induced by discrete AE events or continuous emission, which is a superposition of a series of smaller AE events, such as typically occurs in a cascade of dislocations or small cracks or in the case of increased friction (Ctvrtlik et al., 2019). In any case, the acoustic emission record is actually a series of discrete events lasting tens or hundreds of microseconds, which are called "**hits**." Fig. 11.11C shows hit diagram in full resolution where several amplitude and time parameters can be described. Hit starts as a sudden rise of a periodic signal, which after a certain "**rise time**" reaches the maximum amplitude. The rise time can be used to estimate the lifetime of a physical event generating elastic deformation waves. It should be only considered as an upper bound of the event lifetime, as it is affected by a variety of factors such as sensor characteristics, signal reflection. After reaching the maximum amplitude, the attenuation of the signal occurs more slowly until it falls below the detection limit. This determines the "**hit length**." The amplitude of the signal is of course related to the intensity of the event that generates the AE (cracking, plastic deformation, phase transition, delamination, debonding, friction, etc.). In addition to the maximum amplitude, the intensity of the event can also be evaluated using the "**hit energy**," which is calculated as the area below the signal level. Similarly, the signal RMS (root mean square) is also often used providing direct information about the energy of AE signal (Ctvrtlik et al., 2019).

A Fourier transformation can be used to obtain "**hit spectrum**" that can be regarded along with time — amplitude parameters as a possible fingerprint of a particular type of failure mode in the film — substrate system. The spectrum can be displayed for each specific intervention or even in the form of a spectrogram showing spectral intensities for the full length (duration) of the test (scratch test, indentation, etc.) (see Fig. 11.11D). In addition to the total "**number of hits**," the simplified number of "**counts**" is also a suitable quantification parameter tracking the incidence of individual events and thus the intensity of destructive phenomena during the scratch test. While hits are defined by a higher number of parameters (exceed of amplitude threshold, rise time, hit length, dead time between hits), the "counts" are defined only by the exceedance of the amplitude threshold. Thus, the number of hits determines the number of specific events (whose characteristics may vary according to the types of initiating physical phenomenon) while the number of counts determines the cumulative number of times the amplitude threshold is exceeded (Ctvrtlik et al., 2019; Tomastik, Ctvrtlik, Drab, et al., 2018).

Another huge leap in the advanced AE record evaluation is regarded an employment of neural networks or other artificial intelligence (AI) algorithms, which can be used to analyze individual types of hits and their characteristics in statistically significant numbers and thus differentiate and determine the accompanying physical phenomena (Bi et al., 2019; Bonaccorsi, 2012; Daugela et al., 2021; Griffin & Chen, 2006, 2008; Plaza et al., 2020).

11.3 Electrical contact resistance measurement

Electrically conductive probes such as Boron-doped diamond indenters or customized metallic probes are required for electrical contact resistance (ECR) measurement. The conductive indenter creates an electrical circuit during a nanoscale contact in any nano- or micromechanical test possible with the instrumentation as illustrated schematically in Fig. 11.12. Tests can be run in constant current or constant voltage modes.

The technique has found application in studying complex phase transformations during nanoindentation of silicon. Using an electrically conductive indenter (e.g., boron-doped diamond) can provide additional information about the phase transformations that occur in loading and unloading. It is well known that lower unloading rate promotes sharp and "kink" pop-out events during unloading due to the rapid growth of high-pressure phases, which cause the sudden volume expansion (e.g., Si-III/Si-XII) while faster unloading favors formation of an a-Si and a more gradual "elbow" pop-out. It has been argued that the pop-out does not represent the onset of a phase transformation, which begins as a gradual phase transformation at higher load at the onset of nonelastic unloading (Chang & Zhang, 2009). ECR measurements provide additional information to investigate this. Mann et al. first studied this using a 1 µm radius VC tip as the conductive probe (Mann et al., 2002). More recent examples with a sharp boron-doped diamond Berkovich showing sharp

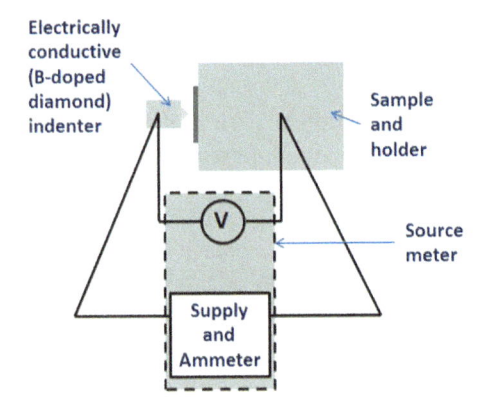

Figure 11.12 Principle of electrical contact resistance measurement. Using an electrically conductive indenter enables an electrical circuit to be created.

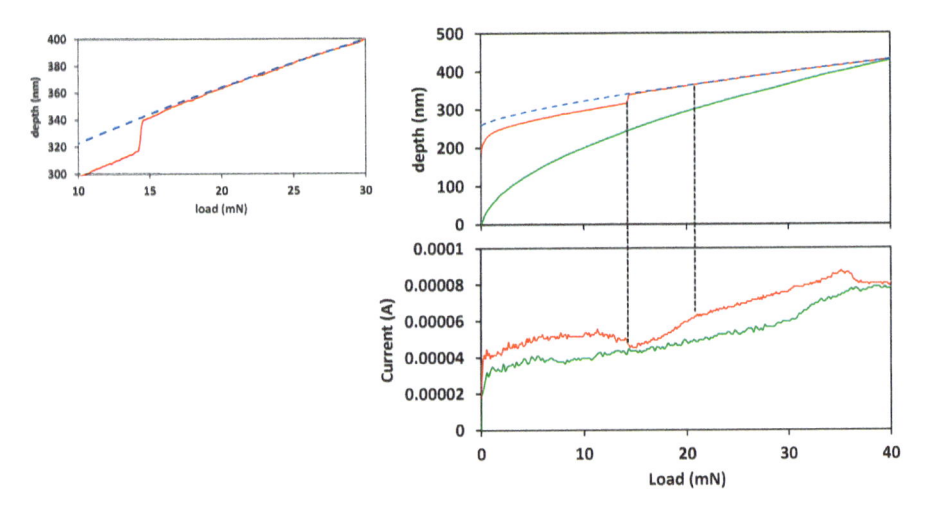

Figure 11.13 Nanoindentation of Si(100) to 40 mN with a Boron-doped diamond Berkovich unloading at 0.667 mN/s. A marked decrease in electrical contact resistance accompanies the sharp pop-out event.

(Fig. 11.13) and kink (Fig. 11.14) pop-outs are shown in Figs. 11.13 and 11.14. The power law fit to the unloading curves is also shown. From closer investigation, it is clear that the initial deviation from elastic unloading associated with phase transformation begins before the main pop-out that occurs with further unloading.

There are also clear differences in the ECR behavior at the pop-out for the two types of response. For the kink pop-out, there is little increase in current, but there is a clearer increase in current for the abrupt pop-out. The initial deviation from elastic unloading was accompanied by a more rapid decrease in electrical current for both cases. Whether this represents a phase transformation from Si-II to a less

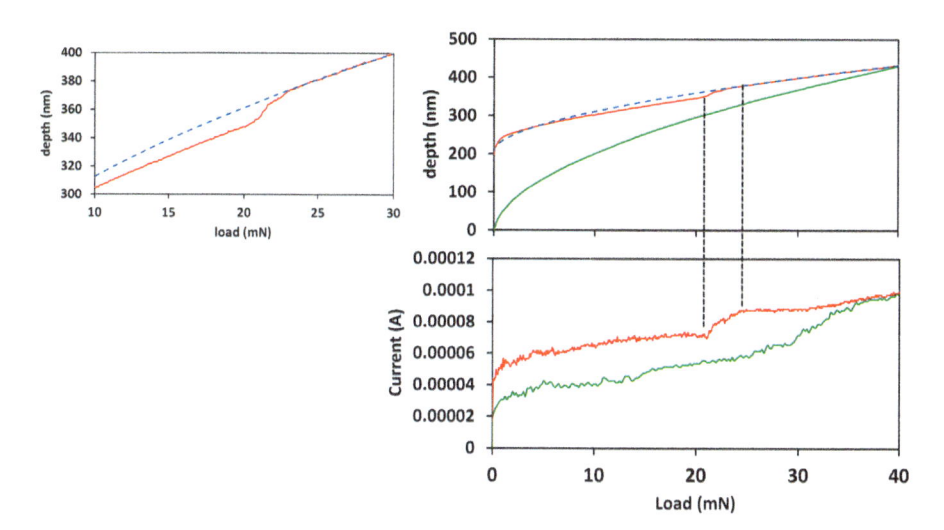

Figure 11.14 Nanoindentation of Si(100) to 40 mN with a Boron-doped diamond Berkovich unloading at 1.33 mN/s. The "kink pop-out" event is associated with little change in electrical contact resistance.

conductive phase (e.g., a-Si) or whether, as Gerbig and coworkers have suggested from Raman measurements, it reflects partial relaxation of the strained Si-II phase, it is clear that the initial deviation from elastic unloading is not due to the development of the same metallic high-pressure phases that occur at the main pop-out since the electrical signals are different.

Kholkhujaev et al. have shown that ECR measurement during indentation provides a way to directly measure the contact area under load, which could provide a method to more accurately measure the properties of metallic materials that display strong pile-up effects that are not taken into account by the standard ISO14577 analysis (Kholkhujaev et al., 2023). They found it could improve the accuracy of the projected area measurement in microindentation of a reference block steel and, therefore, the hardness and elastic modulus of the test material.

ECR measurements have also been employed to better understand the microtribological behavior of novel alloys being investigated as electrical connector materials (Beake et al., 2021). Understanding the nature of the sliding failure that can occur due to frequent mating — unmating contact is an important step in developing improved electrical connector systems. To this end, high cycle reciprocating microwear tests have been performed between silver alloy probes and silver alloy disks and the interrelationships between friction and ECR studied (Beake et al., 2021). Although the spikes in ECR (or friction) did not occur over the entire track, the mean value of these parameters over the 5 mm wear track for each cycle was used to conveniently follow the tribo-contact. Fig. 11.15 shows friction and ECR during (A) run-in and (B and C) two complex multistage failures in a 35,000 cycle 200 mN test between a silver alloy probe and silver alloy disk. Both the probe and

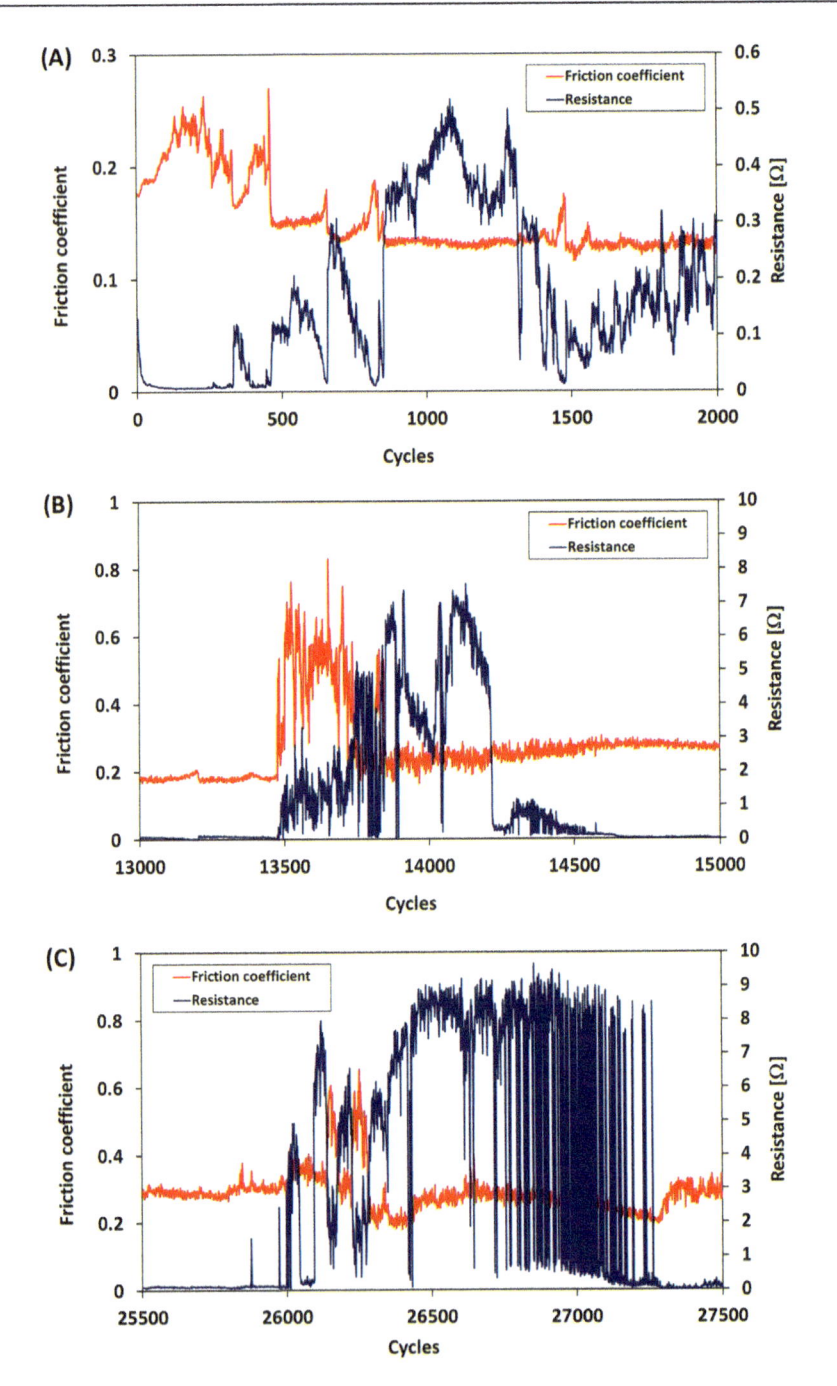

Figure 11.15 Mean friction and mean electrical contact resistance per cycle in a 35,000 cycle test between Ag probe and Ag disk. Variation in (A) friction and contact resistance during run-in (B and C) friction and contact resistance around complex multistage failures.

the disk contained copper so that tribo-formation of copper oxides can occur. Apart from these two events, the friction and ECR remained low.

During the running-in period of ~250 cycles, there is a gradual decrease in contact resistance and accompanying increase in friction as the asperities on the probe and disk are smoothed out increasing the contact area (Fig. 11.15A). A similar contact area/surface topography effect is also likely to explain the two dramatic failures that occurred later in the test. These are consistent with nonconducting oxide debris becoming trapped in the contact increasing the contact area and eventually being ejected. The two spikes in friction and resistance after 13,500 and 26,000 cycles are shown in detail in Fig. 11.15B and C. There is a complex relationship between the friction and resistance within the failure events, with an inverse correlation observed for some regions. Further study is needed to understand this behavior, but it may be related to changing metallic/nonmetallic composition of the contact, as previously buried oxide fragments enter/exit.

11.4 Friction measurement

There are similar benefits in monitoring friction in nanoscratch tests. When failure occurs by dramatic (compressive) fracture in front of the sliding probe, this is very clearly seen as instantaneous apparent decreases in friction due to the abrupt decrease in contact area. A clear example is the extensive brittle fracture of a silicon substrate, which occurs at high load after thin film failure. Fig. 11.16 shows friction and depth data for a region of a nanoscratch test on a physical vapour deposited (PVD) nc-TiN/Si_3N_4 film on Si(100) where the coating has already failed and no longer able to protect the substrate from fracture.

Beake, Harris and Vishnyakov studied the influence of the mechanical properties of ~1.3 μm nc-TiFeN$_x$ coatings on Si(100) on their critical loads and

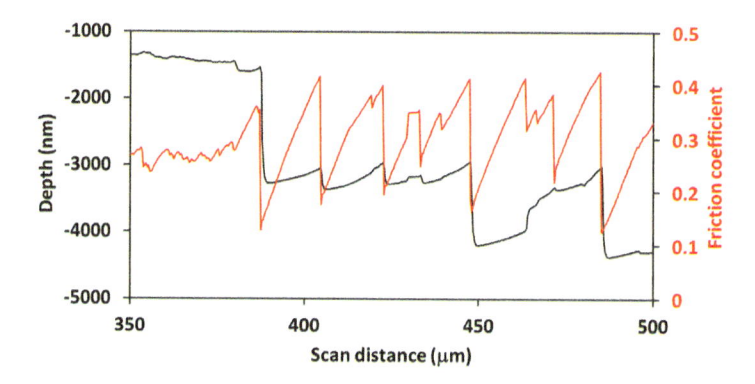

Figure 11.16 Correlation between friction coefficient and on-load depth during a nanoscratch test. Friction and depth during a test on PVD nc-TiN/Si_3N_4 film on Si(100) where the coating has already failed and there is substrate fracture.

failure mechanism in nanoscratch tests with a well-worn Berkovich indenter (end radius ~1 μm) scratching edge-forward (Beake et al., 2011, 2017). The deformation responses and critical loads were a strong function of the coating mechanical properties with ductile behavior for the softest coatings and increasingly brittle behavior for coatings with higher hardness and H^3/E^2. The sensitivity of the friction force to the failure was strongly dependent on where the failure occurs relative to the sliding probe.

Fig. 11.17 shows example friction, and Fig. 11.18 corresponding scanning electron microscopy (SEM) images for nanoscratch tests on four of these coatings whose hardness vary between 10 and 30 GPa. Where the film failure occurs through compressive failure in front of the probe, the decreases in friction are clearly seen, as in (C), a TiFeN coating with $H = 15.8$ GPa, $E = 256$ GPa, and $H^3/E^2 = 0.06$ GPa. There are failures at the side of the scratch track, but these are not seen in the friction. In (A), a TiFeN film with $H = 21$ GPa, $E = 320$ GPa, and $H^3/E^2 = 0.09$ GPa, there was clear film failure with extensive delamination outside the track, but friction coefficient was insensitive to it and remained at ~0.22 throughout the majority of the track. This insensitivity is due to the failure starting behind the probe. FEM has shown maximum tensile stress at the surface located around a half-contact width behind the sliding contact (Holmberg et al., 2007). The maximum tensile stress appearing just behind the trailing edge was also reported by Diao et al. (1994). The location of the tensile peak a small distance behind the contact (rather than at the trailing edge of the contact) provides an explanation why extreme coating failure can occur without being detected in the friction signal. The TiN coating (B) with $H = 30.0$ GPa, $E = 426$ GPa, and $H^3/E^2 = 0.149$ GPa initially fails behind the probe with extensive delamination

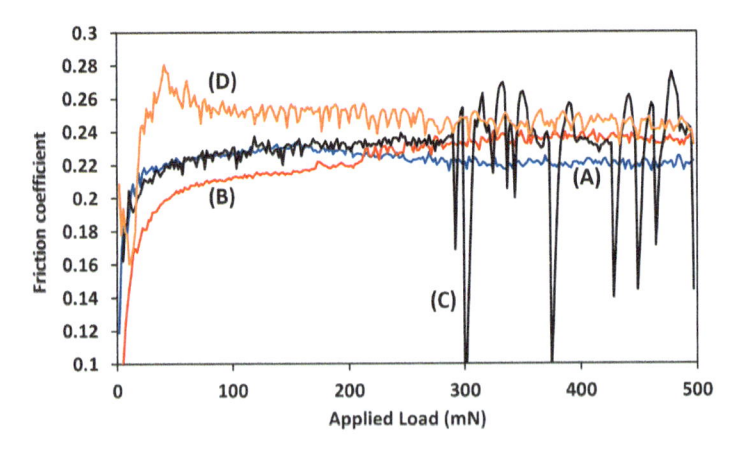

Figure 11.17 Friction coefficient versus load in nano-scratch tests on ~1.3 μm nc-TiFeN$_x$ coatings on Si(100) with blunt Berkovich indenter. Friction coefficient versus load for (A) TiFeN coating with $H = 21$ GPa, $E = 320$ GPa and $H^3/E^2 = 0.09$ GPa (B) TiN coating with $H = 30.0$ GPa, $E = 426$ GPa and $H^3/E^2 = 0.149$ GPa (C) TiFeN coating with $H = 15.8$ GPa, $E = 256$ GPa and $H^3/E^2 = 0.06$ GPa (D) FeN coating with $H = 9.6$ GPa, $E = 229$ GPa and $H^3/E^2 = 0.017$ GPa.

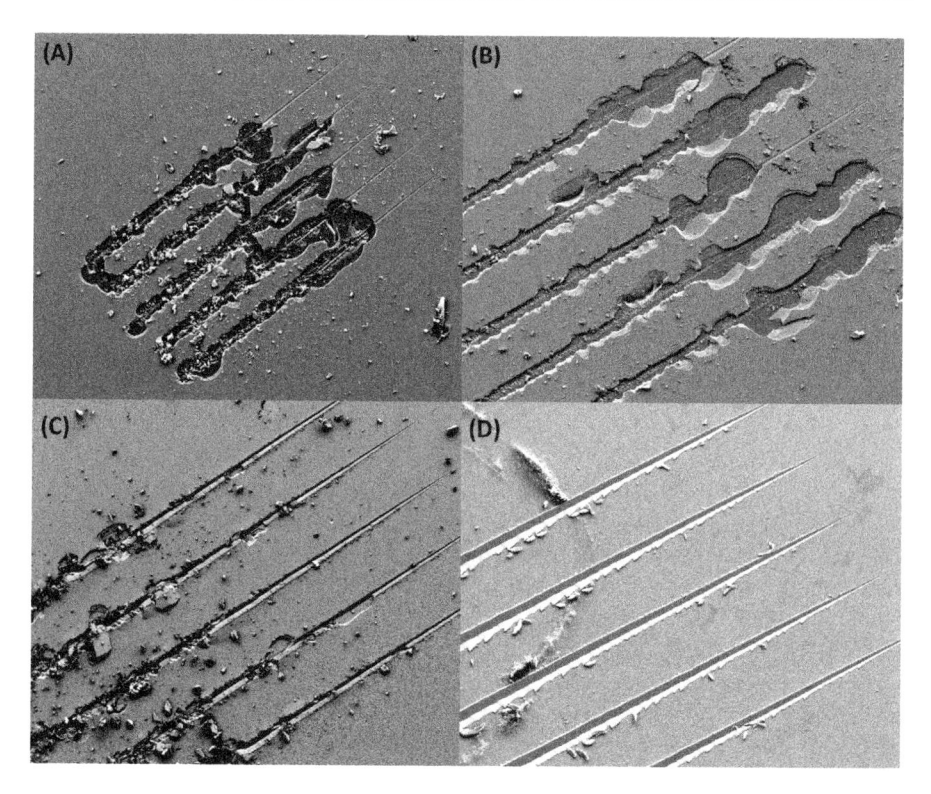

Figure 11.18 SEM images of nanoscratch tests on ~1.3 μm nc-TiFeN$_x$ coatings on Si(100) with blunt Berkovich indenter SEM images of coatings (A–D) from Fig. 11.17.

(not seen in friction) with SEM showing that deformation of the substrate was minimal at this point. At higher load, there is transition to brittle machining damage of the silicon substrate. This transition is seen in a reduction in the extent of delaminated area outside of the track and a small jump in friction. For TiFeN films with even higher H^3/E^2, there was brittle machining failure and/or adhesion failure at low load, but this was barely detected in the friction. An FeN coating (D) with H = 9.6 GPa, E = 229 GPa, and H^3/E^2 = 0.017 GPa displays different behavior showing ductile failure early in the test and higher friction.

11.5 Nanomechanical Raman spectroscopy

Gerbig et al. developed a nanomechanical Raman spectroscopy capability. The setup was used to perform in situ Raman spectroscopic measurements during indentation of Si thin films with a 45 μm probe (Gerbig et al., 2015). They produced a sequence of Raman images to show the evolution of strain fields and changes in phase distributions during the spherical indentation, providing evidence for previous assumptions regarding which phases are dominant at different stages of the indentation. Zhang and coworkers developed a high-temperature integrated nanomechanical Raman spectroscopy measurement platform to enable simultaneous in situ temperature and stress mapping as a function of microstructure during deformation (Zhang et al., 2016, 2018). The platform has been used to perform in situ measurements of the crack tip stress distribution in Inconel 617 under 3-point bending using a flat-ended indenter. A 0.5 μm thick layer of silicon, a Raman-sensitive material, was deposited on the alloy surface. At an average temperature of 200°C under bending loads of 0.5, 2, and 4 N, they observed a correlation between the microstructure-dependent stress distribution and temperature fields at the notch tip. They concluded that stress concentrations influence local hot spot formation, with high stresses being associated with a local reduction in temperature indicating an increase in thermal conductivity.

11.6 Sensors and actuators

Schmidt et al. designed an electrochemically responsive polymer nanocomposite multilayer thin film containing cationic linear poly(ethyleneimine) and 68 vol.% anionic Prussian Blue nanoparticles with actively tunable mechanical properties (Schmidt et al., 2009). The reversible changes from oxidation/reduction in its mechanical properties were evaluated using nanoindentation in an electrochemical cell. -0.2 V (vs Ag/AgCl) reduced the particles to the Prussian White state and $+0.6$ V oxidized the particles back to the Prussian Blue state. The reversible swelling on reduction resulted in a 50% reduction in elastic modulus (measured at an indentation depth of 40 nm with a 5 μm ruby indenter while fully immersed in 0.1 M potassium hydrogen phthalate). Customized electrochemical setups have been used to investigate the interactions between hydrogen and metals to improve our understanding of the hydrogen embrittlement, which decreases fuel cell efficiency (Massone & Kiener, 2022). Duarte et al. have summarized the advantages and disadvantages of "front-side" and "back-side" charging of the test sample with hydrogen (Duarte et al., 2021). In their "back-side" setup, the hydrogen reaches the test surface slowly through diffusion, but the influence of the electrolyte on the nanoindented surface is avoided.

(Pr, Ce)O$_{2-\delta}$ (PCO) has been investigated as a model material for a solid oxide fuel cell (SOFC) cathode. SOFC devices often operate at high temperature and oxygen partial pressure varying from 0.2 (oxidizing in air) to $<10^{-3}$ (reducing) atm (Swallow et al., 2016). The effect of oxygen partial pressure on the elastic modulus of 1 μm PCO films deposited by PVD on yttria-stabilized zirconia (YSZ) was studied by nanoindentation at 600°C (Swallow et al., 2016) in an environmental chamber that enabled the gas concentration to be precisely controlled by flowing nitrogen and oxygen through the chamber. The film elastic modulus decreased from ~ 250 GPa in air at 25°C to ~ 150 GPa in 7.6×10^{-4} pO$_2$ at 600°C. PCO also has potential as a elevated temperature, low voltage electromechanical oxide actuator. For use under extreme conditions, actuator operation requires materials that can reliably sense and actuate at elevated temperatures, and over a range of gas environments. At >500°C PCO, in common with many nonstochiometric oxides, exhibits chemical expansion (i.e., coupling between composition and volume). The oxide expansion at these temperatures is enhanced by electrochemical pumping of oxygen vacancies with an applied electrical bias. Elevated temperature nanoscale electrochemomechanical spectroscopy measurements have been performed on the chemomechanically coupled nonstoichiometric oxide thin film (Swallow et al., 2017). Oxygen activity in 0.3−1 μm PVD PCO films deposited on YSZ was modulated by an alternating bias applied to the working electrode with respect to the reference electrode. This caused oxygen vacancies to be pumped in and out of the film through the YSZ substrate producing a detectable mechanical response due to a combination of film volume change and substrate bending due to chemical expansion of the film. This mechanical response to the applied bias was detected in the NanoTest system through the displacement of a cBN indenter held on the sample surface with the minimum mechanical load to maintain contact at 550°C−650°C. The electrochemically driven "breathing" response has potential advantages for actuation at elevated temperatures, with the possibility to operate at much lower voltages than are needed in elevated temperature piezoelectric actuators (Swallow et al., 2017).

11.7 Multi-sensing in high strain rate contact

In an impact test, the actual impact force is much higher than the actuated force due to inertia effects (Rueda-Ruiz et al., 2020). To a first approximation, the impact process can be modeled as a linearized damped harmonic oscillator. Traditional energy-based methods for determining dynamic hardness from impact data have uncertainties and can overestimate hardness, particularly for materials with high elastic recovery (Zehnder et al., 2018). An alternative approach is to directly measure the impact force. Rueda-Ruiz et al. achieved this with the addition of an integrated piezoelectric force sensor with stiffness comparable with the load frame (∼0.1−1 nm/mN) capable of high-frequency dynamic force measurements in the μN−mN range (Rueda-Ruiz et al., 2020). Contact times are very short in comparison with conventional indentation, and high-resolution monitoring of the load through an additional load cell or the acoustic emission through an additional detector (see earlier) provides additional information about processes that occur rapidly. The load cell provides direct evidence for the greater forces generated in impact. For example, Rueda-Ruiz et al. showed that in a nano-impact test, a 5 mN static load produced an impact force of around 100 mN.

11.8 Electroplastic effect

Electroplastic forming where the electroplastic effect is exploited by passing a current through the workpiece is an interesting approach for deforming materials, such as metallic − intermetallic composites, which are otherwise too brittle. Andre et al. developed an experimental multisensing setup where 100 Ms pulses of electric current (a short enough duration to avoid Joule heating) during nanoindentation with a conductive WC indenter could be used to produce displacement bursts on the scale of the individual microstructural components in alloys (Andre et al., 2019). The approach was used to study the electroplastic effect in individual phases in the metallic − intermetallic composite bulk eutectic Al−Al$_2$Cu. Higher current intensity, higher loading rate, and larger pulsing intervals all resulted in larger displacement bursts. The bursts were interpreted as an electroplastic effect through enhanced dislocation motion, for example, from an "electron wind" de-pinning dislocations from obstacles. A thermal softening effect was ruled out by the observation that the displacement bursts were in the opposite direction during unloading.

References

Andre, D., Burlet, T., Körkemeyer, F., Gerstein, G., Gibson, J. S. K.-L., Sandlöbes-Haut, S., & Korte-Kerzel, S. (2019). Investigation of the electroplastic effect using nanoindentation. *Materials & Design, 183*. Available from https://doi.org/10.1016/j.matdes.2019.108153.

Beake, B. D., Ctvrtlik, R., Harris, A. J., Martin, A. S., Vaclavek, L., Manak, J., & Ranc, V. (2020). High frequency acoustic emission monitoring in nano-impact of alumina and partially stabilised zirconia. *Materials Science and Engineering: A, 780*. Available from https://doi.org/10.1016/j.msea.2020.139159.

Beake, B. D., Harris, A. J., Liskiewicz, T. W., Wagner, J., McMaster, S. J., Goodes, S. R., Neville, A., & Zhang, L. (2021). Friction and electrical contact resistance in reciprocating nano-scale wear testing of metallic materials. *Wear, 474−475*. Available from https://doi.org/10.1016/j.wear.2021.203866, https://www.journals.elsevier.com/wear.

Beake, B. D., Vishnyakov, V. M., & Harris, A. J. (2011). Relationship between mechanical properties of thin nitride-based films and their behaviour in nano-scratch tests. *Tribology International, 44*, 468.

Beake, B. D., Vishnyakov, V. M., & Harris, A. J. (2017). Nano-scratch testing of (Ti, Fe)Nx thin films on silicon. *Surface and Coatings Technology, 309*, 671−679.

Becker, D., Cailleau, B., Dahm, T., Shapiro, S., & Kaiser, D. (2010). Stress triggering and stress memory observed from acoustic emission records in a salt mine. *Geophysical Journal International, 182*(2), 933−948. Available from https://doi.org/10.1111/j.1365-246x.2010.04642.x.

Bi, G., Xu, T., Kang, J., & Peng, Y. (2019). Feature learning method of acoustic emission signal during single-grit scratching on BK7 in brittle regime. *Proceedings of the Institution of Mechanical Engineers, Part C: Journal of Mechanical Engineering Science, 233*(2), 526−538. Available from https://doi.org/10.1177/0954406218762211.

Bonaccorsi, L. (2012). Artificial neural network analyses of AE data during long-term corrosion monitoring of a post-tensioned concrete beam. *Journal of Acoustic Emission, 30*, 40−53.

Chang, L., & Zhang, L. C. (2009). Deformation mechanisms at pop-out in monocrystalline silicon under nanoindentation. *Acta Materialia, 57*(7), 2148−2153. Available from https://doi.org/10.1016/j.actamat.2009.01.008.

Choudhary, R. K., & Mishra, P. (2016). Use of acoustic emission during scratch testing for understanding adhesion behavior of aluminum nitride coatings. *Journal of Materials Engineering and Performance*, *25*(6), 2454−2461. Available from https://doi.org/10.1007/s11665-016-2073-9.

Čtvrtlík, R., Čech, J., Tomáštík, J., Václavek, L., & Haušild, P. (2022). Plastic instabilities explored via acoustic emission during spherical nanoindentation. *Materials Science and Engineering: A*, *841*. Available from https://doi.org/10.1016/j.msea.2022.143019.

Ctvrtlik, R., Tomastik, J., Vaclavek, L., Beake, B. D., Harris, A. J., Martin, A. S., Hanak, M., & Abrham, P. (2019). High-resolution acoustic emission monitoring in nanomechanics. *JOM*, *71*(10), 3358−3367. Available from https://doi.org/10.1007/s11837-019-03700-8.

Daugela, A., Chang, C. H., & Peterson, D. W. (2021). Deep learning based characterization of nanoindentation induced acoustic events. *Materials Science and Engineering: A*, *800*. Available from https://doi.org/10.1016/j.msea.2020.140273, http://www.elsevier.com.

Diao, D. F., Kato, K., & Hayashi, K. (1994). The maximum tensile stress on a hard coating under sliding friction. *Tribology International*, *27*, 267−272.

Duarte, M. J., Fang, X., Rao, J., Krieger, W., Brinckmann, S., & Dehm, G. (2021). In situ nanoindentation during electrochemical hydrogen charging: A comparison between front-side and a novel back-side charging approach. *Journal of Materials Science*, *56* (14), 8732−8744. Available from https://doi.org/10.1007/s10853-020-05749-2.

Gallego, A., Gil, J. F., Vico, J. M., Ruzzante, J. E., & Piotrkowski, R. (2005). Coating adherence in galvanized steel assessed by acoustic emission wavelet analysis. *Scripta Materialia*, *52* (10), 1069−1074. Available from https://doi.org/10.1016/j.scriptamat.2005.01.037.

Gallego, A., Gil, J. F., Castro, E., & Piotrkowski, R. (2007). Identification of coating damage processes in corroded galvanized steel by acoustic emission wavelet analysis. *Surface and Coatings Technology*, *201*(8), 4743−4756. Available from https://doi.org/10.1016/j.surfcoat.2006.10.018.

Gerbig, Y. B., Michaels, C. A., & Cook, R. F. (2015). In situ observation of the spatial distribution of crystalline phases during pressure-induced transformations of indented silicon thin films. *Journal of Materials Research*, *30*, 390−406.

Griffin, J., & Chen, X. (2008). Characteristics of the acoustic emission during horizontal single grit scratch tests: Part 1 characteristics and identification. *International Journal of Abrasive Technology*, *2*(1), 25−42.

Griffin, J., & Chen, X. (2006). Classification of the acoustic emission signals of rubbing, ploughing and cutting during single grit scratch tests. *International Journal of Nanomanufacturing*, *1*(2), 189−209. Available from https://doi.org/10.1504/IJNM.2006.012195.

Hesser, J., Kaiser, D., Schmitz, H., & Spies, T. (2015). *Measurements of acoustic emission and deformation in a repository of nuclear waste in salt rock. Engineering geology for society and territory - Volume 6: Applied geology for major engineering projects* (pp. 551−554). Germany: Springer International Publishing. Available from https://doi.org/10.1007/978-3-319-09060-3, https://doi.org/10.1007/978-3-319-09060-3_99.

Hiroyuki, K. (2012). Analysis of AE signals during scratch test on the coated paperboard. *Journal of Acoustic Emission*, *30*, 1−10.

Holmberg, K., Ronkainen, H., Laukkanen, A., & Wallin, K. (2007). Friction and wear of coated surfaces—scales, modelling and simulation of tribomechanisms. *Surface and Coatings Technology*, *202*(4−7), 1034−1049. Available from https://doi.org/10.1016/j.surfcoat.2007.07.105.

Julia-Schmutz, C., & Hintermann, H. E. (1991). Microscratch testing to characterize the adhesion of thin layers. *Surface and Coatings Technology*, *48*(1), 1−6. Available from https://doi.org/10.1016/0257-8972(91)90121-c.

Kholkhujaev, J., Maculotti, G., Genta, G., & Galetto, M. (2023). Metrological comparison of available methods to correct edge-effect local plasticity in instrumented indentation test. *Materials*, *16*(12). Available from https://doi.org/10.3390/ma16124262.

Mann, A. B., van Heerden, D., Pethica, J. B., Bowes, P., & Weihs, T. P. (2002). Contact resistance and phase transformations during nanoindentation of silicon. *Philosophical Magazine A*, *82*(10), 1921−1929. Available from https://doi.org/10.1080/01418610208235704.

Massone, A., & Kiener, D. (2022). Prospects of enhancing the understanding of material-hydrogen interaction by novel in-situ and in-operando methods. *International Journal of Hydrogen Energy*, *47*(17), 10097−10111. Available from https://doi.org/10.1016/j.ijhydene.2022.01.089.

Nazarchuk, Z., Skalskyi, O., & Serhiyenko, R. (2017). *Acoustic emission − Methodology and application. Foundations of Engineering Mechanics* (11, p. 283). Springer.

Němeček, J., Čtvrtlík, R., Václavek, L., & Němeček, J. (2023). Fracture toughness of cement paste assessed with micro-scratch and acoustic emission. *Powder Metallurgy Progress*, *22*(1), 7−13. Available from https://doi.org/10.2478/pmp-2022-0002, http://www.degruyter.com/view/j/pmp.

Plaza, E. G., Chen, X., & Ouarab, L. A. (2020). Grinding acoustic emission features in relation to abrasive scratch characteristics. *International Journal of Abrasive Technology*, *10*(2), 134−154. Available from https://doi.org/10.1504/IJAT.2020.109874, http://www.inderscience.com/ijat.

Proctor, T. M. (1982). An improved piezoelectric acoustic emission transducer. *Journal of the Acoustical Society of America*, *71*(5), 1163−1168. Available from https://doi.org/10.1121/1.387763.

Rueda-Ruiz, M., Beake, B. D., & Molina-Aldareguia, J. M. (2020). New instrumentation and analysis methodology for nano-impact testing. *Mater Design*, *192*, 108715, 12pp.

Schmidt, D. J., Cebeci, F. C., Kalcioglu, Z. I., Wyman, S. G., Ortiz, C., Van Vliet, K. J., & Hammond, P. T. (2009). Electrochemically controlled swelling and mechanical properties of a polymer nanocomposite. *ACS Nano*, *3*(8), 2207−2216. Available from https://doi.org/10.1021/nn900526c, http://pubs.acs.org/doi/pdfplus/10.1021/nn900526c, United States.

Scruby, C. B. (1987). An introduction to acoustic emission. *Journal of Physics E: Scientific Instruments*, *20*(8), 946−953. Available from https://doi.org/10.1088/0022-3735/20/8/001.

Svoboda, V. (1992). The use of the acoustic emission for the components of the primary circuit of the nuclear power plants. *International Atomic Energy Agency (IAEA)*, 36−38.

Swallow, J. G., Kim, J. J., Kabir, M., Smith, J. F., Tuller, H. L., Bishop, S. R., & Van Vliet, K. J. (2016). Operando reduction of elastic modulus in (Pr, Ce)O$_2$-δ thin films. *Acta Materialia*, *105*, 16−24. Available from https://doi.org/10.1016/j.actamat.2015.12.007, http://www.journals.elsevier.com/acta-materialia/.

Swallow, J. G., Kim, J. J., Maloney, J. M., Chen, D., Smith, J. F., Bishop, S. R., Tuller, H. L., & Van Vliet, K. J. (2017). Dynamic chemical expansion of thin-film non-stoichiometric oxides at extreme temperatures. *Nature Materials*, *16*(7), 749−754. Available from https://doi.org/10.1038/nmat4898, http://www.nature.com/nmat/.

Tomastik, J., Ctvrtlik, R., Bohac, P., Drab, M., Koula, V., Cvrk, K., & Jastrabik, L. (2015). Utilization of acoustic emission in scratch test evaluation. *Key Engineering Materials*, *662*, 119−122. Available from https://doi.org/10.4028/http://www.scientific.net/kem.662.119.

Tomastik, J., Ctvrtlik, R., Drab, M., & Manak, J. (2018). On the importance of combined scratch/acoustic emission test evaluation: SiC and SiCN thin films case study. *Coatings*, *8*(5). Available from https://doi.org/10.3390/coatings8050196.

Tomastik, J., Ctvrtlik, R., Ingr, T., Manak, J., & Opletalova, A. (2018). Effect of nitrogen doping and temperature on mechanical durability of silicon carbide thin films. *Scientific Reports*, *8*(1). Available from https://doi.org/10.1038/s41598-018-28704-3.

von Stebut, J., Lapostolle, F., Bucsa, M., & Vallen, H. (1999). Acoustic emission monitoring of single cracking events and associated damage mechanism analysis in indentation and scratch testing. *Surface and Coatings Technology, 116–119*, 160–171. Available from https://doi.org/10.1016/s0257-8972(99)00211-x.

Zehnder, C., Peltzer, J. N., Gibson, J. S. K. L., & Korte-Kerzel, S. (2018). High strain rate testing at the nano-scale: A proposed methodology for impact nanoindentation. *Materials and Design, 151*, 17–28. Available from https://doi.org/10.1016/j.matdes.2018.04.045.

Zhang, Y., Mohanty, D. P., & Tomar, V. (2016). Visualizing in situ microstructure dependent crack tip stress distribution in IN-617 using nano-mechanical Raman spectroscopy. *JOM, 68*, 2742–2746.

Zhang, Y., Wang, H., & Tomar, V. (2018). Visualizing stress and temperature distribution during elevated temperature deformation of IN-617 using nanomechanical Raman spectroscopy. *JOM, 70*, 464–468.

High-temperature testing

James S. K-L. Gibson
Department of Materials, University of Oxford, Oxford, United Kingdom

12.1 Introduction

Many materials are used at significant fractions of their melting temperature during service. These are not limited to the intuitive selections of superalloys or materials for power generation, but also for low-melting-point materials operating close to room temperature; for magnesium alloys in automotive power trains or lithium in energy storage applications. Understanding how materials behave at their operating temperature as new deformation mechanisms such as creep or other thermally activated deformation processes come into play is key for materials development in these applications.

The motivation for performing such characterisations on the micro- or nanoscale is threefold. First, the material of interest may only exist on these sorts of scales; in the case of protective, deposited coatings, radiation-damaged layers, tribofilms, or microelectronics, for example. Second, by operating at a microstructural length scale, these tests enable the fundamental deformation mechanisms of the material to be elucidated, allowing the study of individual grains, well-characterized grain boundaries, and single precipitates or second phases. Third and finally, small-scale testing complements the growing field of high-throughput testing, allowing rapid testing of easy-to-prepare samples such as plasma-deposited wafers, diffusion couples, or additively manufactured materials.

12.2 Strategies for reliable high-temperature nanomechanical test measurements

There have been a number of recent reviews of variable temperature nanoindentation testing at both high (Chavoshi & Xu, 2018b) and low temperatures (Wang & Zhao, 2020), using modeling insights (Chavoshi & Xu, 2019) and with a focus on irradiated materials (Hosemann et al., 2015). The most comprehensive is likely that from Wheeler et al. (2015) that contains an overview of several commercially available systems, strategies for eliminating thermal drift and an insight into indentation in vacuum. Rather than repeat vast swathes of text regarding the current state of the art within the literature, this section will instead simply focus on current "best practices" for reliable high (or low)-temperature nanoindentation tests.

Being an extension of "regular" nanomechanical testing, the fundamental requirements for reliable measurements at elevated temperatures still apply: flat, damage-free surfaces, mechanically stiff sample mounting, large grains if single-crystal measurements are of interest, and sufficient sample dimensions that the material of interest can be tested without an undue influence of substrates, sample edges, pores, etc. Two further considerations are then needed for testing at elevated temperature: thermal management to ensure isothermal contact and chemical management to prevent sample oxidation or tip-sample chemical reactions.

The latter is perhaps most straightforward to manage. Sample oxidation is best prevented by testing in vacuum ($<10^{-5}$ mbar). While shielding strategies can be effective, for reactive samples such as tungsten, the use of inert atmospheres has been shown to be ineffective in preventing oxidation (Wheeler et al., 2015). It goes without saying that detailed in-situ studies necessarily have to take place in vacuum as well for the operation of the SEM. Wheeler has produced the most detailed study of materials for nanoindenter tips (Wheeler & Michler, 2013), which contains a reference table listing reactivity trends between tip material with sample elements. Many of these trends are necessarily limited in concrete references due to the time and costs involved in systematic studies. However, two main takeaways can be gathered. First, the standard diamond indenters should be avoided above $\cong 300°C$ for both reactivity and tip oxidation concerns, even in vacuum. Second, tungsten carbide (WC) indenters are a good "standard" choice for high-temperature work as they are the least chemically reactive.

Ensuring isothermal contact is critical for high-temperature micromechanics. Especially in nanoindentation, where all the mechanical information is derived from the measurement of indentation depth, having this depth signal unimpacted by thermal drift is vital. This stage is complicated by the fact that isothermal contact needs to take place where the indenter tip meets the sample surface, whereas controlling thermocouples are placed further backward toward the heaters and are thus slightly hotter than the true contact temperature. Often, a system will contain three thermocouples: tip, sample heater, and a third bonded to the surface of the sample to act as a reference of the true indentation temperature. Where this third thermocouple is impractical to mount, the tip thermocouple can be used as a measurement of the indentation temperature. This is best done by indenting a thermocouple junction (Wheeler et al., 2012) to characterize the thermal gradient between the indenter thermocouple and the temperature at the indenter tip. However, a first approximation of the contact temperature can be taken in many systems by simply using the indenter temperature, as the low thermal mass and relatively short conduction distance between the tip and its thermocouple mean that the gradient is of the order of a few degrees or so.

As stated, on the sample side, it is better to have a third thermocouple attached to the sample as an "independent" measure of test temperature, unaffected by the stronger thermal gradients present in the sample and tip heating systems. However, it is then vital that this thermocouple is tightly attached to the sample, particularly in the case where testing takes place in vacuum where conduction is the dominant form of heat transfer. A true mechanical contact using spot welding, or a high-temperature braze assures this, assuming that the sample dimensions are sufficient that any heat-affected zones or similar are small. Failing this (or if the sample material is not conducive to spot welding, e.g., tungsten), then it is recommended that the sample thermocouple be bent in such a way that the tension in the thermocouple wire naturally clamps onto the sample. When the experimentalist is confident in the quality of the contact between thermocouple and sample, this "sample" thermocouple can then be used to control the target temperature during heating, with the "stage" thermocouple reduced to a simple, somewhat irrelevant display. If the contact is solely mechanical, and doubly so in vacuum, it is recommended that this is avoided as a loss of contact will cause rapid overheating and potential failure of the furnace attempting to reach a temperature that is being incorrectly measured.

Heating to target temperature is typically simply a case of specifying a tip and furnace target temperature and letting the system's PID controllers take care of the rest. Depending on the system configuration, it may be that the indenter heater must manually be switched to "fixed power" once the target temperature has been reached. It is worth bearing in mind the thermal gradients and mutual heating of the system when specifying target temperatures; the sample (and therefore indentation) temperature is likely some way below the furnace temperature, with the exact magnitude of the drop depending on sample thickness and thermal conductivity. The furnace temperature should therefore be set slightly higher than the desired indentation temperature. Similarly, the radiative heating from the sample will increase the tip temperature when the two are brought close together. If the sample is being heated over a microscope, some way away from the indenter, then the target indenter temperature should be set slightly below the target indentation temperature. The offset will become noticeable above $\approx 500°C$ as radiative heating increases with T^4 (Lucchesi et al., 2021).

Ensuring isothermal contact essentially involves making repeated contact between the tip and the sample and measuring the temperature change of the indenter tip (as above, due to its low mass the response is much more rapid). For systems where the temperature is not logged at a per second or subsecond resolution, it is generally easiest to do this through a series of test indents and characterizing the thermal drift postindentation. Negative thermal drift indicates an indenter that needs to be cooled (Wheeler et al., 2015). Where the temperature is rapidly logged, this process can be accelerated as the indent does not need to be performed. The tip is stabilized at a constant power output, brought into contact with the sample, and the temperature monitored. A drop in tip temperature simply indicates the indenter needs to be cooled. Locking in a constant power output during this process is important as it means the tip temperature during contact is not influenced by the PID controller attempting to restore the setpoint temperature. This process is illustrated graphically in Fig. 12.1.

The final question is what is "good enough" with regard to drift rate. While the goal of drift rates comparable with those obtained at room temperature, that is <0.05 nm/s, is of course ideal, it must be balanced with the practicality of

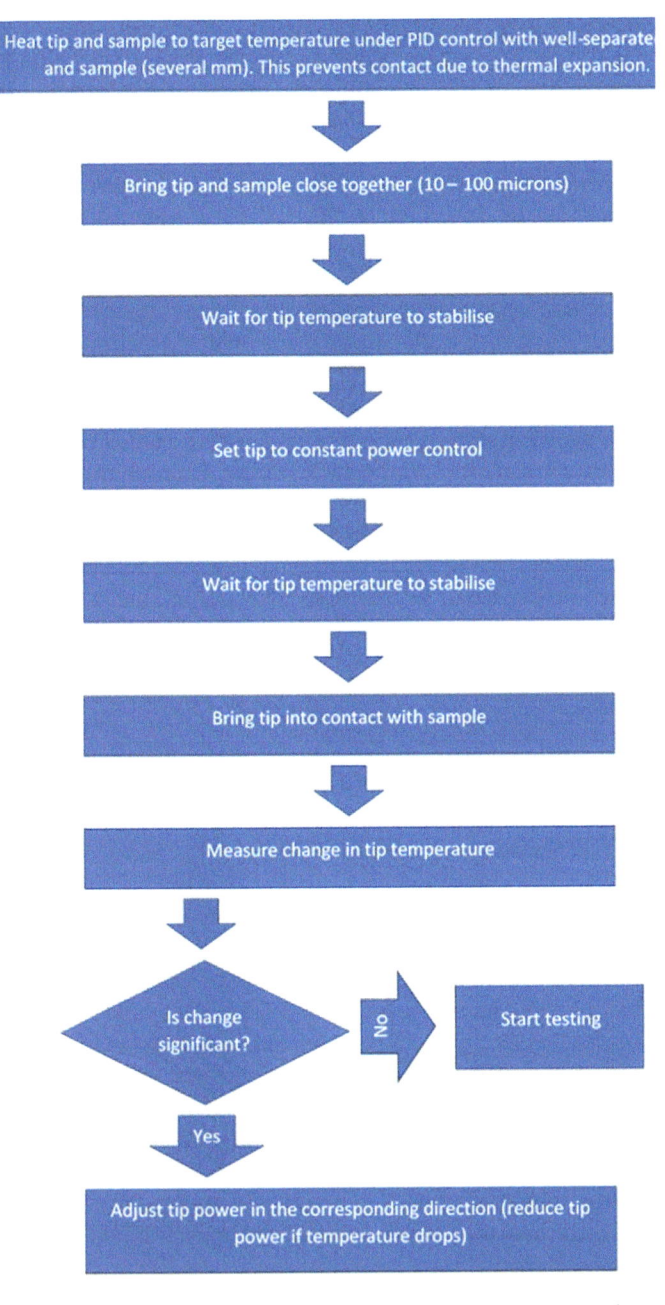

```
┌─────────────────────────────────────────────────────────┐
│ Heat tip and sample to target temperature under PID     │
│ control with well-separated tip and sample (several mm).│
│ This prevents contact due to thermal expansion.         │
└─────────────────────────────────────────────────────────┘
                            ↓
┌─────────────────────────────────────────────────────────┐
│ Bring tip and sample close together (10 – 100 microns)  │
└─────────────────────────────────────────────────────────┘
                            ↓
┌─────────────────────────────────────────────────────────┐
│ Wait for tip temperature to stabilise                   │
└─────────────────────────────────────────────────────────┘
                            ↓
┌─────────────────────────────────────────────────────────┐
│ Set tip to constant power control                       │
└─────────────────────────────────────────────────────────┘
                            ↓
┌─────────────────────────────────────────────────────────┐
│ Wait for tip temperature to stabilise                   │
└─────────────────────────────────────────────────────────┘
                            ↓
┌─────────────────────────────────────────────────────────┐
│ Bring tip into contact with sample                      │
└─────────────────────────────────────────────────────────┘
                            ↓
┌─────────────────────────────────────────────────────────┐
│ Measure change in tip temperature                       │
└─────────────────────────────────────────────────────────┘
                            ↓
                    ⟨Is change        ⟩  No
                    ⟨significant?     ⟩ ─────→  Start testing
                            ↓ Yes
┌─────────────────────────────────────────────────────────┐
│ Adjust tip power in the corresponding direction (reduce │
│ tip power if temperature drops)                         │
└─────────────────────────────────────────────────────────┘
```

Figure 12.1 Flowchart for thermal stabilization in order to achieve isothermal contact between a heated sample and heated indenter.

gradually tuning the contact temperature by fractions of a degree. It should be pointed out that one cannot rely on significant corrections from a thermal drift measurement to be accurate. Once again, as testing (likely) takes place in vacuum, thermal flow will be dominated by conduction. It will therefore change throughout the experiment as the contact area increases with depth, but will be corrected by a constant factor. A suggested guideline is therefore to use the drift rate in combination with the test time and indent depth to calculate a "depth uncertainty," as follows, with the goal of this only being a few percent (Eq. 12.1). For example, a "poor" drift rate of 0.5 nm/s is acceptable, with a <1% error, if a relatively large, 2000 nm deep indent is performed in a very modest 40 seconds.

$$\text{Uncertainty} = \frac{\text{Drift Rate} \times \text{Test Time}}{\text{Indentation Depth}} \quad (12.1)$$

Now that the "hardware" is out of the way, the "software" side of optimum experimental design needs to be considered. Plastic deformation (in the form of pileup around the indent) and creep rates during testing at high temperatures are both likely to be significantly greater than at room temperature. Systems with a dynamic or continuous measurement of contact stiffness are thus potentially better equipped to simultaneously measure the mechanical properties of interest (hardness, elastic modulus and/or indentation creep). Certainly, the capability to monitor the contact stiffness will help account for increased surface pileup effects as materials soften and the deviation from the assumed contact shape increases. However, the potential effects of creep should not be ignored. It has been shown recently by Merle et al. (2019) that at high loading rates, plasticity during an individual CSM cycle will lead to erroneous values of hardness and modulus. It is likely—although as yet unproven—that creep during the CSM loading cycle will induce a similar error.

For systems without a continuous measurement of contact stiffness, the increased creep rates mean that one must optimize the test (or conduct several different experiments), depending on the property of interest. The reason behind this is illustrated in Fig. 12.2, demonstrating a load − displacement curve obtained on a nickel superalloy at 1000°C (Gibson et al., 2017). The experimental parameters are those of a "typical" experiment: Indents were performed to a depth of 1 μm using a an independently heated Berkovich sapphire tip at a constant load rate of 2 mN/s, followed by a 30 seconds dwell period and 5 mN/s unloading rate. This has resulted in a test that is very poorly optimized for measurements of hardness and elastic modulus. It can be seen that the depth increase during the dwell period is significant: an additional 700 nm from an initial indent depth of only 1000 nm. Due to the need to calculate contact area from the unloading slope, this essentially results in a hardness evaluated with a "1000 nm" force at a "1700 nm" depth, reducing the apparent hardness. This will have an enhanced effect in strain-rate-sensitive materials akin to the effect seen by comparing CSM (continuous stiffness measurement, sometimes called dynamic indentation) with the quasistatic load − unload tests illustrated here.

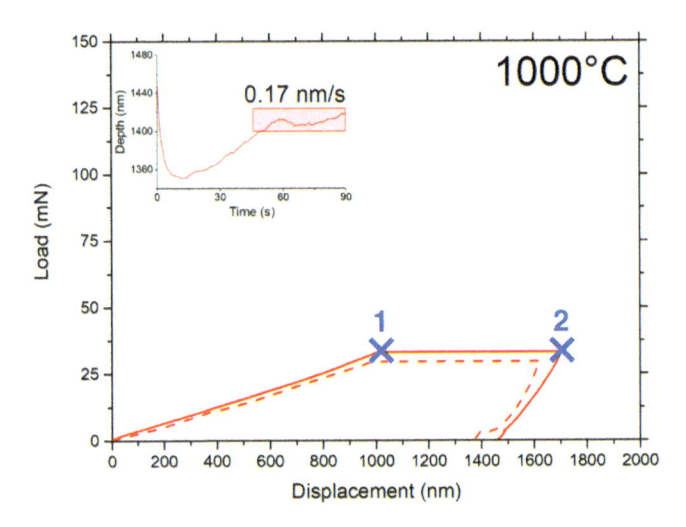

Figure 12.2 Load-displacement curve (Gibson et al., 2017). Load-displacement curve from a nickel superalloy (CMSX-4) tested by nanoindentation at 1000°C. Thermal drift measurement is inset to demonstrate the low achieved drift rate. A significant increase in depth (around 700 nm) can be seen from the point at which loading was stopped (point 1) to where unloading began (point 2). Furthermore, a "nose" in the unloading curve can be seen where residual plastic deformation due to creep has interrupted the (assumed) purely elastic unloading, leading to errors in the calculated modulus.
Source: From Gibson, J. S. K. -L., Schröders, S., Zehnder, C., & Korte-Kerzel, S. (2017). On extracting mechanical properties from nanoindentation at temperatures up to 1000°C. *Extreme Mechanics Letters*, *17*, 43−49, https://doi.org/10.1016/j.eml.2017.09.007.

This is due to the hold period essentially reducing the applied strain rate to the material, therefore also reducing the apparent hardness. (Leitner et al., 2017). As such, if hardness is of interest, the dwell period should be kept as short as possible.

The determination of the unloading slope is also problematic. Residual creep during the "elastic" unloading period has led to a characteristic "nose" in the curve, resulting in a measured Young's modulus around 10% of the true value. While more rapid unloading can be effective, the creep rate here of 16 nm/s would require an unloading rate of 335 mN/s, well beyond what is reasonably achievable with the inertia effects and data acquisition rates of typical nanoindenters. Therefore, if elastic modulus is of interest, the dwell period should be increased until creep rates are neglible. This is then of course thus in direct contradiction to the requirements of reliable hardness measurement.

Finally, for completeness, measurements of indentation creep rates will naturally require a reasonable dwell period for the creep to occur. It is therefore recommended that a series of indents are performed with significant dwell periods to assess modulus and creep behavior, followed by a series of indents with a short or zero dwell period for hardness measurements.

In the earlier discussion of isothermal contact, the drift-free condition requires both instrument drift and sample drift to be near-zero (Wheeler et al., 2015), that is, no thermal gradients within the system, as well as between tip and sample. This necessarily requires some stabilization time once the target temperature has been reached, typically of the order of half an hour or so, for thermal gradients with the instrument to stabilize. The natural consequence of this is that thermally unstable processes such as recovery and recrystallization are impossible to measure. A recent technique, named "high temperature scanning indentation" (Tiphéne et al., 2021) has been proposed to counteract this by minimizing drift during nonisothermal indentation tests. A sister sample is precharacterized such that the individual tip and sample heating ramp rates are known (i.e., if the system were left to stabilize, neither instrument or sample drift would be significant). Additionally, deep, very-high-speed indents are performed—to minimize the effect of drift—with a stepped unloading profile to also minimize the effect of creep on the unloading curve. In combination, these experimental modifications have been shown to be effective in characterizing the hardness evolution during the recovery and recrystallization of an aluminum sample.

12.2.1 Strategies for robust high-temperature measurements on coatings

In discussing testing on coatings, it will implicitly be assumed that testing is taking place perpendicular to the surface of the coating, that is, parallel to the growth direction, such that eventually the indenter would reach the substrate. This therefore confines the indentation experiments to necessarily be taking place at shallow depths. The commonly given rule of thumb is that indents must take place up to a maximum depth of 10% of the coating's thickness, although recent work (Zak et al., 2022) has shown that this should be used with care and perhaps used as an upper limit. This limit becomes even more restrictive at elevated temperature; as the majority of coatings involve a hard coating on a softer substrate, it is likely that the substrate will additionally soften more rapidly with increasing temperature. The composite response of coating and substrate (Ma et al., 2012) will therefore be dominated by that of the substrate unless the indentation depth is correspondingly reduced.

The primary concern when dealing with shallow indents at high temperature is the effect of tip wear and the resultant inaccuracy in mechanical properties that an incorrect area function will introduce. As chemical reactivity is increased at elevated temperatures so too is indenter degradation, of the order of a few indents in extreme cases (Gibson et al., 2017). Tip area function calibrations should therefore be performed more regularly to assess this, before and after an experimental campaign is recommended as a minimum.

Performing calibrations more regularly than this is problematic; as the most common reference material—fused silica—demonstrates an anomalous modulus − temperature curve (Beake & Smith, 2002), it is not very well-suited to in-situ area function calibrations at elevated temperature. A refractory BCC metal, such as tungsten or molybdenum, which has a more straightforward and well-characterized change in modulus with temperature (Harris et al., 2017) could also be installed at the same time at the sample of interest such that an area function calibration could be performed "at will" without having to cool, vent, and change to a calibration sample. In practice, however, this is rarely done, as the calibration sample will likely have different dimensions and a different thermal conductivity to the sample of interest. A full tip-sample temperature calibration would therefore have to be performed to obtain accurate area function data, which is experimentally time-consuming.

Finally, although not directly related to experimental design, it should be noted that the hardness — temperature behavior of materials is not linear, that is, they soften at different rates. An example can be seen in Cu/Nb, Cu/W, and Zr/Nb multilayers, where 10 nm Zr/Nb multilayers are ~ 2 GPa harder than Cu/W multilayers at room temperature, but ~ 1 GPa softer at 200°C (Beake & Harris, 2019).

12.3 High-temperature pillar compression and microcantilever bending

Elevated temperature testing has taken place on focused-ion-beam-manufactured structures since 2009 (Korte & Clegg, 2009; Song et al., 2009), only a few years after the first commercial high-temperature nanoindenters were available. Just as at room temperature, these structures allow various mechanical properties such as strength (Korte & Clegg, 2009), size effects (Abad et al., 2016), twinning (Edwards et al., 2018), or critical resolved shear stresses (Wang et al., 2019) to be measured.

By and large, there are no additional considerations when testing these structures: contact should be isothermal, oxidation should be avoided, substrates should be assumed to be more compliant, and so on. However, there are a handful of factors that make such tests easier than indentation-based experiments. The first is that often only the peak load in the test is of interest, as this characterizes the yield/fracture/twinning stress. As the displacement is not as useful, the need for strictly isothermal contact is less critical as for "traditional," load-controlled systems the thermal drift does not affect the applied load. For piezoelectric-based, intrinsically displacement-controlled systems such as the in-situ systems from Alemnis or Bruker, the load is instead affected by drift and thus strictly isothermal contact is essential.

The second factor that simplifies high-temperature micromechanics is that the stresses and strains in the system are characterized by this applied load and the geometry of the structure in question, that is, they are insensitive to the shape of the indenter. Tip wear is therefore significantly less concerning (likely helped further by the lower loads in micromechanical testing).

The only additional concern is when a gallium-based focussed ion beam has been used to manufacture the structures to be tested. Gallium, as a low-melting-point metal, has the potential to form a low-melting-point alloy with the material to be investigated. Typically the solubility of gallium is extremely low such that the gallium simply forms small metallic balls on the sample surface as can be seen in work on silicon (Armstrong & Tarleton, 2015). However, in sputtered or deposited coatings that have an extremely small grain size, the high grain boundary density can act as sinks for the gallium that then further degrades mechanical properties through grain boundary embrittlement, for example, Best et al. (2016). In these cases, the best solution would be to produce the micromechanical structures through a different method: plasma-FIB, lithography, micro-EDM (electric discharge machining), femtosecond laser, etc. When such methods are not available or not practical, similar sample preparation techniques to TEM foils can be employed (Li et al., 2006), namely low current and/or low-KV polishing steps toward the end of the sample preparation to remove as much FIB damage, and therefore as much gallium, as possible.

12.4 High-temperature scratch and impact/high strain rate testing

There are relatively few works that seek to combine novel scratch or high strain rate testing with high-temperature testing. This is most likely a simple effect of both techniques being relatively specialised in terms of experiment capability of the researcher and their equipment.

Certainly for scratch testing, the effect of frictional loading has obvious applications for the study of coatings for machining, where high-speed cutting applications can reach temperatures of up to 1000°C (Beake, 2022). It can also be imagined that this would be an exciting area of research, given the substantial amount of plastic deformation involved in the scratching process (Kareer et al., 2020) that the changes with temperature can be complex (Salari et al., 2022), and it has been shown—as with softening rates—that wear rates can vary drastically with temperature and initially more wear-resistant coatings can show poorer performance at temperature (Tillmann et al., 2018).

A review of the key concepts in the instrumentation and experimentation of high-temperature microtribology can be found elsewhere (Chavoshi & Xu, 2018a), but by and large, the considerations have already been mentioned: the importance of isothermal contact, the need to prevent oxidation, the strong effects of tip wear, and the increase in substrate compliance.

The latter can have a strong effect if a spherical probe is used. Spheres are often used in microtribology tests because the effect of tip wear is reduced and the scratch direction does not have to be considered; with a sharp indenter, an "edge forward" or "face forward" scratch direction can give differing results. However, there is an additional effect that the depth of the peak stress under the indenter is a function of tip radius. Larger tip radii produce a peak stress at deeper depths and thus allow the interface between substrate and coating to be specifically targeted (Beake et al., 2015). However, as this is again a composite response of coating and substrate, the depth of this peak stress is likely to change as temperature increases and significant substrate softening occurs.

In regard to impact nanoindentation and/or high-speed indentation, it is necessary to differentiate between indents conducted at high strain rates, but where the indenter only makes contact with the surface once (at least with any significant force behind it), termed "dynamic hardness" measurements (Zehnder et al., 2017, 2018), and tests where the tip is repeated driven into the surface, termed "multiple impulse." The former, that is, dynamic hardness testing, has limited application at high temperatures. While high strain rate nanoindentation may occur as a side effect of minimizing test time such that thermal drift is also minimized, this is not the primary focus of the tests. Instead, scientifically the two effects of strain rate and temperature are in opposition: high temperatures encourage the initiation of thermally activated deformation processes, while high strain rates prevent them. As such, to the authors' knowledge, there are no works that perform any dynamic hardness measurements at elevated temperatures.

The multiple impulse technique on the other hand is far better complimented by elevated temperature testing. It has been shown that insights from repeated impact testing can be used to inform the development of wear-resistant coatings (Beake, 2020; Beake & Smith, 2004), and therefore, high-temperature multiple impulse testing (Beake et al., 2019) is also a useful technique in this sector. At the risk of repetition, again the requirements are broadly similar to those listed previously: oxidation should be avoided, tip wear will be significant, and the increased substrate plasticity is likely to play a significant part in the response of the system. The only significant deviation is that of the condition of isothermal contact. As the contact time is so small, and the contact depth not an experimentally significant parameter, the experiment can be run with an unheated indenter tip (Beake et al., 2019) to reduce degradation.

12.5 Summary and conclusions

There are some clear trends in the requirements for reliable high-temperature measurements on the micro- and nanoscale, regardless of the exact tests to be performed:

1. Oxidation of the tip and sample should be avoided, which is best achieved by testing in vacuum and avoiding diamond indenters.
2. Isothermal contact is often essential and can be achieved by careful monitoring of drift rates as the tip is brought into contact with the sample.
3. If the indenter shape is of importance, increased tip wear and pileup mean that area function calibrations should be performed more regularly.

References

Abad, O. T., Wheeler, J. M., Michler, J., Schneider, A. S., & Arzt, E. (2016). Temperature-dependent size effects on the strength of Ta and W micropillars. *Acta Materialia*, *103*, 483−494. Available from https://doi.org/10.1016/j.actamat.2015.10.016.

Armstrong, D. E. J., & Tarleton, E. (2015). Bend testing of silicon microcantilevers from 21°C to 770°C. *Minerals, Metals and Materials Society, United Kingdom JOM*, *67*(12), 2914−2920. Available from https://doi.org/10.1007/s11837-015-1618-y, http://www.springer.com/materials/journal/11837.

Beake, B. D., & Harris, A. J. (2019). Nanomechanics to 1000°C for high temperature mechanical properties of bulk materials and hard coatings. *Vacuum*, *159*, 17−28. Available from https://doi.org/10.1016/j.vacuum.2018.10.011.

Beake, B. D., Liskiewicz, T. W., Vishnyakov, V. M., & Davies, M. I. (2015). Development of DLC coating architectures for demanding functional surface applications through nano- and micro-mechanical testing. *Surface and Coatings Technology*, *284*, 334−343. Available from https://doi.org/10.1016/j.surfcoat.2015.05.050.

Beake, B. D. (2020). Micro-scale impact testing − A new approach to studying fatigue resistance in hard carbon coatings. *Tribology International*, *149*.

Beake, B. D. (2022). Nano- and micro-scale impact testing of hard coatings: a review. *Coatings*, *12*(6).

Beake, B. D., & Smith, J. F. (2004). Nano-impact testing − An effective tool for assessing the resistance of advanced wear-resistant coatings to fatigue failure and delamination. *Surface and Coatings Technology*, *188-189*(1-3), 594−598. Available from https://doi.org/10.1016/j.surfcoat.2004.07.016.

Beake, B. D., Bird, A., Isern, L., Endrino, J. L., & Jiang, F. (2019). Elevated temperature micro-impact testing of TiAlSiN coatings produced by physical vapour deposition. *Thin Solid Films*, *688*. Available from https://doi.org/10.1016/j.tsf.2019.06.008.

Beake, B. D., & Smith, J. F. (2002). High-temperature nanoindentation testing of fused silica and other materials. *Philosophical Magazine A*, *82*(10), 2179−2186. Available from https://doi.org/10.1080/01418610208235727.

Best, J. P., Zechner, J., Wheeler, J. M., Schoeppner, R., Morstein, M., & Michler, J. (2016). Small-scale fracture toughness of ceramic thin films: The effects of specimen geometry, ion beam notching and high temperature on chromium nitride toughness evaluation. *Philosophical Magazine*, *96*(32-34), 3552−3569. Available from https://doi.org/10.1080/14786435.2016.1223891, http://www.informaworld.com/smpp/title~content = t713695589~db = all.

Chavoshi, S. Z., & Xu, S. (2018a). A review on micro- and nanoscratching/tribology at high temperatures: Instrumentation and experimentation. *Journal of Materials Engineering and Performance*, *27*(8), 3844−3858. Available from https://doi.org/10.1007/s11665-018-3493-5, http://www.springerlink.com/content/1059-9495/.

Chavoshi, S. Z., & Xu, S. (2018b). Temperature-dependent nanoindentation response of materials. *MRS Communications*, *8*(1), 15−28. Available from https://doi.org/10.1557/mrc.2018.19, https://www.scopus.com/inward/record.uri?eid = 2-s2.0-85044194476&doi = 10.1557%2fmrc.2018.19&partnerID = 40&md5 = 9cb2b59fa111f0e9a927adbc680514f2.

Chavoshi, S. Z., & Xu, S. (2019). Nanoindentation/scratching at finite temperatures: Insights from atomistic-based modeling. *Progress in Materials Science*, *100*, 1−20. Available from https://doi.org/10.1016/j.pmatsci.2018.09.002, https://www.scopus.com/inward/record.uri?eid = 2-s2.0-85054166978&doi = 10.1016%2fj.pmatsci.2018.09.002&partnerID = 40&md5 = c5d7cd253eb4936438e437732214a3a7.

Edwards, T. E. J., Di Gioacchino, F., Mohanty, G., Wehrs, J., Michler, J., & Clegg, W. J. (2018). Longitudinal twinning in a TiAl alloy at high temperature by in situ microcompression. *Acta Materialia*, *148*, 202−215. Available from https://doi.org/10.1016/j.actamat.2018.01.007, http://www.journals.elsevier.com/acta-materialia/.

Gibson, J. S. K.-L., Schröders, S., Zehnder, C., & Korte-Kerzel, S. (2017). On extracting mechanical properties from nanoindentation at temperatures up to 1000°C. *Extreme Mechanics Letters*, *17*, 43−49. Available from https://doi.org/10.1016/j.eml.2017.09.007.

Harris, A. J., Beake, B. D., Armstrong, D. E. J., & Davies, M. I. (2017). Development of high temperature nanoindentation methodology and its application in the nanoindentation of polycrystalline tungsten in vacuum to 950°C. *Experimental Mechanics*, *57*(7), 1115−1126. Available from https://doi.org/10.1007/s11340-016-0209-3.

Hosemann, P., Shin, C., & Kiener, D. (2015). Small scale mechanical testing of irradiated materials. *Journal of Materials Research*, *30*(9), 1231−1245. Available from https://doi.org/10.1557/jmr.2015.26, https://www.scopus.com/inward/record.uri?eid = 2-s2.0-849-29708093&doi = 10.1557%2fjmr.2015.26&partnerID = 40&md5 = d0ca22360e79ffd07-93434641972593a.

Kareer, A., Tarleton, E., Hardie, C., Hainsworth, S. V., & Wilkinson, A. J. (2020). Scratching the surface: Elastic rotations beneath nanoscratch and nanoindentation tests. *Acta Materialia*, *200*, 116−126. Available from https://doi.org/10.1016/j.actamat.2020.08.051.

Korte, S., & Clegg, W. J. (2009). Micropillar compression of ceramics at elevated temperatures. *Scripta Materialia*, *60*(9), 807−810. Available from https://doi.org/10.1016/j.scriptamat.2009.01.029.

Leitner, A., Maier-Kiener, V., & Kiener, D. (2017). Dynamic nanoindentation testing: Is there an influence on a material's hardness? *Materials Research Letters*, *5*(7), 486−493. Available from https://doi.org/10.1080/21663831.2017.1331384, https://doi.org/10.1080/21663831.2017.1331384.

Li, J., Malis, T., & Dionne, S. (2006). Recent advances in FIB−TEM specimen preparation techniques. *Materials Characterization*, *57*(1), 64−70. Available from https://doi.org/10.1016/j.matchar.2005.12.007.

Lucchesi, C., Vaillon, R., & Chapuis, P.-O. (2021). Temperature dependence of near-field radiative heat transfer above room temperature. *Materials Today Physics*, *21*. Available from https://doi.org/10.1016/j.mtphys.2021.100562.

Ma, Z. S., Zhou, Y. C., Long, S. G., & Lu, C. (2012). On the intrinsic hardness of a metallic film/substrate system: Indentation size and substrate effects. *International Journal of Plasticity*, *34*, 1−11. Available from https://doi.org/10.1016/j.ijplas.2012.01.001.

Merle, B., Higgins, W. H., & Pharr, G. M. (2019). Critical issues in conducting constant strain rate nanoindentation tests at higher strain rates. *Journal of Materials Research*, *34*(20), 3495−3503. Available from https://doi.org/10.1557/jmr.2019.292, https://www.springer.com/journal/43578.

Salari, S., Rahman, M. S., Beheshti, A., & Polycarpou, A. A. (2022). Elevated temperature nanoscratch of inconel 617 superalloy. *Mechanics Research Communications*, *121*. Available from https://doi.org/10.1016/j.mechrescom.2022.103875.

Song, S. X., Lai, Y. H., Huang, J. C., & Nieh, T. G. (2009). Homogeneous deformation of Au-based metallic glass micropillars in compression at elevated temperatures. *Applied Physics Letters*, *94*(6). Available from https://doi.org/10.1063/1.3081111.

Tillmann, W., Kokalj, D., Stangier, D., Paulus, M., Sternemann, C., & Tolan, M. (2018). Investigation on the oxidation behavior of AlCrVxN thin films by means of synchrotron radiation and influence on the high temperature friction. *Applied Surface Science*, *427*, 511−521. Available from https://doi.org/10.1016/j.apsusc.2017.09.029.

Tiphéne, G., Baral, P., Comby-Dassonneville, S., Guillonneau, G., Kermouche, G., Bergheau, J.-M., Oliver, W., & Loubet, J.-L. (2021). High-temperature scanning indentation:

A new method to investigate in situ metallurgical evolution along temperature ramps. *Journal of Materials Research*, *36*(12), 2383−2396. Available from https://doi.org/10.1557/s43578-021-00107-7.

Wang, S., Giuliani, F., & Britton, T. B. (2019). Variable temperature micropillar compression to reveal <a> basal slip properties of Zircaloy-4. *Scripta Materialia*, *162*, 451−455. Available from https://doi.org/10.1016/j.scriptamat.2018.12.014.

Wang, S., & Zhao, H. (2020). Low temperature nanoindentation: Development and applications. *Micromachines*, *11*(4). Available from https://doi.org/10.3390/MI11040407, https://www.scopus.com/inward/record.uri?eid = 2-s2.0-85084429330&doi = 10.3390%2fMI11040407&partnerID = 40&md5 = 18ed016b49d56e711260df67c4ee86cf.

Wheeler, J. M., Armstrong, D. E. J., Heinz, W., & Schwaiger, R. (2015). High temperature nanoindentation: The state of the art and future challenges. *Current Opinion in Solid State and Materials Science*, *19*(6), 354−366. Available from https://doi.org/10.1016/j.cossms.2015.02.002.

Wheeler, J. M., Brodard, P., & Michler, J. (2012). Elevated temperature, in situ indentation with calibrated contact temperatures. *Philosophical Magazine*, *92*(25-27), 3128−3141. Available from https://doi.org/10.1080/14786435.2012.674647.

Wheeler, J. M., & Michler, J. (2013). Invited article: Indenter materials for high temperature nanoindentation. *Review of Scientific Instruments*, *84*(10). Available from https://doi.org/10.1063/1.4824710.

Zak, S., Trost, C. O. W., Kreiml, P., & Cordill, M. J. (2022). Accurate measurement of thin film mechanical properties using nanoindentation. *Journal of Materials Research*, *37*(7), 1373−1389. Available from https://doi.org/10.1557/s43578-022-00541-1.

Zehnder, C., Peltzer, J.-N., Gibson, J. S. K.-L., & Korte-Kerzel, S. (2018). High strain rate testing at the nano-scale: A proposed methodology for impact nanoindentation. *Materials & Design*, *151*, 17−28. Available from https://doi.org/10.1016/j.matdes.2018.04.045.

Zehnder, C., Peltzer, J.-N., Gibson, J. S. K.-L., Möncke, D., & Korte-Kerzel, S. (2017). Non-Newtonian flow to the theoretical strength of glasses via impact nanoindentation at room temperature. *Scientific Reports*, *7*(1). Available from https://doi.org/10.1038/s41598-017-17871-4.

Section III

Modelling approaches

Analytical methods - applied

Troy vom Braucke[1], Norbert Schwarzer[2], Frank Papa[1] and Julius Schwarzer[2]
[1]GP Plasma, Medina, OH, United States, [2]Siomec, Tankow, Ummanz, Germany

13.1 Introduction of analytical modeling of coated systems

About 150 years ago, there was a breakthrough in a field we today know as material science, where scientists introduced the concept of stresses and strains and applied it to technical mechanics. This led to an explosion in technology, just like the Cambrian evolutionary explosion of life on Earth. The massive growth was in all technical fields leading to new weapons, ships (Queen Mary, Titanic), Henry Ford's mass-produced Model T, the Empire State building, weapons developments, and so on. This technology breakthrough never would have been possible without the visualization of stress and strain fields in engineering, allowing engineers to much better understand (and predict) the behavior of materials.

The goal of materials and process engineers remains to have less failures and more success for customers. However, the technical optimization with respect to one aspect of a thin film or a certain coating property can already be very complex. For those interested in an elaboration on the extremely wide and complex field of optimization (nonlinear, linear, or integer), we refer them to the literature (Marthaler, 2013). A seamless connection between deposition process recipe, property measurement and testing, and simulation/refinement optimization cycles provides a holistic methodology to efficiently solve failures. The analytical approach within FilmDoctor Studio (2006) provides this with computational efficiency, allowing for fast-paced development against competitors and quick troubleshooting capability. Fig. 13.1 gives an example for such an optimization procedure regarding the mechanical stability and reliability (Schwarzer, 2014).

Approximately 25 years ago at the University of Chemnitz theoretical physics department, analytical modeling (using FilmDoctor) was compared with a first-principles nontruncated molecular dynamics (MD) orbital model and experimental observations. The results of the analytical approach agreed with both the MD's modeling and reality. Interestingly, although the MD calculations went for a month, the numerical analytical method from (Schwarzer, 2017) needed just a few seconds on a laptop. Similarly, finite element methods have been compared; however, both MD and FE lack invertibility to reverse optimize performance solutions, meaning

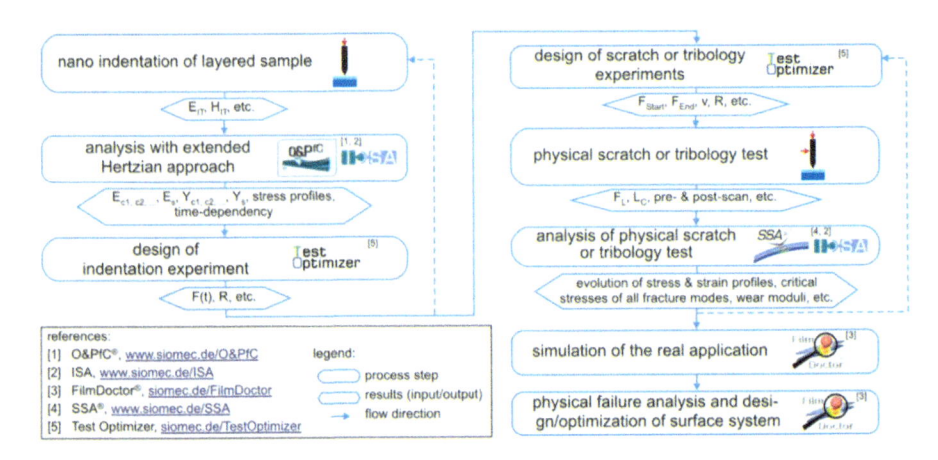

Figure 13.1 A flowchart of the procedure of mechanical characterization and optimization of arbitrary structured surfaces with respect to mechanical stability and reliability. A suite of analytical and simulation modules provide the coating development engineer a variety of options for understanding failure modes and subsequent optimization. *Source*: Courtesy Siomec.

that starting with a performance requirement and calculating back to find suitable materials properties and layer structures is of interest to coating designers.

The quasielastic Hertzian analytical approach (Schwarzer, 2017) allows one to combine all the information collected about the material solution and contact conditions, to both quantifiably and visually see the critical failure points. These can then be addressed by materials selection, test methodology, and/or process adjustment to modify the structure and properties to be more resilient against the expected application contact conditions and importantly to ensure robust coating solutions under manufacturing processes variations—that is, repeatable quality. It is not always the case that an engineer will have the measured properties or the tools at hand to measure and test. The analytical approach is not limited in this case, because it can be used to run thought experiments on contact scenarios, as well as a variety of material properties and layer structures to rapidly search for potential solutions. With the industrial tools of instrumented scratch and indentation, a variety of loading conditions can be simulated to understand the failure modes not accessible by noncontact means.

The chapter from Schwarzer (2017) was summarized as follows "In order to achieve the goal set, namely, the optimization and parameter identification for layered surfaces in a sufficiently generic manner, it is necessary to combine a variety of scientific fields and/or concepts of material science. This chapter covers the some of the following issues: the effective indenter concept made time-dependent; and the physical scratch and/or tribological test and its analysis. It aims to improve the simulation by taking nonlinear and temperature effects into account. The chapter outlines the analytical method that has the big advantage of delivering quasi 'insight-views' about the deformation fields, strains, stresses and so on of the surfaces subjected to such tribology tests. Due to the completely analytical character of the used tools in the chapter, the calculation usually is extremely fast and robust. This way weak spots within the material composition can be found more easily and efficiently."

The following fields and/or concepts bullet pointed are taken from Schwarzer (2017) and were elaborated and referenced within. They outline the main topics covered about the analytical method:

- First principle – based, effective interatomic potential description of mechanical material behavior and its extension into the time domain;
- The effective indenter concept made time-dependent;
- The extension of the Oliver and Pharr method to analyze nanoindentation data to layered materials and time-dependent mechanical behavior;
- The incorporation of intrinsic or residual stresses into the model description.

The interested reader is invited to dig deeper into the analytical methods presented within Schwarzer (2017) and the more fundamental derivations of the theory in Schwarzer (2022) and further explored in Schwarzer (2023).

13.1.1 A digital twin

Having a better understanding of coating stacks and their behavior in the desired application is a crucial point for the optimization of complex systems. To improve the understanding of the system, usually many tests are performed, which are designed to be as close as possible to the application. Better performance in the tests correlates to better performance in the application. The analytical first-principle models described earlier make it possible to improve this process. The idea is to transform as much gathered knowledge as possible to a digital model. With such a model, you will not only know how good the system performed in a specific test, but can use the data and compare the test results to find the initial reason for a certain failure mode.

13.1.1.1 What is needed for such a digital model, or "digital twin"?

You need to know the material parameters for each coating and the substrate. The necessary material parameters are the Young's modulus, the Poisson's ratio, and layer thickness. In addition, the critical values such as yield and tensile strength are needed.

13.1.1.2 How to obtain the material data?

A good way to get the Young's modulus values is to perform nanoindentation measurements. These are relatively simple experiments in which a normal load is applied and the indentation depth is measured. This information allows evaluation of the Young's modulus by using the Oliver and Pharr method. This method is built into the software of commercial measurement devices, but it has limitations for coating systems since it only gives one result for the complete system that does not take the real structure into account. Even if the indentation depth is very low compared with the layer thickness (<10%), there is still some contribution from the substrate, which is more noticeable when there is a large elastic mismatch between the coating and substrate. In contrast, with an analytical approach [termed "extended Oliver & Pharr for Coatings" (Schwarzer, 2017)], the analysis is possible for every coated system as it takes the material structure into account and determines the Young's modulus and yield strength for each part of the system.

13.1.1.3 How to use these data?

Knowing the material parameters also helps to analyze more complex tests such as the scratch test. In scratch tests, a lateral load is added, and so a great variety of contact conditions can be created to reflect the desired application much better. With the Scratch Stress Analyzer from FilmDoctor Studio (2006), for example, such tests can be analyzed and the stress strain field or each traversed position during the scratch can be evaluated. This helps to correlate the observed failure mechanism to a certain field value, for example, the von Mises stress exceeded the yield strength, which indicates the start of plastic flow. Therefore, it is possible to detect the initial reason for a system failure and add the critical values to the digital twin.

13.1.1.4 Why digital?

Having the digital twin allows you to make simulations of real tests in the computer without performing them. This can save time and money and help to narrow down the search path during the optimization process. You can easily transform your system to a new application, by performing calculations with test parameters, which represent another application or test, it's possible to check how good the coating system performs under new conditions. In addition you can play around with some material combinations or material parameters to find out how the system needs to be changed to perform better in a new application; therefore, speeding up the optimization time a lot by choosing the most promising system for a different application using simulations and the digital twins of your coated samples. But an often overlooked advantage of the digital twin is the fact that the visualization of a root cause failure and potential solutions make for a great communication tool to justify research and development paths.

13.2 Working through practical examples

The problem when going down to the nanoscale with layered inhomogeneous materials is that the classical methods become challenged to properly measure and account for uncertainties that the macroscale methods are less prone to, because of the use of monolithic homogeneous half-space solutions. If we are to efficiently solve coating system optimization problems, then the optimal process for the materials design engineer is to understand first the functional application and the design parameters required; from those requirements, to then find the optimal set of properties and material layers to protect the substrate from early failure, thus extending the functional life of the coated system.

However, the perfect material system, property profile, and interface adhesion are not immediately available to the process engineer when beginning such a project. Typically, they must deal with process operating window restrictions from the deposition equipment, substrate geometry and variability, temperature compatibility, tooling, and the deposition machine repeatability as the coating system evolves with deposited films. Additionally, the test equipment needs to be selected and set up in such a way to collect the material property information and, in the case of performance requirements, information about the application contact conditions. This can require setup of lab testing to best replicate contact conditions to extract performance information. The key consideration is to select a suitable load range and scale of contact tip to measure properties at the right scale (see Fig. 13.2) to find the relevant critical failure modes of a coating − substrate system.

For example, if there are pop-ins during an indentation, often those curves are thrown out, but what are the causes, what stress and strain events are occurring that lead to them? The stress combinations leading to them can be explored to find out more about why and how they occur. Regarding coating anisotropy, it is important to understand the contribution of structural effects. The ability to incline the sample to replicate an inclined loading scenario offers a method to understand how microstructure can influence the contact behavior and response of the system under such loading conditions and what stress and strain components may become critical. Lateral loading of an indenter tip can achieve similar information, but they are less commonly available. Alternatively, using instrumented scratch testing can also provide some information on the complex stress fields occurring in a coating system from both the material and structural influences.

Analytical modeling of coated systems when combined with careful data collection from suitable measurement techniques provides efficient methods to navigate the optimization cycle from initial coating system design and failure analysis through to discovering the optimal properties to withstand both the variation in functional use and variation from manufacturing. The optimization cycle utilizes feedback from analytical measurement tools to confirm the properties of what was produced to feedback into the analytical simulations. Final evaluation either from "in application" testing or accelerated testing closely matches the final application conditions provides additional feedback on the design loop as shown in Fig. 13.1.

With functional coatings, stress is a key property to manage in thin-film coating design for function, manufacturing process control and in its interaction with coating quality control. Because stress can quickly destroy coatings when it exceeds the critical von Mises stress of the weakest point in the coating system, it is important to measure and then optimize the stress profile. Determining "why" is important for identifying the root cause and to determine if failure is during an unusual applied use scenario or due to the variability of the coating manufacturing process. To measure property profiles such as stress, it is necessary to find a suitable method to

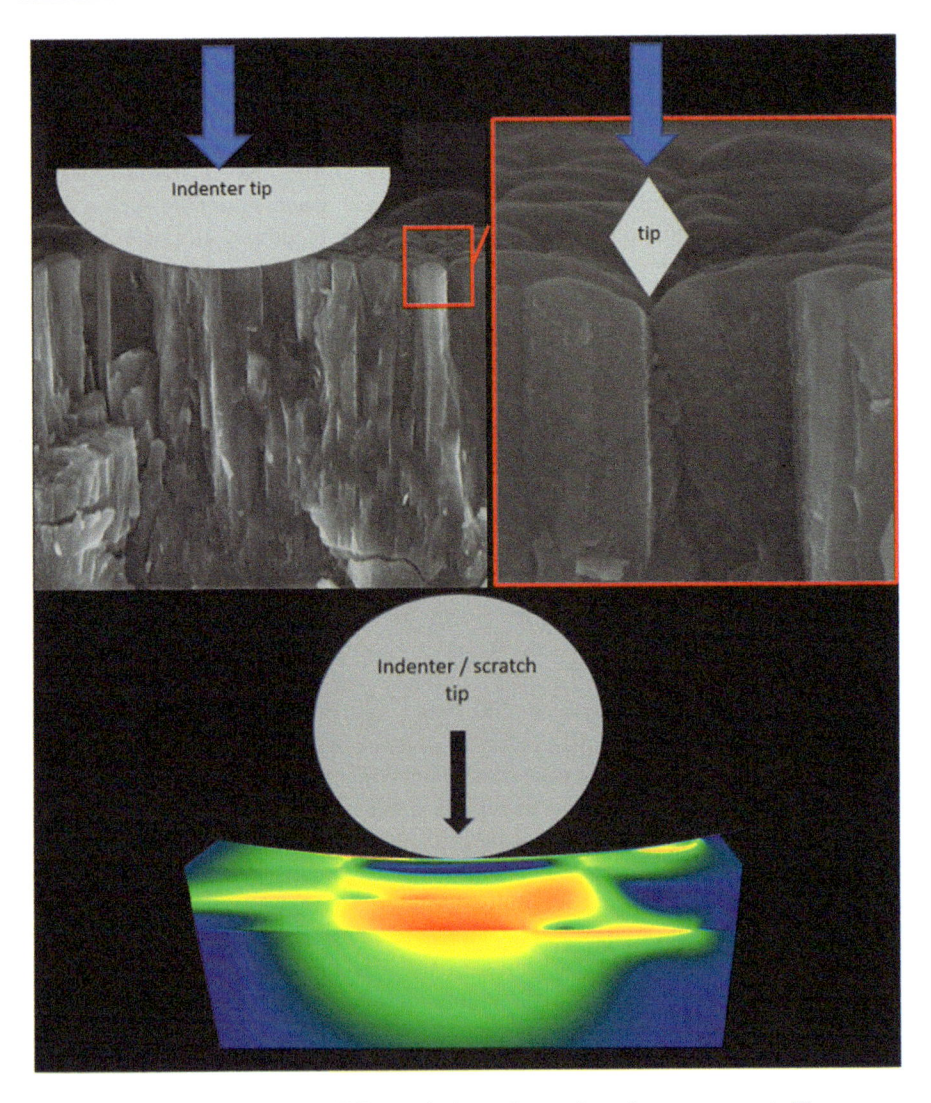

Figure 13.2 Representation of different indentation scales of measurement. The measurement contact "sees" millions of atoms and all are completely summed up into the systems response under loading. We don't need to model every atom, just the ensemble.

section through a coating. Examples are the Calotte ball crater method, which uses a large hardened steel or ceramic ball to grind through the coating stack down to the substrate with increasingly finer abrasives to minimize surface roughness. Alternatively, FIB milling techniques or other sectioning methods may be required to profile through complex geometries or thinner materials. Careful selection, for example, by using Test Optimizer (2013), of the indenter load and tip size and geometry is required to ensure that the effects of surface roughness, polishing damage are largely overcome, and the load is extracting the properties at the right depth within the coating. While it should be noted that the method presented is certainly not the only way to measure coating stress [see Abadias et al. (2018), for an excellent review of alternatives], it is one that is relatively fast for industrial development where understanding the stress evolution in response to directed loading is of importance.

Stress also influences other measured mechanical properties such as the yield strength and Young's modulus. Being able to mathematically extract the influence of stress components on these properties will be important not just for localized measurement information but also for developing material properties with the desired stress, yield strength, and modulus properties through the thickness of the coating. This opens up opportunities to find different material combinations to give the same functional performance and provides options to use more earth abundant materials to replace expensive exotic or supply chain constrained materials (Schwarzer, 2023). Problems then arise to extract the measured properties of materials at the nanoscale, which is most often done with instrumented nanoindentation above several tens of nanometers in film thickness.

We present a series of applied examples to provide an overview of working with an analytical numerical approach.

13.2.1 From geological to the nanoscale

It is known from geology that in cases where there is an incompressible layer, it is usually a water-bearing formation, while the foundation below is made out of rock, the amazed observer will detect an upward bending of the surface even though only normal loads in downward direction are applied. The reason for this apparently paradoxical behavior is to be found in the incompressibility of the water-bearing formation. Being hindered in downward movement by the stiff foundation and not being able to densify due to its large Poisson's ratio, the water-bearing layer—or polymer in our nanoscale example in Fig. 13.3A—deforms elastically sideways and upward, partially even above the original surface (as shown in Fig. 13.3A) and, simultaneously and this way, bends the surface in a concave manner. This finally leads to strongly enforced tensile surface stresses for the hard top layer endangering it to fracture there that have nothing to do with the typical Hertzian cone cracks (Fischer-Cripps, 1997). As a mere by-product, a new mechanical failure mechanism was discovered, which had puzzled and troubled engineers for many years in the coating industry (see Fig. 13.3).

The new crack behavior observed in both geological examples and nanoscale examples was named the double egg-shell effect to distinguish it from the well-known eggshell effect. The latter appears when a stiff top layer is placed over softer substrate materials leading to radial stresses with fractures opening at the interface forming a so-called star-crack when propagating and finally, reaching the surface (Schwarzer, 2015).

A more complete solution to this complex stability problem is to also include intrinsic stresses, consider them as a new degree of freedom, and combine it with

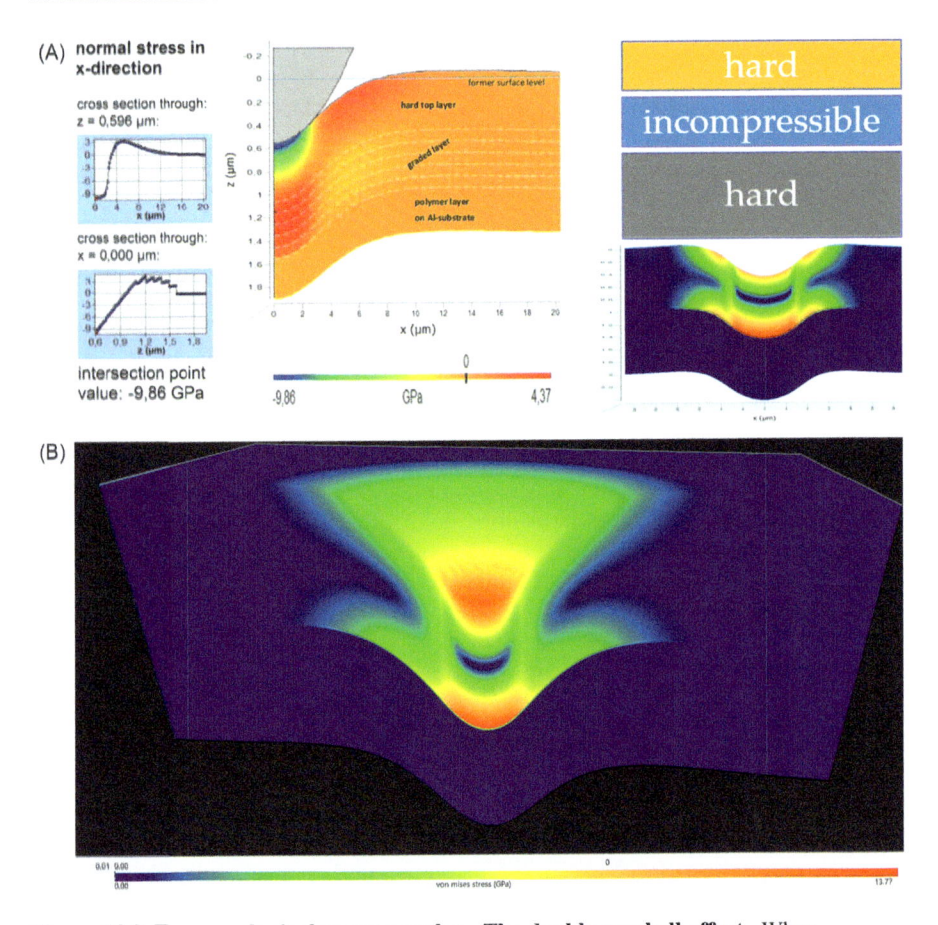

Figure 13.3 From geological to nano scales - The double eggshell effect. When calculating the von Mises stress behavior of a geological uplift formation This same effect was later found to also describe the behavior of a particular geological earthquake formation under study at the University of Ohio where the incompressible layer was a water bearing layer over rock is now called the "double eggshell" effect.

the current structural solution. The latter, namely functionally stressed coatings, was the principle idea of the EU-research project iStress (c.f. http://www.stm.uniroma3.it/iSTRESS) for which an automatic software package was developed, solving the task of designing the intrinsic stresses for a given external or internal load problem respectively worst-case contact scenario (Schwarzer, 2015; iStress-Module, 2014). With a method now available to optimize the intrinsic stress profiles of layer stacks to maximize the load bearing capability, this led to a need to measure the stress and mechanical property profiles so that one can check that the design optimized stress profile was achieved by process development. Some of the following examples explore the measurement of these property profiles as a function of

Analytical methods - applied

coating thickness. The upshot of this example being that the analytical method used is scale-invariant with application from subatomic to geological scales.

13.2.2 Scratch testing of zirconia coatings

For two zirconia coatings deposited on glass with two different thickness, their relative ranking in a scratch test differed from that in a nanoimpact test (i.e., the critical load for coating failure was higher for the thicker sample although its impact resistance was lower). In an attempt to understand the cause of this, the nanoscratch data were analytically modeled with a commercial software package from FilmDoctor Studio (2006).

The input data for the scratch analyzer are mechanical properties of the coating and substrate (H,E) from nanoindentation, thickness, Poisson ratios, and suitable values of constraint factor (H/Y) to determine Y from H, load, scratch depth, and friction from the nanoscratch test, in this case, using a diamond probe with 4.5 μm end radius. Scratch test data are imported to the software (see Fig. 13.4), and the von Mises stress profile analysis is calculated for each position along the scratch path.

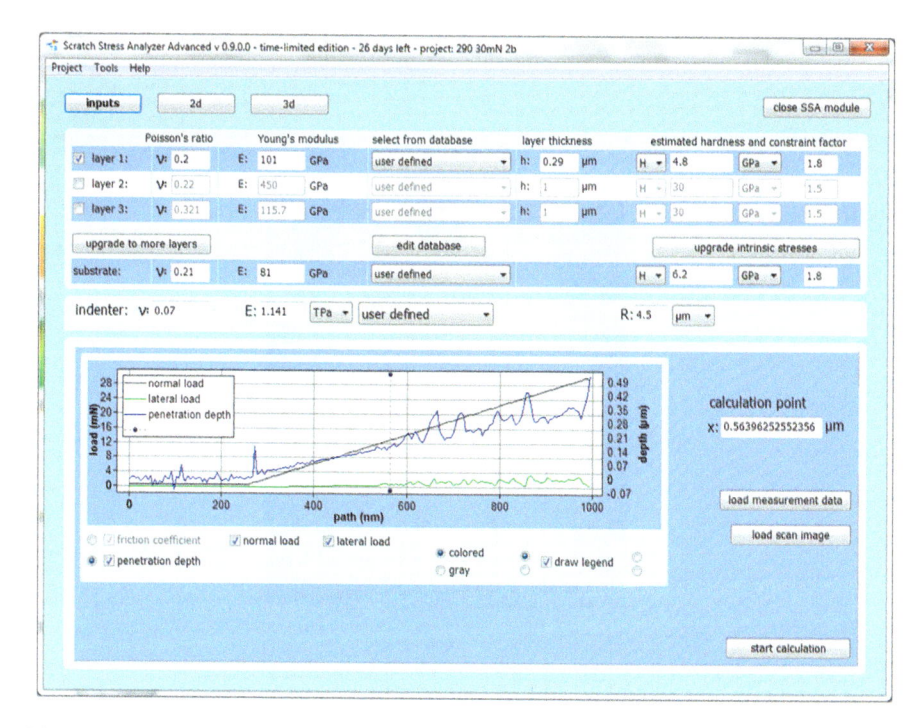

Figure 13.4 Scratch test analysis for 12 mN load on 288 nm ZrO$_2$. Input conditions for the scratch are entered and scratch data imported into the software for analysis.
Source: Courtesy Ben Beake at MML.

The critical load locations were then analyzed. There were differences in plastic flow and interfacial shear stresses, with Fig. 13.5 showing the von Mises stress distributions. The values of a variety of field components (stress, strain, etc.) are calculated together with the simulation plots, allowing quantifiable comparisons between samples. At the critical load for the thinner coating (~ 12 mN), the peak interfacial shear stress driving delamination was 1 GPa but only 0.8 GPa for the thicker

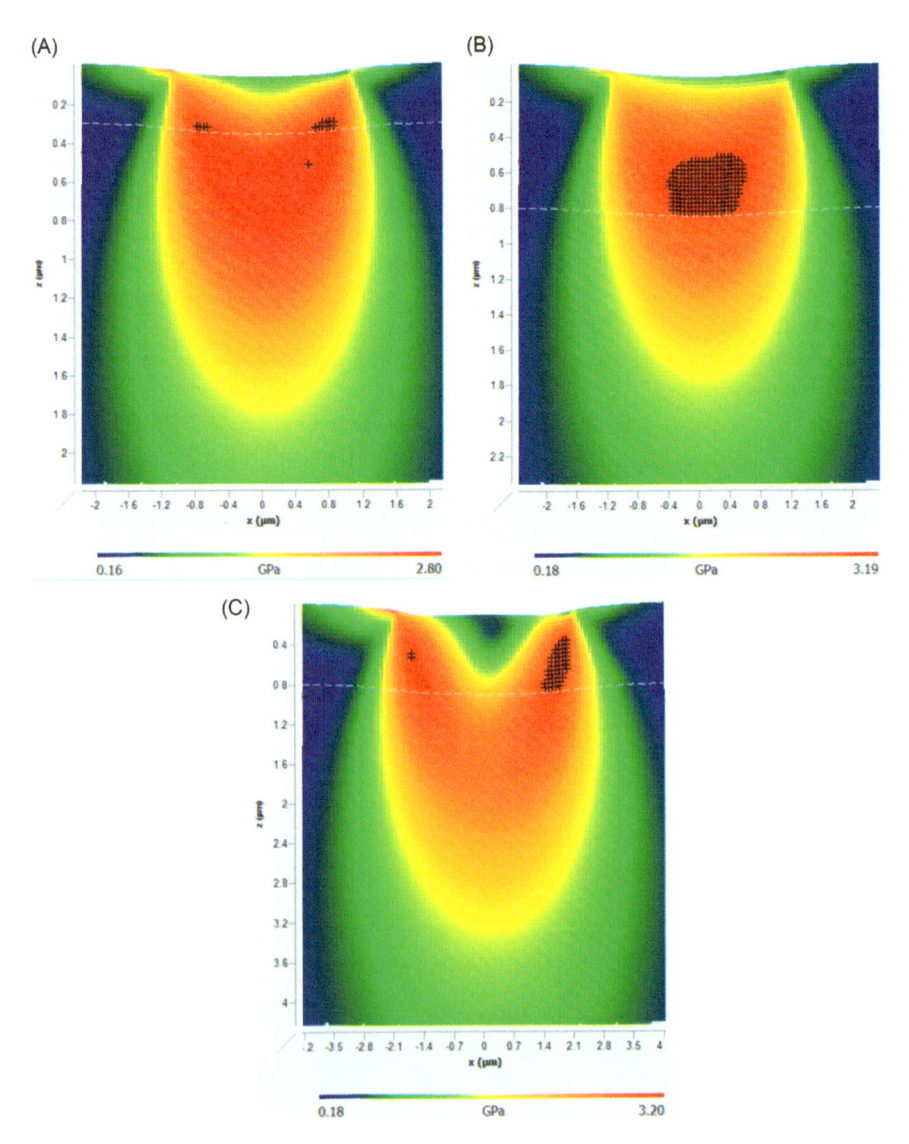

Figure 13.5 Von Mises stress distributions of scratch tests for different zirconia coating thicknesses and scratch loads on glass. The black cross hatching on the image represents the critical von Mises stress region and is the location where failure is most likely to occur.

Analytical methods - applied   353

coating. As the load increases to the critical load on the thicker coating (~ 50 mN), the peak shear stress reaches the same value. This implies that this is the strength of the interface, and under these conditions, the failure is related to adhesion.

Further examples using the scratch analysis software can be found here (Schwarzer, 2022).

13.2.3 Comparing stress and modulus profile tuning for the same cathodic arc deposited coating system

Within a German DFG project consortium, the measurement of stress profiles was undertaken with the Reference Stress Module (FilmDoctor Studio, 2006). Consider the problem of designing coating stress by process recipe changes to minimize coating adhesion failure and maximize load-carrying capability. To better understand the coating system and stress control, three different process recipes were performed with the goal of only changing the stress profile, while coating machine, deposition source parameters, and coating composition were kept the same.

The coating system analyzed was produced using cathodic arc deposition, which inherently has metallic droplets or macroparticles incorporated within the growing film. To provide for reliable statistics, an array of indents was performed around the circumference of the crater edge at the film surface toward the film interface (center) to approximately 60% depth (~ 3 μm of a 5 μm thick film).

The three samples were prepared with a Calotte wear crater by use of a Calotte ball crater test unit. The depth and ball diameter to make the crater were predetermined by use of Test Optimizer software that targets a crater slope of less than 2 degrees for the samples film thickness. Some experimentation may then be required to reveal the substrate slightly to a depth within 5% of the film's total thickness. Polishing techniques are selected to suit the materials being polished so that a balance between wear rate and surface roughness can be found. Surface roughness of less than 50 nm average roughness (Ra) is desirable to minimize indenter measurement errors. Higher roughness requires larger indenter contact sizes to compensate; however, this is at the expense of obtaining measurement resolution in the thinner interface regions of the coated films.

The following schematic shows the layout of the indentation measurements in the Calotte ball crater scar Fig. 13.6. The next image shows an example ball crater (often shortened to "Calo") carefully performed on the coating system Fig. 13.7. The array of indents are performed from crater edge toward center to provide a depth profiling of the coating; however, these indents are not observable on the image without magnification. Once the indentation curve data are imported into the stress profiling software (explored further in the next example), the indent positions are automatically placed within the dimensions measured for the wear scar and their positions refined and checked before moving ahead to calculations (see Fig. 13.8) and (Schwarzer & Schwarzer, 2023a). It must be pointed out that with the reference stress analysis, the calculated stress values are relative to a chosen refence point. This makes it useful for comparative stress profile analysis or if an average stress is

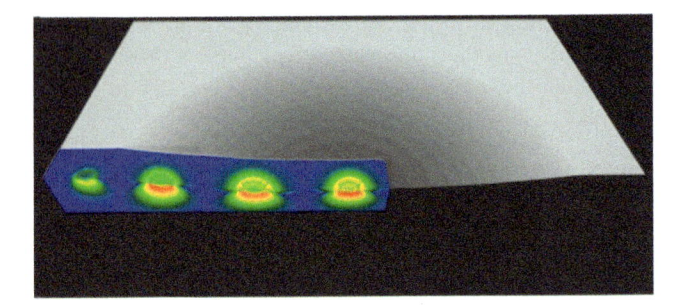

Figure 13.6 Schematic of making indentations inside the wear scar of a Calotte ball crater. Indentations are across the crater profile adjusting the indenter load to lower values as coating thickness reduces.
Source: Courtesy Nick Bierwisch at Siomec.

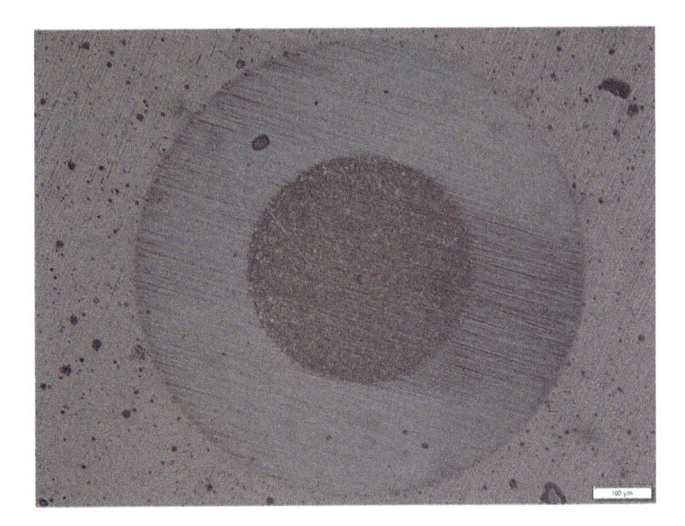

Figure 13.7 Example of a Calotte ball crater wear scar. A wear scar is ground through the coating to provide a shallow slope to make indentation measurements from the top surface down to the coating − substrate interface.
Source: Courtesy GP Plasma LLC.

known for the coating, then the profile's position on the vertical stress axis can be adjusted up or down accordingly to match. Absolute stress measurement can only be made when using an indenter with lateral loading capability to calculate the poisons ratio at each measurement point.

After the indentation curve profiles are checked and outliers identified, the profiles for the reduced hardness, yield strength, and stress are calculated (see Figs. 13.9 and 13.10 respectively). The biaxial stress is calculated with the option to also calculate shear stress or a combination of biaxial and shear stress profiles. This exploration of the different stress components can be useful when analyzing

Analytical methods - applied 355

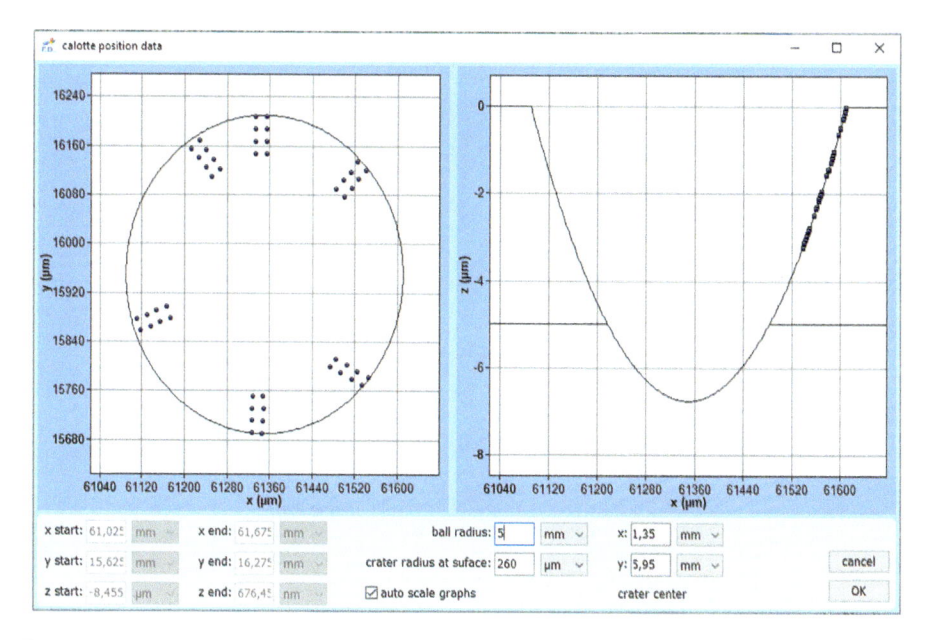

Figure 13.8 Indent positions with the Calotte ball crater wear scar. The indent positions extend from the top edge down through the coating to the substrate, the additional indents at each depth location in this example were made around the circumference.
Source: Courtesy DFG Consortium Project SPP2013 http://www.spp2013.tum.de.

the curves previously identified as outliers. Perhaps they suffered tip sliding on high roughness peaks or landed on a porous or weakened part of the coating. The profile plots place the coating surface as the left side of the chart (zero x value), and the right side represents the coating − substrate interface.

In this particular coating system, the deposited coating material composition remained constant; however, the process engineer adjusted the deposition recipe parameters to control the stress profile of the coating. The profile plots show when the yield strength profile was constant, so was the stress profile. When the recipe was adjusted to increase the stress during the coating growth (substrate is the right side of plots), the yield strength decreased in unison. When the recipe was adjusted to increase the compressive stress of the coating toward the substrate, the yield strength reduced toward the substrate.

Interestingly, it shows that the biaxial stress changes inversely in relation to the yield strength as a function of depth when comparing Figs. 13.9 and 13.10B and C plots. If designing for load-carrying capability, the increase in compressive stress (Fig. 13.10C) leading to a loss in yield strength (Fig. 13.9C) is completely unexpected. Potentially no other approach could have provided such depth of understanding in a low-cost industrial lab capable method. This allows coating engineers to better understand how the recipe process controls impact important coating properties relevant to performance and quality.

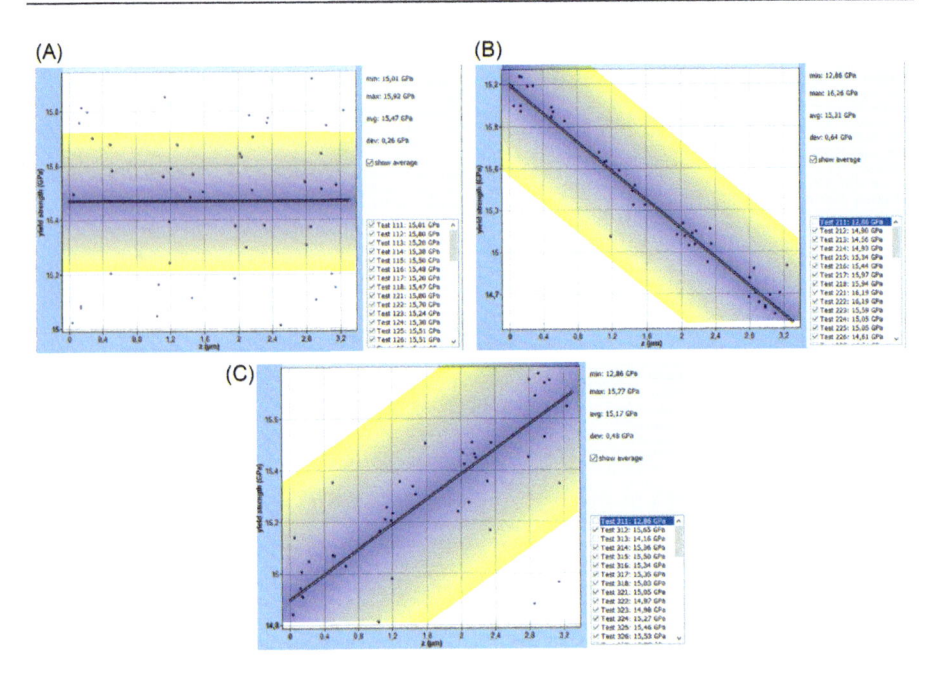

Figure 13.9 Cathodic arc deposited coating showing the Yield strength profile adjustment by via process deposition change. A constant stress recipe (A), followed by (B) a recipe with increasing compressive stress during deposition and then (C) a recipe with decreasing stress during deposition.

13.2.4 Stress profiling of a DLC coating on A2 tool steel

The commercial coating analyzed in this example is provided as a generic diamond-like carbon (DLC) solution for a variety of different substrates and geometries. This has led to many commercially successful applications. However, some customer applications with different requirements led to the coating suddenly failing under certain loading scenarios that lead to coating failure.

To understand the occasional adhesion failure problem, the coatings recipe was repeated on an A2 tool steel substrate sample where both success and failure had been experienced for the same coating thickness and process recipe but under different customer required test conditions. No information was provided about the interface, although it was later discovered that a doped interface was used to promote adhesion of the carbon coating to the steel substrate.

- Average stress measurements by the customer had been done in the past by the curvature method with Stoney's equation showing the coating to have an average compressive stress in the range of −0.5 GPa.
- Produced by a PECVD method, the DLC coating depending on the part geometry and customer requirements would have a film thickness ranging from several tens of nanometers up to several tens of microns.

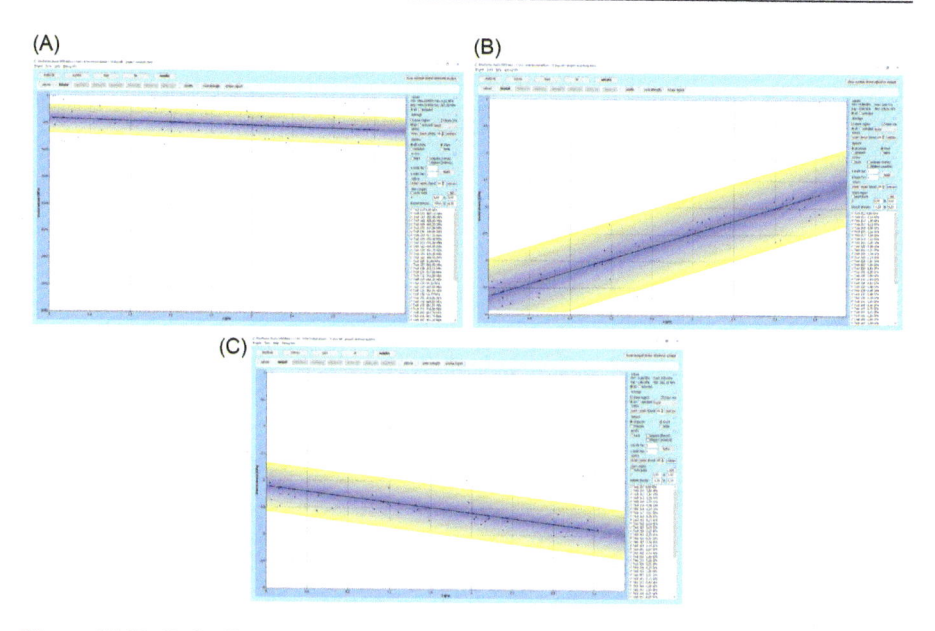

Figure 13.10 Cathodic arc deposited coating showing von Mises stress profile adjustment by via process change. A constant stress recipe (A), followed by (B) a recipe with increasing compressive stress during deposition and then (C) a recipe with decreasing stress during deposition.

The Calo wear scar was made in the DLC coating and an array of indentations made within it for mechanical property profile analysis. Fig. 13.11 shows the first steps to bring the measured data into the stress profile analysis software. Previously measured coating properties from indentation measurements were used as starting points to enter in the material input parameters, then the indentation curves were imported and analyzed for outliers. The position of the indents within the crater was checked, and further analysis of the indentation curves was made. The calculation of each curve takes place including all errors and uncertainties. If uncertainties are too high, a traffic light system stops the calculation at the suspect load − depth curve for guidance on continuing or canceling that dataset.

The results are then presented for the hardness, yield strength, reduced modulus and biaxial stress profiles with option to calculate the shear stress profile. In the case of the DLC coating, the biaxial stress profile was very informative (see Fig. 13.12). The yield strength profile is not shown, but it was inverse to the stress profile with the coating becoming decreasingly lower in yield strength toward the interface as the stress went increasingly tensile toward the interface. For most of the coating thickness, the stress was constant (at the given -0.5 GPa as the reference value). Contact conditions for a given contact size and load may place the maximum critical load within this compressive stress region avoiding putting critical load in the weakened interface region. However, if an application places the critical load within the last few microns, it would be reasonable to assume that a

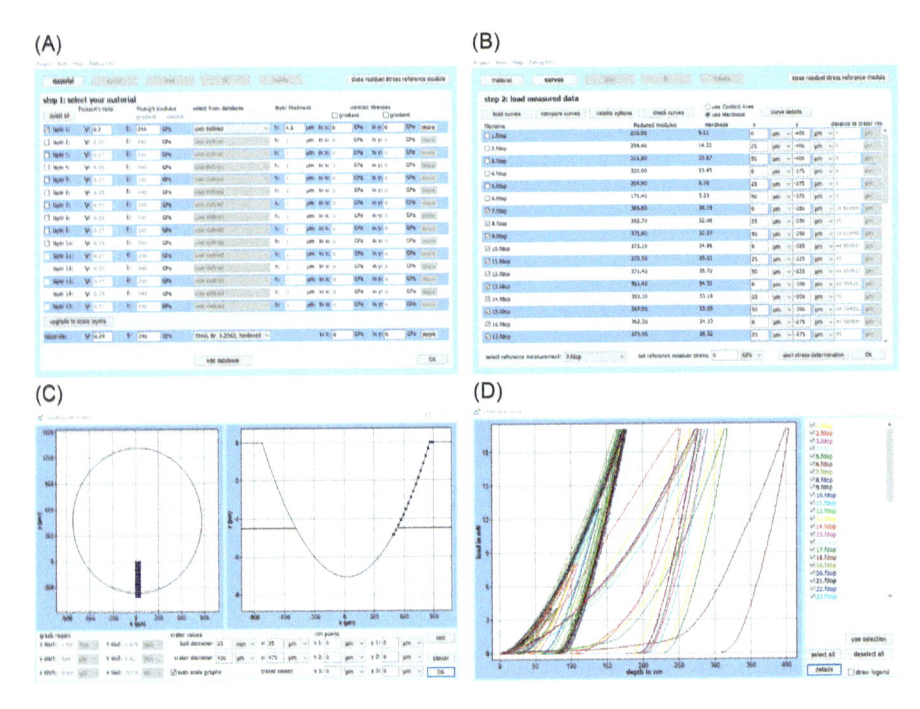

Figure 13.11 Material input page and the curve import page to analyze stress profiles. Importing the load-depth curves from the indentation equipment and the measured mechanical property and thickness of coating and substrate.

sudden failure may occur. The use of the indentation and scratch simulation modules provides a method to explore what the critical contact conditions might be for this coating system for that thickness and substrate combination.

In other circumstances, manufacturing process variations due to preparation, process variation, or batch densities, combined with different substrate material coefficients of thermal expansion and part geometry, may lead to exceeding the yield strength of the coating where it is under tensile load near the interface. As a consequence, coating failure straight after the deposition process might be expected. By being able to measure and see the stress and yield strength profiles, it is much more instructive for the coating designer to solve coating failure problems.

iStress-Module (2014) was developed to find the optimum stress profiles to minimize critical stress situations in the layer stack. The software can start from an assumed stress-free layer stack (if the stress and/or its profile is not yet known) and calculate the ideal stress profile to maximize performance. If a measured stress profile is known, then this can be used to compare with the ideal or used as a starting point for further optimization.

Using FilmDoctor Studio (2006), the critical load conditions can be found that place the interface under a critical stress matching the adhesive failure observed by the application as shown in Fig. 13.13 with the black dots marking the maximum

Analytical methods - applied 359

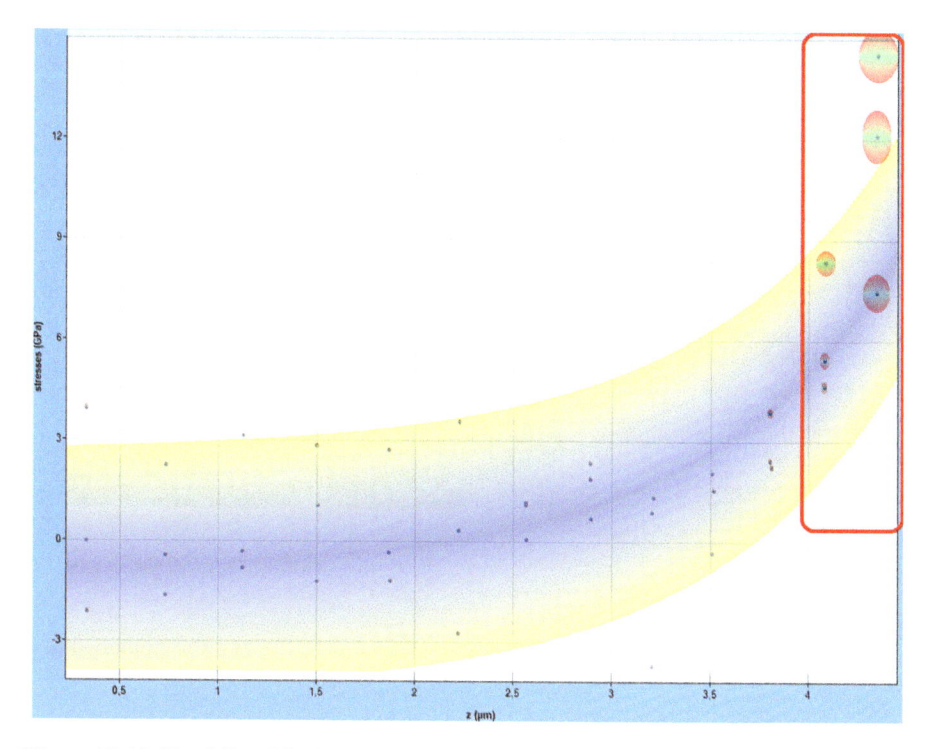

Figure 13.12 Von Mises biaxial stress profile plot. Stress profile plot of DLC coating on A2 tool steel with coating surface on left and substrate on right.
Source: Courtesy Troy vom Braucke at GP Plasma LLC.

von Mises stress locations. The region closest to the substrate acts as a weakened interface only when sufficiently loaded.

The stress profile can then be optimized with (iStress-Module, 2014) for coating designers to adjust the process recipe to achieve a stress profile as close as possible to the simulation. Alternatively, if achieving the calculated stress profile is not possible through the deposition technology, the critical load optimization module in FilmDoctor Studio, 2006 can be used to find a suitable range of property values (i.e., Young's Modulus) to select alternative materials.

13.2.5 Impact simulation on fused silica

Particle impacts have been a challenging topic to analyze and simulate; however, the work of Beake et al. (2024) provides insightful experimental results and simulation is now catching up. Simulation and imaging using a spherical tipped indenter on fused silica showed cone cracking with some chipping, and the imaging under green light revealed subsurface cracks. Green light also showed a vertical line below the surface between the impacts in Fig. 13.14A due to the subsurface interaction between adjacent cone crack systems. Different materials have different

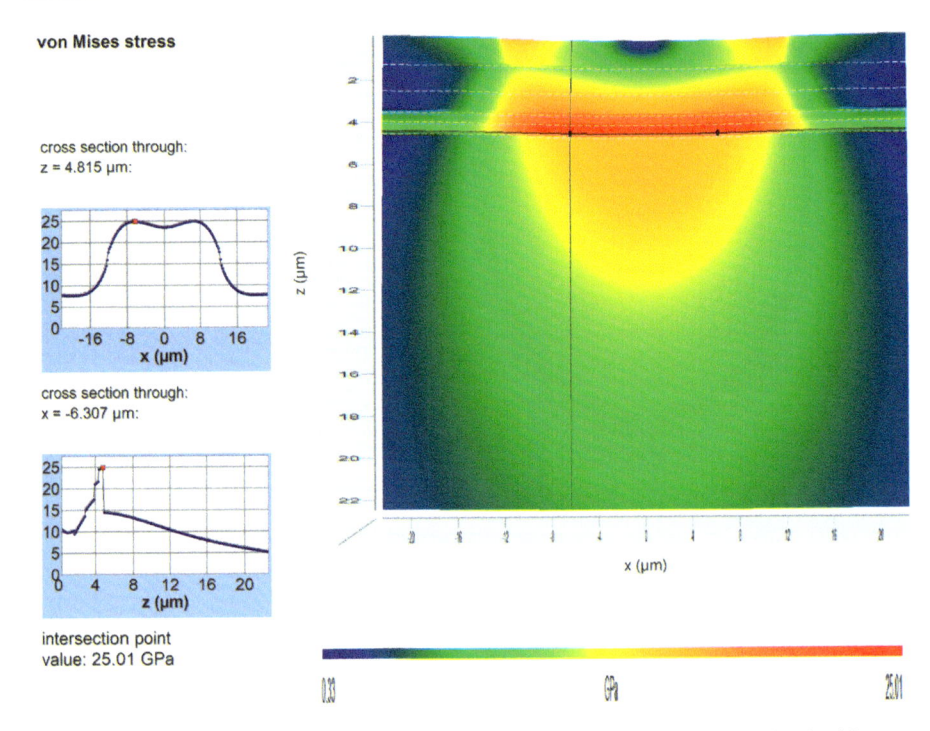

360   Nanomechanics for Coatings and Engineering Surfaces

Figure 13.13 Von Mises stress plot of a DLC coating subject to a contact load with a rotational force 10% of the normal force. The DLC coating property profiles are replicated and analyzed within FilmDoctor to determine the critical load conditions leading to adhesive failure at the interface based on the von Mises stress.
Source: Courtesy GP Plasma.

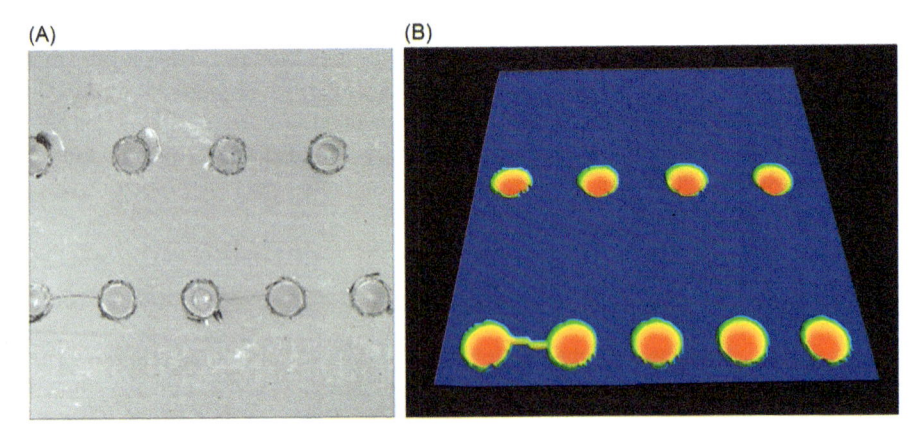

Figure 13.14 Laser scanning confocal microscopy (LSCM) imaging on fused silica compared to impact simulation with FilmDoctor. Critical separation distance exists below which surface cracks between adjacent impacts are observed, with the same behavior shown in the simulation for similar material properties and impact conditions.

Analytical methods - applied 361

"critical interaction thresholds" where damage generated from one impact affects the damage accumulation and subsequent erosion rate.

The image shown in Fig. 13.14A shows a critical separation distance exists below which surface cracks between adjacent impacts were seen, which was 60 μm for fused silica under these impact conditions (bottom row). No cracks were formed when the distance was 70 μm (top row). Fig. 13.14B shows when the same material properties, impact conditions, and spacing are replicated in the FilmDoctor impact module, there is a correlation for the critical separation distance where the surface cracks between adjacent impacts at 60 μm (bottom row). Notably, damage accumulation under repeated impacts can also be incorporated and is briefly discussed in the next section.

13.3 Thought experiments

13.3.1 Impact simulation - mixed droplet and particle impact

A simulation of an impact scenario with multiple impacts hitting the wall of a nuclear fusion device was undertaken as a thought experiment and animated [https://youtu.be/eBWfikGxhuQ]. The simulation considered a base material for a fusion chamber wall Fig. 13.15A, then compared that same material with a protective coating (Fig. 13.15B) and an optimized stress profile using iStress-Module

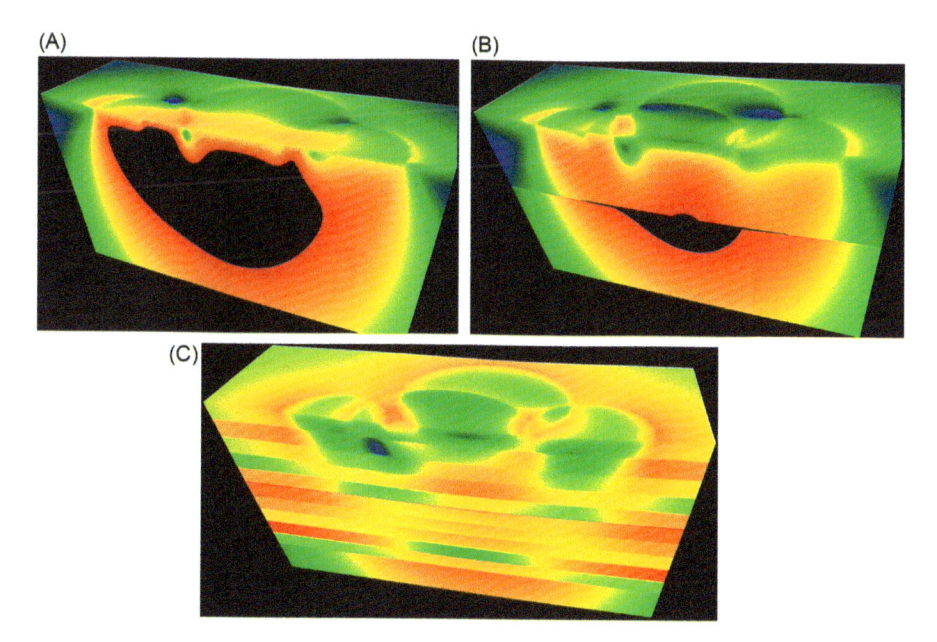

Figure 13.15 Nuclear fusion wall optimization. Multiple solid particle impacts on a fusion chamber wall, comparing uncoated, coated, and coating + stress optimization.

(2014) as shown in Fig. 13.15C. However, the stress and strain evolution of impacting droplets was not considered. However, taking the hot environment regarding such an application, we know that an extension to liquid droplet impact is of essential need.

A mix of particle and droplet impact testing was simulated as a thought experiment shown in Fig. 13.16A to understand the interaction between solid particle-induced stress and the adhesive forces generated from droplet impact. Fig. 13.16B describes the upward adhesive force ranking around the perimeter of the droplet impact zones, while the simulation software quantitively calculates the major properties of interest.

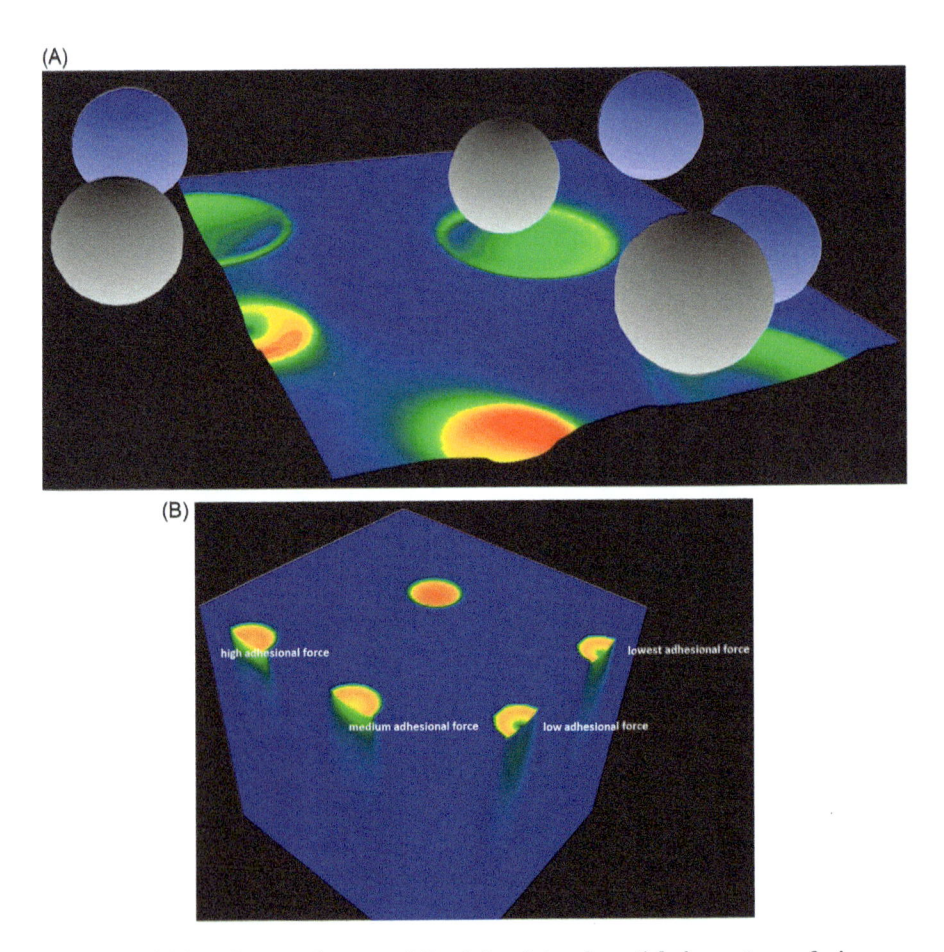

Figure 13.16 Thought experiment — Mixed droplet and particle impact on a fusion reactor wall. The stress and strain evolution of impacting droplets together with particles can be analyzed as a thought experiment to find the failure modes of the coating system. This helps determine research and development paths for promising solutions before spending money on trial and error.

Taking into consideration the effects of damage accumulation with data taken from indentation measurements, the impact module can also be extended to show the damage accumulation within the simulation after each impact Fig. 13.17.

13.3.2 Graphene simulation

Not having measurement data available to explore a material application does not need to stop an exploratory study. By using literature values for single-layer graphene, different "thought" experiments can be run by treating the Pi and Sigma carbon bonds as continuous fields (see Fig. 13.18). This provides opportunities for assessment of interface adhesion and crinkling behavior, dopants, and charge buildup (simulating the charged ions as impacts with corresponding energies and contact size).

This provides opportunities to narrow down a design search field of interest and then using measurement data to better inform the model. Importing AFM data can be one way to analyze the experimental data with the software, rather than rely on literature values used during the thought experiment search phase. Then the effects of localized defects and dopants on the stress and strain can be explored including with inclined or lateral loading to measure the response of the system in different loading directions.

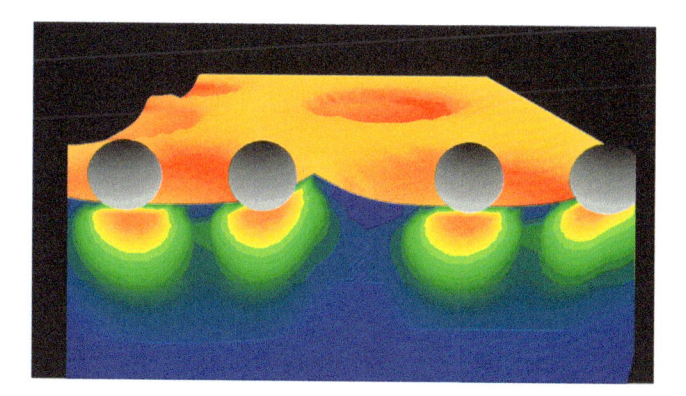

Figure 13.17 Incorporation of damage accumulation for multiple impacts of different sizes and positions. Normally this type of simulation is done with finite element methods, however, using the analytical approach the simulation runs using measured property data as input to aim for as close a match to experiment as possible, limited by the resolution of the measurements made at scales relevant to contact conditions and failure modes. With expert training on the model usage, localized defects can also be incorporated to simulate inhomogeneous materials.
Source: Courtesy Nick Bierwisch at Siomec.

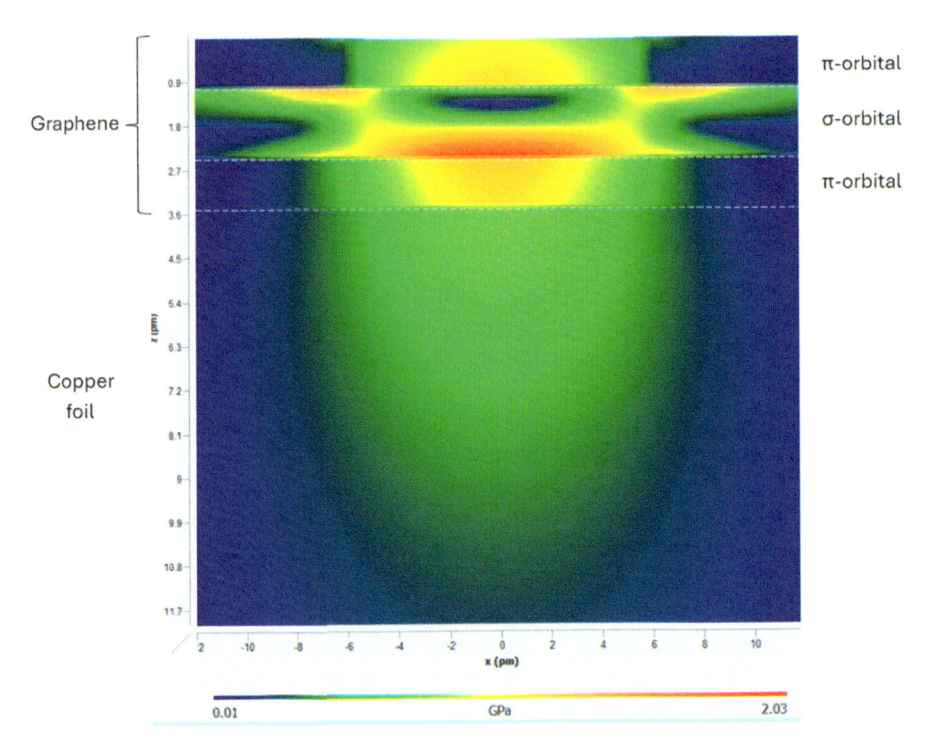

Figure 13.18 Thought experiment – Simulating graphene bonds as continuous fields – going beyond the resolution of nanoindentation, an opportunity for AFM? The von Mises stress distribution for a single sheet of graphene on a copper substrate under contact loading. By simulating each of the Graphene bonds as continuous fields and using literature values for mechanical properties and bond layer thickness, the Pi and Sigma orbital bonds can be represented under a variety of contact loading scenarios such as ion impact or larger contact loads.
Source: Courtesy GP Plasma from: https://gpplasma.com/blog/rapidly-optimize-2d-material-properties-and-become-uncatchable.

13.4 Outlook

- The stress profile measurement method described only allows the stress to be calculated as a profile relative to a datum point such as the coating surface. If the total average stress is known by other measurement methods, then the position of the stress profile can be adjusted accordingly. However, it is possible to measure absolute values of stress for a stress profile. This can be achieved by inclined indentation to measure the properties of the material in addition to normal axis, a more time-efficient method is to use indentations equipment with lateral loading capability. This type of instrument places a lateral load on the indenter tip during the hold cycle of the normal loading. The ability to measure a local Poisson's ratio allows a reduction of the unknowns to calculate the absolute stress at each measured position.

Analytical methods - applied

- Coating designers can use the various modules in commercial analytical packages such as FilmDoctor to also explore nanofretting, scratch, and tribological wear testing as a function of the stress profile. By making these measurements within the Calotte wear crater, the designer can discover the coating performance response for different wear or erosion conditions. Further information can be found in Schwarzer and Schwarzer (2023a, 2023b).
- The analytical digital twin simulation approach can overcome the ISO14577-4 standard's limitations when it comes to handling multilayered coatings or mechanically graded coatings. The ISO standard is based on monolithic homogeneous half-space solutions; therefore, it must fail in all layered systems and can only provide an approximation of the real results for these. This holds for all mechanical tests on layered coating systems. The international standard tries to give you hints, but it can't always steer you in the right direction because it is based on a simplified model that is using approximations. Therefore, opportunity exists to adapt international standards to provide for test setup optimized for the coating system at hand. For example, the Test Optimizer (FilmDoctor Studio, 2006) built from this analytical approach may be able to design a proper test to evaluate an ensemble of loads more effectively than by current ISO14577-4 approaches (Fig. 13.19).
- The graphene example together with renewed interest in novel properties of 2D materials or single atomic layer films are an opportunity to leverage AFM instrumentation data with all the capabilities described in this chapter. The analytical method can allow improved measurement capabilities on optically thin films or MEMs devices to find critical stress values. In addition, weakened interfaces can be explored with higher confidence.

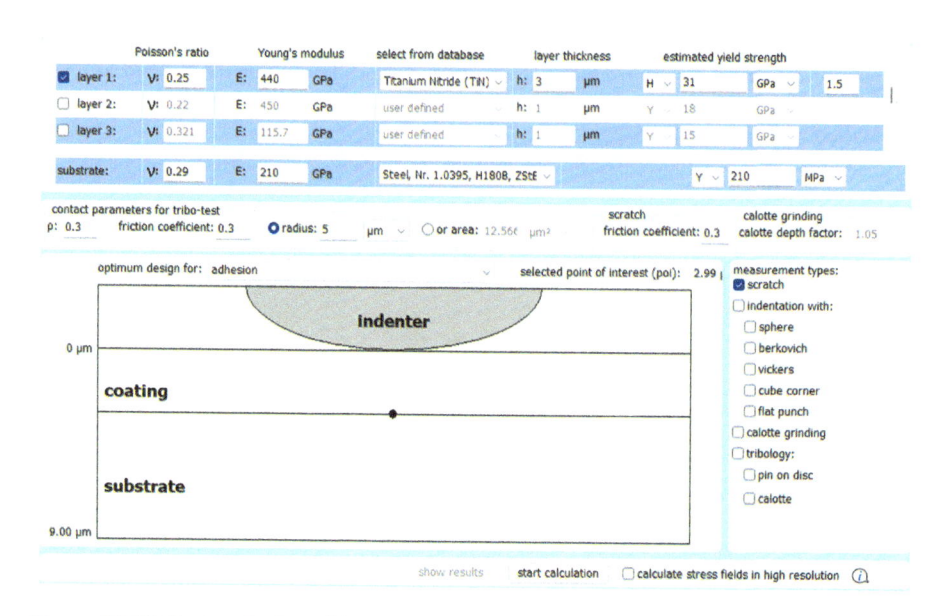

Figure 13.19 Test optimizer allows a user to find the optimum parameters for instrumented scratch or hardness testing. By selecting the Adhesion option, the point of interest is placed at the coating − substrate interface.
Source: Courtesy Siomec.

- Seeing the critical stress in a coating along with the quantified values helps designers understand and improve product development. Whereas the current visualizations cut though the center of the contact zone to see the maximum von Mises stress location, recent developments of the mathematics and graphics software have led to tomographic digital twin visualization. Just like in airport scanners, MRI, and ultrasonic scanners, the data can now be visualized in 3D with adjustable transparent views to see critical regions interacting between multiple contact conditions (see Fig. 13.20). This is particularly useful when there are multiple contact zones or internal defects leading to several critical stress regions that cannot all be sliced through in a single section view. The different regions can be explored by the user with "X-Ray like vision" to see the different stress field interactions.
- With geopolitical risks factors and potential supply chain disruptions, there may come a time when certain valuable materials and rare earths may be in short supply. Tools to find material combinations to replace scarce materials need brute force methods to run through many combinations to find potential candidates. Molecular dynamics (MD), boundary element (BE), finite element (FE) methods are too computationally slow and costly to throw at the problem. In optimization processes or parameter identification tasks, they rely on half empirical sensitivity analysis, but the errors can become too high very quickly—MD ~5% starting error (in dependence on what truncation or tight binding methods are applied), FE/BE ~3% starting error. These methods may not achieve the precision needed and are not invertible for reverse optimization. However, with an analytical solution, it is possible to reverse optimize to find alternative material combinations and structures needed to achieve the desired properties of the now scarce material resource. This provides both supply chain mitigation and potentially lower cost earth abundant alternatives that not only can solve the supply chain problems but may also be applied to find more environmentally friendly material solutions. An exploration of the topic was reported here (vom Braucke, 2023) and the mathematical approach detailed in Schwarzer (2023).

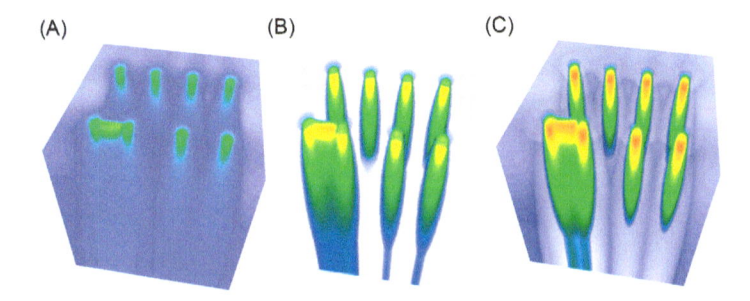

Figure 13.20 Tomographic view of multiple contact loading simulation on Silica. Seeing potential hidden failure modes. When multiple contact conditions occur or there may be contact conditions interacting with localized defects, the tomographic view can make the outer regions transparent to see the stress field interactions within to find the critical regions that are otherwise hidden. The low stress regions around critical stress zones can be made transparent, allowing a more detailed internal inspection of the critical areas.

References

Abadias, G., Chason, E., Keckes, J., Sebastiani, M., Thompson, G. B., Barthel, E., Doll, G. L., Murray, C. E., Stoessel, C. H., & Martinu, L. (2018). Review article: Stress in thin films and coatings: Current status, challenges, and prospects. *Journal of Vacuum Science & Technology A, 36*. Available from https://doi.org/10.1116/1.5011790.

Beake, B. D., Goodes, S. R., Zhang, H., Luis Isern, L., Chalk, C., Nicholls, J. R., & Gee, M. G. (2024). Randomised nano-/micro- impact testing — A novel experimental test method to simulate erosive damage caused by solid particle impacts. *Tribology International, 195*, 109647.

FilmDoctor Studio. (2006). Siomec. Available from https://siomec.com/software/filmdoctor-studio/.

Fischer-Cripps, A. C. (1997). Predicting Hertzian fracture. *Journal of Materials Science, 32*, 1277–1285.

Marthaler, D. E. (2013). An overview of mathematical methods for numerical optimization. *Topics in Applied Physics, 127*, 31–53. Available from https://doi.org/10.1007/978-94-007-6664-8_2, 14370859 03034216.

iStress-Module, (2014), Siomec. https://siomec.com/software/modules/istress/.

Schwarzer, J. (2022). YouTube — Working with SSA in FilmDoctor Studio — Part I. https://youtu.be/KAvQbdDVKRo?.

Schwarzer, J., & Schwarzer F. (2023a). YouTube — How to get the depth profile of a complex coating system. https://youtu.be/EmXfSFiwdVE?.

Schwarzer, J., & Schwarzer F. (2023b). YouTube — Scratch test analysis with the Calotte Module of FilmDoctor. https://youtu.be/sn82LOGMOao?.

Schwarzer, N. (2014). Completely Analytical Tools for the Next Generation of Surface and Coating Optimization. *Coatings, 4*(2). Available from https://doi.org/10.3390/coatings4020253.

Schwarzer, N. (2015). From Hertz via Higgs to a Paradox Failure Mechanism. *SVC Bulletin*, 46–49, Fall issue. Available from https://www.flipsnack.com/svcdigitalpublications/2015-fall-winter-bulletin.html.

Schwarzer, N. (2017). Scale invariant mechanical surface optimization applying analytical time dependent contact mechanics for layered structures. In A. Tiwari, & S. Natarajan (Eds.), *Applied nanoindentation in advanced materials*. Wiley. Available from https://doi.org/10.1002/9781119084501.ch22.

Schwarzer, N. (2022). *The World Formula: A Late Recognition of David Hilbert 's Stroke of Genius*. Jenny Stanford Publishing Available from. Available from https://www.routledge.com/The-World-Formula-A-Late-Recognition-of-David-Hilberts-Stroke-of-Genius/Schwarzer/p/book/9789814877206.

Schwarzer N. 2023 11 23 ASIN: B0CNY7TRH2 Amazon USA Unpublished content Scarce Material Replacements: Project Opportunity to Mitigate Supply Chain Risks, Lower Costs and Accelerate Product Development Timelines https://www.amazon.com/Scarce-Material-Replacements-Opportunity-Development-ebook/dp/B0CNY7TRH2?ref_ = ast_author_dp.

Test Optimizer, (2013), Siomec, https://siomcc.com/software/modules/testoptimizer/.

vom Braucke T. 2023 10 5 News Blog GP Plasma Finding Scarce Material Performance Alternatives to Mitigate Supply Chain Risk and Lower Costs https://gpplasma.com/blog/finding-rare-earth-material-performance-alternatives-to-mitigate-supply-chain-risk-and-lower-costs.

Numerical simulation and finite element analysis

14

Roberto Martins Souza and Newton Kiyoshi Fukumasu
Surface Phenomena Laboratory, Polytechnic School, Universidade de Sao Paulo, Ao Paulo, Brazil

14.1 Introduction

The performance of coatings and engineering surfaces may be significantly affected by the system mechanical behavior, for instance, regarding the response to contact loads typical of tribological situations (Holmberg & Matthews, 2009). However, even in other applications, such as thermal (Wang et al., 2016), magnetic (Sander et al., 2009) or optical (Criado et al., 2006; Praud et al., 2021), mechanical factors may be relevant. Along these lines, independent of the application, system performance may be affected by coating stiffness; excessive tensile stresses may result in coating fracture (Fukumasu et al., 2010; Marx et al., 2015), and excessive compressive residual stresses during processing may have a negative impact on the adhesion of a coating to a substrate (Abadias et al., 2018).

Without neglecting the importance of experimental approaches, several reasons may be presented to justify the use of numerical tools in the analysis of the mechanical behavior of coatings and engineering surfaces. One of these reasons refers to the possibility of the numerical tools to provide data not readily available experimentally, such as the distribution of stresses and strains during a given mechanical loading. Besides, a series of numerical simulations may cover a wide range of input parameters at a time (and a cost) usually significantly lower than those of the experimental procedures. In fact, the range of input parameters may include conditions not yet available experimentally, which may guide system manufacturing toward the development of a given set of mechanical and physical parameters to provide a tailored performance.

The task of presenting a global analysis on the use of numerical tools to study coatings and engineering surfaces is difficult, mainly due to the number of situations that this analysis would have to cover. Starting with the surface modification itself, the literature presents several examples on the use of numerical simulation to study coating deposition (Dollet, 2004; Liu et al., 2021; Smy et al., 1997). Additionally, a broad range of coating applications may be modeled, considering different classes of materials (ceramics, metals, polymer, and composites) and several techniques available for the deposition. Furthermore, different modeling techniques may be used, such as the finite element method (FEM) and molecular dynamics (MD), which frequently correspond to results at different scales.

Nanomechanics for Coatings and Engineering Surfaces. DOI: https://doi.org/10.1016/B978-0-443-13334-3.00014-0
Copyright © 2025 Elsevier Inc. All rights are reserved, including those for text and data mining, AI training, and similar technologies.

Considering this extensive range of options, this chapter will concentrate on contact loads, which may be key factors to analyze system behavior, either as the cause of mechanical failure or as part of the techniques to measure elastic, plastic, and fracture properties. In terms of surface engineering, an emphasis will be placed on hard thin films deposited by physical vapor deposition (PVD) or chemical vapor deposition (CVD) processes.

14.2 Contact stresses

The interest in the modeling of contact stresses is not new. In the description of the fundamentals of contact problems, it is usual to mention the 19th-century approaches of Hertz and Boussinesq (Johnson, 1985), which cover the elastic contact of two nonconforming surfaces and the situation where an elastic half-space is deformed by a rigid punch of arbitrary shape, respectively. The importance of these pioneer works may be noticed, for example, during the relative movement of a sphere with respect to a plane, in a condition where the maximum contact pressure calculated based on Hertz is frequently given in the literature during the description of the experimental procedure. In many of these cases, the contact conditions invalidate assumptions of the Hertz theory, for example, in the presence of nonnegligible plastic deformation or in conditions with tangential loads due to sliding. Thus, the actual contact stresses may be significantly different from those predicted theoretically. Nevertheless, the calculation of maximum contact pressure using the Hertzian approach usually provides a good ranking on the potential severity of the contact.

Contact stresses also play an important role in surface engineering, among other reasons, due to the usual submillimeter (frequently micron or submicron) dimension of surface modifications. To evaluate mechanical properties in regions compatible with this dimension, extensive work has been carried out on instrumented indentation (International Organization for Standardization ISO, 2015). On this topic, classical works from the late 1980s (Doerner & Nix, 1986) and early 1990s (Oliver & Pharr, 1992) were analytical and directed to bulk materials. However, for the reason aforementioned, the technique was later also extensively applied to coated systems (Saha & Nix, 2002). In the beginning, most of the attention was directed to the calculation of hardness and elastic modulus, including the initial works mentioned before. However, over time, the use of the instrumented indentation also expanded to other measurements, such as fracture toughness (Li et al., 1997; Pharr, 1998; Sebastiani et al., 2015) and interfacial fracture toughness (Chicot et al., 1996; Sánchez et al., 1999).

Another evolution observed throughout the years was the incorporation of numerical tools to support the analysis of indentation data (Bolshakov et al., 1996; Han et al., 2022). One usual advantage over analytical methods is the need of less simplifying assumptions, for example, regarding the ability of methods such as FEM to consider plastic deformation. The possibility of analyzing more complex shapes is also a common advantage of numerical tools. However, this advantage may not be significant during the study of indentation problems, due to the geometrical symmetry observed for many indenters, in particular those with axial symmetry, such as spheres and cones.

Over time, numerical approaches have taken advantage of the increase in computational capabilities, and systems with more and more complexity could be simulated. Considering that the finite element method (FEM) is a numerical technique used for solving engineering problems, based on mathematical equations of physical behavior (Barsoum, 1971; Fries & Belytschko, 2010; Pian & Tong, 1969), this tool is particularly useful for problems with complex geometries, irregular boundaries, nonlinear material properties, and discrete system features, including multipoint contact dynamics, multiphysics dynamical coupling (including lubrication (Martinet & Chabrand, 2000) and thermal loads modeling (Seriacopi et al., 2019)), and inhomogeneous anisotropic materials (including coated systems, case hardened surfaces (Lau et al., 1989) and discrete microstructures with precipitates and different phases (Fukumasu et al., 2005; Fukumasu et al., 2017; Seriacopi et al., 2016)). Focusing on contact dynamics, the versatility of the FEM technique allows further understanding of multiscale systems, from atomistic-based surface interactions (Bortoleto et al., 2016; Marques et al., 2017) to full-scale mechanical dynamics (Fukumasu et al., 2016).

14.3 Thin film mechanics and substrate effects during indentation

14.3.1 Initial analyses of indentation problems

Besides its relevance for the measurement of mechanical properties, contact problems are also important to understand the overall mechanical behavior of coated systems. The most simple condition in this case considers a system composed of a single coating layer and a substrate under the action of a normal load. Fig. 14.1 (Souza, 1999) presents a schematic of this configuration, in which $r = 0$ corresponds to a line of axisymmetry. Many works published in the 1990s considered this configuration or similar ones (Djabella & Arnell, 1992; Komvopoulos, 1989; Montmitonnet et al., 1993). In the same decade, additional complexity was incorporated with the consideration of multilayer coatings (Djabella & Arnell, 1993b), and in some cases, a tangential load was also imposed, but without considering a lateral displacement of the indenter (Djabella & Arnell, 1993a).

The condition with one coating layer under a normal load provides valuable insights to understand the situation with more complex coating architectures and loadings. In general, the overall response depends on how the properties of the film relate to those of the substrate, in terms of stiffness and hardness. As an example, it is possible to mention a system with a hard elastic film onto a compliant substrate. In this case, as the load increases, the properties of the substrate become more relevant in determining the amount of deformation. For sufficiently high normal loads

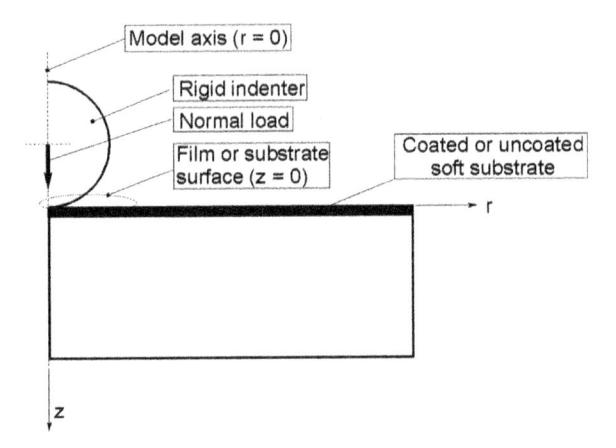

Figure 14.1 Axisymmetric representation of the condition of a spherical indenter applying normal loads onto a system with one coating and a substrate (Souza, 1999).

(P_{max}), and considering perfect film − substrate adhesion, it is possible to consider that the substrate is deformed based on an imposed pressure distribution, and the film, which needs to follow the deformation of the substrate, is deformed based on an imposed displacement.

For this condition of the indentation of a system with a hard film and a compliant substrate, Fig. 14.2 (Souza, 1999) indicates the shape that two hard films (0.6 and 4.6 μm thick) assume after the 50 N indentation with a sphere 1.59 mm in diameter. For this particular situation, Fig. 14.3 presents the distribution of radial stresses (σ_r) along the surface (Fig. 14.3A) and along the film side of the film/substrate interface (Fig. 14.3B) at the maximum value of normal load, that is, before unloading (Souza, 1999). The peak in tensile stresses observed at $r/a_{os} \cong 1.8$, which represents the border of the contact of the sphere with the film, is, in many cases, responsible for the propagation of circular cracks in the film, such as those presented in Fig. 14.4 (Souza, Sinatora, et al., 2001). The situation at the axis of symmetry, $r/a_{os} = 0$, is slightly more complicated. Considering only the "geometry of the deformation," stresses at the top of the film should be compressive, which is not verified in Fig. 14.3A. This result may be explained based on the coefficient of friction between the film and the sphere, which, if low, allows the film to slide with respect to the ball, leading to stretching tensile stresses in the film. Still at the axis of symmetry, the stresses at the interface are even more tensile, since the tensile radial stresses from bending are superimposed to those from stretching. In principle, the stress distribution in Fig. 14.3B should lead to the propagation of cracks that nucleate at the interface and propagate toward the surface. This type of crack is difficult to observe in practice, due to the usual opaque nature of the hard films. However, this trend of cracking was predicted numerically (Abdul-Baqi & Van der Giessen, 2002; Kot et al., 2013) and even observed experimentally (Kot et al., 2013). Note that the earlier analysis does not consider another important source of coating stresses, the residual one, which may reach values compatible with the

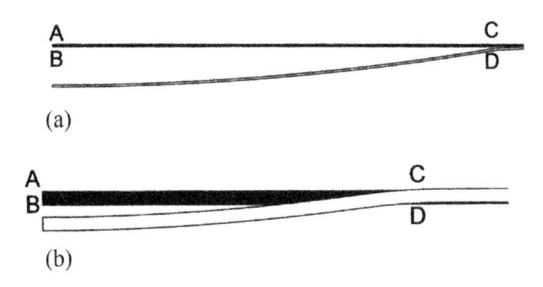

Figure 14.2 FEM results obtained during the 50 N normal indentation of a system with an elastic film and an elastic − plastic substrate. Original (black) and displaced (white) meshes of films with: (a) Thickness $t = 0.6$ μm and (b) Thickness $t = 4.6$ μm (Souza, 1999).

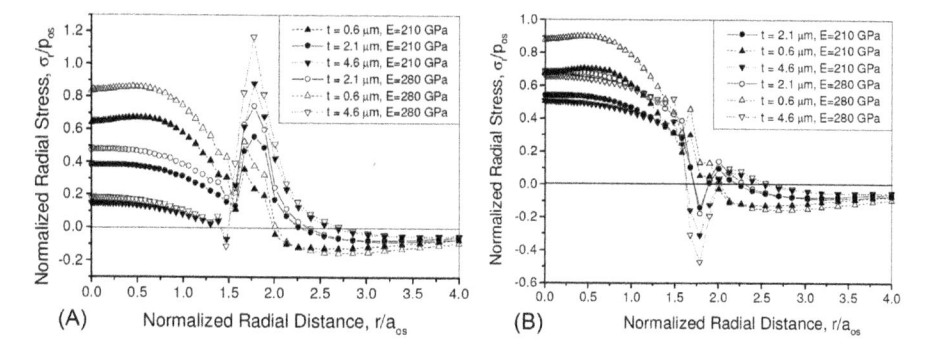

Figure 14.3 Radial stresses (σ σ_r) calculated at the maximum load of 50 N normal indentation of systems with different elastic films and an elastic − plastic substrate. Results for different values of film elastic modulus (E) and film thickness (t): (A) Film surface and (B) film side of the film/substrate interface (Souza, 1999).

contact stresses and are superimposed to them (Holmberg & Ronkainen, Laukkanen, et al., 2009).

The morphology of the cracks observed at the surface also depends on the properties of the substrate. Depending on the indentation conditions, which include geometry, loads, and properties, the evolution of normal load may be associated with a change in the morphology of cracks, for example, as presented in Fig. 14.5. In this case, FEM simulations by Pachler et al. (2007) revealed that the probable cause for the change in crack orientation was the largest normal stress, which changed from being the radial stress (σ_r) at lower loads to become the circular (hoop) stress (σ_θ) during unloading (Fig. 14.6).

Since bending stresses near the contact edge are important regarding cracking in the film (cohesive failure), and considering that the normal load is high enough to induce normal displacements larger than the film thickness, at least at the line of symmetry, system fracture behavior may depend on how the mechanical properties of the substrate affect its deformation. Under indentation, properties such as the

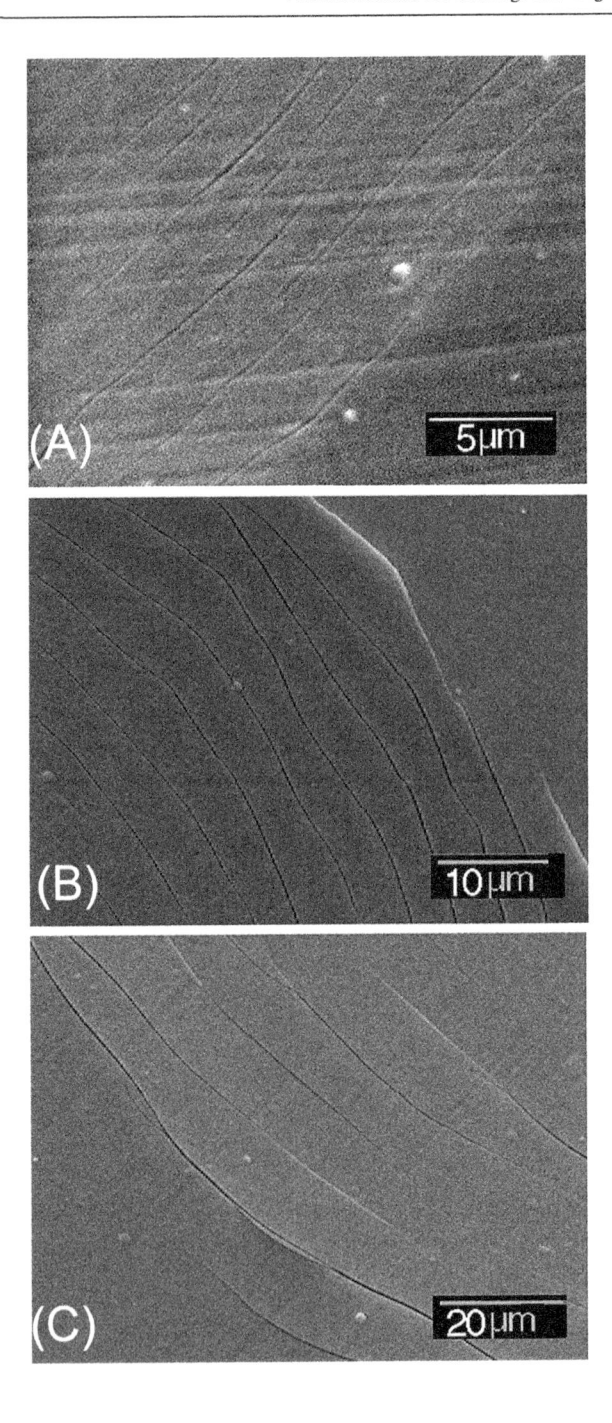

(Continued)

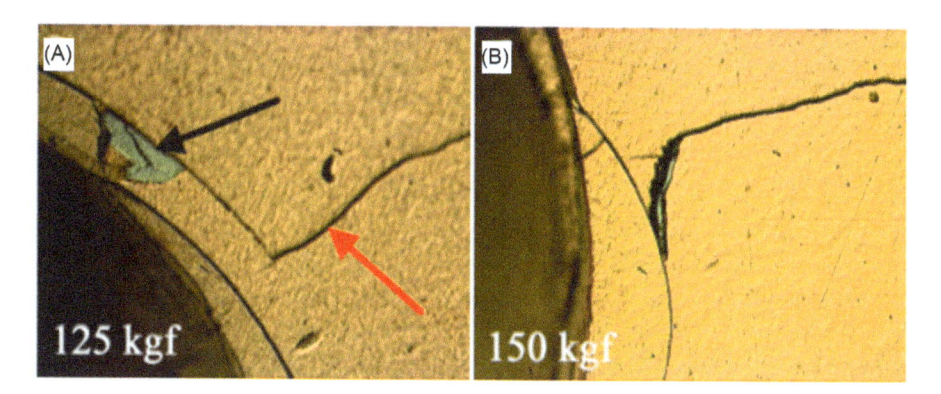

Figure 14.5 Edge of Rockwell C indentations conducted with different loads on a duplex coating (6-h nitriding followed by Titanium nitride coating) onto H13 tool steel. The red arrow indicates a radial crack and the black arrow indicates a film delamination. Loads of: (A) 1226 N (125 kgf) and (B) 1471 N (150 kgf) (Franco Júnior, 2003).

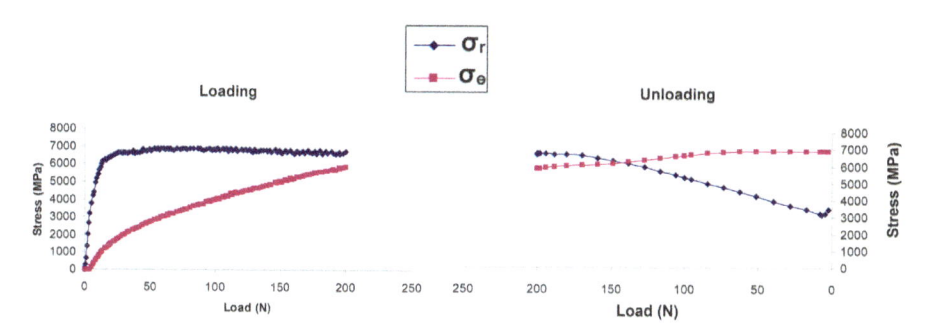

Figure 14.6 Finite element modeling results of the evolution of radial stresses (σ σ_r) and circular stresses (σ $\sigma\theta$) as a function of load during the loading (left) and unloading (right) portion of Rockwell C indentation cycle of a system with an elastic film and an elastic-plastic substrate.

Figure 14.4 Scanning electron microscopy analysis of the circular cracks observed after the 50 N indentation on samples 6061 aluminum substrates coated with silicon carbide (SiC): (A) Film with deposition time of 2 h; (B) film with deposition time of 7 h; (C) film with deposition time of 14 h
Source: From Souza, R. M., Sinatora, A., Mustoe, G. G. W., & Moore, J. J. (2001). Numerical and experimental study of the circular cracks observed at the contact edges of the indentations of coated systems with soft substrates. *Wear, 251*(1), 1337–1346. https://doi.org/10.1016/S0043-1648(01)00778-5.

strain hardening exponent define if the substrate will present pileup or sink-in morphologies (Casals & Alcalá, 2005). Considering these two options, sink-in tends to impose less bending than pileup, and corresponding larger radial stresses are obtained with higher pileups than with lower ones (Pérez R & Souza, 2004).

14.3.2 Effect of substrate anisotropy on coating behavior

Another possibility of controlling system failure based on substrate properties involves anisotropic behavior. The complexity of reproducing material characteristics, such as microstructure and time evolving mechanical properties, tends to promote the assumption of homogeneous and isotropic material properties. However, regular production processes, which include rolling, extrusion, forging, directional solidification, additive manufacturing, shot peening, laser and/or plasma surface treatments, may result in metallic substrates with anisotropic properties. Different methodologies, including analytical modeling (Hill, 1950) and finite element modeling (Fukumasu & Souza, 2006), can be applied to account for the influence of anisotropic properties on system behavior. Focusing on metallic substrates, Hill (1950) developed the theory of anisotropy in yielding that modifies the von Mises equation of equivalent stresses by considering correction factors (F, G, H, L, M, and N) for the anisotropic case. These factors are related to Hill coefficients (R_{xx}, R_{yy}, R_{zz}, R_{xy}, R_{xz}, and R_{yz}), each one associated with normal and shearing directions for a 3D solid system. When those coefficients are set to unity, the correction factors allow the restoration of the original von Mises equivalent stress equation for an isotropic material. When only one direction presents nonisotropic behavior, this may be defined as an orthotropic material, implying that only one Hill coefficient is different from unity, maintaining all other coefficients set to unity. This was the case evaluated by Fukumasu and Souza (2006), in which the substrate presented orthotropic behavior parallel to the loading direction, varying the constant R_{yy} between 0.707 and 3. The lower limit was calculated based on the restriction of a well-defined stress state of the system, in which the F, G, and H parameters must be positive. The upper limit was selected according to technological restrictions, including the possibility of obtaining a material that has yield strength many times larger in one direction than in the other two. For example, values of $R_{yy} > 1$ may be obtained by shot peening of a specimen with initially isotropic properties. Also, the influence of the orthotropic layer thickness (d), compared with the coating thickness, was evaluated, given the importance of the depth of this supporting zone.

Results showed that an increase in R_{yy} promotes a decrease in the ratio between residual penetration depth and the maximum penetration depth (h_f/h_{max}), which impacts the edge morphology of the indented region. Similar to Casals and Alcalá (2005), the change from pileup to sink-in formation was observed by Bolshakov and Pharr (1998), in this case as a result of the decrease in the ratio (E/σ_y). When $R_{yy} > 1$, the substrate plastic properties play an important role in terms of pileup height (h_p); however, for $R_{22} < 1$, film thickness becomes the dominant factor in terms of h_p height. The orthotropic layer thickness affected the minimum value of pileup height, which was found for the intermediate thickness value, and this influence was consistently lower in the case of thicker film.

One important feature of numerical simulations is the possibility to visualize the evolution of results in regions not easily observable in experimental analyses, which is the case for the time evolution of the spatial distribution of the plastic strain in a diametral cross section of a system. In this case, the analyses of such results indicated that, although varying the R_{yy} coefficient, the plastic deformation initiated at similar locations in the substrate. Exceptions were observed only for large values of orthotropic layer thickness ($d = 20\%$) and $R_{yy} > 2$, in which plastic deformation initiated at more than one point of the substrate. The numerical simulations allowed the association of lower values of h_p with lower peaks in radial stresses, which may lead to lower number of indentation circular cracks in systems that present a substrate layer with orthotropic properties and $R_{yy} > 2$, which represents an important technological processing result.

14.3.3 Simulation of cohesive coating failure

Initial simulations of the indentation of coated systems were able to identify regions prone for cohesive (film) or adhesive (coating/substrate interface) fracture, but did not incorporate actual crack propagation. This type of analysis was incorporated in the late 1990s, mainly using criteria available at that time in software packages such as ABAQUS, from HKS at that time and now from Dassault Systémes. One of the early initiatives in this direction included crack propagation based on the calculation of stress states at a given position ahead of the crack tip and comparison with a threshold stress value (Souza, Mustoe, et al., 1999). Some of the limitations of this approach included the need of having an initial crack in the original input model and the need of predefining a path for crack propagation. For indentation problems considering homogeneous coatings, the need of a defined crack path did not represent a serious restriction, since experimental analysis indicated that circular cohesive cracks, due to geometrical factors, would propagate almost entirely in a direction perpendicular to the coating/substrate interface. Despite the limitations, these analyses (Souza, Mustoe, et al., 2001) were able to provide an explanation for the formation of sets of circular cracks nucleating at the coating surface, such as those presented in Fig. 14.4.

To overcome the predefined crack limitation, improved numerical modeling was developed to include arbitrary nucleation sites based on the stress evolution of the system. One technique presenting good results is the eXtended Finite Element Method (XFEM), which is an extension of the traditional FEM that allows for the simulation of problems with complex and evolving geometries, such as cracks, interfaces, and discontinuities (Fries & Belytschko, 2010; Réthoré et al., 2005). The basis of XFEM lies in the enrichment of standard FEM with functions and degrees of freedom to capture numerical singularities associated with these features without the need for remeshing. The use of this technique in indentation analyses allows for the development of crack nucleation sites and propagation path as a direct function of mechanical properties of the material and the evolving stress field, presenting lower influence of mesh size and static parameters (Mariani & Perego, 2003; Sukumar & Prévost, 2003), which is especially important for complex crack patterns and configurations.

Another way to simulate arbitrary crack nucleation and propagation is to use the element erosion technique (Gruben et al., 2013; Liu et al., 2014). This technique relies on the damage modeling of a unit element cell, in which element stiffness is degraded after a critical mechanical property is achieved, such as maximum stress and/or strain of the material. After complete degradation, the element is removed from the mesh by inhibiting the load carrying capacity of this cell, allowing the nucleation and/or propagation of a crack. Given the inherent dependence of the mesh, this modeling technique requires a larger number of elements, with a refined discretization of regions more prone for crack nucleation and propagation, which increases computational costs of the overall model. This technique still presents lower performance compared with other methods (Song et al., 2008), which limits its applications to small domains and phenomena-based analyses.

14.3.4 Simulation of adhesive and cohesive + adhesive failure

A further step toward the increase in the complexity of the analyses involves crack propagation along the film/substrate interface. The stress fields around a crack located at a bimaterial interface are complex (Hutchinson & Suo, 1991), which limited initial attempts to simulate interfacial crack propagation (Souza, Mustoe, et al., 2001). More recently, approaches using cohesive zone models (Barés et al., 2012; Bouvard et al., 2009; Zeng & Li, 2010) became a common way to simulate adhesive problems.

Cohesive zone modeling (CZM) relies on the theory of modeling an extremely thin region to couple two surfaces, allowing the transfer of mechanical or thermal loads considering properties associated to this interface (Elices et al., 2002), including mechanical, thermal, and fracture properties. Usually used for adhesive component modeling, such as double-sided tapes, this region presents lower stiffness compared with the involved materials, being a stress concentrator in the component assembly (Campilho et al., 2011; Khoramishad et al., 2010). Nevertheless, this technique permits the consideration of fracture properties, allowing the decohesion of the interface, considering the gradual reduction on the load-carrying capacity promoted by the damage evolution. The initiation and evolution of the damage condition are associated with fracture properties, including the fracture toughness, for crack nucleation, and energy release rate for crack propagation and failure of the interface (Kubair et al., 2002; Li et al., 2005). Focusing on simulation of indentation conditions, literature reports the use of CZM to predict crack nucleation and propagation for different materials, including thermoviscoelastic materials (Yoon & Allen, 1999), concrete (Toyama et al., 2018), fused silica (Bruns et al., 2017), and brittle coatings (Zahedmanesh et al., 2016).

The previous sections have indicated that the outputs of a simple action of an indenter applying normal loads on a coated system depend on a significant number of variables. Besides the load itself, the response depends on geometrical factors, such as indenter geometry and coating thickness. These factors add to mechanical properties of the system, which include the elastic properties of film and substrate, the plastic properties of both, fracture toughness (especially for brittle materials), and properties of the coating/substrate interface. In an attempt to provide an analysis involving most of these parameters, Fukumasu and Souza conducted 220 simulations of spherical indentation of systems with one coating (Fukumasu & Souza, 2014). The variables considered were the coating thickness (t), the spherical indenter radius (R_i), the elastic modulus of the coating (E_c), the energy release rate for cohesive failure (G_c), the energy release rate for adhesive failure (G_{int}), and the residual stress in the coating (σ_c). Results were placed in a map of coating toughness ($K_c = G_c\, E_c$) as a function of a factor F_1, defined by the following equation, in which P is the applied normal load, and A is the area of contact. For the conditions considered in the analysis, the map indicated that film cracks (cohesive failure) were observed for the majority of the conditions with F_1 lower than 2 and not observed for the majority of the conditions with F_1 higher than 2. Therefore, this parameter was considered as an index to correlate different variables with the resulting failure behavior.

$$F_1 = \frac{P\, R_i^2\, K_c^{\frac{3}{2}}}{A\, \sqrt{t}\, E_c}$$

14.3.5 Other indentation analyses

The simulation of the indentation of coated systems has also allowed the analysis of multilayered systems, either considering a system with a few layers or others with several alternating layers of two materials with a given periodicity. These cases allow exploring other variables, for example, the coating architecture (Tu et al., 2023; Zhao, Xie, & Munroe, 2011), the properties of each of the films and of the interfaces between them (Lin et al., 2017; Rusinowicz et al., 2022; Yuan et al., 2021). The impact of multilayers on the overall coating/substrate adhesion was also explored, for example, to study aluminum and silicon carbide (SiC) multilayers (Jamison & Shen, 2016). A similar system was considered to simulate and understand the formation of shear bands inside the multilayered system (Bigelow & Shen, 2018). In another example, the impact of interface morphology was considered, particularly with respect to the presence of curved interfaces (Verma & Jayaram, 2014).

Another variable explored during the simulation of the indentation of coated systems was roughness. For example, Walter simulated successive indentations on chromium nitride coated systems with roughness (Ra) ranging from 2 to 11.1 nm, either 2D (Walter et al., 2007) or 3D (Walter & Mitterer, 2009), to conclude that roughness can result in an underestimation of elastic modulus calculated from nanoindentation data. The effect of roughness was also analyzed by Xiao (Xiao et al., 2020). In this case, sinusoidal functions were considered, with the amplitude of the function along the interface being always larger than that at the coating surface. In the analyses, the effect of roughness was superimposed to that of coating residual stresses. Conclusions emphasized the difference in having indentation on a peak or in a valley and the preferred positions for cohesive or adhesive failure as a function of the variables considered.

14.4 Simulation of normal loads and lateral indenter displacement

Situations where an indenter applies only normal loads onto a coated system are important, but in many cases are not entirely representative of practical contact situations. An additional condition, which is particularly important for many tribological analyses, involves adding a lateral displacement to the indenter, either with constant or with a gradual increase of the normal load along this displacement. If a single indenter is considered, this situation is commonly referred to as a scratch test. Similar to the pure indentation condition, these tests may either be used to evaluate a given mechanical characteristic of the system, such as coating/substrate adhesion (ASTM International, 2005), or as a tool to understand the mechanical behavior of coated systems in laboratory and field applications.

Fig. 14.7 presents a schematic representation of a scratch test on a coated system, with an indication of the tensile and compressive stresses experienced by the film as a function of indenter displacement (Holmberg et al., 2006). The figure also indicates that contact stresses are superimposed to previous residual stresses in the coating, most often as a result of coating deposition. The figure indicates that the stress evolution is more complex than during indentation. Of particular importance is the "push" stresses indicated in front of the indenter, which may be responsible for adhesive failures during scratch tests.

The literature presents several examples of finite element simulation of scratch tests, which include analyses with only one homogeneous coating (Holmberg et al., 2006; Li & Beres, 2006). Additionally, other features analyzed with indention were also studied during scratches, such as the effect of residual stresses (Holmberg & Ronkainen, Laukkanen, et al., 2009), the analysis of multilayered systems (Ali et al., 2015), and the effect of coating roughness (Feng, 2017). In most of these conditions, the addition of a lateral displacement to the normal load may significantly increase the complexity of the observed phenomena, for example, in terms of the array of cohesive cracks formed during the scratch test of multilayered systems (Fig. 14.8; Araujo et al., 2015).

The simulation of scratch tests was also used as a tool to understand another important aspect of thin films, especially those produced by cathodic arc evaporation, which is the presence of macroparticles in the film structure (Bernardes et al., 2020). In that work, a growth defect was considered in the finite element model, in an attempt to study its effect on the mechanical and tribological behavior of a multilayered system with tungsten carbide and amorphous carbon layers. Numerical results were able to provide a reasonable explanation for the phenomena observed experimentally, which showed that, before the complete loss of coating/substrate adhesion (at normal loads of approximately 20 N), lower loads (below 12 N)

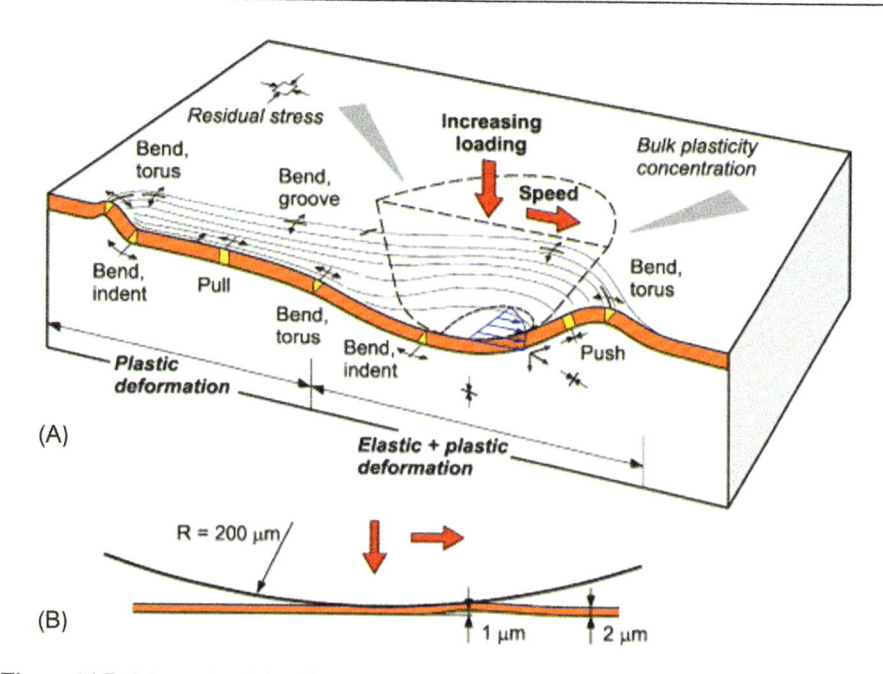

Figure 14.7 Schematic of the sliding of an indenter over a coated surface: (A) loading effects with exaggerated dimensions and deformations and (B) section with correct dimension ratios.
Source: From Holmberg, K., Laukkanen, A., Ronkainen, H., Wallin, K., Varjus, S., & Koskinen, J. (2006). Tribological contact analysis of a rigid ball sliding on a hard coated surface: Part I: Modelling stresses and strains. *Surface and Coatings Technology*, 200(12), 3793–3809. https://doi.org/10.1016/j.surfcoat.2005.03.040.

resulted in defect deformation, in opposition to a gradual increase in the fraction of defect detachment as the load increased from 12 to 20 N.

In another attempt to use finite element analyses to explain experimental phenomena, Fukumasu simulated the scratch test of an hydrogenated amorphous carbon (a-C:H) film deposited onto an AISI H13 steel substrate (Fukumasu et al., 2018). The numerical study calculated highly compressive contact stresses (up to 20 GPa), especially at the coating surface, which could justify a phase transformation of the carbon material at room temperature, in agreement with results observed experimentally from Raman spectroscopy measurements.

14.5 Beyond indentation and scratch analyses for coating/substrate simulation

The versatility of the FEM technique in combining complex geometries, nonlinear mechanical behavior, inhomogeneous and anisotropic materials, and multiscale

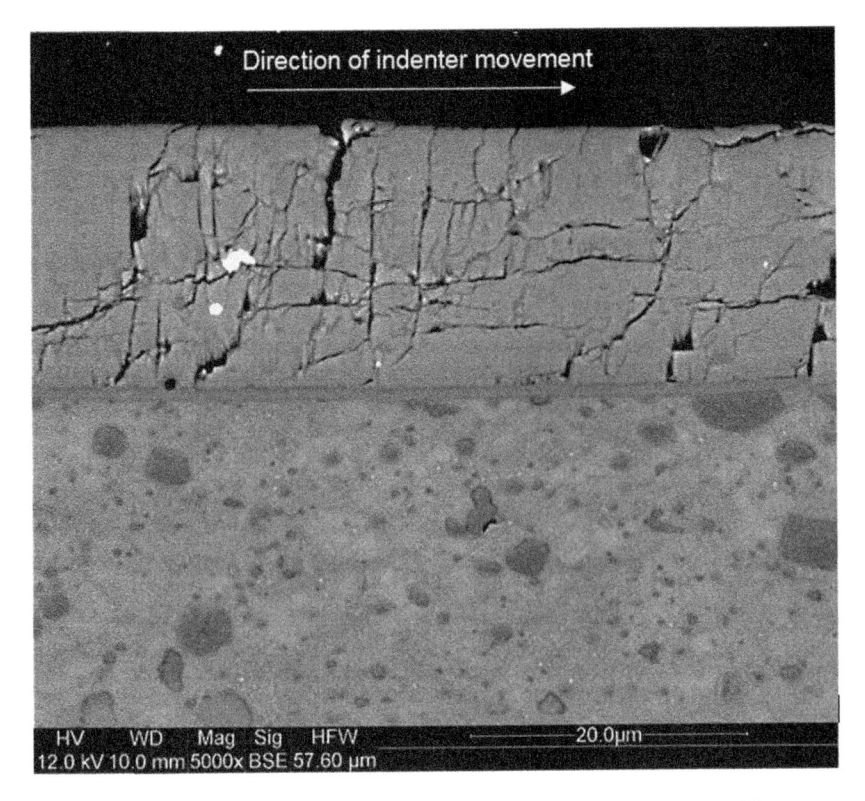

Figure 14.8 Portion of the longitudinal section of a scratch test on a system with gas nitrided martensitic stainless steel (AISI 440B) substrate and a chromium nitride/niobium nitride (CrN/NbN) multilayer coating with 20 nm periodicity.
Source: From Araujo, J. A., Araujo, G. M., Souza, R. M., & Tschiptschin, A. P. (2015). Effect of periodicity on hardness and scratch resistance of CrN/NbN nanoscale multilayer coating deposited by cathodic arc technique. *Wear, 330–331*, 469–477. https://doi.org/10.1016/j.wear.2015.01.051.

features, such as influences of discrete microstructures, allows the prediction of mechanical behavior with evolving stress fields and dynamic loads. Going beyond indentation and scratch analyses of coated systems, numerical prediction of the coated contact conditions enables the design and engineering of systems to present higher performance and durability.

In terms of complex load states promoted by component dynamics, literature reports the use of FEM to analyze the stress distribution induced in given applications, such as in the case of coated piston rings (Lima et al., 2013; Mishra et al., 2022). In this case, macroscale mechanical and thermal loads, obtained from the operation of a real engine, were used to evaluate the behavior of a coated ring, resulting in the distribution of the stress state inside the coating and at the coating/substrate interface. These results allowed the authors to identify the conditions that would improve system durability and performance.

In addition to contact-based FEM models, noncontact conditions may provide important insights on coated system behavior, including the analyses of tensile tests and evaluation of residual stress distribution. The influence of surface singularities, such as the presence of coating cracks, on the evolution of stress state at the coating/substrate interface under tensile tests was reported by Fukumasu et al. (2010) and by Krishnamurthy and Reimanis (2005). These works indicated that crack tip geometry affected the local distribution and intensity of the stress field and lowered the constraints imposed at the crack tip, leading to a stress-relieving phenomenon of the coating. In terms of residual stresses, the simulation of coating topology on nonhomogeneous residual stress distribution was evaluated by Araujo et al. (2016) and Bemporad et al. (2006). In those works, multilayer coatings were modeled with individual layers to evaluate residual stress distribution along the thickness, promoted by the deposition process. Results indicated that lattice parameter, individual layer thickness, and thermal expansion coefficient of layered materials promoted nonhomogeneous residual stress distribution that affect both hardness of the system and fracture properties. Experimental results corroborate numerical data by showing similar trends in mechanical properties and crack propagation through coating thickness influenced by the layer configuration.

Finally, the coupling of algebraic models to govern a variable that affects the evolution of the numerical simulation allows the prediction of macroscopic features based on microscopic properties, such as the wear of a surface. The numerical simulation to predict material removal under dry sliding contact conditions can be predicted by the coupling of the Achard wear model and the FEM simulation. An example of such coupling was reported by Bortoleto et al. (2013), in which the simulation of a pin-on-disk configuration was able to predict wear evolution in agreement to experimental tests, based only on the geometrical arrangement, loading conditions, and mechanical properties. For complex geometries and loading configurations, Sabangban et al. (2016) simulated a strip ironing to predict the wear of die coatings. In this case, the cyclic wear evolution was analyzed, and durability predictions were evaluated based on the wear of the coating thickness, which was calculated again by the coupling of Archard's equation and stress evolution. Results allowed the construction of a wear map to improve the selection of die coatings and process parameters to promote extended die durability.

14.6 Concluding remarks

In this chapter, a portion of the extensive intersection of surface engineering and numerical modeling was emphasized. Contact is often a key issue regarding nanomechanics of coatings and engineering surfaces. Even in noncontact applications, indentation techniques are frequently required to measure properties such as hardness and elastic modulus of surface engineered surfaces. Therefore, a better understanding of contact stresses is usually helpful to more accurate property measurements and to better understand system response, for example, in tribological applications. Not surprisingly, activities in contact modeling are at least almost 150 years old.

Overtime, contact analyses attempted to benefit from the evolution of the numerical techniques, such as the finite element modeling, boundary element modeling, discrete element modeling. This evolution includes not only the increase in computational power but also the development of new methods in the scope of each of these techniques. As a result, more complex geometries were more accurately simulated with an increase in the number of elements and new phenomena could be included in the analyses, such as crack propagation. The indentation of coated systems is a good example of this trend, starting with the response of a single elastic coating on an elastic substrate, moving to the inclusion of plasticity, then to the simulation of cohesive cracks, and finally, to the simulation of both cohesive and adhesive failure. A similar evolution was also observed for the simulation of scratch tests and to other contact or noncontact conditions. Wear simulation represents another step in this evolution, with simulations being able, for example, to alter the position of contacting nodes, allowing the analysis of the evolution of component topography due to wear.

References

Abadias, G., Chason, E., Keckes, J., Sebastiani, M., Thompson, G. B., Barthel, E., Doll, G. L., Murray, C. E., Stoessel, C. H., & Martinu, L. (2018). Review article: Stress in thin films and coatings: Current status, challenges, and prospects. *Journal of Vacuum Science & Technology A*, *36*(2), 020801. Available from https://doi.org/10.1116/1.5011790, https://doi.org/10.1116/1.5011790.

Abdul-Baqi, A., & Van der Giessen, E. (2002). Numerical analysis of indentation-induced cracking of brittle coatings on ductile substrates. *International Journal of Solids and Structures*, *39*(6), 1427–1442. Available from https://doi.org/10.1016/S0020-7683(01)00280-3, https://www.sciencedirect.com/science/article/pii/S0020768301002803.

Ali, R., Sebastiani, M., & Bemporad, E. (2015). Influence of Ti–TiN multilayer PVD-coatings design on residual stresses and adhesion. *Materials & Design*, *75*, 47–56. Available from https://doi.org/10.1016/j.matdes.2015.03.007, https://www.sciencedirect.com/science/article/pii/S0261306915000904.

Araujo, J. A., Araujo, G. M., Souza, R. M., & Paulo Tschiptschin, A. (2015). Effect of periodicity on hardness and scratch resistance of CrN/NbN nanoscale multilayer coating deposited by cathodic arc technique. *Wear*, *330–331*, 469–477. Available from https://doi.org/10.1016/j.wear.2015.01.051, https://www.sciencedirect.com/science/article/pii/S004316481500068X.

Araujo, J. A., Giorjão, R. A. R., Bettini, J., Souza, R. M., & Paulo Tschiptschin, A. (2016). Modeling intrinsic residual stresses built-up during growth of nanostructured multilayer NbN/CrN coatings. *Surface and Coatings Technology*, *308*, 264–272. Available from https://doi.org/10.1016/j.surfcoat.2016.07.108, https://www.sciencedirect.com/science/article/pii/S0257897216309136.

ASTM International. (2005). *ASTM C1624-05 – Standard test method for adhesion strength and mechanical failure modes of ceramic coatings by quantitative single point scratch testing*. ASTM International, 10.1520/C1624-05.

Barsoum, R. S. (1971). Finite element method applied to the problem of stability of a nonconservative system. *International Journal for Numerical Methods in Engineering*, *3*(1), 63−87. Available from https://doi.org/10.1002/nme.1620030110, https://doi.org/10.1002/nme.1620030110.

Barés, J., Gélébart, L., Rupil, J., & Vincent, L. (2012). A joined finite element based method to simulate 3D crack network initiation and propagation in mechanical and thermal fatigue. *International Journal of Fatigue*, *44*, 279−291. Available from https://doi.org/10.1016/j.ijfatigue.2012.04.005, https://www.sciencedirect.com/science/article/pii/S0142112312001326.

Bemporad, E., Sebastiani, M., Pecchio, C., & De Rossi, S. (2006). High thickness Ti/TiN multilayer thin coatings for wear resistant applications. *Surface and Coatings Technology*, *201*(6), 2155−2165. Available from https://doi.org/10.1016/j.surfcoat.2006.03.042, https://www.sciencedirect.com/science/article/pii/S0257897206002829.

Bernardes, C. F., Fukumasu, N. K., Lima, A. O., Souza, R. M., & Machado, I. F. (2020). Influence of growth defects on the running-in behavior of an a-C:H:W coating under pure sliding contact conditions. *Surface and Coatings Technology*, *402*, 126278. Available from https://doi.org/10.1016/j.surfcoat.2020.126278, https://www.sciencedirect.com/science/article/pii/S0257897220309476.

Bigelow, S., & Shen, Y.-L. (2018). Parametric computational analysis of indentation-induced shear band formation in metal-ceramic multilayer coatings. *Surface and Coatings Technology*, *350*, 779−787. Available from https://doi.org/10.1016/j.surfcoat.2018.04.055, https://www.sciencedirect.com/science/article/pii/S0257897218304316.

Bolshakov, A., Oliver, W. C., & Pharr, G. M. (1996). Influences of stress on the measurement of mechanical properties using nanoindentation: Part II. Finite element simulations. *Journal of Materials Research*, *11*(3), 760−768. Available from https://doi.org/10.1557/JMR.1996.0092, https://doi.org/10.1557/JMR.1996.0092.

Bolshakov, A., & Pharr, G. M. (1998). Influences of pileup on the measurement of mechanical properties by load and depth sensing indentation techniques. *Journal of Materials Research*, *13*(4), 1049−1058. Available from https://doi.org/10.1557/JMR.1998.0146, https://doi.org/10.1557/JMR.1998.0146.

Bortoleto, E. M., Rovani, A. C., Seriacopi, V., Profito, F. J., Zachariadis, D. C., Machado, I. F., Sinatora, A., & Souza, R. M. (2013). Experimental and numerical analysis of dry contact in the pin on disc test. *Wear*, *301*(1), 19−26. Available from https://doi.org/10.1016/j.wear.2012.12.005, https://www.sciencedirect.com/science/article/pii/S0043164812004334.

Bortoleto, E. M., Prados, E. F., Seriacopi, V., Fukumasu, N. K., Luiz, G. D. B., Lima, da S., Machado, I. F., & Souza, R. M. (2016). Numerical modeling of adhesion and adhesive failure during unidirectional contact between metallic surfaces. *Friction*, *4*(3), 217−227. Available from https://doi.org/10.1007/s40544-016-0119-5, https://doi.org/10.1007/s40544-016-0119-5.

Bouvard, J. L., Chaboche, J. L., Feyel, F., & Gallerneau, F. (2009). A cohesive zone model for fatigue and creep−fatigue crack growth in single crystal superalloys. *International Journal of Fatigue*, *31*(5), 868−879. Available from https://doi.org/10.1016/j.ijfatigue.2008.11.002, https://www.sciencedirect.com/science/article/pii/S0142112308002521.

Bruns, S., Johanns, K. E., Rehman, H. U. R., Pharr, G. M., & Durst, K. (2017). Constitutive modeling of indentation cracking in fused silica. *Journal of the American Ceramic Society*, *100*(5), 1928−1940. Available from https://doi.org/10.1111/jace.14734, https://doi.org/10.1111/jace.14734.

Campilho, R. D. S. G., Banea, M. D., Pinto, A. M. G., da Silva, L. F. M., & de Jesus, A. M. P. (2011). Strength prediction of single- and double-lap joints by standard and extended finite element modelling. *International Journal of Adhesion and Adhesives*, *31*

(5), 363−372. Available from https://doi.org/10.1016/j.ijadhadh.2010.09.008, https://www.sciencedirect.com/science/article/pii/S0143749611000273.

Casals, O., & Alcalá, J. (2005). The duality in mechanical property extractions from Vickers and Berkovich instrumented indentation experiments. *Acta Materialia, 53*(13), 3545−3561. Available from https://doi.org/10.1016/j.actamat.2005.03.051, https://www.sciencedirect.com/science/article/pii/S1359645405002181.

Chicot, D., Démarécaux, P., & Lesage, J. (1996). Apparent interface toughness of substrate and coating couples from indentation tests. *Thin Solid Films, 283*(1), 151−157. Available from https://doi.org/10.1016/0040-6090(96)08763-9, https://www.sciencedirect.com/science/article/pii/0040609096087639.

Criado, D., Alayo, M. I., Fantini, M. C. A., & Pereyra, I. (2006). Study of the mechanical and structural properties of silicon oxynitride films for optical applications. *Advances in Optical Materials, 352*(23), 2319−2323. Available from https://doi.org/10.1016/j.jnoncrysol.2006.03.012, https://www.sciencedirect.com/science/article/pii/S0022309306004492.

Djabella, H., & Arnell, R. D. (1992). Finite element analysis of the contract stresses in an elastic coating on an elastic substrate. *Thin Solid Films, 213*(2), 205−219. Available from https://doi.org/10.1016/0040-6090(92)90284-I, https://www.sciencedirect.com/science/article/pii/004060909290284I.

Djabella, H., & Arnell, R. D. (1993a). Finite element comparative study of elastic stresses in single, double layer and multilayered coated systems. *Thin Solid Films, 235*(1), 156−162. Available from https://doi.org/10.1016/0040-6090(93)90259-R, https://www.sciencedirect.com/science/article/pii/004060909390259R.

Djabella, H., & Arnell, R. D. (1993b). Finite element analysis of the contact stresses in elastic coating/substrate under normal and tangential load. *Thin Solid Films, 223*(1), 87−97. Available from https://doi.org/10.1016/0040-6090(93)90731-4, https://www.sciencedirect.com/science/article/pii/0040609093907314.

Doerner, M. F., & Nix, W. D. (1986). A method for interpreting the data from depth-sensing indentation instruments. *Journal of Materials Research, 1*(4), 601−609. Available from https://doi.org/10.1557/JMR.1986.0601, https://doi.org/10.1557/JMR.1986.0601.

Dollet, A. (2004). Multiscale modeling of CVD film growth—A review of recent works. *Surface and Coatings Technology, 177−178*, 245−251. Available from https://doi.org/10.1016/j.surfcoat.2003.09.040, https://www.sciencedirect.com/science/article/pii/S0257897203010491.

Elices, M., Guinea, G. V., Gómez, J., & Planas, J. (2002). The cohesive zone model: Advantages, limitations and challenges. *Engineering Fracture Mechanics, 69*(2), 137−163. Available from https://doi.org/10.1016/S0013-7944(01)00083-2, https://www.sciencedirect.com/science/article/pii/S0013794401000832.

Feng, B. (2017). Effects of surface roughness on scratch resistance and stress-strain fields during scratch tests. *AIP Advances, 7*(3), 035217. Available from https://doi.org/10.1063/1.4979332, https://doi.org/10.1063/1.4979332.

Franco Júnior, A.R., (2003). Obtenção de revestimentos dúplex por nitretação a plasma e PVD-TiN em aços ferramenta AISI D2 e AISI H13, *Doctoral Thesis*, Escola Politécnica, University of São Paulo, São Paulo. doi: https://doi.org/10.11606/T.3.2003.tde-02102003-114623.

Fries, T.-P., & Belytschko, T. (2010). The extended/generalized finite element method: An overview of the method and its applications. *International Journal for Numerical Methods in Engineering, 84*(3), 253−304. Available from https://doi.org/10.1002/nme.2914, https://doi.org/10.1002/nme.2914.

Fukumasu, N. K., Angelo, C. M., Ignat, M., & Souza, R. M. (2010). Numerical study of tensile tests conducted on systems with elastic-plastic films deposited onto elastic-plastic substrates. *Surface and Coatings Technology*, 205(5), 1415−1419. Available from https://doi.org/10.1016/j.surfcoat.2010.07.104, https://www.sciencedirect.com/science/article/pii/S0257897210006262.

Fukumasu, N. K., Bernardes, C. F., Ramirez, M. A., Trava-Airoldi, V. J., Souza, R. M., & Machado, I. F. (2018). Local transformation of amorphous hydrogenated carbon coating induced by high contact pressure. *Tribology International*, 124, 200−208. Available from https://doi.org/10.1016/j.triboint.2018.04.006, https://www.sciencedirect.com/science/article/pii/S0301679X18301932.

Fukumasu, N. K., Boidi, G., Seriacopi, V., Machado, G. A. A., Souza, R. M., & Machado, I. F. (2017). Numerical analyses of stress induced damage during a reciprocating lubricated test of fecmo sps sintered alloy. *Tribology International*, 113, 443−447. Available from https://doi.org/10.1016/j.triboint.2016.12.025, https://www.sciencedirect.com/science/article/pii/S0301679X16304947.

Fukumasu, N. K., Machado, G. A. A., Souza, R. M., & Machado, I. F. (2016). Stress analysis to improve pitting resistance in gear teeth. *Procedia CIRP*, 45, 255−258. Available from https://doi.org/10.1016/j.procir.2016.02.349, https://www.sciencedirect.com/science/article/pii/S2212827116006442.

Fukumasu, N. K., Pelegrino, P. L., Cueva, G., Souza, R. M., & Sinatora, A. (2005). Numerical analysis of the stresses developed during the sliding of a cylinder over compact graphite iron. *Wear*, 259(7), 1400−1407. Available from https://doi.org/10.1016/j.wear.2005.01.014, https://www.sciencedirect.com/science/article/pii/S0043164805000359.

Fukumasu, N. K., & Souza, R. M. (2006). Numerical analysis of the contact stresses developed during the indentation of coated systems with substrates with orthotropic properties. *Surface and Coatings Technology*, 201(7), 4294−4299. Available from https://doi.org/10.1016/j.surfcoat.2006.08.072, https://www.sciencedirect.com/science/article/pii/S0257897206008553.

Fukumasu, N. K., & Souza, R. M. (2014). Numerical evaluation of cohesive and adhesive failure modes during the indentation of coated systems with compliant substrates. *Surface and Coatings Technology*, 260, 266−271. Available from https://doi.org/10.1016/j.surfcoat.2014.07.093, https://www.sciencedirect.com/science/article/pii/S0257897214007130.

Gruben, G., Hopperstad, O. S., & Børvik, T. (2013). Simulation of ductile crack propagation in dual-phase steel. *International Journal of Fracture*, 180(1), 1−22. Available from https://doi.org/10.1007/s10704-012-9791-2, https://doi.org/10.1007/s10704-012-9791-2.

Han, G., Marimuthu, K. P., & Lee, H. (2022). Evaluation of thin film material properties using a deep nanoindentation and ANN. *Materials & Design*, 221, 111000. Available from https://doi.org/10.1016/j.matdes.2022.111000, https://www.sciencedirect.com/science/article/pii/S0264127522006220.

Hill, R. (1950). *The mathematical theory of plasticity oxford classic texts in the physical sciences*. Oxford: Oxford Clarendon Press.

Holmberg, K., Laukkanen, A., Ronkainen, H., Wallin, K., Varjus, S., & Koskinen, J. (2006). Tribological contact analysis of a rigid ball sliding on a hard coated surface: Part I: Modelling stresses and strains. *Surface and Coatings Technology*, 200(12), 3793−3809. Available from https://doi.org/10.1016/j.surfcoat.2005.03.040, https://www.sciencedirect.com/science/article/pii/S0257897205004755.

Holmberg, K., & Matthews, A. (2009). *Coatings tribology: Properties, mechanisms, techniques and applications in surface engineering tribology and interface engineering* (2nd Edition). Elsevier.

Holmberg, K., Ronkainen, H., Laukkanen, A., Wallin, K., Hogmark, S., Jacobson, S., Wiklund, U., Souza, R. M., & åStåhle, P. S. (2009). Residual stresses in TiN, DLC and

MoS$_2$ coated surfaces with regard to their tribological fracture behaviour. *Wear, 267* (12), 2142−2156. Available from https://doi.org/10.1016/j.wear.2009.01.004, https://www.sciencedirect.com/science/article/pii/S0043164809000064.

Hutchinson, J. W., & Suo, Z. (1991). Mixed mode cracking in layered materials mixed mode cracking in layered materials. *Advances in Applied Mechanics, 29*, 63−191. Available from https://doi.org/10.1016/S0065-2156(08)70164-9, https://www.sciencedirect.com/science/article/pii/S0065215608701649.

International Organization for Standardization (ISO). (2015). ISO 14577-1:2015, *Metallic materials Instrumented indentation test for hardness and materials parameters. Part 1: Test method.*

Jamison, R. D., & Shen, Y.-L. (2016). Delamination analysis of metal−ceramic multilayer coatings subject to nanoindentation. *Surface and Coatings Technology, 303*, 3−11. Available from https://doi.org/10.1016/j.surfcoat.2016.01.038, https://www.sciencedirect.com/science/article/pii/S025789721630038X.

Johnson, K. L. (1985). *Normal contact of elastic solids − Hertz theory contact mechanics* (pp. 84−106). Cambridge: Cambridge University Press. Available from: https://www.cambridge.org/core/product/5CEDA95E252B9CFE3FDDA16AB328A744, https://doi.org/10.1017/CBO9781139171731.005.

Khoramishad, H., Crocombe, A. D., Katnam, K. B., & Ashcroft, I. A. (2010). Predicting fatigue damage in adhesively bonded joints using a cohesive zone model. *International Journal of Fatigue, 32*(7), 1146−1158. Available from https://doi.org/10.1016/j.ijfatigue.2009.12.013. Available from:, https://www.sciencedirect.com/science/article/pii/S0142112309003648.

Komvopoulos, K. (1989). Elastic-plastic finite element analysis of indented layered media. *Journal of Tribology, 111*(3), 430−439. Available from https://doi.org/10.1115/1.3261943, https://doi.org/10.1115/1.3261943.

Kot, M., Rakowski, W., Lackner, J. M., & Major, Ł. (2013). Analysis of spherical indentations of coating-substrate systems: Experiments and finite element modeling. *Materials & Design, 43*, 99−111. Available from https://doi.org/10.1016/j.matdes.2012.06.040, https://www.sciencedirect.com/science/article/pii/S026130691200427X.

Krishnamurthy, S., & Reimanis, I. (2005). Multiple cracking in CrN and Cr$_2$N films on brass. *Surface and Coatings Technology, 192*(2), 291−298. Available from https://doi.org/10.1016/j.surfcoat.2004.06.025, https://www.sciencedirect.com/science/article/pii/S0257897204004761.

Kubair, D. V., Geubelle, P. H., & Huang, Y. Y. (2002). Analysis of a rate-dependent cohesive model for dynamic crack propagation. *Engineering Fracture Mechanics, 70*(5), 685−704. Available from https://doi.org/10.1016/S0013-7944(02)00042-5, https://www.sciencedirect.com/science/article/pii/S0013794402000425.

Lau, A. C. W., Shivpuri, R., & Chou, P. C. (1989). An explicit time integration elastic-plastic finite element algorithm for analysis of high speed rolling. *International Journal of Mechanical Sciences, 31*(7), 483−497. Available from https://doi.org/10.1016/0020-7403(89)90098-2, https://www.sciencedirect.com/science/article/pii/0020740389900982.

Li, J., & Beres, W. (2006). Three-dimensional finite element modelling of the scratch test for a TiN coated titanium alloy substrate. *Wear, 260*(11), 1232−1242. Available from https://doi.org/10.1016/j.wear.2005.08.008, https://www.sciencedirect.com/science/article/pii/S0043164805004357.

Li, S., Thouless, M. D., Waas, A. M., Schroeder, J. A., & Zavattieri, P. D. (2005). Use of mode-I cohesive-zone models to describe the fracture of an adhesively-bonded polymer-matrix composite. *Composites Science and Technology, 65*(2), 281−293. Available from https://doi.org/10.1016/j.compscitech.2004.07.009, https://www.sciencedirect.com/science/article/pii/S0266353804001800.

Li, X., Diao, D., & Bhushan, B. (1997). Fracture mechanisms of thin amorphous carbon films in nanoindentation. *Acta Materialia*, *45*(11), 4453−4461. Available from https://doi.org/10.1016/S1359-6454(97)00143-2, https://www.sciencedirect.com/science/article/pii/S1359645497001432.

Lima, L. G. D. B. S., Nunes, L. C. S., Souza, R. M., Fukumasu, N. K., & Ferrarese, A. (2013). Numerical analysis of the influence of film thickness and properties on the stress state of thin film-coated piston rings under contact loads. *Surface and Coatings Technology*, *215*, 327−333. Available from https://doi.org/10.1016/j.surfcoat.2012.04.102, https://www.sciencedirect.com/science/article/pii/S0257897212010912.

Lin, P., Shen, F., Yeo, A., Liu, B., Xue, M., Xu, H., & Zhou, K. (2017). Characterization of interfacial delamination in multi-layered integrated circuit packaging. *Surface and Coatings Technology*, *320*, 349−356. Available from https://doi.org/10.1016/j.surfcoat.2016.12.050, https://www.sciencedirect.com/science/article/pii/S025789721631338X.

Liu, S.-H., Trelles, J. P., Murphy, A. B., He, W.-T., Shi, J., Li, S., Li, C.-J., Li, C.-X., & Guo, H.-B. (2021). Low-pressure plasma-induced physical vapor deposition of advanced thermal barrier coatings: Microstructures, modelling and mechanisms. *Materials Today Physics*, *21*, 100481. Available from https://doi.org/10.1016/j.mtphys.2021.100481, https://www.sciencedirect.com/science/article/pii/S2542529321001425.

Liu, Y., Filonova, V., Hu, N., Yuan, Z., Fish, J., Yuan, Z., & Belytschko, T. (2014). A regularized phenomenological multiscale damage model. *International Journal for Numerical Methods in Engineering*, *99*(12), 867−887. Available from https://doi.org/10.1002/nme.4705, https://doi.org/10.1002/nme.4705.

Mariani, S., & Perego, U. (2003). Extended finite element method for quasi-brittle fracture. *International Journal for Numerical Methods in Engineering*, *58*(1), 103−126. Available from https://doi.org/10.1002/nme.761, https://doi.org/10.1002/nme.761.

Marques, F. P., Scandian, C., Bozzi, A. C., Fukumasu, N. K., & Tschiptschin, A. P. (2017). Formation of a nanocrystalline recrystallized layer during microabrasive wear of a cobalt-chromium based alloy (Co-30Cr-19Fe). *Tribology International*, *116*, 105−112. Available from https://doi.org/10.1016/j.triboint.2017.07.006, https://www.sciencedirect.com/science/article/pii/S0301679X17303493.

Martinet, F., & Chabrand, P. (2000). Application of ALE finite elements method to a lubricated friction model in sheet metal forming. *International Journal of Solids and Structures*, *37*(29), 4005−4031. Available from https://doi.org/10.1016/S0020-7683(99)00205-X, https://www.sciencedirect.com/science/article/pii/S002076839900205X.

Marx, V. M., Toth, F., Wiesinger, A., Berger, J., Kirchlechner, C., Cordill, M. J., Fischer, F. D., Rammerstorfer, F. G., & Dehm, G. (2015). The influence of a brittle Cr interlayer on the deformation behavior of thin Cu films on flexible substrates: Experiment and model. *Acta Materialia*, *89*, 278−289. Available from https://doi.org/10.1016/j.actamat.2015.01.047, https://www.sciencedirect.com/science/article/pii/S1359645415000609.

Mishra, P. C., Tiwari, P., & Khoshnaw, F. (2022). Finite element modelling for failure prevention of coated piston compression ring. *International Journal of Manufacturing, Materials, and Mechanical Engineering*, *12*(1), 1−15. Available from https://doi.org/10.4018/IJMMME.299057, http://doi.org/10.4018/IJMMME.299057.

Montmitonnet, P., Edlinger, M. L., & Felder, E. (1993). Finite element analysis of elastoplastic indentation: Part II—Application to hard coatings. *Journal of Tribology*, *115*(1), 15−19. Available from https://doi.org/10.1115/1.2920970, https://doi.org/10.1115/1.2920970.

Oliver, W. C., & Pharr, G. M. (1992). An improved technique for determining hardness and elastic modulus using load and displacement sensing indentation experiments. *Journal of Materials Research, 7*(6), 1564−1583. Available from https://doi.org/10.1557/JMR.1992.1564, https://doi.org/10.1557/JMR.1992.1564.

Pachler, T., Souza, R. M., & Tschiptschin, A. P. (2007). Finite element analysis of peak stresses developed during indentation of ceramic coated steels. *Surface and Coatings Technology, 202*(4), 1098−1102. Available from https://doi.org/10.1016/j.surfcoat.2007.07.041, https://www.sciencedirect.com/science/article/pii/S0257897207008171.

Pharr, G. M. (1998). Measurement of mechanical properties by ultra-low load indentation. *Materials Science and Engineering: A, 253*(1), 151−159. Available from https://doi.org/10.1016/S0921-5093(98)00724-2, https://www.sciencedirect.com/science/article/pii/S0921509398007242.

Pian, T. H. H., & Tong, P. (1969). Basis of finite element methods for solid continua. *International Journal for Numerical Methods in Engineering, 1*(1), 3−28. Available from https://doi.org/10.1002/nme.1620010103, https://doi.org/10.1002/nme.1620010103.

Praud, F., Schmitt, T., Zabeida, O., Maïza, S., Martinu, L., & Lévesque, M. (2021). Phase field fracture models to predict crack initiation and propagation in anti-reflective coatings. *Thin Solid Films, 736*, 138920. Available from https://doi.org/10.1016/j.tsf.2021.138920, https://www.sciencedirect.com/science/article/pii/S004060902100403X.

Pérez R, E. A., & Souza, R. M. (2004). Finite element analysis on the effect of indenter diameter and load on the contact stresses during indentation of coated systems. *Journal of Metastable and Nanocrystalline Materials, 20−21*, 763−768. Available from https://doi.org/10.4028/http://www.scientific.net/JMNM.20-21.763, https://www.scientific.net/JMNM.20-21.763.

Rusinowicz, M., Parry, G., Volpi, F., Mercier, D., Eve, S., Lüders, U., Lallemand, F., Choquet, M., Braccini, M., Boujrouf, C., Hug, E., Coq Germanicus, R., & Verdier, M. (2022). Failure of a brittle layer on a ductile substrate: Nanoindentation experiments and FEM simulations. *Journal of the Mechanics and Physics of Solids, 163*, 104859. Available from https://doi.org/10.1016/j.jmps.2022.104859, https://www.sciencedirect.com/science/article/pii/S0022509622000710.

Réthoré, J., Gravouil, A., & Combescure, A. (2005). A combined space−time extended finite element method. *International Journal for Numerical Methods in Engineering, 64*(2), 260−284. Available from https://doi.org/10.1002/nme.1368, https://doi.org/10.1002/nme.1368.

Sabangban, N., Mahayotsanun, N., Sucharitpwatskul, S., & Mahabunphachai, S. (2016). Wear prediction of die coatings in strip ironing by finite element simulation. *Transactions of the IMF, 94*(4), 199−203. Available from https://doi.org/10.1080/00202967.2016.1180813, https://doi.org/10.1080/00202967.2016.1180813.

Saha, R., & Nix, W. D. (2002). Effects of the substrate on the determination of thin film mechanical properties by nanoindentation. *Acta Materialia, 50*(1), 23−38. Available from https://doi.org/10.1016/S1359-6454(01)00328-7, https://www.sciencedirect.com/science/article/pii/S1359645401003287.

Sander, D., Tian, Z., & Kirschner, J. (2009). The role of surface stress in structural transitions, epitaxial growth and magnetism on the nanoscale. *Journal of Physics: Condensed Matter, 21*, 134015. Available from https://doi.org/10.1088/0953-8984/21/13/134015.

Sebastiani, M., Johanns, K. E., Herbert, E. G., & Pharr, G. M. (2015). Measurement of fracture toughness by nanoindentation methods: Recent advances and future challenges. *Current Opinion in Solid State and Materials Science, 19*(6), 324−333. Available from https://doi.org/10.1016/j.cossms.2015.04.003, https://www.sciencedirect.com/science/article/pii/S1359028615000285.

Seriacopi, V., Fukumasu, N. K., Souza, R. M., & Machado, I. F. (2016). Analysis of abrasion mechanisms in the AISI 303 stainless steel: Effect of deformed layer. *Procedia CIRP*, *45*, 187−190. Available from https://doi.org/10.1016/j.procir.2016.02.326, https://www.sciencedirect.com/science/article/pii/S2212827116006211.

Seriacopi, V., Fukumasu, N. K., Souza, R. M., & Machado, I. F. (2019). Finite element analysis of the effects of thermo-mechanical loadings on a tool steel microstructure. *Engineering Failure Analysis*, *97*, 383−398. Available from https://doi.org/10.1016/j.engfailanal.2019.01.006, https://www.sciencedirect.com/science/article/pii/S1350630718308185.

Smy, T., Tan, L., Winterton, S. S., Dew, S. K., & Brett, M. J. (1997). Simulation of sputter deposition at high pressures. *Journal of Vacuum Science & Technology A*, *15*(6), 2847−2853. Available from https://doi.org/10.1116/1.580838, https://doi.org/10.1116/1.580838.

Song, J.-H., Wang, H., & Belytschko, T. (2008). A comparative study on finite element methods for dynamic fracture. *Computational Mechanics*, *42*(2), 239−250. Available from https://doi.org/10.1007/s00466-007-0210-x, https://doi.org/10.1007/s00466-007-0210-x.

Souza R.M. (1999). Finite element modeling of contact stresses during the indentation of wear resistant coatings on soft substrates. *Ph.D. Thesis*, Colorado School of Mines.

Souza, R. M., Mustoe, G. G. W., & Moore, J. J. (1999). Finite-element modeling of the stresses and fracture during the indentation of hard elastic films on elastic-plastic aluminum substrates. *Thin Solid Films*, *355-356*, 303−310. Available from https://doi.org/10.1016/S0040-6090(99)00505-2, https://www.sciencedirect.com/science/article/pii/S0040609099005052.

Souza, R. M., Mustoe, G. G. W., & Moore, J. J. (2001). Finite element modeling of the stresses, fracture and delamination during the indentation of hard elastic films on elastic−plastic soft substrates. *Thin Solid Films*, *392*(1), 65−74. Available from https://doi.org/10.1016/S0040-6090(01)00959-2, https://www.sciencedirect.com/science/article/pii/S0040609001009592.

Souza, R. M., Sinatora, A., Mustoe, G. G. W., & Moore, J. J. (2001). Numerical and experimental study of the circular cracks observed at the contact edges of the indentations of coated systems with soft substrates. *Wear*, *251*(1), 1337−1346. Available from https://doi.org/10.1016/S0043-1648(01)00778-5, https://www.sciencedirect.com/science/article/pii/S0043164801007785.

Sukumar, N., & Prévost, J.-H. (2003). Modeling quasi-static crack growth with the extended finite element method Part I: Computer implementation. *International Journal of Solids and Structures*, *40*(26), 7513−7537. Available from https://doi.org/10.1016/j.ijsolstr.2003.08.002, https://www.sciencedirect.com/science/article/pii/S0020768303004499.

Sánchez, J. M., El-Mansy, S., Sun, B., Scherban, T., Fang, N., Pantuso, D., Ford, W., Elizalde, M. R., Martínez-Esnaola, J. M., Martín-Meizoso, A., Gil-Sevillano, J., Fuentes, M., & Maiz, J. (1999). Cross-sectional nanoindentation: A new technique for thin film interfacial adhesion characterization. *Acta Materialia*, *47*(17), 4405−4413. Available from https://doi.org/10.1016/S1359-6454(99)00254-2, https://www.sciencedirect.com/science/article/pii/S1359645499002542.

Toyama, H., Kishida, H., & Yonezu, A. (2018). Characterization of fatigue crack growth of concrete mortar under cyclic indentation loading. *Engineering Failure Analysis*, *83*, 156−166. Available from https://doi.org/10.1016/j.engfailanal.2017.10.004, https://www.sciencedirect.com/science/article/pii/S135063071730907X.

Tu, R., Jiang, M., Yang, M., Ji, B., Gao, T., Zhang, S., & Zhang, L. (2023). Effects of gradient structure and modulation period on mechanical performance and thermal stress of TiN/TiSiN multilayer hard coatings. *Materials Science and Engineering: A, 866*, 144696. Available from https://doi.org/10.1016/j.msea.2023.144696, https://www.sciencedirect.com/science/article/pii/S092150932300120X.

Verma, N., & Jayaram, V. (2014). Role of interface curvature on stress distribution under indentation for ZrN/Zr multilayer coating. *Thin Solid Films, 571*, 283–289. Available from https://doi.org/10.1016/j.tsf.2014.06.001, https://www.sciencedirect.com/science/article/pii/S0040609014006440.

Walter, C., Antretter, T., Daniel, R., & Mitterer, C. (2007). Finite element simulation of the effect of surface roughness on nanoindentation of thin films with spherical indenters. *Surface and Coatings Technology, 202*(4), 1103–1107. Available from https://doi.org/10.1016/j.surfcoat.2007.07.038, https://www.sciencedirect.com/science/article/pii/S0257897207008018.

Walter, C., & Mitterer, C. (2009). 3D versus 2D finite element simulation of the effect of surface roughness on nanoindentation of hard coatings. *Surface and Coatings Technology, 203*(20), 3286–3290. Available from https://doi.org/10.1016/j.surfcoat.2009.04.006, https://www.sciencedirect.com/science/article/pii/S0257897209003521.

Wang, L., Li, D. C., Yang, J. S., Shao, F., Zhong, X. H., Zhao, H. Y., Yang, K., Tao, S. Y., & Wang, Y. (2016). Modeling of thermal properties and failure of thermal barrier coatings with the use of finite element methods: A review. *Journal of the European Ceramic Society, 36*(6), 1313–1331. Available from https://doi.org/10.1016/j.jeurceramsoc.2015.12.038, https://www.sciencedirect.com/science/article/pii/S0955221915302971.

Xiao, Y., Wu, L., Luo, J., & Zhou, L. (2020). Mechanical response of thin hard coatings under indentation considering rough surface and residual stress. *Diamond and Related Materials, 108*, 107991. Available from https://doi.org/10.1016/j.diamond.2020.107991, https://www.sciencedirect.com/science/article/pii/S0925963520305446.

Yoon, C., & Allen, D. H. (1999). Damage dependent constitutive behavior and energy release rate for a cohesive zone in a thermoviscoelastic solid. *International Journal of Fracture, 96*(1), 55–74. Available from https://doi.org/10.1023/A:1018601004565, https://doi.org/10.1023/A:1018601004565.

Yuan, Z., Han, Y., Zang, S., Chen, J., He, G., Chai, Y., Yang, Z., & Fu, Q. (2021). Damage evolution behavior of TiN/Ti multilayer coatings under high-speed impact conditions. *Surface and Coatings Technology, 426*, 127807. Available from https://doi.org/10.1016/j.surfcoat.2021.127807, https://www.sciencedirect.com/science/article/pii/S0257897221009816.

Zahedmanesh, H., Vanstreels, K., & Gonzalez, M. (2016). A numerical study on nano-indentation induced fracture of low dielectric constant brittle thin films using cube corner probes. *Microelectronic Engineering, 156*, 108–115. Available from https://doi.org/10.1016/j.mee.2016.01.006, https://www.sciencedirect.com/science/article/pii/S0167931716300065.

Zeng, X., & Li, S. (2010). A multiscale cohesive zone model and simulations of fractures. *Computer Methods in Applied Mechanics and Engineering, 199*(9), 547–556. Available from https://doi.org/10.1016/j.cma.2009.10.008, https://www.sciencedirect.com/science/article/pii/S0045782509003533.

Zhao, X., Xie, Z., & Munroe, P. (2011). Nanoindentation of hard multilayer coatings: Finite element modelling. *Materials Science and Engineering: A, 528*(3), 1111–1116. Available from https://doi.org/10.1016/j.msea.2010.09.073, https://www.sciencedirect.com/science/article/pii/S0921509310011032.

High-performance molecular dynamics simulations to investigate nanoindentation of advanced engineering materials

15

Saurav Goel[1,2] and Pengfei Fan[3]
[1]School of Engineering, London South Bank University, London, United Kingdom,
[2]Department of Mechanical Engineering, University of Petroleum and Energy Studies, Dehradun, Uttarakhand, India, [3]Centre for Nanoscience and Nanotechnology, University of Bath, Bath, United Kingdom

Abbreviations

ABOP	Analytical bond order potential
BCC	Body-centered cubic
CAT	Crystal analysis tool
COMB	Charge-optimized many body potential
DFT	Density functional theory
DXA	Dislocation extraction algorithm
EAM	Embedded-atom-method
FCC	Face-centered cubic
GB	Grain boundary
MD	Molecular dynamics
NVE	Microcanonical ensemble
OVITO	Open Visualization tool (software program)
PBC	Periodic boundary condition
TEM	Transmission electron microscope
VMD	Visual molecular dynamics

Nomenclatures

A	Contact radius of the spherical indenter
A	Projected area
B	Direction of Burgers vector
B	Bulk modulus
DI	Simulation using a rigid diamond indenter
TI	Rigid tungsten indenter
RI	A purely repulsive indenter
C_{ij}	Elastic constants

DI-NA	Diamond indenter with no pairwise attraction
E	Elastic modulus of the material
E_r	Reduced elastic modulus of the material system
F or P	Normal force or load on the indenter
$F(r)$	Repulsive force potential
G	Shear modulus (GPa)
h	Instantaneous displacement of the indenter
h_f	Residual depth of indentation
h_{max}	Maximum depth of indentation
H	Hardness of the material
K	Force constant
L	Total length of dislocations in Å
P-h	Load – displacement curve
p_m	Contact pressure
R	Radius of the indenter
R_{pl}	Radius of the plastic zone
r	Distance between two dimers
r_0	Cutoff radius between indenter and the substrate
S	Slope of the indentation unloading curve
W	Tungsten
$\sigma_{hydrostatic}$	Hydrostatic stress
$\sigma_1, \sigma_2,$ **and** σ_3	Principal stresses
ε	Strain
τ	Shear stress

Nanoindentation technology has become crucial to investigate the nanoscale mechanics of advanced engineering materials. In light of the recent experimental efforts in material developments (Longbottom & Lanham, 2005), for example, auxetic materials, metamaterials, 2D materials, scintillating, and doped materials in heterogenous ratios, computational modeling and simulation to predict material behavior are gaining rapid momentum. In this spirit, various simulation tools have emerged over the past decades, including advances in macroscopic combinatorial coupling schemes such as finite element analysis (FEA) and computation fluid dynamics (CFD) (Pervaiz et al., 2014) as well as methods such as homogenization in time (Gracie & Belytschko, 2011; Oskay & Fish, 2004), model reduction techniques (Kerfriden et al., 2012), movable cellular automaton (Psakhie et al., 1995), the discrete element method (Tan et al., 2009), and coupling of finite element method (FEM) with molecular dynamics (MD) simulation (Pen et al., 2011). However, while these methods have solved the problem of the size scale, they have not succeeded in mitigating the problem of the timescale. The fundamental quest to probe the atomistic origins of materials deformation triggering dislocation movements continues to excite the computational mechanists to use molecular dynamics to investigate material deformation at the atomic scale. This chapter provides up-to-date knowledge on advances made in the field of high-performance molecular dynamics simulations (MDS) to study nanoindentation of advanced materials, such as metals, semiconductor, and alloys including the newly developed high-entropy alloy. The scope of this chapter includes a discussion on the working principle of

MD simulation, boundary condition, and model setup to study the nanoindentation process, postprocessing of the MD data, and various types of deformation mechanisms observed in a variety of advanced engineering materials to date. It is hoped that this chapter can guide the new researchers in the field of MD to appreciate the scientific information that the virtual experiments can provide in a time-efficient manner to better understand the deformation mechanics at the nanoscale.

15.1 Ingredients of a trustworthy molecular dynamics simulation

15.1.1 Considerations for simulation of the nanoindentation process

While there are many codes available these days, this chapter was built on the use of a "Large-scale atomic/molecular massively parallel simulator" (Plimpton, 1995) software developed by Sandia Labs in the USA.

15.1.2 Boundary conditions and ensemble

The MD simulation software LAMMPS can either directly be used to generate the CAD geometry of the simulation model or external software such as "Materials Studio" or "Atomsk" can conveniently be used for this task. The output files generated by the external software can very easily be imported to LAMMPS to perform the MD simulations. An example representative model of the MD simulation model after equilibration is shown in Fig. 15.1.

Once the model is imported in the LAMMPS file, one needs to apply the appropriate boundary conditions. This is usually done by following the traditional scheme of partition by having Newton region, thermostat region, and a fixed region. The atoms in the Newton region directly affected by the chemical interactions were allowed to follow Newtonian dynamics (LAMMPS NVE dynamics), while atoms in a thin boundary layer were subjected to a thermostat (LAMMPS NVT dynamics) to dissipate the heat generated in the artificial volume, which would have otherwise taken away by the air during nanoindentation or lubricant (in cutting). The fixed region is prescribed to maintain the crystal symmetry and anchor substrate to resist the applied load without moving.

The next choice is the selection of the indenter geometry. Normally, pyramidal indenters, such as Berkovich or cube corner, are classified as "sharp" indenters while spherical indenters are referred to as "blunt" indenters (Goel, Yan, et al., 2014). In practice, almost all indenters have some finite edge radius (despite being referred to as extremely sharp), and therefore, a spherical shaped indenter for small depths of indentation does represent a typical Berkovich indenter (Fig. 15.1). The atoms in the indenter can be made to follow fixed boundary conditions under the assumption that the indenter does not wear out during the process of indentation.

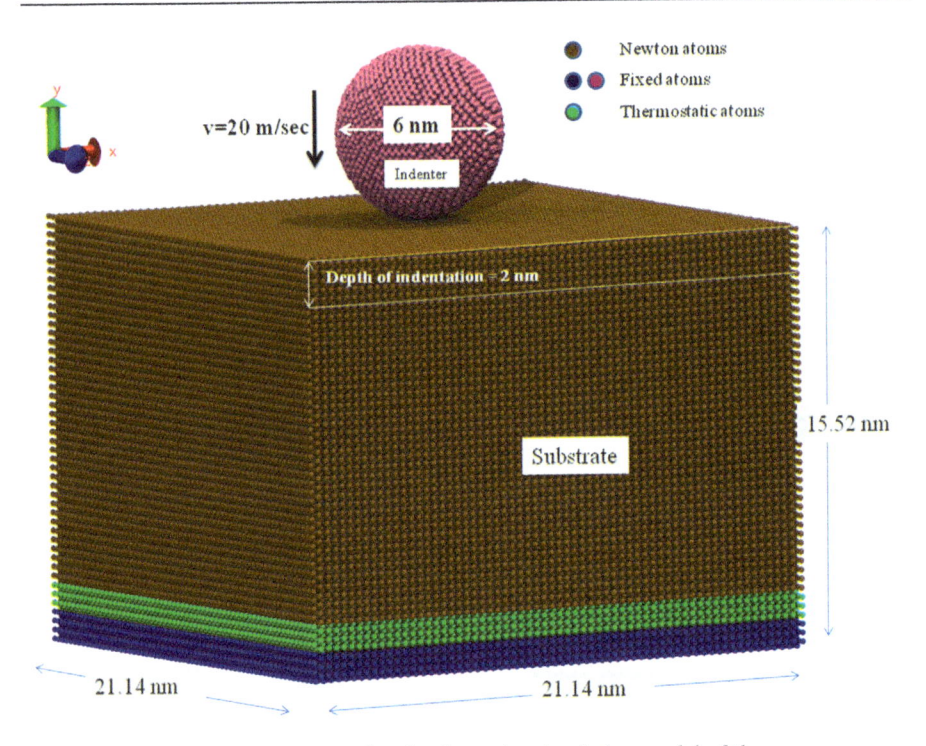

Figure 15.1 Schematic diagram molecular dynamics simulation model of the nanoindentation.
Source: From Goel, S., Beake, B., Chan, C. W., Haque Faisal, N., & Dunne, N. (2015). Twinning anisotropy of tantalum during nanoindentation. *Materials Science and Engineering: A*, *627*, 249–261. https://doi.org/10.1016/j.msea.2014.12.075.

Fig. 15.2 highlights the nanoindentation model of polycrystalline alloy materials with a spherical rigid indenter. Nanoindentation simulation studies need to consider an appropriate indenter. In previous works, the role of tip − sample interaction perturbations on continuum contact mechanics concepts using state-of-the-art, fully atomistic molecular dynamics modeling was examined (Goel et al., 2018). This exploration revealed the importance of adhesion (clinging between two different atomic species) and cohesion (clinging between the same atomic species) and how it affects the initial contact and its consequential effects on the contact mechanics at the meso- and macroscales.

15.1.3 *Potential energy function or a force field*

MD simulation requires a constitutive description of the terms with which the atoms in a simulation interact. This interaction is governed by a potential energy function often referred to as a force field that roughly approximates the quantum level details of the atoms being simulated, so that the force field can replicate bulk macroscopic properties of the material being simulated. Most potential functions or

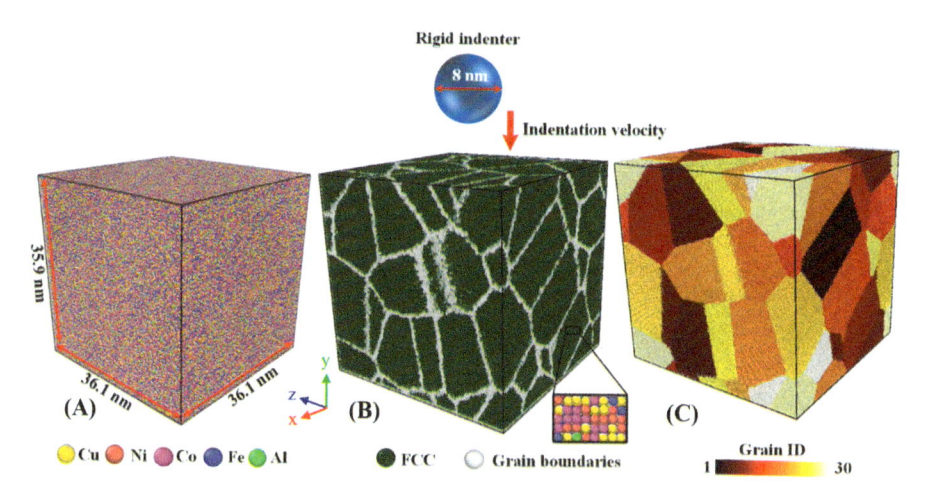

Figure 15.2 (A) Monocrystalline $Ni_{25}Cu_{18.75}Fe_{25}Co_{25}Al_{6.25}$ HEA model, (B) polycrystalline $Ni_{25}Cu_{18.75}Fe_{25}Co_{25}Al_{6.25}$ HEA model, (C) Identification of individual grains in the polycrystalline HEA sample.
Source: From Fan, P., Nirmal, K., Arshad, M., Bai, M., Mao, H., & Goel, S. (2024). Anisotropic plasticity mechanisms in a newly synthesised high entropy alloy investigated using atomic simulations and nanoindentation experiments. *Journal of Alloys and Compounds*, 970. https://doi.org/10.1016/j.jallcom.2023.172541.

force fields are empirical and consist of a summation of forces associated with chemical bonds, bond angles, dihedrals, nonbonding forces associated with van der Waals forces and electrostatic forces. Silicon and carbon by far are two of the most commonly studied materials using atomic simulations. In recent times, Balamane et al. (1992) presented a comprehensive review of the potential energy functions that have been used to simulate silicon. Similarly, Tomas et al. reviewed (de Tomas et al., 2016, 2019) similar developments for the carbon force field and their applicability to study amorphous carbon atoms.

It is worth noting that a potential function is usually parametrized for a fixed and known (albeit wide ranging) combination of coordination and bonding topology. This parametrization strategy for bulk fitting inevitably introduces potential energy barriers (e.g., the Albe barrier), thereby limiting the potential to accurately describe near-surface phenomena: a trade-off between the strain energy and number of dangling bonds leads to a nonphysical handling of bond-bending forces. A researcher therefore must be aware that an underestimation of either the strain energy or the number of dangling bonds may lead to an observed nonphysical behavior in the scenario of contact mechanics. This problem is well known in the MD community and led Tersoff to develop separate parametrisations to study surface and bulk silicon (Fan et al., 2021; Tersoff, 1988a, 1988b). Despite this, while the newly developed force fields are able to achieve improve accuracy over the old potential functions, they tend to be computationally expensive. A summary of this development is shown in Table 15.1.

Table 15.1 List of potential functions with respect to the time of introduction (Goel et al., 2015).

S. no.	Year	Name of the potential function	Materials suited
1	1984	EAM: embedded-atom method (Daw & Baskes, 1984)	Cu
2	1985	Stillinger–Weber potential (Stillinger & Weber, 1985)	Si
3	1987	SPC: simple point charge (Berendsen et al., 1987)	H_2O
4	1988	BOP: bond–order potential	Si
	1988	Tersoff-1 variant for silicon (Tersoff, 1988b)	Si
	1989	Tersoff-2 for better elastic properties of silicon (Tersoff, 1988a)	Si, Ge, and C
	1990	Tersoff-3 for Si, C and germanium (Tersoff, 1989, 1990b)	Si and C
	1994	Tersoff-4 for silicon and carbon (Tersoff, 1990a)	SiC
		Tersoff-5 for amorphous silicon carbide (Tersoff, 1994)	Si and C
		Refinements in Tersoff potential function (Agrawal et al., 2005; Devanathan et al., 1998; Kumagai et al., 2007)	Si and C
5	1989	EDIP (Bazant et al., 1997; Lucas et al., 2010)	Universal
6	1990	MEAM: modified embedded-atom method (Baskes et al., 1989)	
7	2000	REBO: reactive empirical bond order (Brenner, 1990)	Carbon
		AIREBO: adaptive intermolecular reactive empirical bond order (Stuart et al., 2000) (4 body potential function)	Hydrocarbons and carbon
8	2001	ReaxFF: reactive force field (Van Duin et al., 2001) (capable of bond breaking and bond-formation during the simulation)	Universal
9	2005	ABOP: analytical bond order potential (Erhart & Albe, 2005) (3 body potential function)	Si and C
10	2007	COMB: charge optimized many-body (Yu et al., 2007)	SiO_2, Cu, and Ti
11	2008	EIM: embedded-ion method (Zhou & Doty, 2008)	Ionic, for example, NaCl
12	2010	GAP: Gaussian approximation potential (Bartók et al., 2010)	Universal
13	1998–2001	Other important potential functions relevant in contact loading problems (de Brito Mota et al., 1998; Matsunaga & Iwamoto, 2001; Matsunaga et al. 2000)	Si; B, and N
14	2013	Screened potential functions (Pastewka et al., 2008, 2013)	Range of materials

Source: From Goel, S. Luo, X., Agrawal, A. & Reuben, R.L. (2015). Diamond machining of silicon: A review of advances in molecular dynamics simulation. International Journal of Machine Tools and Manufacture, 88, 131–164. https://doi.org/10.1016/j.ijmachtools.2014.09.013.

15.2 Postprocessing of the molecular dynamics data to extract the nanomechanical properties

Nanoindentation of materials is carried out to probe the nanomechanical properties of materials such as nanoindentation hardness and elastic modulus and to understand the origins of plasticity through tracking of the movement of dislocations, all of which are influenced by the crystallographic structure and orientation of the workpiece. The most basic thing to be first understood during the nanoindentation process is the load − displacement ($P-h$) behavior.

15.2.1 Estimation of the load − displacement (P−h) curve

The basic principle of nanoindentation is to apply load on the surface of the unknown material by a nanoindenter tool usually made from diamond due to its high hardness. This in turn results in the displacement or advancement of the indenter into the material. During the process of loading and retraction, a piezo attached to the indentor records the forces in real time, which are plotted against the displacement made by the indenter. This curve is popularly called as a $P-h$ curve, which highlights loading and unloading events during the indentation run. Usually when the indenter is retracted, the unloading curve does not follow the loading curve, and thus, hysteresis originates from the plastic deformation of the indented material. Fig. 15.3 show a typical load − displacement curve obtained from the simulated nanoindentation test under four different material setups, namely the high entropy alloy indented on the (100), (110), (111) orientations and the polycrystalline HEA sample. Clearly, the $P-h$ plots in all cases can be seen to indicate differences in the material behavior due to the differences in the microstructure and most specifically the crystal orientation. During the loading process, elastic deformation can be seen to take place in all cases. With an increase of the load, plastic deformation begins to appear. The h_{max} signifies maximum depth of indentation. The value of h_f/h_{max} was estimated from the simulation in all four cases. It can be seen that the polycrystalline HEA showed the minimum h_f/h_{max} ratio of 0.398 while the (100), (110), and (111) monocrystalline HEA substrate showed this ratio to be 0.733, 0.493, and 0.721, respectively. This ratio signifies the extent of residual work or how significant has been the recovery after the nanoindentation process.

Hertzian contact theory applied to the $P-h$ plot continues to remain a popular way to precisely mark the elastic − plastic transition point on the loading curve. This can be easily plotted by using the equation Force $= 4/3 E_r R^{1/2} h^{3/2}$. where E_r is the reduced elastic modulus, R is the radius of the indenter, and h is the instantaneous displacement of the indenter. Fig. 15.4 show a Hertzian curve fitted to identify the elastic − plastic transition point. This point coincides at an indentation depth of ∼8000 nm, similarly to that observed at smaller scale for the nanoindentation of HEA while using a spherical indenter with a diameter of 8 nm.

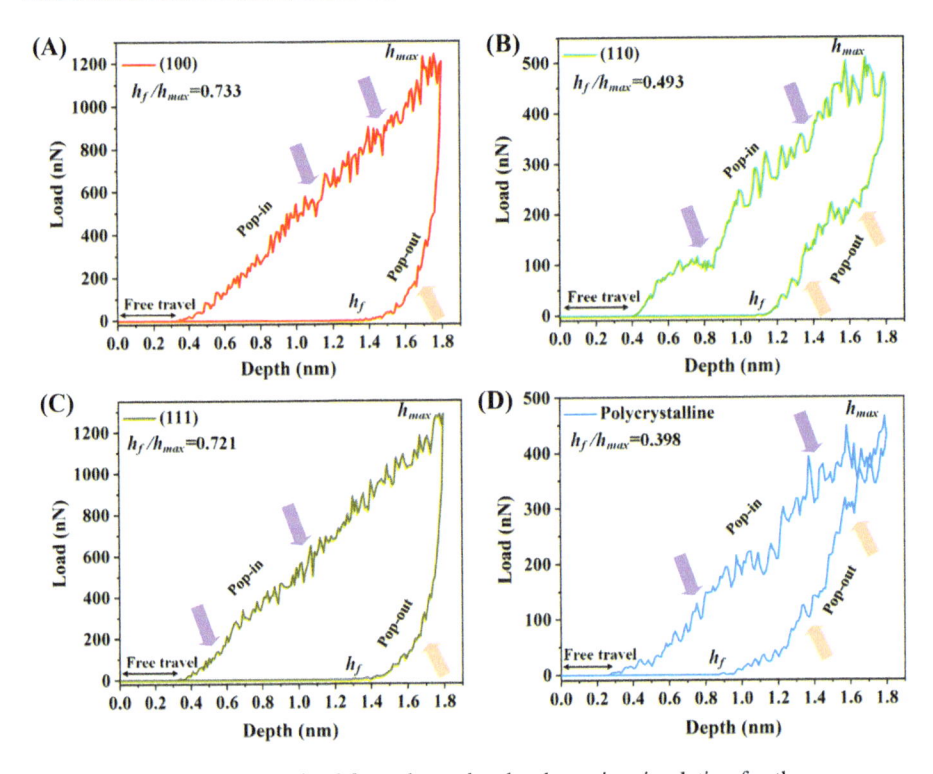

Figure 15.3 P-h curves obtained from the molecular dynamics simulation for the Ni$_{25}$Cu$_{18.75}$Fe$_{25}$Co$_{25}$Al$_{6.25}$ HEA (A) (100) orientation, (B) (110) orientation, (C) (111) orientation, and (D) polycrystalline HEA. Violet arrows and orange arrows refer to Pop-in and Pop-out events, respectively (Fan et al., 2024).
Source: From Fan, P., Nirmal, K., Arshad, M., Bai, M., Mao, H., & Goel, S. (2024). Anisotropic plasticity mechanisms in a newly synthesised high entropy alloy investigated using atomic simulations and nanoindentation experiments. *Journal of Alloys and Compounds*, 970. https://doi.org/10.1016/j.jallcom.2023.172541.

15.2.2 *Nanoindentation hardness and elastic modulus*

The nanoindentation results in the form of $P-h$ curve can be analyzed with respect to the contact pressure (p_m) underneath the indenter. Contact pressure (p_m) is defined as the ratio of the peak indentation force (F) on the indenter and the projected contact area (A) made in the workpiece postindentation. Building on recent practice (Mo et al., 2009; Ziegenhain et al., 2010), the contact area can be determined by the size of the contiguous region of the sample atoms involved in a chemical interaction with the indenter at a given instance or using the physical geometrical area using Oliver – Pharr approach. To estimate the atomic contact area, a chemical interaction cutoff (defined by the potential function) is required. The cutoff distances as per the ABOP potential for the Carbon – Tungsten contract can be 3.0 Å, for instance. This

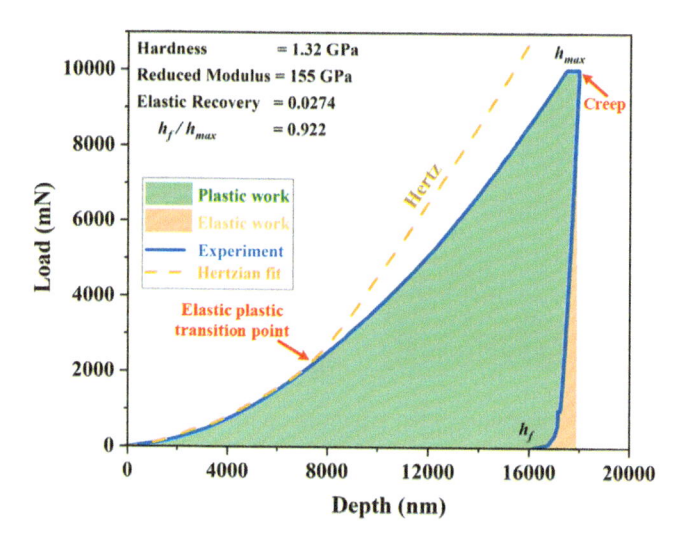

Figure 15.4 Experimental P–h plot and Hertzian fit on the Ni$_{25}$Cu$_{18.75}$Fe$_{25}$Co$_{25}$Al$_{6.25}$ polycrystalline substrate highlighting elastic and plastic work involved in the HEA indentation. *Source*: From Fan, P., Nirmal, K., Arshad, M., Bai, M., Mao, H., & Goel, S. (2024). Anisotropic plasticity mechanisms in a newly synthesised high entropy alloy investigated using atomic simulations and nanoindentation experiments. *Journal of Alloys and Compounds*, *970*. https://doi.org/10.1016/j.jallcom.2023.172541.

in turn gives us a three-dimensional elliptical contact area underneath the indenter as shown by the yellow atoms highlighted in Fig. 15.5.

After mapping the number of contacts, the projected contact diameter 2a, and hence, πa^2, the chemical contact area can be obtained. Recently, this contact area was compared with the purely continuum geometric approach of the Oliver and Pharr method (Oliver & Pharr, 1992) where the projected contact area for a spherical indenter is expressed as ($\pi \times a^2$) (here $a = \sqrt{h(2R - h)}$ is the contact radius of the spherical indenter, R is the radius of the indenter, and h is the instantaneous displacement of the indenter) (Fischer-Cripps, 2011). Interestingly, in the geometric estimation approach, the contact area remains unchanged with the indenter type, especially when the indentation force changes with the change of the atomic species in the indenter. In the chemical approach, both the contact area and the indentation force change with the change in the material and the indenter's boundary condition, and this helps increase the accuracy of the estimation. For a given indenter shape, the projected area of a residual impression (i.e., plastic deformation) divided by the peak indentation load used to form it is known as the hardness H of the material (for that shape). The contact pressure (p_m) or true stress can be plotted against true strain during the nanoindentation (Herbert et al., 2001) to calculate the elastic modulus of the material and determine the elastic − plastic limit of the substrate material. An alternative method for calculating the hardness from a spherical indenter in the MD simulation

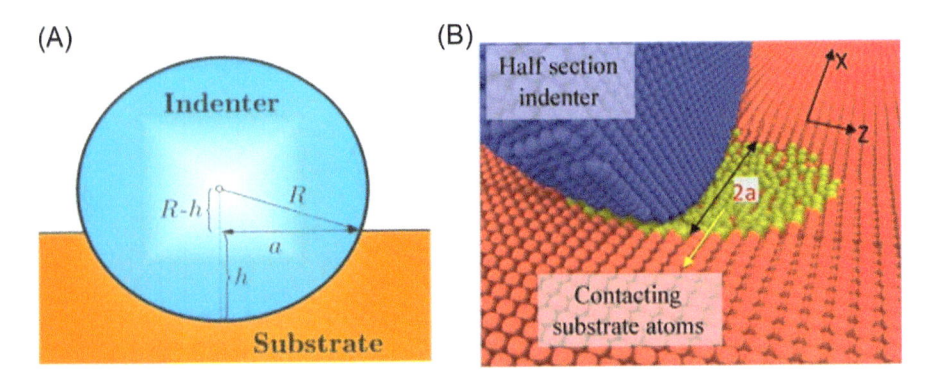

Figure 15.5 Schematic illustration of the scheme to estimate the projected contact area by using (A) geometry based area function approach, that is, $(\pi \times a^2)$, where $a =$ is the projected contact radius, so that R is the radius of the indenter and h is the instantaneous displacement of the indenter and (B) screening the contact atoms using fixed cutoff as per the ABOP potential function and then using $(\pi \times a^2)$ formulation where $2a$ is the contact diameter taken as the largest distance determined from the $(X_{high} - X_{low})$ and $(Z_{high} - Z_{low})$ values with X and Z being the atomic coordinates.
Source: From Goel, S., Cross, G., Stukowski, A., Gamsjäger, E., Beake, B., & Agrawal, A. (2018). Designing nanoindentation simulation studies by appropriate indenter choices: Case study on single crystal tungsten. *Computational Materials Science*, 152, 196–210. https://doi.org/10.1016/j.commatsci.2018.04.044.

framework also exists (Chen & Ke, 2004) but is only applicable to a rigid indenter, while the approach discussed earlier can be used for any type of the indentor.

In addition to hardness, nanoindentation can be used to characterize the elasticity of the material. The most popular way of calculating elastic modulus (E) of the material from the load–displacement (P–h) plot was proposed by Oliver and Pharr (O&P) (Oliver & Pharr, 1992). The method relies on the calculation of the projected contact area from a power law fitted to the unloading curve of the P–h plot. The slope of this curve (S) enables one to obtain the reduced elastic modulus of the material, which can then be used to obtain E using the following equation:

$$\frac{1-\vartheta^2}{E} = \frac{1}{E_r} - \frac{1-\vartheta_{indenter}^2}{E_{indenter}} \tag{15.1}$$

where E_r is the reduced modulus while υ is the Poisson's ratio.

15.2.3 Estimation of the strain rate sensitivity

The calculations of stress–strain can be used to further analyze strain rate sensitivity (only during loading). Stress (σ) during the indentation can be calculated as the ratio of the instantaneous load and the projected area (hardness) while the indentation strain rate $\dot{\varepsilon}$ (instantaneous descent rate of the indenter divided by that depth) and m (strain rate sensitivity) are two other important parameters expressed as (Mayo et al., 1990):

$$\sigma = \frac{P}{A} \tag{15.2}$$

$$\dot{\varepsilon} = \frac{1}{h_i}\left(\frac{dh}{dt}\right) \tag{15.3}$$

$$m = \frac{d\ln\sigma}{d\ln\dot{\varepsilon}} \tag{15.4}$$

where h_i is the displacement of the indenter at i th time step, dh/dt is the velocity of the indenter (since here it is a constant displacement indentation), A is the projected area ($2\pi \times R \times h_i$ for spherical indenter), and m is the strain rate sensitivity.

Strain rate sensitivity is an important indicator to assert if a large variation in the indentation speed will lead to significant changes in the outcome of the indentation results. Figs. 15.6 and 15.7 show the plots of the log stress versus log strain rate. These plots were fitted with a linear trend line to estimate the slope of each plot. The value of the slope (strain rate sensitivity) obtained from Fig. 15.6 for Fe_3C was in the

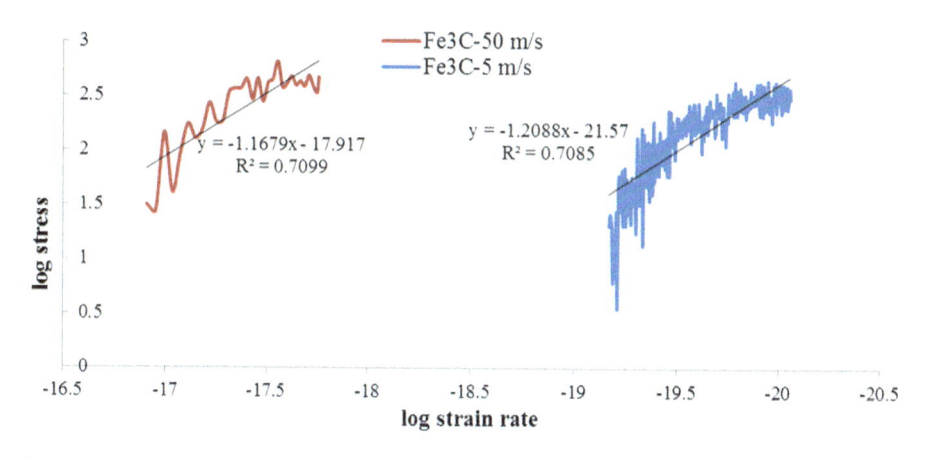

Figure 15.6 Comparison of log (stress) and log (strain rate) for Fe_3C obtained from the molecular dynamics simulation with respect to change in the indentation speed (Goel et al., 2014).
Source: From Goel, S., Joshi, S.S., Abdelal, G., & Agrawal, A. (2014). Molecular dynamics simulation of nanoindentation of Fe_3C and Fe_4C. *Materials Science and Engineering: A, 597*, 331−341. https://doi.org/10.1016/j.msea.2013.12.091.

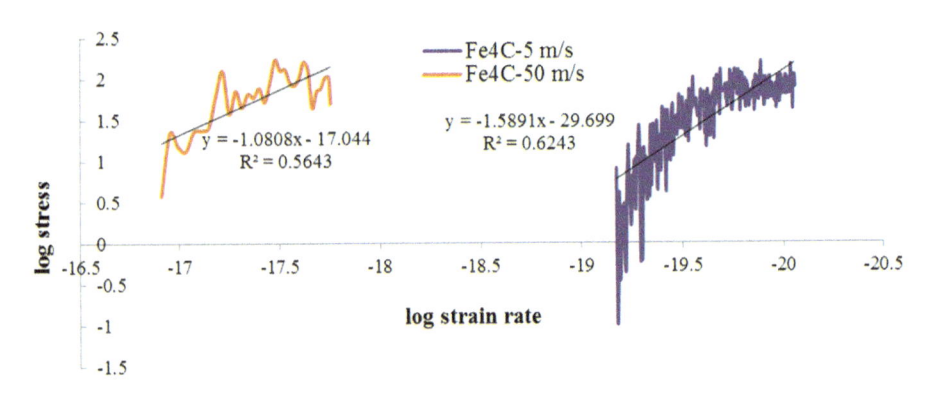

Figure 15.7 Comparison of log (stress) and log (strain rate) for Fe$_4$C obtained from the MD simulation with respect to change in the indentation speed.
Source: From Goel, S., Joshi, S.S., Abdelal, G., & Agrawal, A. (2014). Molecular dynamics simulation of nanoindentation of Fe$_3$C and Fe$_4$C. *Materials Science and Engineering: A, 597*, 331–341. https://doi.org/10.1016/j.msea.2013.12.091.

range of -1.16–-1.2, whereas that for Fe$_4$C (Fig. 15.7) was observed to vary between -1.08 and -1.589 across the speed of indentation of 5–50 m s^{-1}. The variations in the magnitude of strain rate sensitivity are marginal and therefore signify that the MD simulation results shown here remain insensitive to the speed of the indenter.

15.2.4 Estimation of the stresses causing deformation during nanoindentation

The state of stress acting on a small volume of material undergoing three-dimensional deformation during the process of nanoindentation is schematically shown in Fig. 15.8. To estimate this, MD is a very good tool; however, the magnitude of the instantaneous value of the stress calculated from the MD simulation should always be time-averaged (Fan et al., 2020; Wang et al., 2021). One fundamental problem with the computation of atomic stress is that the volume of an atom does not remain fixed during deformation. To mitigate this problem, the best method is to plot the stresses on the fly by considering an elemental atomic volume in the cutting zone. The total stresses acting on that element could be computed and divided by the precalculated total volume of that element to obtain the physical stress tensor (engineering stress). When a stress tensor from the simulation is available, the following equations can readily be used to obtain the Tresca stress, von Mises stress, octahedral shear stress, and hydrostatic stress:

$$\text{Stress tensor} = \begin{pmatrix} \sigma_{xx} & \tau_{xy} & \tau_{xz} \\ \tau_{xy} & \sigma_{yy} & \tau_{yz} \\ \tau_{xz} & \tau_{xz} & \sigma_{zz} \end{pmatrix} \quad (15.5)$$

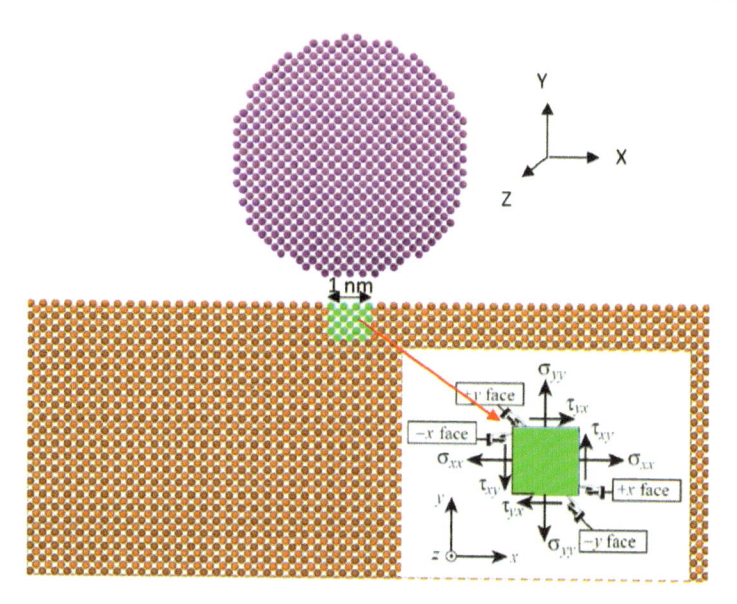

Figure 15.8 Volume of material (1 × 1 × 1 nm) considered for stress computation (only 2D representation of 3D volume in XY plane is shown here).
Source: From Goel, S., Beake, B., Chan, C.W., Haque Faisal, N., & Dunne, N. (2015). Twinning anisotropy of tantalum during nanoindentation. *Materials Science and Engineering: A, 627*, 249–261. https://doi.org/10.1016/j.msea.2014.12.075.

Invariants:

$$I_1 = \sigma_{xx} + \sigma_{yy} + \sigma_{zz} \tag{15.6}$$

$$I_2 = \sigma_{xx}\sigma_{yy} + \sigma_{yy}\sigma_{zz} + \sigma_{zz}\sigma_{xx} - \tau_{xy}^2 - \tau_{xz}^2 - \tau_{yz}^2 \tag{15.7}$$

$$I_3 = \sigma_{xx}\sigma_{yy}\sigma_{zz} + 2(\tau_{xy}\tau_{yz}\tau_{xz}) - \tau_{xz}^2\sigma_{yy} - \tau_{yz}^2\sigma_{xx} - \tau_{xy}^2\sigma_{zz} \tag{15.8}$$

$$A_1 = -I_1; A_2 = I_2; A_3 = -I_3; \tag{15.9}$$

$$Q = \frac{3A_2 - A_1^2}{9} \tag{15.10}$$

$$R = \frac{9A_1A_2 - 27A_3 - 2A_1^3}{54} \tag{15.11}$$

$$D = Q^3 + R^2 \tag{15.12}$$

If $D<0$, then as follows: else the condition is 2D stress

$$\theta = \cos\left(\frac{R}{\sqrt{-Q^3}}\right)^{-1} \tag{15.13}$$

$$R_1 = 2\sqrt{-Q} \times \cos\left(\frac{\theta}{3}\right) - \frac{A_1}{3} \tag{15.14}$$

$$R_2 = 2\sqrt{-Q} \times \cos\left(\frac{\theta + 4\pi}{3}\right) - \frac{A_1}{3} \tag{15.15}$$

$$R_3 = 2\sqrt{-Q} \times \cos\left(\frac{\theta + 2\pi}{3}\right) - \frac{A_1}{3} \tag{15.16}$$

Major principal stress $(\sigma_1) = \max(R_1, R_2, R_3);$ \qquad (15.17)

Major principal stress $(\sigma_3) = \min(R_1, R_2, R_3);$ \qquad (15.18)

$$\sigma_{tresca} = \frac{\sigma_1 - \sigma_3}{2} \tag{15.19}$$

$$\sigma_{VONMises} = \sqrt{\frac{(\sigma_{xx}-\sigma_{yy})^2 + (\sigma_{yy}-\sigma_{zz})^2 + (\sigma_{zz}-\sigma_{xx})^2 + 6(\tau_{xy}^2 + \tau_{yz}^2 + \tau_{xz}^2)}{2}} \tag{15.20}$$

$$\sigma_{octahedral} = \sqrt{\frac{(\sigma_{xx}-\sigma_{yy})^2 + (\sigma_{yy}-\sigma_{zz})^2 + (\sigma_{zz}-\sigma_{xx})^2 + 6(\tau_{xy}^2 + \tau_{yz}^2 + \tau_{xz}^2)}{2}}$$

$$= \frac{\sqrt{2}}{3}\sigma_{VONMises} \tag{15.21}$$

15.2.5 Subsurface deformation mechanisms, dislocations, and crystal defects

Powerful analysis and visualization techniques play a key role in the data analysis as the simulated material systems become complex. Without the right software tool, key information would remain undiscovered, inaccessible, and unused. OVITO (Stukowski, 2010) is a visualization and analysis software for output data generated in molecular dynamics, which can be used to visualize and analyze the atomistic simulation data while other tools such as the automated "dislocation extraction algorithm" (DXA) (Stukowski et al., 2012) and crystal analysis tool (CAT) (Stukowski &

Albe, 2010) can be used for automated identification of crystal defects, dislocation lines, and their Burgers Vector from the output of the MD data. An example of subsurface dislocations extracted during indentation of tungsten with different types of indenter materials is shown in Fig. 15.9. It highlights the differences in the subsurface mechanisms observed when using different indenter types.

Similarly, the evolution of microstructure during indentation of $Ni_{25}Cu_{18.75}Fe_{25}Co_{25}Al_{6.25}$HEA is shown by Figs. 15.10−15.12 revealing various atomic movements indicating mechanisms of reversible and inelastic behavior of HEA at indentation depths of 0.5 nm and 1.5 nm as various orientations were indented. In particular FCC phase HEA, several dislocations such as the 1/2 <110> perfect dislocation, 1/6 <112> Shockley dislocation, 1/6 <110> Stair-rod dislocation, 1/3 <100> Hirth dislocation, and 1/3 <111> Frank dislocation were seen to emerge that can clearly be seen in Figs. 15.10 and 15.11. These were accompanied by multiple shear dislocation loops leading to a forest of dislocations after unloading the indentation load as also reported experimentally through the TEM observations (Lee, 2020). On indentation of the (110) orientation, a "lasso" type dislocation loop can be seen to emerge, which was not evidenced in any other case of indentation (Remington et al., 2014; Xiang et al., 2017). The (110) orientation showed multiple shear loops, which were seen to advance under the stress caused by the indenter into the HEA material by the advancement of their edge components. The dislocations for the (111) orientation were mostly 1/6 <112> Shockley, 1/6 <110> Stair-rod, and 1/3 <100> Hirth. Note that the Hirth dislocation can result from reactions between Shockley dislocations as $1/6[\bar{1}2\bar{1}] + 1/6[\bar{1}2\bar{1}] = 1/3[\bar{1}00]$ and Frank dislocation can result from reactions of Stair-rod and Shockley, for example, $1/6[011] + 1/6[211] = 1/3[111]$. Consequently, Hirth dislocations are also known as Hirth locks (Fan et al., 2023; Paulauskas et al., 2015; Wang, Fan et al., 2021), and Frank dislocations are known as Lomer−Cottrell (LC) locks (Bulatov et al., 1998; Fan et al., 2022, 2024).

A typical FCC structure has 12 nonequivalent partial dislocation slip systems $\langle 110 \rangle/2 - \{111\}$. However, not all these slip systems are operative during nanoindentation. Based on tensor rotation, we can convert the applied shear stress ($\sim 5-6$ GPa) to each of the slip systems. The slip systems with the maximum conversion factors (analogous to Schmid factor) are operative. It was observed that the (010) orientation would have eight most likely operative slip systems $[101]/2 - (\bar{1}11)$, $[0\bar{1}1]/2 - (\bar{1}11)$, $[101]/2 - (11\bar{1})$, $[011]/2 - (11\bar{1})$, $[011]/2 - (\bar{1}1\bar{1})$, $[10\bar{1}]/2 - (\bar{1}1\bar{1})$, $[0\bar{1}1]/2 - (111)$, $[\bar{1}01]/2 - (111)$ with a conversion factor of 0.41, the (110) orientation has four most likely operative slip systems $[110]/2 - (1\bar{1}\bar{1})$, $[1\bar{1}0]/2 - (11\bar{1})$, $[110]/2 - (\bar{1}1\bar{1})$, $[1\bar{1}0]/2 - (111)$ with a conversion factor of 0.82, and the (111) orientation has one most likely operative slip systems $[1\bar{1}0]/2 - (111)$ with a conversion factor of 1.00, two second likely operative slip systems $[110]/2 - (\bar{1}11)$, $[110]/2 - (1\bar{1}1)$ with a conversion factor of 0.87, and four most likely operative slip systems $[101]/2 - (1\bar{1}\bar{1})$, $[011]/2 - (\bar{1}1\bar{1})$,

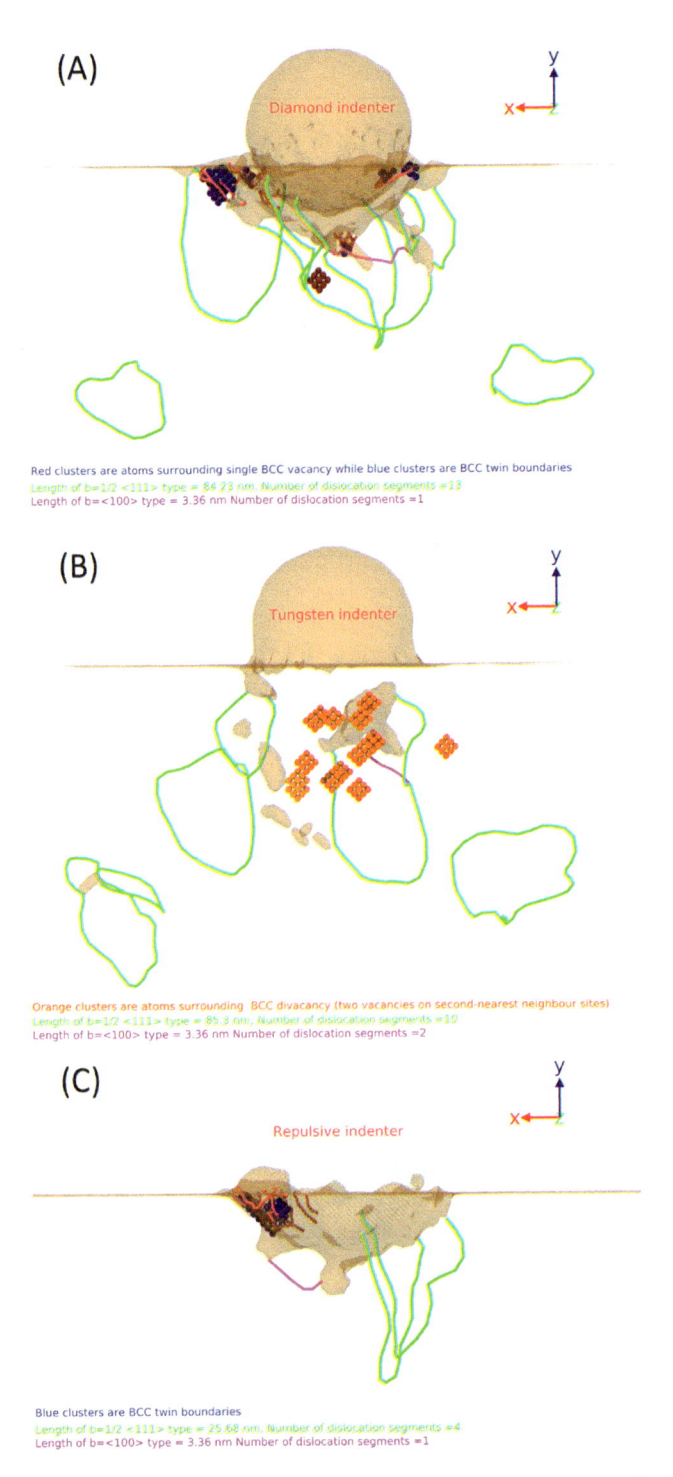

Figure 15.9 Views in the XY planes obtained from DXA processing shows plastic deformation zone, crystal defects and free surfaces comparing sample defects created for diamond (DI-full), tungsten (TI) and smooth repulsive (RI) at a penetration depth of 2 nm.

(Continued)

$[0\bar{1}1]/2-$ (111), $[\bar{1}01]/2-$ (111) with a conversion factor of 0.50. Note that during the (010) indentation, every slip vector is activated on two slip planes, for instance, the [101]/2 vector is activated on the $(\bar{1}11)$ and $(11\bar{1})$ planes in the $[101]/2-(\bar{1}11)$ and $[101]/2-(11\bar{1})$ slip systems. This feature is not present on the other orientations.

In terms of the crystal defects, they were seen to consist of FCC intrinsic stacking faults (ISF) resembling two atomic layers of hcp coordination, adjacent intrinsic stacking faults (quad faults and resembles as four atomic layers of hcp coordination), coherent $\sum 3$ twin boundary, and a coherent twin boundary next to an intrinsic stacking fault (triple fault) resembling three atomic layers of hcp coordination. The evolution of microstructure of monocrystalline HEA and polycrystalline HEA is demonstrated, where green color atoms refer to the FCC microstructure, white color atoms are disordered, and the red color atoms highlight stacking faults. At an indentation depth of 0.5 nm, FCC microstructure becomes disordered and stacking faults appeared in the deformation zone. At a deeper indent depth of 1.5 nm, stacking faults were seen to grow in numbers. The stacking faults were formed on both sides of the deformed area for the (100) HEA substrate while for the (110) and (111) orientation, the stacking faults extend deeply along the indent direction. As the material recovers during unloading, some of the stacking faults reverse elastically. Furthermore, it highlights that the presence of grain boundary obstructs the growth and propagation of shear loops leading to form a complex and concentrated junction of stacking faults in the indented area. Thus, in the case of (110) and polycrystalline HEA substrates, the elastic work in the indentation hysteresis loop was seen to be larger than the (100) and the (111) orientations.

Another typical example of dislocations and crystal defects was studied in tantalum. Fig. 15.13 shows the (convoluted looking) dislocation structures in the plastic zone of tantalum post indentation by a depth of 2 nm on three crystallographic orientations. The bottom portion of Fig. 15.13 shows the magnified view both at the peak indentation depth and after the indenter was retracted. Further details of these dislocations are provided in Table 15.2. These data were used to estimate the dislocation density (m^2) in all the three simulation cases (Gao et al., 2014; Wang et al., 2014) using Density of dislocations = Length of dislocations/$\left(\frac{2\pi R_{pl}^3}{3}\right) - (2\pi h^3/3)$, where R_{pl} is the radius of the plastic zone (largest distance of a dislocation from the indentation

◀ Dislocations with b = 1/2 <111> are shown in green and b = <100> are shown in pink. Common neighbor analysis was used to identify and remove the crystalline atoms to highlight crystalline defects. Red atoms represent atoms surrounding a single BCC vacancy while orange clusters are atoms surrounding a BCC divacancy (two vacancies on second-nearest neighbor sites) and blue atoms represent BCC twin boundaries. The smooth RI indenter is represented for visualization, but it was not physically present in the simulation. (Goel et al., 2018).

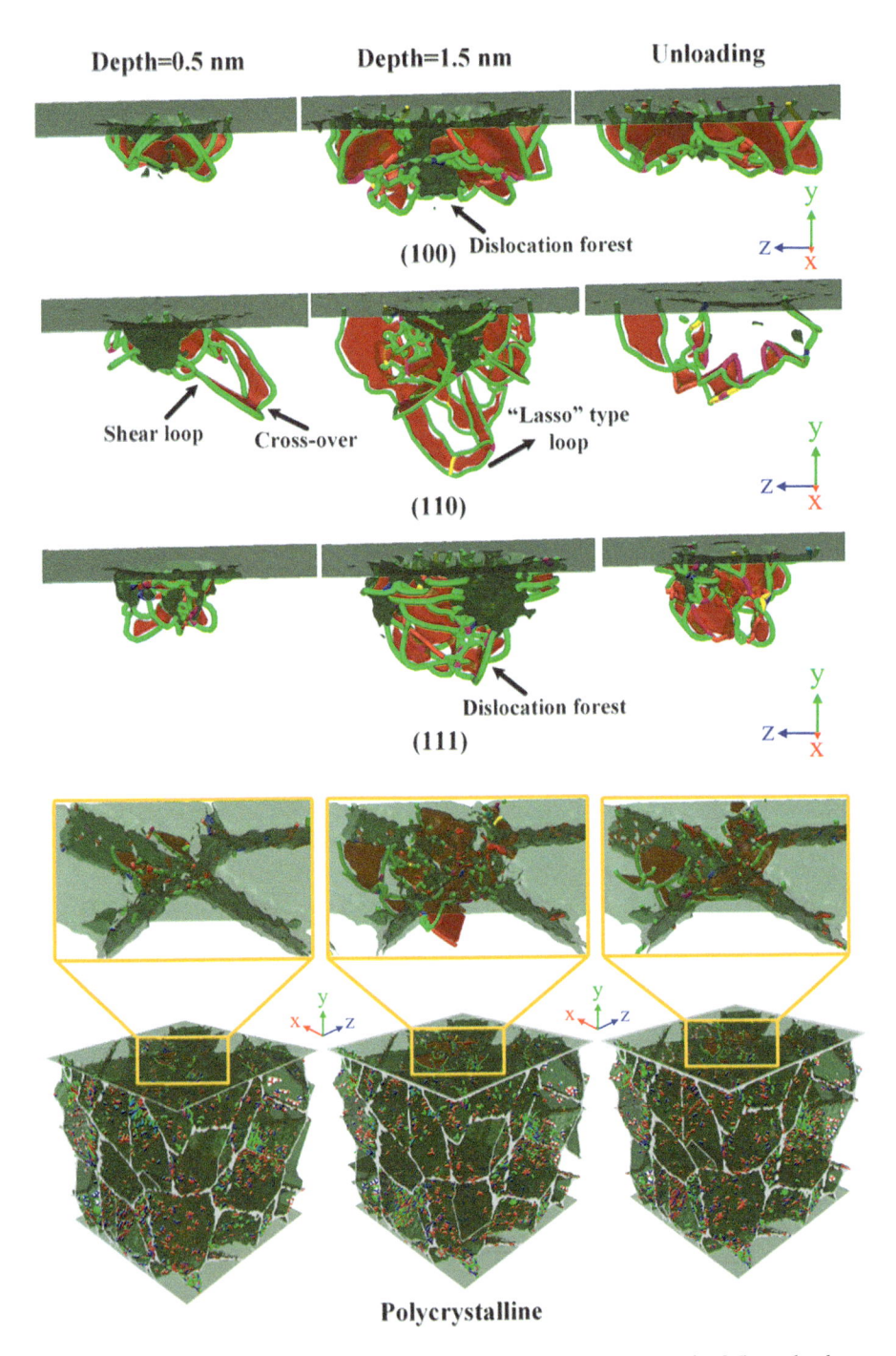

Figure 15.10 The evolution of dislocation nucleation and movement at the 0.5 nm depths, 1.5 nm depths and the unloading stage for (100) monocrystalline, (110) monocrystalline, (111) monocrystalline and polycrystalline structure.

(Continued)

point), which is assumed to be hemispherical, and for an indentation depth of R, the indented volume is also considered hemispherical. Consequently, the dislocation density obtained from MD during indentation of Ta by 2 nm depth was found to be of the order of 1.6×10^{14} m^2.

Fig. 15.13 also revealed that the prismatic dislocation loops were observed to transport the material downward under the wake of the indenter. Furthermore, Burgers vector of the dislocation lines (arrows shown in Fig. 15.13) were found perpendicular to the dislocation line, clearly indicating the edge nature of the dislocations rather than screw dislocations observed during tensile pulling (Smith et al., 2014) of Ta. There were two major types of dislocations captured with b = 1/2 <111> (blue color) and b = <100> (red color). On some instances, the <100> dislocation was found to form because of the interaction between two 1/2 <111> dislocation loops. On unloading of the indenter on the (110) orientation, the dislocations were seen to be completely reversible in contrast to the (100) and (111) orientations meaning that the plasticity on this orientation in particular is not only driven by dislocation nucleation.

Apart from the prismatic dislocation loops, crystal defects in the form of twin boundaries below the indenter were also observed on the (110) and (111) oriented Ta, which are highlighted in Fig. 15.14. In Fig. 15.14, green-colored atoms refer to twin boundaries. It is noticeable that the twin boundaries (twin planes (112) planes) (well known to be responsible for plasticity in BCC metals) were noticed only during indentation on the (110) and (111) plane and not on the (100) plane when the indentation was performed with the rigid tantalum indenter. However, with a purely repulsive indenter, the twin boundaries were observed on the (Goel et al., 2014) plane as well. This is one of those gray area where the use of a diamond indenter can verify the presence of twin boundaries during indentation on the (Goel et al., 2014) orientation of Ta; however, to date, a reasonable interaction potential of diamond and Ta is not available.

15.3 High-order postprocessing analysis of the molecular dynamics simulation files to correlate with experiments

15.3.1 Anisotropic mechanisms during nanoindentation

A notable feature of plasticity in BCC metals is their asymmetry in tension-compression behavior (Healy & Ackland, 2014), and this asymmetry has been

---

◂ *Source*: From Fan, P., Nirmal Kumar, K., Arshad, M., Bai, M., Mao, H., & Goel, S. (2024). Anisotropic plasticity mechanisms in a newly synthesised high entropy alloy investigated using atomic simulations and nanoindentation experiments. *Journal of Alloys and Compounds*, 970. https://doi.org/10.1016/j.jallcom.2023.172541.

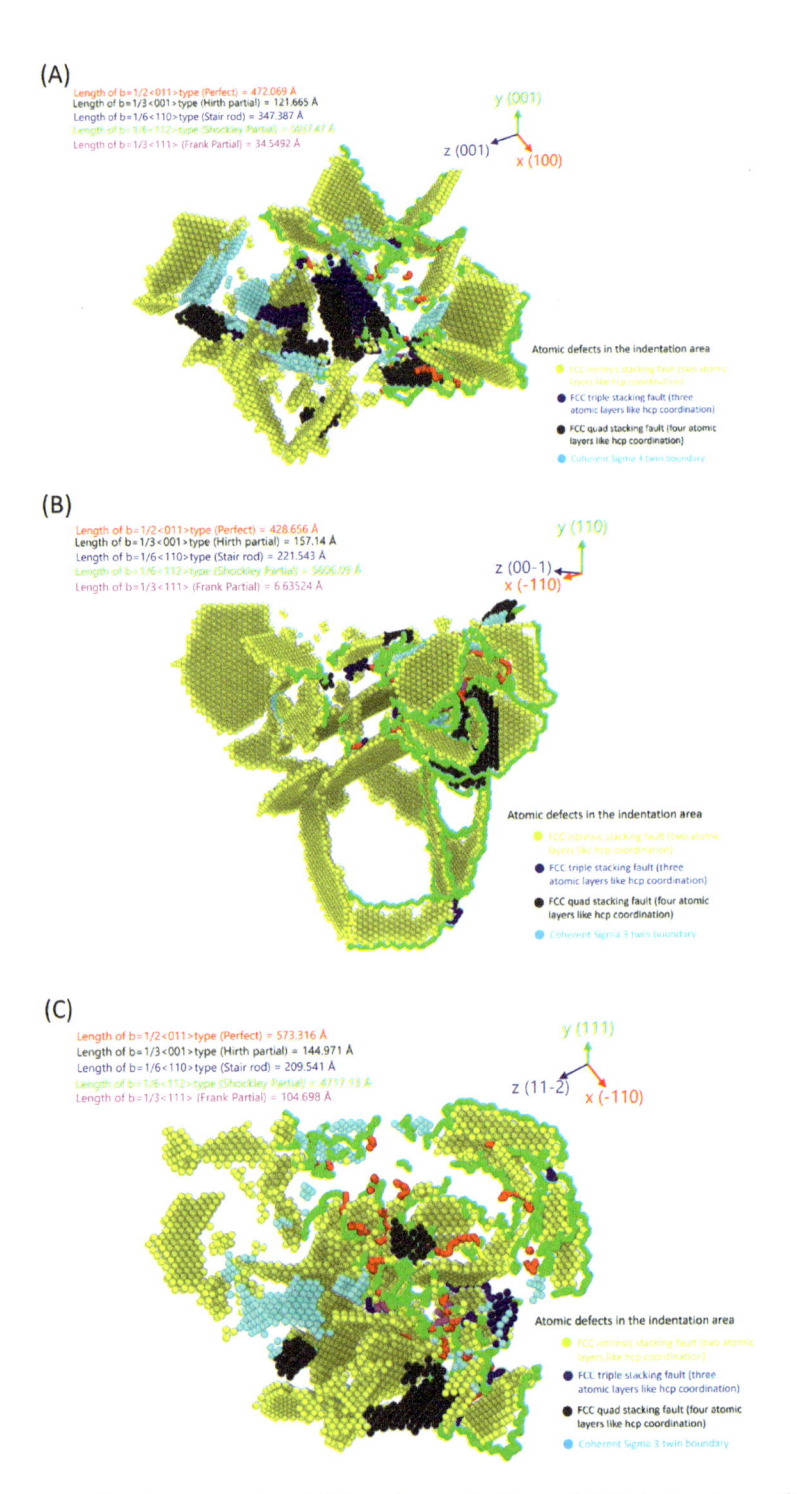

Figure 15.11 The demonstration of dislocation nucleation and defect atom types after indentation for **(A) (100) monocrystalline, (B) (110) monocrystalline, (C) (111) monocrystalline** (Fan et al., 2024). (A) Dislocation and defect atom types after indentation

(Continued)

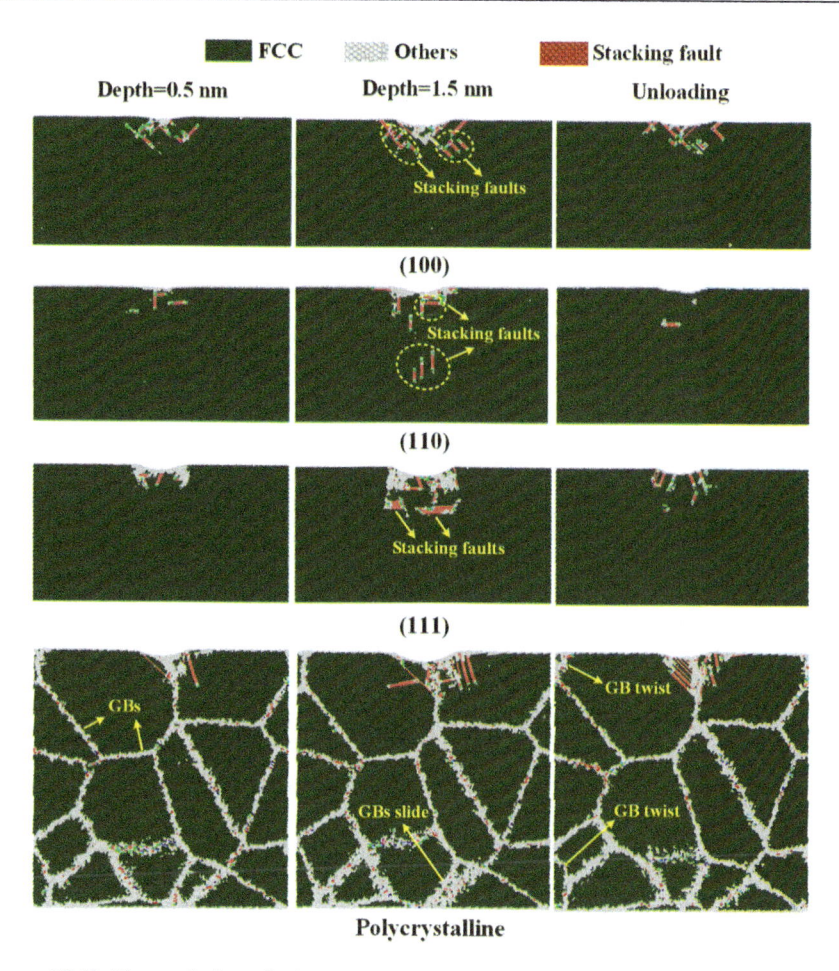

Figure 15.12 The evolution of microstructure at depths of 0.5 nm, 1.5 nm and after unloading.
Source: From Fan, P., Nirmal Kumar, K., Arshad, M., Bai, M., Mao, H., & Goel, S. (2024). Anisotropic plasticity mechanisms in a newly synthesised high entropy alloy investigated using atomic simulations and nanoindentation experiments. *Journal of Alloys and Compounds*, *970*. https://doi.org/10.1016/j.jallcom.2023.172541.

◀ on (100) monocrystalline HEA showing several nested "shear loops" (B) Dislocation and defect atom types after indentation on (110) monocrystalline HEA showing "lasso-like" dislocation loop (C) Dislocation and defect atom types after indentation on (111) monocrystalline HEA.

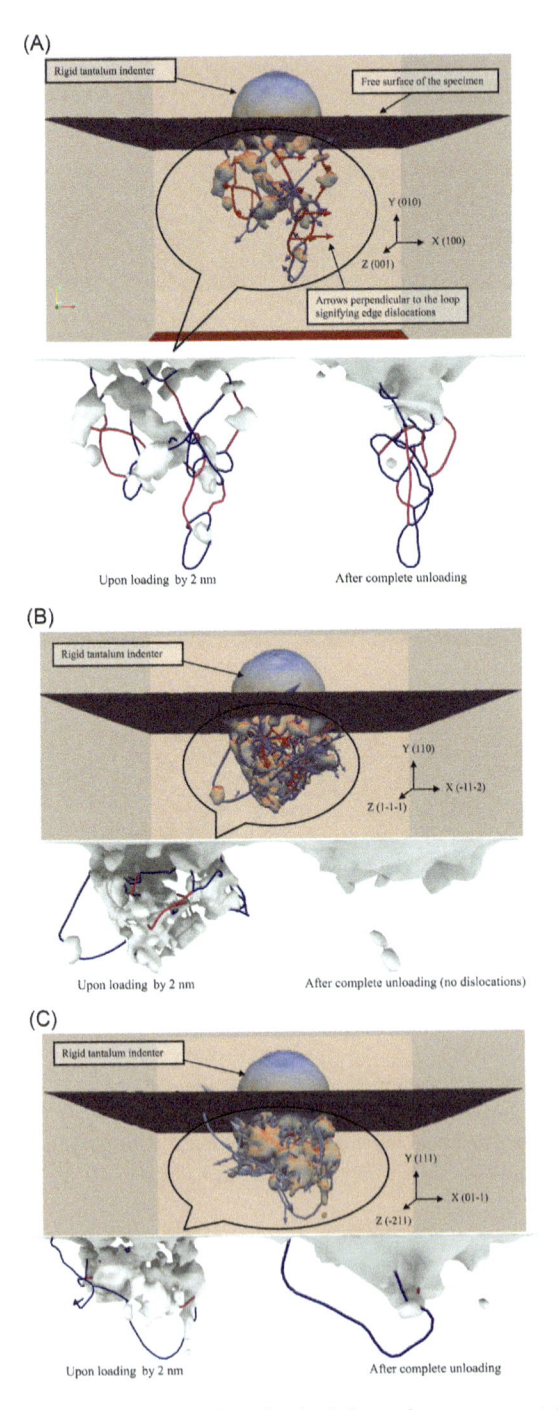

Figure 15.13 Output of the DXA showing plastic deformation zone, crystal defects, free surfaces and dislocation lines during nanoindentation of tantalum on three orientations (A) (Goel et al., 2014) orientation, (B) (110) orientation, and (C) (111) orientation.

(Continued)

Table 15.2 Characteristics of the dislocations obtained using the three simulation cases with rigid tantalum indenter. L is the total length of dislocations in Å (Goel et al., 2015).

	Goel et al. (2014) orientation	(110) orientation	(111) orientation
Burgers Vector (b = 1/2 <111>)			
L (total length) of dislocations	663.914 Å	763.899 Å	993.06 Å
Burgers Vector (b = <100>)			
L (total length) of dislocations	96.255 Å	149.3 Å	36.44 Å
Total length of dislocations	760.17 Å	913.19 Å	1029.5 Å
Radius of the plastic zone (R_{pl})	131 Å	140 Å	145 Å
h (displacement of the indenter)	20 Å	20 Å	20 Å
Dislocation density obtained from the MD simulation	1.62×10^{14} m^2	1.59×10^{14} m^2	1.615×10^{14} m^2

Source: From Goel, S., Beake, B., Chan, C.W., Haque Faisal, N., & Dunne, N. (2015). Twinning anisotropy of tantalum during nanoindentation. *Materials Science and Engineering: A, 627*, 249−261. https://doi.org/10.1016/j.msea.2014.12.075.

attributed to the differences in the different deformation mechanisms, that is, dislocation glide prevails in compression while twinning is prominent during tensile pulling of BCC metals. An exception to this observation has recently come from recent experiments (Wang et al., 2005), where unlike other BCC metals, the plastic deformation of Ta during its nanoindentation was found to be dominated by the deformation twinning. Another area where Ta has shown deviation from the classical knowledge is that its elastic − plastic transition on the (110) crystallographic direction occurs beyond the theoretical critical value of shear stress (Guerrero & Marucho, 2013), thereby pointing to the fact that the parameter critical resolved shear stress may not be applicable at atomic scale where the role of dislocation mechanics is more prevalent. A fundamental criterion for nucleation of plastic deformation could be that the work done over the displaced surface should be greater than the line

◄ The geometric boundaries of tantalum are shown, while the geometric boundaries of the disordered phases are not visible in these visualizations. The top part shows the bulk view at peak loading conditions while the bottom part shows the magnified view of the plastic zone at peak loading and upon unloading. Dislocations with **b** = 1/2⟨111⟩ are shown in blue and **b** = ⟨100⟩ are shown in red. Arrows indicate the direction of **b** (Burgers Vector) with respect to the dislocation loop. Here these arrows are in a direction perpendicular to the dislocation lines signifying that the dislocations are pure edge dislocations.
Source: From Goel, S., Beake, B., Chan, C.W., Haque Faisal, N., & Dunne, N. (2015). Twinning anisotropy of tantalum during nanoindentation. *Materials Science and Engineering: A, 627*, 249−261. https://doi.org/10.1016/j.msea.2014.12.075.

Figure 15.14 Spliced cross-sectional views in the *XY* plane showing twinning on the (110) and (111) crystal orientations. No twin boundaries were detected on the (Goel et al., 2014) orientation. Common neighbor analysis (CNA) is used to identify and distinguish the crystalline atoms (shown in brown color), atoms with crystalline defects are shown in yellow color and rigid indenter is shown in pink color. Green color is used to mark atoms that form the twin boundaries in the entire substrate.
Source: From Goel, S., Beake, B., Chan, C. W., Haque Faisal, N., & Dunne, N. (2015). Twinning anisotropy of tantalum during nanoindentation. *Materials Science and Engineering: A*, *627*, 249−261. https://doi.org/10.1016/j.msea.2014.12.075.

energy of the new dislocation loop (Kelchner et al., 1998). These observations should not be surprising since Ta does not obey Schmidt law. Anisotropy in deformation of Ta can be observed in terms of 3D stress analysis, maximum shear stress, and pileup formation. Fig. 15.15A shows the sectional view in the XY plane representing the distribution of the octahedral shear stress during the deformation of tantalum on each of the three orientations when the indenter has moved into the substrate by 2 nm. Fig. 15.15B on the other hand shows the spliced view in the XZ plane (top view) of the substrate in conjunction with Fig. 15.15A, and compare the topography of the lattice deformation and pileup with Fig. 15.15C obtained from the experiments (Biener et al., 2007). Interestingly, a very distinct distribution of octahedral shear stress is apparent, which varies with its location during indentation on each plane. This could certainly be attributed to the differences in the BCC crystal structure, and this difference leads to the deformation patterns to resemble as to what has been shown in Fig. 15.15B. In Fig. 15.15B, the simulations performed on the (Goel et al., 2014) orientation of Ta showed the pileup pattern (lattice deformation) along the closed packed direction of the BCC metal. Interestingly, in the case of the (110) orientation, where both in-plane and out-of-plane slip directions are available, the pileup is predominantly found along the in-plane slip directions resembling closer with the microscale experimental findings (Biener et al., 2007).

These results are incorporated in Table 15.3, and it can be seen that the repulsive indenter overestimated the deformation stresses by a slight margin. The classical Hertzian contact theory suggests a multiplicative factor of 0.465 with the mean pressure to arrive at the maximum shear stress underneath the indenter (Grau et al., 2002). This value is taken as a conventional wisdom in a variety of indentation studies and has somewhat become a common lore particularly since the direct measurement of maximum shear stress during the experiments is difficult. MD simulation results (Table 15.4) clarify that this multiplicative factor from the classical Hertzian contact may be incorrect at the atomic scale. Table 15.4 obtained from the MD results shows (1) this ratio to be about 0.4 for the (110) and (111) orientation and ~0.8 for the (100) orientation of tantalum and (2) the maximum shear (Tresca) stress was found to exceed the theoretical shear strength of tantalum on the (Goel et al., 2014) orientation but was well within for the (110) and (111) orientations. This unique observation on the (Goel et al., 2014) plane seems to be in accord with a recent study (Guerrero & Marucho, 2013) where the elastic − plastic limit of Ta was observed to go past the theoretically predicted critical shear stress.

Another example of nanoindentation is that of the high entropy alloy. In the MDS results, the residual indentation showed strong anisotropy on the (100), (110), (111) orientations as well as on the polycrystalline HEA shown in Fig. 15.16. The (100) HEA showed clear fourfold symmetry atomic pileup (see Fig. 15.16A) along the $\langle 011 \rangle$, $\langle 0\bar{1}\bar{1} \rangle$, $\langle 0\bar{1}1 \rangle$, and $\langle 01\bar{1} \rangle$ directions, which is consistent with the previous report for the Fe-Ni-Cr-Co-Cu high-entropy alloy (Ruestes & Farkas, 2022). The (110) monocrystalline setup showed evenly distributed deformation shown in Fig. 15.16B with somewhat a fourfold symmetry. In the case of (111) orientation, material flow and pileup shown in Fig. 15.16C did not indicate any clear preferential direction of deformation. Interestingly, for the polycrystalline HEA model, the

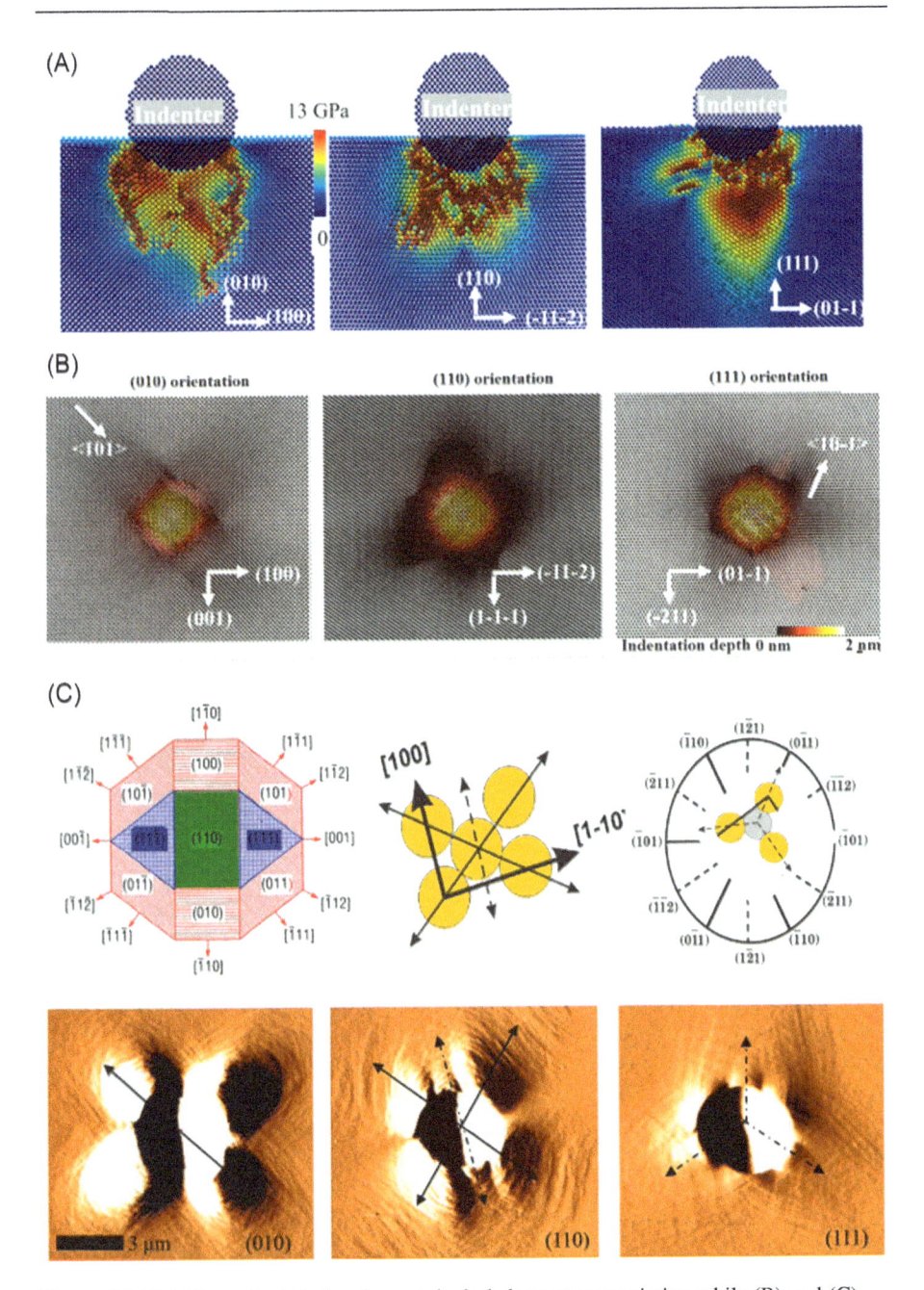

Figure 15.15 MD results (A) showing octahedral shear stress variation while (B) and (C) showing pileup and lattice deformation along the closed packed directions. (A) View on the XY Plane (front view) showing variation in the octahedral shear stress on the three orientations of tantalum underneath the indenter obtained from the MD simulation. (B) View

(Continued)

Table 15.3 Nanoscale yielding stresses of tantalum obtained from the molecular dynamics simulation (Goel et al., 2015).

Critical value of maximum stress (GPa) in the deformation zone of tantalum	Goel et al. (2014) orientation		(110) orientation	(111) orientation
	With a pure repulsive indenter	With a rigid tantalum indenter	With a rigid tantalum indenter	With a rigid tantalum indenter
Von Mises stress	28.69	21.94	10.05	12.51
Octahedral shear stress	13.52	10.34	4.74	5.9
Tresca stress	14.4	11.78	5.43	6.86
Major principal stress	−36.53	−30.66	−22.24	−30.05
Minor principal stress	−7.74	−7.09	−11.85	−16.85
Hydrostatic stress	−20.56	−25.74	−17.18	−24.52

Source: From Goel, S., Beake, B., Chan, C.W., Haque Faisal, N., & Dunne, N. (2015). Twinning anisotropy of tantalum during nanoindentation. *Materials Science and Engineering: A, 627*, 249–261. https://doi.org/10.1016/j.msea.2014.12.075.

fourfold or threefold symmetry atomic pileup was less clear due to the interaction of several grains, which can be seen from Fig. 15.16D. To further study this, the polycrystalline dump file was postprocessed by removing the top lattice layers using the polyhedral template matching algorithm within OVITO, which leaves a clear visibility of the individual grains as shown in Fig. 15.17.

For ease of understanding, the polycrystalline configuration has been compared pre and postindentation deformation by bringing an EBSD equivalent color scheme. It can be seen from Fig. 15.17 that the indenter made contact in the region, which is at the intersection of several grains with orientations (310), (111), (101), and (210). As opposed to a perfect spherical symmetry, the lines of plastic deformation can be seen to be traversing through these grains carrying different propensity of deformation. Of those grains, the (210) orientation can be seen to be affected the most showing that the extent of dislocation propagation has affected a good half part of the grain, thus confirming that this orientation is most amenable compared with the other three grains in the ease of its plastic deformation.

◀ in the XZ plane (top view) showing atomic displacements (lattice deformation) on all three crystal planes of tantalum obtained from the MD simulation. (C) Experimental results (Biener et al., 2007) on the nanoindentation on Ta on three different orientations.
Source: From Goel, S., Beake, B., Chan, C. W., Haque Faisal, N., & Dunne, N. (2015). Twinning anisotropy of tantalum during nanoindentation. *Materials Science and Engineering: A, 627*, 249–261. https://doi.org/10.1016/j.msea.2014.12.075.

Table 15.4 Ratio of mean pressure (p_m) to the Tresca stress or maximum shear stress (Goel et al., 2015).

Value of sress (GPa)	Goel et al. (2014) orientation		(110) orientation	(111) orientation
	With a pure repulsive indenter	With a rigid tantalum indenter	With a rigid tantalum indenter	With a rigid tantalum indenter
Shear strength of Ta ($G/2\pi$)	10.03	10.03	10.03	10.03
Tresca (T) or maximum shear stress in the deformation zone	14.4	11.78	5.43	6.86
Mean pressure (p_m) (Instant force/projected contact area)	14.48	14.17	14.13	16.13
Ratio of (T/p_m)	1	0.83	0.38	0.42

Source: From Goel, S., Beake, B., Chan, C.W., Haque Faisal, N., & Dunne, N. (2015). Twinning anisotropy of tantalum during nanoindentation. *Materials Science and Engineering: A*, *627*, 249−261. https://doi.org/10.1016/j.msea.2014.12.075.

15.3.2 Influence of the indentation velocity

A drawback of the MD simulation is that the computational time and resource limitation do not permit scaling of the simulation parameters to the experimental scale. An MDS example about indenting silicon using spherical indenter is demonstrated in this chapter (Goel, Haque Faisal, et al., 2014; Wang, Yan et al., 2021). The length scales in this regard are restricted to up to few nanometers (or few millions of atoms). In addition, the velocity of indentation, or velocity of cutting, such as that used in the current simulation (50 ms^{-1}) against an experimental speed of upto ~ 1 ms^{-1} in nanometric cutting or $0.1-10$ μ ms^{-1} during nanoindentation, is a factor that might cause unexpected spurious effects. It may, however, be anticipated that a higher loading rate, such as the one used to demonstrate here, leads to high strain rate deformation. It is, therefore, necessary to explore the effect of such strain rates on the MD simulation results. This section details the investigation of the effect of loading rate on Si-I in Si-II phase transformation. Several MD trials were carried out on single crystal silicon at different indentation and retraction speeds to explore such effects, and the results of these trials are listed in Table 15.5 and plotted in Fig. 15.18.

Fig. 15.18 shows that the loading curve remains unaffected due to the differences in the indentation speeds as much as the unloading curve, which leads to the appearance of different phases of silicon during unloading, as is well documented in the literature. To analyze these differences, a parameter Coefficient of Variation

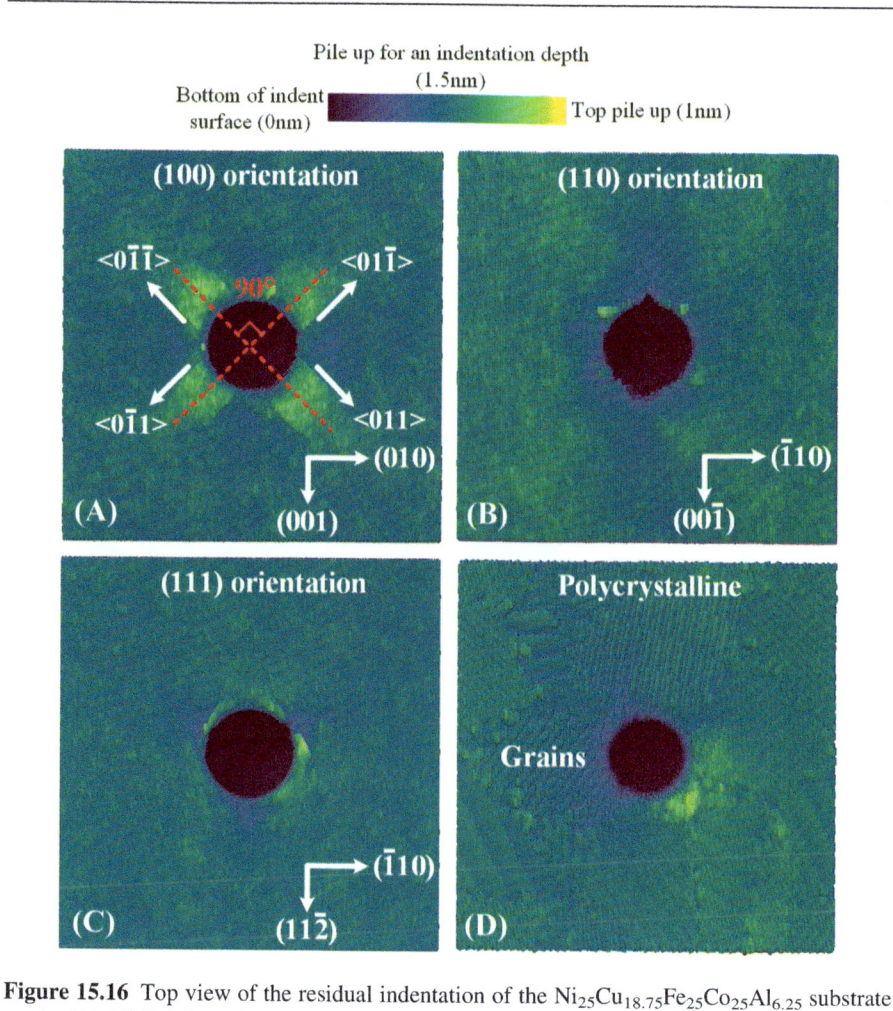

Figure 15.16 Top view of the residual indentation of the $Ni_{25}Cu_{18.75}Fe_{25}Co_{25}Al_{6.25}$ substrate on the (A) (100) orientation, (B) (110) orientation, (C) (111) orientation, and (D) polycrystalline HEA (see Fig. 15.17 for further analysis).
Source: From Fan, P., Nirmal Kumar, K., Arshad, M., Bai, M., Mao, H., & Goel, S. (2024). Anisotropic plasticity mechanisms in a newly synthesised high entropy alloy investigated using atomic simulations and nanoindentation experiments. *Journal of Alloys and Compounds*, *970*. https://doi.org/10.1016/j.jallcom.2023.172541.

(CV), defined as the ratio of standard deviation to the arithmetic mean, which eliminates the effect of sample population since it is a normalized measure of dispersion of a probability distribution was used. It may be seen from Table 15.5 that even when the indentations were performed at varying speeds, the CV by contrast is much higher for the peak temperature (0.1891) than the peak von Mises stress (0.042). The variation in the stress and temperature is plotted in Fig. 15.19, which

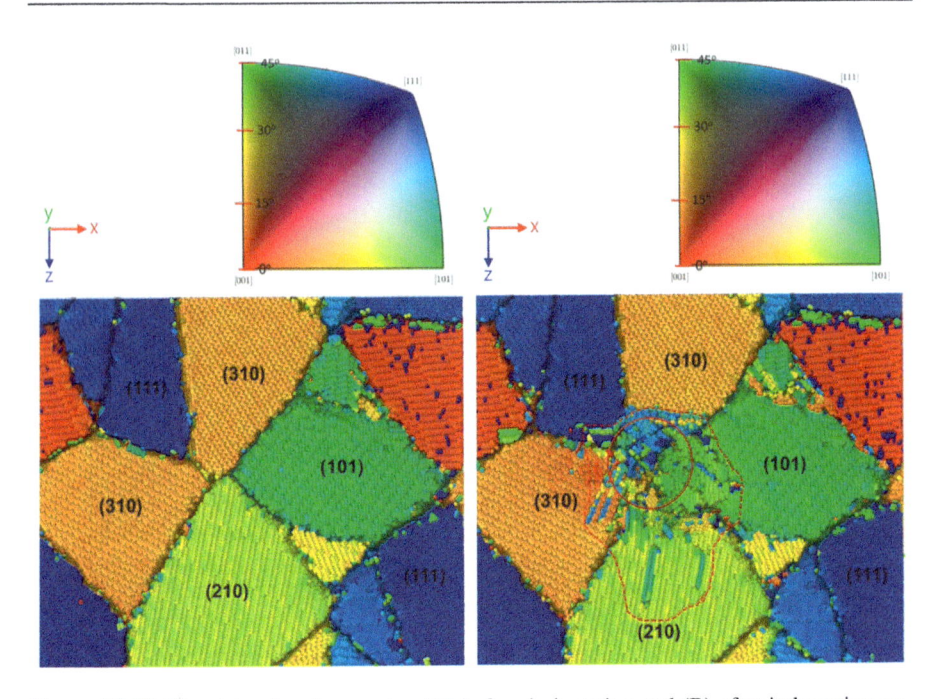

Figure 15.17 The state of various grains (A) before indentation and (B) after indentation on the polycrystalline HEA.
Source: From Fan, P., Nirmal Kumar, K., Arshad, M., Bai, M., Mao, H., & Goel, S. (2024). Anisotropic plasticity mechanisms in a newly synthesised high entropy alloy investigated using atomic simulations and nanoindentation experiments. *Journal of Alloys and Compounds*, 970. https://doi.org/10.1016/j.jallcom.2023.172541.

shows that an increase in the indentation speed causes only a marginal change in the peak stress, whereas the temperature rise is more significant.

This observation strengthens the earlier observation of Niihara (1979), who asserted that the influence of large hydrostatic stresses (such as those existing during nanoindentation) could lead to the plastic deformation of almost any material (including superhard substances such as diamond), even at low temperatures. Especially in brittle materials such as silicon, HPPT is known to arouse ductility. Finally, to explore whether Si-I to Si-II transformation is a stress-driven or a temperature-driven process, the stress and temperature values as a function of indent speed were mapped on the pressure − temperature phase diagram of silicon (see Fig. 15.20). By extrapolating this line along a lower temperature, it can be seen that at a sufficiently lowered indentation speed, the formation of Si-II is achievable by the virtue of stress alone and is not due to the temperature, and hence, the observation of HPPT being a stress-driven process appears to be true.

Table 15.5 Additional exploratory trials on single crystal silicon with spherical indenter.

Indentation speed (dh/dt)	5 m s^{-1}	10 m s^{-1}	50 m s^{-1}	100 m s^{-1}	CV (Standard deviation/mean)
Retraction speed	10 m s^{-1}	50 m s^{-1}	50 m s^{-1}	50 m s^{-1}	
Plastic depth of indentation (h_f)	1.13 nm	1.14 nm	1.16 nm	1.21 nm	0.027
Ratio of plastic depth and maximum indentation depth (h_f/h_{max})	0.543	0.548	0.558	0.582	0.027
Peak indentation force (P)	1155 nN	1088 nN	1116.2 nN	1119 nN	0.0212
Reduced elastic modulus (E_r)	145.50 GPa	138.52 GPa	145.2 GPa	153.93 GPa	0.0375
Young's modulus (E_s)	134.9 GPa	128.42 GPa	134.62 GPa	142.71 GPa	0.0375
Peak average temperature	552K	681K	725K	932K	0.1891
Peak average von Mises stress	9.74 GPa	9.93 GPa	9.21 GPa	10.35 GPa	0.042
Peak average shear stress	3.7 GPa	3.05 GPa	3.56 GPa	2.91 GPa	0.101
Peak compressive stress	22.56 GPa	22.95 GPa	23.4 GPa	23.06 GPa	0.013

Source: From Goel, S., Haque Faisal, N., Luo, X., Yan, J., & Agrawal, A. (2014). Nanoindentation of polysilicon and single crystal silicon: Molecular dynamics simulation and experimental validation. *Journal of Physics D: Applied Physics*, 47(27). https://doi.org/10.1088/0022-3727/47/27/275304.

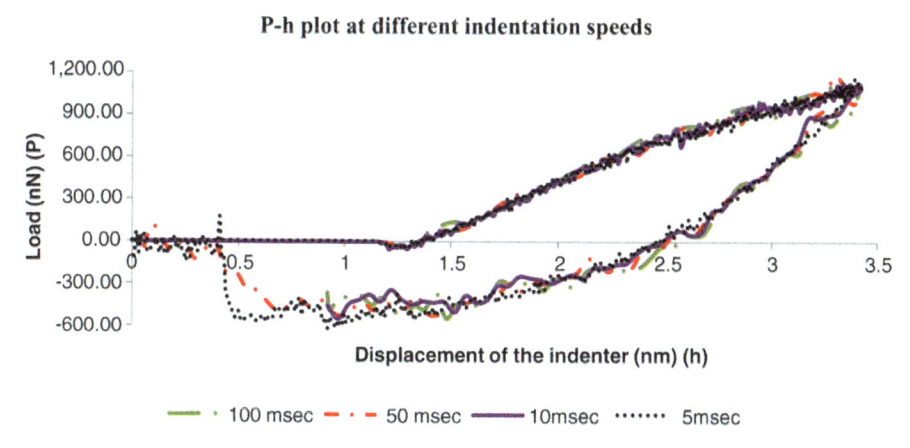

Figure 15.18 Variation in the $P-h$ plot obtained from the molecular dynamics simulation at different speeds of indentation while indenting silicon using spherical indenter.
Source: From Goel, S., Haque Faisal, N., Luo, X., Yan, J., & Agrawal, A. (2014). Nanoindentation of polysilicon and single crystal silicon: Molecular dynamics simulation and experimental validation. *Journal of Physics D: Applied Physics*, 47(27). https://doi.org/10.1088/0022-3727/47/27/275304.

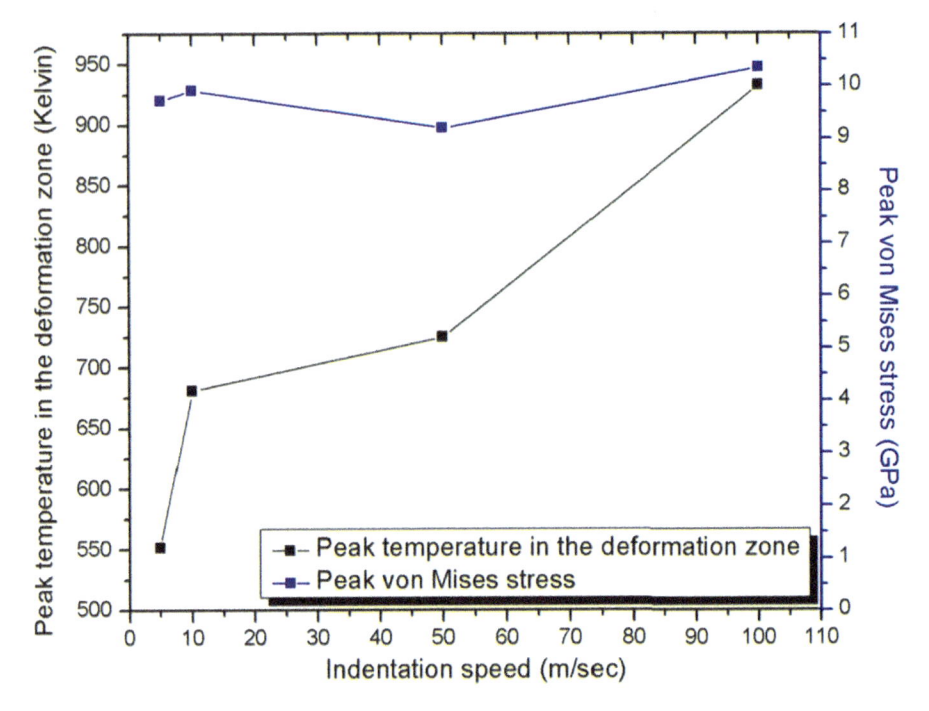

Figure 15.19 Variation in the stress and temperature obtained from molecular dynamics simulation with respect to the speed of indentation.
Source: From Goel, S., Haque Faisal, N., Luo, X., Yan, J., & Agrawal, A. (2014). Nanoindentation of polysilicon and single crystal silicon: Molecular dynamics simulation and experimental validation. *Journal of Physics D: Applied Physics*, *47*(27). https://doi.org/10.1088/0022−3727/47/27/275304.

15.3.3 Cyclic nanoindentation

A diamond indenter was used to perform cyclic nanoindentation on single crystal tungsten at different depths of 0.5, 1, 1.5, and 2 nm at a temperature of 10K. Fig. 15.21 shows the variation in the $P-h$ profiles. The overlapping in the loading profile provides a strong robustness check for the MD simulation results. Even at a small indentation depth of 0.5 nm, the $P-h$ profile shows some residual plasticity. Four indentation depths were chosen to evaluate the unloading slopes at different unloading depths, which helped to obtain the nanoindentation hardness and reduced elastic modulus of the tungsten substrate at various indentation depths as shown in Fig. 15.22. Fig. 15.21 also shows that the unloading slope (stiffness) obtained for the different $P-h$ plots differs significantly, suggesting that the same value of slope should be used to calculate the elastic modulus at different indentation depths.

For an indentation depth of 2 nm, the unloading slope (stiffness) was obtained as 2392 N m^{-1} as opposed to 1059 N m^{-1} for an indentation depth of 0.5 nm. Also,

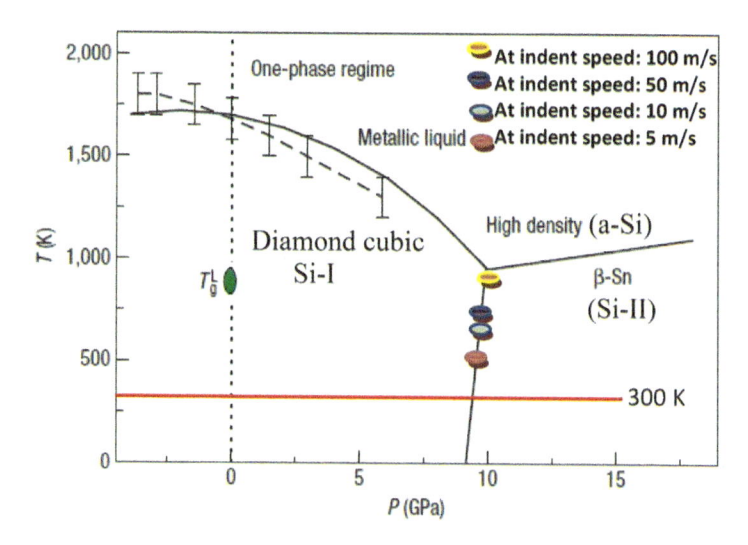

Figure 15.20 Peak stress and peak temperature in the indentation zone obtained from the MD simulation has been fitted to the experimentally obtained phase diagram of silicon reflecting Si-I to Si-II phase transformation as a function of indentation speed (Mcmillan et al., 2005). The dashed line, with error bars represent the uncertainty in the melting point determination using Stillinger − Weber potential function while indicates the LDA polymorph transition and its details can be had from its respective reference.
Source: Reproduced with permission from Mcmillan, P. F. Wilson, M., Daisenberger, D., & Machon, D. (2005). A density-driven phase transition between semiconducting and metallic polyamorphs of silicon. *Nature Materials*, 4(9), 680 −684, Available from: http://www.nature.com/nmat/. doi:10.1038/nmat1458; Goel, S., Haque Faisal, N., Luo, X., Yan, J., & Agrawal, A. (2014). Nanoindentation of polysilicon and single crystal silicon: Molecular dynamics simulation and experimental validation. *Journal of Physics D: Applied Physics*, 47 (27). https://doi.org/10.1088/0022−3727/47/27/275304.

the peak load and the projected contact area differed for different indentation depths. These data were used as an input to obtain the reduced modulus and hardness shown in Fig. 15.22. Fig. 15.22 shows that a strong extrinsic size effect is present. For instance, at a shallow indentation depth of 0.5 nm, the reduced modulus was very low (~320 GPa), but as the indentation depth increased, the value saturated to about 420 GPa. Similarly, the value of nanoindentation hardness fluctuated between 40 and 44 GPa except for the indentation depth of 1.5 nm where the nanoindentation hardness was of the order of 49 GPa.

Fig. 15.22 also shows graphical variation in the von Mises stress with the indentation depth in the atomic framework. At a small indentation depth of 1 nm, the von Mises stress remained concentrated underneath the indenter (signifying compression) and then moves on to become tensile at the periphery of the indenter leading to pileup of the material.

Finally, stress tensor information was also used to obtain information on the nanoscale yielding in tungsten. This was done by converting the atomic stress tensor to a

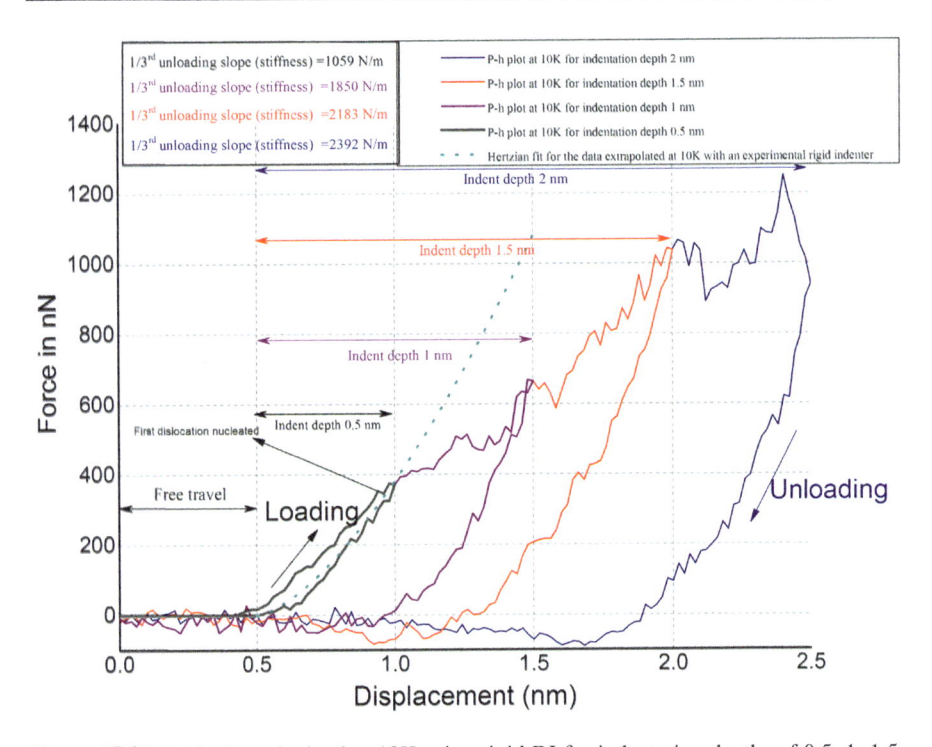

Figure 15.21 $P-h$ plots obtained at 10K using rigid DI for indentation depths of 0.5, 1, 1.5 and 2 nm obtained by performing cyclic nanoindentation tests fitted with experimental equivalent Hertz model.
Source: From Goel, S., Cross, G., Stukowski, A., Gamsjäger, E., Beake, B., & Agrawal, A. (2018). Designing nanoindentation simulation studies by appropriate indenter choices: Case study on single crystal tungsten. *Computational Materials Science*, *152*, 196−210. https://doi.org/10.1016/j.commatsci.2018.04.044.

physical stress tensor, which was then used to calculate Major Principal stress, Minor Principal stress, Tresca stress, von Mises stress, and Octahedral shear stress (Goel et al., 2016). Table 15.6 shows the yielding stress indicators of tungsten in the deformation zone in all three cases of crystallographic orientations. It can be seen that the Tresca stress and von-Mises stress (more appropriate for predicting the yielding of ductile materials) were similar for the DI and TI but differed for the RI, suggesting that the nanoindentation performed by an RI provides an overestimated value of the nanoscale yielding stress for tungsten. This implies the need of more healthy assessment of the indenter choice while performing the MD simulations. Especially for materials where diamond interaction is not available, the indentation study made from a hypothetical indenter can show some artifacts. Experimentally, the mean contact pressure (p_m) is obtained as 0.667 times the maximum contact pressure in Hertzian spherical contact and for the smallest R values of the indenter, a value of

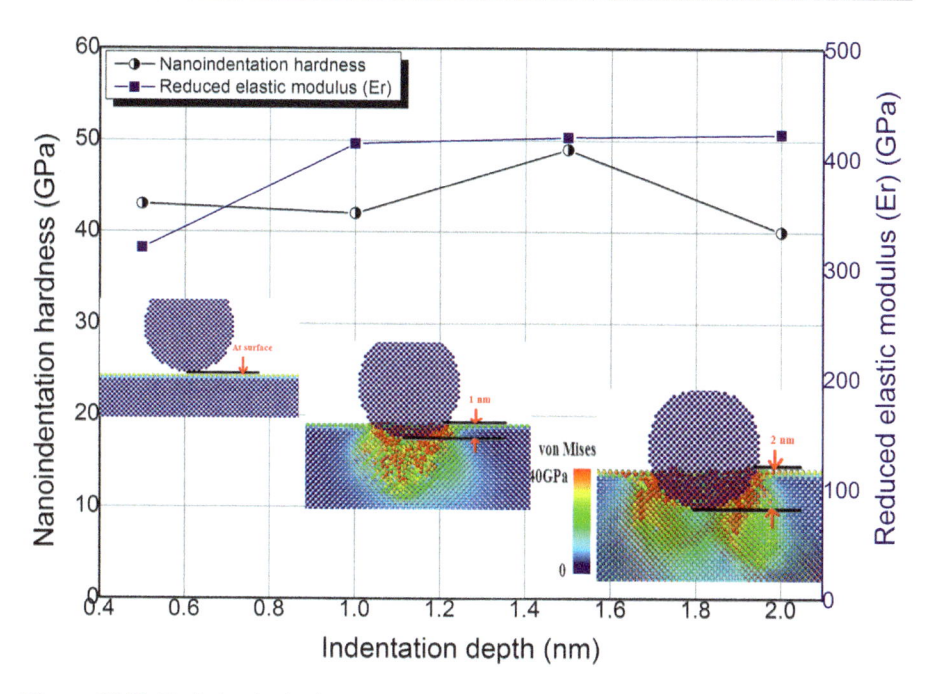

Figure 15.22 Variation in the instantaneous nanoindentation hardness (GPa) and reduced elastic modulus (*Er*) of the tungsten substrate obtained by applying O&P method in conjunction with the *P−h* plots obtained at depths 0.5, 1, 1.5, and 2 nm, respectively. *Source*: From Goel, S., Cross, G., Stukowski, A., Gamsjäger, E., Beake, B., & Agrawal, A. (2018). Designing nanoindentation simulation studies by appropriate indenter choices: Case study on single crystal tungsten. *Computational Materials Science*, *152*, 196−210. https://doi.org/10.1016/j.commatsci.2018.04.044.

Table 15.6 Nanoscale yielding stresses of tungsten obtained from the molecular dynamics simulation.

Stress peak (GPa) causing nanoscale yielding of tungsten	Rigid DI	Rigid TI	Rigid RI
Von-Mises stress	30	30	42
Octahedral shear stress	14	14	20
Tresca stress	17	17	21
Major principal stress	−81	−53	−77
Minor principal stress	−47	−19	−34
Hydrostatic stress	−62	−35	−49

Source: From Goel, S., Cross, G., Stukowski, A., Gamsjäger, E., Beake, B., & Agrawal, A. (2018). Designing nanoindentation simulation studies by appropriate indenter choices: Case study on single crystal tungsten. *Computational Materials Science*, *152*, 196−210. https://doi.org/10.1016/j.commatsci.2018.04.044.

(33 ± 5) GPa for the (100) oriented W was obtained, close to what is obtained from the MD for DI and TI and to an extent with RI too.

A substantive variation was observed in the Major, Minor, and hydrostatic stress between the three cases, and the immediate reason for these differences was not clear. The large variations in the principal stresses could possibly be attributed to the cohesive dynamics between the two kinds of atoms. It may be recalled that a purely repulsive indenter repels all the atoms away so there is no cohesive force between the tungsten atoms. In the TI case, this anomaly arises from Van der Waals forces between the two kinds of atoms. It is already known that the surface forces are dominated by Van der Waals forces (Delrio et al., 2005). At interfaces (during approach), the adhesion between the diamond and the tungsten atoms is mainly due to Van der Waals forces, whereas in the case of TI, this adhesion is replaced by cohesion, which might be responsible for the observed difference. It may also be noted that while the difference in the critical stresses for DI and TI may look quite large at the first glance, RI being a repulsive indenter exerts a purely repulsive force (along the radial direction), and hence, the friction between the indenter and the substrate surface is disregarded. These differences indicate that the von-Mises and Tresca criteria are more suitable for prediction of nanoscale yielding of metals than the principal stress criterion, which was observed to be more appropriate for brittle semiconductive material such as silicon (Goel et al., 2016).

15.4 Scale differences between molecular dynamics simulations and nanoindentation experiments

Scale differences restrict commercial use of MD for advancing our understanding about the indentation of materials. The world of MD simulations is dependent on present-day computers. This presents a limitation in that real-world scale simulation models are yet to be developed. The simulated length and timescales are far shorter than the experimental scales. Also, simulations of nanoindentation are typically done in the speed range of a few hundreds of m/s against the experimental speeds of typically about 1 m sec^{-1}. This presents a level of difficult in validation of results, but there are indirect means to achieve this that are discussed here.

15.4.1 Means to validate molecular dynamics simulation studies

To validate the nanoindentation simulation results, the Oliver and Pharr method can be used to estimate the Young's modulus from the $P-h$ curve. As for test case indentations made on silicon, the MD simulation $P-h$ plots revealed values of about 135 and 165 GPa as Young's modulus of single crystal and polycrystalline silicon, respectively, when a spherical indenter was used.

To compare these values, a displacement controlled quasistatic nanoindentation experiment on a single crystal silicon specimen using a three-sided pyramidal Berkovich indenter at extremely fine indentation depths of several nanometers was performed. The nanoindentation tests were performed on a TI 900 Hysitron TriboIndenter at a room temperature of about 293K. The specimen used was a single crystal silicon wafer with crystal orientation (0 0 1), with a diameter of 50 mm and thickness of 5 mm. It was polished on both sides. The tip of the indenter was noted to be blunt, having an edge radius of 300 nm as opposed to a newly procured tip radius of 150 nm. However, this blunt nature turned out to be a benefit rather than an experimental difficulty because the blunted geometry of the nanoindenter can often be approximated as spherical.

During the experiment, a displacement control feedback system was chosen over a load-controlled feedback system so as to achieve a finite indentation depth (Kruzic et al., 2009). The time allowed for reaching maximum displacement in all the cases was 10 s and the indenter was retracted immediately after attaining the peak indentation depth in the same duration of 10 s. Each indentation was conducted using a quick approach method so as to ensure the accuracy of depth measurements sensed by the indentation probe. Since the modulus of the material is a fundamental property, it can be compared even at a relatively higher indentation depth. Therefore, an indentation at a depth of 15 nm was made so that the substrate size effects and sensitivity effects of the ambience on the results can be eliminated.

The indentation result obtained from the instrument is shown in Fig. 15.23. The reduced elastic modulus obtained from the instrument was $E_r = 132.8$ GPa, which is in excellent agreement with the value obtained from the MD simulation during indentation of single crystal silicon from a spherical indenter using MD simulation. The small difference plausibly originated from several factors, such as sample roughness, air lubrication, sensitivity of the equipments, purity of the material, and accuracy of the measurements, etc. Moreover, the experimental value of Young's modulus of silicon reported in the literature is in the range of 120–170 GPa (Luo et al., 2012), depending on the orientation of the wafer, the parameters used for the indentation, and the sensitivity of the measurements. Thus, an excellent agreement in the values of elastic modulus of silicon obtained from MD simulation and the experiment provides confidence in the simulation results.

15.4.2 Restrictions and limitations of molecular dynamics simulations

MD scales differ in the timescale probed in nanoindentation experiments and, with some exceptions, the length scale. As such, MD simulation results are incomparable to the state-of-the-art experimental tools. Besides differing in the timescale (MD observations are in pico-nano seconds as opposed to the experimental observations seen at the microsecond scale), the presence of crystal defects in the real-world materials is always inherent, whereas a perfect crystal model as in an MD simulation always assumes an activation energy barrier. Also, at an experimental scale, the size of the indenter is substantially larger, leading to activation of several dislocations to create a dislocation burst at yield. In practice, the stress at which this happens is dependent on the preexisting

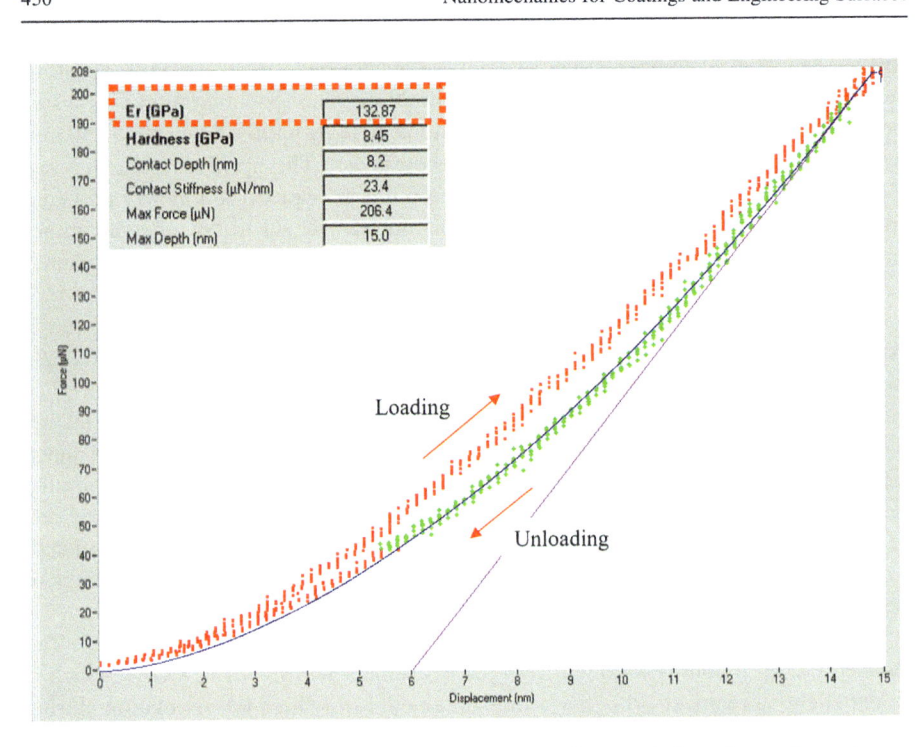

Er (GPa)	132.87
Hardness (GPa)	8.45
Contact Depth (nm)	8.2
Contact Stiffness (µN/nm)	23.4
Max Force (µN)	206.4
Max Depth (nm)	15.0

Figure 15.23 $P-h$ plot obtained from the experiments during nanoindentation performed on silicon at a depth of 15 nm. (The dashed line represents loading and unloading, while the solid line represents fitting curve on the unloading line to evaluate the mechanical properties). *Source*: From Goel, S., Haque Faisal, N., Luo, X., Yan, J., & Agrawal, A. (2014). Nanoindentation of polysilicon and single crystal silicon: Molecular dynamics simulation and experimental validation. *Journal of Physics D: Applied Physics*, *47*(27). https://doi.org/10.1088/0022-3727/47/27/275304.

dislocation density, nanoscale surface roughness, residual stress, the testing environment, and indenter size, all of which can affect the probability of encountering dislocations in a given volume. At the MD scale, a highly localized yield is affected rather than a burst and a dislocation free ideal crystal is the starting configuration in the presence of vacuum. Overall, MD simulations are based on materials having a specific theoretical strength, whereas in experiments, this can vary stochastically.

Furthermore, two key assumptions of the O&P method are: (1) Unloading is purely elastic, and (2) the elastic unloading relaxation involves only vertical displacements (Mott assumption). Sometimes the simulation condition may involve an indentation tip consisting of an atomic structure that is chemically active in terms of interaction with the substrate, but the positions of the atoms are completely fixed and the elastic or plastic relaxation of the tip is not allowed. However, the substrate (tungsten) atoms are both chemically active and mobile under all interaction forces and can exhibit elastic and plastic relaxation. In principle, they can also diffuse, but this is unlikely in tungsten at 10K and over short simulation timescales. All these aspects count toward making the MD simulation findings not directly comparable with the experiments, but complementary.

15.5 Concluding remarks

The indentation behavior of materials is characterized by combination of plastic, elastic − plastic deformation, and fracture. However, there are two main approaches to the mechanics of indentation depending upon the whether the accommodation is by plastic deformation or by fracture. Considering the MD simulation, and if the contact force is applied to a material, then constituent atoms will undergo reconfiguration from its original state, deforming the initial contact or near contact bonds in the process. If the deformation state reverses to the original configuration as the applied load is unloaded, then the deformation is elastic. If, however, the state shows residual deformation, after unloading, then the deformation could be described as plastic. Thus, plastic deformation can be characterized by permanent displacements of atoms. However, if those few broken bonds do not organize back to the original state, a fracture leading to a crack formation can also occur. The bulk material response under indentation is a function of the individual deformations of the bonds and can therefore be combination of elastic, elastic − plastic deformation, and fracture. Since MD simulation is an important theoretical technique to understand the surface and subsurface changes in materials, the current study has been motivated partly by the identification of initial stages of yielding and fracture mechanism, crystal defects, dislocation monitoring and differentiate the types of dislocation nucleation and its motion. This chapter sheds light on the imminent opportunities available for scholars to leverage high-performance computing simulation tools to address various nanoindentation issues through atomic simulations in advanced engineering materials. Specifically, this chapter demonstrates that the application of the alloy materials, semiconductors, and metals can all be examined conveniently using the MD modeling approaches, leading to observe new insights, hitherto unseen.

Acknowledgments

SG would like to acknowledge the financial support provided by the UKRI via Grants No. EP/S036180/1 and EP/T024607/1, Hubert Curien Partnership Programme from the British Counsil and the International exchange Cost Share award by the Royal Society (IEC\NSFC\223536). This work also accessed the Isambard Bristol, UK supercomputing service via Resource Allocation Panel (RAP), Kittrick HPC at LSBU, and Param Kamrupa HPC based at Guwahati, India.

References

Agrawal, P. M., Raff, L. M., & Komanduri, R. (2005). Monte Carlo simulations of void-nucleated melting of silicon via modification in the Tersoff potential parameters. *Physical Review B − Condensed Matter and Materials Physics*, *72*(12). Available from https://doi.

org/10.1103/PhysRevB.72.125206, http://oai.aps.org/oai/?verb = ListRecords&metadataPrefix = oai_apsmeta_2&set = journal:PRB:72.
Balamane, H., Halicioglu, T., & Tiller, W. A. (1992). Comparative study of silicon empirical interatomic potentials. *Physical Review B*, *46*(4), 2250−2279. Available from https://doi.org/10.1103/physrevb.46.2250.
Bartók, A. P., Payne, M. C., Kondor, R., & Csányi, G. (2010). Gaussian approximation potentials: The accuracy of quantum mechanics, without the electrons. *Physical Review Letters*, *104*(13). Available from https://doi.org/10.1103/PhysRevLett.104.136403, http://oai.aps.org/oai?verb = GetRecord&Identifier = oai:aps.org:PhysRevLett.104.136403&metadataPrefix = oai_apsmeta_2.
Baskes, M. I., Nelson, J. S., & Wright, A. F. (1989). Semiempirical modified embedded-atom potentials for silicon and germanium. *Physical Review B*, *40*(9), 6085−6100. Available from https://doi.org/10.1103/physrevb.40.6085.
Bazant, M. Z., Kaxiras, E., & Justo, J. (1997). Environment-dependent interatomic potential for bulk silicon. *Physical Review B − Condensed Matter and Materials Physics*, *56*(14), 8542−8552. Available from https://doi.org/10.1103/PhysRevB.56.8542.
Berendsen, H. J. C., Grigera, J. R., & Straatsma, T. P. (1987). The missing term in effective pair potentials. *The Journal of Physical Chemistry*, *91*(24), 6269−6271. Available from https://doi.org/10.1021/j100308a038.
Biener, M. M., Biener, J., Hodge, A. M., & Hamza, A. V. (2007). Dislocation nucleation in bcc Ta single crystals studied by nanoindentation. *Physical Review B − Condensed Matter and Materials Physics*, *76*(16). Available from https://doi.org/10.1103/PhysRevB.76.165422, http://oai.aps.org/oai?verb = GetRecord&Identifier = oai:aps.org:PhysRevB.76.165422&metadataPrefix = oai_apsmeta_2.
Brenner, D. W. (1990). Empirical potential for hydrocarbons for use in simulating the chemical vapor deposition of diamond films. *Physical Review B*, *42*(15), 9458−9471. Available from https://doi.org/10.1103/PhysRevB.42.9458.
de Brito Mota, F., Justo, J. F., & Fazzio, A. (1998). Structural properties of amorphous silicon nitride. *Physical Review B*, *58*(13), 8323−8328. Available from https://doi.org/10.1103/physrevb.58.8323.
Delrio, F. W., De Boer, M. P., Knapp, J. A., Reedy, E. D., Clews, P. J., & Dunn, M. L. (2005). The role of van der Waals forces in adhesion of micromachined surfaces. *Nature Materials*, *4*(8), 629−634. Available from https://doi.org/10.1038/nmat1431, http://www.nature.com/nmat/.
Bulatov, V., Abraham, F. F., Kubin, L., Devincre, B., & Yip, S. (1998). Connecting atomistic and mesoscale simulations of crystal plasticity. *Nature*, *391*(6668), 669−672. Available from https://doi.org/10.1038/35577.
Chen, S., & Ke, F. (2004). MD simulation of the effect of contact area and tip radius on nanoindentation. *Science in China, Series G: Physics Astronomy*, *47*(1), 101−112. Available from https://doi.org/10.1360/03yw0163, http://www.springer.com/physics/journal/11433.
Daw, M. S., & Baskes, M. I. (1984). Embedded-atom method: Derivation and application to impurities, surfaces, and other defects in metals. *Physical Review B*, *29*(12), 6443−6453. Available from https://doi.org/10.1103/PhysRevB.29.6443.
Devanathan, R., Diaz de la Rubia, T., & Weber, W. J. (1998). Displacement threshold energies in β-SiC. *Journal of Nuclear Materials*, *253*(1−3), 47−52. Available from https://doi.org/10.1016/s0022-3115(97)00304-8.
Erhart, P., & Albe, K. (2005). Analytical potential for atomistic simulations of silicon, carbon, and silicon carbide. *Physical Review B − Condensed Matter and Materials Physics*, *71*(3). Available from https://doi.org/10.1103/PhysRevB.71.035211.

Fan, P., Ding, F., Luo, X., Yan, Y., Geng, Y., & Wang, Y. (2020). A simulated investigation of ductile response of GaAs in single-point diamond turning and experimental validation. *Nanomanufacturing and Metrology, 3*(4), 239−250. Available from https://doi.org/10.1007/s41871-020-00080-5, http://springer.com/journal/41871.

Fan, P., Goel, S., Luo, X., Yan, Y., Geng, Y., & Wang, Y. (2021). An atomistic investigation on the wear of diamond during atomic force microscope tip-based nanomachining of gallium arsenide. *Computational Materials Science, 187*. Available from https://doi.org/10.1016/j.commatsci.2020.110115.

Fan, P., Katiyar, N. K., Arshad, M., Bai, M., Mao, H., & Goel, S. (2024). Anisotropic plasticity mechanisms in a newly synthesised high entropy alloy investigated using atomic simulations and nanoindentation experiments. *Journal of Alloys and Compounds, 970*. Available from https://doi.org/10.1016/j.jallcom.2023.172541.

Fan, P., Katiyar, N. K., Goel, S., He, Y., Geng, Y., Yan, Y., Mao, H., & Luo, X. (2023). Oblique nanomachining of gallium arsenide explained using AFM experiments and MD simulations. *Journal of Manufacturing Processes, 90*, 125−138. Available from https://doi.org/10.1016/j.jmapro.2023.01.002.

Fan, P., Katiyar, N. K., Zhou, X., & Goel, S. (2022). Uniaxial pulling and nano-scratching of a newly synthesized high entropy alloy. *APL Materials, 10*(11). Available from https://doi.org/10.1063/5.0128135, http://scitation.aip.org/content/aip/journal/aplmater.

Fischer-Cripps, A. C. (2011). *Nanoindentation testing* (pp. 21−37). Springer Science and Business Media LLC. Available from http://doi.org/10.1007/978-1-4419-9872-9_2.

Gao, Y., Ruestes, C. J., & Urbassek, H. M. (2014). Nanoindentation and nanoscratching of iron: Atomistic simulation of dislocation generation and reactions. *Computational Materials Science, 90*, 232−240. Available from https://doi.org/10.1016/j.commatsci.2014.04.027.

Goel, S., Beake, B., Chan, C. W., Haque Faisal, N., & Dunne, N. (2015). Twinning anisotropy of tantalum during nanoindentation. *Materials Science and Engineering: A, 627*, 249−261. Available from https://doi.org/10.1016/j.msea.2014.12.075, http://www.elsevier.com.

Goel, S., Cross, G., Stukowski, A., Gamsjäger, E., Beake, B., & Agrawal, A. (2018). Designing nanoindentation simulation studies by appropriate indenter choices: Case study on single crystal tungsten. *Computational Materials Science, 152*, 196−210. Available from https://doi.org/10.1016/j.commatsci.2018.04.044.

Goel, S., Haque Faisal, N., Luo, X., Yan, J., & Agrawal, A. (2014). Nanoindentation of polysilicon and single crystal silicon: Molecular dynamics simulation and experimental validation. *Journal of Physics D: Applied Physics, 47*(27). Available from https://doi.org/10.1088/0022-3727/47/27/275304, http://iopscience.iop.org/0022-3727/47/27/275304/pdf/0022-3727_47_27_275304.pdf.

Goel, S., Joshi, S. S., Abdelal, G., & Agrawal, A. (2014). Molecular dynamics simulation of nanoindentation of Fe$_3$C and Fe$_4$C. *Materials Science and Engineering: A, 597*, 331−341. Available from https://doi.org/10.1016/j.msea.2013.12.091, http://www.elsevier.com.

Goel, S., Kovalchenko, A., Stukowski, A., & Cross, G. (2016). Influence of microstructure on the cutting behaviour of silicon. *Acta Materialia, 105*, 464−478. Available from https://doi.org/10.1016/j.actamat.2015.11.046, http://www.journals.elsevier.com/acta-materialia/.

Goel, S., Luo, X., Agrawal, A., & Reuben, R. L. (2015). Diamond machining of silicon: A review of advances in molecular dynamics simulation. *International Journal of Machine Tools and Manufacture, 88*, 131−164. Available from https://doi.org/10.1016/j.

ijmachtools.2014.09.013, http://www.journals.elsevier.com/international-journal-of-machine-tools-and-manufacture/.

Goel, S., Yan, J., Luo, X., & Agrawal, A. (2014). Incipient plasticity in 4H-SiC during quasi-static nanoindentation. *Journal of the Mechanical Behavior of Biomedical Materials, 34*, 330−337. Available from https://doi.org/10.1016/j.jmbbm.2013.12.005, http://www.elsevier.com/wps/find/journaldescription.cws_home/711005/description#description.

Gracie, R., & Belytschko, T. (2011). An adaptive concurrent multiscale method for the dynamic simulation of dislocations. *International Journal for Numerical Methods in Engineering, 86*(4-5), 575−597. Available from https://doi.org/10.1002/nme.3112.

Grau, P., Lorenz, D., & Zeckzer, A. (2002). Fundamentals of dislocation nucleation at nanoindentation. *Radiation Effects and Defects in Solids, 157*(6-12), 863−869. Available from https://doi.org/10.1080/10420150215829.

Guerrero, O., & Marucho, M. (2013). Elastic-plastic transition under uniaxial stress BCC tantalum. *Journal of Material Sciences and Engineering, 3*, 153−160.

Healy, C. J., & Ackland, G. J. (2014). Molecular dynamics simulations of compression-tension asymmetry in plasticity of Fe nanopillars. *Acta Materialia, 70*, 105−112. Available from https://doi.org/10.1016/j.actamat.2014.02.021.

Herbert, E. G., Pharr, G. M., Oliver, W. C., Lucas, B. N., & Hay, J. L. (2001). On the measurement of stress-strain curves by spherical indentation. *Materials Research Society Symposium − Proceedings Q3.4.6*, United States, 649.

Kelchner, C. L., Plimpton, S. J., & Hamilton, J. C. (1998). Dislocation nucleation and defect structure during surface indentation. *Physical Review B, 58*(17), 11085−11088. Available from https://doi.org/10.1103/physrevb.58.11085.

Kerfriden, P., Passieux, J. C., & Bordas, S. P. A. (2012). Local/global model order reduction strategy for the simulation of quasi-brittle fracture. *International Journal for Numerical Methods in Engineering, 89*(2), 154−179. Available from https://doi.org/10.1002/nme.3234.

Kruzic, J. J., Kim, D. K., Koester, K. J., & Ritchie, R. O. (2009). Indentation techniques for evaluating the fracture toughness of biomaterials and hard tissues. *Journal of the Mechanical Behavior of Biomedical Materials, 2*(4), 384−395. Available from https://doi.org/10.1016/j.jmbbm.2008.10.008.

Kumagai, T., Izumi, S., Hara, S., & Sakai, S. (2007). Development of bond-order potentials that can reproduce the elastic constants and melting point of silicon for classical molecular dynamics simulation. *Computational Materials Science, 39*(2), 457−464. Available from https://doi.org/10.1016/j.commatsci.2006.07.013.

Lee, S. (2020). In-situ observation of the initiation of plasticity by nucleation of prismatic dislocation loops. *Nature Communications, 11*(1), 1−11.

Longbottom, J. M., & Lanham, J. D. (2005). Cutting temperature measurement while machining − A review. *Aircraft Engineering and Aerospace Technology, 77*(2), 122−130. Available from https://doi.org/10.1108/00022660510585956.

Lucas, G., Bertolus, M., & Pizzagalli, L. (2010). An environment-dependent interatomic potential for silicon carbide: Calculation of bulk properties, high-pressure phases, point and extended defects, and amorphous structures. *Journal of Physics: Condensed Matter, 22*(3). Available from https://doi.org/10.1088/0953-8984/22/3/035802.

Luo, X., Goel, S., & Reuben, R. L. (2012). A quantitative assessment of nanometric machinability of major polytypes of single crystal silicon carbide. *Journal of the European Ceramic Society, 32*(12), 3423−3434. Available from https://doi.org/10.1016/j.jeurceramsoc.2012.04.016.

Matsunaga, K., Fisher, C., & Matsubara, H. (2000). Tersoff potential parameters for simulating cubic boron carbonitrides. *Japanese Journal of Applied Physics, Part 2: Letters, 39* (1), L48−L51. Available from https://doi.org/10.1143/jjap.39.l48.

Matsunaga, K., & Iwamoto, Y. (2001). Molecular dynamics study of atomic structure and diffusion behavior in amorphous silicon nitride containing boron. *Journal of the American Ceramic Society, 84*(10), 2213−2219.

Mayo, M. J., Siegel, R. W., Narayanasamy, A., & Nix, W. D. (1990). Mechanical properties of nanophase TiO_2 as determined by nanoindentation. *Journal of Materials Research, 5* (5), 1073−1082. Available from https://doi.org/10.1557/jmr.1990.1073.

Mcmillan, P. F., Wilson, M., Daisenberger, D., & Machon, D. (2005). A density-driven phase transition between semiconducting and metallic polyamorphs of silicon. *Nature Materials, 4*(9), 680−684. Available from https://doi.org/10.1038/nmat1458, http://www.nature.com/nmat/.

Mo, Y., Turner, K. T., & Szlufarska, I. (2009). Friction laws at the nanoscale. *Nature, 457* (7233), 1116−1119. Available from https://doi.org/10.1038/nature07748.

Niihara, K. (1979). Slip systems and plastic deformation of silicon carbide single crystals at high temperatures. *Journal of The Less-Common Metals, 65*(1), 155−166. Available from https://doi.org/10.1016/0022-5088(79)90161-9.

Oliver, W. C., & Pharr, G. M. (1992). An improved technique for determining hardness and elastic modulus using load and displacement sensing indentation experiments. *Journal of Materials Research, 7*(6), 1564−1583. Available from https://doi.org/10.1557/jmr.1992.1564.

Oskay, C., & Fish, J. (2004). Fatigue life prediction using 2-scale temporal asymptotic homogenization. *International Journal for Numerical Methods in Engineering, 61*(3), 329−359. Available from https://doi.org/10.1002/nme.1069.

Pastewka, L., Klemenz, A., Gumbsch, P., & Moseler, M. (2013). Screened empirical bond-order potentials for Si-C. *Physical Review B − Condensed Matter and Materials Physics, 87*(20). Available from https://doi.org/10.1103/PhysRevB.87.205410, http://oai.aps.org/filefetch?identifier = 10.1103/PhysRevB.87.205410&component = fulltext &description = markup&format = xml.

Pastewka, L., Pou, P., Pérez, R., Gumbsch, P., & Moseler, M. (2008). Describing bond-breaking processes by reactive potentials: Importance of an environment-dependent interaction range. *Physical Review B − Condensed Matter and Materials Physics, 78* (16). Available from https://doi.org/10.1103/PhysRevB.78.161402, http://oai.aps.org/oai?verb = GetRecord&Identifier = oai:aps.org: PhysRevB.78.161402&metadataPrefix = oai_apsmeta_2.

Paulauskas, T., Buurma, C., Stafford, B., Sun, C., Chan, M., Sivalingham, S., Kim, M., & Klie, R. F. (2015). Atomic scale study of Lomer-Cottrell and Hirth lock dislocations in CdTe. *Microscopy and Microanalysis, 21*(S3), 2087−2088. Available from https://doi.org/10.1017/S1431927615011216.

Pen, H. M., Liang, Y. C., Luo, X. C., Bai, Q. S., Goel, S., & Ritchie, J. M. (2011). Multiscale simulation of nanometric cutting of single crystal copper and its experimental validation. *Computational Materials Science, 50*(12), 3431−3441. Available from https://doi.org/10.1016/j.commatsci.2011.07.005.

Pervaiz, S., Deiab, I., Ibrahim, E. M., Rashid, A., & Nicolescua, M. (2014). A coupled FE and CFD approach to predict the cutting tool temperature profile in machining. *Procedia CIRP, 17*, 750−754. Available from https://doi.org/10.1016/j.procir.2014.01.104, http://www.sciencedirect.com/science/journal/22128271.

Plimpton, S. (1995). Fast parallel algorithms for short-range molecular dynamics. *Journal of Computational Physics*, *117*(1), 1−19. Available from https://doi.org/10.1006/jcph.1995.1039.

Psakhie, S. G., Horie, Y., Korostelev, S. Y., Smolin, A. Y., Dmitriev, A. I., Shilko, E. V., & Alekseev, S. V. (1995). Method of movable cellular automata as a tool for simulation within the framework of mesomechanics. *Russian Physics Journal*, *38*(11), 1157−1168. Available from https://doi.org/10.1007/bf00559396.

Remington, T. P., Ruestes, C. J., Bringa, E. M., Remington, B. A., Lu, C. H., Kad, B., & Meyers, M. A. (2014). Plastic deformation in nanoindentation of tantalum: A new mechanism for prismatic loop formation. *Acta Materialia*, *78*, 378−393. Available from https://doi.org/10.1016/j.actamat.2014.06.058.

Ruestes, C. J., & Farkas, D. (2022). Dislocation emission and propagation under a nanoindenter in a model high entropy alloy. *Computational Materials Science*, *205*. Available from https://doi.org/10.1016/j.commatsci.2022.111218, https://www.journals.elsevier.com/computational-materials-science.

Smith, L., Zimmerman, J. A., Hale, L. M., & Farkas, D. (2014). Molecular dynamics study of deformation and fracture in a tantalum nano-crystalline thin film. *Modelling and Simulation in Materials Science and Engineering*, *22*(4). Available from https://doi.org/10.1088/0965-0393/22/4/045010, http://iopscience.iop.org/0965-0393/22/4/045010/pdf/0965-0393_22_4_045010.pdf.

Stillinger, F. H., & Weber, T. A. (1985). Computer simulation of local order in condensed phases of silicon. *Physical Review B*, *31*(8), 5262−5271. Available from https://doi.org/10.1103/PhysRevB.31.5262.

Stuart, S. J., Tutein, A. B., & Harrison, J. A. (2000). A reactive potential for hydrocarbons with intermolecular interactions. *Journal of Chemical Physics*, *112*(14), 6472−6486. Available from https://doi.org/10.1063/1.481208, http://scitation.aip.org/content/aip/journal/jcp.

Stukowski, A. (2010). Visualization and analysis of atomistic simulation data with OVITO- the open visualization tool. *Modelling and Simulation in Materials Science and Engineering*, *18*(1). Available from https://doi.org/10.1088/0965-0393/18/1/015012.

Stukowski, A., & Albe, K. (2010). Extracting dislocations and non-dislocation crystal defects from atomistic simulation data. *Modelling and Simulation in Materials Science and Engineering*, *18*(8). Available from https://doi.org/10.1088/0965-0393/18/8/085001.

Stukowski, A., Bulatov, V. V., & Arsenlis, A. (2012). Automated identification and indexing of dislocations in crystal interfaces. *Modelling and Simulation in Materials Science and Engineering*, *20*(8). Available from https://doi.org/10.1088/0965-0393/20/8/085007, http://iopscience.iop.org/0965-0393/20/8/085007/pdf/0965-0393_20_8_085007.pdf.

Tan, Y., Yang, D., & Sheng, Y. (2009). Discrete element method (DEM) modeling of fracture and damage in the machining process of polycrystalline SiC. *Journal of the European Ceramic Society*, *29*(6), 1029−1037. Available from https://doi.org/10.1016/j.jeurceramsoc.2008.07.060.

Tersoff, J. (1988a). Empirical interatomic potential for silicon with improved elastic properties. *Physical Review B*, *38*(14), 9902−9905. Available from https://doi.org/10.1103/physrevb.38.9902.

Tersoff, J. (1988b). New empirical approach for the structure and energy of covalent systems. *Physical Review B*, *37*(12), 6991−7000. Available from https://doi.org/10.1103/physrevb.37.6991.

Tersoff, J. (1989). Modeling solid-state chemistry: Interatomic potentials for multicomponent systems. *Physical Review B*, *39*(8), 5566−5568. Available from https://doi.org/10.1103/physrevb.39.5566.

Tersoff, J. (1990a). Erratum: Modeling solid-state chemistry: Interatomic potentials for multicomponent systems. *Physical Review B, 41*(5), 3248. Available from https://doi.org/10.1103/physrevb.41.3248.2.

Tersoff, J. (1990b). Carbon defects and defect reactions in silicon. *Physical Review Letters, 64*(15), 1757–1760. Available from https://doi.org/10.1103/physrevlett.64.1757.

Tersoff, J. (1994). Chemical order in amorphous silicon carbide. *Physical Review B, 49*(23), 16349–16352. Available from https://doi.org/10.1103/physrevb.49.16349.

de Tomas, C., Aghajamali, A., Jones, J. L., Lim, D. J., López, M. J., Suarez-Martinez, I., & Marks, N. A. (2019). Transferability in interatomic potentials for carbon. *Carbon, 155*, 624–634. Available from https://doi.org/10.1016/j.carbon.2019.07.074, http://www.journals.elsevier.com/carbon/.

de Tomas, C., Suarez-Martinez, I., & Marks, N. A. (2016). Graphitization of amorphous carbons: A comparative study of interatomic potentials. *Australia Carbon, 109*, 681–693. Available from http://www.journals.elsevier.com/carbon/, https://doi.org/10.1016/j.carbon.2016.08.024.

Van Duin, A. C. T., Dasgupta, S., Lorant, F., & Goddard, W. A. (2001). ReaxFF: A reactive force field for hydrocarbons. *Journal of Physical Chemistry A, 105*(41), 9396–9409. Available from https://doi.org/10.1021/jp004368u.

Wang, Y., Fan, P., Luo, X., Geng, Y., Goel, S., Wu, W., Li, G., & Yan, Y. (2021). Fabrication of three-dimensional sin-shaped ripples using a multi-tip diamond tool based on the force modulation approach. *Journal of Manufacturing Processes, 72*, 262–273. Available from https://doi.org/10.1016/j.jmapro.2021.10.032, http://www.elsevier.com/wps/find/journaldescription.cws_home/620379/description#description.

Wang, B., Gao, Y., & Urbassek, H. M. (2014). Microstructure and magnetic disorder induced by nanoindentation in single-crystalline Fe. *Physical Review B — Condensed Matter and Materials Physics, 89*(10). Available from https://doi.org/10.1103/PhysRevB.89.104105, http://harvest.aps.org/bagit/articles/10.1103/PhysRevB.89.104105/apsxml.

Wang, Y. M., Hodge, A. M., Biener, J., Hamza, A. V., Barnes, D. E., Liu, K., & Nieh, T. G. (2005). Deformation twinning during nanoindentation of nanocrystalline Ta. *Applied Physics Letters, 86*(10), 1–3. Available from https://doi.org/10.1063/1.1883335.

Wang, J., Yan, Y., Li, Z., Geng, Y., Luo, X., & Fan, P. (2021). Processing outcomes of atomic force microscope tip-based nanomilling with different trajectories on single-crystal silicon. *Precision Engineering, 72*, 480–490. Available from https://doi.org/10.1016/j.precisioneng.2021.06.009, https://www.journals.elsevier.com/precision-engineering.

Xiang, H., Li, H., Fu, T., Huang, C., & Peng, X. (2017). Formation of prismatic loops in AlN and GaN under nanoindentation. *Acta Materialia, 138*, 131–139. Available from https://doi.org/10.1016/j.actamat.2017.06.045, http://www.journals.elsevier.com/acta-materialia/.

Yu, J., Sinnott, S. B., & Phillpot, S. R. (2007). Charge optimized many-body potential for the Si/SiO_2 system. *Physical Review B — Condensed Matter and Materials Physics, 75*(8). Available from https://doi.org/10.1103/PhysRevB.75.085311, http://oai.aps.org/oai?verb = GetRecord&Identifier = oai:aps:PhysRevB.75.085311&metadataPrefix = oai_apsmeta_2.

Zhou, X. W., & Doty, F. P. (2008). Embedded-ion method: An analytical energy-conserving charge-transfer interatomic potential and its application to the La-Br system. *Physical Review B — Condensed Matter and Materials Physics, 78*(22). Available from https://doi.org/10.1103/PhysRevB.78.224307, http://oai.aps.org/oai?verb = GetRecord&Identifier = oai:aps.org:PhysRevB.78.224307&metadataPrefix = oai_apsmeta_2.

Ziegenhain, G., Urbassek, H. M., & Hartmaier, A. (2010). Influence of crystal anisotropy on elastic deformation and onset of plasticity in nanoindentation: A simulational study. *Journal of Applied Physics, 107*, 6. Available from https://doi.org/10.1063/1.3340523.

Section IV

Coatings and engineering surfaces: design strategies and industrial applications

The significance of hardness and elastic modulus in the design of engineered surfaces

16

Adrian Leyland[1] and Allan Matthews[2]
[1]Department of Materials Science and Engineering, The University of Sheffield, Sheffield, United Kingdom, [2]Department of Materials, University of Manchester, Manchester, United Kingdom

16.1 Introduction

In recent decades, the benefits in designing appropriate engineered surfaces to improve the practical performance and durability of engineering components in their intended applications and environments have become increasingly recognized. Similarly, the availability of an increasing range of advanced surface modification technologies—alongside the continuous improvement of existing surface engineering treatments and processes—has facilitated broader scientific study and increased application in manufacturing industry.

Surface engineering is now commonly accepted as a scientific discipline in its own right by researchers, designers, and practitioners alike, particularly in the fields of Materials Science and Mechanical Engineering. However, the terminology only emerged quite recently—and was barely recognized as such before the 1980s—although a number of pioneering researchers around the world were, in the 1960s and 1970s, developing the concepts and technologies that would lead to its inception (Bunshah & Raghuram, 1972; Halling & Arnell, 1984; Matthews & Teer, 1980; Mattox, 2003; Muenz et al., 1982; Schiller et al., 1976). Also, there are many definitions of what surface engineering actually entails but, particularly for the purpose and intent of this chapter, it is appropriate to use (with a little para-phrasing) a definition promoted primarily by the late Professor Tom Bell of the University of Birmingham, United Kingdom (Bell, 1991), namely, "*the design of a coating/treatment and a substrate, as a system, to provide a cost-effective performance enhancement of which neither is capable alone.*" Furthermore, the wider systematic study of engineering surfaces in moving contact (i.e. Tribology) only started to emerge as a distinctly recognizable discipline in the 1940s and 1950s, due primarily to the pioneering work of a few visionary researchers (Archard, 1953; Bowden & Leben, 1939; Tabor, 1951 - and see also review by Hutchings, 2009). It was brought into wider prominence in the 1960s by the UK Government Report of Peter Jost (Jost, 1966), detailing the high cost to industry/society of wear and corrosion and promoting the concept of Tribology as a discipline worthy of financial support. This led to

Nanomechanics for Coatings and Engineering Surfaces. DOI: https://doi.org/10.1016/B978-0-443-13334-3.00016-4
Copyright © 2025 Elsevier Inc. All rights reserved, including those for text and data mining, AI training, and similar technologies.

the establishment in 1968 of three centers of tribology in the United Kingdom (Williams, 2024) (at the Universities of Leeds and Swansea, and at the UK Atomic Energy Authority site at Risley, Cheshire), and to the development and/or expansion of similar activities in other industrialized countries worldwide—such as Germany (BMFT, 1980), the United States (Pinkus & Wilcock, 1977), France, and Japan (see also, for further examples, Holmberg & Erdemir (2017)—and Refs. therein). Recent attempts to revisit the Jost Report's findings have only confirmed the continuing importance for the global economy—and for sustainability—in addressing the issues that it raised (Ciulli, 2024; Holmberg & Erdemir, 2017). It remains clear that the main life-limiting factor for the vast majority of engineering components, and particularly for those in loaded/moving contact, is degradation of the surface condition—and the engineering consequences thereof—and that cost-effective measures should be taken to mitigate against surface degradation of engineering components. However, against the backdrop of more recent rapid technological advances in Tribology and Surface Engineering, it should be mentioned that the application of these disciplines to the mitigation of wear and corrosion in engineering has been a factor of major concern since the beginning of the first Industrial Revolution (c.1760), and was studied extensively by Renaissance Engineers and Scholars—see, e.g., Hutchings (2024), and references therein, for details. The importance of "surface engineering" can actually be traced back into deeper history, through the weapons and regalia of the "terracotta army" of the Qin dynasty in China—and through engineering artifacts from the early Roman Empire —through to ancient Greece, and possibly yet further back (Bell, 1991; Holmberg & Erdemir, 2017; Hutchings, 2024; Mayrhofer et al., 2024).

Returning to the modern era, however, there are a number of interacting physical, mechanical, and chemical properties of materials that inevitably influence strongly their degradation behavior in engineering applications. It is also no coincidence that the increased recognition and accelerated use of surface modification technologies since the Second World War have occurred in parallel with the rapid development (and/or increased availability) of a wide array of materials characterization techniques. These include, for example, electron (and "atom-probe") microscopy, X-ray diffraction, other spectroscopic analysis techniques, instrumented indentation methods (particularly nanoindentation)—and an ever-increasing range of other, sophisticated mechanical, tribological, corrosion (and tribocorrosion) test capabilities. Such capabilities allow us to investigate the finest details of materials' structure and properties, dramatically improving our materials design and processing capabilities—and allowing us to push the performance of engineering materials to their fundamental scientific limits. On the other hand, it can be argued that, as a result of this enhanced capability, we are currently "running out of materials"—that is, the Grand Challenge, going forward, is to accelerate the development of new engineering materials (while taking account also of the environmental, ethical, and various "geopolitical" pressures that modern society faces)—and to perform this faster than we exhaust the capabilities of our current materials inventory. Regardless of whether we can achieve this aim, in whole or in part, the need for advanced materials characterization methods—the data from which can be related directly to industrial applications performance—will not diminish. Furthermore, the intensity of demand for surface coatings and treatments to enhance the performance of our existing (as well as newly emerging) engineering materials can only increase.

16.2 The importance of hardness and elastic modulus in tribology and surface engineering

We turn now to the specifics of the physical, mechanical, and chemical properties of engineering materials—and particularly of surface engineered materials (and how to evaluate them). The interactions between different measurable material parameters and the extraction of meaningful values (depending on the prior processing path—and resulting macro/micro/nanostructural features) are immensely complex and challenging to assess, to quantify, and to contextualize. This is particularly true in terms of the relevance of laboratory data to real-life applications (the reader could consult, for example, Fig. 5.23, p. 356, in Holmberg and Matthews (1994), and the pertaining narrative). Furthermore, the parameters that define the material properties, both of the substrate bulk and of the engineered surface (and of any graded/modulated layers that lie between the two), are often significantly different. Various researchers (including the authors of this chapter) have made concerted attempts to develop basic design rules and guidelines (Farrow, 1986; Holmberg & Matthews, 1994; Luo et al., 2011; Matthews & Swift, 1983; Matthews et al., 1998; Matthews et al., 2007), to assist the design engineer to navigate through what can (on first sight) seem to be an intractable series of problems—but the "rules" are inevitably somewhat generalized and open to (mis)interpretation. Various other aspects of the (nano)measurement of coated/treated surface properties are addressed in other chapters of this book; however, here we focus mainly on considerations of Hardness (H) and Elastic modulus (E) —and the significance and meaning of ratios between the two parameters. Nonetheless, such "mechanical" (H) and "physical" (E) properties cannot be discussed in isolation since, in Tribology and Surface Engineering, the chemical behavior (specifically, the *tribo*chemistry) of the contacting surfaces is often a dominant factor in determining the resulting levels of surface attrition, depending on both application and environment.

The aforementioned distinction between the property "definitions" of H and E is intentional in the sense that, from a Physics (rather than Mechanics) perspective, E generally has a precise and unique value for a single (crystalline) material phase of a specific chemical composition. This can be related directly to basic *physical* parameters, such as the length, strength, and coordination number of the interatomic bonds, which define the material's behavior; that is, for polycrystalline materials particularly, E is essentially an intrinsic material property (notwithstanding the elastic anisotropy that the individual grains of many crystallographic phases possess). However, there is a frequent need—particularly in instrumented indentation measurements—to convert some measured "reduced modulus" value into the required quantity (assuming that the Poisson's ratios of the contacting materials are known, or can be estimated with reasonable precision). On the other hand, unlike E, hardness cannot be described as an intrinsic material property—and the numerical value of H obtained will inevitably depend on a variety of extrinsic factors, such as material grain size, residual stress state, applied indentation load/indenter contact geometry (and whether the latter two generate significant work hardening of the indented material).

Contemporaneously, in the measurement of Hardness—by any means (but generally by some kind of indentation method)—and of wear resistance, it has often been assumed that the "laws" proposed tentatively by Archard in the 1950's (e.g. Archard, 1953; Archard & Hirst, 1956), which led to the development of a "simple" wear equation (subsequently adopted widely by tribologists), are universally applicable in quantifying the wear of materials. In particular, the assumption that the rate of wear will be inversely proportional to the hardness of the material being worn has become a widespread belief—with it also being said that the counterface needs to possess some minimum, higher hardness value (typically $\geq 25\%-40\%$ higher; we discuss this later). These assumed "facts" remain in use today, for example, in the calculation of dimensionless wear coefficients from tribological testing data (typically for bulk materials, sometimes lubricated—and often metals). However, over a similar timeframe to Archard's early work, questions were also raised, as to the universal applicability of the "accepted" approach. For example, Oberle (1951) published work based on many practical experiments carried out at Caterpillar in the United States, on a range of metals of varying hardness, and also some engineering ceramics. His findings showed that Hardness alone often did not correlate directly with the wear performance of real engineering materials/components. In fact, if an increase in Hardness was also accompanied by an increase in Elastic modulus, the wear resistance appeared to decrease. Kragelskii (1965) also proposed that, to avoid severe wear, it could be appropriate that the near-surface of a material be less hard than the interior. Furthermore, Lancaster and coworkers (e.g. Evans & Lancaster, 1979; Lancaster, 1978) showed that the wear behavior of polymers also rarely follows the expected trend prescribed by the wear equation attributed to Archard. In most of the aforementioned cases—and in others, e.g., Halling (1983); Smart (1978) —it has been clearly shown that, either directly or indirectly, a ratio between the Yield strength and the Elastic modulus (i.e., a Y/E value) often provides a more consistent correlation to the wear rates of a wide range of different materials.

Yield *strength*, σ_Y (i.e., some critical *tensile* normal force, applied over a measured cross-sectional area; F/A) can, with careful consideration, be correlated to Yield Pressure, P_C (i.e., a critical *compressive* normal Force, also applied over a measured contact area) (Johnson, 1985; Marshall et al., 1982). As originally reported by Tabor (1951), and subsequently summarized elsewhere (Tabor, 1996; Wyatt & Dew-Hughes, 1974), the indentation hardness (of metals, particularly) is generally found to be of the order of three times the equivalent tensile Yield stress of the same material—but with several provisos regarding the precise indentation behavior of the pertaining material (Tabor, 1951). Hence, with a resulting assumption that there is a direct (and consistent) linear relationship between σ_Y or, in fact, *flow pressure*, p_m—as originally defined in Archard (1953) and Archard & Hirst (1956)—and indentation Hardness, H, the simple "Archard Wear Equation" (that we will return to later) was widely adopted.

On the other hand, the ratio between Hardness and Elastic modulus (H/E)—as well as being investigated by indentation hardness measurement (e.g. Marshall et al., 1982)—also emerged in recent decades as a convenient measure of the *elastic* strain to failure of a range of engineering materials, For example, bearing designers have for many years incorporated an E/H parameter in the "plasticity index" (Greenwood & Williamson, 1966; Halling, 1983), used often to assess the effect of surface roughness on the load-carrying capacity of bearings (and thereby to avoid plastic yield in use). Furthermore, the increasingly widespread availability of instrumented indentation methods since the early 1990s has facilitated easy measurement of both H and E, for bulk materials and surface coatings/treatments alike. However, as Tabor also pointed out in his seminal work on *"The Hardness and Strength of Metals"* (Tabor, 1951), the precise relationship between Y and H in metal alloys, from which many engineering components are manufactured, varies from metal to metal and depends on heat treatment path (and resulting microstructure). Furthermore, it is also influenced by the work-hardening exponent of the alloy (including any prior work of deformation—as well as indenter-induced work hardening). Tabor proposed that the latter points could be accommodated by the use of an averaged "effective" yield strength (Y_e) at low-to moderate levels of plastic strain. The Poisson's ratio (v) of a material also influences the behavior of different materials—particularly under compressive indentation loading, where the elastic strain field in the material that surrounds the indent under load constrains the local yielding behavior (often described as "Plastic Constraint").

Many engineering metals possess very similar Poisson's ratios, but the values of v for ceramics and polymers (and indeed for composite materials) can vary over a wide range—possibly an order of magnitude, or more (particularly when auxetic materials are also considered, see e.g. Surjadi et al. (2019). Thus, in nanoindentation, the "reduced" modulus, arising from the contact mechanics between two materials of significantly different stiffness and/or Poisson's ratio (i.e., between indenter and substrate), needs to be accounted for—as does the "real" contact area/depth. It is now well known that the latter can vary significantly due to substrate pile-up/sink-in—and depends also on the ratio between elastic/plastic work of indentation (and whether the indented material is metal, ceramic, polymer, or composite). Furthermore, any attempts to use indentation methods to assess coating fracture toughness require careful consideration of the many factors involved. Methods that measure indentation cracking in brittle materials to assess bulk fracture toughness have been available for many decades now [e.g., Anstis et al. (1981) and Chantikul et al. (1981)], but similar methods applied to thin coatings show varying degrees of success [e.g., Chen (2012), Chen and Bull (2008), Xia et al. (2004), and Zhang et al. (2005)]—not least because it is difficult to deconvolute coating and substrate deformation/fracture behavior. There is also an apparent lack of awareness of the specific conditions of Mode I opening, under plain strain (that define textbook "K_{IC}" fracture toughness values)—and that such conditions are virtually impossible to obtain when evaluating thin films.

Furthermore, as rightly pointed out more recently by Chen et al. (2019), despite the incorporation of *H/E* as an elastic-strain parameter in such fracture toughness calculations—and the use of H^3/E^2 as an indicator of P_C (Johnson, 1985; Tsui et al., 1995)—both of these parameters are essentially considered based on Hookean, linear-elastic behavior (and the limits thereof). As such, neither can be reasonably expected to describe with any accuracy significant plastic strain behavior in ductile substrates and/or metallic (or metal-containing) wear-resistant coatings—even taking account of the reality that indentation hardness is in most cases measured by evaluating the dimensions of a permanently (and predominantly *plastically*) deformed area. Thus the frequent attempts over the last 20 years or so by coatings researchers, to correlate coating wear performance to Hardness-to-Modulus ratios (using data extracted primarily from nanoindentation measurements), are often of questionable merit. This is due primarily to the fact that the tribochemistry of the contacting surfaces, rather than any tribo*mechanical* consideration, is frequently the determining factor in the degradation and wear behavior of many surface engineered systems (and of tribological systems in general). Factors such as the tribochemical-adhesive (rather than mechanical-ploughing) component of friction, the development of passive (or active) tribofilms on either counterface, and local plastic yielding under load, can (1) alter the real load-bearing area in a nonlinear fashion (with work-hardening effects often also involved), and can (2) have a much more significant effect on actual wear behavior than the Hardnesses (or indeed Elastic moduli) of the contacting surfaces alone. However, the now widespread use in Tribology and Surface Engineering of thin ceramic or, increasingly, nanostructured/nanocomposite coatings (with ceramic/metallic, crystalline/amorphous phase constituents) continues to yield many exciting opportunities to improve the performance of engineering materials. Nevertheless, such opportunities clearly also bring threats—in terms of the many more levels of complexity to accurate assessment of hardness, toughness, wear, and corrosion. The linking of such generally nonintrinsic (often somewhat subjective) parameters to expected/required load-carrying capacity—and/or to application-specific environment challenges—for commercial engineering components and systems remains a significant, sometimes intractable, problem for researchers and design engineers alike to solve, to expand the performance envelope (and extending "durability") in service.

Discussing specifically now, the ratio of Hardness to Elastic modulus, the authors of this chapter published, nearly 25 years previously, a widely cited paper on the significance of the *H/E* ratio in wear control (Leyland & Matthews, 2000), followed a few years later by a review of potentially suitable design criteria for nanocomposite coatings (Leyland & Matthews, 2004). The latter focused particularly on the potential to use metallic and/or amorphous phases to control (and, where expedient, to minimize) coating elastic modulus values, to the potential benefit of wear performance. In both papers, the influence of material stiffness was highlighted, from a "global" coating/substrate system design perspective. We emphasized—as several other authors had also previously articulated (e.g., Almond (1984), Kramer (1983), Ramalingam (1984))—that an important aspect of successful Surface Engineering design is to consider the elastic properties of both coating *and* substrate, preferably finding a way to minimize (or, at least, integrate) differences between the two and, if possible, eliminate interfacial elastic mismatch—thereby reducing the propensity for coating/layer disbondment under mechanical loading. We also pointed out in (Leyland & Matthews, 2004) that the ultrahigh hardness values being pursued (and occasionally achieved) at that time by PVD coating researchers, although impressive, were probably of secondary importance to the achievement of high wear resistance—even in abrasion and cutting since, as alluded to earlier in this chapter, the hardness need only be "sufficiently high" to suppress tribomechanical wear (Eyre, 1976, 1978; Khruschov, 1974; Khruschov & Babichev, 1953; Richardson, 1967, 1968).

However, it is expedient to point out that high hardness tends also to be closely correlated with strong ionic or covalent bonding (as may be seen commonly in ceramic materials), which is in turn responsible for the high Elastic moduli of many bulk engineering ceramics, and of most ceramic coatings. It is perhaps unfortunate that the interatomic bonding characteristics that promote high hardness are also largely responsible for the brittleness (and poor toughness) of many ceramic materials. On the other hand, another consequential characteristic for many ceramic materials is a very low surface energy—which can be beneficial in promoting benign tribo*chemical* behavior. Thus an ongoing challenge for tribological-coating designers is to find the right combinations of hardness, toughness, resilience, and (tribo)chemical inertness for an intended coating application—taking account of both the loading conditions and the environment—to optimize the mechanical behavior of the coating/substrate system as a whole, without compromising surface tribochemical requirements.

16.3 The Archard Wear Equation

Revisiting now the commonly used "Archard Wear Equation" (and the wear coefficients derived from it), we should first remind ourselves of its origins and of the assumptions made in its implementation. As previously mentioned, Archard (1953) published a (widely cited) paper, reviewing and extending previous seminal work by Swedish electrical engineer, Ragnar Holm (1922, 1967), which examined the nature of metallic electrical contacts and how the electrical conductivity between stationary plates of copper or nickel increased when two similar, nominally flat, surfaces were pressed together. Archard proceeded to consider the effects of relative movement of contacting metal surfaces and the wear which resulted. The main conclusion drawn was that, due to microscopic roughness (at the asperity level) of the nominally flat surfaces, the real contact area between the plates was initially only a fraction of that which was apparent from the measured area. However, the real area appeared to increase, in proportion to an increasing applied load, gradually approaching the apparent contact area—with a supposition made, that surface asperity contacts increased in number (but not in size). Bowden and Tabor (1939) had previously drawn similar conclusions for metal surfaces sliding against each other, noting also an increase in frictional heating as the contact load was raised (and the real contact area increased).

Archard's initial approach was to consider asperities as hemispheres of ideally plastic material pressed against a "nondeformable" counterface, to estimate a load-dependent asperity contact area (Archard, 1953). Although he did not directly reference the original work of Hertz (1882) on nonadhesive, *elastic* contact mechanics (or any subsequent derivatives thereof), Archard did, however, acknowledge that elastic considerations may alter the actual contact area by a factor of 2. When proceeding to describe wear resulting from asperity contacts, other assumptions were also made; that is, (1) each asperity contact created a hemispherical wear particle of dimensions equivalent to the assumed circular contact area, and (2) the asperity contact area remained the same with increasing load (and only the number of contacts increased; not their size). Nevertheless, despite the uncertainties that the earlier assumptions introduce, Archard was able to plausibly adapt Holm's originally proposed relationship of $W = Z.P/p_m$ (Eq. 1)—where W is the worn volume per unit sliding distance (and P the applied load), with p_m being the (previously mentioned) flow pressure, and Z the number of atoms removed per atomic encounter—to develop a revised formula that was based more upon the physical realities of the surface topography of engineering materials. It should, however, be mentioned at this stage that the p_m parameter, as defined by Holm (1922) (and subsequently reiterated by Archard), is initially introduced in (Archard, 1953) as a term defining the flow pressure for asperity deformation in a softer, deformable (and ideally plastic) material, rubbing against a "nondeformable" (and perfectly smooth) hard counterface—i.e. very similar to the model originally envisaged by Holm.

However, Archard then proceeded to reconsider Holm's "atoms-removed-per-atomic-encounter" probability scenario, to introduce a revised "probability factor" K, based on adhesive contact encounters between asperities of an idealized shape and distribution. These were shown conceptually in (Archard, 1953) as having an identical, flat surface area (on both contacting surfaces)—implying a pairing of materials of similar roughness and deformability (and therefore probably also hardness). Notwithstanding the uncertainties that these numerous assumptions may introduce, particularly in regard to what p_m might physically represent, this approach allowed Archard to consider the (more likely) scenario that only some smaller fraction of the total asperity-to-asperity encounters between two "real," microscopically rough, contacting surfaces in sliding motion would generate wear—and to propose that: $W = K.P/3a$ (Eq. 2). Here, the probability factor K is related to the likelihood of an asperity encounter producing a wear particle, and a is the wear particle radius (with an assumption that this is similar to the asperity contact radius). In his paper, (Archard, 1953), two main "rules" (out of four tentatively proposed) arose from the theoretical considerations—and from some compiled load-dependent wear-rate test data (for brass and Stellite rubbing against tool steel)—namely, that (1) wear rate is proportional to load, and (2) wear rate is independent of the apparent contact area. Archard and Hirst (1956) subsequently reported a further iteration of (Eq. 2) above (more directly related to Holm's original equation, Eq. (1)), specifically that: $W = K.s.P/p_m$ (Eq. 3), with s being the sliding distance. They also demonstrated with test data that the two rules stated above appeared to hold firm for a range of metal-on-metal dry sliding pairs (Archard & Hirst, 1956).

In terms of a direct linear relationship between applied load and wear rate (or between Hardness and wear resistance), further assumptions are required for Eq. (3) to remain valid. The most significant of these were "that p_m and K remain constant" (this was the basis of "rule (c)" in Archard (1953), which was not discussed further at that time)— but neither of these assumptions is true in practice. Even if the correct p_m-value of the softer material can be successfully identified (see again comments above, about the highly idealized origins of this parameter (Archard, 1953; Holm, 1922)—and considering also that the first point of yielding in compression occurs below the surface (Hertz, 1882; Johnson, 1985), and is very difficult to measure—then, as has been pointed out by other authors (e.g., Eyre (1976, 1978)), the dimensionless wear coefficient K also has little or no meaning. It is often the case in real engineering applications that the wear *regime* changes from mild to severe (Lim & Ashby, 1987) and/or (likewise) the wear *mechanism* itself changes as wear progresses; see for instance Farrow & Gleave (1984) (for an early example of a "wear interactions" schematic). Another valid argument against the effectiveness of the Archard-attributed wear equation is that, despite being formulated with the intention to describe *adhesive* sliding wear, the equation does not actually contain any relevant term to account for the adhesive strength of a "welded" asperity interface, and/or the work required to break this interface apart (the consequences of which may be the production of wear debris). The dimensionless wear coefficient K indirectly takes account of this aspect but—since K can in practice vary by many orders of magnitude for different engineering material pairings—it is often difficult to attribute a relevant meaning to K for practical applications.

On the other hand, Rabinowicz (1958)—who was looking for an improved understanding of how and why wear debris is created in sliding wear—proposed a simple equation, describing the volumetric energy expenditure required to generate wear particles of a particular critical size, based on the release of stored elastic strain energy to generate the new surfaces caused by asperity detachment. In many ways, his approach was analogous to the thermodynamic "energy balance" considered by Griffith (1921), in developing his fracture mechanics equation (with the toughness— or "strain energy release rate"—being assigned the parameter "G"). The key function used by both Griffith and Rabinowicz to quantify the energy released from a volume of elastically-strained material during the formation of new fracture surfaces was "σ^2/E" (a parameter that we shall return to, later on in this chapter). Although Rabinowicz himself seems to have subsequently endorsed Archard's approach (again, we shall return to this in more detail in the following sections), there seems to be some merit—but also a lack of universality—in the approaches of both researchers. Thus, despite the widespread acceptance of the Archard Wear Equation, some researchers have suggested that, if the "energy" approach of Rabinowicz and the "hardness" approach of Archard could somehow be combined, then a more universally applicable formula for wear prediction might emerge (see, e.g., Popov (2019)). Furthermore, as our computational materials modeling capabilities continue to develop and strengthen, some researchers have recently started to revisit the fundamental arguments (developed in the 1940s and 1950s) around the nature of friction and wear, to try to model the nanoscale material interactions whose precise nature remains so elusive (e.g., Garcia-Suarez et al. (2023)).

Notwithstanding the many assumptions, uncertainties and contradictions highlighted earlier, the fact is that Archard and Hirst themselves did not actually develop, either in Archard (1953) or in Archard and Hirst (1956), any equations (or "laws") based on the direct use of Hardness. They even went so far in Archard and Hirst (1956)—whilst briefly mentioning an apparent correlation between *abrasive* wear and Hardness (seen from the contemporary research of others, including Rabinowicz)—as to question the validity of trying to correlate Hardness to wear rate. Nevertheless, Eq. (3) was subsequently adapted to incorporate a Hardness parameter, in place of Holm's "flow pressure." Archard did, on the other hand, briefly suggest in his 1959 paper on *"The Temperature of Rubbing Surfaces"*(Archard, 1959) that Holm's asperity "flow pressure," p_m, could also be considered to be the hardness, H, but no explanation was given as to why this was a valid assumption, nor did Archard attempt to associate this consideration in any way to the above-mentioned Eq. (3), introduced through his earlier 1956 work with Hirst (on the unlubricated wear of metals) (Archard & Hirst, 1956)—although other researchers are likely to have done so. The association between p_m and H may also have occurred via Tabor's contemporaneous assertion that $H \approx 3\sigma_Y$ (for metals) (Tabor, 1951)—and thus an assumption that p_m (a theoretical compressive stress) and σ_Y (a measurable tensile stress) should be directly interchangeable —to essentially state that: $W/s = K.P/H$ (Eq. 3a). This provided what is recognizably the Archard Wear Equation in common use today—with the "probability factor" K reinterpreted as a (dimensionless) wear coefficient, and with W/s redefined as a wear rate Q, and the p_m parameter interpreted as being a direct proxy for the measured hardness H (of the material being worn). Slightly confusingly, "W" is now quite often used instead to describe the normal load (formerly "P"). Thus the equation (and socalled "simple theory of wear") now attributed to Archard is frequently expressed in tribology textbooks as $Q = K.W/H$ (Eq. 4) (Hutchings & Shipway, 2017). In practice however, most tribologists (and particularly *coating* tribologists) harness the principles behind the Archard wear equation to conveniently plot graphically, and compare (in journal publications), a *specific* wear rate—but in this case not using the supposedly "dimensionless" wear coefficient (i.e., probability factor) K. In fact, K is typically divided by the Hardness H to generate a *dimensional* wear coefficient, which was initially designated as little "k" (Hutchings & Shipway, 2017); that is, $(K/H =) k = V/w.s$ (Eq. 5), with V being the wear volume; note also that the load (big "W") is, in this version of the equation, reduced to a normalized "unit load"—little "w" (e.g., Holmberg & Matthews, 1994)—and that also, in many journal publications, authors sometimes now ascribe big "K" to the *dimensional* wear coefficient, k.

Thus, through a number of translational steps, the original wear rate equation (Eq. 3) has been reformulated in terms of the total volume, V, of material removed (typically measured quite conveniently by stylus, or optical, profilometry in a wear test scar/crater)—per unit of normal load applied, per unit of sliding distance (both also being easily measurable)—with appropriate measurement units of k typically being of the order of 10^{-3}–10^{-6} mm^3 N^{-1} m^{-1}. Although other eminent researchers such as Ashby (2011) have, not unreasonably, described such "normalized" wear rates in the standard SI units of "$m^2 N^{-1}$," the aforementioned mixed-unit "$mm^3 N^{-1} m^{-1}$" parameter has proved to be a convenient (and widely adopted) measurement tool for describing and comparing the wear rates of materials and coatings. However, it should be pointed out that, although dimensional analysis of little-k (the dimensional wear coefficient - or specific wear rate) reveals units of 'reciprocal pressure' (ie. $1/Nm^{-2}$), these units of (normalised) wear rate under a moving contact should not be confused with those of (quasi-static) indentation pressure - ie. Hardness. Again, it is rarely possible in practice to correlate a static contact pressure to that which is found under the dynamic conditions of sliding wear - even if the contact geometries in each case are notionally similar. Perhaps most significant therefore, is that this commonly used *dimensional* wear coefficient actually requires no direct consideration of the Hardness (of either of the two counterfacing materials) to be made. From a modern coatings tribology perspective, such an approach appears to have been accepted as being of general validity, based on the approach now taken by most tribology researchers in presenting wear data acquired from a broad range of (predominantly) metallic substrates, and a multitude of (mainly) ceramic wear-resistant coatings deposited thereon—using the panoply of instrumented laboratory wear tests that are now readily available to the research community. However, one might conversely argue that, although convenient (and universally understood), a wear equation which considers neither H nor E—although providing useful comparative data through which to down-select "better" Surface Engineering solutions—can in itself be no more than an iterative tool, with no "predictive" insights provided directly. On the other hand, the wear test scars and measurement data resulting from simple, instrumented laboratory-based tribological tests, *do* provide an indirect "window" of access for the increasingly powerful microscopic, and spectroscopic, analytical techniques now at our disposal—through which researchers can more confidently deduce precisely *how* and *why* wear occurred (rather than just "what" occurred).

In specific regard to the assumed "translate-ability" of p_m and H, which seems to have led to the widespread adoption of the "Archard Wear Equation"—and, indirectly, to a further assumption that the inverse proportionality between Hardness and wear holds true not just for abrasion of metals, but for adhesive (and other forms of) wear of other materials—the present authors have a number of concerns. Firstly, about the highly idealized origins of p_m, that (as we mentioned previously) arise from static-loading considerations of an ideally plastic softer, hemispherical asperity, pressed against an (again idealized) "nondeformable," perfectly smooth counterface. In a dynamic (sliding) scenario, it is then assumed that the asperity has adhered/welded to the counterface —and that the tangential force required to initiate motion is equivalent to the shear strength σ_S, of the softer material (Archard, 1953; Bowden & Tabor, 1939; Rabinowicz, 1958). Archard further assumed (Archard, 1953) that the circular contact area of the deformed asperity was now "encountering," in sliding, deformed asperities of identical contact area on the counterface—and that, with increasing load, the number of asperity contacts increased, but their average radius did not (Archard, 1953; Burwell & Strang, 1952; Rabinowicz & Tabor, 1951); this average radius was apparently assumed to be of the order of several tens of microns. From these considerations, it follows naturally that the friction coefficient should be $\mu = \sigma_S/p_m$, but the values of μ that such idealized considerations yield are typically between three and seven times smaller than those seen in practice (Barwell, 1979). Also, the size, shape, and volume of wear debris actually found in sliding contacts rarely bears any resemblance to the mechanisms proposed above.

Secondly, the aforementioned theory is also based on having an intimate knowledge of the *real* contact area. For many static, *elastic* contacts this area can probably be estimated with reasonable accuracy (Hertz, 1882; Johnson, 1985)—but for plastic deformation (or fracture) and detachment of asperities in sliding wear, determining what this area is (and/or how it might change with sliding distance) remains a challenging topic. Early work by Bowden and Tabor (1939) showed data for contacting steel-on-steel flats of 21 cm² area where, between loads of 3 and 300 kg (equivalent to contact pressures of approximately 140 MPa–14 GPa), the real contact area—presumably measured (or estimated) under static loading (but not clearly stated)—was apparently very small, varying from <0.0006% to >0.6% [described in Bowden and Tabor (1939) as: "1/170,000–1/130"] of the apparent area under the loads applied. However, no specific guidance on how the area might change with time, under sliding, was provided. At the other limit of a spectrum of possibilities, Lim and Ashby, whilst developing wear maps for steel-on-steel contacts (Lim & Ashby, 1987), suggested that "seizure" would occur as the real contact area in pin-on-disk sliding approached 100% of the apparent area—but, clearly, the point-of-seizure (in dry, unlubricated sliding of steels) will (1) probably occur at much less than 100% contact, and (2) vary significantly, depending on both frictional heating (i.e., velocity, at a given load) and a multitude of other, tribochemical, factors related to the precise details of the material pairings and their surface finish. In regard to surface finish, Greenwood and Williamson's well-known 1966 paper on the "Contact of nominally flat surfaces" (Greenwood & Williamson, 1966) examined in some detail the effect of asperity size/shape on contact pressure (outside of the earlier, idealized hemispherical asperity considerations). They developed a "plasticity index" to guide bearing designers on what level of polishing may be required, to achieve predominantly elastic contact conditions, where: $\psi = E'/H \cdot (\alpha/\beta)^{1/2}$, with α and β representing the asperity tip height and radius, respectively.

Thus a formula combining measurable material and topographical parameters was introduced (n.b. also that the included "material parameters" of significance are in fact H and E—*with E' being the contact modulus*). Of more relevance to our immediate discussion however is that Greenwood and Williamson also systematically tabulated and plotted contact area versus applied load (Greenwood & Williamson, 1966)—albeit again for a static contact—revealing approximate contact areas of 0.01%, 0.1%, and 1.0% for loads of 1, 10, and 100 kg. They also showed largely overlapping slopes for nominal contact areas of both 1 and 10 cm², demonstrating the independence of the real contact area from the chosen nominal area. Taking however the 10 cm² example—that is closest to the 21 cm² value described by Bowden and Tabor (1939)—the contact pressures involved are of the order of 10 MPa, 100 MPa, and 1 GPa, respectively. Comparing these results with those of Bowden and Tabor, there seems to be a potential discrepancy of around an order of magnitude. Other authors—for example, Eyre (1987)—state a typical real area in wear contacts of " < 1/1000" (i.e., less than 0.1%), that seems to fall within the range identified by Greenwood and Williamson—but no mention of "how much less than 0.1%" is given, nor what materials were being considered (Eyre, 1987).

Over the last one or two decades, researchers have attempted to measure the real contact area for sliding wear in situ, using transparent counterfaces (such as glass or acrylic)—see for example Krick et al. (2012). Furthermore, as laser surface processing technologies (and associated laser-interferometry optical imaging capabilities) have advanced, other authors have recently attempted to monitor the wear behavior of more relevant engineering tribological contacts; for example, using a laser patterned tungsten substrate and monitoring its wear evolution in reciprocating-sliding against an alumina ball (pass-by-pass), using 3D-imaging digital microscopy, based on twin-laser-beam interferometry (Lechthaler et al., 2019). A recent example of such work by Lechthaler et al. (2019) reveals a fractional contact area rising from less than 0.05 after 20 passes (the authors state that the wear scar could not be reliably detected below this number), to around 0.1 after 100 passes, then fluctuating in a range between 0.1 and 0.15 until the test was terminated after 500 passes. Extrapolating visually the plotted data in Lechthaler et al. (2019) (Fig. 5 therein, specifically), it seems likely that the initial contact area fraction was certainly <0.01 (i.e., substantially less than 1%); thus, in percentage terms, Greenwood and Williamson's original (unworn) contact area data seems to be of the correct order of magnitude and, after a short running-in period, the real fractional contact area apparently rises above 0.1 (i.e. >10%), and then settles in a range between 10% and 15%. Another recent paper by Li et al. (2021) provides a fairly comprehensive historical overview of other key developments in this area to date.

16.4 A brief historical perspective on hardness, hardness-to-modulus ratio, and wear-rate determination

Notwithstanding the extensive contributions of Bowden and Tabor to the measurement of friction and hardness, dating back to the late 1930s (e.g., Bowden and Tabor (1939), Hutchings (2009), Rabinowicz and Tabor (1951), and Tabor (1951, 1996)), the early work of Archard and colleagues at Associated Electrical Industries (AEI) Ltd. (located near Aldermaston, Berkshire, United Kingdom), from the mid-1950s to early 1960s, collectively made a significant (and somewhat underacknowledged) contribution to the soon-to-emerge discipline of Tribology. At this time, a group of researchers who were mainly Physicists by discipline, led by Wallace Hirst (a former student of Bowden at Cambridge)—and including John Frederick ("Jack") Archard, John Lancaster, and several others (Hirst, 1995)—had been gathered together in the recently-established AEI Fundamental Laboratory at Aldermaston Court. From the late 1940s onward, Hirst and his team performed a lot of groundbreaking work on the sliding wear of metals, on the abrasive wear of polymers, and in the lubrication tribology of materials in general. Of particular importance to the purpose of the present authors (i.e., primarily in discussing the ratio of Hardness to Elastic modulus, and pertaining topics in tribology and wear), was the work of John Lancaster—together with Hirst in the early years at AEI, and then later (jointly and individually) when both researchers subsequently took up Academic positions at the University of Reading. Around the same time that Archard published with Hirst his data on metallic wear in dry sliding (Archard & Hirst, 1956), that led to the "simple theory of wear" (and, indirectly, to what many now know of as the "Archard Wear Equation"), Lancaster and Hirst also published in 1956 an article investigating the effects of oxide tribofilm formation on the wear of metals (Hirst & Lancaster, 1956), already noting several circumstances where "rules" described by (or later indirectly attributed to) Archard were not necessarily obeyed.

In the years that followed, Lancaster went on to investigate the wear of nonmetallic materials—and particularly of polymers—demonstrating wear relationships that showed closer correlations with the Elastic modulus of the worn material (Lancaster, 1963), and/or correlated not at all with Hardness, but with the stress−strain product of the material under test (Lancaster, 1969, 1978). That is, a relationship of wear to the energy consumed (work done) defined by the load − displacement behavior of the (mainly viscoelastic-polymer) materials investigated and, as we will show in the following section, exhibiting a direct relationship to H/E via the measured strain-to-failure. His approach of assessing wear in terms of the work done (energy expended/dissipated), rather than some probability of wear particle generation, seems to be more closely aligned with the original, much earlier (mid-1800s), "pivot friction" theory of German Scientist and Mathematician, Theodor Reye (1860), than with the Holm/Archard approach (Archard, 1953; Holm, 1922). Lancaster also proceeded to compare the wear behavior of a wide range of metals and polymers (amongst themselves, and in combination) (Lancaster, 1978)—pointing out the challenges in translating acquired laboratory wear data to real-life performance in each case. Lancaster found that, although metals tended—under abrasion—to follow a fairly linear relationship of wear to hardness, and that this relationship also scaled linearly with surface roughness, neither of these factors held true for *polymer* wear (the reader might find Figs. 9−11 in Lancaster, 1978) illuminating in this regard). For the wear of polymers (particularly under dry abrasion), it was found that the reciprocal product of ultimate stress and ultimate strain ($1/\sigma_U.\varepsilon_U$) —that is, work done in (visco)elastic strain to failure—gave a much more accurate representation of wear rate than did $1/H$ [see again Lancaster, 1978, Evans & Lancaster, 1979, and Lancaster (1969)]— in contradiction to the earlier work on abrasion of metals (Khruschov & Babichev, 1953; Rabinowicz et al., 1961; Spurr & Newcomb, 1957). However, what Lancaster did not go so far as to elucidate at the time, was that this "energy approach" could be used more generally, to describe also the wear behavior of metals, ceramics, and composite materials—including (nano)composite coatings.

In fact, as alluded to previously, this approach concurs with the earlier practical findings of Oberle (1951), who discussed that, for a range of metals (but also for chrome hard plating, and for ceramic/cermet materials such as Al_2O_3/WC-Co), "surface stability" (in resisting wear) could be correlated to the area under the stress − strain (or, by analogy, load − extension) *curve*. This is an interesting (and mathematically valid, but possibly accidental) choice of wording—since his conversation is focused entirely on the linear-elastic *slope*—although it was mentioned also that the (Brinell) Hardness squared, divided by (2×) the Elastic modulus, could be considered as a "modulus of resilience" (with possible 'nonlinear-deformation' implications). However, it should be pointed out that the nature of the Brinell indentation method (i.e., compression loading by a "blunt" (hemi)sphere, rather than the higher strain contact of a "sharp" cone or pyramid) does lead to much more extensive work hardening and, as Tabor (1951) pointed out, Brinell indentation hardness values therefore correlate very well with the ultimate tensile strength (UTS) of metals, but (unlike conical/pyramidal indenters) less so with yield strength. It is not clear if Oberle was aware of this indenter-geometry distinction at the time that he published his work; however, he did also consider, more generally, $S^2/2E$ (where S is some "maximum allowable stress") in the same context—mentioning the idea that the energy stored (per unit volume) should be a good indicator of resilience. Note here, the similarity—in terms of a "stored elastic strain energy" approach—to the previously mentioned 1950s work by Rabinowicz on wear particle detachment (Rabinowicz, 1958).

Although not explicitly stated by Oberle, the "$S^2/2E$" consideration implies that *plastic* deformation work above the yield strength (flow pressure) may also be considered in the same way, including work-hardening behavior—something that, incidentally, does not appear to have been considered by Rabinowicz (1958). Furthermore, Oberle also mentioned the significance of rubber (as a low-modulus material that can distribute applied loads over a larger deformation volume), in avoiding/delaying the attainment of some critical contact stress for damage/failure. For simplicity of argument, Oberle focused on his so-called "Modell Factor"—essentially a measure of H/E, but not in SI units. Again, it should also be borne in mind that significant work-hardening tends to occur in Brinell indentation of metals specifically (but not of polymers or ceramics), and that Oberle may not have fully considered this in his measurement interpretations. Nevertheless, the aforementioned points, in regard to the area under the load-extension curve—and the stored (or expended) elastic (or plastic) strain energies that this area represents—are aligned with Lancaster's findings in abrasive wear of both metals and polymers, and arguably provide the key to a more comprehensive understanding of both the significance and the *limitations* of H/E in the measurement/prediction of wear behavior.

Despite not directly including a parameter for asperity contact adhesion strength (and subsequent detachment as a wear particle), the "simple" Archard Equation for the adhesive, sliding wear of metals nevertheless provided a relatively straightforward theoretical approach to wear prediction (assuming that a specific wear regime and mechanism could be maintained in practice)—but, more importantly for contemporary use, it yielded also a basic equation with which to conveniently measure and compare *specific* wear rates (without necessarily taking the hardness of the materials involved into direct consideration). However, as initially theorized (at least from an elastic deformation perspective) by Rabinowicz (1958), an accurate assessment of work done—via the stored/recovered (elastic) *and* the expended/unrecovered (*plastic*) strain energies involved—in the wear and failure of engineering components is clearly a more universal approach, and an important goal in permitting both researchers and design engineers/practitioners to correlate measured wear data to likely performance in service. As alluded to above, the approach of assessing work done (due to frictional energy expended during the wear process) predates the Hardness-oriented work of Tabor, and the asperity deformation/adhesion-oriented work of Archard, by nearly a century.

Although less widely acknowledged in modern tribology, the work of Theodor Reye in the mid-1800s was arguably the first to devise a useable theorem upon which to measure (and predict) friction and wear. His work on "The Theory of Pivot Friction" (Reye, 1860) remains quite widely studied in mainland Europe, but has been largely neglected in the United Kingdom, United States, and elsewhere. Reye established that the sliding wear of articulating joints appeared to show a linear relationship between the tangential work of friction and the volume of wear debris produced. Researchers continue to contest the wider validity of both the Reye approach (to which Rabinowicz's early arguments also bear some resemblance) and the Archard approach in wear rate prediction and attempt to model nanoscale asperity interactions, to elucidate fundamental mechanisms (see e.g., Garcia-Suarez et al. (2023)). Furthermore, the previously mentioned work of Khruschov and coworkers (Khruschov, 1974; Khruschov & Babichev, 1953) took a work of deformation approach to abrasive wear which, whilst also arguably closer in conception to Reye's work (in terms of wear debris generation due to tangential forces), nevertheless reached a similar conclusion to "the Archard approach," in terms of a basic equation that describes a wear rate versus hardness proportionality —albeit for *tangential* abrasive ploughing forces, rather than for asperity deformation/adhesion due to the applied *normal* load.

At this point, it should be reiterated that many researchers still continue to associate the Archard wear equation with the concept that wear rate is inversely proportional to hardness. However, as we pointed out above, none of the "rules" initially proposed by Archard (Archard, 1953; Archard & Hirst, 1956) give any mention of Hardness as a key property consideration. For example, Archard considers (in the two papers cited above) wear rate primarily as a function of sliding distance or applied load. However, what *is* stated in the discussion section of the 1956 paper by Archard and Hirst is that: "…. *the wear per unit load per unit sliding distance…. may be described as the coefficient of wear….*" – which seems to be a rather prescient statement, in terms of how we typically measure wear rates today. Furthermore, contemporaneous abrasive wear work (by Khruschov and others) is also briefly mentioned by Archard and Hirst—and, in particular, the apparent finding in some cases of an inverse relationship between wear and hardness. However, the authors then proceed to say that, in their opinion, such results are in contradiction to most of the other research literature available at that time, and comment also that—since both K and p_m are highly material dependent—a direct relationship between hardness and wear rate cannot be inferred from the theoretical equations presented (Archard and Hirst, 1956). Thus, the association between wear and hardness seems most likely to have arisen from the work of Khruschov and Babichev (1953) on abrasive wear, that was occurring in Russia at around the same time as Archard's work—and this was probably made more noticeable to a Western audience by the work of Rabinowicz at MIT (early development of a three-body abrasion test) that followed closely in 1961, (Rabinowicz et al., 1961).

Rabinowicz (with industrial coworkers Dunn and Russell) showed an inverse relationship between wear volume and hardness in a range of metals, where alumina grit was introduced between two metal annuli (made of the same alloy), one of which was floating (but static), the other being motor-driven and rotating (Rabinowicz et al., 1961). Rabinowicz discussed the recent work in Russia by Khruschov and Babichev (1953) (and by others, such as Spurr and Newcomb at Ferodo UK, (Spurr & Newcomb, 1957), and Burwell and Strang (1952) (the latter, like Rabinowicz, being based at MIT, USA), reporting similar outcomes. However, it was highlighted that, in the Russian work particularly, the results presented were generally comparative data (i.e., measured proportionately against a known, relatively soft, standard)—rather than giving absolute values of wear. Rabinowicz proceeded to present a "simple theory of *abrasive* wear" (Rabinowicz et al., 1961), that mathematically bore a very close resemblance to Archard's theoretical considerations in Archard (1953) and Archard and Hirst (1956). Rabinowicz readily acknowledged this similarity and, furthermore, suggested that—although Archard's approach was based on asperity adhesion and detachment in metallic wear—the same equation might be used also to describe material cutting and displacement in abrasive wear. Conversely, Rabinowicz appeared at this time to be searching for a test that could establish more accurately what the actual dimensionless K-values might be for different metal pairings, that were proving difficult to establish with certainty in adhesive sliding wear (due to the variabilities caused by surface contamination/oxidation, generation of entrained wear debris of different size/morphology, and cross-transfer of wear products between the two surfaces), and he proposed that an abrasion test largely avoided these complexities—and might therefore be useful to establish values of K for the design and selection of metal-on-metal pairings for tribology applications (including for sliding wear).

In the years that followed, Rabinowicz promoted the idea that the Archard equation could be used to predict both abrasive and sliding wear behavior—as may be seen in, for example, Rabinowicz (1980, p. 475–478), with H (or $3H$) directly replacing p_m (depending on whether the softer or the harder of the two counterbodies was being considered)—and this wider endorsement of Archard's equation, together with Archard's unsubstantiated comment in Archard (1959) (suggesting that p_m and H might be interchangeable in some circumstances), may have reinforced the notion that the statically-loaded "flow pressure" parameter, p_m, in Archard's equation from Archard and Hirst (1956) could be directly replaced by "H" for the "dynamic" asperity contact conditions of adhesive sliding. Rabinowicz also commented in (Rabinowicz et al., 1961)—in concurrence with Archard and Hirst's tentative suggestions regarding wear measurement units in Archard and Hirst (1956)—that ".... *the wear rate per unit length of sliding will be directly proportional to the load*....," but also that (unlike Archard and Hirst (1956)) the wear rate "*.... will vary inversely as the hardness of the surface*...." However, it is important to note that the latter of these two comments by Rabinowicz does relate specifically to *abrasive* wear—and what was also shown clearly in Rabinowicz's 1961 paper, in this specific context, was that such an inverse correlation does *not* appear to hold true for *sliding* wear of metals (the reader might wish to compare Figs. 3 and 10 in Rabinowicz et al. (1961), to confirm this point). Furthermore, Rabinowicz (1980) questioned the usefulness of an "inverse stress" parameter (i.e., m^2/N, or equivalent). Although the careful consideration of dimensionless wear coefficient values promoted by Rabinowicz (and others) has clearly benefited design engineers in guiding their materials selection criteria for (predominantly metallic) sliding wear contact pairs over the years, it should be pointed out that, in contrast to his comments (Rabinowicz, 1980) about the *"rather cumbersome units of volume/force-distance"*, the m^2/N pressure reciprocal—and particularly the mixed-unit $mm^3/N.m$ parameter—has proved to be of immense use to the tribological research community (and particularly to coatings tribologists), in the post-1970s Surface Engineering era. We shall return to this point in the following section, when discussing the "inverse stress–*strain*" parameter of Lancaster (1978), and its relationship to *H/E*.

As we mentioned previously, it seems that many researchers have adopted Rabinowicz's suggestion that the Archard equation could provide correlations between wear and hardness in both abrasion and sliding tests—but did not consider the important point that, despite some evidence to the contrary, it may not work very well for most sliding (adhesive) wear situations. In contrast, for three-body abrasion particularly (Rabinowicz et al., 1961)—and probably for many two-body situations, as Khruschov and Babichev (1953) and Spurr and Newcomb (1957) mainly considered—there is arguably a case for the Archard wear equation having a more universal applicability in evaluating and comparing abrasion data for metals and ceramics but, as Lancaster has clearly illustrated, not for polymers (or polymer composites) (Evans & Lancaster, 1979; Lancaster, 1969, 1978). Intriguingly, Archard himself later pointed out (Archard, 1980) that: *"a high value of Hardness is usually helpful. However, the probable factor of significance is not hardness, but the ratio (H/E) which represents the yield or proof strain of the material."*—that is, Archard appears to have recognized himself, that Hardness alone was unlikely to be the most important factor in determining rates of wear. For completeness, it should be pointed out finally that, although wear mechanisms are complex, and tend to fluctuate and change in many practical applications (Farrow, 1986; Farrow & Gleave, 1984; Holmberg & Matthews, 1994; Lim & Ashby, 1987), abrasive wear is widely considered to be the dominant wear regime in at least 50% of industrial applications (Eyre, 1976, 1978; Farrow & Gleave, 1984). However, there also remain inbuilt limitations in abrasive wear theory vs. the reality (see, e.g., in scratch test modeling (Williams, 1996), of material being displaced ahead and to the side of the wear track—and also significantly work-hardened, in many practical cases (Lindroos et al., 2015)—rather than simply being detached as wear debris) and with accumulated wear debris influencing the wear process itself—these factors all being challenging to incorporate into constitutive models (Lindroos et al., 2015; Williams, 1996).

16.5 Practical use of hardness-to-modulus ratio based parameters

As we pointed out earlier in this chapter, the use of hardness-to-modulus ratios (such as H/E and/or H^3/E^2) traditionally considers the *elastic* response of a tribological system, and little or no account is taken of the (often unavoidable) *plastic* work involved in many engineering applications. However, it should also be remembered that hardness is generally measured using some form of indentation method (to create a permanent material displacement), and that, to establish what the *elastic* surface contact pressure limit is (that largely defines the hardness value), significant *plastic* work needs to be done and—depending on the indenter geometry (and the indenter/substrate frictional interactions)—substantial additional energy expenditure may occur, such as that of work-hardening. Returning now to Oberle's considerations (Oberle, 1951), the contention that $S^2/2E$ is of importance—as a measure of work done per-unit-volume to cause wear—has wider implications, for the establishment of an H/E-based parameter that can provide some measure of the total energy input (both elastic *and* plastic) needed to cause failure. Considering, on a superficial level, the deformation behavior of metals then, as Oberle essentially described, $H^2/2E$ and $S^2/2E$ are somewhat interchangeable—primarily through Tabor's well-established assertion that $H \approx 3\sigma_Y$ (where, up to the elastic limit of Hookean stress − strain behavior), S and σ can be considered as interchangeable parameters.

First, let us define the work done (per unit volume of material) in a tensile test when elastically displacing a material through a distance x by the application of an applied load P. Here the elastic work expended is: $W_e = \frac{1}{2}P.x$ (i.e., the area under the slope on a graphical plot of P vs x)—and the same load-displacement linear extension (slope) could also be considered in terms of a stress−strain ($\sigma - \varepsilon$) plot, where the slope defines the Young's modulus, E. Since stress, $\sigma = P/A$ (load/area), and strain, $\varepsilon = x/l$ (i.e., extension/length), it is clear that $\frac{1}{2}P.x = \frac{1}{2}\sigma A.\varepsilon l = \frac{1}{2}\sigma \varepsilon.V$. Furthermore, since by Hooke's law strain, $\varepsilon = \sigma/E$, it can also be stated that (elastic) work, $W_e = (\sigma^2/2E).V$, meaning that the work done per unit volume of material is $\sigma^2/2E$. As mentioned previously, both Rabinowicz (1958) and Griffith (1921) made similar considerations.

Since the factor of 2 in '2E' is an artifact of the *elastic* work done, it is not unreasonable to conveniently neglect this factor, for the purposes of further discussion—noting also that Oberle's designation of $(\frac{1}{2})S^2/E$ (Oberle, 1951) could be used to signify a "maximum permitted stress" that may be reached above σ_Y (yet where the load-extension arguments for "area under slope/curve" defining the work done still hold true), but clearly the "real" *hardness* between σ_Y and the UTS will likely increase significantly for a moderately ductile material—typically by some material-specific power law, depending on the work-hardening characteristics of the material (Wyatt & Dew-Hughes, 1974).

Despite the "hardness uncertainty" outlined above, this leads us to consider further (in terms of both elastic and *plastic* deformation) the parameter H^2/E—something that the present authors did not previously publish in the scientific literature, but have mentioned frequently in conference presentations over the last 20 + years

—e.g., Leyland & Matthews, 2007; Leyland & Matthews, 2003 and Matthews & Leyland, 2002. The H^2/E parameter is important in the context of an energy approach to the deformation of materials in general—and in the use of H/E-based ratios as mechanical design parameters. Oberle's interpretation of the importance of $(½)S^2/E$ appears to have been a linear-elastic one; however, his choice of the word "curve," rather than slope, (Oberle, 1951)—although mathematically precise—seems strangely prescient. As alluded to above, the work term $(½)P.x$ is not conditional on linear elastic behavior; it does indeed relate also to the (total) area under a displacement *curve*—and the energy expenditure this represents (whatever form the curve might take). In the case of Lancaster's polymer abrasion, for example Evans & Lancaster (1979) and Lancaster (1969), and bearing in mind our discussion earlier about units of modern wear measurement essentially comprising a (reciprocal) stress value, the stress–*strain* product, $S.\varepsilon$ (or $\sigma.\varepsilon$, which—since $\varepsilon = \sigma/E$—could also, in a linear-elastic sense, be designated as σ^2/E) will, in most cases, define (for polymers) an approximately s-shaped visco-elastic curve, leading to the point at which tensile fracture occurs. Lancaster (with Evans) (Evans & Lancaster, 1979; Lancaster, 1963, 1969, 1978) plotted wear resistance against a reciprocal stress–strain product—initially designated "$1/s.e$" (Lancaster, 1969) (note "little s" for stress) but, for polymers, essentially being a reciprocal product of ultimate stress times ultimate strain (i.e., $1/\sigma_U.\varepsilon_U$)—finding an excellent correlation between this parameter and increasing abrasive wear rate for a wide range of polymer-based materials; that is, a high "$s.e$" value correlates to low wear. This parameter (to describe the abrasion of polymers and elastomers) appears to originate from work by Ratner and coworkers (Ratner et al., 1965, 1967) in Russia, who tried to correlate the impact strength (i.e., fracture toughness) of polymers to abrasive wear resistance—finding that H was of negligible importance, but that the work done (i.e., the product of "s" and "e") was a significant factor. In a metal, the slope will by comparison be entirely linear until the yield stress (or, in many cases, flow stress) is reached, and will then deviate sublinearly with further increase in applied stress—at a rate which depends primarily on the work-hardening characteristics of the material. In an ideally brittle ceramic, the slope will also be linear up to the fracture stress, σ_F but, in reality, any bulk engineering ceramic (or indeed a ceramic coating) is likely to exhibit at least some sublinear yield above σ_F, albeit—from an energy expenditure viewpoint—a vanishingly small fraction of the total energy input required to cause failure. In the common tribological case of a high-modulus, brittle ceramic coating on a relatively low-modulus metallic substrate, the ceramic coating will in practice typically fracture (and/or disbond), when (1) the coating/substrate elastic strain mismatch exceeds the fracture strength of the coating (or the interfacial adhesion strength) or, if neither of the abovementioned phenomena has yet occurred, when (2) the substrate yield strength is significantly exceeded, and extensive plastic strain occurs below the coating/substrate interface.

Turning now to the specific example of impact wear of metals, Eyre (1976, 1978) reported that, in cavitation erosion, there is inevitably significant plastic deformation and work-hardening of the impacted metal surface prior to material removal—and that, beyond an initial incubation period, the wear rates of different alloys can be correlated to the square of the UTS, divided by the Elastic modulus of the alloy in question. Thus Oberle's $(\frac{1}{2})S^2/E$ "modulus of resilience" parameter does indeed appear to be relevant for both elastic (as Oberle primarily considered, Oberle, 1951) and *plastic* deformation, if the "maximum permitted stress" is in fact not the yield strength, but the UTS. Smart closely followed Eyre's bulk material comments on this matter (Smart, 1978; Smart & Moore, 1979) with a discussion in the context of *coated* materials, finding similar correlations and introducing the terminology of "ultimate resilience," as a ranking parameter for cavitation erosion resistance of coating/substrate systems. However, the likelihood of such correlations in erosive wear was not unknown before the mid-late 1970s work of Eyre/Smart. In 1963 Thiruvengadam (1963) published a "Unified Theory of Cavitation Damage," based on work being undertaken on behalf of the US Office of Naval Research (Thiruvengadam & Waring, 1964), which was subsequently incorporated into an ASTM Standard Technical Procedure (Thiruvengadam, 1967). Furthermore, such considerations can to some extent be traced even further back—particularly in the context of impact damage due to water-droplet erosion (WDE)—via the early work of Cook (1928) on the "water-hammer" effect in droplet impact.

Surprisingly there has been little or no further work since that of Smart (Smart, 1978; Smart & Moore, 1979), to try to correlate S^2/E—and/or H^2/E—to the cavitation erosion of coated surfaces, apart from one or two isolated comments (see, e.g., Godoy et al. (2009)); nor to find correlations from the reversible/irreversible work of (nano) indentation (i.e., from $W_e + W_p = W_t$, or from the indentation "resilience," in terms of W_e / W_t), that has been extensively studied in recent years (e.g. Bull, 2006; Chen, 2012; Chen & Bull, 2006a, 2006b, 2008). In the Authors' opinion, this is probably to a large extent due to the difficulties in accurately determining the variable (and often uncertain) parameters that should be considered above the yield/flow stress; that is: what are the true values of H, that might be input to an H^2/E parameter in the non-Hookean region between σ_Y and the UTS...? This consideration may be of less significance for brittle, monolithic ceramic coatings on ductile/compliant metallic substrates, but is clearly quite relevant to the multilayered, (nano)composite and duplex/hybrid/functionally-graded Surface Engineering solutions increasingly being researched (and now implemented industrially) in the early part of the 21st century.

Considering further the issue of how to address H/E-based ratios in the context of the *plastic* contribution to total energy expenditure (and the work-hardening aspects of this, between the bounds of the yield/flow stress and the UTS), many authors—particularly Cheng/Cheng (Cheng & Cheng, 1998), Malzbender/de With et al. (den Toonder et al., 2002; Malzbender & de With, 2000a, 2000b), Bull/Chen (Bull, 2006; Chen, 2012; Chen & Bull, 2006a, 2006b, 2008), and Yang et al. (2010) —have discussed (and extensively modeled) different elastic-plastic work-of-deformation scenarios, largely in the context of linear, or nonlinear, relationships between H/E and W_p/W_t (i.e., the ratio of irreversible Plastic-Work to Total-Work of indentation). It is interesting to note that these authors seem not to have taken account of previous work showing that S^2/E (Eyre, 1978; Oberle, 1951; Smart & Moore, 1979) (and thus possibly H^2/E (Beake, 2022; Godoy et al., 2009; Oberle, 1951; Leyland & Matthews, 2007; Leyland & Matthews, 2003; Matthews & Leyland, 2002) might be a more appropriate parameter to incorporate into their deliberations. Since S^2/E can (1) be related to the energy input per unit volume (whether that be elastic or plastic), and (2) is shown to correlate well to deformation/wear processes that involve significant plastic yielding, such as cavitation erosion, this omission is rather surprising. Furthermore, it seems to the present authors that analogies should also be drawn to the field of Fracture Mechanics—and not just to Griffith (1921), but particularly to Elastic–Plastic Fracture Mechanics (EPFM) and the *J*-integral theory, developed by Rice, Hutchinson (and coworkers) (Hutchinson, 1968; Hutchinson & Suo, 1991; Rice & Rosengren, 1968; Rice, 1968). Here, a "line-integral" approach was taken to the summing of work done (albeit in this case by the displacements caused in the tensile extension of a growing/moving crack, rather than the predominantly compressive displacements caused by a mechanical indenter), leading to a mathematical solution that is valid for "nonlinear-elastic" deflections (i.e., *plastic* work), but also provides a valid solution for linear-elastic work (that, as mentioned earlier, Griffith defined as "*G*" in his original energy balance considerations for fast fracture, Griffith, 1921). Thus, the nonlinear energy expenditure term, *J* converges with *G* (in the case that plastic work is negligible)—and both can be related to the fracture toughness K_C via the Elastic modulus: i.e., $K_C = [E.(G_C; J_C)]^{\frac{1}{2}}$.

Although the 3D mathematical complexity of computing the precise form/magnitude of stress/strain fields around an indenter remains as a rather intractable problem (particularly for coating/substrate systems with often intricate and nonlinear changes in mechanical properties with depth), the present authors believe that the use of H^2/E (as an elastic/plastic "resilience" parameter), together with *J*-integral-related mathematical approaches to addressing the "hardness uncertainty" above σ_Y, are worthy of further study—and might facilitate significant additional advances in our understanding of coating deformation behavior and fracture toughness, leading to better predictive tools for coating design and selection. Returning to a specific comment we made previously on this last point, many authors in the coatings community, when trying to determine coating fracture toughness values, obtain (by various means) $MPa\sqrt{m}$ values, which they define in terms of "K_{IC}," perhaps not realizing that the "*I*" (in simple, "*Mode I*" crack opening, under plane *strain*) has some very specific implications for how textbook Fracture Toughness data is collected/presented for bulk materials—and that any value computed for a thin film will, almost inevitably, have been acquired under predominantly plane *stress* conditions (and is thus likely to be an overestimate of whatever the "real" K_{IC} value might be, by a factor of the order of 1.5–2.0). It is in our opinion rather inappropriate therefore, to use the term "K_{IC}" when presenting *coating* fracture toughness data.

16.6 Appropriate use of hardness-to-modulus ratio as a design tool—fundamental considerations

Returning now to the broader topic of the appropriate use of *H*-to-*E* ratios in coating design; as mentioned previously, although *H/E* can be (and is, widely) used as a coating development, screening and performance evaluation criterion, it is simply a *mechanical* performance consideration. Despite being a likely parameter of significance in erosion and abrasion wear situations, where the surface chemical interactions are often (but not always) negligible—and possibly also under certain low-amplitude fretting conditions (where repeated elastic deflections of the contacting surfaces contribute significantly to the ensuing wear process)—*H/E* is of little or no use in the sliding wear and contact adhesion of smooth surfaces; that is: it can in no way represent the tribochemistry of a tribological contact. Any correlations between *H/E* and wear resistance that are found in such circumstances may relate in part to the mechanical properties of (thermochemically directed) stable tribofilms, once formed—but are, in the authors' opinion, likely to be purely coincidental in many cases. Czichos has probably framed this issue most succinctly in Czichos (2001) (see, e.g., Fig. 10 therein), by categorizing tribological interactions, that may lead to wear (and debris formation), into two main regimes; "Stress Interactions" (involving both loading regime and frictional forces) and "Material Interactions" (relating predominantly to interatomic bonding). Although frictional heating may have an influence on material behavior in either category, Czichos proposed that Stress Interactions comprise mainly *Abrasion* and surface *Fatigue*, whilst Material Interactions comprise mainly *Adhesion* and *Tribochemical* reactions. For the purposes of considering what *H/E*-based mechanical/physical parameters can, or can't, tell us (and indeed why an Archard-type wear equation—and/or trying to correlate hardness alone to wear performance—may not work as effectively for adhesive wear as for abrasion), this simple categorization by Czichos is rather illuminating.

As discussed in Section 16.2, the primary purpose of *H/E* as a design tool should be in the holistic consideration of the coating/substrate system mechanics—including any interlayers and/or substrate pretreatment prior to application of the coating top-layer. As a number of researchers considering coating design have commented over many years, as well as the need for sufficient mechanical strength (hardness) and load-bearing capacity (Almond, 1984; Bell, 1991; Kramer, 1983; Leyland et al., 1991; Matthews & Leyland, 1995; Ramalingam, 1984), coating elastic properties (and, for some applications, also their thermal expansion coefficient and thermal conductivity) need to be given careful consideration (Almond, 1984; Kramer, 1983; Leyland & Matthews, 2000, 2004; Ramalingam, 1984)—together with the interplay of such properties with the substrate material under mechanical and/or thermal load. Any significant property mismatch between coating and substrate is likely to compromise the overall system performance and although mechanical aspects are often a primary consideration for many applications at or near ambient temperature, *H/E* will again not yield any direct benefits in the optimization of coating thermal (or thermo*chemical*) properties. Referring back to the early work of authors such as Khruschov and Babichev (1953) and Richardson (1967, 1968), it is clear that, although a certain minimum level of hardness is required to resist mechanical wear (and particularly abrasion), increasing the hardness beyond some optimal level (which seems to be of the order of only 1.25 − 1.4 times the hardness of the abradant,

Khruschov, 1974; Khruschov & Babichev, 1953; Richardson, 1967, 1968) leads to a plateauing of the wear rate, with little additional benefit found in increasing hardness further. However, the arguments for chemical (adhesive) wear are of course more nuanced, in the sense that high hardness tends also to arise as a consequence of strong interatomic bonding, and a commensurately low surface reactivity. On the other hand, a certain level of reactivity is often desirable in sliding wear of smoother, hardened surfaces—if this leads to the development of a stable tribofilm on one, or both, of the mating faces. Thus the pursuit of "ultrahigh" Hardness is of questionable merit, particularly since very hard, chemically stable materials tend also, through the nature of their interatomic bonding, to exhibit very high values of Elastic modulus—often many times higher than the preferred substrate material (leading to very high levels of coating/substrate elastic mismatch).

As the authors previously observed in Leyland and Matthews (2004), the enhanced capabilities of modern Surface Engineering deposition processes now permit us to step outside of the rather limited range of bulk-material H and E values historically available to the design engineer, and to develop composite, layered and functionally graded surfaces, where (with improved understanding of the likely performance needs), we can tailor the system mechanical properties to achieve sufficient (but not unnecessarily high) surface hardness values. At the same time, the coating elastic properties can be adjusted to more closely match those of candidate substrates or, where the coating/substrate modulus ratio is very high, to include some degree of functional grading—through duplex/hybrid treatments involving substrate pretreatments or interlayers (Matthews & Leyland, 1995).

16.7 A case study

A suitable "vehicle" with which to elaborate on a few practical design aspects, that pertain to many of the above points, is based on the Authors' own experience in the development and use of PVD Titanium Nitride, TiN—a coating material which revolutionized the cutting tool industry in the early mid 1980s (and was arguably the main reason for the emergence of "Surface Engineering" as a recognizable discipline in its own right at around the same time). In particular, if one considers TiN together with Titanium (and its alloys), as a candidate coating/substrate system for both light weight and high strength, then a discussion around the mechanical and tribological properties of this system can be enlightening. A range of Hardness values are quoted in the scientific literature for both TiN coatings and Ti-alloy substrates, but they typically fall in the range of $H = 20-25$ GPa, for TiN, and 1.5–7.5 GPa, for Ti and its various alloys. Similarly with Elastic moduli, $E = 450-520$ GPa is not an untypical range quoted for TiN, with near-alpha / alpha-beta Ti alloys typically quoted around 105–120 GPa, and near-beta / metastable-beta alloy values tending to be somewhat lower—in a range that, in some cases, reaches below $E = 70$ GPa. Thus it is realistic to say, for a notional TiN/Ti coating/substrate combination, that the Hardness ratio will be at least 3 (and possibly as high as 15 or 16), and the modulus ratio could range between 4 and $\sim 7½$. Even taking the minimum H and E "mismatch" values, of 3 and 4, respectively, it is clear that neither the substrate load-bearing capacity, nor the substrate/coating interfacial behavior, will be adequate for optimized performance in the basic Ti/TiN material pairing.

Having previously examined the load-bearing capabilities of Ti/TiN multilayer coatings in abrasion and erosion (Leyland & Matthews, 1994), the authors more recently carried out extensive work to optimize a Ti/TiN substrate/coating system (Cassar et al., 2010, 2011, 2012a, 2012b, 2012c), as part of a UK Industrial Collaborative Research and Development project to develop lightweight, exclusively titanium-based, bearing materials for applications in passenger aircraft (Wallwork HT/Airbus UK, 2013). In this case, the substrate of choice was a Ti-6Al-4V "ELI" (i.e., controlled low oxygen content) material, with Elastic modulus of ~ 115 GPa and heat-treated hardness of ~ 3.8 GPa. The broader project objective was to be able to replace typical solid, uncoated aerospace spherical bearings—comprising steel ball and bronze housing—with treated/coated titanium alloy components (of like-for-like dimensions). The intention was that this would lead to a weight reduction in excess of 40% per bearing (relative densities [$\times 10^{-3}$ kg m^{-3}] of ~ 4.5 vs $\sim 7.7-7.8$), with the potential to save hundreds of kg per aircraft. The main technical issue to address was the achievement of sufficient load-bearing capacity in the titanium-based bearing; in this regard, preliminary trials with PVD-TiN on Ti-6Al-4V had indicated a maximum permitted contact pressure of around 80 MPa, for a coating to survive without cracking/flaking, when the design specification of the conventional bearing was validated up to around 220 MPa—thus a nearly threefold increase in load-bearing capacity was required. As the present authors (and others) have described, load support is an important mechanical design consideration for practical applications involving surface engineering (Almond, 1984; Bell, 1991; Kramer, 1983; Leyland et al., 1991; Matthews & Leyland, 1995; Ni et al., 2004; Ramalingam, 1984; Ramalingam & Zheng, 1995; Wang et al., 2016). Furthermore, it was quite clear in this case that, with the elastic modulus of the Ti-alloy substrate being barely half that of the conventional materials in current use, the elastic deflections of the bearing under load would inevitably be significantly higher, and that this could potentially put a TiN tribological coating (of around four times the stiffness of the substrate) under immense interfacial shear stress, when loaded.

It thus became apparent that an appropriate duplex combination of a plasma-thermochemical diffusion treatment and a plasma-assisted PVD ceramic coating was required, to satisfy the mechanical loading demands. Preferentially, this duplex treatment should be implemented as a "hybrid" process (Leyland et al., 1991; Matthews & Leyland, 1995), in which both diffusion and coating treatment stages could be performed in the same equipment, in a commercially upscaleable manner, and at a maximum treatment temperature no higher than 700°C to maintain substrate core strength. The chosen candidate solution was a Triode Plasma Nitriding (TPN) substrate pretreatment, in combination with electron-beam (EB) evaporative PVD of a TiN ceramic coating, the reasoning behind which is detailed below. The "triode" terminology relates to the inclusion of a negatively-biased, electron-emitting third electrode—namely, a heated tungsten filament (generating thermionic electron emission at energies above the ionization potentials of nitrogen and argon)—to intensify the plasma, and sustain it at vacuum chamber pressures much lower than those conventionally used for traditional "diode" plasma nitriding, (i.e., <1 Pa; c.f. >100 Pa; Roliński, 2015). This type of configuration has been used routinely for EB-PVD of TiN—via direct, reactive evaporation of molten Ti—for many years (Bunshah & Raghuram, 1972; Matthews & Teer, 1980), preceding by over a decade the now more common approach of magnetron sputter PVD (that was challenging to implement commercially until the advent of closed-field, unbalanced-magnetron, multitarget systems in the late 1980s). Unlike magnetron sputtering (or indeed cathodic-arc evaporation, CAE—that, historically, preceded both "plasma-assisted" EB-PVD and sputtering) (Mattox, 2003), electron-beam PVD is fundamentally a thermally driven vapor deposition process, with no "plasma assistance" inherently available to the vapourised deposition species—unlike sputtering or CAE, where the deposition species are created (and energized) by localized plasma bombardment at the vapor source. Thus, the need for an independent plasma creation/enhancement device, in plasma-assisted EB-PVD.

This brings technical complexities and challenges for duplex process integration, but also potential advantages, since the thermionic-electron enhancement provides fully-independent control of the deposition plasma, with particular flexibility to control the energy distributions (and/or degree of ionization) of the active plasma species, as they arrive at the substrate being treated. As the authors have previously demonstrated (Leyland et al., 1991), this configuration can be adapted to perform plasma nitriding treatments on steel substrates—as well as PVD coating deposition—in the same treatment chamber, bringing several technical advantages for the former process.

Firstly, the use of a low-pressure intensified plasma designed for plasma-assisted EBPVD (with ionization levels at several percent—compared with barely 0.01% in a traditional diode plasma) allows a very thin (and virtually collision-less) cathode sheath to be created at the substrate surface, providing almost mono-energetic ionized N-species and high treatment uniformity (Leyland et al., 1991)—even on complex-shaped components with significant sharp corners and/or recesses. Secondly, the intense (but fully adjustable) plasma bombardment permits nascent surface oxide layers to be physically removed prior to, and during, diffusion treatment—preventing poor treatment thickness uniformity and optimizing the rate of inward diffusion of the interstitial species. This circumvents the need for hydrogen (and/or ammonia) gas—both commonly used in diode plasma nitriding—to chemically clean oxides the surface (and act as a diluent, to control the nitriding potential). While generally true for the treatment of both steels and Ti-alloys, an ability to preclude hydrogen is particularly important for Ti-alloys—since the susceptibility of titanium to hydrogen embrittlement is high.

Thirdly, the intensity (and adjustability) of the triode plasma allows the nitride-ceramic compound layer, that typically forms on both steels and Ti-alloys, to be suppressed as/when required. Again, this is particularly relevant for Ti-alloys, since the early formation of a Ti_xN compound layer can be particularly damaging to treatment efficiency—being known to provide a nitrogen diffusion rate at least 30 times lower than that of pure titanium (Kashaev et al., 2004; Wood & Paasche, 1977) in the typical treatment temperature range of interest (c.700°C–950°C). In an Argon/Nitrogen triode plasma, if the nitrogen content is kept relatively low (e.g., ≤30%), and the substrate voltage also maintained at a low level (unlike diode plasma nitriding, a substrate negative bias voltage of much less that −500 V can be easily maintained)—typically ∼200 V is routinely possible (Cassar et al., 2012; Leyland et al., 1991)—then titanium nitride compound layer formation can be avoided. This allows rapid nitriding to occur to a depth of several tens of microns in only a few hours, even at 700°C [which is sufficient to provide a useful increase in near-surface load-bearing capacity, whilst avoiding significant bulk grain growth (Cassar, 2011) that tends to rapidly degrade the substrate core strength in most commonly used Ti-based structural alloys, at temperatures of 800°C and above]. Conversely, an adjustment of the nitriding parameters to a higher nitrogen plasma partial pressure (e.g., from 30% to 70%), and/or an increase in substrate negative bias voltage, from ∼200 to ∼1000 V, can be deployed in the later stages of any TPN diffusion treatment, to "seal" the diffusion-hardened case with a thin nitride compound layer (Cassar et al., 2012a)—should this provide additional mechanical and/or tribological benefits.

As Fig. 16.1 shows, the depth-dependent microstructure of plasma-nitrided Ti-6Al-4V (and indeed of most thermochemically-treated titanium alloys) is quite complex but, with correct implementation, can provide a near-surface region that is functionally graded in both hardness and modulus. Across the nitrogen

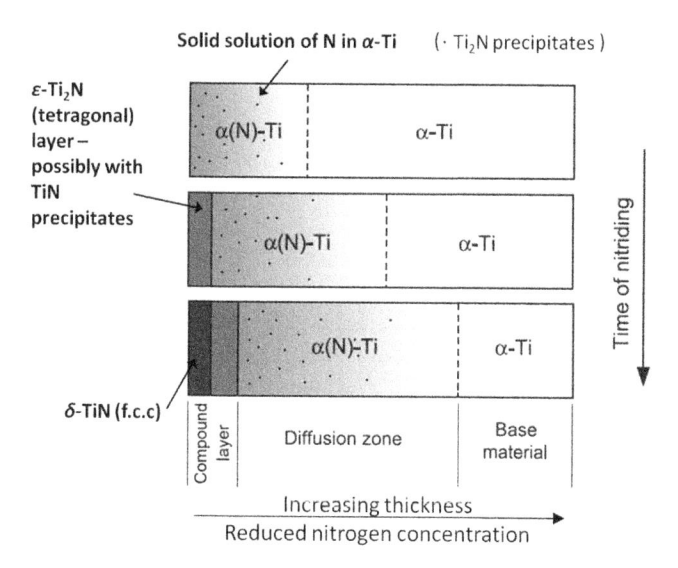

Figure 16.1 Triode-plasma nitrided surface layers. A schematic representation of the kinetics of formation and growth of triode-plasma nitrided surface layers of alpha-titanium. Note: For dual-phase Ti-alloys such as Ti-6Al-4V, a layer of nitrogen-stabilized α-Ti is created between the α(N)-Ti/Ti$_2$N precipitation zone and the nitrogen-containing (α + β)-Ti base material (Cassar, 2011).

interstitially-strengthened and nitride precipitation-hardened regions, a two- to three-fold increase in hardness can be achieved, with the elastic modulus also increasing by a similar (but somewhat smaller) amount across the precipitation-hardened zone. Furthermore, this diffusion-treated region does not need to be particularly thick, to yield an acceptable performance enhancement. In conventional (or plasma-enhanced) nitriding of ferritic steels, a layer thickness of at least 150 μm (and preferably ≥250 μm) is generally considered to be "sufficient" for many applications (Leyland et al., 1991; Roliński, 2015) but, depending on the level of substrate alloying, this can typically take up to 24 hours to achieve (at ∼500°C treatment temperature). In contrast, the low-temperature (≤450°C) nitriding of *austenitic* stainless steels, to produce so-called "expanded austenite" (also named S-phase) (Ichii et al., 1986; Leyland et al., 1993; Sun et al., 1999; Tao et al., 2019, 2021) typically develops a hardened case of only around 20–30 μm thick over a similar timeframe—but still with significant benefits for load-bearing capacity and wear resistance. In the case of low-temperature TPN of Ti-alloys (where in this context "low" means ∼700°C, rather than 900°C or above), a 25–40 μm thick case, with functionally graded properties, can be developed in only 4–8 hours of treatment, if the correct treatment parameters are applied. In practice, this has proved to be "sufficient" in our development work to achieve a load-bearing capacity of >200 MPa, once a 2–4 μm thick PVD-TiN top-coat is added to the nitrided surface (Cassar et al., 2010).

Discussing the aforementioned in terms of hardness and elastic modulus, the *H/E* characteristics of the PVD-TiN coating are of rather minor importance. Referring again to Fig. 16.1, and to Cassar et al. (2012a), Cassar (2011), the gradual increase in hardness from substrate to near-surface (over a distance of 25 μm or more, after a 4 hour nitriding treatment at 700°C) is sufficient to increase the static load-bearing capacity of the substrate material by a factor of ∼2.5×, that is approximately proportional to the increase in permitted design stress enabled by the diffusion zone (i.e., from ∼80 to >200 MPa, with a concomitant hardness increase from <4 to ∼10 GPa). However, it is important to note that, although the *H/E* ratio is also increasing, the factor involved is somewhat less than 2.5. An increase in stiffness from the baseline value of *E* for Ti-6Al-4V of ∼115 GPa is created by the Ti_2N precipitates (that increase in number in the diffusion zone as the surface is approached), the "composite" elastic modulus is raised by only 20%–30%. The moderate increase in *E* effectively means that the *H/E* ratio of the treated substrate surface is less than 2x that of the untreated bulk. This factor is nevertheless sufficient to beneficially reduce the elastic mismatch to any Ti_xN compound layer formed above the diffusion zone—and indeed the formation of such a compound layer, with hardness typically in the range of 18–20 GPa, toward the end of the diffusion treatment (by increasing the substrate negative bias from 200 to 1000 V for the last hour of the 4 hours treatment) is found to provide a significant additional boost to the sliding wear performance of a subsequently-deposited PVD Ti/TiN coating on top. The reasons for this are not entirely clear—but the hardness differential between the PVD coating (∼25 GPa) and the compound layer ≤20 GPa points to a possible nitrogen substoichiometry in the upper

TiN region of the layer. Importantly, a low nitriding temperature "locks-in" a compressive stress state to the diffusion-treatment layers, that significantly reduces the risk of fracture/delamination of the higher-modulus PVD-TiN coating under cyclic, fatigue loading (Cassar, Avelar-Batista Wilson, et al., 2011).

Certainly, in reciprocating sliding tests, significant differences in sliding wear performance can be seen between (1) nitride compound layer but no PVD coating, (2) PVD-TiN but no compound layer, and (3) nitride compound layer + PVD-TiN —with the latter clearly being superior (Cassar, 2011). However, in the spherical bearing application itself, despite the *H/E*-based mechanical-property considerations of the duplex nitriding/coating treatment, it is in fact the tribochemistry of the contact that ultimately determines its durability. As is well known, the ability of typical engineering lubricants (historically designed for steels) to adhere to titanium alloys is very limited, with few (if any) satisfactory solutions available. Despite its acknowledged tribological inertness against steel and other (but not all) metal pairings TiN, although exhibiting ceramic hardness, also possesses a high degree of metallic bonding character (Pierson, 1996; Yu et al., 2015)—and this tribochemical factor appears to be important for the functionality of extreme pressure (EP) additives found in many commercial grease lubricants already qualified and used for spherical bearings in aircraft applications.

16.8 Summarizing remarks

Having discussed briefly the emergence of Tribology as a recognized scientific discipline—and the subsequent recognition of Surface Engineering as a distinct and important subset of the former—we then proceeded to discuss the significance of *H/E* ratios in the design of engineered surfaces, why hardness alone is rarely a reliable indicator of wear resistance, and also why the substitution of "yield" or "flow" pressure (whether considered in tension or, particularly, in compression) by a measured hardness value requires careful consideration. When using H/E as an indicator of "strain to failure" under load (or indeed H^3/E^2, to signify some limiting contact pressure) in tribological contacts, the primary goal for the design engineer should, firstly, be to make "holistic" assessments—based on system requirements involving both coating *and* substrate selection—and, where necessary, any pretreatment of the latter.

For the substrate, hardness *does* matter greatly (in terms of a close correlation between mechanical strength and load-bearing capacity), but any surface treatment deployed to facilitate a higher load-bearing capacity (such as plasma-thermochemical processing) should not be achieved at the expense of excessive reductions in ductility and toughness. However, *coating* hardness again requires particularly careful consideration, since the "*super-*" (≥ 40 GPa) and, particularly, "*ultra-*" (≥ 70 GPa) high hardness values—that many PVD ceramic coating researchers have attributed such importance to over the last 3–4 decades—are, in themselves, rarely of any major significance in achieving tribological performance benefits on real industrial components. Such coatings tend also to exhibit high values of Elastic modulus that are often dangerously mis-matched to typical (metallic-alloy) engineering substrates and (as alluded to in our earlier discussions) the benefits that chemically inert ceramic coatings bring to tribo*chemical* performance need to be weighed against any adverse effects on system integrity (and coating tribo*mechanical* durability).

By way of a general example, many authors (including ourselves) have described PVD coatings—both ceramic and ceramic/metallic "nanocomposite" in nature— that exhibit high to "super-high" hardness values (in the c.25−40+GPa range), but also possess high *H/E* ratios approaching or exceeding 0.1, which is a value that is typically at least twice that of the most "resilient" conventional bulk engineering materials. In a tribomechanical context, such highly resilient coatings are desirable to create—but it should be noted that, even achieving an arbitrary "0.1" target value, implies coating *E*-values of around 250−400 GPa, and that steels, titanium-, aluminum- and magnesium-alloys (i.e., commonly used engineering substrates) possess values of only ∼40−200 GPa. Thus, the major mechanical risk to such coatings is substrate/coating elastic mismatch, even if the load-bearing capacity of the substrate is sufficiently high—and it is difficult to cost-effectively produce commercially-viable engineering substrates with a hardness much above ∼1.3 GPa (i.e., compressive strength ∼4.0 GPa), for serial production applications.

As our short case study in Section 16.7 illustrated, careful functional grading of substrate mechanical properties (both Hardness and, where possible, Elastic modulus), together with a "benign" stress state (i.e., moderately compressive, ideally)—to offset the negative effects of any remaining coating/substrate mechanical property mismatch—yields excellent practical results, but ultimately it is the tribochemistry between the treated/coated surface and its intended counterface that primarily determines successful implementation. In this regard, *H/E* ratios are a useful *mechanical* property optimization tool, but (1) the *H/E* ratio (and its variants) is just one tool amongst many, (2) optimization and alignment of *substrate* mechanical properties to the intended coating (and not vice versa) will always be more beneficial, rather than focusing on coating properties alone, (3) in sliding wear particularly, neither of the previous two points is of much relevance if they are considered ahead of the tribochemical (and/or tribocorrosion) demands of the intended application.

Presenting finally a *specific* coating example, the proliferation and commercialization over the last 25 years of a multitude of diamond-like carbon (DLC)-based coating systems in a range of industrial sector applications—particularly automotive, but also aerospace, nuclear and biomedical—relates to multiple factors, but stems from a flexible combination of highly adjustable properties, both mechanical and chemical. For instance, a DLC coating with a hardness of ∼15 GPa (i.e., "sufficiently" hard for most mechanical wear purposes) and an elastic modulus of ∼150 GPa (i.e., *H/E* ratio of ∼0.1) that is elastically compatible with most engineering alloy substrates, is clearly likely to be a promising candidate as a protective, low-friction coating for many sliding wear applications. That the coating chemistry can also be adjusted, to be either tribochemically inert (i.e., be particularly "diamond-like") or to be tribochemically active, and readily form a lubricious tribofilm (i.e., be particularly "carbon-like"), only serves to extend the range of potential applications. Most other coating systems are significantly less flexible in their range of possibilities—but the wider principle, of the balanced matching of mechanical *and* chemical coating properties (to both substrate material *and* application environment) is pragmatic. However, the appropriate use of *H/E* ratios to assist in optimizing the mechanical property aspects of a synergistic, "Surface Engineered" solution (i.e., considering both coating *and* substrate) can be highly beneficial, so long as such considerations are not prioritized over and above the tribochemical requirements and environmental demands of the specific application.

Acknowledgment

Many thanks to Prof. Glenn Cassar (University of Malta) for assistance with initial proofreading of this Chapter, and for technical comments/suggestions.

References

Almond, E. A. (1984). Aspects of various processes for coating and surface hardening. *Vacuum*, *34*(10−11), 835−842. Available from https://doi.org/10.1016/0042-207x(84)90160-x.

Anstis, G. R., Chantikul, P., Lawn, B. R., & Marshall, D. B. (1981). A critical evaluation of indentation techniques for measuring fracture toughness: I, direct crack measurements. *Journal of the American Ceramic Society*, *64*(9), 533−538. Available from https://doi.org/10.1111/j.1151-2916.1981.tb10320.x.

Archard, J. F. (1953). Contact and rubbing of flat surfaces. *Journal of Applied Physics*, *24*(8), 981−988. Available from https://doi.org/10.1063/1.1721448.

Archard, J. F., & Hirst, W. (1956). The wear of metals under unlubricated conditions. *Proceedings of the Royal Society of London, Series A: Mathematical and Physical Sciences*, *236*. Available from https://doi.org/10.1098/rspa.1956.0144.

Archard, J. F. (1959). The temperature of rubbing surfaces. *Wear*, *2*, 438−455.

Archard, J. F. (1980). Wear theory and mechanisms., in: *Wear control handbook; (eds. M.B. Peterson & W.O. Winer)* (pp. 35−80). pub. American Society of Mechanical Engineers (ASME), USA. Available from https://www.asme.org/publications-submissions/books/find-book/wear-control-handbook/1980/print-book.

Ashby, M. F. (2011). Chapter 3., in: *Materials selection in mechanical design* (4th Edition); ISBN: 978-1-85617-663-7 (pp. 31−56). Elsevier.

Barwell, F. T. (1979). *Theories of wear and their significance for engineering practice, Treatise on Materials Science and Technology*, *13*, 1−83. Elsevier. Available from https://doi.org/10.1016/S0161-9160(13)70065-6

Beake, B. D. (2022). The influence of the H/E ratio on wear resistance of coating systems − Insights from small-scale testing. *Surface and Coatings Technology*, *442*, 128272. Available from https://doi.org/10.1016/j.surfcoat.2022.128272.

Bell, T. (1991). Towards designer surfaces. *Metals and Materials*, *7*(8), 478−485.

Bowden, F. P., & Leben, L. (1939). The nature of sliding and the analysis of friction. *Proceedings of the Royal Society of London, Series A: Mathematical and Physical Sciences*, *169*, 371−391. Available from https://doi.org/10.1098/rspa.1939.0004.

Bowden, F. P., & Tabor, D. (1939). The area of contact between stationary and between moving surfaces. *Proceedings of the Royal Society of London, Series A: Mathematical and Physical Sciences, 169*, 391−413. Available from https://doi.org/10.1098/rspa.1939.0005.

Bull, S. J. (2006). Using work of indentation to predict erosion behavior in bulk materials and coatings. *Journal of Physics D: Applied Physics, 39*(8), 1626. Available from https://doi.org/10.1088/0022-3727/39/8/023.

Bunshah, R. F., & Raghuram, A. C. (1972). Activated reactive evaporation process for high rate deposition of compounds. *Journal of Vacuum Science and Technology, 9*(6), 1385−1388. Available from https://doi.org/10.1116/1.1317045.

Burwell, J. T., & Strang, C. D. (1952). On the empirical law of adhesive wear. *Journal of Applied Physics, 23*(1), 18−28. Available from https://doi.org/10.1063/1.1701970.

BMFT. (1980). *Tribology − High losses due to friction and wear.* (in English). (translated from BundesMinisterium für Forschung und Technologie (BMFT) Newsletter: Report FB-T76-26; 1976), Bonn, FDR.

Cassar, G. (2011). Improvement in the tribological characteristics of titanium alloys using duplex intensified plasma treatments. *PhD Thesis.* The University of Sheffield, UK.

Cassar, G., Avelar-Batista Wilson, J. C., Banfield, S., Housden, J., Matthews, A., & Leyland, A. (2010). A study of the reciprocating-sliding wear performance of plasma surface treated titanium alloy. *Wear, 269*(1−2), 60−70. Available from https://doi.org/10.1016/j.wear.2010.03.008.

Cassar, G., Avelar-Batista Wilson, J. C., Banfield, S., Housden, J., Fenech, M., Matthews, A., & Leyland, A. (2011). Evaluating the effects of plasma diffusion processing and duplex diffusion/PVD-coating on the fatigue performance of Ti−6Al−4V alloy. *International Journal of Fatigue, 33*(9), 1313−1323. Available from https://doi.org/10.1016/j.ijfatigue.2011.04.004.

Cassar, G., Banfield, S., Avelar-Batista Wilson, J. C., Housden, J., Matthews, A., & Leyland, A. (2012a). Micro-abrasion wear testing of triode plasma diffusion and duplex treated Ti−6Al−4V alloy. *Wear, 274−275*, 377−387. Available from https://doi.org/10.1016/j.wear.2011.10.002.

Cassar, G., Banfield, S., Avelar-Batista Wilson, J. C., Housden, J., Matthews, A., & Leyland, A. (2012b). Impact wear resistance of plasma diffusion treated and duplex treated/PVD-coated Ti−6Al−4V alloy. *Surface and Coatings Technology, 206*(10), 2645−2654. Available from https://doi.org/10.1016/j.surfcoat.2011.10.054.

Cassar, G., Matthews, A., & Leyland, A. (2012c). Triode plasma diffusion treatment of titanium alloys. *Surface and Coatings Technology, 212*, 20−31. Available from https://doi.org/10.1016/j.surfcoat.2012.09.006.

Chantikul, P., Anstis, G. R., Lawn, B. R., & Marshall, D. B. (1981). A critical evaluation of indentation techniques for measuring fracture toughness: II, strength method. *Journal of the American Ceramic Society, 64*(9), 539−543. Available from https://doi.org/10.1111/j.1151-2916.1981.tb10321.x.

Chen, J. (2012). On the determination of coating toughness during nanoindentation. *Surface and Coatings Technology, 206*(13), 3064−3068. Available from https://doi.org/10.1016/j.surfcoat.2011.12.006.

Chen, J., & Bull, S. J. (2006a). Assessment of the toughness of thin coatings using nanoindentation under displacement control. *Thin Solid Films, 494*(1−2), 1−7. Available from https://doi.org/10.1016/j.tsf.2005.08.176.

Chen, J., & Bull, S. J. (2006b). A critical examination of the relationship between plastic deformation zone size and Young's modulus to hardness ratio in indentation testing. *Journal of Materials Research, 21*(10), 2617−2627. Available from https://doi.org/10.1557/jmr.2006.0323.

Chen, J., & Bull, S. J. (2008). A modified model to determine limiting values of coating toughness by nanoindentation. *Tribology — Materials, Surfaces and Interfaces*, *2*(4), 219–224. Available from https://doi.org/10.1179/175158308X394628.

Chen, X., Du, Y., & Chung, Y-. W. (2019). Commentary on using H/E and H^3/E^2 as proxies for fracture toughness of hard coatings. *Thin Solid Films*, *688*, 137265. Available from https://doi.org/10.1016/j.tsf.2019.04.040.

Cheng, Y. T., & Cheng, C. M. (1998). Relationships between hardness, elastic modulus, and the work of indentation. *Applied Physics Letters*, *73*(5), 614–616. Available from https://doi.org/10.1063/1.121873.

Ciulli, E. (2024). Vastness of tribology research fields and their contribution to sustainable development. *Lubricants*, *12*(2), 33. Available from https://doi.org/10.3390/lubricants12020033.

Cook, S. S. (1928). Erosion by water hammer. *Proceedings of the Royal Society of London, Series A: Containing Papers of a Mathematical and Physical Character*, *119*, 481–488. Available from https://doi.org/10.1098/rspa.1928.0107.

Czichos, H. (2001). Tribology and its many facets: From macroscopic to microscopic and nano-scale phenomena. *Meccanica*, *36*(6), 605–615. Available from https://doi.org/10.1023/A:1016388517893.

den Toonder, J., Malzbender, J., de With, G., & Balkenende, R. (2002). Fracture toughness and adhesion energy of sol-gel coatings on glass. *Journal of Materials Research*, *17*(1), 224–233. Available from https://doi.org/10.1557/JMR.2002.0032.

Evans, D. C., & Lancaster, J. K. (1979). The Wear of Polymers, Treatise on Materials Science and Technology, 13, 85–139. https//doi.org/10.1016/s0161-9160(13)70066-8

Eyre, T. S. (1976). Wear characteristics of metals. *Tribology International*, *9*(5), 203–212. Available from https://doi.org/10.1016/0301-679x(76)90077-3.

Eyre, T. S. (1978). The mechanisms of wear. *Tribology International*, *11*(2), 91–96. Available from https://doi.org/10.1016/0301-679x(78)90135-4.

Eyre T. S. (1987). Selecting the optimum surfacing technique for wear resistance. *Proceedings of the 2nd International Conference on Surface Engineering, Stratford-upon-Avon, 15th–18th June 1987 (ed. I.M. Bucklow), pub. The Welding Institute, Abingdon, UK. (Conference Paper #43)*.

Farrow M. (1986). Selecting wear resistant surface treatments. *Proceedings of the 13th International Conference on Metallurgical Coatings, San Diego, USA, 11th-16th April (1986); pub. National Centre of Tribology, Risley, UK; Feb. 1986*.

Farrow, M., & Gleave, C. (1984). Wear resistant coatings. *Transactions of the IMF*, *62*(1), 74–80. Available from https://doi.org/10.1080/00202967.1984.11870676.

Garcia-Suarez, J., Brink, T., & Molinari, J.-F. (2023). Breakdown of Reye's theory in nanoscale wear. *Journal of the Mechanics and Physics of Solids*, *173*, 105236. Available from https://doi.org/10.1016/j.jmps.2023.105236.

Godoy, C., Mancosu, R. D., Machado, R. R., Modenesi, P. J., & Avelar-Batista, J. C. (2009). Which hardness (nano or macrohardness) should be evaluated in cavitation? *Tribology International*, *42*(7), 1021–1028. Available from https://doi.org/10.1016/j.triboint.2008.09.007.

Greenwood, J. A., & Williamson, J. B. P. (1966). Contact of nominally flat surfaces. *Proceedings of the Royal Society of London, Series A: Mathematical and Physical Sciences*, *295*. Available from https://doi.org/10.1098/rspa.1966.0242.

Griffith, A. A. (1921). The phenomena of rupture and flow in solids. *Philosophical Transactions of the Royal Society A: Containing Papers of a Mathematical and Physical Character*, *221*, 163–198. Available from https://doi.org/10.1098/rsta.1921.0006.

Halling, J. (1983). The tribology of surface films. *Thin Solid Films*, *108*(2), 103−115. Available from https://doi.org/10.1016/0040-6090(83)90496-0.

Halling, J., & Arnell, R. D. (1984). Ceramic coatings in the war on wear. *Wear*, *100*(1−3), 367−380. Available from https://doi.org/10.1016/0043-1648(84)90022-x.

Hertz, H. (1882). Ueber die Berührung fester elastischer Körper. *Journal für die Reine und Angewandte Mathematik*, *1882*(92). Available from https://doi.org/10.1515/crll.1882.92.156.

Hirst, W. (1995). Twenty-five years of tribology. *Tribology International*, *28*(1), 23−27. Available from https://doi.org/10.1016/0301-679x(95)99489-8.

Hirst, W., & Lancaster, J. K. (1956). Surface film formation and metallic wear. *Journal of Applied Physics*, *27*(9), 1057−1065. Available from https://doi.org/10.1063/1.1722541.

Holm, R. (1922). *Zeitschrift für Technische Physik*, 3, Articles published on pages: 290-294; 320-327; 349-357.

Holm, R. (1967). *Electrical contacts: Theory and application* (4th Edition; ISBN: 978-3-642-05708-3). pub. Springer-Verlag GmbH, Germany (originally published in 1946; later re-written in English).

Holmberg, K., & Matthews, A. (1994), in: Dowson, D. (Ed.), Tribology Series: 28. *Coatings tribology − Properties, techniques and applications in surface engineering*. Elsevier. Available from https://www.sciencedirect.com/bookseries/tribology-series/vol/28/.

Holmberg, K., & Erdemir, A. (2017). Influence of tribology on global energy consumption, costs and emissions. *Friction*, *5*(3), 263−284. Available from https://doi.org/10.1007/s40544-017-0183-5, http://www.springer.com/engineering/mechanical + engineering/journal/40544.

Hutchings, I., & Shipway, P. (2017). *Tribology: Friction and wear of engineering materials* (Second Edition). United Kingdom: Elsevier Inc. Available from https://www.sciencedirect.com/science/book/9780081009109.

Hutchings, I. M. (2009). The contributions of David Tabor to the science of indentation hardness. *Journal of Materials Research*, *24*(3), 581−589. Available from https://doi.org/10.1557/jmr.2009.0085.

Hutchings, I. M. (2024). Leonardo da Vinci's writings on sliding bearings, lubrication and wear. *Proceedings of the Institution of Mechanical Engineers, Part J: Journal of Engineering Tribology*, *238*, 761−771. Available from https://doi.org/10.1177/13506501241232036.

Hutchinson, J. W. (1968). Singular behaviour at the end of a tensile crack in a hardening material. *Journal of the Mechanics and Physics of Solids*, *16*(1), 13−31. Available from https://doi.org/10.1016/0022-5096(68)90014-8.

Hutchinson, J. W., & Suo, Z. (1991). Mixed mode cracking in layered materials. *Advances in Applied Mechanics*, *29*(C), 63−191. Available from https://doi.org/10.1016/S0065-2156(08)70164-9.

Ichii, K., Fujimora, K., & Takase, T. (1986). Structure of the ion-nitrided layer of 18-8 stainless steel. *Technology Reports of Kansai University*, *27*, 135−144.

Johnson, K. L. (1985). *Contact mechanics*. Cambridge University Press (published online 2012). Available from https://doi.org/10.1017/CBO9781139171731.

Jost, H. P. (1966). Lubrication (tribology), education and research: A report on the present position and industry's needs. (*report presented to the British Department of Trade & Industry Committee*; August 1966). pub. HMSO, UK.

Kashaev, N., Stock, H.-R., & Mayr, P. (2004). Nitriding of Ti−6% Al−4%V alloy in the plasma of an intensified glow discharge. *Metal Science and Heat Treatment*, *46*(7/8), 294−298. Available from https://doi.org/10.1023/B:MSAT.0000048837.39784.e2.

Khruschov, M. M. (1974). Principles of abrasive wear. *Wear*, *28*(1). Available from https://doi.org/10.1016/0043-1648(74)90102-1.

Khruschov, M. M., & Babichev, M. A. (1953). Resistance to abrasive wear and the hardness of metals. *Doklady Akademii Nauk SSSR (in Russian)*, *88*, 445−448.

Kragelskii, I. V. (ed.). (1965). *Friction and wear. (translated from Russian by L. Ronson)*. ISBN-10: 0408200006: pub. Butterworth & Co. (Publishers) Ltd, UK.
Kramer, B. M. (1983). Requirements for wear-resistant coatings. *Thin Solid Films, 108*(2), 117−125. Available from https://doi.org/10.1016/0040-6090(83)90497-2.
Krick, B. A., Vail, J. R., Persson, B. N. J., & Sawyer, W. G. (2012). Optical in situ micro tribometer for analysis of real contact area for contact mechanics, adhesion, and sliding experiments. *Tribology Letters, 45*(1), 185−194. Available from https://doi.org/10.1007/s11249-011-9870-y.
Lancaster, J. K. (1963). The relationship between the wear of carbon brush materials and their elastic moduli. *British Journal of Applied Physics, 14*(8), 497. Available from https://doi.org/10.1088/0508-3443/14/8/311.
Lancaster, J. K. (1969). Abrasive wear of polymers. *Wear, 14*(4), 223−239. Available from https://doi.org/10.1016/0043-1648(69)90047-7.
Lancaster, J. K. (1978). Wear mechanisms of metals and polymers. *Transactions of the IMF, 56*(1), 145−153. Available from https://doi.org/10.1080/00202967.1978.11870471.
Lechthaler, B., Ochs, G., Mücklich, F., & Dienwiebel, M. (2019). Evolution of the true contact area of laser textured tungsten under dry sliding conditions. *Frontiers in Mechanical Engineering*, 5. Available from https://doi.org/10.3389/fmech.2019.00003, http://www.frontiersin.org/journals/mechanical-engineering#.
Leyland, A., Fancey, K. S., & Matthews, A. (1991). Plasma nitriding in a low pressure triode discharge to provide improvements in adhesion and load support for wear resistant coatings. *Surface Engineering, 7*(3), 207−215. Available from https://doi.org/10.1179/sur.1991.7.3.207.
Leyland, A., Lewis, D. B., Stevenson, P. R., & Matthews, A. (1993). Low temperature plasma diffusion treatment of stainless steels for improved wear resistance. *Surface and Coatings Technology, 62*(1−3), 608−617. Available from https://doi.org/10.1016/0257-8972(93)90307-a.
Leyland, A., & Matthews, A. (1994). Thick Ti/TiN multilayered coatings for abrasive and erosive wear resistance. *Surface and Coatings Technology, 70*(1), 19−25. Available from https://doi.org/10.1016/0257-8972(94)90069-8.
Leyland, A., & Matthews, A. (2000). On the significance of the H/E ratio in wear control: A nanocomposite coating approach to optimised tribological behaviour. *Wear, 246*(1−2), 1−11. Available from https://doi.org/10.1016/s0043-1648(00)00488-9.
Leyland, A. & Matthews, A. (2003). The role of hardness and elastic modulus in determining the wear behaviour of nanocomposite coatings. *30th International Conference on Metallurgical Coatings and Thin Films*; 28th April - 2nd May 2003, San Diego, USA.
Leyland, A., & Matthews, A. (2004). Design criteria for wear-resistant nanostructured and glassy-metal coatings. *Surface and Coatings Technology, 177−178*, 317−324. Available from https://doi.org/10.1016/j.surfcoat.2003.09.011.
Leyland, A. & Matthews, A. (2007). Nanostructured coatings for enhanced tribological performance. *Proceedings of the 16th International Federation for Heat Treatment and Surface Engineetring Congress (IFHTSE2007) − Thermal processing and surface engineering: Key activities in the global knowledge economy*. Brisbane, Australia (30th October - 1st November 2007).
Li, L. T., Liang, X. M., Xing, Y. Z., Yan, D., & Wang, G. F. (2021). Measurement of real contact area for rough metal surfaces and the distinction of contribution from elasticity and plasticity. *Journal of Tribology, 143*(7), 071051. Available from https://doi.org/10.1115/1.4048728.

Lim, S. C., & Ashby, M. F. (1987). Overview no. 55 wear-mechanism maps. *Acta Metallurgica*, *35*(1), 1−24. Available from https://doi.org/10.1016/0001-6160(87)90209-4.

Lindroos, M., Valtonen, K., Kemppainen, A., Laukkanen, A., Holmberg, K., & Kuokkala, V. T. (2015). Wear behavior and work hardening of high strength steels in high stress abrasion. *Wear*, *322−323*, 32−40. Available from https://doi.org/10.1016/j.wear.2014.10.018.

Luo, D. B., Fridrici, V., & Kapsa, Ph. (2011). A systematic approach for the selection of tribological coatings. *Wear*, *271*(9−10), 2132−2143. Available from https://doi.org/10.1016/j.wear.2010.11.049.

Malzbender, J., & de With, G. (2000a). The use of the loading curve to assess soft coatings. *Surface and Coatings Technology*, *127*(2−3), 265−272. Available from https://doi.org/10.1016/s0257-8972(00)00640-x.

Malzbender, J., & de With, G. (2000b). Energy dissipation, fracture toughness and the indentation load−displacement curve of coated materials. *Surface and Coatings Technology*, *135*(1), 60−68. Available from https://doi.org/10.1016/s0257-8972(00)00906-3.

Marshall, D. B., Noma, T., & Evans, A. G. (1982). A simple method for determining elastic-modulus-to-hardness ratios using Knoop indentation methods. *Journal of the American Ceramics Society*, *65*(10), c175−c176. Available from https://doi.org/10.1111/j.1151-2916.1982.tb10357.x.

Matthews, A., Franklin, S., & Holmberg, K. (2007). Tribological coatings: Contact mechanisms and selection. *Journal of Physics D: Applied Physics*, *40*(18), 5463. Available from https://doi.org/10.1088/0022-3727/40/18/S07.

Matthews, A., & Leyland, A. (1995). Hybrid techniques in surface engineering. *Surface and Coatings Technology*, *71*(2), 88−92. Available from https://doi.org/10.1016/0257-8972(94)01004-3.

Matthews, A., Leyland, A., Holmberg, K., & Ronkainen, H. (1998). Design aspects for advanced tribological surface coatings. *Surface and Coatings Technology*, *100−101*(1−3), 1−6. Available from https://doi.org/10.1016/s0257-8972(97)00578-1.

Matthews, A., & Swift, K. G. (1983). Intelligent knowledge-based systems for tribological coating selection. *Thin Solid Films*, *109*(4), 305−311. Available from https://doi.org/10.1016/0040-6090(83)90183-9.

Matthews, A., & Teer, D. G. (1980). Deposition of Ti-N compounds by thermionically assisted triode reactive ion plating. *Thin Solid Films*, *72*(3), 541−549. Available from https://doi.org/10.1016/0040-6090(80)90545-3.

Matthews, A. & Leyland, A. (2002). The importance of the ratio of hardness to elastic modulus in determining the wear performance of coatings. *36th IUVSTA workshop*, 20th - 24th October 2002, Plzen, Czech Republic.

Mattox, D. M. (2003). Foundations of vacuum coating technology: The stories behind the facts. *Proceedings, 46th Annual Technical Conference − Society of Vacuum Coaters, 3rd - 8th May 2003, San Francisco, USA*, 11−20. Available from https://www.svc.org/index.php?xsearch%5B0%5D = mattox&xsearch%5B1%5D = Foundations&xsearch%5B2%5D = &submit = Search&xsearch_id = resource_library_search&src = directory &srctype = resource_library_lister&view = resource_library&submenu = DigitalLibrary.

Mayrhofer, P. H., Clemens, H., & Fischer, F. D. (2024). Materials science-based guidelines to develop robust hard thin film materials. *Progress in Materials Science*, *146*, 101323. Available from https://doi.org/10.1016/j.pmatsci.2024.101323, https://www.sciencedirect.com/science/journal/00796425.

Muenz, W. D., Hofmann, D., & Hartig, K. (1982). High rate sputter process for the formation of hard, friction reducing TiN coatings on tools, in: *Proceedings, conference on ion-assisted surface treatments, techniques and processes. The University of Warwick, UK; 14th – 16th September 1982.* (pp. 17.1–17.6). pub. The Metals Society, London, UK. (ISBN: 9780904357486).

Ni, W., Cheng, Y. T., Lukitsch, M. J., Weiner, A. M., Lev, L. C., & Grummon, D. S. (2004). Effects of the ratio of hardness to Young's modulus on the friction and wear behavior of bilayer coatings. *Applied Physics Letters, 85*(18), 4028–4030. Available from https://doi.org/10.1063/1.1811377.

Oberle, T. L. (1951). Properties influencing wear of metals. *Journal of Metals, 3,* 438; 439A-G.

Pierson, H. O. (1996). *Handbook of refractory carbides and nitrides: Properties, characteristics, processing, and applications; Chapters 9–11.* (pp. 156–208). pub. Elsevier. (ISBN: 978-0-8155-1392-6).

Pinkus, O., & Wilcock, D. F. (1977). Strategy for energy conservation through tribology. *US Department of Energy Report: TID-28175.* pub. American Society of Mechanical Engineers (ASME), New York, USA.

Popov, V. L. (2019). Generalized Archard law of wear based on Rabinowicz criterion of wear particle formation. *Facta Universitatis, Series: Mechanical Engineering, 17*(1). Available from https://doi.org/10.22190/FUME190112007P.

Rabinowicz, E. (1958). The effect of size on the looseness of wear fragments. *Wear, 2*(1), 4–8. Available from https://doi.org/10.1016/0043-1648(58)90335-1.

Rabinowicz, E. (1980). Wear coefficients – Metals, in: *Wear control handbook; (eds. M.B. Peterson & W.O. Winer)* (pp. 475–506). pub. American Society of Mechanical Engineers (ASME), USA.

Rabinowicz, E., Dunn, L. A., & Russell, P. G. (1961). A study of abrasive wear under three-body conditions. *Wear, 4*(5), 345–355. Available from https://doi.org/10.1016/0043-1648(61)90002-3.

Rabinowicz, E., & Tabor, D. (1951). Metallic transfer between sliding metals: An autoradiographic study. *Proceedings of the Royal Society of London, Series A: Mathematical and Physical Sciences, 208,* 455–475. Available from https://doi.org/10.1098/rspa.1951.0174.

Ramalingam, S. (1984). Tribological characteristics of thin films and applications of thin film technology for friction and wear reduction. *Thin Solid Films, 118*(3), 335–349. Available from https://doi.org/10.1016/0040-6090(84)90204-9.

Ramalingam, S., & Zheng, L. (1995). Film-substrate interface stresses and their role in the tribological performance of surface coatings. *Tribology International, 28*(3), 145–161. Available from https://doi.org/10.1016/0301-679x(95)98963-e.

Ratner, S. B. (1965). Wear of polymers as a process of fatigue failure. In S. B. Ratner, G. S. Klitenik, & E. G. Lure (Eds.), *Theory of friction and wear* (pp. 156–159). Moscow, USSR: Nauka.

Ratner, S. B., Farberova, I. I., Radyukevich, O. V., & Lure, E. G. (1967). Connection between the abrasion resistance of plastics and other properties. *Abrasion of Rubber (ed. D.I. James; translated from Russian by M.E. Jolley)* (p. 145) pub. MacLaren Press, London, UK. (ISBN: 0853340021).

Reye, K. T. (1860). Zur Theorie der Zapfenreibung. *Der Civilingenieur - Zeitschrift für das Ingenieurwesen, 6,* 236–254. Available from https://www.digitale-sammlungen.de/de/view/bsb10709353.

Rice, J. R. (1968). A path independent integral and the approximate analysis of strain concentration by notches and cracks. *Journal of Applied Mechanics, 35*(2), 379–386. Available from https://doi.org/10.1115/1.3601206.

Rice, J. R., & Rosengren, G. F. (1968). Plane strain deformation near a crack tip in a power-law hardening material. *Journal of the Mechanics and Physics of Solids*, *16*(1), 1−12. Available from https://doi.org/10.1016/0022-5096(68)90013-6.
Richardson, R. C. D. (1967). The wear of metals by hard abrasives. *Wear*, *10*(4), 291−309. Available from https://doi.org/10.1016/0043-1648(67)90093-2.
Richardson, R. C. D. (1968). The wear of metals by relatively soft abrasives. *Wear*, *11*(4), 245−275. Available from https://doi.org/10.1016/0043-1648(68)90175-0.
Roliński, E. (2015). Plasma-assisted nitriding and nitrocarburizing of steel and other ferrous alloys. In Mittermeijer E. J. & Somers M. A. J.). (Eds.), *Thermochemical surface engineering of steels* (pp. 413−457). Elsevier BV. Available from https://doi.org/10.1533/9780857096524.3.413.
Schiller, S., Heisig, U., & Goedicke, K. (1976). Ionenplattieren - ein neues verfahren der vakuumbeschichtung. *Vakuumtechnik*, *25*(3), 672−695.
Smart, R. F. (1978). Selection of surfacing treatments. *Tribology International*, *11*(2), 97−104. Available from https://doi.org/10.1016/0301-679X(78)90136-6.
Smart, R. F., & Moore, J. C. (1979). Materials selection for wear resistance. *Wear*, *56*(1), 55−67. Available from https://doi.org/10.1016/0043-1648(79)90006-1.
Spurr, R. T., & Newcomb, T. P. (1957). The adhesion theory of friction. *Proceedings of the Physical Society. Section B*, *70*(1), 98. Available from https://doi.org/10.1088/0370-1301/70/1/314.
Sun, Y., Bell, T., Kolosvary, Z., & Flis, J. (1999). The response of austenitic stainless steels to low-temperature plasma nitriding. *Heat Treatment of Metals*, *26*(1), 9−16.
Surjadi, J. U., Gao, L., Du, H., Li, X., Xiong, X., Fang, N. X., Lu, Y., et al. (2019). Mechanical metamaterials and their engineering applications. *Advanced Engineering Materials*, *21*. Available from https://doi.org/10.1002/adem.201800864.
Tabor, D. (1951). The hardness and strength of metals. *Journal of the Institute of Metals*, *79*, 1−18.
Tabor, D. (1996). Indentation hardness: Fifty years on a personal view. *Philosophical Magazine A*, *74*(5), 1207−1212. Available from https://doi.org/10.1080/01418619608239720.
Tao, X., Li, X., Dong, H., Matthews, A., & Leyland, A. (2021). Evaluation of the sliding wear and corrosion performance of triode-plasma nitrided Fe-17Cr-20Mn-0.5N high-manganese and Fe-19Cr-35Ni-1.2Si high-nickel austenitic stainless steels. *Surface and Coatings Technology*, *409*, 126890. Available from https://doi.org/10.1016/j.surfcoat.2021.126890.
Tao, X., Liu, X., Matthews, A., & Leyland, A. (2019). The influence of stacking fault energy on plasticity mechanisms in triode-plasma nitrided austenitic stainless steels: Implications for the structure and stability of nitrogen-expanded austenite. *Acta Materialia*, *164*, 60−75. Available from https://doi.org/10.1016/j.actamat.2018.10.019, http://www.journals.elsevier.com/acta-materialia/.
Thiruvengadam, A. (1963). A unified theory of cavitation damage. *Journal of Basic Engineering*, *85*(3), 365−373. Available from https://doi.org/10.1115/1.3656610.
Thiruvengadam, A. (1967). The concept of erosion strength. *ASTM International*, *STP408*. Available from https://doi.org/10.1520/stp46044s.
Thiruvengadam, A., & Waring, S. (1964). Mechanical properties of metals and their cavitation damage resistance. *US Office of Naval Research (Technical Report 233-5, June 1964)*. Available from https://www.osti.gov/biblio/4689001.
Tsui, T. Y., Pharr, G. M., Oliver, W. C., Bhatia, C. S., White, R. L., Anders, S., Anders, A., & Brown, I. G. (1995). Nanoindentation and nanoscratching of hard carbon coatings for magnetic disks. *Materials Research Society Symposium − Proceedings*, *383*, 447−452. Available from https://doi.org/10.1557/proc-383-447.

Wallwork HT/Airbus UK. (2013). "A Lighter Way to Fly" <https://www.wallworkht.co.uk/content/posters/> (accessed: 01-08-2024).

Wang, C. T., Hakala, T. J., Laukkanen, A., Ronkainen, H., Holmberg, K., Gao, N., Wood, R. J. K., & Langdon, T. G. (2016). An investigation into the effect of substrate on the load-bearing capacity of thin hard coatings. *Journal of Materials Science, 51*(9), 4390−4398. Available from https://doi.org/10.1007/s10853-016-9751-8.

Williams, J. (2024). The invention of tribology: Peter Jost's contribution. *Lubricants, 12*(3), 65. Available from https://doi.org/10.3390/lubricants12030065.

Williams, J. A. (1996). Analytical models of scratch hardness. *Tribology International, 29*(8), 675−694. Available from https://doi.org/10.1016/0301-679x(96)00014-x.

Wood, F. W., & Paasche, O. G. (1977). Dubious details of nitrogen diffusion in nitrided titanium. *Thin Solid Films, 40*(C), 131−137. Available from https://doi.org/10.1016/0040-6090(77)90111-0.

Wyatt, O. H., & Dew-Hughes, D. (1974). *Metals ceramics and polymers.* pub. Cambridge University Press, London, UK, 1974.

Xia, Z., Curtin, W. A., & Sheldon, B. W. (2004). A new method to evaluate the fracture toughness of thin films. *Acta Materialia, 52*(12), 3507−3517. Available from https://doi.org/10.1016/j.actamat.2004.04.004.

Yang, R., Zhang, T., & Feng, Y. (2010). Theoretical analysis of the relationships between hardness, elastic modulus, and the work of indentation for work-hardening materials. *Journal of Materials Research, 25*(11), 2072−2077. Available from https://doi.org/10.1557/jmr.2010.0267.

Yu, S., Zeng, Q., Oganov, A. R., Frappere, G., & Zhang, L. (2015). Phase stability, chemical bonding and mechanical properties of titanium nitrides: A first-principles study. *Physical Chemistry Chemical Physics: PCCP, 17,* 11763−11769. Available from https://doi.org/10.1039/C5CP00156K.

Zhang, S., Sun, D., Fu, Y., & Du, H. (2005). Toughness measurement of thin films: A critical review. *Surface and Coatings Technology, 198*(1−3), 74−84. Available from https://doi.org/10.1016/j.surfcoat.2004.10.021, http://www.journals.elsevier.com/surface-and-coatings-technology/.

Thin protective coatings on silicon for microelectromechanical systems

17

Ben Beake[1] and Tomasz Liskiewicz[2]

[1]Micro Materials Ltd., Wrexham, United Kingdom, [2]Department of Engineering, Manchester Metropolitan University, Manchester, United Kingdom

17.1 Introduction

The high surface-to-volume ratio makes interfacial interactions a dominant factor in the wear and lifetime of MEMS (microelectromechanical system) devices. While MEMS can be fabricated out from several different materials, silicon is the common choice due to the large infrastructure for Si-based devices of the microelectronics industry. At elevated temperature, silicon can show some ductility (Schaffar et al., 2023), but at room temperature it is a brittle material with little or no conventional plasticity, low fracture toughness, and sensitivity to shock loading. It exhibits highly complex mechanical and tribological behavior with phase transformations and lateral cracking observed in indentation and brittle fracture in a wide range of mechanical contacts (Bhowmick et al., 2009; Cook, 2006; Domnich & Gogotsi, 2002). For these reasons, the reliability of silicon-based MEMS when/if mechanical contact occurs is limited by high wear and stiction forces (Ku et al., 2011; Tanner et al., 1998; Williams & Le, 2006). Typically, commercial applications have therefore limited device motion, but if stiction and wear issues could be tackled successfully, many more applications could be developed (Kim et al., 2007). Coating strategies for protective low-friction overcoats for Si-based MEMS include solid lubrication with self-assembled monolayers, atomic layer deposition, and the deposition of thin hard carbon films such as tetrahedral amorphous carbon (ta-C) films (Maboudian et al., 2000; Scharf et al., 2006; Shi et al., 1996; Smallwood et al., 2006). Ta-C films deposited by filtered cathodic vacuum arc (FCVA) technology have been developed for MEMS applications including capacitive sensors and protective coatings for micromachined components. The mechanical and interfacial behavior of the contacting silicon surfaces is modified by these thin, low surface energy films. Ta-C films deposited by FCVA have high hardness, but the films can be highly stressed (Sheeja, Tay, Lau, et al., 2002; Sheeja, Tay, Leong, et al., 2002; Shi et al., 1996), which limits their thickness.

This chapter focuses on the nanomechanical behavior and nano/microtribological performance of 5–80 nm ta-C films deposited as protective overcoats on Si(100) wafers. It is organized by first discussing the mechanical and tribological behavior of uncoated silicon and then showing how the behavior of the thin film systems is influenced by the underlying complex deformation of the silicon substrate. Despite being extensively studied in indentation, much less is understood about its behavior under more complex loading geometries that occur in practical tribological situations. To address this deficiency, the behavior of the thin ta-C films and also the uncoated Si(100) has been investigated in different tribocontacts and with different probe geometries. Multifunctional nanomechanical NanoTest systems were used for all the nanoindentation, nano-scratch, nanofretting, and nano-impact tests. This approach has enabled, for example, the influence of tangential loading in more complex tribocontact situations to be directly compared with deformation in the idealized loading in the nanoindentation test using the same spheroconical indenter in all the tests (Beake et al., 2011).

17.2 Silicon (100)—phase transformation and cracking

17.2.1 Indentation

To more deeply understand the indentation behavior of ultrathin ta-C films on silicon, it is first necessary to obtain data on uncoated silicon wafers with the same probes as the films are to be tested with. This approach enables the influence of the substrate on the ta-C/Si system to be determined later. Phase transformations and lateral cracking have been observed in silicon in indentation and brittle fracture in many different mechanical contacts. Its behavior under the idealized quasistatic loading conditions of a nanoindentation experiment has been the subject of many studies using Berkovich and spherical indenters (Chang & Zhang, 2009a; 2009b; Juliano, Domnich, et al., 2004; Juliano, Gogotsi et al., 2003; Oliver et al., 2008). It is now well established that a phase transformation from semiconductor to metallic behavior takes place beneath the contact site and that on unloading the characteristic "pop-out" event is a consequence of another phase transformation and its accompanying volumetric expansion.

Nanoindentation curves on Si(100) with a spheroconical diamond probe with end radius of 4.6 μm probe showed excellent reproducibility above and below the first pop-in event, with some variation in the critical load required (40 ± 5 mN). The critical pop-in load is in good agreement with previous work with similar-sized indenters; 28 mN for a 4.2 μm indenter (Bradby et al., 2001) and (37 ± 4) mN for a 5 μm indenter (Weppelmann et al., 1995). With the 4.6 μm probe, more small pop-ins were observed at higher loads. Pop-outs during unloading were observed after loading to ≥ 60 mN, with some sensitivity to unloading rate. Pop-outs occurred at lower loads when the unloading rate is higher. Strong kink pop-outs during unloading were more common at higher unloading rate. The load required for pop-outs varied with unloading rate. With sharp and large radius (13.5 μm) spherical indenters, several events (pop-outs, kink pop-outs, and elbow events) can be observed during unloading depending on the unloading rate and the peak load (Domnich & Gogotsi, 2002; Juliano et al., 2003, 2004). Pop-outs and kink pop-outs were favored by higher peak load and slower unloading rate. The dependence of the critical load on loading rate found with the 4.6 μm probe was very similar to that observed at 200–700 mN using a 13.5 μm spherical indenter (Juliano et al., 2003). At higher load, brittle deformation predominates. With the 4.6 μm probe cracking around the indentation site occurs at ≥ 200 mN with a radial-lateral crack system developing at >300 mN with more extensive lateral cracking requiring >400 mN and a penetration depth of >2 μm. The large pop-ins found when silicon is indented at high load with spherical indenters is highly stochastic (Oliver et al., 2008). When indented with a 4.3 μm, probe Oliver and coworkers found typically (350 ± 100) mN was required and larger pop-ins (>1 μm) were more likely to occur at >500 mN.

17.2.2 When does phase transformation start?

Chang and Zhang have argued that, using data from sharp indentation, the occurrence of a pop-out does not represent the onset of a phase transformation, which begins as a gradual phase transformation at higher load at the onset of nonelastic unloading (Chang & Zhang, 2009a, 2009b). During unloading, as the load reduces from the peak load, along with this gradual phase transformation, rapid growth of high-pressure phases within the deformation zone occurs and causes the sudden volume expansion and observed pop-out. The observed differences with load and loading rates are the result of different evolution processes in the phase transformation of Si (100).

To investigate these in greater detail, in situ electrical contact resistance measurements using a Boron-doped diamond Berkovich indenter as the electrically conductive probe can provide additional information about the phase transformations that occur in loading and unloading (Beake & Jochum, 2022). In this case study, mirror polished Si(100) was indented with Boron-doped diamond Berkovich to 40 mN at 2 mN s^{-1}, with different unloading rates = 0.267–20 mN s^{-1} to investigate rate dependence in phase transformation. A typical example with a slow unloading rate of 0.267 mN s^{-1} is shown in Fig. 17.1.

By including the power law fit in Fig. 17.1A, it becomes clear that initial deviation from elastic unloading associated with phase transformation begins (at point 1 in Fig. 17.1A and B) well before the well-known "pop-out" event (at point 2 in Fig. 17.1A and B) that occurs with further unloading. ECR and depth data show changes during unloading due to phase transformation, which begin before the main pop-out event. The initial deviation from elastic unloading was accompanied by a more rapid decrease in electrical current. The subsequent pop-out was accompanied by an abrupt increase in electrical current. It is not immediately clear what causes this behavior. On possibility is that on unloading Si-II initially transforms to a less conductive phase (e.g., a-Si) with a later transition to more electrically conductive high-pressure phases such as Si-III/Si-XII at the pop-out. Gerbig et al. have performed in situ Raman spectroscopic measurements during spherical indentation of Si thin films with a 45 μm probe (Gerbig et al., 2015). They also reported that changes to the Raman spectra occur between initial unloading and the start of the

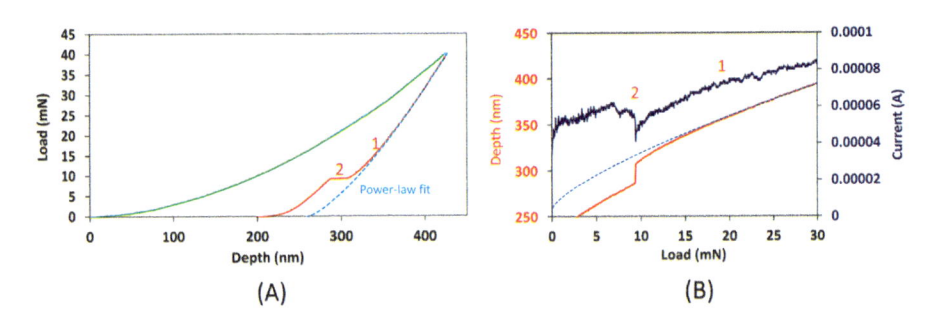

Figure 17.1 Nanoindentation of Si with a boron-doped diamond Berkovich. Nanoindentation of Si(100) to 40 mN with a boron-doped diamond Berkovich unloading at 0.267 mN s^{-1}. (A) Load — displacement curve and power-law fit; (B) variation in depth and contact resistance during unloading. Point 1 marks the deviation from elastic unloading and Point 2 marks the pop-out.

deviation from elastic behavior, which was interpreted as a partial relaxation of the strained Si-II phase. Evidence of both high pressure and a-Si phases was found in spectra before the elbow pop-out.

17.2.3 Nano-scratch of Si(100)

Si(100) undergoes a predominantly brittle response in scratch testing. In dry sliding at the macroscale, brittle wear was considered the major mechanism with limited pressure-induced metallization resulting in plastic flow of the metallic phase (Kovalchenko et al., 2002). In nano-scratch tests with small spherical probes (e.g., ~5 μm radius), several critical loads can be determined as the severity of the contact increases. These are:- (1) L_y—yield-onset of nonelastic deformation; (2) L_{c1}—pop-in is observed in the residual depth data; at the side of the scratch track ductile chipping was found above this; (3) L_{c2} marks the onset of more pronounced chipping at the sides of the track and edge cracking; (4) L_{c3} is lateral cracking (brittle chipping) and is marked by a more jagged scratch track beyond this point due to probe twisting due to the highly uneven load support of the cracked surface. The scratch tracks are completely straight up to this point due to the high lateral stiffness of the instrumentation used. Ductile chipping, edge cracking, and brittle chipping were also observed on Si(111) in a single point diamond turning test with a spherical probe of ill-defined radius (Gogotsi et al., 2001).

The first pop-in marks the onset of nonelastic behavior in indentation and scratch testing. In both of these tests, the behavior just after the pop-in is almost completely elastic since the pop-in is usually only a few nanometers and the residual scratch depth is only a few nm. Analysis of the critical pressure required for the first yield/pop-in to occur in the nano-scratch test gives 12–13 GPa, very similar to the hardness determined from analysis of nanoindentation unloading curves. The friction coefficient at yield is sufficiently low that the stress distributions remain very similar.

In indentation with a 4.6 μm probe, cracking is observed from >200 mN around the contact, with the radial-lateral crack system developing at >300 mN (Beake et al., 2011), and more extensive lateral cracking starting from >400 mN at a penetration depth of >2 μm. In contrast, when the same probe is used in nano-scratch testing, lateral cracking (L_{c3}) required only ~150–170 mN, due to the different stress distribution.

The kinetics of the pop-in and pop-out behavior in indentation similarly complicate the nano-scratch response of Si(100). Although the L_{c2} and L_{c3} events varied with dL/dx with maximum values at $dL/dx = 1.25$ mN mm^{-1}, there was relatively little variation across the entire range of dL/dx studied. This small dependence on dL/dx was considered to be due to the competing influence of the effects of scan speed and loading rate on the relative importance of phase transformation and brittle machining mechanisms. The critical load for the generation of significant debris and edge cracking (L_{c2}) is important as it results in the delamination when ≤ 80 nm ta-C films are deposited on Si, with L_{c2} on Si and on the 80 nm ta-C showing exactly the same trends with dL/dx (Beake et al., 2011).

17.2.4 Nanofretting of Si(100)

In contrast to its behavior in the nano-scratch and nanoindentation tests where it showed relatively little rate sensitivity, in the nanofretting test, wear of Si(100) was more strongly dependent on the rate of initial loading. This appears to be due to the generation of more extensive cracking when loaded more rapidly. Subsequent crack propagation and material removal under constant load nanofretting appear to be influenced by the type of cracks that are generated during the initial loading stage. Nanofretting tests were performed at 30–300 mN with 10,000 cycles at peak load where the load was applied either abruptly (<0.3 s) or slowly (10 s) (Beake, Liskiewicz, Pickford, et al., 2012; Beake et al., 2011). Other conditions were: 4.6 μm probe; 2 μm track length, 10 Hz oscillation frequency; ~12–13 GPa contact pressure. Over most of the load range, this resulted in gross slip (track length <2 × contact radius) and elongated wear scar geometry. With abrupt loading, there was radial and lateral cracking with large displacement jumps (pop-ins) during the oscillation at constant load. The crack morphology was similar to that in repetitive nano-impact tests and nanoindentation at >300 mN load with the same probe. With slow loading, radial cracking was not observed, but the crack morphology and wear rate showed a marked load dependence. At 90 mN normal load, damage was minimal, but there was evidence of slip with an elongated wear scar ~2 μm in length, which corresponded to the amplitude of the applied displacement. The wear volume measured by confocal microscopy (material removed from the crater-shape wear scar) was only 2 μm^3 at this load. The wear debris appeared to be ductile with characteristic ribbon-like features that have been observed in reciprocating AFM nanowear (Zhao & Bhushan, 1998) and defined as an agglomeration of high surface energy fine Si particles. With only a small increase of load to 100–110 mN, there was a dramatic transition to a more severe wear regime. For example, an experiment at 110 mN generated a wear volume that was over 20 times higher than at

90 mN. The debris was fragmented with microchipping damage outside the wear scar. Increasing the load to 200 mN reduced the track length leading to energy dissipation on a smaller contact area and a further dramatic increase in wear volume to >400 μm^3 (i.e., over two orders of magnitude greater than at <100 mN). Under these conditions, silicon appears to show a brittle/ductile response to nanofretting, characterized by the presence of brittle microscale chips around the wear track and more ductile wear debris in the center of the contact. Increasing the load of the nanofretting tests resulted in reduced slip ratio and smaller track length, but also produced greater penetration depth leading to higher wear rates. The shorter wear track under higher load is consistent with classical macrofretting experiments and typical of a gross-slip to stick-slip transition (Vingsbo & Söderberg, 1988).

17.2.5 Si(100): tangential loading—influence on yield and wear mechanisms

A novel method has been introduced allowing quantitative comparison of deformation during loading in the nanoindentation, nano-scratch, and nanofretting tests (Beake et al., 2011). As an example, Fig. 17.2 shows typical loading curves from indentation, scratch, and fretting tests on Si(100) to 200 mN using the same diamond probe with a 4.6 μm end radius. The loading curves in all three tests were almost identical at very low load. Differences become apparent as the load increases. The tangential loading in the nano-scratch and nanofretting tests promoted yielding. This earlier yield onset resulted in (1) more significant damage occurring at lower loads (and depths) and (2) greater penetration depths at higher load, than in comparative indentation tests.

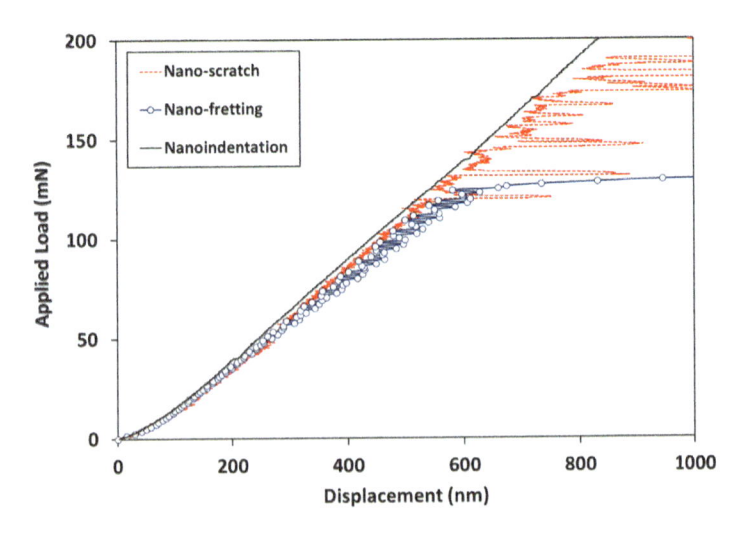

Figure 17.2 Nanoindentation, nano-scratch, and nanofretting tests of silicon. Comparison of loading curves in nanoindentation, nano-scratch, and nanofretting tests of Si(100) with the same 4.6 μm end radius spheroconical probe.

This behavior is in agreement with analytical results, which have shown that in comparison with indentation, tangential loading with friction facilitates yielding (Zok & Miserez, 2007). Zok and Miserez proposed that the magnitude of the decrease in critical load is related to the friction coefficient by (Eq. 17.1)

$$P_y/P_y^0 = 1 - 2.9\,\mu^2 \tag{17.1}$$

Where P_y^0 is the yield load in the absence of friction, P_y is the yield load with friction, and μ is the friction coefficient. Eq. (17.1) is valid for friction coefficients <0.3, which Hamilton and Goodman showed is the point at which the maximum stress occurs on the surface at the edge of the contact and surface yield occurs (Hamilton & Goodman, 1966). Blees et al. reported that the critical load in scratching sol − gel coatings on polypropylene decreased by an order of magnitude when the friction coefficient increased from 0.5 to 3 (Blees et al., 2000). However, on Si at yield, the friction coefficient was only 0.07 so the effect was rather small, so that with the same 4.6 μm probe $L_y = (40 \pm 5)$ mN in nanoindentation, (37 ± 5) mN in nano-scratch, and ∼30 mN in nanofretting (Beake et al., 2011). Wu et al. also noted a reduction in the critical load for phase transformation in nano-scratching compared with nanoindentation when using sharper probes (Wu et al., 2010). Hsu et al. reported low critical loads of around 0.5 and 2.3 mN when scratching with diamond probes with 0.5 and 1.2 μm end radii (Hsu et al., 2018). The friction coefficient in the elastic regime was 0.04 for both probes.

The pronounced lateral-radial cracking observed in high load nano-scratch and nanoindentation tests was largely absent in the nanofretting tests, indicative of some differences in the deformation mechanism and consistent with the fretting wear process minimizing strain accumulation. Despite this, there are clear similarities in the behavior of the Si(100) in the nano-scratch and nanofretting tests. In both tests, the damage became more pronounced at ∼600 nm, and the friction coefficient at failure was around 0.2. Microchipping damage observed at the edge of the scratch track above L_{c1} appears similar to the microchipping debris observed at the edge of the fretting scars. Similar damage morphology has been found in larger-scale tests; in radial fretting with a 1.6 mm probe and a contact zone of ∼250 μm, multiple ring cracking and chipping and debris were observed at the edge of the contact zone (Bhowmick et al., 2009). In repetitive sliding contacts, there is also a transition to more severe wear with extensive lateral cracking and the generation of debris. These include repetitive scratches in same track and statistically distributed scratch tests that involve multiple parallel scratches. In tests on Si(100) with a 5 μm probe, the loads required for the transition (125 mN) were similar to those required in the nanofretting test.

Smaller-scale nanofretting tests of Si have been performed using SiO_2 probes. Varenberg et al. studied partial and gross slip fretting with a 1.55 μm radius SiO_2 probe, finding higher friction at the transition between the partial and gross slip regimes accompanied by changing wear scar appearance for the two regimes (Varenberg et al., 2005). Yu et al. studied the behavior over displacement amplitudes between 0.5 and 250 nm using SiO_2 AFM tips with 0.15−0.9 μm end radii

(Yu et al., 2009, 2010, 2012; Yu, Qian, et al., 2009a, 2009b). The energy ratio related to the transition from partial to gross slip regime was 0.32−0.64, which was higher than the 0.2 observed in classic macroscale fretting. Depending on the applied load and test environment, either "hillocks" (uplifted material in the scar) or a typical wear trench was observed (Yu et al., 2009a, 2012). At lower contact pressure, hillocks formed, but grooves were observed when the Hertzian contact pressure approached the hardness of silicon (Yu et al., 2009a).

17.3 Coatings to protect silicon − thin ta-C films

ta-C films of 5, 20, and 80 nm thickness were deposited on Si(100) substrates by (Beake & Lau, 2005; Beake, Vishnyakov, et al., 2004), an industrial filtered cathodic vacuum arc system (Nanofilm Technologies Pte. Ltd., Singapore). To improve adhesion, the silicon surface was argon ion cleaned to remove its native oxide before film deposition. The R_a surface roughness of 80 nm ta-C over 100 mm line profiles determined with a 3.8 μm probe was 0.6 nm, which was the same as measured over the same track length on Si(100) without any coating, that is, in agreement with other reports that deposition of DLC coatings typically does not significantly increase the surface roughness of the substrate.

Nanoindentation, nano-scratch, nanofretting, and nano-impact tests of 5, 20, and 80 nm ta-C films deposited on Si(100) have been performed using spheroconical indenters with a range of end radii to investigate the role of film thickness, tangential loading, contact pressure, and deformation mechanism in the different contact situations (Beake et al., 2013; Shi et al., 2019). The influence of the mechanical properties and phase transformation behavior of the silicon substrate in determining the tribological performance (critical loads, damage mechanism) of the ta-C coated samples was investigated.

17.3.1 Challenges in measuring film hardness in ultrathin films

For reliable hardness measurements, it is necessary to have a fully developed plastic zone. On ultrathin hard carbon films such as the ta-C films, the hardness measured by nanoindentation may be less than the true film hardness, either at very shallow penetration depths where the indenter is more spherical in shape and the contact is essentially elastic (so that the mean pressure is less than the hardness) or at greater depths where plastic yield in the substrate occurs before the plastic zone is fully developed within the coating, as has been described in the ISO standard.

Chudoba has investigated the minimum depth (in relation to the tip radius) where constant hardness could be expected for a conical indenter with a spherical end cap (Chudoba, 2006). The plateau corresponding to the coating hardness is typically from a depth of $0.2\,R$ (where R is the indenter radius) to a relative indentation depth $(h_c/t_c) = 0.1$. For hardness measurements on coatings of different thicknesses to be directly comparable, it follows that the minimum thickness be at least $2\,R$.

Since the tip radius of a Berkovich indenter is often around 100 nm, it follows that the minimum thickness is ~200 nm. Clearly, some care must be taken in comparing hardness measurements for coatings thinner than this.

For an ultrathin ta-C film tested with a Berkovich indenter, it is necessary to consider the combined influence of the substrate and indenter blunting on the measured data. Fig. 17.3 shows that the measured hardness of 60 and 80 nm ta-C films first increases, then decreases with increasing contact depth. The hardness measured in the intermediate depth region (equivalent to the plateau region found in thicker films) is often taken as the film hardness since in this region it is less influenced by tip geometry and substrate behavior (Lemoine et al., 2000), that is, to employ the standard ISO14577-4 methodology. Alternatively, to extract the film-only hardness from data including some substrate contribution, the system can be treated as a two-layer composite system where the overall hardness is the sum of volume law of mixtures contributions (Burnett & Rickerby, 1987). Bhattacharya and Nix (1988) developed an analytical equation for a hard film on a softer substrate, which Lemoine et al. modified for use on ultrathin films by accounting for tip blunting (Eq. 17.2), where H_m is the measured hardness (Lemoine et al., 2007).

$$H_m = H_s(h_c) + (H_f(h_c) - H_s(h_c))\exp(-k_1 * \text{RID}) \tag{17.2}$$

Using this approach, these authors reported values of 29 and 38 GPa for 10 and 50 nm ta-C on Si, respectively, which were larger than those otherwise determined without accounting for tip blunting. Hardness >80 GPa has been reported for thin ta-C films (Chartidis, 2010) but at very low penetration depths, the accuracy of the indenter area function, and the method used to determine it may also affect the accuracy of these results. Fisher-Cripps et al. have also questioned the accuracy of extremely high hardness measurements in thin films (Fischer-Cripps et al., 2012). Another argument against these is that the elastic modulus determined by

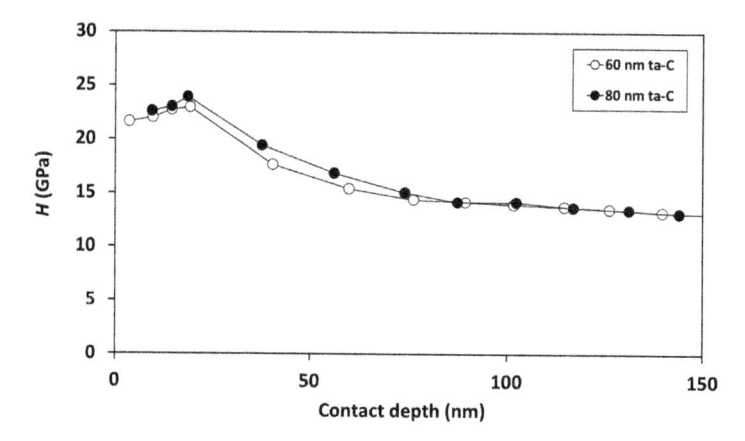

Figure 17.3 Hardness of carbon-based films. Hardness versus contact depth for 60 and 80 nm ta-C films on Si.

ISO14577-4 methods (i.e., extrapolation to zero depth to give the "film-only" response) is simply inconsistent with the films being that hard.

Sharper indenters (i.e., brand new Berkovich indenters) and indenters with smaller apical angles such as the cube corner have been used to measure coating-only hardness of thin films since they promote plastic deformation in the coating before substrate yield. When ta-C films were deposited on a substrate with lower yield stress (glass), the hardness measured with a Berkovich indenter was lower than with a sharper cube corner indenter, due to substrate yield before the mean pressure in the contact reaches the film hardness. However, when cube corner indenters are used on brittle coatings with high H/E ratios, the possibility of cracking should be considered. Jungk et al. reported that when a 110 nm ta-C film was indented with a cube corner indenter, there was a completely elastic response below 1.5 mN, but multiple pop-ins due to cracking were observed above this (Jungk et al., 2006). Applying Hertz's theory to the indentation of elastic materials shows that hardness is dependent on indenter angle (Martinez & Esteve, 2001). It is therefore not possible to directly compare hardness results obtained with cube corner indenters to with those obtained with more commonly used Berkovich indenters. At shallow depths, hardness measured with pyramidal indenters is a function of indentation depth due to the nonideal shape of the indenter (Borodich et al., 2003). For example, Lemoine et al. reported that $H(h) \propto h^{0.38}$ for the first 20 nm when using a Berkovich to test ta-C films (Lemoine et al., 2000). Hertzian contact mechanics shows that the measured hardness increases with the depth in an asymptotic manner $H(h) \propto h^{0.5}$ until it reaches the value, which would be measured with an ideally sharp indenter.

The hardness and elastic modulus of the 80 nm ta-C film were determined by ISO14577-4 methodology as 23.9 and 331 GPa, respectively, so that $H/E = 0.072$ and $H^3/E^2 = 0.125$ GPa (Beake et al., 2022). The relatively high hardness is due to >70% sp^3. The thickness of protective coatings deposited by FCVA may be limited to not much more than 80 nm since FCVA ta-C films can be highly stressed and can spontaneously delaminate if deposited to too large thickness (Sheeja, Tay, Lau, et al., 2002; Sheeja, Tay, Leong, et al., 2002).

17.3.2 Spherical indentation of ta-C films

Due to their low thickness, they cannot completely eliminate phase transformation in the underlying silicon substrate and pop-outs occur during unloading of ta-C films deposited on silicon. In indentation tests with a Berkovich indenter, there was some stochastic variability in the pop-out load so repeat tests were used to investigate the variation with thickness (Beake & Lau, 2005). Forty repeat tests to 20 mN were performed with 0.5 mN s^{-1} loading and unloading rate. The pop-out load varied from (4.7 ± 1.7) mN on Si, (4.8 ± 1.3) mN on 5 nm ta-C, (4.0 ± 1.2) mN on 20 nm ta-C and (3.3 ± 0.6) mN on 80 nm ta-C. Spherical indentation tests were performed on the ta-C films with the same 4.6 μm indenter and conditions that were used on the uncoated Si enabling direct comparison. Even though they are very thin, the ta-C films were able to alter the phase transformation behavior of the Si by providing load support, reducing the effective load reaching the substrate, and

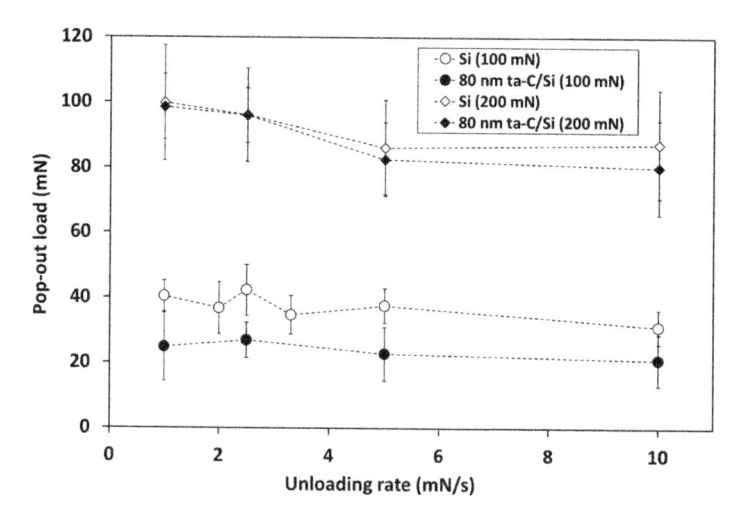

Figure 17.4 Pop-out load on unloading. Dependence of pop-out load on unloading rate when unloading from 100 or 200 mN.

spreading the deformation out over a wider area. Thicker coatings have enhanced load carrying ability in comparison with thinner coatings, and consequently, coating thickness influences the stress fields generated in the substrate and alters the loads/depths at which stress-induced phase transformations occur. As well as reducing the load transferred to the substrate, hard coatings also distribute the applied load over a wider area consequently resulting in phase transformation and plastic deformation in the Si over a wider area. Hence, for ta-C films on Si pop-ins are bigger on than on uncoated Si.

Fig. 17.4 shows pop-outs on 80 nm ta-C and Si(100) when unloading from 100 or 200 mN. The figure shows that the pop-outs were clearly lower when unloading from 100 mN, but the influence of the film on the tests to 200 mN was minimal. Although the critical loads at which pop-ins and pop-outs occurred were modified by the presence of the ta-C overlayer, their rate dependence was not influenced by their presence confirming that they are due to transitions in the underlying Si rather than the film. In 500 mN nanoindentation tests, the thin hard ta-C films showed no obvious protection against lateral cracking, which occurred only occasionally, irrespective of the ta-C film thickness.

17.3.3 Nano-scratch of ta-C films

Very small radius probes (e.g., ~100 nm) are progressively worn away on repeated loaded sliding against hard films (Sundararajan & Bhushan, 1999). Molecular dynamics simulations have investigated this diamond wear in cutting silicon wafers (Goel et al., 2014). Nano-scratch tests were performed on the ta-C films with diamond probes with end radii of 1.1, 3.1, 4.6, and 9.0 μm (Beake & Lau, 2005; Beake et al., 2009, 2013). Using several probes enabled the variation in measured critical load with the radius of the spherical probe to be investigated. Being less sharp, these spherical probes were less likely to blunt during the testing than probes with ~100 nm end radii and had well defined end radii that could be accurately calibrated over a range of depths.

Nano-scratch tests with a 4.6 μm probe were performed on the 80 nm ta-C film over a wide range of loading rates ($dL/dt = 0.1-12$ mN s^{-1}) and scan speeds ($dx/dt = 0.1-40$ μm s^{-1}) to provide critical load data over a range of $dL/dx = 0.1-100$ mN μm^{-1}. An example with loading rate = 1 mN s^{-1} and scan speed = 2 μm s^{-1} is shown in Fig. 17.5. Friction coefficients were ~0.09 at yield, rising to ~0.14-0.15 at L_{c2} film failure and ~0.21-0.24 at the onset of lateral cracking.

The nano-scratch and nanowear resistance of FCVA ta-C films has also been studied with spherical probes of 100-1200 nm. Critical loads were between 0.1 and 7.5 mN depending on the sharpness of the probe and the thickness of the films (Beake & Lau, 2005; Lemoine et al., 2004; Li & Bhushan, 1999; Sundararajan & Bhushan, 1999). In one study with a nominal 25 μm tip, a closer examination of the on-load depth at failure and the scratch track widths determined that the effective radius of the probe (Beake & Lau, 2005) was only ~1.1 μm at the shallow depths required to fracture the ta-C films. In this case, the probe had a sharp asperity, which made it effectively much sharper than its nominal radius. The nano-scratch data using the 1.1 μm probe on the 20 nm ta-C film are in good agreement with data by Lemoine and coworkers using a ~1200 nm indenter on 10 nm ta-C films also produced by FCVA (Lemoine et al., 2004) and 20 nm FCVA carbon films tested by Li and Bhushan with a ~1 μm probe (Li & Bhushan, 1999). Despite the different radii, the friction coefficients at failure were consistent across the different studies.

Increasing the thickness of the ta-C film from 5 to 80 nm was found to increase the critical load for film delamination by a factor of 2 when a 3.1 μm probe was used, but smaller dependencies were found for L_y and L_{c1} (Fig. 17.6). Above the

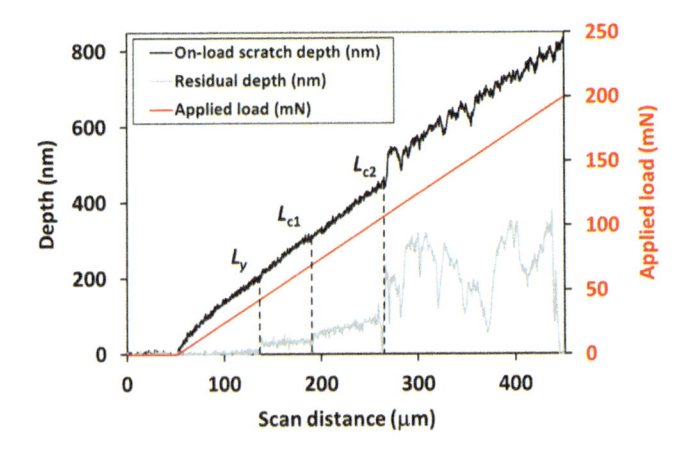

Figure 17.5 Nano-scratch test. Nano-scratch test on 80 nm ta-C with a 4.6 μm probe.

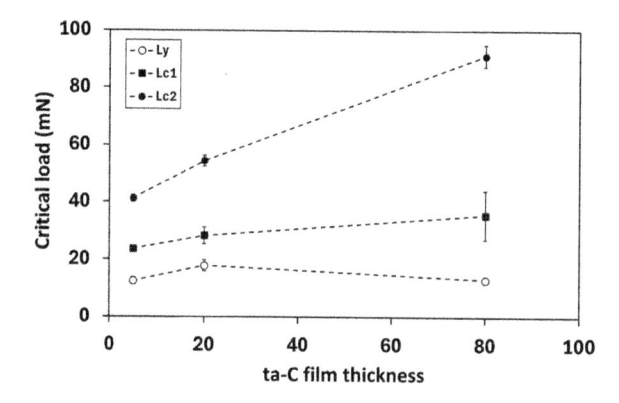

Figure 17.6 Critical loads in nano-scratch experiment. Variation in critical loads in tests with 3.1 μm probe with ta-C film thickness.

Figure 17.7 SEM image of nano-scratch. SEM image of a scratch test on 80 nm ta-C with 4.6 μm probe showing cracking before μ SEM image of a scratch test on 80 nm ta-C with 4.6 μm probe showing cracking before L_{c2} failure. The corresponding back-scattered image is shown in the inset (The authors express their gratitude to Vlad Vishnyakov for his assistance in acquiring the SEM image).

critical load for edge cracking, L_{c1}, but before the total film failure, some isolated small delamination events could occasionally be seen in the scratch track. SEM images show periodic cracking across the scratch track between L_{c1} and L_{c2} in ta-C. For example, Fig. 17.7 shows SEM of a test with a loading rate of 0.5 mN s^{-1} and scan speed of 2 μm s^{-1} showing clear tensile cracking behind the probe before the L_{c2} failure. This tensile cracking is more likely for coatings with large elastic modulus mismatch to the substrate (i.e., $E_c/E_s \gg 1$), as is the case for the 80 nm ta-C here. At film failure, the abrupt change in depth—for example, as shown in Fig. 17.7 with a 4.6 μm probe—is very close to the film thickness.

In nano-scratch tests with a 3.1 μm end radius diamond probe, a contact pressure of about 12 GPa was required for yield on the 5, 20, and 80 nm films. Although the critical load is strongly dependent on probe sharpness (see further), the mean pressure is relatively unchanged. For example, on the 80 nm film, the calculated mean pressure (P_y) at yield was 12.0 GPa with $R \sim 1.1$ μm; 12.9 GPa with $\mu \sim 1.1$ μm; 12.9 GPa with $R = 3.1$ μm; 14.8 GPa with $\mu = 3.1$ μm; 14.8 GPa with $R = 4.6$ μm and 11.4 GPa with $\mu = 4.6$ μm and 11.4 GPa with $R = 9.0$ μm. The Hertzian method of analyzing the nano-scratch data indicates the onset of non-elastic deformation in the nano-scratch test is due to substrate yield rather than film deformation. Maximum stresses are generated at a distance of $\sim 0.5\ a$ under the surface in a Hertzian indentation contact, which for these very thin films is deep into the silicon substrate, consistent with substrate yield. At the onset of nonelastic deformation, $P_m = 1.1\ Y$. As yield stress of silicon is 11.3 GPa, so $1.1\ Y = 12.4$ GPa, which is almost the same as the pressure required for phase transformation (~ 12 GPa) and hardness (12.5 GPa) emphasizing the importance of yield by phase transformation in the silicon substrate. The closeness of the mean pressures at yield to this with the thin films is consistent with substrate yield. Although the presence of a tangential load would produce a less uniform pressure distribution and result in the maximum stress shifting toward the surface (Arnell, 1990; Djabella & Arnell, 1993; Hamilton & Goodman, 1966), the friction coefficients were (only) 0.1−0.15 at yield on the 5−80 nm hard films. As the load increases, the maximum von Mises stress moves further into the substrate. At the critical loads for cracking and film failure, the maximum von Mises stress is in the substrate so that film failure is driven by the deformation of the Si substrate due to plasticity and predominantly through phase transformation.

On the 80 nm ta-C, the variation in L_{c1} with probe radius is shown in the following. This dependency has been modeled by fitting the data to a relationship of the form

$$L_C = xR^m \qquad (17.3)$$

Where m is a best fitting parameter. Assuming that plastic yielding (or cracking) occurs at a critical pressure on a bulk material leads to the critical load depending on R^2. For coating systems with different mechanical properties, thickness, etc., the situation is more complex and m is usually <2 in practice. For the ta-C, the onset of yield is in the silicon substrate when using $R = 1.1-9.0$ μm probes and $m \sim 1.8$. For the critical loads for cracking and film failure, the values of m are close to this, with $m \sim 1.7$ being reported for the 5−80 nm ta-C films (Beake et al., 2009). Tests with the 4.6 μm probe with loading rate of 12 mN s^{-1} and scanning speed 10 μm s^{-1} were performed to 300 mN to study the lateral cracking (L_{c3} failure) that occurs on Si at higher load and investigate whether the ta-C films could protect the substrate or not. On Si(100) lateral cracking occurred at (130 ± 10) mN. Slightly higher values were found for the 5 nm ta-C film (152 ± 35) mN and the 20 nm ta-C (163 ± 45) mN. However, the critical load for the L_{c3} failure increased dramatically to (276 ± 20) mN on the 80 nm ta-C, and the lateral cracking was less pronounced.

The nano-scratch test data show that tangential loading promotes the formation of large lateral cracks on all the ta-C film samples at much lower load than in nanoindentation as was found for uncoated Si(100). In 500 mN nanoindentation tests, lateral cracking occurred only occasionally, though in the nano-scratch test, the tangential loading promoted dramatic film failure at much lower load (e.g., ∼115 mN for 80 nm ta-C) with extensive lateral cracking at <300 mN.

On failure, the removal of 5–80 nm ta-C thin films is restricted to the scratch track. It appears that when the thin ta-C films failed in the scratch track, the intact ta-C film outside the scratch track is still capable of maintaining a measure of load support and providing some protection, particularly for the 80 nm film. The increasing critical load for thicker films is due to their ability to better protect the Si substrate. Nano-scratch studies on a-C films deposited on Si by closed field unbalanced magnetron sputtering (CFUBMS) and Si:a-C:H films deposited on glass by plasma enhanced chemical vapor deposited (PECVD) have also shown that the critical load for total film failure in the nano-scratch test can be strongly correlated with film thickness reflecting enhanced load support and substrate protection. However, thin PVD films can be highly stressed, and these high internal stresses coupled with shear stresses in the scratch test can lead to poor adhesion and dramatic coating spallation. In nano-scratch testing of 450 and 962 nm ECR-CVD DLC films deposited on Si without interlayers, it was found that the thicker film failed at around 60% of the critical load of the thinner film to due high stress. Substrate bias during deposition can significantly alter film stress (Zeng et al., 2002). Shi et al. reported nano-scratch data for 200 and 1000 nm a-C films deposited with varying substrate bias voltage (-20 to -140 V) by CFUBMS (Shi et al., 2008). At 200 nm thickness, the highest H/E films performed best since the intrinsic stress in the film was relatively low. For ∼1 μm a-C films, they observed a general relationship between H/E and the scratch test critical loads. When H/E is ∼0.08–0.09, there was a tendency of the a-C films to delaminate behind the moving probe at low critical load. Nano-scratch testing of ta-C films has shown that they are sufficiently thin to not show large area delamination and scratch resistance increases with H/E. Nano-scratch tests on the 80 nm ta-C film showed a small rate dependence. The critical load for film failure was approximately constant (113 ± 15) mN over a 100-fold variation in dL/dx from 0.1–10 mN μm^{-1}. This very small rate-sensitivity in the scratch response is consistent with previous studies of a-C:H and a-C:H/Si films on glass, which showed no obvious rate sensitivity over a 20-fold range of dL/dx from 0.05 to 1 mN μm^{-1} (Beake et al., 2006). The similarity of critical loads, and their rate sensitivity, on the coated and uncoated Si strongly suggests that deformation of the silicon substrate is responsible for the failure in the coated system. The critical load for the onset of more extreme debris and edge cracking on silicon (L_{c2}) is important as it results in the delamination of the ta-C films deposited on Si. The presence of the very thin hard coating merely acts to provide some additional load support, which slightly alters the stress distribution.

As the friction coefficient at yield of the ta-C films when using the 4.6 μm probe was only 0.09, it is unsurprising that there was little difference in the critical load for yield in the nano-scratch test being generally no lower than in nanoindentation.

The increase in friction force with load follows closely the behavior observed on uncoated Si. In both cases, this is due to increasing plowing contribution to the friction force as the severity of the contact increases. Similar behavior was observed by Sundararajan and Bhushan in AFM-nano-scratch tests, where friction coefficients on 3.5–20 nm carbon films were initially 0.04–0.06, which was the same as Si (100) tested under the same conditions, increasing to ~0.1 at the critical load (Sundararajan & Bhushan, 1999, 2001). In tests on the 80 nm ta-C with the 1.1 μm probe, the friction coefficient was 0.1 at yield, 0.17 at L_{c1} and 0.23 at L_{c2} (Beake & Lau, 2005). Despite being at quite different values of critical load, the friction coefficients at the transitions were close to those measured with the 4.6 μm probe—further supporting the idea that when the "deformation level" is taken into account that the friction forces can be consistent across different contact sizes.

17.3.4 Nanofretting of ta-C films

A range of probe radii were used in the first studies of nanofretting of thin and ultrathin carbon films ranging from ruby sphere of radius 150 μm (Wilson et al., 2008, 2009; Wilson & Sullivan, 2009) to 10 μm and Berkovich (~100 nm) diamond probes (Liskiewicz et al., 2010). Larger probes were employed to perform tests with more fretting cycles, with up to 216,000 cycles being used in tests with the 150 μm ruby probe. Wilson and coworkers investigated 10–2000 nm sputtered Cr-doped a-C films. They studied (1) the influence of film thickness on specific wear rate and (2) the transition from fretting/partial slip to gross slip by changing track length and applied load, in nanofretting tests. In 10 Hz tests with 14 μm track length, the specific wear rate (defined as worn volume of the coating per unit load per unit slid distance) reduced exponentially with decreasing film thickness and with increasing applied load from 100 μN to 10 mN. A rapid reduction of specific wear rate was observed during the first 3000 oscillation cycles. Higher wear for the thicker films may relate to them being more stressed than the thinner films. The transition from fretting/partial slip to gross slip was shown by changing track length (2–14 μm) and applied load (10–200 mN). Two distinct fretting wear regimes were found, namely a W-shaped wear scar under low oscillation amplitude and a U-shaped wear scar produced at larger amplitudes.

Liskiewicz and coworkers studied the behavior of soft (70 and 150 nm MoST) and hard (70 nm a-C:H and 80 nm ta-C) thin coatings deposited on Silicon in 10,000 cycle nanofretting tests with 25 μm spheroconical probes and 20 Hz oscillation frequency (Liskiewicz et al., 2010). In nanofretting tests at 10 mN, no abrupt changes in the probe depth under load were observed on any of the coatings, and the lowest wear depth was found for the 80 nm ta-C, which was the hardest film. A gradual wear process without any abrupt changes during the tests due to discrete fractures was reported in tests on 70 nm a-C:H across a wide load range from 0.2 to 100 mN. Sharper probes promoted distinct failure events. In nanofretting tests with a sharper pyramidal Berkovich test probe, abrupt increases in probe depth were observed at low load, for example, on a 150 nm a-C:H abrupt depth-steps consistent with multiple failure events were observed even at 1 mN.

Nanofretting tests on 5, 20, and 80 nm ta-C films at 5 Hz with 10 μm track length using 37 μm end radius diamond probes were at significantly lower contact pressure than was required for plastic deformation and phase transformation in nanoindentation and nano-scratch testing so that fatigue could be studied. The initial Hertzian mean contact pressures were ∼3–4 GPa at 10 mN to ∼10 GPa at 200 mN. By performing tests for varying numbers of fretting cycles and combining friction and on-load depth signals during the tests with SEM/EDX analysis, it was possible to investigate differences in the durability of the films with film thickness. For example, comparison of tests at 50 mN showed that the 20 nm film was worn through after a smaller number of cycles than the 80 nm film, and at 10 mN similarly, the 5 nm film was worn away more rapidly than the 20 nm film. EDX elemental line profiles across fretting scars revealed thinning and complete removal of the films depending on the film thickness, fretting load, and number of cycles. The fretting wear on the 5 and 20 nm films was accompanied by more oxygen incorporation within the track, which may limit their durability. Molecular dynamics simulations have indicated oxidation as a possible mechanism in the sliding wear of ta-C films (Moras et al., 2011). The 5 nm film at 10 mN showed oxygen incorporation and film thinning within 1500 cycles with complete removal of the film from the wear scar region within 3000 cycles. For the 20 nm film, similar behavior occurred at 50 mN, with thinning and oxygen incorporation even in 300 cycle tests, and complete film removal (EDX profiling showing Si but no C in the track) after 3000 cycles.

Fig. 17.8 shows the friction and on-load depth during a 6000 cycle test on the 80 nm ta-C film with the 37 μm probe at 200 mN. Close examination shows that the slope in the on-load fretting depth shows some correlation with the measured friction, as changes in depth wear rate were observed when friction changes, most noticeably after ∼3200 cycles. Wear scars from 18,000 cycle tests on the 80 nm

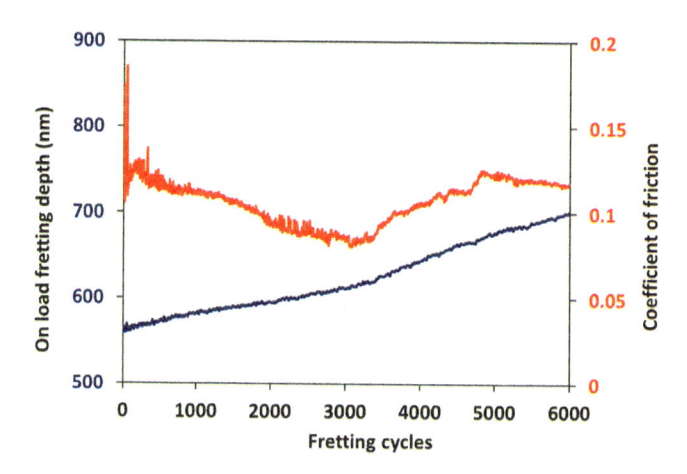

Figure 17.8 Nanofretting test on carbon-based coating. 200 mN nano-fretting test on the 80 nm ta-C film with a 37 μm probe.

ta-C showed a reduction in track length at 200 mN compared with 50 mN. On the 80 nm ta-C film, nanofretting tests with a sharper 5 μm probe showed wear at only 10 mN (mean initial Hertzian contact pressure ~6 GPa). The maximum von Mises stress is well into the substrate (at a depth ~260 nm) and the applied load below that required for Si phase transformation. One possibility is that fretting damage may be initiated by micro-slip in the substrate.

Typically, there was initial variability in friction in nanofretting tests on the ta-C, which was followed by a reduction and subsequent increase in friction after the film was worn away. In the nanofretting tests on ta-C (e.g., Fig. 17.3), friction and wear rate are correlated, which appears to be a contact area effect. The higher friction after failure could partly relate to changing interfacial component since the interfacial friction between Si and diamond is higher. Effects including a transfer film formation and/or breaking/reforming < 1 nm native oxide layer may also play a role (Liu et al., 2019; Wang et al., 2020). The adhesion term dominates friction when there is high surface conformity and friction scales with contact area (Hsia et al., 2020). The on-load and residual depth data enable contact area to be estimated. Abrupt and gradual changes in probe depth have different influences on the contact area and hence, the adhesion contribution to the measured frictional force. Typically, a gradual increase in depth is due to material removal that gradually increases the area in contact, whereas an abrupt increase in probe depth is associated with a (transitory) decrease in contact area due to fracture.

Analysis of the SEM images of fretting wear scars revealed distinctive gross-slip type of damage with scratches generated during sliding present within the wear scar area. The damage mechanics of ta-C films seems to be progressive wear by debris generation and debris aggregation outside the contact area. Deformation in the nanofretting tests of the ta-C films proceeded by a fatigue mechanism with a gradual wearing away of the film, as shown by the EDX profiles across scars, and the absence of any abrupt changes in depth or friction. The fatigue wear process is different from the film behavior in the nano-scratch test where the contact pressure is greater and distinct critical loads are observed, with abrupt increases in probe depth and friction. Appreciable substrate deformation and bending at the film − substrate interface is necessary before the thin films fail. The probe depths under load at failure are significantly lower in the nanofretting test due to lower substrate deformation, and consequently, the friction force at failure is lower in the nanofretting test due to less contribution from plowing. Li and Bhushan proposed a mechanism for the fatigue failure of 3.5−20 nm ta-C films in reciprocating sliding with 3 mm sapphire sliders. In their mechanism, wear fatigue cracks form within the coating and on thinner (3.5, 5 nm ta-C) propagate to the interface resulting in delamination but for 20 nm ta-C films do not reach the interface. Since this mechanism does not take into account substrate deformation and interface bending, therefore it is not directly applicable to the situation in the nano-scratch test where appreciable elastic and plastic substrate deformation occurs and failure does not occur until the probe depth is greater than the film thickness. However, the mechanism of film failure in the nanofretting tests is closer to that proposed by Li and Bhushan, although some substrate deformation does occur.

Despite the differences in contact pressure and failure mechanism in the nano-fretting and nano-scratch tests, there was a correlation in coating performance with higher film thickness providing more load support and protection of the Si substrate in both tests. Thinner films offered significantly less protection, failing at lower load in the scratch test and more rapidly and/or at lower load in the fretting test. Ma et al. studied the influence of film thickness in AFM nano-wear of 5−85 nm sputtered a-C films on Si(100) with a 200 nm end radius Berkovich indenter (Ma et al., 2003), finding that thinner films had reduced load carrying capacity. Chen et al. studied AFM nanowear of 2 and 5 nm DLC films deposited by filtered cathodic arc on Si(100) against either 1 μm SiO_2 or 0.5 μm diamond tips (Chen et al., 2011). Although the thicker DLC showed the best wear resistance, the thinner film also was more wear resistant than the Si(100). Interestingly, Liu et al. (2021) discovered that the wear resistance of fluorinated graphene nanosheets deposited on Si in AFM nanowear tests improved as the coating thickness decreased from 4.2 to 0.8 nm. With support from density function theory calculations, these authors suggested that this was due to differences in interfacial charge transfer between the fluorinated graphene and the Si, which affected the adhesion strength. Although it may seem surprising that such thin films can be effective, it should be noted that that thin native oxide layers can offer a similar protective effect, for example, in protecting Ti alloys, which undergo dramatic wear in nanofrettting and microscale reciprocating wear once the layer is compromised, as discussed in Chapter 7.

17.3.5 Impact—influence of probe geometry

Silicon is a brittle material and is susceptible to impact damage. Reliability of MEMS devices under severe shock conditions is an active research area, which has been reviewed by Peng and You (2021). Kishimura et al. showed that under shock loading, Si undergoes transition to high-pressure phases at ∼13 GPa (Kishimura et al., 2015).

The behavior of 5−80 nm FCVA ta-C on Si (Beake et al., 2004; Shi et al., 2019) and 100 nm sputtered DLC on Si (Faisal et al., 2014) under repetitive impact loading has been studied in nano-impact tests with different impact probe geometry and applied load. Repetitive impact of the 100 nm DLC with a 10 μm end radius spheroconical indenter produced delamination and uplift at low load. Finite element modeling showed the uplift (blistering) and delamination occurred when maximum von Mises stresses were close to the interface with the silicon substrate. In lower strain rate indentation fatigue tests (where the probe did not leave the sample between contacts) coating failure occurred after longer much fatigue cycles.

In nano-impact tests of ta-C coatings using a well-worn Berkovich indenter with 100−300 mN applied load (Beake, Lau, et al., 2004), it was found that the 60 and 80 nm ta-C coatings failed clearly after only a few impacts. These films were less resistant to impact induced damage than the underlying Si under these conditions. The on-load probe depths at the end of the test (including elastic deformation) therefore represent differences in film thickness (so that higher depths were found on thicker coatings). Under the low impact forces, the impact-induced stresses were not high enough to cause phase transformation or lateral cracking in the Si substrate, but the fatigue process causes film failure. Goel et al. reported that in MD simulation of very thin coatings, they were able to resist nano-impact by reducing the contact pressure in the Si substrate (Goel et al., 2014) to below that required for phase transformation.

Single impact and repetitive (cyclic) nano-impact tests with a $R = 4.6$ μm spherical diamond probe were performed over a range of loads on the 5 and 80 nm ta-C coatings and uncoated Si(100) to investigate how damage tolerance of silicon was modified by the presence of the ultrathin coatings (Shi et al., 2019). The impact resistance of different coatings can be assessed by the number of impacts required for failure to occur in 50% of the tests. Rebound impacts are essentially elastic (Wheeler et al., 2019) so that only the initial impact in each cycle is counted. Failure probability can be estimated by ranking the number of impacts-to-failure events in order of increasing fatigue resistance and then assigning a probability of failure to the n th ranked failure event in a total sample size of N, according to Eq. (17.4), in an analogous approach to the treatment of distributions of failure stresses in Weibull statistics.

$$P(f) = n/(N+1) \tag{17.4}$$

By combining failure probability data at different loads, it is possible to produce a plot of the number of impacts required for failure to occur in 50% of tests versus the impact force. Failure in the nano-impact test can be strongly load-dependent. Fig. 17.9 illustrates how the failure probability changes with load and number of impacts for a 80 nm ta-C coating on Si when impacted by a spherical indenter with 4.6 μm end radius (Shi et al., 2019).

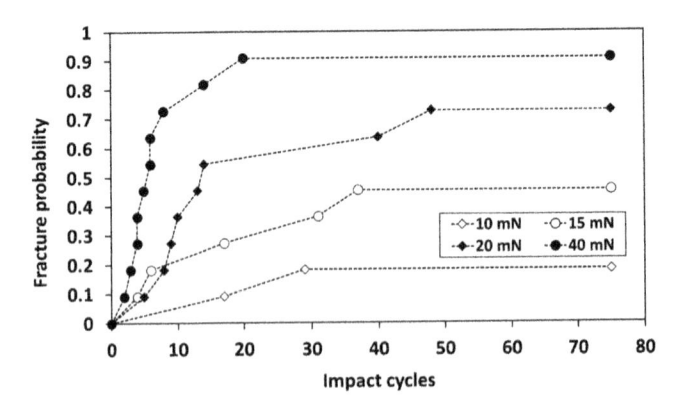

Figure 17.9 Failure probability in the nano-impact test. Relationship between failure probability, impact load, and number of impacts for an 80 nm ta-C coating on Si when impacted by a spherical indenter with 4.6 μm end radius.

At low impact load, the deformation mechanisms involved coating damage with minimal permanent substrate damage, with delamination outside of the impact crater for the 80 nm coating, but this did not occur for the 5 nm coating. Fracture of the silicon substrate occurred at higher loads through a failure mechanism involving initially plastic deformation/phase transformation during the first few impact cycles, with subsequent brittle fracture after the completed plastic deformation. Changing contact pressure during the test provided further support for the degradation mechanism and the influence of phase transformation in the silicon substrate. For the ta-C coated systems, there was lower impact depth and more impacts required before substrate fracture than on the uncoated silicon, with the thicker coating showing a better protective role than the thinner one. This appears to be related to their enhanced load support, which affects phase transformation in the substrate, with potentially delamination providing an additional impact energy dissipation mechanism.

17.4 Conclusions

The nanomechanical behavior and nano/microtribological performance of 5–80 nm ta-C films deposited as protective overcoats on Si(100) wafers are strongly influenced by the underlying phase transformation and brittle deformation of the silicon substrate. By investigating the response of the thin ta-C films and the uncoated Si in different tribocontacts (nanoindentation, nano-scratch, nanofretting, and nano-impact tests) and with different probe geometries (Berkovich and spherical indenters with radii 1–37 μm), it has been possible to study (1) rate sensitivity, (2) influence of tangential loading, (3) differences in deformation mechanisms between nanofretting and nano-scratch, (4) the role of lateral cracking of the Si, and (5) the influence of ta-C film thickness on their ability to protect the Si substrate. A quantitative comparison of the deformation during loading in the nanoindentation, nano-scratch, and nanofretting tests using the same spherical probe showed that loading curves on Si(100) were very similar at very low load, but the tangential loading in the nano-scratch and nanofretting tests promoted earlier yield and at higher load severe cracking at lower load than in nanoindentation. The nanofretting tests on Si showed strong load sensitivity, with a clear transition to more severe wear at a critical load.

Tests on the ta-C films showed that despite their limited thickness, they were able to provide some protection to the underlying Si by providing load support, reducing the effective load reaching the substrate, and spreading the deformation out over a wider area, which all affect phase transformation in the Si substrate. The thin films modify the critical loads for pop-ins and pop-outs commonly observed on Si, but their rate dependence was not affected by the presence of the ta-C confirming that they originate in the substrate. Although tangential loading in the nano-scratch test resulted in the formation of large lateral cracks at much lower load than in nanoindentation the 80 nm ta-C film provided a significantly higher critical load for this. Film fatigue wear in nanofretting occurred at significantly lower contact pressure than is required for plastic deformation and phase transformation in nanoindentation and nano-scratch testing. In contrast to the abrupt ta-C film failures in the nano-scratch tests, in the nanofretting tests. there was a more gradual increase in depth, which was associated with higher contact area and higher friction. The longer, higher cycle and lower contact pressure nanofretting tests can be an effective method to study damage tolerance under more application-relevant conditions. The 80 nm ta-C film provided more load support and protection of the Si substrate than for thinner films, which failed at lower load in the nano-scratch test and more rapidly and/or at lower load in nanofretting and nano-impact tests.

References

Arnell, R. D. (1990). The mechanics of the tribology of thin film systems. *Surface and Coatings Technology, 43−44*(1−3), 674−687. Available from https://doi.org/10.1016/0257-8972(90)90011-z.

Beake, B. D., Davies, M. I., Liskiewicz, T. W., Vishnyakov, V. M., & Goodes, S. R. (2013). Nano-scratch, nanoindentation and fretting tests of 5−80 nm ta-C films on Si(100). *Wear, 301*(1−2), 575−582. Available from https://doi.org/10.1016/j.wear.2013.01.073.

Beake, B. D., Goodes, S. R., & Shi, B. (2009). Nanomechanical and nanotribological testing of ultra-thin carbon-based and MoST films for increased MEMS durability. *Journal of Physics D: Applied Physics, 42*, 065301.

Beake, B. D. & Jochum, T. (2022). Correlation between electrical contact resistance, deviation from elastic unloading and phase transformation in silicon, *Symposium: 30 Years of Nanoindentation with the Oliver-Pharr Method and Beyond*, TMS.

Beake, B. D., & Lau, S. P. (2005). Nanotribological and nanomechanical properties of 5−80 nm tetrahedral amorphous carbon films on silicon. *Diamond and Related Materials, 14*(9), 1535−1542. Available from https://doi.org/10.1016/j.diamond.2005.04.002.

Beake, B. D., Lau, S. P., & Smith, J. F. (2004). Evaluating the fracture properties and fatigue wear of tetrahedral amorphous carbon films on silicon by nano-impact testing. *Surface and Coatings Technology, 177−178*, 611−615.

Beake, B. D., Liskiewicz, T. W., Pickford, N. J., & Smith, J. F. (2012). Accelerated nanofretting testing of Si(100). *Tribology International, 46*, 114−118.

Beake, B. D., Liskiewicz, T. W., & Smith, J. F. (2011). Deformation of Si(100) in spherical contacts—Comparison of nano-fretting and nano-scratch tests with nano-indentation. *Surface and Coatings Technology, 206*, 1921−1926.

Beake, B. D., McMaster, S. J., & Liskiewicz, T. W. (2022). Contact size effects on the friction and wear of amorphous carbon films. *Applied Surface Science Advances, 9*, 100248.

Beake, B. D., Ogwu, A. A., & Wagner, T. (2006). Influence of experimental factors and film thickness on the measured critical load in the nanoscratch test. *Materials Science and Engineering: A, 423*, 70−73.

Beake, B. D., Vishnyakov, V. M., Goodes, S. R., & Rahmati, A. T. (2004). Statistically distributed nano-scratch testing of AlFeMnNb, AlFeMnNi and TiN/Si$_3$N$_4$ thin films on silicon. *Journal of Vacuum Science & Technology A, 42*.

Bhattacharya, A. K., & Nix, W. D. (1988). Analysis of elastic and plastic deformation associated with indentation testing of thin films on substrates. *International Journal of Solids and Structures*, 24(12), 1287−1298. Available from https://doi.org/10.1016/0020-7683(88)90091-1.

Bhowmick, S., Cha, H., Jung, Y.-G., & Lawn, B. R. (2009). Fatigue and debris generation at indentation-induced cracks in silicon. *Acta Materialia*, 57, 582−589.

Blees, M. H., Winkelman, G. B., Balkenende, A. R., & den Toonder, J. M. J. (2000). The effect of friction on scratch adhesion testing: Application to a sol−gel coating on polypropylene. *Thin Solid Films*, 359(1), 1−13. Available from https://doi.org/10.1016/s0040-6090(99)00729-4.

Borodich, F. M., Keer, L. M., & Korach, C. S. (2003). Analytical study of fundamental nanoindentation test relations for indenters of non-ideal shapes. *Nanotechnology*, 14, 803−808.

Bradby, J. E., Williams, J. S., Wong-Leung, J., Swain, M. V., & Munroe, P. (2001). Mechanical deformation in silicon by micro-indentation. *Journal of Materials Research*, 16(5), 1500−1507. Available from https://doi.org/10.1557/jmr.2001.0209.

Burnett, P. J., & Rickerby, D. S. (1987). The relationship between hardness and scratch adhesion. *Thin Solid Films*, 154(1−2), 403−416. Available from https://doi.org/10.1016/0040-6090(87)90382-8.

Chang, L., & Zhang, L. C. (2009a). Deformation mechanisms at pop-out in monocrystalline silicon under nanoindentation. *Acta Materialia*, 57, 2148−2153.

Chang, L., & Zhang, L. C. (2009b). Mechanical behaviour characterisation of silicon and effect of loading rate on pop-in: A nanoindentation study under ultra-low loads. *Materials Science and Engineering A*, 506, 125−129.

Chartidis, C. A. (2010). Nanomechanical and nanotribological properties of carbon-based thin films: A review. *International Journal of Refractory Metals and Hard Materials*, 28, 51−70.

Chen, L., Yang, M., Yu, J., Qian, L., & Zhou, Z. (2011). Nanofretting behaviours of ultrathin DLC coating on Si(100) substrate. *Wear*, 271(9−10), 1980−1986. Available from https://doi.org/10.1016/j.wear.2010.11.016.

Chudoba, T. (2006). *Measurement of hardness and Young's modulus of coatings by nanoindentation* (pp. 216−260). Springer.

Cook, R. F. (2006). Strength and sharp contact fracture of silicon. *Journal of Materials Science*, 41(3), 841−872. Available from https://doi.org/10.1007/s10853-006-6567-y.

Djabella, H., & Arnell, R. D. (1993). Finite element analysis of the contact stresses in elastic coating/substrate under normal and tangential load. *Thin Solid Films*, 223(1), 87−97. Available from https://doi.org/10.1016/0040-6090(93)90731-4.

Domnich, V., & Gogotsi, Y. (2002). Phase transformations in silicon under contact loading. *Reviews on Advanced Materials Science*, 3, 1−36.

Faisal, N. H., Ahmed, R., Goel, S., & Fu, Y. Q. (2014). Influence of test methodology and probe geometry on nanoscale fatigue failure of diamond-like carbon film. *Surface and Coatings Technology*, 242, 42−53. Available from https://doi.org/10.1016/j.surfcoat.2014.01.015.

Fischer-Cripps, A. C., Bull, S. J., & Schwarzer, N. (2012). Critical review of claims for ultra-hardness in nanocomposite coatings. *Philosophical Magazine*, 92, 1601−1630.

Gerbig, Y. B., Michaels, C. A., & Cook, R. F. (2015). In situ observation of the spatial distribution of crystalline phases during pressure-induced transformations of indented silicon thin films. *Journal of Materials Research*, 30, 390−406.

Goel, S., Agrawal, A., & Faisal, N. H. (2014). Can a carbon nano-coating resist metallic phase transformation in silicon substrate during nanoimpact? *Wear*, 315, 38−41.

Gogotsi, Y., Zhou, G., Ku, S. S., & Cetinkunt, S. (2001). Raman microspectroscopy analysis of pressure-induced metallization in scratching of silicon. *Semiconductor Science and Technology*, *16*(5), 345−352. Available from https://doi.org/10.1088/0268-1242/16/5/311.

Hamilton, G. M., & Goodman, L. E. (1966). The stress field created by a circular sliding contact. *Journal of Applied Mechanics*, *33*(2), 371−376. Available from https://doi.org/10.1115/1.3625051.

Hsia, F. C., Elam, F. M., Bonn, D., Weber, B., & Franklin, S. E. (2020). Wear particle dynamics drive the difference between repeated and non-repeated reciprocated sliding. *Tribology International*, *142*. Available from https://doi.org/10.1016/j.triboint.2019.105983, http://www.elsevier.com/inca/publications/store/3/0/4/7/4.

Hsu, S., Patakamuri, G., & Li, L. (2018). Surface and interface designs for friction control. *Tribology Online*, *13*, 178−187.

Juliano, T., Domnich, V., & Gogotsi, Y. (2004). Examining pressure-induced phase transformations in silicon by spherical indentation and raman spectroscopy: A statistical study. *Journal of Materials Research*, *19*, 3099−3108.

Juliano, T., Gogotsi, Y., & Domnich, V. (2003). Effect of indentation unloading conditions on phase transformation induced events in silicon. *Journal of Materials Research*, *18*, 1192−1201.

Jungk, J., Boyce, B., Buchheit, T., Friedmann, T., Yang, D., & Gerberich, W. (2006). Indentation fracture toughness and acoustic energy release in tetrahedral amorphous carbon diamond-like thin films. *Acta Materialia*, *54*(15), 4043−4052. Available from https://doi.org/10.1016/j.actamat.2006.05.003.

Kim, S. H., Asay, D. B., & Dugger, M. T. (2007). Nanotribology and MEMS. *Nano Today*, *2*, 22−29.

Kishimura, H., Matsumoto, H., & Thadhani, N. N. (2015). Effect of shock compression on single crystalline silicon. *Journal of Physics: Conference Series*, *215*, 012145.

Kovalchenko, A., Gogotsi, Y., Domnich, V., & Erdemir, A. (2002). Phase transformations in silicon under dry and lubricated sliding. *Tribology Transactions*, *45*, 372−380.

Ku, I. S. Y., Reddyhoff, T., Holmes, A. S., & Spikes, H. A. (2011). Wear of silicon surfaces in MEMS. *Wear*, *271*(7−8), 1050−1058. Available from https://doi.org/10.1016/j.wear.2011.04.005.

Lemoine, P., Quinn, J. P., Maguire, P., & McLaughlin, J. A. (2004). Comparing hardness and wear data for tetrahedral amorphous carbon and hydrogenated amorphous carbon thin films. *Wear*, *257*(5−6), 509−522. Available from https://doi.org/10.1016/j.wear.2004.01.010.

Lemoine, P., Quinn, J. P., Maguire, P. D., Zhao, J. F., & McLaughlin, J. A. (2007). Intrinsic mechanical properties of ultra-thin amorphous carbon layers. *Applied Surface Science*, *253*(14), 6165−6175. Available from https://doi.org/10.1016/j.apsusc.2007.01.028.

Lemoine, P., Zhao, J. F., Quinn, J. P., McLaughlin, J. A., & Maguire, P. (2000). Hardness measurements at shallow depths on ultra-thin amorphous carbon films deposited onto silicon and Al_2O_3−TiC substrates. *Thin Solid Films*, *379*(1−2), 166−172. Available from https://doi.org/10.1016/s0040-6090(00)01543-1.

Li, X., & Bhushan, B. (1999). Micro/nanomechanical and tribological characterization of ultrathin amorphous carbon coatings. *Journal of Materials Research*, *14*(6), 2328−2337. Available from https://doi.org/10.1557/jmr.1999.0309.

Liu, Y., Jiang, Y., Sun, J., Wang, Y., Qian, L., Kim, S. H., & Chen, L. (2021). Inverse relationship between friction and wear of fluorinated graphene: "Thinner is better". *Nano Letters*.

Liskiewicz, T. W., Beake, B. D., & Smith, J. F. (2010). In situ accelerated micro-wear - a new technique to fill the measurement gap. *Surface and Coatings Technology, 205*, 1455−1459.

Liu, Y., Chen, L., Zhang, B., Cao, Z., Shi, P., Peng, Y., Zhou, N., Zhang, J., & Qian, L. (2019). Key role of transfer layer in load dependence of friction on hydrogenated diamond-like carbon films in humid air and vacuum. *Materials, 12*(9). Available from https://doi.org/10.3390/ma12091550.

Ma, X.-G., Komvopoulos, K., Wan, D., Bogy, D. B., & Kim, Y.-S. (2003). Effects of film thickness and contact load on nanotribological properties of sputtered amorphous carbon thin films. *Wear, 254*(10), 1010−1018. Available from https://doi.org/10.1016/s0043-1648(03)00307-7.

Maboudian, R., Ashhurst, W. R., & Carraro, C. (2000). Self-assembled monolayers as anti-stiction coatings for MEMS: Characterisation and recent developments. *Sensors Actuators, 82*, 219−223.

Martinez, E., & Esteve, J. (2001). Nanoindentation hardness measurements using real-shape indenters: Application to extremely hard and elastic materials. *Applied Physics A, 72*, 319−324.

Moras, G., Pastewka, L., Gumbsch, P., & Moseler, M. (2011). Formation and oxidation of linear carbon chains and their role in the wear of carbon materials. *Tribology Letters, 44*(3), 355−365. Available from https://doi.org/10.1007/s11249-011-9864-9.

Oliver, D. J., Lawn, B. R., Cook, R. F., Reitsma, M. G., Bradby, J. E., Williams, J. S., & Munroe, P. (2008). Giant pop-ins in nanoindented silicon and germanium caused by lateral cracking. *Journal of Materials Research, 23*(2), 297−301. Available from https://doi.org/10.1557/jmr.2008.0070.

Peng, T., & You, Z. (2021). Reliability of MEMS in shock environments: 2000−2020. *Micromachines, 12*(11). Available from https://doi.org/10.3390/mi12111275.

Schaffar, G. J. K., Tscharnuter, D., & Maier-Kiener, V. (2023). Exploring the high-temperature deformation behavior of monocrystalline silicon − An advanced nanoindentation study. *Materials & Design, 233*, 112198.

Scharf, T. W., Prasad, S. V., Dugger, M. T., Kotula, P. G., Goeke, R. S., & Grubbs, R. K. (2006). Growth, structure, and tribological behavior of atomic layer-deposited tungsten disulphide solid lubricant coatings with applications to MEMS. *Acta Materialia., 54*(18), 4731−4743. Available from https://doi.org/10.1016/j.actamat.2006.06.009.

Sheeja, D., Tay, B. K., Lau, S. P., Leong, K. W., & Lee, C. H. (2002). An empirical relation for the critical load of DLC coatings prepared on silicon substrates. *International Journal of Modern Physics B, 16*(2002), 958−962.

Sheeja, D., Tay, B. K., Leong, K. W., & Lee, C. H. (2002). Effect of film thickness on the stress and adhesion of diamond-like carbon coatings. *Diamond and Related Materials, 11*, 1643−1647.

Shi, B., Sullivan, J. L., & Beake, B. D. (2008). An investigation into which factors control the nanotribological behaviour of thin sputtered carbon films. *Journal of Physics D: Applied Physics, 41*, 045303.

Shi, X., Beake, B. D., Liskiewicz, T. W., Chen, J., & Sun, Z. (2019). Failure mechanism and protective role of ultrathin ta-C films on Si (100) during cyclic nano-impact. *Surface and Coatings Technology, 364*, 32−42. Available from https://doi.org/10.1016/j.surfcoat.2019.02.082, http://www.journals.elsevier.com/surface-and-coatings-technology/.

Shi, X., Tay, B. K., Tan, H. S., Zhong, L., Tu, Y. Q., Silva, S. R. P., & Milne, W. I. (1996). Properties of carbon ion deposited tetrahedral amorphous carbon films as a function of ion energy. *Journal of Applied Physics*, *79*(9), 7234−7240. Available from https://doi.org/10.1063/1.361440, http://scitation.aip.org/content/aip/journal/jap.

Smallwood, S. A., Eapen, K. C., Patton, S. T., & Zabinski, J. S. (2006). Performance results of MEMS coated with a conformal DLC. *Wear*, *260*(11−12), 1179−1189. Available from https://doi.org/10.1016/j.wear.2005.07.019.

Sundararajan, S., & Bhushan, B. (2001). Development of a continuous microscratch technique in an atomic force microscope and its application to study scratch resistance of ultrathin hard amorphous carbon coatings. *Journal of Materials Research*, *16*, 437−445.

Sundararajan, S., & Bhushan, B. (1999). Micro/nanotribology of ultra-thin hard amorphous carbon coatings using atomic force/friction force microscopy. *Wear*, *225−229*(I), 678−689. Available from https://doi.org/10.1016/s0043-1648(99)00024-1.

Tanner, D. M., Miller, W. M., Eaton, W. P., Irwin, L. W., Peterson, K. A., Dugger, M. T., Senft, D. C., Smith, N. F., Tangyunyong, P., & Miller, S. L. (1998). Effect of frequency on the lifetime of a surface micromachined microengine driving a load. *1998/01 Annual proceedings − Reliability Physics (Symposium) IEEE*, United States. https://doi.org/10.1109/relphy.1998.670438, 00999512, 26−35.

Varenberg, M., Etsion, I., & Halperin, G. (2005). Nanoscale fretting wear study by scanning probe microscopy. *Tribology Letters*, *18*, 493.

Vingsbo, O., & Söderberg, S. (1988). On fretting maps. *Wear*, *126*, 131−147.

Wang, K., Zhang, J., Ma, T., Liu, Y., Song, A., Chen, X., Hu, Y., Carpick, R. W., & Luo, J. (2020). Unraveling the friction evolution mechanism of diamond-like carbon film during nanoscale running-in process toward superlubricity. *Small (Weinheim an der Bergstrasse, Germany)*, *17*, 2005607.

Weppelmann, E. R., Field, J. S., & Swain, M. V. (1995). Influence of spherical indentor radius on the indentation-induced transformation behaviour of silicon. *Journal of Materials Science*, *30*, 2455−2462.

Wheeler, J. M., Dean, J., & Clyne, T. W. (2019). Nano-impact indentation for high strain rate testing: The influence of rebound impacts. *Extreme Mechanics Letters*, *26*, 35−39.

Williams, J. A., & Le, H. R. (2006). Tribology and MEMS. *Journal of Physics D: Applied Physics*, *39*(12), R201−R214. Available from https://doi.org/10.1088/0022-3727/39/12/R01.

Wilson, G. M., Smith, J. F., & Sullivan, J. L. (2008). A nanotribological study of thin amorphous C and Cr doped amorphous C coatings. *Wear*, *265*, 1633−1641.

Wilson, G. M., Smith, J. F., & Sullivan, J. L. (2009). A DOE nano-tribological study of thin amorphous carbon-based films. *Tribology International*, *42*, 220−228.

Wilson, G. M., & Sullivan, J. L. (2009). An investigation into the effect of film thickness on nanowear with amorphous carbon-based coatings. *Wear*, *266*, 1039−1043.

Wu, Y. Q., Huang, H., Zou, J., Zhang, L. C., & Dell, J. M. (2010). Nanoscratch-induced phase transformation of monocrystalline Si. *Scripta Materialia*, *63*(8), 847−850. Available from https://doi.org/10.1016/j.scriptamat.2010.06.034.

Yu, B., Dong, H., Qian, L., Chen, Y., Yu, J., & Zhou, Z. (2009). Friction-induced nanofabrication on monocrystalline silicon. *Nanotechnology*, *20*(465303), 8.

Yu, B., Li, X., Dong, H., Chen, Y., Qian, L., & Zhou, Z. (2012). Towards a deeper understanding of the formation of friction-induced hillocks on monocrystalline silicon. *Journal of Physics D: Applied Physics*, *45*(145301), 6.

Yu, J., Qian, L., Yu, B., & Zhou, Z. (2010). Effect of surface hydrophilicity on the nanofretting behavior of Si(100) in atmosphere and vacuum. *Journal of Applied Physics*, *108*, 034314.

Yu, J. X., Qian, L. M., Yu, B. J., & Zhou, Z. R. (2009a). Nanofretting behaviors of monocrystalline silicon (100) against diamond tips in atmosphere and vacuum. *Wear, 267*, 322−329.

Yu, J., Qian, L., Yu, B., & Zhou, Z. (2009b). Nanofretting behavior of monocrystalline silicon (100) against SiO$_2$ microsphere in vacuum. *Tribology Letters, 34*, 31−40.

Zeng, X. T., Zhang, S., Ding, X. Z., & Teer, D. G. (2002). Comparison of three types of carbon composite coatings with exceptional load-bearing capacity and high wear resistance. *Thin solid films, 420−421*, 366−370. Available from https://doi.org/10.1016/s0040-6090(02)00840-4.

Zhao, X., & Bhushan, B. (1998). Material removal mechanisms of single-crystal silicon on nanoscale and at ultralow loads. *Wear, 223*(1−2), 66−78. Available from https://doi.org/10.1016/s0043-1648(98)00302-0.

Zok, F. W., & Miserez, A. (2007). Property maps for abrasion resistance of materials. *Acta Materialia, 55*(18), 6365−6371. Available from https://doi.org/10.1016/j.actamat.2007.07.042.

Diamond-like carbon coatings on steel for automotive applications

18

Samuel James McMaster[1,2]
[1]Research Centre for Manufacturing and Materials (CMM), The Institute for Advanced Manufacturing and Engineering (AME), Coventry, United Kingdom, [2]Pillarhouse International Ltd. Chelmsford, Essex, United Kingdom

18.1 Introduction to diamond-like carbon coatings

Diamond-like carbon (DLC) coatings are an important class of coatings for multiple industries. DLCs are mechanically hard, amorphous, metastable carbonaceous materials (Grill, 1999; Hainsworth & Uhure, 2007; Lifshitz, 1999; Robertson, 1992, 2002). They are applied widely for increased performance of components such as pistons, tappets, and bearings in the automotive field. This chapter will present an overview of DLC coatings, methodologies for their deposition, and mechanical and structural characterization. After these fundamentals have been reviewed, specific case studies on multiscale scratch testing, fretting testing, and impact testing will be presented to evaluate the suitability of different DLC coatings for automotive applications.

18.1.1 History of diamond-like carbons and types of diamond-like carbons

A review of DLCs would not be complete without a short aside on their history. Bewilogua and Hofmann (2014) reviewed the history of DLCs and collated the number of publications referring to DLC from 1970 to 2012. DLC had a surge in popularity in the 1990s with annual publications exceeding 500 from 2004 to 2009. As such, it is immediately apparent that DLCs are an important and useful coating. A great strength of DLCs is the ability to tailor their properties to different application areas; this can be achieved by changing the deposition conditions or using dopant elements (Hainsworth & Uhure, 2007; Lifshitz, 1999; Robertson, 1992, 2002; Sánchez-López & Fernández, 2008). This will be explored later in Section 18.2. Examples of doping elements are fluorine used for altering surface energy to affect wettability properties, silicon for thermal stability, and titanium for enhanced adhesion (Bewilogua et al., 2000; Choi et al., 2007; Hasselbruch et al., 2015; Lanigan et al., 2016; Sánchez-López & Fernández, 2008).

The initial developments of DLC films were accidental. In 1953, Schmellenmier was the first to note the formation of a hard black film on the cathode of his glow discharge apparatus during the study of the influence of ionized acetylene gas (C_2H_2) on the surfaces of tungsten − cobalt alloys (Schmellenmeier, 1953). Following this, he noted that these micron thickness films were amorphous as determined by X-ray diffraction (XRD); however, some regions were diamond crystallites (Schmellenmeier, 1956). 1971 heralded the next major advance with the work of Aisenberg and Chabot (1971). They coined the term "diamond-like carbon" and utilized ion-beam deposition to form carbon films.

To understand more about the terminology and types of DLCs, we should review some of the essentials of carbon chemistry. Carbon's ability to exist in a variety of crystalline and disordered structures is due to its three-bond hybridisations. Hybridization is the overlapping of electron orbitals (Robertson, 2002). The different physical forms (crystalline structures) of carbon are known as allotropes. Carbon is not the only element able to be assembled thus, other group 14 elements in the periodic table are able to do this also (Pierson, 1993). We will now review carbon's bonding types that allow for these configurations. In sp^3 bonding (found in diamond), the carbon atom's four valence electrons are assigned to a tetrahedrally directed sp^3 orbital. These strong bonds are referred to as a σ bond. Whereas for graphite, carbon has a threefold configuration. Three of the four valence electrons enter trigonally arranged sp^2 orbitals forming σ bonds in the plane. The fourth electron is in a $p\pi$ orbital, orienting itself normal to the σ bonding plane. As such, carbon bonded in the sp π orbitals forming σ bonds in the plane. The fourth electron is in a $p\pi$ orbital, orienting itself normal to the σ bonding plane. As such, carbon bonded in the sp^2 configuration has strong bonding within the plane but only weak Van der Waals bonding between layers (Brian & Kelly, 1981; Dresselhaus et al., 1996). Therefore, graphite has strongly directional mechanical properties; the layers can be easily sheared (Hainsworth & Uhure, 2007). This contributes to graphite's excellent tribological properties. Diamond, on the other hand, does not have this directional dependence and has incredibly strong omnidirectional bonding (Angus & Hayman, 1988; Hainsworth & Uhure, 2007). The result of which is its high hardness ($H = 100$ GPa) and stiffness ($E = 1144$ GPa) (Hainsworth & Uhure, 2007).

DLC is a generalized term for amorphous carbon films, but more specific terminology exists to classify the various families of amorphous carbon compounds (Casiraghi et al., 2005; Hainsworth & Uhure, 2007; Robertson, 2008a):

- a-C − Hydrogen-free amorphous carbon films with a greater proportion of sp^2 bonding. The "a" donates an amorphous coating.
- a-C:H − Amorphous hydrogenated films containing a modest sp^3 fraction. "H" denotes an amount of hydrogenation in the coating. Several subtypes of this coating exist:
- a-C:H films with the highest H content (40%−50%) can have up to 60% sp^3; however, the bonds are hydrogen-terminated resulting in a soft material with low density; these films are known as polymer-like a-C:H (PLCH).
- a-C:H films with intermediate H content (20%−40%) have lower sp^3 content compared with polymer-like a-C:H (PLCH) but more C−C sp^3 bonds; this results in better mechanical properties. These films are called diamond-like a-C:H (DLCH).

- a-C:H with low H content (<20%). These films have high sp^2 content and sp^2 clustering and are commonly known as graphite-like a-C:H (GLCH).
- ta-C:H — Tetrahedral amorphous hydrogenated films possessing a significant sp^3 bonding fraction (>70%). These films can have ~25% hydrogen content. These properties result in a high density (up to 2.4 g cm^{-1}) and bonding fraction (>70%). These films can have ~25% hydrogen content. These properties result in a high density (up to 2.4 g cm^{-3}) and higher Young's modulus (up to 300 GPa).
- ta-C — Tetrahedral amorphous carbon. These films lack hydrogenation and have high sp^3 content giving them a high hardness (40–90 GPa).

Hainsworth and Uhure (2007) summarized the typical properties of the most common subtypes of DLCs described above. One of the most well-known figures related to DLCs is from Robertson (2002); the ternary phase diagram shows the proportions of the different hybridized bonding in DLC types and the levels of hydrogen therein.

Metallic elements doped into DLC structures can be noted as such a-C:H:X (where X is the metallic element such as W). Nonmetals are instead denoted as Y-a-C:H (where Y is the nonmetallic element such as Si). Additional terminology that is present in the literature is graphitic-like carbon (GLC) (Lin et al., 2023), referring to higher sp^2 content. As the naming conventions aren't standardized, one may encounter different formats of naming, but this represents reasonable conventions found throughout academic literature. More novel amorphous carbon structures exist such as onion-like carbon (OLC) (Chen & Li, 2020). Berman has explored their use in achieving superlubricity (Berman et al., 2013, 2018, 2015). These studies utilized DLC as a counterface sliding against graphene. The proposed mechanism for achieving superlubricious conditions was the formation of nanoscrolls of graphene around nanodiamonds. Graphene has also integrated into DLC coating structures as in the work of Brittain et al. (2023), to further decrease the friction and wear experienced in the coating system.

18.2 Performance of diamond-like carbons

Before delving into the known performance of DLC coatings for applications in the automotive sector, we will first explore the various components that DLCs have been applied to and additionally what the requirements are for these coatings. Robertson (2008b) compared the performance of DLC with diamond; it would be beneficial to review why amorphous diamond-like carbon would be chosen for automotive tribological applications based instead of diamond. DLC has many more options in terms of deposition methodologies (see Section 18.3); however, industrial diamond is grown by epitaxy, thereby reducing its deposition rate. Though extremely hard, making it very wear-resistant for sliding applications, diamond will generally have a higher coefficient of friction due to the lack of hydrogen (Hainsworth & Uhure, 2007). Its surface interaction is dominated by carbon — carbon dangling bond interactions (Donnet & Erdemir, 2008).

It should be noted that for the majority of academic publications, tribological studies will be performed on smaller-scale tribometers or fretting test rigs. Depending on the equipment, wear assessments can be made in-situ, but a proper assessment will require some form of profilometry with some of the most effective for measuring wear volume being white light interferometry or confocal/infinite focus microscopy. Depending on the optical properties of the DLC coating, white light interferometry can have trouble picking up the required interference patterns, and for this reason, other forms of profilometry may be preferred. This equipment, as with all lab-scale assessments, will provide an estimation of the field performance of a prospective coating. A more complete tribological assessment should always be performed to provide a complete performance assessment.

As this chapter is tailored toward automotive applications, we should note that no lab-scale test can completely replace engine test bays or full-scale car tests. Running an entire engine assembly or an entire vehicle in a dynamometer allows for the evaluation of whole system performance that most lab-scale tests cannot capture.

18.2.1 Common automotive applications for diamond-like carbons

Al Mahmud et al. (2015) reviewied the application of DLC coatings to engine parts and found that they were commonly applied to vales, bearings, gears, piston rings, piston pins, and direct injection fuel systems. The substrates identified in this study were varied but included CrMo steel, M2 tool steel, stainless steel, and 52,100 steel. In addition to standard DLCs (undoped a-C:H), tungsten-doped DLC was utilized for gears. Though the coatings would be tailored for the environment, softer coatings were utilized as they could more easily from a transfer layer (Al Mahmud et al., 2015). Vinoth et al. (2019) conducted a study of automotive piston rings coated with diamond-like carbon. They noted that the coating mechanical properties were altered depending upon RF power. 150 W resulted in a film with a hardness of 15 GPa, as measured by microindentation, compared with 8.21 GPa at 200 W. The DLC produced with 150 W of RF power had superior wear and tribological properties. Aldeeb et al. (2019) observed the resultant difference in $sp^3\%$ across a range of bias voltages. Robertson noted this also with the change of density and $sp^3\%$ depending upon the impinging ion energy and bias voltage (Robertson, 1992, 1994).

Kano (2014) reviewed the applications of diamond-like carbon for the Japanese society of tribologists and identified further application areas such as SUV 4WD torque controlled coupling clutches (Si-a-C:H), SUV differential gears (a-C:H:WC), and motorbike engine piston rings (a-C:H:WC). Tetrahedral amorphous carbon was also utilized in engine valve lifters and piston rings. Friction coefficients of 0.02 were noted for this coating under oil lubricated conditions. Renondeau et al. (2009) reviewed DLC coatings tribological performance under lubricated automotive applications. A TE-77 was used to produce ball-on-plate sliding, and MTM was utilized for ball-on-disk rolling/sliding. Importantly, it was summarized that no "universal" solution was identified in this study; low friction and wear could be achieved on the coating, but the steel counterface experienced significantly higher wear despite the presence of tribofilms.

Lawes et al. (2010) performed impact wear testing of diamond-like carbon films to assess their suitability for valve-tappet surfaces. To mimic the angled impact of the cam on the tappet surface, a bespoke test setup was created to impact with an Al_2O_3 stylus removing the sliding and rolling mechanisms that are also present in normal operations. Though simplified, this type of testing allows for the effect of impact alone to be studied. Tests were conducted dry and with SAW 5W30 engine oil. Sliding and rolling were also studied with a bespoke cam-tappet test rig (Lawes et al., 2007). Overall, DLC was found to perform well in valve train applications though a-C:H was found to be particularly susceptible under lubricated impact wear (<50 impacts to failure) possibly due to the formation of an abrasive slurry of hard particulate wear debris. The a-C coating performed better in this case (>450 impacts to failure). The a-C:H coating was superior in sliding and rolling tests, remaining intact after the testing regime.

Impact studies of DLCs are detailed in Section 18.7; however, the tests were performed on to a sample surface that is normal to the indenter. An angled sample holder can be employed for micro and nanoscale impact tests, but this was not used in this case.

Lawes et al. (2007) performed a tribological evaluation of a-C:H and WC/C a-C:H for engine applications. Nanoindentation determined that the a-C:H was harder with a greater degree of elastic recovery compared with the much more plastic response of the WC/C a-C:H. In scratch testing, high critical loads were observed with WC/C a-C:H (similar to Section 18.5); however, larger-scale spallation was found after failure. In sliding tests, a-C:H was superior with mild scuffing wear compared with the abrasive wear and pitting in WC/C a-C:H. Finally, the authors noted that, for optimum DLC selection, a complete mechanical and chemical characterization of the coatings is required to lead to an improved understanding of how coatings perform in an engine. Though beyond the scope of this chapter, it is interesting to note that diamond-like carbon has been applied to aluminum alloys for automotive applications (Malaczynski et al., 1997).

Though the applications for DLC coatings are many in the automotive industry, their high-temperature performance must be managed by dopants. The thermal stability of DLC is dependent upon structure with nonhydrogenated and more sp^3 rich ta-C stable up to 600°C (Erdemir & Donnet, 2006). Whereas a-C:H film only remain stable up to 400°C (Erdemir & Donnet, 2006). Titanium doping (Konkhunthot et al., 2019) and light elements (Sánchez-López & Fernández, 2008) have been found to be effective for this purpose.

18.2.2 Tribology of diamond-like carbon

Two common metrics, derived for physical parameters, are often used to predict the tribological performance of coatings. Those being *H/E* and H^3/E^2, Leyland and

Matthews (2000) were instrumental in popularizing *H/E* for use in the prediction and ranking of coating wear performance. The ratio itself relates to the elastic strain to break (or strain to failure) (Austin, 2014; Leyland & Matthews, 2000). *H/E* is present in the plasticity index as defined by Greenwood and Williamson (Greenwood & Williamson, 1966):

$$\varphi = \frac{E}{H}\left(\frac{\sigma}{\beta}\right)^{\frac{1}{2}}$$

where σ is the surface roughness, and β is the asperity radius. The plasticity index is used as a measure of the limit of elastic behavior in the design of sliding and rolling element bearings, making it important in the minimization of wear (Leyland & Matthews, 2000). Contacts with small plasticity indices must have a larger contact stress to induce significant plastic deformation. At a constant surface roughness and asperity radius, *H/E* is inversely proportional to the plasticity index. Therefore, a coating with a larger *H/E* less likely to deform plastically at a given stress and will have a higher toughness where there is no plastic deformation (Chen et al., 2019). A dimensionless plasticity index (PI) also exists from the Cheng and Cheng relating the plastic work (W_p) and the elastic work (W_e) during indentation (Beake et al., 2009; Cheng & Cheng, 2004):

$$\mathrm{PI} = \frac{W_p}{(W_p + W_e)} = 1 - x\left(\frac{H}{E_r}\right)$$

where x is a constant.

The use of H^3/E^2 is derived from Hertzian contact mechanics. For a flat surface in elastic/plastic contact with a rigid radius (R), the yield pressure (P_y) is dependent upon H^3/E^2 (Beake, 2022a; Chen et al., 2019; Johnson, 1985):

$$P_y = 0.78 R^2 \left(\frac{H^3}{E^2}\right)$$

Therefore, at a given contact pressure, the contact is more likely to be elastic for a surface with high H^3/E^2. As such, H^3/E^2 can be used as a measure to plastic deformation or load-carrying capacity in a material.

Beyond these parameters, H^2/E has been explored by Eyre (1976, 1978) and Smart (1978) as a correlation between erosion resistance and the ultimate resilience of a material as the ratio relates to the amount of energy a material can absorb before cracking occurs.

H/E can also relate to frictional performance (Ni et al., 2004) by assuming that multiasperity contacts can be modeled locally by either conical or spherical indentations. The deformed surface around the indenter can exhibit pileup or sink-in dependent upon the mechanical properties of the surface and indenter geometry (Cheng & Cheng, 2004). For materials with a large ratio of yield strength over elastic modulus, sinking in is expected, and the tendency of sink in increases with H/E (Ni et al., 2004). As friction is widely considered to consist of two main components: an interfacial/adhesive component and plowing component (Beake et al., 2013); the plowing component will deform materials in the direction of sliding. The plowing force is expected to be smaller for sinking in than for piling up; therefore, the plowing contribution to friction is expected to decrease with increasing H/E (Ni et al., 2004), thereby reducing friction. However, this is a simplification. Surface interactions and geometric considerations will change the frictional response as the wear track is modified (Meine et al., 2002a, 2002b; Santner et al., 2006). Graphitization and transfer layer formation will also change the frictional response of DLCs throughout testing (Grill, 1993; Angus & Hayman, 1988; Blanpain et al., 1993; Chen, Yang, et al., 2011; Field et al., 2004; Liu et al., 1996, 2019; Ronkainen & Holmberg, 2008).

DLCs are well known for having low coefficients of friction and low wear. There is a range of performance achievable with carbonaceous coatings that depends upon their structure (the deposition of which is detailed in Section 18.3.1) and the environmental conditions. Humidity plays a role in the coefficient of friction that can be achieved.

At low humidity, hydrogen-free DLC has a relatively high friction coefficient of approximately 0.8. This high friction is due to carbon − carbon dangling bond interactions. Graphitization leads to the reduction of friction. If the humidity were to be increased, then the dangling bond interactions will change (the presence of water in the tribosystem), thereby lowering friction. Hydrogenated DLC has a lower coefficient of friction of approximately 0.15. This is due to the interaction of hydrogen-terminated dangling bonds. Increasing humidity for this DLC type will increase the coefficient of friction (Donnet & Erdemir, 2008).

The formation of a carbonaceous transfer layer is an essential component in the low-friction performance of DLC. The formation of this layer reduces the friction coefficient; a high normal load and high sliding speed can generate a thick and compact layer. The layer is graphitic in nature and has lower values of hardness and elastic modulus compared with the as-deposited coating (Grill, 1993; Field et al., 2004; Liu et al., 1996, 2019; Ronkainen & Holmberg, 2008). Nanoindentation mapping could be used to analyze the mechanical properties of the transformed structure (Fouvry et al., 2002). This can be correlated with the structural transformation determined by Raman spectroscopy as utilized by McMaster et al. (2020) and Shi et al. (2020).

Zahid et al. (2016) conducted a review of the intrinsic and extrinsic conditions affecting the tribological performance of diamond-like carbon. They found that ultralow friction can be achieved with hydrogenated DLCs; however, harder hydrogen-free DLCs are more wear-resistant. Adhesion increased interlayers such as chromium or titanium and increased the service life of the coating. By tuning the deposition parameters, the hardness of the coating can be adjusted by altering the sp^2/sp^3 ratio; however, care must be taken as higher hardness coatings can suffer from delamination and brittle fracture. Increased temperature (beyond 100°C) led to a decrease in the friction coefficient as the rate of graphitization increased. Surface modification such as modification or texturing can improve tribological performance, but better entraining lubricant and ensuring the contacts aren't starved. Finally, though additives such as ZDDP and MoDTC are designed for interaction with ferrous surfaces, they can also aid effective operation of DLCs.

The use of DLC coatings for electric vehicles will require the characterization of electrical properties alongside tribological behavior as in the work of Farfan-Cabrera et al. (2023). Additional wear mechanisms such as fluting, electrical pitting, tribo-oxidation and -corrosion, frosting, and corrugation may be caused by the electrostatic charge buildups due to triboelectrotrification. The effects of electrification on friction were marginal on uncoated steel pairs during dry sliding with a slight increase, but the opposite was observed with lubricated testing. The opposite effects were observed with DLC pairs. Electrification had an extreme effect on the wear of uncoated and DLC-coated pairs with a substantial in wear. H-free DLC experienced catastrophic wear due to the removal of coating from the sliding surfaces. Hydrogenated DLC fared much better with little change in the friction and wear behaviors, making it the most effective option in suppressing tribological failures in tribological elements subjected to electrical discharges in electric vehicles.

Farfan-Cabrera (2019) conducted a review of the tribology of electric vehicles in 2019 finding that coatings alongside efficient greases can be used to advance the efficiency of electric motor systems. The elements to be optimized are rolling bearings and brushes/sliding rings. In-wheel motor systems and multispeed transmission systems can be further improved by the application of optimized coatings pairing with efficient oils and surface texturing.

18.2.3 Lubricant interactions of diamond-like carbons

Though DLCs are often used in dry conditions due to their ability to graphitize and self-lubricate (Angus & Hayman, 1988; Blanpain et al., 1993; Chen, Yang, et al., 2011), it would be remiss of this chapter to exclude their known lubricant interactions. DLC has been found to have an adverse reaction with oils that contain MoDTC. Kosarieh et al. (2016) found that the formation of a MoDTC tribofilm reduced friction but increased wear in the presence of iron. Over longer time periods, this may cause issues for DLC parts in automotive engines.

De Barros Bouchet and coworkers (De Barros Bouchet et al., 2005; De Feo et al., 2016) have conducted multiple studies of the inclusion of MoDTC additives in the lubrication of DLC contacts. In their 2005 study (De Barros Bouchet et al., 2005), it was found that MoDTC and ZDDP react directly on amorphous carbon surfaces, being seemingly more active with selected hydrogenated amorphous carbon. MoS_2 was found to be formed in the contact area only. On amorphous carbon, the role of the ZDDP antiwear additive was to enhance the formation of MoS_2 sheets leading to lower friction. Tribofilms were of similar composition to that formed on steel, and the reactions to form them were able to occur with the iron catalyst. Their 2016 work, De Feo et al. (2016) further explored the surface chemical modification of lubricated DLC coatings in the presence of MoDTC additive. In boundary lubrication conditions, the MoDTC additive showed reduced friction overall, but the silicon-doped DLC had relatively high friction combined with a massive increase in wear. This work provides evidence that the formation of a molybdenum carbide compound is an important step in the wear mechanism of DLC coatings rubbing against a steel counterface. A further work by De Barros Bouchet et al. (2017) explored superlubricous behavior of ta-C pairs under oleic acid conditions. This reduction of friction was due to the triboformation of a graphene-like structure at the top surface (thickness lower than 1 nm). The formation mechanism is proposed to be the enrichment of the top surface of ta-C with sp^2 carbon due to relatively high shear stress leading to a preferential formation of the 2D structure at the top of colliding asperities. Finally, the graphene can be oxidized by the oleic acid decomposition.

Many other authors have analyzed the interaction of MoDTC with DLC films. Okubo and Sasaki (2017) used in situ Raman observations to observe the structural transformation of DLC finding that wear acceleration of a-C:H was present with a steel counterface. The I_D/I_G ratio of the coating decreased gradually over time indicating that it was not graphitization but transformation to other carbon compounds such as Mo-carbides. These carbides contribute to the wear as they are displaced into the lubricant solution. A later work by Okubo et al. (2019) further investigated the structural transformation of carbon films in lubricants containing MoDTC. Nanoindentation of the wear scar found that the tribofilm at the interface was much harder. This was due to the carburization of Mo and, as before, this contributed to additional abrasive wear. Wear acceleration was not present in harder ta-C films. ZDDP was found to inhibit the carburization of the Mo source.

Espejo et al. (2019) analyzed the differences between model lubricants and fully formulated engine oils using in-situ Raman microscopy. Wear was increased in steel/DLC systems for the steel counterface in comparison to steel/steel systems due to surface chemical interactions and hardness differences with the coating. The addition of MoDTC increased the wear of a-C:H coating systems similar to other studies. MoS_2 was formed as well as large carbonaceous deposits on the steel counterface. Once again, ta-C systems exhibited a low friction coefficient and little to no wear in MoDTC containing lubricant. This study identified the presence of an MoS_2 containing tribolayer on the DLC coating contributing to a low friction coefficient of below 0.04. With a fully formulated oil, MoDTC instead reacts with other additives and therefore does not produce an MoS_2 containing tribolayer. This reduced the excessive wear of a-C:H that would otherwise occur in the presence of MoDTC. Yoshida and Kunitsugu (2018) also studied the interaction of DLCs with MoDTC finding that higher sp^2/sp^3 coatings experienced a higher wear coefficient when immersed in MoDTC containing oil. The mechanism identified in this case was the oxidative cleavage of C—C π bonds due to interactions MoO π coatings experienced a higher wear coefficient when immersed in MoDTC containing oil. The mechanism identified in this case was the oxidative cleavage of C—C π bonds due to interactions with MoO_3.

The interaction of doped DLCs with MoDTC is also of interest due to the increased performance from the doping process. Bae et al. (2023) performed a tribochemical investigation of Cr-doped DLC interacting with MoDTC containing engine oil in boundary lubrication conditions. I_D/I_G ratio was increased with the introduction of Cr due to embedded CrC_x nanoparticles, which were able to act as catalysts promoting the formation of sp^2 sites. Friction is reduced due to the formation of MoS_2 in this doped coating also, but wear is still deleteriously affected with a maximum increase in wear rate of 14 times compared with non-MoDTC containing oil. MoO_3 was confirmed to be the cause by XPS analysis. The decrease in friction coefficient from standard DLC to chromium-doped DLC was attributed to the decreasing contribution of $MoS_{2-x}O_x$.

Yue et al. (2015) studied the effects of tungsten doping on the tribological behaviors of DLCs lubricated with oils containing MoDTC. Additive containing oils were able to reduce friction to a greater degree than base oils alone. The coefficient of friction was stable at 0.06, which was not affected by varying levels of W content. Graphitization and the formation of a sulfide containing tribofilm were the cause of this increased performance. Wear rate was, however, affected with MoDTC-containing lubricants experiencing higher wear due to MoO_x formation and graphitization. The lowest wear rate was reported with the W containing DLC with the higher *H/E* ratio.

Komori and Umehara (2017) performed a study on the tribological properties of tetrahedral silicon doped hydrogenated carbon under lubricated conditions with ZnDTP and MoDTC. The coatings had a relatively high hardness of 20–26 GPa due to a highly sp^3-bonded structure. Secondary ion mass spectroscopy analysis showed the presence of multiple Mo-oxysulfides ions and Mo-oxides on the worn tribosurfaces. A higher surface roughness Si containing DLC and its counterface contained a higher ratio of these ions. MoS_2 formation was promoted through the Si-containing DLC leading to a lower friction coefficient. A high thermal load test showed that compared with undoped hydrogenated DLC and hydrogen-free DLC, the Si-doped DLC had no significant wear and lower friction coefficient in oil containing MoDTC. A further study was performed in 2020 by Kassim et al. (2020) focusing on the MoDTC derived particle on Si-doped and hydrogenated DLC at room temperature. Wear acceleration as found to be due to Mo_2C formation from MoDTC degradation. MoO_3 was found in the lubricants; however, the wear acceleration was deemed to be due to a catalytic effect due to the hardness differences between the MoDTC (1 GPa) and the coatings (18 GPa for a-C:H and 25 GPa for Si-DLC).

The need for specialized formulation of lubricants of was highlighted in 2009, Jarry et al. (2009) identified the future development need for lubricant formulations specifically tailored for DLC coatings. Furthermore, lubricants must be crafted especially for electric vehicle applications. Cañellas et al. (2023) found that PAO as a base stock with additives such as esters and MoDTP had great efficacy. ZDDP was unactive without high temperature and high pressure. DLC lubricant compatibility was highlighted as part of the study of trimethylolpropane (TMP) esters as an additive in a PAO formulation containing glycerol mono-oleate (GMO), MoDTC, or ZDDP.

Khanmohammadi et al. (2022) investigated the ability of ionic liquids as friction modifiers in water-based lubricants for metal-doped DLCs on stainless steel substrates. Tungsten and silver doping reduced the hardness and elastic modulus of the coatings. The undoped DLC had the worst adhesion of all the tested coatings, which contributed to its failure in all tribological tests. Under lubricated conditions, W-doped DLC showed the lowest friction wherein the friction evolution was purported to be controlled by an electron transfer mechanism. The Ag-doped DLC possessed higher friction than WDLC, and the controlling mechanism was electrochemistry. Oxidation of the silver was revealed in the anodic branch of the polarization curve, and higher anodic potentials are required to activate the surface adsorption sites with ionic liquid additive lubricants.

18.3 Deposition of diamond-like carbon coatings

Both physical vapor deposition (PVD) and chemical vapor deposition (CVD) are used to deposit DLC coatings; however, it should be noted that the deposition method chosen will generate a different type of carbon film. The usage of hydrocarbon precursors introduces a certain level of hydrogenation into the DLC structure (Hainsworth & Uhure, 2007; Robertson, 2002). PECVD deposited films can have hydrogen contacts as high as 60% (Grill, 1993; Hainsworth & Uhure, 2007). The character of the film is affected by the ion energy (dependent on the bias voltage and precursor gas) of the species utilized in the coating process (Robertson, 1994, 2008a).

Zia and Birkett (2021) reviewed the deposition of diamond-like carbon coatings for conventional and nonconventional markets in 2021. They identified the need to explore further methods of DLC deposition to allow for the use of the coating in nonconventional environments where the deposition temperatures, vacuum requirements, or limitations in bulk-scale deposition may limit applications. Microplasma, dielectric barrier discharge, microwave resonators, and plasma guns are ideals methods to use for specialized purposes.

18.3.1 Common deposition methods

The common types of atomistic vaporization technology used to produce DLC films are radio-frequency (RF) or direct-current (DC) chemical vapor deposition, magnetron sputtering, vacuum arc, and ion beam deposition (Robertson, 2002).

The introduction of dopants into carbon structures will generally involve the use of additional deposition technologies besides the primary one chosen for the carbon film. As an example, if we were depositing a chromium-doped hydrogenated amorphous carbon (a-C:H:Cr), we would use a hydrocarbon as the precursor with plasma-enhanced chemical vapor deposition but could also utilize magnetron sputtering from a chromium target to codeposit chromium.

Interlayers are key to an optimal DLC coating. Interlayers are used to overcome DLC's high residual compressive stresses, leading to adhesion issues. The measurement of the residual internal stress is typically performed by measuring the curvature of the substrate materials by applying the Stoney equation (Stoney, 1909). A surface profilometer is generally the best device to employ for this task (Baba et al., 2020; Bonetti et al., 2006; Ren et al., 2015; Shi et al., 2020).

Metallic and ceramic interlayers are common in these coating structures. The common metals that are used as interlayers are Ti, Zr, W, and Cr. Ceramics such as WC or CrN are common. Silicon is also sometimes used as an interlayer material (Pauleau, 2008). Though single interlayers can be used, it's often more beneficial to adhesion to have several layers; for example, a Cr layer deposited by magnetron sputtering followed by either W and WC, also magnetron sputtered. Ti/TiN and Ti/TiN/TnCN are also commonly used and effective (Pauleau, 2008). In 2019, Tyagi et al. (2019) conducted a review of the deposition of DLC coatings across numerous academic publications noting the substrate, type of coating, deposition technique, deposition parameters, and layer structure.

The growth of coatings in atomistic deposition and the nature of said coatings are determined by an ion subplantation process. The range of ion energy (partially determined by the choice of precursor hydrocarbon in a chemical vapor deposition process) varies; however, an optimum range exists for maximum sp^3 content. The hydrogen content of an a-C:H film is always lower than the precursor hydrocarbon molecule as hydrogen is lost due to ion bombardment. This hydrogen is lost from C–H bonds forming H_2 molecules that leave the growing carbon network. The initial ion subplantation will generally create sp^3 sites at first. Following this, diffusion will occur toward a lower energy configuration. If an ion has enough energy, it will pass through the outer layer of the growing film and cause subsurface growth. This increases density in this local area. Robertson (2002) reviewed this process in detail, and it is recommended to consult his texts for further information.

For deposition of DLC coatings by commercial suppliers, it is recommended to speak to their technical team to advise on the surface finish requirements, geometric requirements, and mounting requirements for their process. Many of the coating structural parameters will not be able to given, but it's likely that the DLCs will have been tested to ASTM or BSI standards to report comparable coefficients of friction. Many commercial DLC suppliers will also produce a coating tailored to esthetic requirements, as such the process will be optimized to prioritize the surface finish rather than the mechanical properties or tribological properties.

18.3.2 Deposition of diamond-like carbons for this study

For research, if the coatings are produced commercially, then it's natural for knowledge of certain parameters to not be made available to the academics. Those coating parameters are the intellectual property of the supplier. It's common for research on industrial coatings (whether it be DLC or others) to reference the specific supplier of coatings as that allows for cross-referencing of known properties (Austin, 2014; Beake, 2022b; Bromark et al., 1995; Chen et al., 2016; Field et al., 2004; Yang et al., 2001; Zeng et al., 2002).

This case study will focus on the three DLCs (a-C:H, Si-a-C:H, and a-C:H:W). The substrates chosen for deposition were 316L stainless steel (SS) and hardenend

M2 tool steel (HTS). The notation used to simplify reference to these coatings is as follows:

- a-C:H (Coating A)
- Si-a-C:H (Coating B)
- a-C:H:W (Coating C)

Coating deposition was carried out with the Hauzer Flexicoat 850 physical vapor deposition (PVD) and plasma-assisted chemical vapor deposition (PACVD) system located in the University of Leeds. Substrates were polished to a roughness of 0.01 μm R_a. Table 18.1 shows a summary of all the deposition parameters of the DLC coatings utilized for the case study (McMaster, 2020; McMaster et al., 2020, 2023).

PVD was utilized to deposit the adhesive and gradient layers on the substrates (see Table 18.2). PACVD was used with an acetylene (C_2H_2) precursor for the top layer DLC. Hexamethyldisiloxane (HMDSO), vaporized in the chamber, was utilized to achieve the doping for Coating B. A WC magnetron sputtering target was engaged to provide the doping for Coating C. The same coating procedures were used for both substrates (McMaster, 2020; McMaster et al., 2020, 2023). Differences in the coating thickness can be attributed to the increased hardness of the tool steel (Table 18.3) benefitting coating growth as less ion subplantation will occur before true film growth (Lifshitz et al., 1995). The coating thicknesses of the top layers shown in Table 18.2 were found by calotesting and corroborated by FIB-SEM and TEM cross sections. Adhesion and gradient layer thicknesses were determined by FIB-SEM and TEM cross sections alone (McMaster, 2020).

Fig. 18.1 shows the FIB-SEM cross sections of unworn areas of the coatings. This cross-sectional SEM was obtained utilizing an FEI Helios G4 CX Dualbeam SEM in LEMAS (Leeds Electron Microscopy and Spectroscopy Centre). Platinum was deposited on the surface of the same to protect it during the ion-beam milling operation. The sample was tilted, and a beam of gallium (Ga) ions was used to mill into the surface to reveal the microstructure of the coating. An Oxford Instruments Aztec electron diffraction X-ray spectroscopy (EDX) system was used to identify the composition of the cross section (McMaster, 2020; McMaster et al., 2020). Any globular structures in the foreground of the images are the result of redeposition of material from the ion beam. In Coating B, we can see Mo and V concentrations at the surface of the steel and proceeding into the microstructure. We can see the distinct layer structures of each coating, of note is the gradient layer transitioning into the top layer of the DLC in Coating C.

18.4 Mechanical and structural characterization

Typically, mechanical characterization of DLC coatings (or any thin film) would be to the ISO14577 standard (Metallic materials − Instrumented indentation test for hardness & materials parameters − Part 1: Test method, 2015), but this is not the

Table 18.1 Coating deposition parameters.

Deposition step/ conditions	Temperature (°C)	Pressure (× 10^{-5} mbar)	Bias voltage (V)	Cr target power (kW)	WC target power (kW)	Ar flow rate (sccm)	C$_2$H$_2$ flow rate (sccm)	HMDSO flow rate (sccm)	Table rotation speed (rpm)	Time (mins)
Chamber heating	200	4	–	–	–	–	–	–	1	60
Target cleaning	–	–	500 (DC)	6	3	130	–	–	2	20
Plasma surface etching	150	–	200 (PLS low)	–	–	50	–	–	2	45
Cr deposition	–	–	–	3	–	130	–	–	3	25
Cr/WC deposition	–	–	–	3	3	110	–	–	3	30
a-C:H:W deposition	–	–	–	–	3	90	8–30 (30 mins)	–	3	75 (A&B) 120 (C)
a-C:H deposition	–	–	740 (PLS high)	–	–	–	380–270 (8 mins)	–	1.5	150
Si-a-C:H deposition	–	–	740 (PLS high)	–	–	–	200–120 (8 mins)	18–12 (8 mins)	1.5	120

Source: From McMaster, S. J. (2020). Nanomechanical characterisation of diamond-like carbon coatings for tribological performance. https://doi.org/10.13140/RG.2.2.13315.02083 (Original work published 2020); McMaster, S. J., Kosarieh, S., Liskiewicz, T. W., Neville, A., & Beake, B. D. (2023). Utilising H/E to predict fretting wear performance of DLC coating systems. *Tribology International, 185*. https://doi.org/10.1016/j.triboint.2023.108524.

Table 18.2 Coating architecture and layer thicknesses.

Substrate	DLC Recipe	Substrate roughness (μm R_a)	Adhesion (Cr) layer (μm)	Gradient layer (μm)	DLC layer (μm)	Total coating thickness (μm)
Stainless Steel (316L)	a-C:H (A)	0.026 ± 0.004	0.29 ± 0.03	0.89 ± 0.08	1.60 ± 0.17	2.78 ± 0.19
	Si:a-C:H (B)	0.026 ± 0.004		0.89 ± 0.08	1.16 ± 0.24	2.34 ± 0.25
	a-C:H:W (C)	0.026 ± 0.004		0.29 ± 0.06	1.10 ± 0.22	1.62 ± 0.23
Hardened tool steel (M2)	a-C:H (A)	0.026 ± 0.004	0.29 ± 0.03	0.89 ± 0.08	2.20 ± 0.20	3.38 ± 0.21
	Si:a-C:H (B)	0.026 ± 0.004		0.89 ± 0.08	2.17 ± 0.16	3.35 ± 0.18
	a-C:H:W (C)	0.026 ± 0.004		0.29 ± 0.06	1.17 ± 0.12	1.75 ± 0.13

Source: From McMaster, S. J. (2020). Nanomechanical characterisation of diamond-like carbon coatings for tribological performance. https://doi.org/10.13140/RG.2.2.13315.02083 (Original work published 2020); McMaster, S. J., Kosarieh, S., Liskiewicz, T. W., Neville, A., & Beake, B. D. (2023). Utilising H/E to predict fretting wear performance of DLC coating systems. *Tribology International*, 185. https://doi.org/10.1016/j.triboint.2023.108524.

Table 18.3 Mechanical properties of coatings and substrates. Single indentations were used to analyse the substrates and partial load-unload indentations was used for coatings.

Substrate	DLC recipe	DLC coating thickness (μm)	Surface roughness (nm R_a)	H (GPa)	E (GPa)	H/E	H^3/E^2 (GPa)
Stainless Steel (316 L)	Uncoated	–	26 ± 4	2.6 ± 0.1	223 ± 5	0.01 ± 0.001	0.0004 ± 0.00003
	A	1.60 ± 0.17	22.2 ± 9.4	19.4	215	0.090	0.158
	B	1.16 ± 0.24	25.7 ± 1.5	18.3	187	0.098	0.176
	C	1.10 ± 0.22	40.5 ± 8.0	16.2	235	0.069	0.077
Hardened tool steel (M2)	Uncoated	–	26 ± 4	10.0 ± 0.3	239 ± 7	0.042 ± 0.002	0.017 ± 0.001
	A	2.20 ± 0.20	26.0 ± 10.3	20.2	199	0.101	0.207
	B	2.17 ± 0.16	48.0 ± 12.4	13.1	164	0.080	0.083
	C	1.17 ± 0.12	26.2 ± 1.9	13.9	218	0.064	0.056

Source: From McMaster, S. J. (2020). Nanomechanical characterisation of diamond-like carbon coatings for tribological performance. https://doi.org/10.13140/RG.2.2.13315.02083 (Original work published 2020); McMaster, S. J., Kosarieh, S., Liskiewicz, T. W., Neville, A., & Beake, B. D. (2023). Utilising H/E to predict fretting wear performance of DLC coating systems. *Tribology International, 185*. https://doi.org/10.1016/j.triboint.2023.108524.

only option for coated structures. If we were to indent to the ISO standards to determine the mechanical properties of the coating alone, we would need to determine the thickness of the coating. This could be by either calotesting, cross-sectional electron microscopy, magnetic flux, eddy current or capacitance measurements, and ultrasonic measurements (Holmberg et al., 1998). Additionally, the surface roughness is important in following the ISO standard as the ideal R_a value should be less than 5% of the maximum penetration depth (Metallic materials − Instrumented indentation test for hardness & materials parameters − Part 1: Test method, 2015). For real surfaces where the surface roughness cannot be changed, performing a large number of indentations and performing statistical analysis are also viable.

Nanoindentation following the ISO 14577 standard was used to examine the mechanical properties of DLC. A depth-controlled indentation with a maximum of 10% of coating thickness was used with the depth set for each coating. A maximum of 100 mN was allowed to reach this depth. An initial contact load of 0.01 mN was used with a loading and unloading rate of 0.2 mN s^{-1}. A dwell period of 60 s was used for thermal drift correction postindentation. Typically, a large number of indents would be utilized for a testing methodology such as this (≥ 100). Each indent will generate a single data value of hardness and elastic modulus requiring an indentation grid to be programmed. This, however, can be useful to map mechanical properties across a sample surface. For crystalline materials, this can allow for the correlation of mechanical properties with grain structure or phase (Hu et al., 2017). Liskiewicz et al. (2017) also explored nanomechanical property mapping to measure the tribologically transformed layers in fretting wear scars.

The area function of the indenter was found by indentation into a fused silica reference sample. Hardness (H) and elastic modulus (E) were calculated by applying Oliver-Pharranalysis (Oliver & Pharr, 1992; Pharr & Oliver, 1992). E and η (0.2) are Young's modulus and Poisson's ratio for the coating. The Poisson's ratio of 0.2 was taken as an approximation from the collected values of Hainsworth and Uhure (2007). E_i (1140 GPa) and η η_i (0.07) are the same quantities for diamond, respectively (Hainsworth & Uhure, 2007; Oliver & Pharr, 1992).

Fig. 18.2 shows a model ISO standard single indent on Coating A. The contact depth can be seen to not exceed 170 nm to adhere to the specification of indents being no more than 10% of the total coating thickness. A pop-in event can be seen to occur at approximately 150 nm of contact depth. The hardness measured for the coating using this indentation method was 21.9 ± 2.3 GPa compared with 19.4 GPa for the partial load − unload method (Table 18.3).

18.4.1 Partial-load indentation

Partial loading nanoindentation under load control was carried out using a Nanotest Vantage nanoindentation system (Micro Materials, UK) with a Berkovich diamond indenter. A total of 10 indentations with 40 loading points, in a range of 0−500 mN, per sample, were used to characterize the change in mechanical properties with depth in the coatingsystem. A time of 2 s was used for the loading and unloading. A dwell time of 1 s was used at maximum load. Thermal drift correction

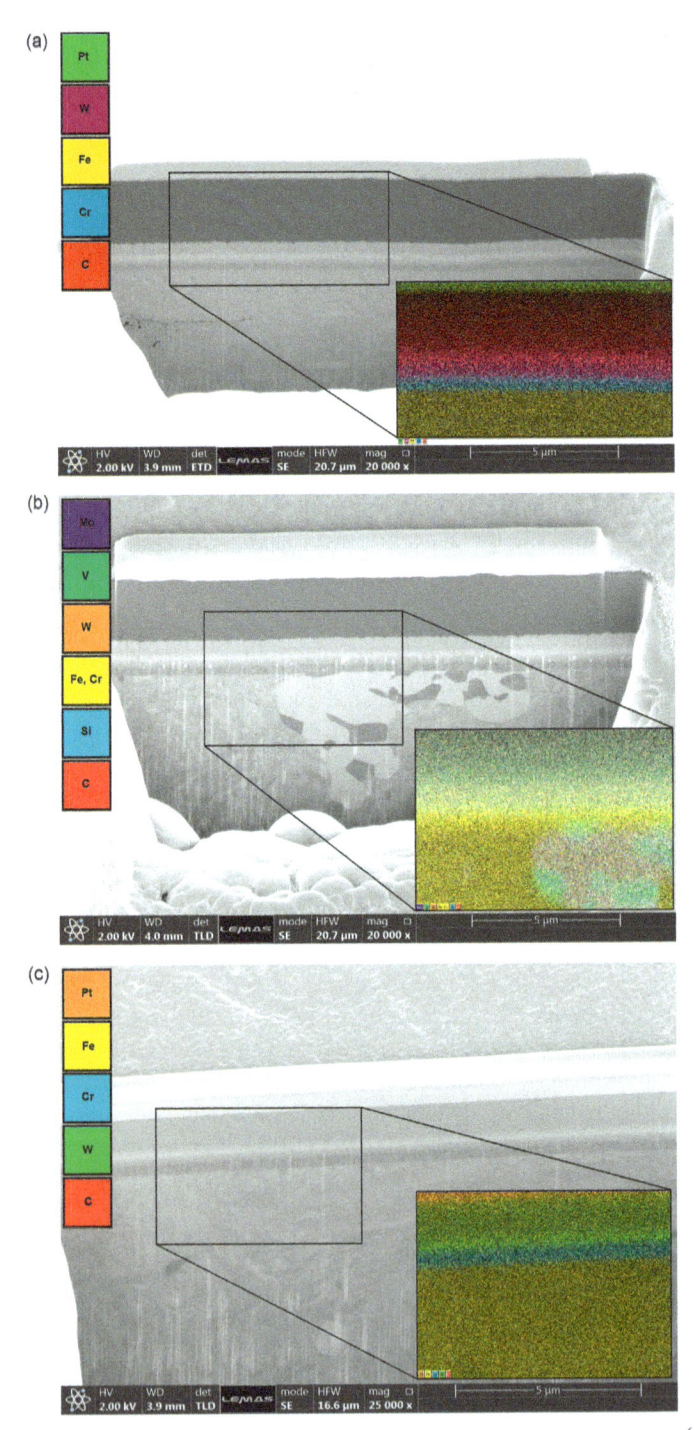

(*Continued*)

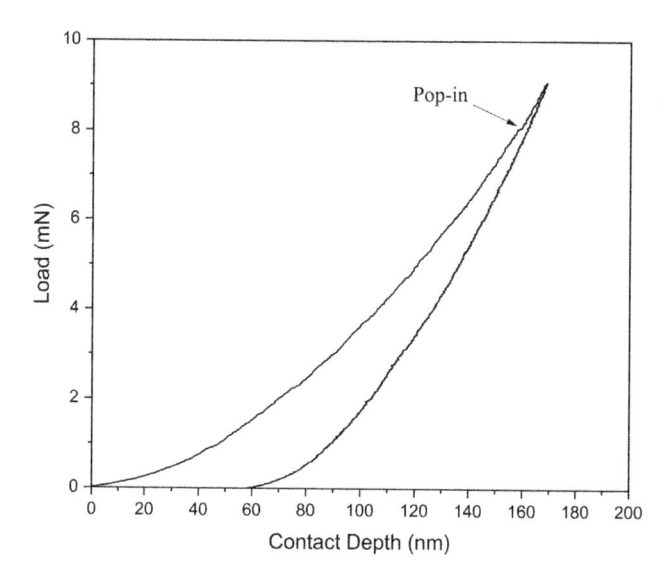

Figure 18.2 Load against depth hysteresis for Coating A on 316L stainless steel performed according to the ISO 14577 standard.
Source: From McMaster, S. J. (2020). Nanomechanical characterisation of diamond-like carbon coatings for tribological performance. https://doi.org/10.13140/RG.2.2.13315.02083 (Original work published 2020).

was performed by a 60 s hold in the final unload step. The indenter contact velocity was set to 0.50 μM s^{-1}. The same analysis parameters for E are used as mentioned above.

Fig. 18.3 shows a model partial load − unload on Coating A. The same broad hysteresis pattern is seen as in Fig. 18.2; however, the "load" section of the graph consists of multiple load and unload steps. Oliver − Pharr analysis (Oliver & Pharr, 1992; Pharr & Oliver, 1992) is performed on each of these steps to obtain the mechanical properties at the maximum contact depth reached for that individual step. The thermal drift correction in the final unload portion of the graph can be seen.

18.4.2 *Multilayer structures and their mechanical properties*

Partial load − unload indentation allows for the characterization of mechanical properties with respect to depth as seen in Fig. 18.4; however, the analysis is more

Figure 18.1 Cross section of coatings (a) A, (b) B, and (c) C on hardened tool steel. Insert is an EDX map of the exposed cross section.
Source: From McMaster, S. J. (2020). Nanomechanical characterisation of diamond-like carbon coatings for tribological performance. https://doi.org/10.13140/RG.2.2.13315.02083 (Original work published 2020); McMaster, S. J., Kosarieh, S., Liskiewicz, T. W., Neville, A., & Beake, B. D. (2023). Utilising H/E to predict fretting wear performance of DLC coating systems. *Tribology International*, 185. https://doi.org/10.1016/j.triboint.2023.108524.

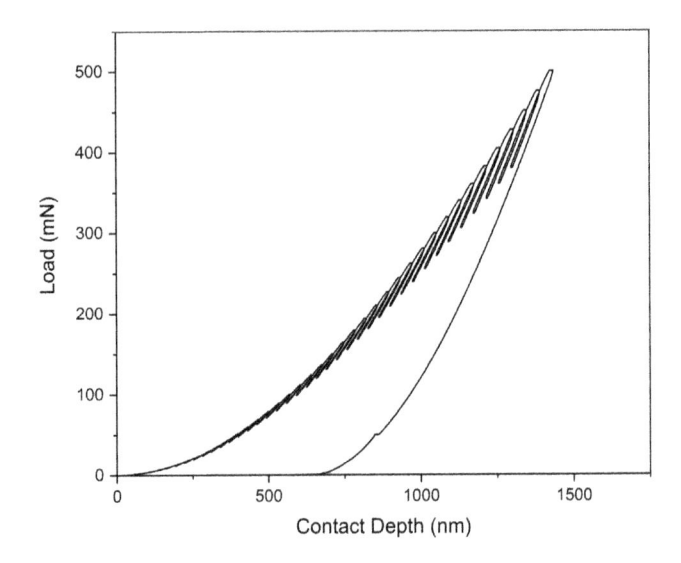

Figure 18.3 Load against contact depth hysteresis graph for coating A on 316L stainless steel using the partial load − unload method.
Source: From McMaster, S. J. (2020). Nanomechanical characterisation of diamond-like carbon coatings for tribological performance. https://doi.org/10.13140/RG.2.2.13315.02083 (Original work published 2020).

complex. Instead of simply reading one hardness or elastic modulus value per indentation, extrapolations must be used. Hardness (H) is found by extrapolating the maximum hardness to the y-axis to give the surface hardness (Fischer-Cripps, 2006). The elastic modulus (E) is determined by taking the mean of the maximum range to negate the surface contact effects reducing modulus at low contact depths (Fischer-Cripps, 2006).

The differences in the progression of hardness with depth allow us to clearly see the difference in load support provided by the HTS substrate as well as the overall lower difference in surface hardness of the coating compared with the substrate underneath (McMaster, 2020).

The average mechanical properties of the substrates and coatings are shown in Table 18.3. As multiple indentations were used for the substrates, we can report an error in the hardness and elastic modulus values. The extrapolation method leaves us without an error value to report. The HTS substrate is much harder with a value of 10 GPa compared with 2.6 GPa. They have similar values of elastic modulus. Overall coating systems, Coating A is hardest (19−20 GPa); however, the highest elastic modulus belongs to Coating C (218−235 GPa). For the HTS substrate, H/E is highest (0.101) with Coating A; however, with SS the similar H values and lower E result in it having a higher H/E. H^3/E^2 follows the same trend as H/E. The results are typical of coatings of this type as per the work of Hainsworth and Uhure (2007), Lanigan et al. (2016), and Yue et al. (2015).

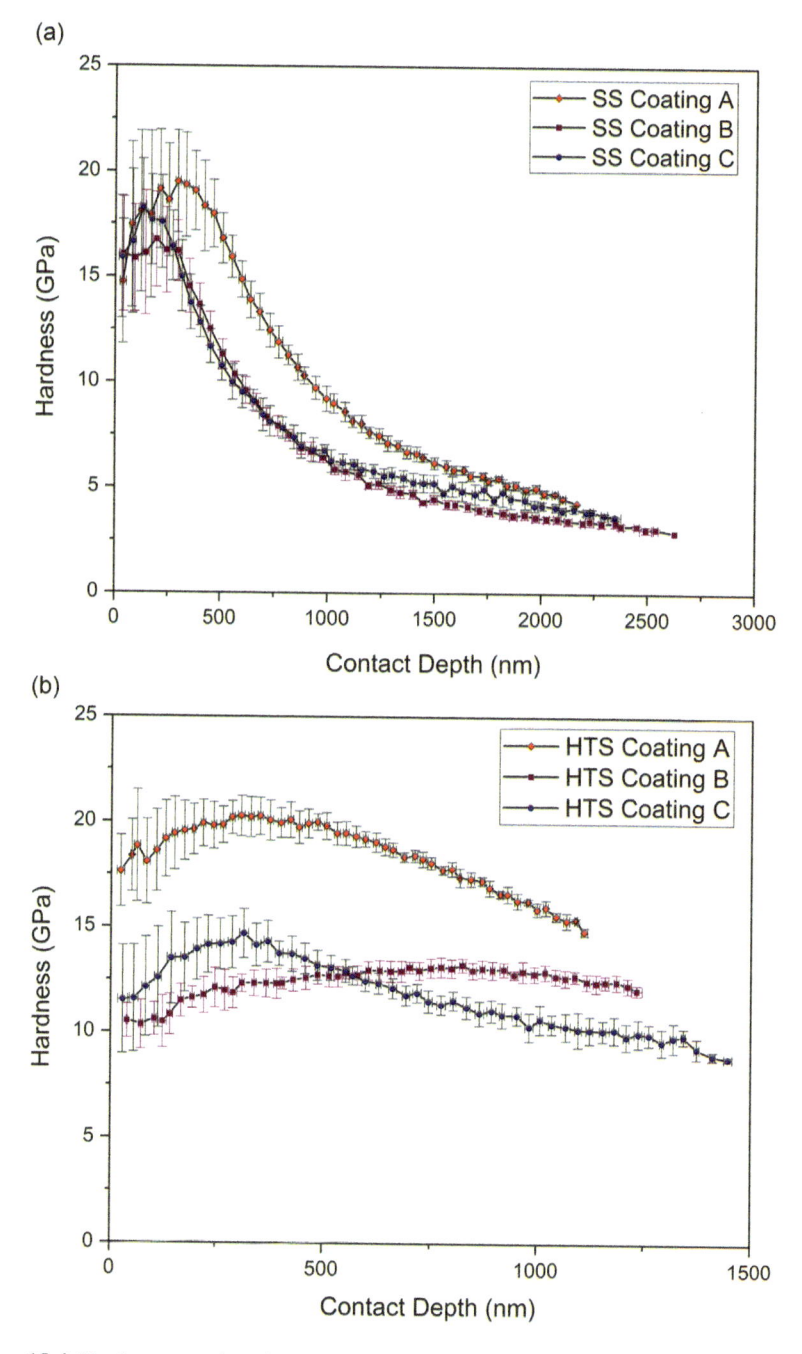

Figure 18.4 Hardness as a function of depth for Coatings A, B and C on (a) 316L stainless steel and (b) hardened M2 tool steel.
Source: From McMaster, S. J. (2020). Nanomechanical characterisation of diamond-like carbon coatings for tribological performance. https://doi.org/10.13140/RG.2.2.13315.02083 (Original work published 2020).

Surface roughness measurements were made using a topographic scan on the NanoTest platform. A 4.5 μm radius spheroconical probe was used with a 200 μm scan distance. The probe velocity was 10 μm s^{-1}. The load applied during the topographic scan was 0.1 mN, and a total of five scans were performed each with a spacing of 500 μm from the previous scan (McMaster, 2020; McMaster et al., 2023).

18.4.3 Structural characterization of doped diamond-like carbons

Structural characterization by Raman spectroscopy allows us to measure the level of amorphization in coatings (via the I_D/I_G ratio) (Filik et al., 2003; Okubo & Sasaki, 2017; Tai et al., 2006). Electron energy loss spectroscopy (EELS) carried out in a transmission electron microscope (TEM) allows for direct measurement of sp^2/sp^3 ratio (Arenal, 2017). X-ray photoelectron spectroscopy (XPS) is another method that can be used to analyze the sp^2/sp^3 ratios of DLC coatings (Filik et al., 2003; Lomon et al., 2018; Tai et al., 2006).

18.4.3.1 Raman spectroscopy

A Renishaw Raman microscope was used with a 488 nm laser targeting a Raman shift range of 1000–2000 cm^{-1}. In this frequency range, the D peak is around 1350 cm^{-1}, and the G peak is around 1580–1600 cm^{-1} (Ferrari & Robertson, 2000; Rincón et al., 2001). A Gaussian fitting with the ratios of the peak area was used for these spectra as opposed to the Lorentzian with full width half maxima (Ferrari & Robertson, 2000). Therefore, we refer to the intensities of the D and G peaks (areas beneath each peak) as I_D and I_G, respectively. Baseline subtraction and peak fitting were performed using OriginPro.

Table 18.4 displays the relative levels of amorphization of the coatings as well as D and G peak positions.

The lowest levels of amorphization are found on Coating B on both substrates (lowest level of I_D/I_G) indicating the lowest levels of sp^2 carbon present in rings (Ferrari, 2002). The highest amorphization is found in Coating C attributed to a high level of sp^2 carbon (Yong et al., 2016). These properties relate to the mechanical properties as high amorphization due to high sp^2 content leads to lower hardness (Table 18.3). G and D peak positions can also tell us about the structural transformation of the coating due to the introduction of dopants. An increase in G peak wavenumber indicates increased chains and clustering, which is typical of an increase in sp^2 bonding (Ferrari & Robertson, 2000). G peak position inversely correlates with hydrogen content in a-C:H (Schwan et al., 1996).

18.4.3.2 Electron energy loss spectroscopy

An FEI Titan Themis TEM was used for the EELS results generation (Fig. 18.5). The proportions of sp^2 and sp^3 carbon were found from the carbon K edge. The sp^2 fraction is found by calculating the area of a Gaussian fitting of the 285 eV peak (p states) and

Table 18.4 Raman spectroscopy characterisation of diamond-like carbon coatings A, B and C on 316L and hardened M2 tool steel substrates.

Substrate	Coating	I_D/I_G	D peak position (cm^{-1})	G peak position (cm^{-1})
316L stainless steel (SS)	A	0.79	1367.68	1552.40
	B	0.68	1353.94	1504.84
	C	3.5	1391.95	1567.04
Hardened M2 tool steel (HTS)	A	0.48	1344.21	1545.03
	B	0.29	1444.26	1505.92
	C	2.96	1392.26	1570.06

Source: From McMaster, S. J. (2020). Nanomechanical characterisation of diamond-like carbon coatings for tribological performance. https://doi.org/10.13140/RG.2.2.13315.02083 (Original work published 2020); McMaster, S. J., Kosarieh, S., Liskiewicz, T. W., Neville, A., & Beake, B. D. (2023). Utilising H/E to predict fretting wear performance of DLC coating systems. *Tribology International, 185.* https://doi.org/10.1016/j.triboint.2023.108524.

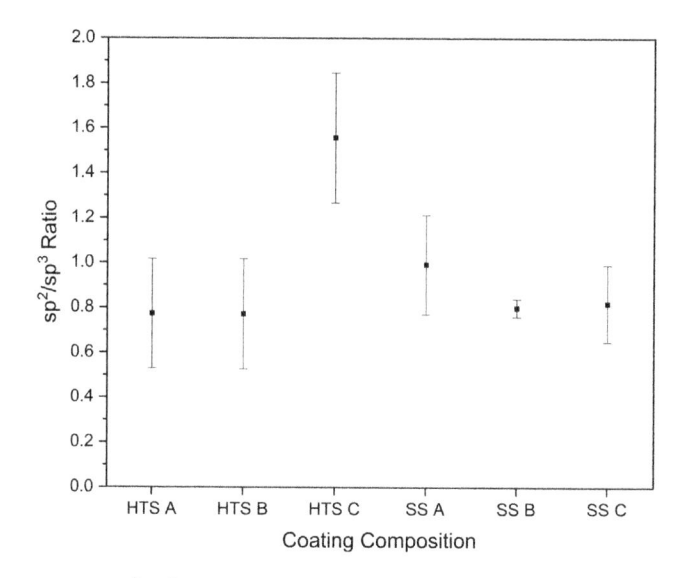

Figure 18.5 Average sp^2/sp^3 ratios of diamond-like carbon coatings.
Source: From McMaster, S. J. (2020). Nanomechanical characterisation of diamond-like carbon coatings for tribological performance. https://doi.org/10.13140/RG.2.2.13315.02083 (Original work published 2020).

290 eV due to the *s* states. This was compared with highly orientated polycrystalline graphite (HOPG), which is 100% sp^2 (Daniels et al., 2007; Ferrari et al., 2000). EELS data processing was performed with Gatan GMS 3 software.

The highest level of sp^2 bonding was found in Coating C on HTS. The others do not vary to a great degree. Hainsworth and Uhure (2007) note that for a hard a-C:H around 40% sp^3 content is expected. For W doping, this rises to 50%. It is typical for Si DLCs to be in the range of 60%–84%. The softening of Coating C (Table 18.3) is partly explainable by the differences in the proportions of carbon bonding in the coating but also by the presence of the doping element.

18.5 Scratch testing for adhesion testing of diamond-like carbons

Scratch testing for the determination of adhesion was used by Friedrich Mohs for ranking of material hardness (Bhushan, 1999; Fischer-Cripps, 2004). It has now progressed into a more analytical study of failure phenomena in progressive loading tests, which can inform on interfacial stresses using measurement methods such as acoustic emission (Bull, 1991; Burnett & Rickerby, 1987; Valli et al., 1985, 1986).

18.5.1 Macroscratch

Scratch crack propagation resistance ($L_{C1}(L_{C2} - L_{C1})$) (sometimes referred to as scratch toughness) is a useful term for analyzing the critical load failures of progressive loading scratch tests. It represents the resistance to crack initiation and propagation throughout the scratch test thereby giving a metric for the toughness of the film (Fox-Rabinovich et al., 2006; Zhang, Sun, et al., 2004). This parameter is, however, dependent upon coating thickness and radius of the indenter used; therefore, it can only compare like-for-like testing conditions.

Progressive loading scratch testing was performed with a Tribotechnic Millennium 200 scratch tester to assess the coating adhesion. A load of 0–50 N with a loading speed of 100 N min^{-1} and a scratching speed of 10 mm min^{-1} were applied. A 200 μm radius Rockwell C diamond indenter was used for testing. L_{C1} and L_{C2} failure points were analyzed. The larger-scale scratch test was performed to assess the cracking resistance of the total coating structure. Fig. 18.6 shows microscopy image (taken with the Keyence VHX-6000) of a typical macroscratch with the critical load positions annotated. Spallation is seen at L_{C1} with chevron cracking appearing before and continuing after indicating a brittle failure. Tensile

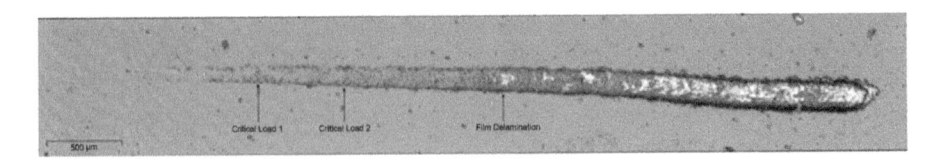

Figure 18.6 Optical micrograph of a macroscratch of diamond-like carbon A on SS. Critical loads 1 and 2 and film delamination are annotated.
Source: From McMaster, S. J. (2020). Nanomechanical characterisation of diamond-like carbon coatings for tribological performance. https://doi.org/10.13140/RG.2.2.13315.02083 (Original work published 2020); McMaster, S. J., Kosarieh, S., Liskiewicz, T. W., Neville, A., & Beake, B. D. (2023). Utilising H/E to predict fretting wear performance of DLC coating systems. *Tribology International, 185*. https://doi.org/10.1016/j.triboint.2023.108524.

cracking is seen at L_{C2}. Gross spallation is seen near the maximum load of the scratch (Bull, 1991; Burnett & Rickerby, 1987). Lateral stiffness of the scratch test platform is seen to be lacking at higher loads with the track drift observed.

Fig. 18.7 shows critical loads and calculated scratch crack propagation (CPR$_S$) resistances of all coatings on both substrates. Generally, HTS has a higher CPR$_S$ level across all coatings. Despite having a thinner top layer than the other coatings (Table 18.2), Coating C is very well adhered with high CPR$_S$ values, especially on SS where it is the highest with a value of 241 N^2.

Coating adhesion follows a similar trend as reported by Hasselbruch et al. (2015) on almost identical coatings. The excellent adhesion of coating C can be attributed to its low H^3/E^2 value (Table 18.3), allowing it to deform more elastically prior to L_{C2} failure. Zhang and coworkers (Wang & Zhang, 2014; Zhang, Bui, Fu, Butler, et al., 2004; Zhang, Bui, Fu, & Du, 2004) showed that increased scratch toughness and adhesion were at the expense of hardness and that reduced compressive residual stresses improved adhesion.

18.5.2 Nanoscratch

Nanoscratch testing allows for greater analytical measurements of the critical load failures of nanometer-scale films and the isolation of adhesion testing to the upper portions of coatings. Smaller proves can be used to increase maximum Hertzian contact pressure and induce film failure (Beake et al., 2015). Furthermore, the friction force during scratch testing can be deconvoluted into interfacial and plowing components to analyze the fundamentals of the interaction between surfaces (Beake et al., 2013). This case study will, however, focus on the determination of adhesion. As of 2021, the nanoscratch testing method has been standardized (NANOINDENT-PLUS project Standardising the nano-scratch test, NMP-2012-CSA-6-319208. at, 2012; CEN/TS. TS CEN/TS 17629, 2021). It should be noted that, as the length scale is different, the critical loads and scratch crack propagation resistances cannot be compared between macro- and nanoscales.

The nanoscratch module consists of the normal loading head as part of the NanoTest Platform with the addition of a friction probe (in the form of a Wheatstone Bridge strain gage). The typical methodology for this type of testing is a three-pass technique known as topography − scratch − topography (Beake et al., 2013; Beake, Shi, et al., 2011). In the topography steps, an extremely low load is used to ensure that no surface deformation takes place. Constant load scratch or progressive load scratches can be performed depending on whether cracking/topography changes, or critical load failures are to be investigated.

A probe scanning velocity of 10 µm s^{-1} and scan length of 1000 µm were used with all passes. The topography steps use a load of 0.10 mN. The scratch step had a maximum load of 500 mN applied after 200 µm with a loading rate of 8 mN s^{-1}. A total scan length of 1000 µm was used. Five repeated scratches were performed with a separation of 200 µm. Fig. 18.8 shows the evolution of coefficient of friction (COF) with applied load for all coatings on the HTS substrate.

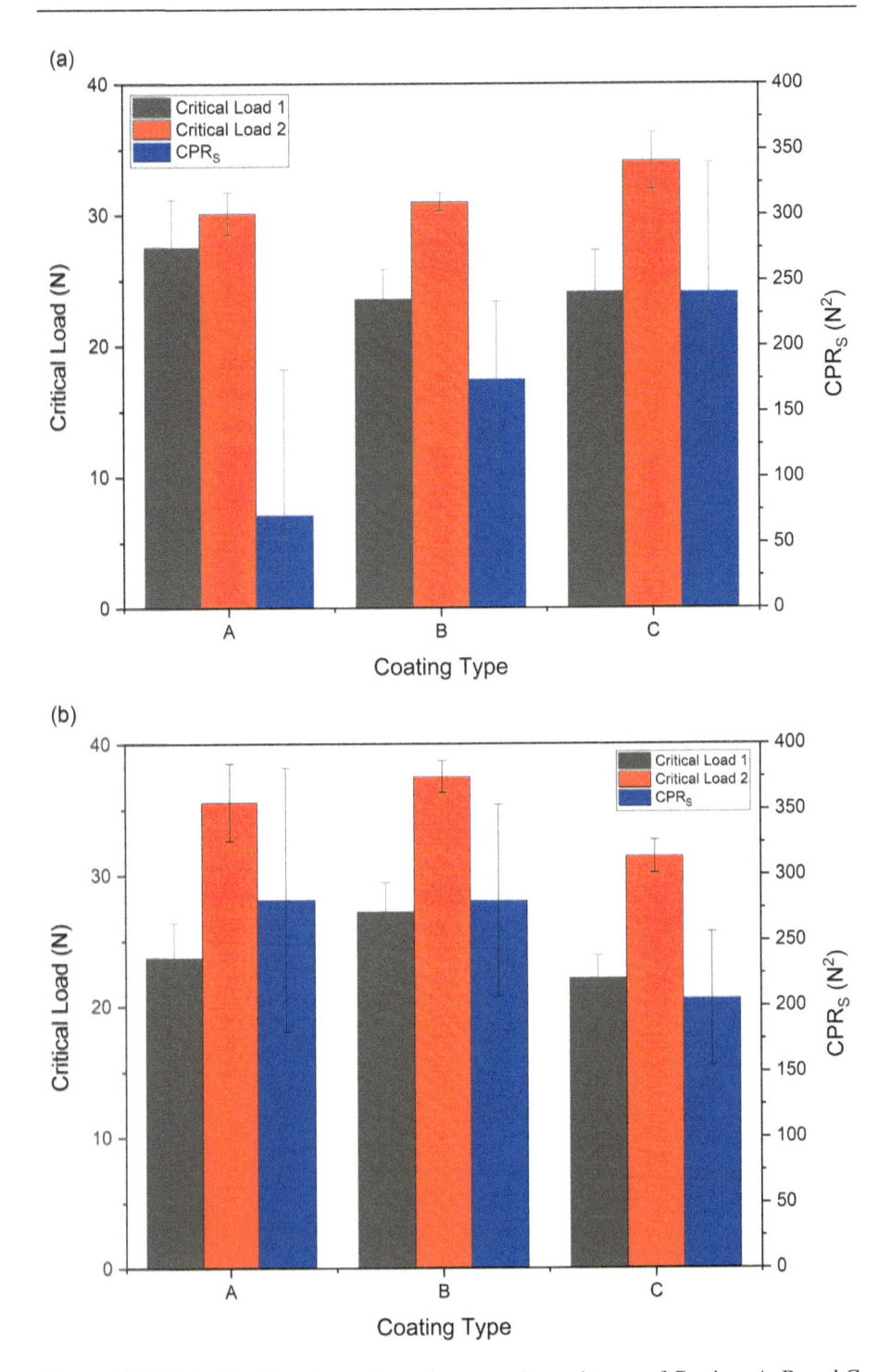

Figure 18.7 Critical Loads and scratch crack propagation resistance of Coatings A, B, and C on (a) 316L SS and (b) HTS determined via macroscratch testing.

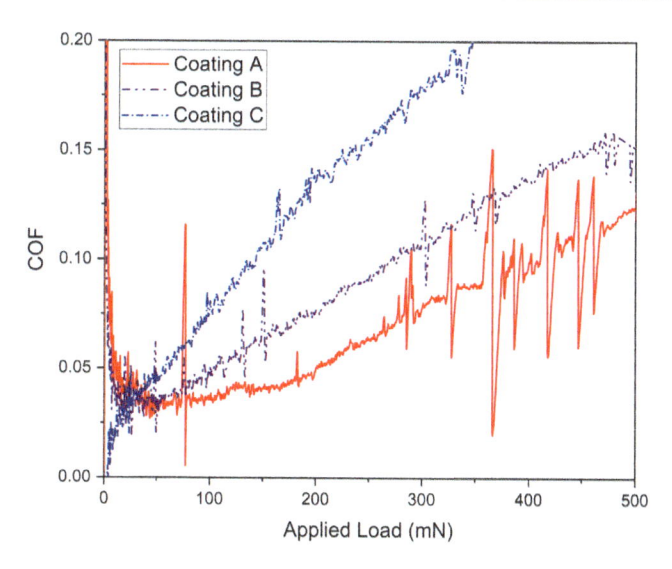

Figure 18.8 Friction coefficient against applied load for diamond-like carbons A, B, and C on hardened M2 tool steel during nanoscratch testing.
Source: From McMaster, S. J. (2020). Nanomechanical characterisation of diamond-like carbon coatings for tribological performance. https://doi.org/10.13140/RG.2.2.13315.02083 (Original work published 2020).

As well as the cracking observed through microscopy, spikes in the coefficient of friction (as in the trace of Coating A in Fig. 18.8) and rapid changes in depth can be used to precisely indicate film failure.

Fig. 18.9 shows the yield loads, critical loads, and calculated CPR_S values for Coatings A, B, and C on both substrates. The highest CPR_S value was with Coating A on SS (0.049 N^2) while the lowest was on Coating B (0.025 N^2). Though at a different length scale, Coating C recorded relatively high CPR_S values, thereby giving it good adhesion (Beake et al., 2015; Wang & Zhang, 2014; Zhang, Bui, Fu, Butler, et al., 2004; Zhang, Bui, Fu, & Du, 2004). The lack of failures on Coatings A and B on HTS is evidence of a higher contact pressure requirement to induce film failure (Beake et al., 2015).

This is similar to the work of Beake et al. (2015), wherein failure could not be observed with nanoscratch testing, and high load (up to 5 N) microscratch testing had to be used with a larger probe (radius of 25 μm). The location of critical yield stress is not at a point where decohesion or cracking would occur. Though a sharper probe could be used, these tend to be more sensitive to blunting and could inhibit the application of Hertzian analysis for yield stress calculations.

18.6 Fretting testing of diamond-like carbons

Fretting wear is small-amplitude oscillatory wear between two surfaces in contact, generally under vibration (Waterhouse, 1976, 1984). This study was performed to

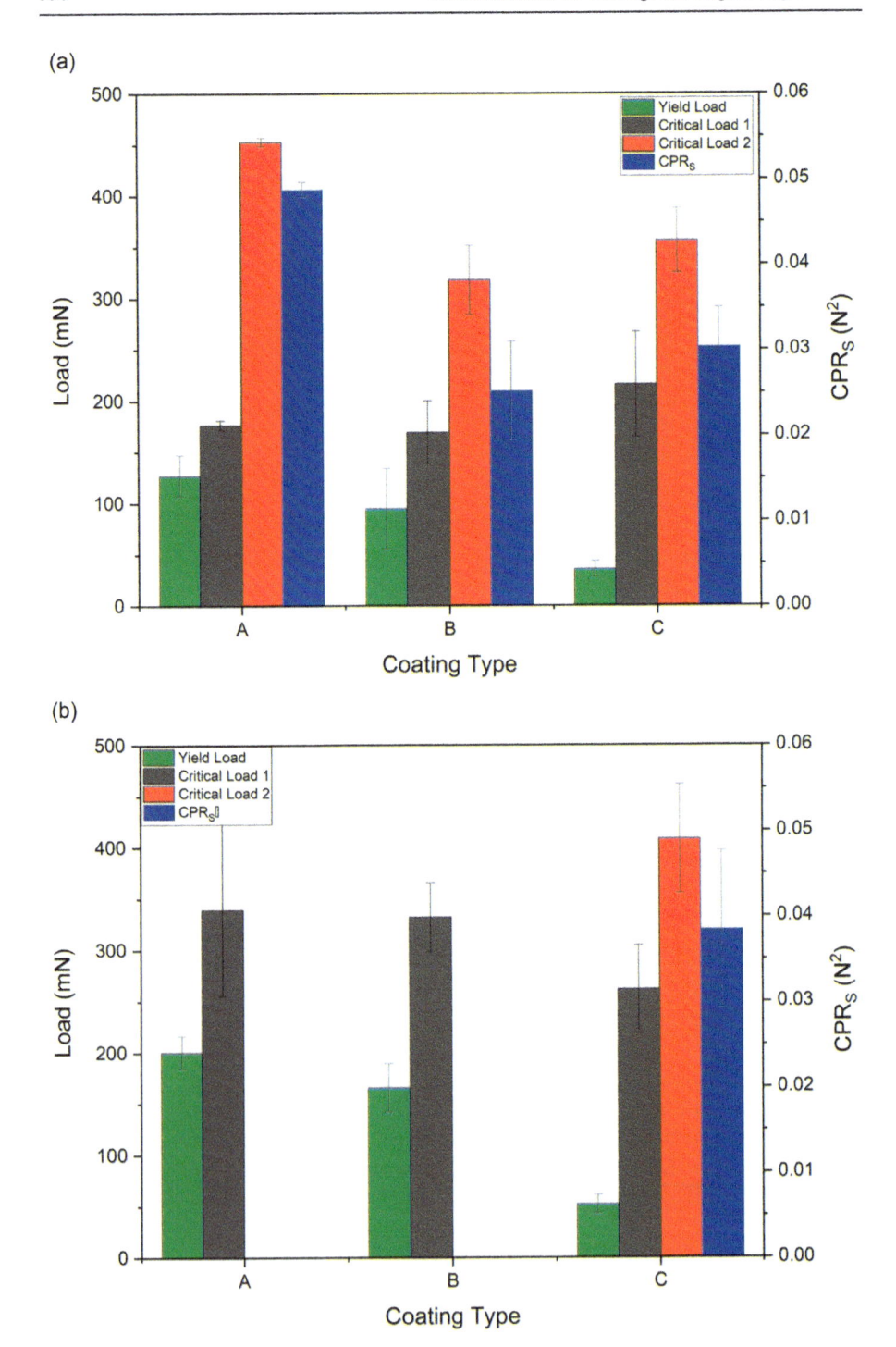

Figure 18.9 Critical loads and scratch crack propagation resistance of Coatings A, B, and C on (a) 316L SS and (b) HTS determined via nanoscratch testing.

assess the capacity to use *H/E* ratio to predict both frictional and wear performance of DLC coating in gross slip fretting. Liskiewicz et al. (2005) found that, for hard PVD coatings, a lower Young's modulus (corresponding to higher *H/E*) promoted low wear. The application of coatings to protect against fretting damage is widely used; coatings with high residual compressive stresses (such as DLC) can be particularly useful as this provides protection against cracking (Fouvry et al., 2006).

Fretting tests were performed with a bespoke electrodynamic shaker driven fretting rig as shown in (Wade et al., 2020). The system is controlled with a bespoke LabVIEW program. Tangential force during testing is monitored with a load cell. The optical displacement sensor adjusts the displacement of the electrodynamic shaker through a feedback loop to maintain it within the parameters specified. All fretting tests were performed in laboratory conditions with a temperature of 22°C and a humidity of 34%.

The counterfaces for the testing were 10 mm diameter 52,100 steel balls, this material was chosen as it is a common bearing steel (Azom, 2012; Fouvry et al., 1994; Siu & Li, 2000). Testing was performed in dry conditions with 15,000 wear cycles at a frequency of 5 Hz. This results in a test time of 50 min. The amplitude was set to ± 50 μm giving wear in the gross slip fretting regime (as determined using sliding ratio (Fouvry et al., 1995)). Sliding ratio is defined as the ratio between the displacement amplitude at zero Q and the maximum displacement amplitude ($\frac{\delta_0}{\delta^*}$) (Fouvry et al., 1995). The effective velocity was 250 μM s^{-1}. Three tests were performed to assess variability in the fretting friction. Tests were performed at a load of 20 N. Coatings deposited on the HTS substrate only were evaluated for this study.

Fig. 18.10 shows the evolution of coefficient of friction (COF) with cycles. Coating C has the highest COF, which rises throughout the test due to wear through of the coating. Coating B had a long running in period (up to 7000 cycles) but eventually reduced to the lowest average COF. Coating A's running in period was extremely short after which it had the most stable friction (average of 0.18). The uncoated substrate (displayed in the inset) had extremely high fretting friction as is typical of metal-on-metal contacts in gross slip fretting (Kubiak et al., 2011).

Due to the stability of Coating A (highest *H/E* ratio as seen in) and its relatively low friction, we can conclude that this one performs best. The wear through in Coating C is partially due to its lower hardness, and also as it has approximately 1 μm less of the top coating (Table 18.2) (McMaster et al., 2023).

Fig. 18.11 shows the relation between *H/E* and wear volume. Wear volumes were analyzed with a Bruker Alicona (Institute for Advanced Manufacturing and Engineering, Coventry University) using a 50× lens. MountainsMap software was used to process the scan morphological data to generate the wear volume. We can see a relatively good inverse correlation between the *H/E* ratio and wear volume. Coating A performs best with the lowest wear volume; this relation gives initial indications that *H/E* can be used a predictor for fretting wear performance in DLC coatings.

18.7 Impact for fatigue determination

Impact testing on the micro- and nanoscale is typically a less utilized characterization technique but can yield lots of information. It can be used to assess multiple

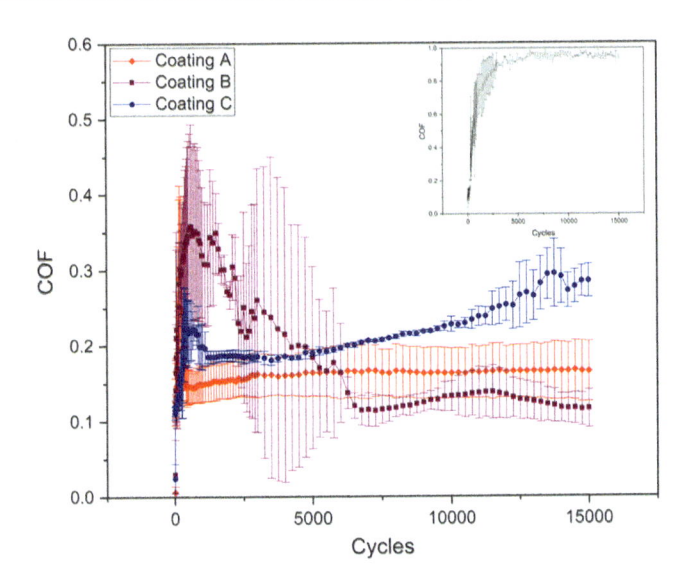

Figure 18.10 Coefficient of friction evolution of diamond-like carbon coatings A, B, and C on hardened M2 tool steel. Inset shows the friction evolution of the uncoated substrate. *Source*: From McMaster, S. J., Kosarieh, S., Liskiewicz, T. W., Neville, A., & Beake, B. D. (2023). Utilising H/E to predict fretting wear performance of DLC coating systems. *Tribology International*, *185*. https://doi.org/10.1016/j.triboint.2023.108524.

parameters such as fatigue strength, creep, adhesion, residual stresses, toughness, erosive wear resistance, and dynamic hardness of coatings (Beake et al., 2001, 2002, 2004; Rueda-Ruiz et al., 2020; Skordaris et al., 2020).

Sharp and blunt probes can be utilized for this type of testing depending on the desired fracture or fatigue conditions being assessed (Beake et al., 2019). Decreasing the radius of the probe will increase the stresses imparted to the surface, and sharper probes will force failure of the coating system faster. This allows customization of the testing regime to mimic a specific environment or simply to function as a comparative metric for a series of coatings. Coating performance is dependent upon architecture with ordered coatings benefitting from a higher resistance to plastic deformation (H^3/E^2) (Beake, Vishnyakov, et al., 2011), whereas amorphous coatings such as DLC benefit from higher toughness (E/H) (Beake, 2005; McMaster et al., 2020).

Micro-impact testing was used with an impact load of 750 mN with a total testing time of 300 seconds. A solenoid connected to a timed relay was used to produce repetitive impacts on the coating surface. Computer control ensured that each impact was in the same location for each load and occurred every 4 s (Beake et al., 2019; Beake, 2005). Maximum testing time was 300 s resulting in 75 impacts in the test duration. A sphero-conical indenter of 12–15 μm radius (dependent on depth from apex) was used. A Micro Materials Vantage system with a microloading head (0.4–5 N) was used for impact testing. Three repeats, where possible, were used in different locations on the sample. In all impacts, the indenter was retracted 40 μm from the surface.

Diamond-like carbon coatings on steel for automotive applications 539

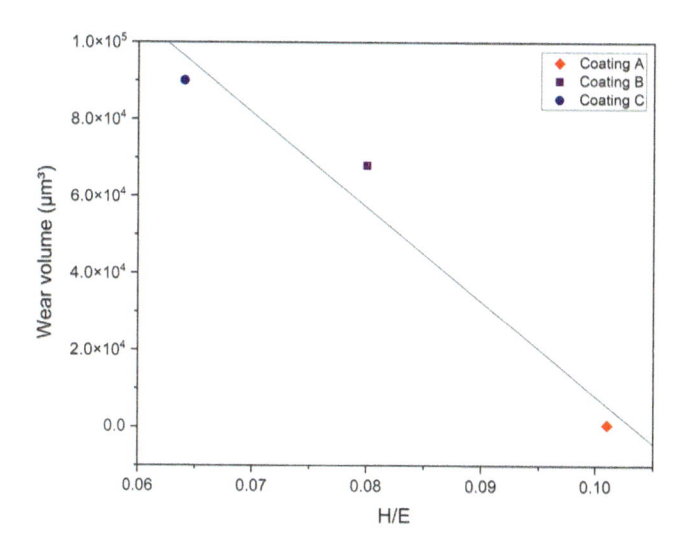

Figure 18.11 Relation between *H/E* and wear volume for diamond-like carbons A, B, and C. Pearson's $r = -0.978721$. $R^2 = 0.91576$.
Source: From McMaster, S. J., Kosarieh, S., Liskiewicz, T. W., Neville, A., & Beake, B. D. (2023). Utilising H/E to predict fretting wear performance of DLC coating systems. *Tribology International*, 185. https://doi.org/10.1016/j.triboint.2023.108524.

Nano-impact at 100 mN load with 75 impacts (corresponding to 300 s) was used to probe the initial cracking behavior of the coatings. Three repeats were performed to ensure repeatability. The same indenter geometry of 12–15 μm was utilized for this testing. A retraction distance of 15 μm is used for nano-impact tests. The nano-loading head has a range of 10–100 mN when used in impact mode.

Fig. 18.12 shows micro- and nano-impact results for the DLC coating systems defined for this case study.

Assessment of the impact and fatigue resistance can be undertaken by simple comparison of the depths reached within a certain number of impacts, but parametric analysis can be carried out with such parameters as (McMaster et al., 2020):

- I_0 — quasistatic depth
- I_1 — the depth of the first true impact.
- I_f — depth of the final impact.
- I_δ — the ratio of final depth to initial depth normalized by the initial impact depth. This parameter shows the relative level of fatigue (depth increase due to crack formation) between each loading step.

I_δ is defined as:

$$I_\delta = \frac{I_f - I_1}{I_1}$$

(a)

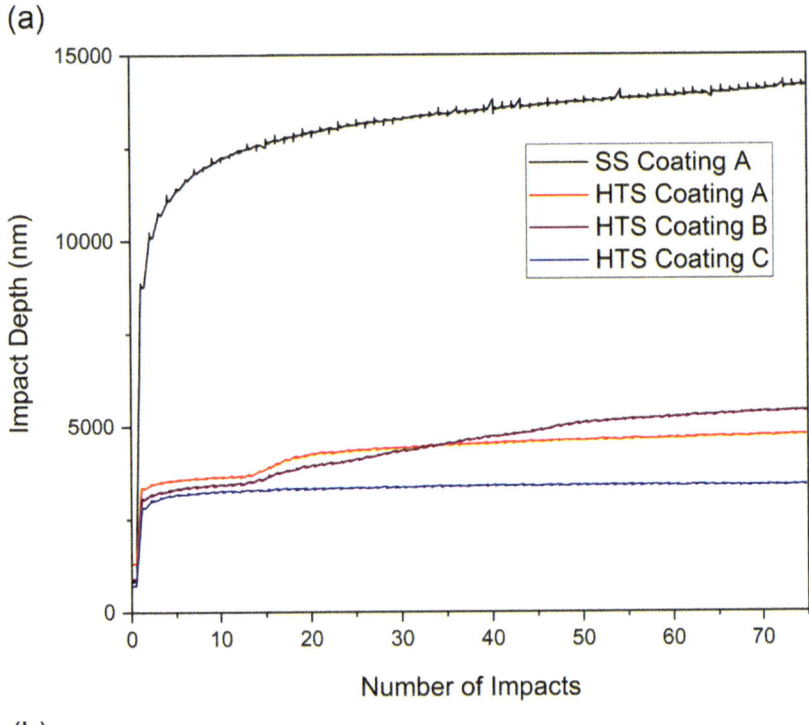

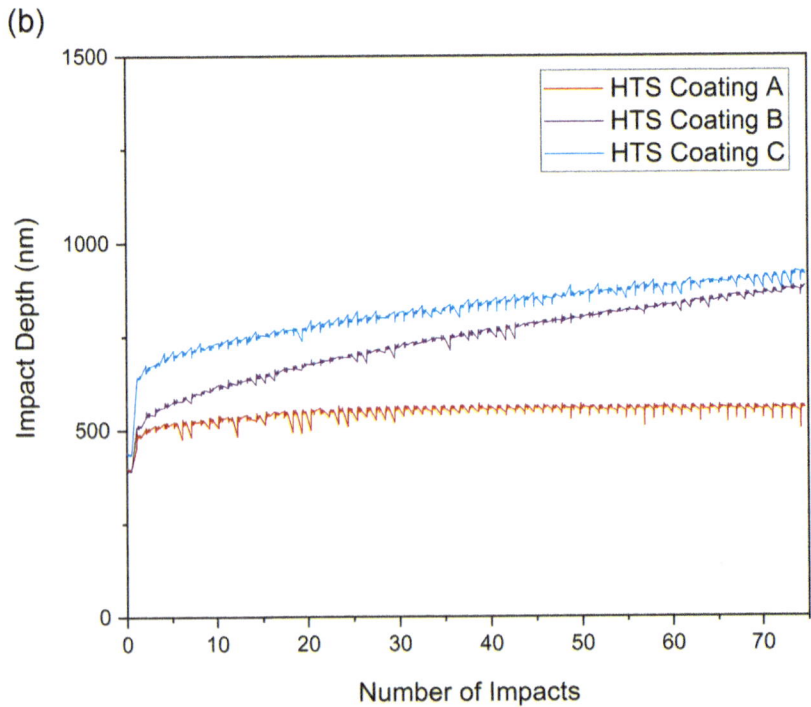

(b)

(Continued)

In micro-impact (Fig. 18.12A), we can see that Coating A on SS reaches an initial depth of approximately 8000 nm indicating the poor performance of this coating system. We can attribute the failure of the system in this case to the poor load support provided by the stainless steel substrate (McMaster et al., 2020). In contrast, all coatings on HTS performed better with the maximum depth of approximately 5000 nm reached by Coating B. Both coatings A and B experienced a cracking event at 12 impacts (signified by the increase in depth). Coating B continued to increase in depth indicating the worst performance, but coating A reached a plateau of depth.

The nano-impact results (Fig. 18.12B) tell a different story as a result of the lower impact load and retraction distance, thereby imparting a lower amount of energy to the surface with each impact. No dramatic cracking events occur as each coating either has a steady increase in depth or maintains a plateau. In this lower-energy scenario, Coating A performs best with a constant depth of 500 nm through the testing regime. Coating C being softer has a larger amount of initial plastic deformation and therefore increased impact depth. Coating C's depth gradient is less severe than B showing its overall better performance.

The almost instantaneous failure of the coating system on the stainless steel substrate is due to egg-shelling (crème brûlée effect) where impact causes coating structural collapse as a result of yield and plastic deformation of the substrate beneath the coating (Bentzon et al., 1994; Hu et al., 2012; Wright & Page, 1992). This is corroborated by the similar elastic moduli of the substrates but markedly different hardness. In nano-impact testing, the dramatic eggshell type failures are not observed; however, the lack of load support is still evident as the depths reached are at minimum 1000 nm higher (McMaster et al., 2020; McMaster, 2020).

Toughness (*E/H*) is the defining parameter for impact resistance for amorphous coatings such as DLC. As such, Coating C is the superior coating in this loading regime. As per Lawn and coworkers and Pharr, this is due to the increased fracture toughness of the coating (Anstis et al., 1981; Beake, 2005; Chantikul et al., 1981; Leyland & Matthews, 2000). The structure of the coating plays a part also. W doping in DLC films increases the sp^2 fraction, thereby giving a more graphitic structure and softening the coating. The sp^2 rings and presence of tungsten in its microstructure increase the elastic modulus; furthermore, it's likely that

Figure 18.12 (a) Representative micro-(impact load = 750 mN) and (b) nano-impact (Impact load = 100 mN) depth against the number of impacts of three coatings on coatings A, B, and C on HTS. Coating A on SS substrate included for comparison. Maximum number of impacts = 75 (300 s of testing).
Source: From McMaster, S. J. (2020). Nanomechanical characterisation of diamond-like carbon coatings for tribological performance. https://doi.org/10.13140/RG.2.2.13315.02083 (Original work published 2020); McMaster, S. J., Kosarieh, S., Liskiewicz, T. W., Neville, A., & Beake, B. D. (2023). Utilising H/E to predict fretting wear performance of DLC coating systems. *Tribology International, 185*. https://doi.org/10.1016/j.triboint.2023.108524.

nanocrystalline WC or W_2C is present (Chen, Peng, et al., 2011; Pei, Galvan, & De Hosson, 2005; Rincón et al., 2001; Yong et al., 2016; Yue et al., 2015).

The newest advancement to instrumented impact testing is the introduction of randomised impact capability allowing for the definition of a distributed area in which impacts will occur. Energy density for a specific area can be better defined, and cracking dynamics between impact craters can be studied. This will allow for more realistic mimicry of erosive conditions.

18.8 Conclusion

In summary, we can see that DLCs are often subjected to challenging environments. Though the performance of this coating is generally suited to those environments, it must be tailored for the environment to achieve optimal performance. This involves the study of the performance of the substrate and how it integrates into the coating system, interlayer structures of the coating, and, of course, the performance of the top layer. Multiple authors have studied the effect of adhesion (Bernal et al., 2018; Cemin et al. 2015, 2016; Crespi et al., 2019; Konkhunthot et al., 2019) and residual stress (Corbella et al., 2004; Li et al., 2015; Santiago et al., 2019; Solis et al., 2016; Zhang et al., 1999; Zou et al., 2013) on DLC as this is one of the predominant issues affecting the deployment of the coating into further applications in the automotive field and beyond.

The tunability of DLCs will allow them to be useful for vehicles utilizing alternative propulsion sources. The electrical properties of the coating will need to be considered for electric vehicles much more than for traditional internal combustion engines, but there are now multiple studies evaluating their performance for this new use case. Humphrey et al. (2023) analyzed the potential of DLC coatings for energy and range savings in electric vehicles in comparison to a superfinished steel. It was found that the use of DLC coatings could save 1.1 km of vehicle range during urban driving. It was also finally remarked that further investigation into surface coating technology could provide further benefits by optimizing the surface lubricant system by maximizing range by reducing losses. Holmberg and Erdemir (2019) reviewed of the impact of tribology on energy use and CO_2 emissions in internal combustion engine (ICE) and electric vehicles. They concluded that 20% (103 EJ) of world energy consumption is used to overcome friction. In total, 18%–30% of that can be saved by the application of new technologies. For an internal combustion engine passenger car, approximately 30% of the fuel consumption is used to overcome friction; for an electric car, the frictional losses are about half that of an ICE vehicle. This chapter further expands upon their 2017 review (Holmberg & Erdemir, 2017).

Using a multitechnique approach is best to build a complete picture of the coating performance. No one technique can tell us everything about a coating's performance. By changing the analytical technique applied to the coating, we can mimic the tribological scenario being targeted by the application of different stresses and loading conditions.

References

Aisenberg, S., & Chabot, R. (1971). Ion-beam deposition of thin films of diamondlike carbon. *Journal of Applied Physics*, *42*(7), 2953–2958. Available from https://doi.org/10.1063/1.1660654.

Al Mahmud, K. A. H., Kalam, M. A., Masjuki, H. H., Mobarak, H. M., & Zulkifli, N. W. M. (2015). An updated overview of diamond-like carbon coating in tribology. *Critical Reviews in Solid State and Materials Sciences*, *40*(2), 90–118. Available from https://doi.org/10.1080/10408436.2014.940441, http://www.tandf.co.uk/journals/titles/10408436.asp.

Aldeeb, M. A., Morgan, N., Abouelsayed, A., Amin, K. M., & Hassablla, S. (2019). Correlation of acetylene plasma discharge environment and the optical and electronic properties of the hydrogenated amorphous carbon films. *Diamond and Related Materials*, *96*, 74–84. Available from https://doi.org/10.1016/j.diamond.2019.04.021.

Angus, J. C., & Hayman, C. C. (1988). Low-pressure, metastable growth of diamond and "diamondlike" phases. *Science (New York, N.Y.)*, *241*(4868), 913–921. Available from https://doi.org/10.1126/science.241.4868.913.

Anstis, G. R., Chantikul, P., Lawn, B. R., & Marshall, D. B. (1981). A critical evaluation of indentation techniques for measuring fracture toughness: I, direct crack measurements. *Journal of the American Ceramic Society*, *64*(9), 533–538. Available from https://doi.org/10.1111/j.1151-2916.1981.tb10320.x.

Arenal, R. (2017). EELS studies on nanodiamonds and amorphous diamond-like carbon materials. *Microscopy and Microanalysis*, *23*(S1), 2274–2275. Available from https://doi.org/10.1017/s143192761701203x.

Austin, L.B. (2014) Evaluation and optimisation of diamond-like carbon for tribological applications. PhD thesis, University of Leeds.

Azom. (2012). AISI 52100 Alloy Steel (UNS G52986). 1–2. https://www.azom.com/article.aspx?ArticleID = 6704 (accessed July 27, 2018)

Baba, K., Hatada, R., Flege, S., & Ensinger, W. (2020). Diamond-like carbon films with low internal stress by a simple bilayer approach. *Coatings*, *10*(7). Available from https://doi.org/10.3390/coatings10070696, https://res.mdpi.com/d_attachment/coatings/coatings-10-00696/article_deploy/coatings-10-00696.pdf.

Bae, S. M., Horibata, S., Miyauchi, Y., & Choi, J. (2023). Tribochemical investigation of Cr-doped diamond-like carbon with a MoDTC-containing engine oil under boundary lubricated condition. *Tribology International*, *188*. Available from https://doi.org/10.1016/j.triboint.2023.108849, http://www.elsevier.com/inca/publications/store/3/0/4/7/4.

Beake, B. D., Fox-Rabinovich, G. S., Veldhuis, S. C., & Goodes, S. R. (2009). Coating optimisation for high speed machining with advanced nanomechanical test methods. *Surface and Coatings Technology*, *203*(13), 1919–1925. Available from https://doi.org/10.1016/j.surfcoat.2009.01.025.

Beake, B. D., Goodes, S. R., Smith, J. F., Madani, R., Rego, C. A., Cherry, R. I., & Wagner, T. (2002). Investigating the fracture resistance and adhesion of DLC films with micro-impact testing. *Diamond and Related Materials*, *11*(8), 1606–1609. Available from https://doi.org/10.1016/s0925-9635(02)00107-3.

Beake, B. D., Goodes, S. R., & Smith, J. F. (2001). Micro-impact testing: A new technique for investigating thin film toughness, adhesion, erosive wear resistance, and dynamic hardness. *Surface Engineering*, *17*(3), 187–192. Available from https://doi.org/10.1179/026708401101517755.

Beake, B. D., Harris, A. J., & Liskiewicz, T. W. (2013). Review of recent progress in nano-scratch testing. *Tribology — Materials, Surfaces & Interfaces, 7*(2), 87–96. Available from https://doi.org/10.1179/1751584X13Y.0000000037.

Beake, B. D., Isern, L., Endrino, J. L., & Fox-Rabinovich, G. S. (2019). Micro-impact testing of AlTiN and TiAlCrN coatings. *Wear, 418-419*, 102–110. Available from https://doi.org/10.1016/j.wear.2018.11.010, https://www.journals.elsevier.com/wear.

Beake, B. D., Lau, S. P., & Smith, J. F. (2004). Evaluating the fracture properties and fatigue wear of tetrahedral amorphous carbon films on silicon by nano-impact testing. *Surface and Coatings Technology, 177-178*, 611–615. Available from https://doi.org/10.1016/S0257-8972(03)00934-4, http://www.journals.elsevier.com/surface-and-coatings-technology/.

Beake, B. D., Liskiewicz, T. W., Vishnyakov, V. M., & Davies, M. I. (2015). Development of DLC coating architectures for demanding functional surface applications through nano- and micro-mechanical testing. *Surface and Coatings Technology, 284*, 334–343. Available from https://doi.org/10.1016/j.surfcoat.2015.05.050.

Beake, B. D., Shi, B., & Sullivan, J. L. (2011). Nanoscratch and nanowear testing of TiN coatings on M42 steel. *Tribology — Materials, Surfaces and Interfaces, 5*(4), 141–147. Available from https://doi.org/10.1179/1751584X11Y.0000000017, http://www.ingentaconnect.com/search/download?pub = infobike%3a%2f%2fmaney%2ftrb%2f2011%2f00-000005%2f00000004%2fart00003&mimetype = application%2fpdf&exitTargetId = 132-4732727881, United Kingdom.

Beake, B. D., Vishnyakov, V. M., & Colligon, J. S. (2011). Nano-impact testing of TiFeN and TiFeMoN films for dynamic toughness evaluation. *Journal of Physics D: Applied Physics, 44*(8). Available from https://doi.org/10.1088/0022-3727/44/8/085301.

Beake, B. D. (2005). Evaluation of the fracture resistance of DLC coatings on tool steel under dynamic loading. *Surface and Coatings Technology, 198*(1-3), 90–93. Available from https://doi.org/10.1016/j.surfcoat.2004.10.048.

Beake, B. D. (2022a). The influence of the *H/E* ratio on wear resistance of coating systems — Insights from small-scale testing. *Surface and Coatings Technology, 442*. Available from https://doi.org/10.1016/j.surfcoat.2022.128272.

Beake, B. D. (2022b). Nano- and micro-scale impact testing of hard coatings: A review. *Coatings, 12*(6). Available from https://doi.org/10.3390/coatings12060793.

Bentzon, M. D., Mogensen, K., Hansen, J. B., Barholm-Hansen, C., Træholt, C., Holiday, P., & Eskildsen, S. S. (1994). Metallic interlayers between steel and diamond-like carbon. *Surface and Coatings Technology, 68-69*(C), 651–655. Available from https://doi.org/10.1016/0257-8972(94)90232-1.

Berman, D., Deshmukh, S. A., Sankaranarayanan, S. K. R. S., Erdemir, A., & Sumant, A. V. (2015). Macroscale superlubricity enabled by graphene nanoscroll formation. *Science (New York, N.Y.), 348*(6239), 1118–1122. Available from https://doi.org/10.1126/science.1262024, http://www.sciencemag.org/content/348/6239/1118.full.pdf.

Berman, D., Erdemir, A., & Sumant, A. V. (2013). Reduced wear and friction enabled by graphene layers on sliding steel surfaces in dry nitrogen. *Carbon, 59*, 167–175. Available from https://doi.org/10.1016/j.carbon.2013.03.006.

Berman, D., Erdemir, A., & Sumant, A. V. (2018). Approaches for achieving superlubricity in two-dimensional materials. *ACS Nano, 12*(3), 2122–2137. Available from https://doi.org/10.1021/acsnano.7b09046, http://pubs.acs.org/journal/ancac3.

Bernal, R. A., Chen, P., Schall, J. D., Harrison, J. A., Jeng, Y. R., & Carpick, R. W. (2018). Influence of chemical bonding on the variability of diamond-like carbon nanoscale

adhesion. *Carbon*, *128*, 267−276. Available from https://doi.org/10.1016/j.carbon.2017.11.040, http://www.journals.elsevier.com/carbon/.
Bewilogua, K., Cooper, C. V., Specht, C., Schröder, J., Wittorf, R., & Grischke, M. (2000). Effect of target material on deposition and properties of metal-containing DLC (Me-DLC) coatings. *Surface and Coatings Technology*, *127*(2−3), 223−231. Available from https://doi.org/10.1016/s0257-8972(00)00666-6.
Bewilogua, K., & Hofmann, D. (2014). History of diamond-like carbon films − From first experiments to worldwide applications. *Surface and Coatings Technology*, *242*, 214−225. Available from https://doi.org/10.1016/j.surfcoat.2014.01.031.
Bhushan, B. (1999). *Handbook of Micro/Nano Tribology*. Boca Raton: CRC Press.
Blanpain, B., Celis, J. P., Roos, J. R., Ebberink, J., & Smeets, J. (1993). A comparative study of the fretting wear of hard carbon coatings. *Thin Solid Films*, *223*(1), 65−71. Available from https://doi.org/10.1016/0040-6090(93)90728-8.
Bonetti, L. F., Capote, G., Santos, L. V., Corat, E. J., & Trava-Airoldi, V. J. (2006). Adhesion studies of diamond-like carbon films deposited on Ti$_6$Al$_4$V substrate with a silicon interlayer. *Thin Solid Films*, *515*(1), 375−379. Available from https://doi.org/10.1016/j.tsf.2005.12.154.
Brian, B. T., & Kelly, T. (1981). *Physics of graphite* (17). Springer.
Brittain, R., Liskiewicz, T., Morina, A., Neville, A., & Yang, L. (2023). Diamond-like carbon graphene nanoplatelet nanocomposites for lubricated environments. *Carbon*, *205*, 485−498. Available from https://doi.org/10.1016/j.carbon.2023.01.061, http://www.journals.elsevier.com/carbon/.
Bromark, M., Hedenqvist, P., & Hogmark, S. (1995). The influence of substrate material on the erosion resistance of TiN coated tool steels. *Wear*, *186-187*(1), 189−194. Available from https://doi.org/10.1016/0043-1648(95)07163-6.
Bull, S. J. (1991). Failure modes in scratch adhesion testing. *Surface and Coatings Technology*, *50*(1), 25−32. Available from https://doi.org/10.1016/0257-8972(91)90188-3.
Burnett, P. J., & Rickerby, D. S. (1987). The relationship between hardness and scratch adhesion. *Thin Solid Films*, *154*(1-2), 403−416. Available from https://doi.org/10.1016/0040-6090(87)90382-8.
Casiraghi, C., Piazza, F., Ferrari, A. C., Grambole, D., & Robertson, J. (2005). Bonding in hydrogenated diamond-like carbon by Raman spectroscopy. *Diamond and Related Materials*, *14*(3-7), 1098−1102. Available from https://doi.org/10.1016/j.diamond.2004.10.030.
Cañellas, G., Emeric, A., Combarros, M., Navarro, A., Beltran, L., Vilaseca, M., & Vives, J. (2023). Tribological performance of esters, friction modifier and antiwear additives for electric vehicle applications. *Lubricants*, *11*(3). Available from https://doi.org/10.3390/lubricants11030109, http://www.mdpi.com/journal/lubricants.
Cemin, F., Bim, L. T., Leidens, L. M., Morales, M., Baumvol, I. J. R., Alvarez, F., & Figueroa, C. A. (2015). Identification of the chemical bonding prompting adhesion of a-C:H thin films on ferrous alloy intermediated by a SiC$_x$:H Buffer Layer. *ACS Applied Materials & Interfaces*, *7*(29), 15909−15917. Available from https://doi.org/10.1021/acsami.5b03554.
Cemin, F., Boeira, C. D., & Figueroa, C. A. (2016). On the understanding of the silicon-containing adhesion interlayer in DLC deposited on steel. *Tribology International*, *94*, 464−469. Available from https://doi.org/10.1016/j.triboint.2015.09.044.
CEN/TS. TS CEN/TS 17629:2021 Nanotechnologies — Nano- and micro- scale scratch testing. at https://standards.iteh.ai/catalog/standards/cen/86dfcdee-bb2f-4a81-8069-3ac68d-7d2071/cen-ts-17629-2021 (2021).

Chantikul, P., Anstis, G. R., Lawn, B. R., & Marshall, D. B. (1981). A critical evaluation of indentation techniques for measuring fracture toughness: II, Strength method. *Journal of the American Ceramic Society, 64*(9), 539−543. Available from https://doi.org/10.1111/j.1151-2916.1981.tb10321.x.

Chen, L., Yang, M., Yu, J., Qian, L., & Zhou, Z. (2011). Nanofretting behaviours of ultrathin DLC coating on Si(100) substrate. *Wear, 271*(9-10), 1980−1986. Available from https://doi.org/10.1016/j.wear.2010.11.016.

Chen, X., Du, Y., & Chung, Y. W. (2019). Commentary on using H/E and H_3/E^2 as proxies for fracture toughness of hard coatings. *Thin Solid Films, 688*. Available from https://doi.org/10.1016/j.tsf.2019.04.040, http://www.journals.elsevier.com/journal-of-the-energy-institute.

Chen, X., & Li, J. (2020). Superlubricity of carbon nanostructures. *Carbon, 158*, 1−23. Available from https://doi.org/10.1016/j.carbon.2019.11.077, http://www.journals.elsevier.com/carbon/.

Chen, X., Peng, Z., Fu, Z., Wu, S., Yue, W., & Wang, C. (2011). Microstructural, mechanical and tribological properties of tungsten-gradually doped diamond-like carbon films with functionally graded interlayers. *Surface and Coatings Technology, 205*(12), 3631−3638. Available from https://doi.org/10.1016/j.surfcoat.2011.01.004.

Chen, Y., Wu, J. M., Nie, X., & Yu, S. (2016). Study on failure mechanisms of DLC coated Ti_6Al_4V and CoCr under cyclic high combined contact stress. *Journal of Alloys and Compounds, 688*, 964−973. Available from https://doi.org/10.1016/j.jallcom.2016.07.254.

Cheng, Y. T., & Cheng, C. M. (2004). Scaling, dimensional analysis, and indentation measurements. *Materials Science and Engineering R: Reports, 44*(4-5), 91−149. Available from https://doi.org/10.1016/j.mser.2004.05.001.

Choi, J., Kawaguchi, M., Kato, T., & Ikeyama, M. (2007). Deposition of Si-DLC film and its microstructural, tribological and corrosion properties. *Microsystem Technologies, 13*(8-10), 1353−1358. Available from https://doi.org/10.1007/s00542-006-0368-8.

Corbella, C., Vives, M., Oncins, G., Canal, C., Andújar, J. L., & Bertran, E. (2004). Characterization of DLC films obtained at room temperature by pulsed-dc PECVD. *Diamond and Related Materials, 13*(4-8), 1494−1499. Available from https://doi.org/10.1016/j.diamond.2003.10.079.

Crespi, Â. E., Leidens, L. M., Antunes, V., Perotti, B. L., Michels, A. F., Alvarez, F., & Figueroa, C. A. (2019). Substrate bias voltage tailoring the interfacial chemistry of a-SiC x:H: A surprising improvement in adhesion of a-C:H thin films deposited on ferrous alloys controlled by oxygen. *ACS Applied Materials and Interfaces, 11*(19), 18024−18033. Available from https://doi.org/10.1021/acsami.9b03597, http://pubs.acs.org/journal/aamick.

Daniels, H., Brydson, R., Rand, B., & Brown, A. (2007). Investigating carbonization and graphitization using electron energy loss spectroscopy (EELS) in the transmission electron microscope (TEM). *Philosophical Magazine, 87*(27), 4073−4092. Available from https://doi.org/10.1080/14786430701394041.

Grill, A. (1999). Diamond-like carbon: State of the art. *Diamond and Related Materials, 8*, 262−263. Available from https://doi.org/10.1016/S0925-9635(98.

De Barros Bouchet, M. I., Martin, J. M., Avila, J., Kano, M., Yoshida, K., Tsuruda, T., Bai, S., Higuchi, Y., Ozawa, N., Kubo, M., & Asensio, M. C. (2017). Diamond-like carbon coating under oleic acid lubrication: Evidence for graphene oxide formation in superlow friction. *Scientific Reports, 7*. Available from https://doi.org/10.1038/srep46394, http://www.nature.com/srep/index.html.

De Barros Bouchet, M. I., Martin, J. M., Le-Mogne, T., & Vacher, B. (2005). Boundary lubrication mechanisms of carbon coatings by MoDTC and ZDDP additives. *Tribology*

International, *38*(3), 257−264. Available from https://doi.org/10.1016/j.triboint.2004.08.009.

De Feo, M., De Barros Bouchet, M. I., Minfray, C., Le Mogne, T., Meunier, F., Yang, L., Thiebaut, B., & Martin, J. M. (2016). MoDTC lubrication of DLC-involving contacts. Impact of MoDTC degradation. *Wear*, *348-349*, 116−125. Available from https://doi.org/10.1016/j.wear.2015.12.001.

Donnet, C., & Erdemir, A. (2008). *Tribology of diamond-like carbon films: Fundamentals and applications* (pp. 1−664). France: Springer US. Available from http://www.springerlink.com/openurl.asp?genre = book&isbn = 978-0-387-30264-5, 10.1007/978-0-387-49891-1.

Dresselhaus, M. S., Dresselhaus, G., & Eklund, P. C. (1996). *Science of fullerenes and carbon nanotubes*. Academic Press. Available from doi:10.1016/B978-0-12-221820-0.X5000-X.

Erdemir, A., & Donnet, C. (2006). Tribology of diamond-like carbon films: Recent progress and future prospects. *Journal of Physics D: Applied Physics*, *39*(18), R311−R327. Available from https://doi.org/10.1088/0022-3727/39/18/R01.

Espejo, C., Thiébaut, B., Jarnias, F., Wang, C., Neville, A., & Morina, A. (2019). MoDTC tribochemistry in steel/steel and steel/diamond-like-carbon systems lubricated with model lubricants and fully formulated engine oils. *Journal of Tribology*, *141*(1). Available from https://doi.org/10.1115/1.4041017, http://tribology.asmedigitalcollection.asme.org/journal.aspx.

Eyre, T. S. (1976). Wear characteristics of metals. *Tribology International*, *9*(5), 203−212. Available from https://doi.org/10.1016/0301-679x(76)90077-3.

Eyre, T. S. (1978). The mechanisms of wear. *Tribology International*, *11*(2), 91−96. Available from https://doi.org/10.1016/0301-679x(78)90135-4.

Farfan-Cabrera, L. I. (2019). Tribology of electric vehicles: A review of critical components, current state and future improvement trends. *Tribology International*, *138*, 473−486. Available from https://doi.org/10.1016/j.triboint.2019.06.029, http://www.elsevier.com/inca/publications/store/3/0/4/7/4.

Farfan-Cabrera, L. I., Cao-Romero-Gallegos, J. A., Lee, S., Komurlu, M. U., & Erdemir, A. (2023). Tribological behavior of H-DLC and H-free DLC coatings on bearing materials under the influence of DC electric current discharges. *Wear*, *522*. Available from https://doi.org/10.1016/j.wear.2023.204709, https://www.journals.elsevier.com/wear.

Ferrari, A. C., Kleinsorge, B., Adamopoulos, G., Robertson, J., Milne, W. I., Stolojan, V., Brown, L. M., LiBassi, A., & Tanner, B. K. (2000). Determination of bonding in amorphous carbons by electron energy loss spectroscopy, Raman scattering and X-ray reflectivity. *Journal of Non-Crystalline Solids*, *266-269*, 765−768. Available from https://doi.org/10.1016/s0022-3093(00)00035-1.

Ferrari, A., & Robertson, J. (2000). Interpretation of Raman spectra of disordered and amorphous carbon. *Physical Review B − Condensed Matter and Materials Physics*, *61*(20), 14095−14107. Available from https://doi.org/10.1103/PhysRevB.61.14095.

Ferrari, A. C. (2002). Determination of bonding in diamond-like carbon by Raman spectroscopy. *Diamond and Related Materials*, *11*(3-6), 1053−1061. Available from https://doi.org/10.1016/S0925-9635(01)00730-0.

Field, S. K., Jarratt, M., & Teer, D. G. (2004). Tribological properties of graphite-like and diamond-like carbon coatings. *Tribology International*, *37*(11-12), 949−956. Available from https://doi.org/10.1016/j.triboint.2004.07.012.

Filik, J., May, P. W., Pearce, S. R. J., Wild, R. K., & Hallam, K. R. (2003). XPS and laser Raman analysis of hydrogenated amorphous carbon films. *Diamond and Related*

Materials, *12*(3-7), 974−978. Available from https://doi.org/10.1016/s0925-9635(02)00374-6.
Fischer-Cripps, A. C. (2004). *Nanoindentation*. New York: Springer. Available from https://doi.org/10.1007/978-1-4757-5943-3.
Fischer-Cripps, A. C. (2006). Critical review of analysis and interpretation of nanoindentation test data. *Surface and Coatings Technology*, *200*(14-15), 4153−4165. Available from https://doi.org/10.1016/j.surfcoat.2005.03.018.
Fouvry, S., Fridrici, V., Langlade, C., Kapsa, P., & Vincent, L. (2006). Palliatives in fretting: A dynamical approach. *Tribology International*, *39*(10), 1005−1015. Available from https://doi.org/10.1016/j.triboint.2006.02.038.
Fouvry, S., Kapsa, P., Sauger, E., Martin, J. M., Ponsonnet, L., & Vincent, L. (2002). Tribologically transformed structure in fretting. *Wear*, *245*. Available from https://doi.org/10.1016/s0043-1648(00, 464 −6.
Fouvry, S., Kapsa, P., & Vincent, L. (1994). Fretting behaviour of hard coatings under high normal load. *Surface and Coatings Technology*, *68-69*(C), 494−499. Available from https://doi.org/10.1016/0257-8972(94)90207-0.
Fouvry, S., Kapsa, P., & Vincent, L. (1995). Analysis of sliding behaviour for fretting loadings: Determination of transition criteria. *Wear*, *185*(1-2), 35−46. Available from https://doi.org/10.1016/0043-1648(94)06582-9.
Fox-Rabinovich, G. S., Beake, B. D., Endrino, J. L., Veldhuis, S. C., Parkinson, R., Shuster, L. S., & Migranov, M. S. (2006). Effect of mechanical properties measured at room and elevated temperatures on the wear resistance of cutting tools with TiAlN and AlCrN coatings. *Surface and Coatings Technology*, *200*(20-21), 5738−5742. Available from https://doi.org/10.1016/j.surfcoat.2005.08.132.
Greenwood, J. A., & Williamson, J. B. P. (1966). Contact of Nominally Flat Surfaces. *Proceedings of the Royal Society A: Mathematical, Physical and Engineering Sciences*, *295*(1442), 300−319. Available from https://doi.org/10.1098/rspa.1966.0242.
Grill, A. (1993). Review of the tribology of diamond-like carbon. *Wear*, *168*(1-2), 143−153. Available from https://doi.org/10.1016/0043-1648(93)90210-D.
Hainsworth, S. V., & Uhure, N. J. (2007). Diamond like carbon coatings for tribology: Production techniques, characterisation methods and applications. *International Materials Reviews*, *52*(3), 153−174. Available from https://doi.org/10.1179/174328007X160272, http://www.tandfonline.com/loi/yimr20#.VwHbh01f1Qs.
Hasselbruch, H., Herrmann, M., Mehner, A., Zoch, H.W., & Kuhfuss, B. (2015). Development, characterization and testing of tungsten doped DLC coatings for dry rotary swaging. *MATEC Web of Conferences*. 21 EDP Sciences Germany. http://www.matec-conferences.org/ https://doi.org/10.1051/matecconf/20152108012.
Holmberg, K., & Erdemir, A. (2017). Influence of tribology on global energy consumption, costs and emissions. *Friction*, *5*(3), 263−284. Available from https://doi.org/10.1007/s40544-017-0183-5, http://www.springer.com/engineering/mechanical+engineering/journal/40544.
Holmberg, K., & Erdemir, A. (2019). The impact of tribology on energy use and CO_2 emission globally and in combustion engine and electric cars. *Tribology International*, *135*, 389−396. Available from https://doi.org/10.1016/j.triboint.2019.03.024, http://www.elsevier.com/inca/publications/store/3/0/4/7/4.
Holmberg, K., Matthews, A., & Ronkainen, H. (1998). Coatings tribology − Contact mechanisms and surface design. *Tribology International*, *31*(1-3), 107−120. Available from https://doi.org/10.1016/S0301-679X(98)00013-9.

Hu, J., Sun, W., Jiang, Z., Zhang, W., Lu, J., Huo, W., Zhang, Y., & Zhang, P. (2017). Indentation size effect on hardness in the body-centered cubic coarse-grained and nanocrystalline tantalum. *Materials Science and Engineering: A, 686,* 19−25. Available from https://doi.org/10.1016/j.msea.2017.01.033, http://www.elsevier.com.

Hu, Z., Schubnov, A., & Vollertsen, F. (2012). Tribological behaviour of DLC-films and their application in micro deep drawing. *Journal of Materials Processing Technology, 212*(3), 647−652. Available from https://doi.org/10.1016/j.jmatprotec.2011.10.012.

Humphrey, E., Elisaus, V., Rahmani, R., Mohammadpour, M., Theodossiades, S., & Morris, N. (2023). Diamond like-carbon coatings for electric vehicle transmission efficiency. *Tribology International, 189.* Available from https://doi.org/10.1016/j.triboint.2023.108916.

Jarry, O., Jaoul, C., Tristant, P., Merle-Méjean, T., Colas, M., Dublanche-Tixier, C., Ageorges, H., Lory, C., & Jacquet, J. M. (2009). Tribological behaviour of diamond-like carbon films used in automotive application: A comparison. *Plasma Processes and Polymers, 6*(1), S478−S482. Available from https://doi.org/10.1002/ppap.200931007, http://www3.interscience.wiley.com/cgi-bin/fulltext/122406437/PDFSTART.

Johnson, K. L. (1985). *Contact Mechanics.* Cambridge University Press.

Kano, M. (2014). Diamond-like carbon coating applied to automotive engine components. *Tribology Online, 9*(3), 135−142. Available from https://doi.org/10.2474/trol.9.135, https://www.jstage.jst.go.jp/article/trol/9/3/9_135/_pdf.

Kassim, K. A. M., Tokoroyama, T., Murashima, M., & Umehara, N. (2020). The wear classification of MoDTC-derived particles on silicon and hydrogenated diamond-like carbon at room temperature. *Tribology International, 147.* Available from https://doi.org/10.1016/j.triboint.2020.106176.

Khanmohammadi, H., Wijanarko, W., Cruz, S., Evaristo, M., & Espallargas, N. (2022). Triboelectrochemical friction control of W- and Ag-doped DLC coatings in water-glycol with ionic liquids as lubricant additives. *RSC Advances, 12*(6), 3573−3583. Available from https://doi.org/10.1039/d1ra08814a, http://pubs.rsc.org/en/journals/journal/ra.

Komori, K., & Umehara, N. (2017). Friction and wear properties of tetrahedral si-containing hydrogenated diamond-like carbon coating under lubricated condition with engine-oil containing zndtp and modtc. *Tribology Online, 12*(3), 123−134. Available from https://doi.org/10.2474/trol.12.123, https://www.jstage.jst.go.jp/article/trol/12/3/12_123/_pdf.

Konkhunthot, N., Photongkam, P., & Wongpanya, P. (2019). Improvement of thermal stability, adhesion strength and corrosion performance of diamond-like carbon films with titanium doping. *Applied Surface Science, 469,* 471−486. Available from https://doi.org/10.1016/j.apsusc.2018.11.028, http://www.journals.elsevier.com/applied-surface-science/.

Kosarieh, S., Morina, A., Flemming, J., Lainé, E., & Neville, A. (2016). Wear mechanisms of hydrogenated DLC in oils containing MoDTC. *Tribology Letters, 64*(1). Available from https://doi.org/10.1007/s11249-016-0737-0, http://www.springerlink.com/(snp-xut45gxflnr45vb2gia45)/app/home/journal.asp?referrer = parent&backto = searchpublicationsresults,1,2.

Kubiak, K. J., Liskiewicz, T. W., & Mathia, T. G. (2011). Surface morphology in engineering applications: Influence of roughness on sliding and wear in dry fretting. *Tribology International, 44*(11), 1427−1432. Available from https://doi.org/10.1016/j.triboint.2011.04.020.

Lanigan, J. L., Wang, C., Morina, A., & Neville, A. (2016). Repressing oxidative wear within Si doped DLCs. *Tribology International, 93,* 651−659. Available from https://doi.org/10.1016/j.triboint.2014.11.004.

Lawes, S. D. A., Fitzpatrick, M. E., & Hainsworth, S. V. (2007). Evaluation of the tribological properties of DLC for engine applications. *Journal of Physics D: Applied Physics, 40*(18), 5427−5437. Available from https://doi.org/10.1088/0022-3727/40/18/S03.

Lawes, S. D. A., Hainsworth, S. V., & Fitzpatrick, M. E. (2010). Impact wear testing of diamond-like carbon films for engine valve-tappet surfaces. *Wear*, *268*(11-12), 1303–1308. Available from https://doi.org/10.1016/j.wear.2010.02.011.

Leyland, A., & Matthews, A. (2000). On the significance of the *H/E* ratio in wear control: A nanocomposite coating approach to optimised tribological behaviour. *Wear*, *246*(1-2), 1–11. Available from https://doi.org/10.1016/s0043-1648(00)00488-9.

Li, X., Ke, P., & Wang, A. (2015). Probing the stress reduction mechanism of diamond-like carbon films by incorporating Ti, Cr, or W carbide-forming metals: Ab initio molecular dynamics simulation. *Journal of Physical Chemistry C*, *119*(11), 6086–6093. Available from https://doi.org/10.1021/acs.jpcc.5b00058, http://pubs.acs.org/journal/jpccck.

Lifshitz, Y. (1999). Diamond-like carbon – Present status. *Diamond and Related Materials*, *8*(8-9), 1659–1676. Available from https://doi.org/10.1016/s0925-9635(99)00087-4, https://www.journals.elsevier.com/diamond-and-related-materials.

Lifshitz, Y., Lempert, G. D., Grossman, E., Avigal, I., Uzan-Saguy, C., Kalish, R., Kulik, J., Marton, D., & Rabalais, J. W. (1995). Growth mechanisms of DLC films from C$^+$ ions: experimental studies. *Diamond and Related Materials*, *4*(4), 318–323. Available from https://doi.org/10.1016/0925-9635(94)05205-0.

Lin, H., Dai, R., Shi, Y., Yuan, J., & Alfano, M. (2023). Effect of the GLC coating thickness on the mechanical and tribological properties of the CrN/GLC coatings. *Tribology Letters*, *71*(3). Available from https://doi.org/10.1007/s11249-023-01768-7, https://www.springer.com/journal/11249.

Liskiewicz, T., Fouvry, S., & Wendler, B. (2005). Hard coatings durability under fretting-wear. *Tribology and Interface Engineering Series*, *48*, 657–665. Available from https://doi.org/10.1016/s0167-8922(05)80067-7, http://www.elsevier.com.

Liskiewicz, T., Kubiak, K., & Comyn, T. (2017). Nano-indentation mapping of fretting-induced surface layers. *Tribology International*, *108*, 186–193. Available from https://doi.org/10.1016/j.triboint.2016.10.018.

Liu, Y., Chen, L., Zhang, B., Cao, Z., Shi, P., Peng, Y., Zhou, N., Zhang, J., & Qian, L. (2019). Key role of transfer layer in load dependence of friction on hydrogenated diamond-like carbon films in humid air and vacuum. *Materials*, *12*(9). Available from https://doi.org/10.3390/ma12091550, https://www.mdpi.com/1996-1944/12/9/1550/pdf.

Liu, Y., Erdemir, A., & Meletis, E. I. (1996). A study of the wear mechanism of diamond-like carbon films. *Surface and Coatings Technology*, *82*(1-2), 48–56. Available from https://doi.org/10.1016/0257-8972(95)02623-1.

Lomon, J., Chaiyabin, P., Saisopa, T., Seawsakul, K., Saowiang, N., Promsakha, K., Poolcharuansin, P., Pasaja, N., Chingsungnoen, A., Supruangnet, R., Chanlek, N., Nakajima, H., & Songsiriritthigul, P. (2018). XPS and XAS preliminary studies of diamond-like carbon films prepared by HiPIMS technique. *Journal of Physics: Conference Series*, *1144*(1). Available from https://doi.org/10.1088/1742-6596/1144/1/012048.

Malaczynski, G. W., Hamdi, A. H., Elmoursi, A. A., & Qiu, X. (1997). Diamond-like carbon coating for aluminum 390 alloy – Automotive applications. *Surface and Coatings Technology*, *93*(2-3), 280–286. Available from https://doi.org/10.1016/S0257-8972(97)00061-3, http://www.journals.elsevier.com/surface-and-coatings-technology/.

McMaster, S. J. (2020) Nanomechanical Characterisation of Diamond-Like Carbon Coatings for Tribological Performance. PhD thesis, University of Leeds.

McMaster, S. J., Kosarieh, S., Liskiewicz, T. W., Neville, A., & Beake, B. D. (2023). Utilising *H/E* to predict fretting wear performance of DLC coating systems. *Tribology*

International, *185*. Available from https://doi.org/10.1016/j.triboint.2023.108524, http://www.elsevier.com/inca/publications/store/3/0/4/7/4.

McMaster, S. J., Liskiewicz, T. W., Neville, A., & Beake, B. D. (2020). Probing fatigue resistance in multi-layer DLC coatings by micro- and nano-impact: Correlation to erosion tests. *Surface and Coatings Technology*, *402*. Available from https://doi.org/10.1016/j.surfcoat.2020.126319, http://www.journals.elsevier.com/surface-and-coatings-technology/.

Meine, K., Schneider, T., Spaltmann, D., & Santner, E. (2002a). The influence of roughness on friction Part I: The influence of a single step. *Wear*, *253*(7-8), 725–732. Available from https://doi.org/10.1016/S0043-1648(02)00159-X.

Meine, K., Schneider, T., Spaltmann, D., & Santner, E. (2002b). The influence of roughness on friction Part II. The influence of multiple steps. *Wear*, *253*(7-8), 733–738. Available from https://doi.org/10.1016/S0043-1648(02)00160-6.

Metallic materials − Instrumented indentation test for hardness and materials parameters − Part 1: Test method. (2015).

NANOINDENT-PLUS project Standardising the nano-scratch test, NMP-2012-CSA-6-319208. at. (2012). https://cordis.europa.eu/project/id/319208

Ni, W., Cheng, Y. T., Lukitsch, M. J., Weiner, A. M., Lev, L. C., & Grummon, D. S. (2004). Effects of the ratio of hardness to Young's modulus on the friction and wear behavior of bilayer coatings. *Applied Physics Letters*, *85*(18), 4028–4030. Available from https://doi.org/10.1063/1.1811377.

Okubo, H., & Sasaki, S. (2017). In situ Raman observation of structural transformation of diamond-like carbon films lubricated with MoDTC solution: Mechanism of wear acceleration of DLC films lubricated with MoDTC solution. *Tribology International*, *113*, 399–410. Available from https://doi.org/10.1016/j.triboint.2016.10.009, http://www.elsevier.com/inca/publications/store/3/0/4/7/4.

Okubo, H., Tadokoro, C., Sumi, T., Tanaka, N., & Sasaki, S. (2019). Wear acceleration mechanism of diamond-like carbon (DLC) films lubricated with MoDTC solution: Roles of tribofilm formation and structural transformation in wear acceleration of DLC films lubricated with MoDTC solution. *Tribology International*, *133*, 271–287. Available from https://doi.org/10.1016/j.triboint.2018.12.029, http://www.elsevier.com/inca/publications/store/3/0/4/7/4.

Oliver, W. C., & Pharr, G. M. (1992). An improved technique for determining hardness and elastic modulus using load and displacement sensing indentation experiments. *Journal of Materials Research*, *7*(6), 1564–1583. Available from https://doi.org/10.1557/jmr.1992.1564.

Pauleau, Y. (2008). *Residual stresses in DLC films and adhesion to various substrates. Tribology of diamond-like carbon films: Fundamentals and applications* (pp. 102–136). France: Springer US. Available from http://www.springerlink.com/openurl.asp?genre = book&isbn = 978-0-387-30264-5, https://doi.org/10.1007/978-0-387-49891-1_4.

Pei, Y. T., Galvan, D., & De Hosson, J. T. M. (2005). Nanostructure and properties of TiC/a-C:H composite coatings. *Acta Materialia*, *53*(17), 4505–4521. Available from https://doi.org/10.1016/j.actamat.2005.05.045.

Pharr, G. M., & Oliver, W. C. (1992). Measurement of thin film mechanical properties using nanoindentation. *MRS Bulletin*, *17*(7), 28–33. Available from https://doi.org/10.1557/s0883769400041634.

Pierson, H. O. (1993). *Handbook of carbon, graphite, diamonds and fullerenes: Processing, properties and applications (Materials science and process technology)*. Noyes Publications. Available from https://doi.org/10.1016/B978-0-8155-1339-1.50008-6.

Ren, S., Zheng, S., Pu, J., Lu, Z., & Zhang, G. (2015). Study of tribological mechanisms of carbon-based coatings in antiwear additive containing lubricants under high temperature. *RSC Advances*, *5*(81), 66426−66437. Available from https://doi.org/10.1039/c5ra08879h, http://pubs.rsc.org/en/journals/journalissues.

Renondeau, H., Papke, B. L., Pozebanchukz, M., & Parthasarathy, P. P. (2009). Tribological properties of diamond-like carbon coatings in lubricated automotive applications. *Proceedings of the Institution of Mechanical Engineers, Part J: Journal of Engineering Tribology*, *223*(3), 405−412. Available from https://doi.org/10.1243/13506501JET548.

Rincón, C., Zambrano, G., Carvajal, A., Prieto, P., Galindo, H., Martínez, E., Lousa, A., & Esteve, J. (2001). Tungsten carbide/diamond-like carbon multilayer coating on steel for tribological applications. *Surface and Coatings Technology*, *148*(2-3), 277−283. Available from https://doi.org/10.1016/S0257-8972(01)01360-3.

Robertson, J. (1992). Properties of diamond-like carbon. *Surface and Coatings Technology*, *50*(3), 185−203. Available from https://doi.org/10.1016/0257-8972(92)90001-q.

Robertson, J. (1994). The deposition mechanism of diamond-like a-C and a-C:H. *Diamond and Related Materials*, *3*(4-6), 361−368. Available from https://doi.org/10.1016/0925-9635(94)90186-4.

Robertson, J. (2002). Diamond-like amorphous carbon. *Materials Science and Engineering: R: Reports*, *37*(4-6), 129−281. Available from https://doi.org/10.1016/s0927-796x(02)00005-0.

Robertson, J. (2008a). Comparison of diamond-like carbon to diamond for applications. *Physica Status Solidi (A)*, *205*(9), 2233−2244. Available from https://doi.org/10.1002/pssa.200879720.

Robertson, J. (2008b). *Classification of diamond-like carbons. Tribology of diamond-like carbon films: Fundamentals and applications* (pp. 13−24). United Kingdom: Springer US. Available from http://www.springerlink.com/openurl.asp?genre = book&isbn = 978-0-387-30264-5, 10.1007/978-0-387-49891-1_1.

Ronkainen, H., & Holmberg, K. (2008). *Environmental and thermal effects on the tribological performance of DLC coatings. Tribology of diamond-like carbon films: Fundamentals and applications* (pp. 155−200). Finland: Springer US. Available from http://www.springerlink.com/openurl.asp?genre = book&isbn = 978-0-387-30264-5, 10.1007/978-0-387-49891-1_6.

Rueda-Ruiz, M., Beake, B. D., & Molina-Aldareguia, J. M. (2020). New instrumentation and analysis methodology for nano-impact testing. *Materials and Design*, *192*. Available from https://doi.org/10.1016/j.matdes.2020.108715, https://www.journals.elsevier.com/materials-and-design.

Santiago, J. A., Fernández-Martínez, I., Kozák, T., Capek, J., Wennberg, A., Molina-Aldareguia, J. M., Bellido-González, V., González-Arrabal, R., & Monclús, M. A. (2019). The influence of positive pulses on HiPIMS deposition of hard DLC coatings. *Surface and Coatings Technology*, *358*, 43−49. Available from https://doi.org/10.1016/j.surfcoat.2018.11.001.

Santner, E., Klaffke, D., Meine, K., Polaczyk, C., & Spaltmann, D. (2006). Effects of friction on topography and vice versa. *Wear*, *261*(1), 101−106. Available from https://doi.org/10.1016/j.wear.2005.09.028.

Schmellenmeier, H. (1953). Die Beeinflussung von festen Oberflachen durch eine ionisierte. *Experimentelle Technik der Physik*, *1*, 49−68.

Schmellenmeier, H. (1956). Kohlenstoffschichten mit Diamantstruktur. *Zeitschrift für Physikalische Chemie*, *205O*(1), 349−350. Available from https://doi.org/10.1515/zpch-1956-20541.

Schwan, J., Ulrich, S., Batori, V., Ehrhardt, H., & Silva, S. R. P. (1996). Raman spectroscopy on amorphous carbon films. *Journal of Applied Physics*, *80*(1), 440−447. Available from https://doi.org/10.1063/1.362745.

Shi, X., Liskiewicz, T. W., Beake, B. D., Chen, J., & Wang, C. (2020). Tribological performance of graphite-like carbon films with varied thickness. *Tribology International*, *149*. Available from https://doi.org/10.1016/j.triboint.2019.01.045, http://www.elsevier.com/inca/publications/store/3/0/4/7/4.

Siu, J. H. W., & Li, L. K. Y. (2000). An investigation of the effect of surface roughness and coating thickness on the friction andwear behaviour of a commercial MoS_2-metal coating on AISI 400C steel. *Wear*, *237*(2), 283−287. Available from https://doi.org/10.1016/S0043-1648(99)00349-X.

Skordaris, G., Bouzakis, A., & Bouzakis, K.-D. (2020). Impact test applications supported by FEA models in surface engineering for coating characterization. *Materials Proceedings*, *2*. Available from https://doi.org/10.3390/ciwc2020-06809.

Smart, R. F. (1978). Selection of surfacing treatments. *Tribology International*, *11*(2), 97−104. Available from https://doi.org/10.1016/0301-679x(78)90136-6.

Solis, J., Zhao, H., Wang, C., Verduzco, J. A., Bueno, A. S., & Neville, A. (2016). Tribological performance of an H-DLC coating prepared by PECVD. *Applied Surface Science*, *383*, 222−232. Available from https://doi.org/10.1016/j.apsusc.2016.04.184.

Sánchez-López, J. C., & Fernández, A. (2008). *Doping and alloying effects on DLC coatings. Tribology of diamond-like carbon films: Fundamentals and applications* (pp. 311−328). Spain: Springer US. Available from http://www.springerlink.com/openurl.asp?genre = book&isbn = 978-0-387-30264-5, 10.1007/978-0-387-49891-1_12.

Tai, F. C., Lee, S. C., Wei, C. H., & Tyan, S. L. (2006). Correlation between I_D/I_G ratio from visible Raman spectra and sp^2/sp^3 ratio from XPS spectra of annealed hydrogenated DLC film. *Materials Transactions*, *47*(7), 1847−1852. Available from https://doi.org/10.2320/matertrans.47.1847.

Stoney, G. G. (1909). The tension of metallic films deposited by electrolysis. *Proceedings of the Royal Society of London. Series A, Containing Papers of a Mathematical and Physical Character*, *82*(553), 172−175. Available from https://doi.org/10.1098/rspa.1909.0021.

Tyagi, A., Walia, R. S., Murtaza, Q., Pandey, S. M., Tyagi, P. K., & Bajaj, B. (2019). A critical review of diamond like carbon coating for wear resistance applications. *International Journal of Refractory Metals and Hard Materials*, *78*, 107−122. Available from https://doi.org/10.1016/j.ijrmhm.2018.09.006, http://www.journals.elsevier.com/international-journal-of-refractory-metals-and-hard-materials/.

Valli, J., Makela, U., & Matthews, A. (1986). Assessment of coating adhesion. *Surface Engineering*, *2*(1), 49−54. Available from https://doi.org/10.1179/sur.1986.2.1.49.

Valli, J., Mäkelä, U., Matthews, A., & Murawa, V. (1985). TiN coating adhesion studies using the scratch test method. *Journal of Vacuum Science & Technology A: Vacuum, Surfaces, and Films*, *3*(6), 2411−2414. Available from https://doi.org/10.1116/1.572848.

Vinoth, I. S., Detwal, S., Umasankar, V., & Sarma, A. (2019). Tribological studies of automotive piston ring by diamond-like carbon coating. *Tribology − Materials, Surfaces and Interfaces*, *13*(1), 31−38. Available from https://doi.org/10.1080/17515831.2019.1569852, http://www.tandfonline.com/loi/ytrb20#.VwHe3E1f1Qs.

Wade, A., Copley, R., Alsheikh Omar, A., Clarke, B., Liskiewicz, T., & Bryant, M. (2020). Novel numerical method for parameterising fretting contacts. *Tribology International*, *149*. Available from https://doi.org/10.1016/j.triboint.2019.06.019.

Wang, Y. X., & Zhang, S. (2014). Toward hard yet tough ceramic coatings. *Surface and Coatings Technology*, *258*, 1−16. Available from https://doi.org/10.1016/j.surfcoat.2014.07.007, http://www.journals.elsevier.com/surface-and-coatings-technology/.
Waterhouse, R. B. (1976). Fretting fatigue. *Materials Science and Engineering*, *25*(C), 201−206. Available from https://doi.org/10.1016/0025-5416(76)90071-9.
Waterhouse, R. B. (1984). Fretting wear. *Wear*, *100*(1-3), 107−118. Available from https://doi.org/10.1016/0043-1648(84)90008-5.
Wright, T., & Page, T. F. (1992). Nanoindentation and microindentation studies of hard carbon on 304 stainless steel. *Surface and Coatings Technology*, *54-55*, 557−562. Available from https://doi.org/10.1016/S0257-8972(07)80082-X.
Yang, S., Li, X., Renevier, N. M., & Teer, D. G. (2001). Tribological properties and wear mechanism of sputtered C/Cr coating. *Surface and Coatings Technology*, *142-144*, 85−93. Available from https://doi.org/10.1016/S0257-8972(01)01147-1.
Yong, Q., Ma, G., Wang, H., Chen, S., & Xu, B. (2016). Influence of tungsten content on microstructure and properties of tungsten-doped graphite-like carbon films. *Journal of Materials Research*, *31*(23), 3766−3776. Available from https://doi.org/10.1557/jmr.2016.433, http://journals.cambridge.org/action/displayJournal?jid = JMR.
Yoshida, Y., & Kunitsugu, S. (2018). Friction wear characteristics of diamond-like carbon coatings in oils containing molybdenum dialkyldithiocarbamate additive. *Wear*, *414-415*, 118−125. Available from https://doi.org/10.1016/j.wear.2018.08.004, https://www.journals.elsevier.com/wear.
Yue, W., Liu, C., Fu, Z., Wang, C., Huang, H., & Liu, J. (2015). Effects of tungsten doping contents on tribological behaviors of tungsten-doped diamond-like carbon coatings lubricated by MoDTC. *Tribology Letters*, *58*(2). Available from https://doi.org/10.1007/s11249-015-0508-3, http://www.springerlink.com/(snpxut45gxflnr45vb2gia45)/app/home/journal.asp?referrer = parent&backto = searchpublicationsresults,1,2.
Zahid, R., Masjuki, H. H., Varman, M., Kalam, M. A., Mufti, R. A., Mohd Zulkifli, N. W. B., Gulzar, M., & Nor Azman, S. S. B. (2016). Influence of intrinsic and extrinsic conditions on the tribological characteristics of diamond-like carbon coatings: A review. *Journal of Materials Research*, *31*(13), 1814−1836. Available from https://doi.org/10.1557/jmr.2016.31, http://journals.cambridge.org/action/displayJournal?jid = JMR.
Zeng, X. T., Zhang, S., Ding, X. Z., & Teer, D. G. (2002). Comparison of three types of carbon composite coatings with exceptional load-bearing capacity and high wear resistance. *Thin Solid Films*, *420-421*, 366−370. Available from https://doi.org/10.1016/s0040-6090(02)00840-4.
Zhang, S., Bui, X. L., Fu, Y., Butler, D. L., & Du, H. (2004). Bias-graded deposition of diamond-like carbon for tribological applications. *Diamond and Related Materials*, *13*(4-8), 867−871. Available from https://doi.org/10.1016/j.diamond.2003.10.043.
Zhang, S., Bui, X. L., Fu, Y., & Du, H. (2004). Development of carbon-based coating of extremely high toughness with good hardness. *International Journal of Nanoscience*, *3*(4-5), 571−578. Available from https://doi.org/10.1142/s0219581x04002395, http://www.worldscinet.com/ijn/ijn.shtml.
Zhang, S., Sun, D., Fu, Y., & Du, H. (2004). Effect of sputtering target power on microstructure and mechanical properties of nanocomposite nc-TiN/a-SiN$_x$ thin films. *Thin Solid Films*, *447-448*, 462−467. Available from https://doi.org/10.1016/S0040-6090(03)01125-8.
Zhang, S., Xie, H., Zeng, X., & Hing, P. (1999). Residual stress characterization of diamond-like carbon coatings by an X-ray diffraction method. *Surface and Coatings Technology*, *122*(2-3), 219−224. Available from https://doi.org/10.1016/S0257-8972(99)00298-4.

Zia, A. W., & Birkett, M. (2021). Deposition of diamond-like carbon coatings: Conventional to non-conventional approaches for emerging markets. *Ceramics International*, *47*(20), 28075−28085. Available from https://doi.org/10.1016/j.ceramint.2021.07.005, https://www.journals.elsevier.com/ceramics-international.

Zou, C. W., Wang, H. J., Feng, L., & Xue, S. W. (2013). Effects of Cr concentrations on the microstructure, hardness, and temperature-dependent tribological properties of Cr-DLC coatings. *Applied Surface Science*, *286*, 137−141. Available from https://doi.org/10.1016/j.apsusc.2013.09.036.

Machining of difficult materials

Jose Luis Endrino[1], German Fox-Rabinovich[2] and Ben Beake[3]
[1]Universidad Loyola Andalucia, Dos Hermanas, Seville, Spain, [2]MMRI, McMaster University, Hamilton, ON, Canada, [3]Micro Materials Ltd., Wrexham, United Kingdom

19.1 Hard coatings for cutting processes

The evaluation of hard coatings can encompass damage, analysis of wear mechanisms, and nanomechanical surface studies (Schalk et al., 2022). In metal cutting scenarios, substantial forces exert themselves in the tool-to-workpiece contact area. The chip formation process induces plastic deformation in shear zones, giving rise to heat and contact stress within the tool, which, in turn, experiences abrasive and adhesive wear. When a softer material moves across the surface of a harder material, it may carry a significant concentration of hard particles. These hard particles function analogous to small cutting edges and to a grinding wheel. The presence of these hard particles contributes to the wear and tear of the tool material. Sometimes, particles of the hard material emerge from the surface and are dragged along the contact surface, further affecting the wear on the tool. Mostly, the role of hard coatings is to offer protection against wear and thermal effects on the tool substrate. The rapid commercial growth of physical vapor deposition (PVD) coatings for cutting tools has been boosted by advancements in manufacturing technologies such as increased cutting speeds and dry machining (Kalss et al., 2006).

Large manufacturers of cutting tools and coatings prioritize productivity. While achieving a 30% reduction in tool costs or a 50% increase in tool lifetime only leads to a modest 1% reduction in manufacturing costs; on the other hand, a 20% increase in cutting speed can result in a significant 15% reduction in manufacturing costs (Kalss et al., 2006). To enhance productivity, different strategies, such as high-performance cutting (HPC) and high-speed cutting (HSC), can be employed. Nevertheless, the specific criteria for hard coatings are contingent upon the nature of the cutting process and the material of the workpiece. Progress in coating technologies, including TiCN, TiAlXN, AlTiXN, AlCrXN, and their related nanocomposite coatings, has facilitated improvements in multiple manufacturing processes (Beake et al., 2017; Endrino et al., 2007; Fox-Rabinovich et al., 2006; Endrino & Derflinger, 2005). The success story of PVD coatings began 43 years ago with TiN and continued its evolution with coatings based on the Al-Ti-N and Al-Cr-N systems, which also maintain a NaCl (Rocksalt) structure. These coatings exhibit superior abrasive wear resistance and enhanced oxidation resistance and show excellent results in cutting tool applications (Kalss et al., 2006). Oxidation and nanomechanical tests conducted at elevated temperatures have been able to assess their chemical stability in such environments and subsequent correlations to coated tool performance in various metal cutting tests have been reported in the literature (Beake & Fox-Rabinovich, 2014; Beake et al., 2017; Beake, Fox-Rabinovich, et al., 2009; Beake, Smith et al., 2007; Fox-Rabinovich et al., 2006). Besides oxidation resistance and excellent mechanical properties at high temperature, it is acknowledged that other coating physical properties such thermal conductivity can play a critical role. For instance, studies have shown that crater wear stems from elevated thermal loads on the tool material. Hence, to prevent premature failure, either a low thermal conductivity or a substantial coating thickness is necessary. Also, studies on the thermal conductivity of hard coatings indicate that coatings based on AlCrN exhibit a low thermal conductivity in comparison to AlTiN and TiAlN coatings, with a key aspect being the lack of increase in thermal conductivity at high temperature and over time (Kalss et al., 2006). This makes AlCrN coatings one of the key players in dry cutting applications. For example, AlCrN-coated tools have performed better than $Ti_{0.5}Al_{0.5}N$-coated tools in end milling of AISI 1040 steel, interrupted turning 42CrMo4V steel and deep hole drilling of hardened structural steel (Beake et al., 2017; Fox-Rabinovich et al., 2006; Kalss et al., 2006).

AlCrN has also shown high resistance to fracture in cyclic nano-impact testing. These rapid nano- and microscale impact tests performed with the NanoTest system have shown a direct correlation to coating performance in applications (Beake & Fox-Rabinovich, 2014; Beake et al., 2009; Beake, Isern, et al., 2019; Bouzakis, Klocke, et al., 2011; Bouzakis et al., 2012; Skordaris, Bouzakis, Charalampous, et al., 2014; Skordaris, Bouzakis, Kotsanis, et al., 2017).

Fig. 19.1 shows comparative behavior of AlCrN and TiAlN coatings in (A) cyclic nano-impact tests, (B) cyclic micro-impact testing at 2 N, and (C) end milling tests of AISI 1040 structural steel.

Fig. 19.1 shows that in the impact and cutting tests after the run-in, there is a period of relatively low damage before a transition to a more severe wear mode and ultimately tool failure. In the cutting tests, the transition occurred after around 40 m on TiAlN and around 280 m for AlCrN. Kalss et al. (2006) similarly showed that in high-speed milling tests of AISI 1045 carbon steel that although initially the flank wear of AlCrN coated mills was worse than TiCN or TiAlN-coated mills, the AlCrN coating enabled a more effective transition to a low wear rate, so that ultimately tool life was $\times 2$ greater than TiAlN and $\times 3$ greater than TiCN.

In the nano-impact test, the transition occurred after (6 ± 2) impacts on TiAlN and (14 ± 3) impacts on AlCrN. As with the cutting test, the increase in wear rate after the transition was much more gradual on AlCrN. Micro-impact tests were performed at 2 and 2.5 N, with 15 repeats at each load. At 2 N, the AlCrN did not fail in any of the 15 tests, but the TiAlN coating failed after (24 ± 10) impacts. At 2.5 N, it failed after (23 ± 6) impacts. Under these more severe conditions, the AlCrN failed in 9/15 tests, with mean failure after (55 ± 22) impacts. In some cases, AlCrN coating optimization has introduced a TiAlN sublayer to improve adhesion (Beake et al., 2015).

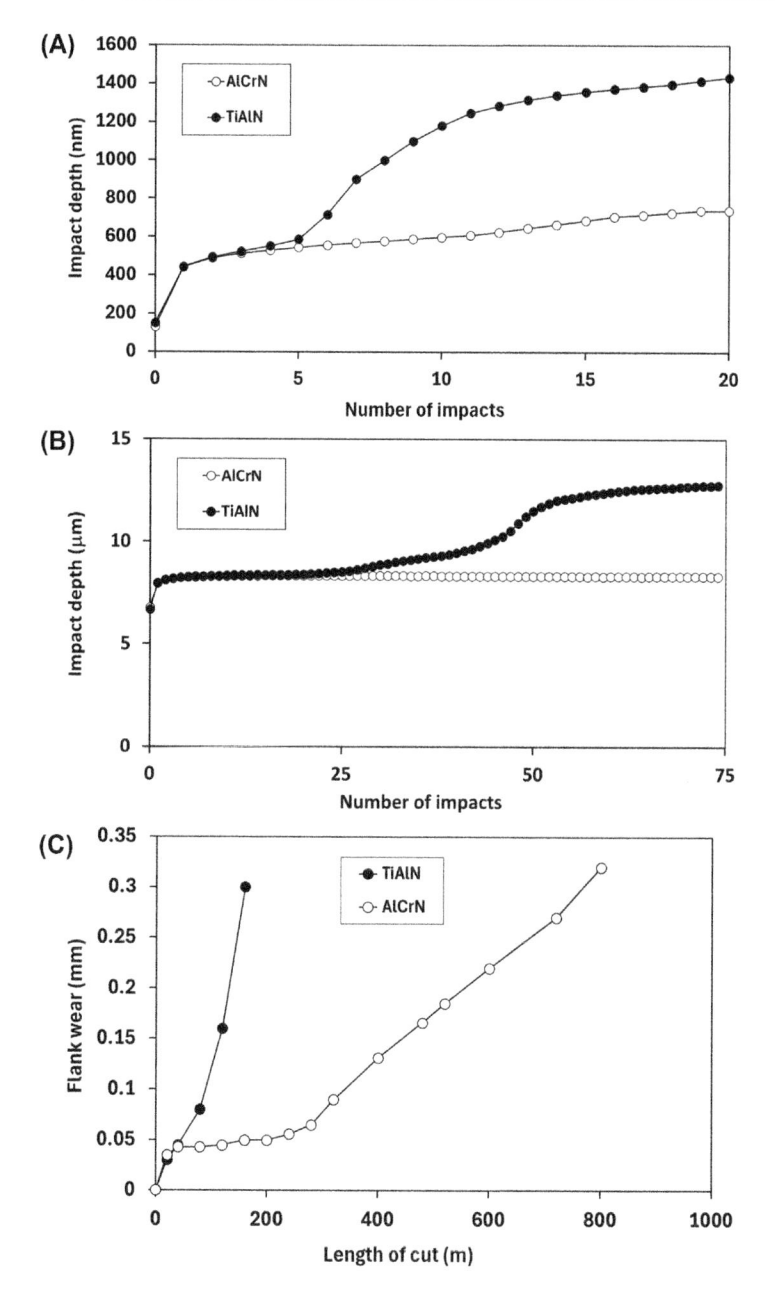

Figure 19.1 Comparative performance of AlCrN and TiAlN coatings in (A) cyclic nano-impact tests, (B) cyclic micro-impact testing at 2 N, and (C) end milling tests of AISI 1040 structural steel.

Tools for machining Ti alloys normally experience rapid tool wear limiting cutting speeds to around 60 m min^{-1}. AlCrN coatings have been used in high-speed machining of Ti6Al4V at 100 m min^{-1} (He et al., 2024). Chipping wear on the rake face was the major factor responsible for tool failure. AlCrN coating design with moderate stress showed more gradual wear in nano-impact tests and longer tool life. The high fracture resistance of AlCrN in nano-impact tests has also been observed in tests at elevated temperature. In nano-impact tests of TiAlN, AlTiN (67%AlTi33%) and AlCrN coatings at 50–200 mN at room temperature, the AlCrN was more resistant to fracture than either of the (Ti,Al)N coatings. At 500°C, all the coatings were more resistant to fracture (Beake et al., 2007), which may be in part a result of the lower developed stresses due to coating softening. Under the test conditions, the AlCrN and AlTiN coatings did not fracture at elevated temperature.

19.2 Hard nitride coatings for machining ductile materials

In one of our previous investigations, stainless-steel plates were subjected to machining using cemented carbide finishing end mills coated with four different PVD coatings, each containing a high aluminum content and a B1-NaCl crystalline structure (Endrino et al., 2005, 2006). The combination of high toughness and ductility in austenitic stainless steel contributes to the generation of elongated continuous chips and substantial adhesion of the workpiece material to the cutting tool surface. This, in turn, increases adhesive wear. Furthermore, elevated temperatures at the tool – chip interface promote increased diffusion and chemical wear. The formation of a buildup edge, typical of machining a high ductile material, and its subsequent tearing during cutting can produce a high instability in machining forces, ultimately leading to chipping of the cutting edge. The selected coatings for this machining process included AlCrN, AlCrNbN, fine-grained (fg) AlTiN, and nanocrystalline (nc) AlTiN. Since both AlTiN and AlCrN-based coatings exhibit excellent oxidation resistance attributed to the formation of aluminum oxide surface layers, they were considered as hard coating candidates for this machining operation. The machinability of austenitic stainless steels can be significantly improved by employing coated cutting tools and the use of hard PVD coatings with low thermal conductivity and improved surface finishes. Thus, the coating design approach should be to improve frictional characteristics at the tool/workpiece interface by selecting a coating with very high oxidation resistance and to achieve an efficient chip evacuation process through the use of an appropriate PVD coating posttreatment (Endrino et al., 2006).

The XRD patterns of AlCrNbN and AlCrN coatings are shown in Fig. 19.2A, representing two examples of AlCr-based coatings. Both coatings exhibit a cubic B1 (NaCl) structure, with their major peaks located in proximity to the corresponding cubic AlN peaks. However, despite these similarities, the two coatings have different textures: the AlCrN has a (111) preferential orientation and the AlCrNbN has a (200) to a (111)-diffraction ratio of 4.2. Such texture change could be caused by

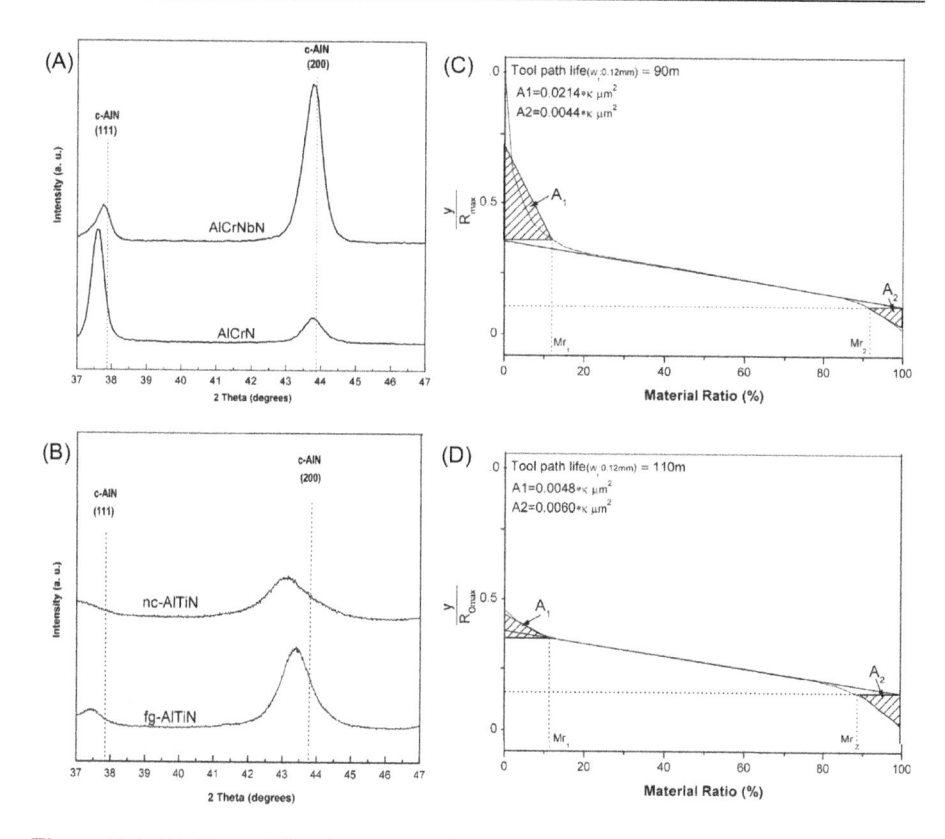

Figure 19.2 (A) X-ray diffraction patterns for AlCrNbN compared with AlCrN. (B) Nanocrystalline AlTiN compared with fine-grained AlTiN. (C) Material ratio curve for "as deposited" AlCrN. (D) Material ratio curve for AlCrN after surface treatment.
Source: With permission from Endrino, J. L., Fox-Rabinovich, G. S., & Gey, C. 2006. Hard AlTiN, AlCrN PVD coatings for machining of austenitic stainless steel. *Surface and Coatings Technology*, 200, 6840–6845. https://doi.org/10.1016/j.surfcoat.2005.100.030, Copyright (2006).

the lower solubility of the aluminum nitride cubic phase in NbAlN in comparison to AlCrN, causing a change of preferential orientation in the grains with crystal growth. The textural change resulted in a lifetime increase of ~20%, which could be due to a smoother surface of (200) planes. Similarly, Fig. 19.2B shows the XRD patterns for fine grained-AlTiN and nanocrystalline-AlTiN with both coatings showing a cubic B1 (NaCl) and a (200) texture, the estimated grain sizes for these two hard tool coatings were ~20 nm for the nc-TiAlN in comparison to ~35 nm for the fg-AlTiN. In this case, the lifetime increase was ~80% higher for the nc-AlTiN, which underscores the importance of coating grain size in machining of ductile materials (Endrino et al., 2005, 2006).

Another crucial factor to consider is the influence of postdeposition treatment on tool wear intensity. In cathodic arc evaporation processes, the presence of macroparticles on the coating surface adversely affects the chip flow. A surface posttreatment, which polishes the surface and removes these deposited macroparticles, can be used to enhance cutting performance. Fig. 19.2C and D show the quantification of peak and valley areas of two material ratio curves corresponding to hard AlCrN PVD coating before and after posttreatment. In this case, the peak area was significantly reduced, the valley area increased around 25%, and the increase in tool path life was ~20%. It was speculated that the slight increase in valley area could also be beneficial in machining ductile materials due to the presence of lubricant reservoirs (Endrino et al., 2005).

19.3 Self-organization processes during wear of cutting tools

In friction and wear processes, coating materials in the tribological contact can experience self-adaptation, often resulting in substantial structural alterations within the surface layer (Fox-Rabinovich et al., 2007). The adaptability of any material, including hard coatings, is intricately linked to the deposition process or to changes in its properties and triggered by external stimuli, leading to a diminishing impact of their influence (Fox-Rabinovich et al., 2007, 2023; Pardo et al., 2013). Adaptation and self-organization manifest during friction and wear, causing significant evolution in the characteristics of both the surface and underlying layers (Fox-Rabinovich et al., 2014). This self-organization phenomenon is particularly evident in the extreme tribological conditions of high-performance dry machining of hardened tool steels, where the friction surface temperature reaches $1000°C-1200°C$, and stresses range around $3-5$ GPa (Chowdhury et al., 2018). Fig. 19.3 shows a diagram of the evolutionary self-organization mechanism for nanostructured hard PVD coatings under extreme tribological conditions. The first part of the evolution process corresponds to the running-in stage characterized by the initiation of the formation of self-organized adaptive secondary structures, which strongly decrease the wear and the entropy production. The second part of the evolution is called post running-in stage. In this later stage, entropy production is minimized, wear rate is low, and the friction coefficient is minimum.

As an example of the aforementioned mechanism, an advanced nanomultilayer coating of TiAlCrSiYN/TiAlCrN, characterized by a thermodynamic state significantly far from equilibrium, acts as an effective catalyst for favorable tribological physical − chemical processes in which the multilayer forms mullite triboceramics first and sapphire/mullite triboceramics (along with various tribological low friction layers) during the post running-in stage. This triboprocess of self-organization provides the engineered friction surface with self-protective properties. The outcomes of partial X-ray emission (XES) experiments on worn cutting tools with the multilayer coating are illustrated in Fig. 19.3(B). Notably, there is a significant increase in Si-K line intensity compared with Al K after high-vacuum annealing at 600°C. Despite a substantial rise in Si emission intensity postannealing at 600°C, further increase in the intensity is not observed at higher temperatures (as seen in spectra at

	Running-in Stage of wear (15m)	Post-running-in Stage of wear (30m)
	The self-organization process is initiated. Strong reduction in wear rate and entropy production under extreme frictional conditions.	The self-organization process is finalized. Wear rate and entropy production is minimized, leading to milder frictional conditions.
Nanolayer of the tribo-films	Functional hierarchy is established. Mullite triboceramics predominantly form with the best termal properties.	Formation of high amount of termal barrier sapphire/mullite tribo-ceramics and some amount of various tribofilms
Layer of the coating	Complex columnar/nano multilayer structure with non-equilibrium state.	

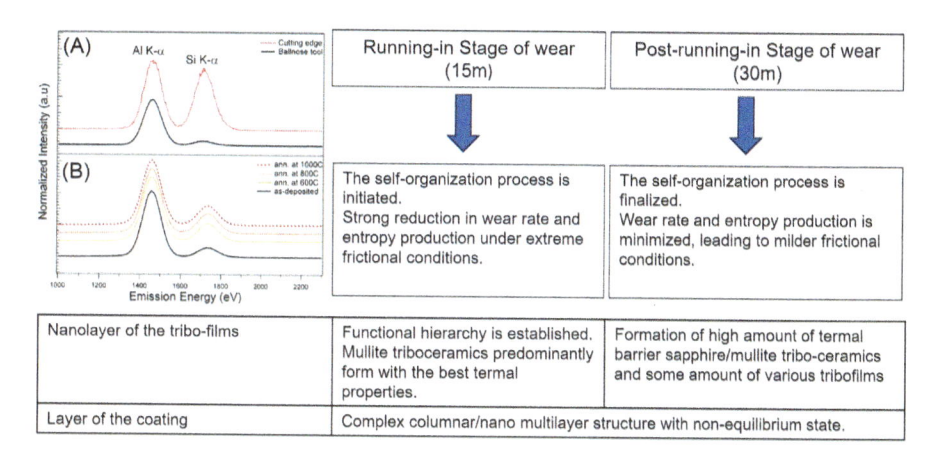

Figure 19.3 Diagram of the evolutionary self-organization mechanism in nanostructured TiAlCrSiYN/TiAlCrN PVD coatings under extreme tribological conditions. Inset (A): Partial X-ray emission scan measured at incident energy of 1860 eV in the tool with multilayer coating, comparison of Al-K and Si-K lines in different spots of the surface of the worn ball nose end mill; (B) comparison of Al K and Si K curves in as-deposited coating and after annealing during 30 min in vacuum at various temperatures.

800°C and 1000°C), indicating saturation in Si mass transfer to the surface. On the contrary, at the cutting edge, the Si K spectrum intensity is significantly higher, indicating the correct functioning of the self-organization due to the extreme tribological conditions. The tribooxidation analysis indicated the replacement of nitrogen by oxygen atoms on the friction surface, leading to the later release of Si from the multilayer coating. X-ray absorption spectroscopy (XAS) was also employed to investigate the chemical state of silicon at the cutting edge, the XAS fingerprint differed significantly from the one of pure Si, which is another evidence of the self-organization process in the multilayer (Fox-Rabinovich et al., 2014).

19.4 Influence of fracture resistance and high-temperature mechanical properties in self-adaptive behavior of TiAlCrSiYN-based coatings

Effective tribofilm formation and replenishment in tribologically extreme conditions are aided by having a low-wear environment without excessive coating fracture. The optimization of the coating architecture for improved performance under tribologically extreme conditions of dry ball nose high-speed milling of hardened H13 steel at 600 m min^{-1} has focused on optimization of the residual stress and interlayer thickness.

Chowdhury et al. (2018) compared the performance of (1) 3 μm monolayer TiAlCrN and TiAlCrSiYN, (2) 2 μm nanomultilayered TiAlCrSiYN/TiAlCrN, (3) 2 and 3 μm nanomultilayered TiAlCrSiYN/TiAlCrN with 100 nm TiAlCrN interlayer, investigating their micromechanical properties, triboceramic films, and cutting data in dry ball nose high-speed milling of hardened H13 steel. The best tool life was shown by the 3 μm thick nanomultilayered TiAlCrSiYN/TiAlCrN with 100 nm TiAlCrN interlayer, with optimized residual stress. There was more gradual wear evolution on the nanomultilayered coatings with the interlayer in the cutting tests and in the impact tests. The nanolaminate structure in the nanomultilayer coatings provides an "effective toughening" mechanism where its resistance to through-thickness cracking is enhanced by crack deflection, for example, along interfaces. A greater proportion of protective sapphire and mullite triboceramic thermal barrier films were formed for 3 μm nanomultilayered TiAlCrSiYN/TiAlCrN with 100 nm TiAlCrN interlayer. The 2 μm nanomultilayer coating with the interlayer also showed higher scratch toughness, through a relatively low load for cracking combined with a larger load for total failure, which reflects its initially high flank and rake wear combined with good durability thereafter seen in the cutting tests. Subsequently, the influence of varying TiAlCrN interlayer thickness between 100 and 500 nm on the performance of 2 μm nanomultilayered TiAlCrSiYN/TiAlCrN in the same application has been studied (Chowdhury et al., 2019). The best performance was with a 300 nm interlayer. In nano-impact tests, this coating was also more resistant to spallation than the coatings with 100 or 500 nm interlayers.

Coating optimization based on room-temperature properties can be misleading if the relative ranking changes with temperature, for example, due to a more significant change in mechanical properties (Inspektor & Salvador, 2014) or differences in an adaptive mechanism that is not operative at lower temperature. For this reason, alongside room temperature tests, elevated temperature nanomechanical and microtribological measurements can provide more application-relevant, useful characterization data. Coating (and substrate) mechanical properties of the coatings decrease with increasing temperature, although there are differences in how much. Fig. 19.4 shows the variation in (A) hardness (B) H/E and (C) H^3/E^2 with temperature for monolayer TiAlN, AlCrN, TiAlCrN, TiAlCrSiYN, and nanomultilayered TiAlCrSiYN/TiAlCrN coatings on cemented carbide from 25°C to 600°C. The elastic modulus of the coatings is less strongly influenced by temperature than hardness so that changes in H/E are of similar magnitude to changes in hardness. As the temperature increases, there is a much stronger decrease in H^3/E^2.

The cutting temperatures may be >1000°C in continuous high-speed turning operations while in ultrahigh-speed milling of hardened steels, temperatures can reach 1100°C–1200°C at a cutting speed of 600 m min^{-1} (Chowdhury et al., 2019). Retaining high coating resistance to plastic deformation at the operating temperature (i.e., hot hardness) is important to protect against substrate softening. In some interrupted cutting operations of hard-to-cut materials, operating speeds are lower and the contact temperatures developing through friction can be lower. An example is in the interrupted cutting of Ti6Al4V where temperatures well below 500°C have been reported. Under these conditions, longer tool life has been shown for AlTiN, compared with TiAlCrN or TiAlN. Nanomechanical measurements show that the AlTiN coating had relatively low high temperature hardness over

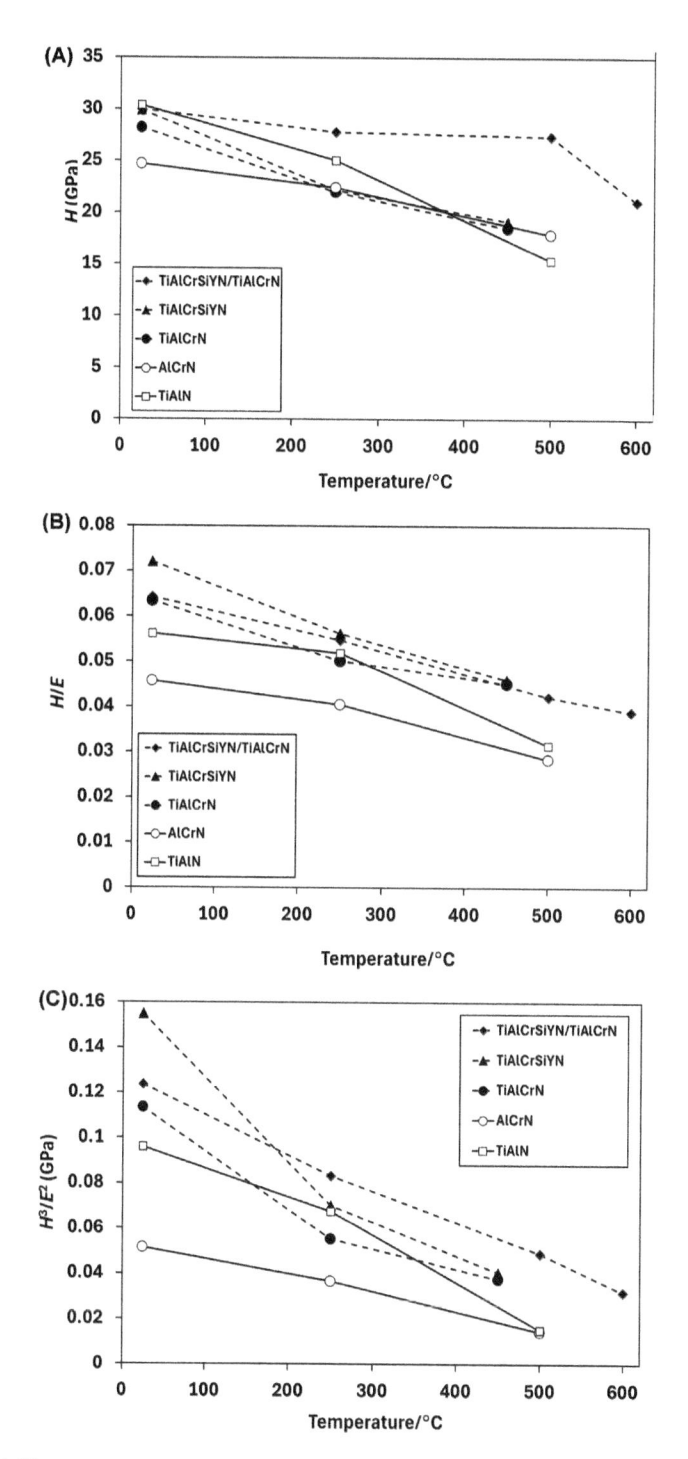

Figure 19.4 Temperature dependency of mechanical properties of selected hard coatings. (A) Hardness versus temperature; (B) H/E versus temperature; (C) H^3/E^2 versus temperature.

250°C but excellent fracture resistance at elevated temperature, shown in nano-impact and high critical loads for cracking in microscratch tests at 500°C.

The relative importance of the high-temperature hardness (or yield stress) and fracture resistance of the coating in determining cutting tool life vary with the cutting test. As mentioned earlier, high hot hardness is paramount in high-speed continuous turning and ultrahigh-speed milling of hardened steels. For example, nanomultilayered TiAlCrSiYN/TiAlCrN had a longer tool life than TiAlCrN and TiAlCrSiYN in turning the hard-to-cut Ni-base superalloys Inconel 718 and ME16, as it can combine high hot hardness and fracture resistance (Fox-Rabinovich et al., 2010). The TiAlCrSiYN/TiAlCrN family of nanomultilayer coatings combine a range of desirable properties that work together to achieve extended tool life in the tribologically extreme conditions through self-organized adaptation with protective triboceramic thermal barrier films. In comparison to monolayered columnar coatings that can perform very well under less severe conditions (e.g., lower cutting speeds), the advantages of the nanomultilayer coatings include (1) improved high temperature mechanical properties, with higher H^3/E^2 providing (a) enhanced load carrying capability and (b) resistance to crack initiation; (2) nano/microstructural advantages of (a) dense microstructure—eliminating/minimizing weak columnar boundaries that act as defects for through-thickness cracks and (b) nanomultilayer structure providing "effective toughening" mechanism where resistance to through-thickness cracking is also enhanced by crack deflection, for example, along interfaces. These advantages enable the TiAlCrSiYN/TiAlCrN family of nanomultilayer coatings to perform at higher cutting speeds.

19.5 Summary

PVD hard coatings display complex adaptive behavior that extends tool life when machining difficult to cut materials and in ultrahigh-speed cutting. In machining applications, coated components are subjected to high contact stresses under normal and tangential loads (Gsellmann et al., 2020), with high strain rates that can approach 10^5 s^{-1}. Nanomechanical and nano/microtribological characterizations under conditions that can more effectively replicate key features of the contact (e.g., stresses, strain rates, temperature, nature of contact) than quasistatic tests at room temperature have proved extremely useful in helping to understand some of the reasons for their improved performance. Nano- and microscale scratch tests can replicate the deformation in highly loaded sliding/abrasive contact (e.g., continuous turning) (Beake et al., 2017), but to date, more attention has focused on impact testing. Cyclic nano- and micro-impact tests can effectively simulate the interrupted, high strain rate, highly loaded mechanical contacts in milling operations. The cyclic nano- and micro-impact tests are accelerated, providing a more severe test of coating behavior than in larger-scale impact tests with mm-sized probes. A potential advantage of testing at smaller scale is that it may offer closer simulation with highly loaded high cutting speed conditions, where finite element modeling of the cutting process has shown that coatings are overstressed and plastically deform. This situation is not replicated in cyclic impact tests at larger scale where the substrate deforms elastoplastically and the coating deforms elastically. However, with the much smaller contact size and higher pressures in the nano- and microscale impact tests, there is some initial coating plasticity and higher bending stresses. Skordaris et al. reported perfect agreement between nano-impact tests and cutting behavior, but agreement with larger-scale impact tests was less good (Skordaris et al., 2017).

Coatings with improved mechanical properties at high temperatures protect the tool by providing a low-wear environment without excessive coating fracture enabling effective formation and replenishment of protective triboceramic films in tribologically extreme conditions.

References

Beake, B. D., Endrino, J. L., Kimpton, C., Fox-Rabinovich, G. S., & Veldhuis, S. C. (2017). Elevated temperature repetitive micro-scratch testing of AlCrN, TiAlN and AlTiN PVD coatings. *International Journal of Refractory Metals and Hard Materials*, 69, 215−226. Available from https://doi.org/10.1016/j.ijrmhm.2017.08.017.

Beake, B. D., & Fox-Rabinovich, G. S. (2014). Progress in high temperature nanomechanical testing of coatings for optimising their performance in high speed machining. *Surface and Coatings Technology*, 255, 1021115.

Beake, B. D., Fox-Rabinovich, G. S., Veldhuis, S. C., & Goodes, S. R. (2009). Coating optimisation for high speed machining with advanced nanomechanical test methods. *Surface and Coatings Technology*, 203(13), 1919−1925. Available from https://doi.org/10.1016/j.surfcoat.2009.01.025.

Beake, B. D., Isern, L., Endrino, J. L., & Fox-Rabinovich, G. S. (2019). Micro-impact testing of AlTiN and TiAlCrN coatings. *Wear*, 418−419, 102−110.

Beake, B. D., Ning, L., Gey, C., Veldhuis, S. C., Komarov, A., Weaver, A., Khanna, M., & Fox-Rabinovich, G. S. (2015). Wear performance of different PVD coatings during hard wet end milling of H13 tool steel. *Surface and Coatings Technology*, 279, 118−125. Available from https://doi.org/10.1016/j.surfcoat.2015.08.038, http://www.journals.elsevier.com/surface-and-coatings-technology/.

Beake, B. D., Smith, J. F., Gray, A., Fox-Rabinovich, G. S., Veldhuis, S. C., & Endrino, J. L. (2007). Investigating the correlation between nano-impact fracture resistance and hardness/modulus ratio from nanoindentation at $25-500°C$ and the fracture resistance and lifetime of cutting tools with Ti1 − xAlxN (x = 0.5 and 0.67) PVD coatings in milling operations. *Surface and Coatings Technology*, 201(8), 4585−4593. Available from https://doi.org/10.1016/j.surfcoat.2006.09.118.

Bouzakis, K.-D., Klocke, F., Skordaris, G., Bouzakis, E., Gerardis, S., Katirtzoglou, G., & Makrimallakis, S. (2011). Influence of dry micro-blasting grain quality on wear behaviour of TiAlN coated tools. *Wear*, 271(5−6), 783−791. Available from https://doi.org/10.1016/j.wear.2011.03.010.

Bouzakis, K.-D., Michailidis, N., Skordaris, G., Bouzakis, E., Biermann, D., & M'Saoubi, R. (2012). Cutting with coated tools: coating technologies, characterization methods and performance optimisation. *CIRP Annals − Manufacturing Technology*, 61, 703−723.

Chowdhury, S., Beake, B. D., Yamamoto, K., Bose, B., Aguirre, M., Fox-Rabinovich, G. S., & Veldhuis, S. C. (2018). Improvement of wear performance of nano-multilayer PVD coatings under dry hard end milling conditions based on their architectural development. *Coatings*, *8*(2). Available from https://doi.org/10.3390/coatings8020059, https://res.mdpi.com/coatings/coatings-08-00059/article_deploy/coatings-08-00059-v2.pdf?filename = &attachment = 1.

Chowdhury, S., Bose, B., Yamamoto, K., & Veldhuis, S. C. (2019). Effect of interlayer thickness on nano-multilayer coating performance during high speed dry milling of H13 tool steel. *Coatings*, *9*(11). Available from https://doi.org/10.3390/coatings9110737.

Endrino, J. L., & Derflinger, V. (2005). The influence of alloying elements on the phase stability and mechanical properties of AlCrN coatings. *Surface and Coatings Technology*, *200*, 988–992. Available from https://doi.org/10.1016/j.surfcoat.2005.02.196.

Endrino, J. L., Fox-Rabinovich, G. S., & Gey, C. (2006). Hard AlTiN, AlCrN PVD coatings for machining of austenitic stainless steel. *Surface and Coatings Technology*, *200*, 6840–6845. Available from https://doi.org/10.1016/j.surfcoat.2005.100.030.

Endrino, J. L., Fox-Rabinovich, G. S., Reiter, A., Veldhuis, S. C., Escobar Galindo, R., Albella, J. M., et al. (2007). Oxidation tuning in AlCrN coatings. *Surface and Coatings Technology*, *201*, 4505–4511. Available from https://doi.org/10.1016/j.surfcoat.2006.090.089.

Endrino, J.L., Wachter, A., Kuhnt, E., Mettler, T., Neuhaus, J., & Gey, C. (2005). Hard PVD coatings for austenitic stainless steel machining: New developments, *Materials Research Society Symposium Proceedings, Liechtenstein*, *843*, 19–24.

Fox-Rabinovich, G., Gershman, I., Goel, S., & Endrino, J. L. (2023). Control over multi-scale self-organization-based processes under the extreme tribological conditions of cutting through the application of complex adaptive surface-engineered systems. *Lubricants*, *11*(3), 106. Available from https://doi.org/10.3390/lubricants11030106.

Fox-Rabinovich, G., Kovalev, A., Aguirre, M. H., Yamamoto, K., Veldhuis, S., Gershman, I., Rashkovskiy, A., Endrino, J. L., Beake, B., Dosbaeva, G., Wainstein, D., Junifeng Yuan, J. W., & Bunting. (2014). Evolution of self-organization in nano-structured PVD coatings under extreme tribological conditions. *Applied Surface Science*, *297*, 22–32. Available from https://doi.org/10.1016/j.apsusc.2014.01.052.

Fox-Rabinovich, G., Veldhuis, S. C., Kovalev, A. I., Wainstein, D. L., Gershman, I. S., Korshunov, S., Shuster, L. S., & Endrino, J. L. (2007). Features of self-organization in ion modified nanocrystalline plasma vapor deposited AlTiN coatings under severe tribological conditions. *Journal of Applied Physics*, *102*(7). Available from https://doi.org/10.1063/1.2785947.

Fox-Rabinovich, G. S., Beake, B. D., Endrino, J. L., Veldhuis, S. C., Parkinson, R., Shuster, L. S., et al. (2006). Effect of mechanical properties measured at room and elevated temperatures on the wear resistance of cutting tools with TiAlN and AlCrN coatings. *Surface and Coatings Technology*, *200*, 5738–5742. Available from https://doi.org/10.1016/j.surfcoat.2005.080.132.

Fox-Rabinovich, G. S., Beake, B. D., Yamamoto, K., Aguirre, M. H., Veldhuis, S. C., Dosbaeva, G., Elfizy, A., Biksa, A., & Shuster, L. S. (2010). Structure, properties and wear performance of nano-multilayered TiAlCrSiYN/TiAlCrN coatings during machining of Ni-based aerospace superalloys. *Surface and Coatings Technology*, *204*(21–22), 3698–3706. Available from https://doi.org/10.1016/j.surfcoat.2010.04.050.

Fox-Rabinovich, G. S., Kovalev, A. I., Endrino, J. L., Veldhuis, S. C., Shuster, L. S., & Gershman, I. S. (2006). Surface-engineered tool materials for high-performance machining. In Fox-Rabinovich, & G. E. Totten (Eds.), *Self-organization during friction* (pp. 231–296). Boca Raton, Florida, USA: CRC Press.

Gsellmann, M., Klünsner, T. K., Mitterer, C., Marsoner, S., Skordaris, G., Bouzakis, K., Leitner, H., & Ressel, G. (2020). Near-interface cracking in a TiN coated high speed steel due to combined shear and compression under cyclic impact loading. *Surface and Coatings Technology, 394*. Available from https://doi.org/10.1016/j.surfcoat.2020.125854.

He, Q., Saciotto, V., DePaiva, J. M., Guimaraes, M. C., Kohlscheen, J., Martins, M. M., & Veldhuis, S. C. (2024). Enhancing tool performance in high-speed end milling of Ti-6Al-4V alloy: The role of AlCrN PVD coatings and resistance to chipping wear. *Journal of Manufacturing and Materials Processing, 8*(2). Available from https://doi.org/10.3390/jmmp8020068.

Inspektor, A., & Salvador, P. A. (2014). Architecture of PVD coatings for metalcutting applications: A review. *Surface and Coatings Technology, 257*, 138−153. Available from https://doi.org/10.1016/j.surfcoat.2014.08.068, http://www.journals.elsevier.com/surface-and-coatings-technology/.

Kalss, W., Reiter, A., Derflinger, V., Gey, C., & Endrino, J. L. (2006). Modern coatings in high performance cutting applications. *International Journal of Refractory Metals and Hard Materials, 24*(5), 399−404. Available from https://doi.org/10.1016/j.ijrmhm.2005.11.005.

Pardo, A., Buijnsters, J. G., Endrino, J. L., Gómez-Aleixandre, C., Gómez-Aleixandre., Abrasonis, G., Bonet, R., & Caro, J. (2013). Effect of the metal concentration on the structural, mechanical and tribological properties of self-organized a-C:Cu hard nanocomposite coatings. *Applied Surface Science, 280*, 791−798. Available from https://doi.org/10.1016/j.apsusc.2013.05.063.

Schalk, N., Tkadletz, M., & Mitterer, C. (2022). Hard coatings for cutting applications: Physical vs. chemical vapor deposition and future challenges for the coatings community. *Surface and Coatings Technology, 429*, 127949. Available from https://doi.org/10.1016/j.surfcoat.2021.127949.

Skordaris, G., Bouzakis, K. D., Charalampous, P., Bouzakis, E., Paraskevopoulou, R., Lemmer, O., & Bolz, S. (2014). Brittleness and fatigue effect of mono- and multi-layer PVD films on the cutting performance of coated cemented carbide inserts. *CIRP Annals − Manufacturing Technology, 63*(1), 93−96. Available from https://doi.org/10.1016/j.cirp.2014.03.081, http://www.elsevier.com/wps/find/journaldescription.cws_home/709764/description#description.

Skordaris, G., Bouzakis, K. D., Kotsanis, T., Charalampous, P., Bouzakis, E., Breidenstein, B., Bergmann, B., & Denkena, B. (2017). Effect of PVD film's residual stresses on their mechanical properties, brittleness, adhesion and cutting performance of coated tools. *CIRP Journal of Manufacturing Science and Technology, 18*, 145−151. Available from https://doi.org/10.1016/j.cirpj.2016.11.003.

Nanoindentation-based techniques for evaluating irradiated fuel and structural materials

20

David Frazer[1] and Peter Hosemann[2]
[1]General Atomics Electromagnetic Systems, Nuclear Materials & Technology Division, San Diego, CA, United States, [2]University of California Berkeley, Department of Nuclear Engineering, Berkeley

20.1 Introduction

The miniaturization of ASTM standard test specimens holds a rich history within the nuclear materials community, primarily driven by spatial constraints in neutron irradiation facilities and the volume reduction on highly irradiated materials. Irradiation facilities often face challenges as irradiations can span from years to decades, further limiting the availability of materials for evaluation. To address these constraints, various testing techniques have been developed, including shear punch (Lucas et al., 1986; Maloy et al., 2011) of TEM disks, SS J-2 tensile specimens (Alam et al., 2016), Vickers hardness (Cappia et al., 2016), and an array of other methods.

In addition to space challenges in neutron irradiation facilities, there are radiation exposure concerns to workers, which can hamper examination of the irradiated materials. To mitigate radiation exposure and facilitate testing of neutron-irradiated materials outside of hot cells, small-scale mechanical testing (SSMT) emerges as a crucial solution (Hosemann, 2018; Kiener et al., 2012). By enabling the measurement of mechanical properties on minute volumes of material, SSMT offers several advantages.

Furthermore, the reduction in sample size enables efficient use of the material while also minimizing radiation exposure to workers by enabling testing to be conducted outside of hot cells (Hosemann, 2018). This shift not only saves time and costs associated with the postirradiation examination (PIE) process but also improves overall safety protocols.

In addition, SSMT allows for testing the same retrieved sample under various conditions, such as low and high temperatures or within fluid cells. This versatility enables researchers to gather a wealth of information efficiently from small material volumes. For example, utilizing the undeformed head of an SS-J2 tensile specimen, multiple microtensile specimens could be manufactured and tested across a range of temperatures. This approach eliminates the need to choose between testing at more temperatures/conditions or increasing the number of samples at a single temperature, offering the opportunity for both improved statistical analysis and a more comprehensive understanding of material behavior under different conditions. These are a rather important consideration for PIE of materials as there is usually only one or two specimens per a testing condition, greatly limiting statistics during testing (Schulthess, 2011).

The assessment of mechanical properties of irradiated materials extends beyond laboratory testing to include in-service components, particularly relevant for aging materials in light water reactors (LWRs) (Naziris et al., 2021). These materials are scrutinized to determine their remaining service life while ensuring compliance with licensing requirements. Consequently, small-scale mechanical testing remains pertinent for evaluating the integrity of in-service components for extending LWRs' operational lifetimes (Naziris et al., 2021).

The challenge lies in understanding how these materials behave under prolonged exposure to the harsh operational conditions. Small-scale mechanical testing provides insights into the material's ability to withstand stress, fatigue, and other degradation mechanisms over time to inform predictions of remaining life (Barnoush et al., 2019; Connolley et al., 2005; Dehm et al., 2018; Gianola et al., 2023; Islam et al., 2021; Jaya et al., 2022)

Microhardness testing emerges as a pivotal miniaturization technique applied to both structural materials and fuel. Vickers testing, in particular, enjoys popularity due to its suitability for use in hot cell environments and its spatial resolution, facilitating the measurement of heterogeneity in irradiated fuels and structural materials (Cappia et al., 2016; Lucas et al., 1986; Zacharie-Aubrun et al., 2021).

However, Vickers testing has its drawbacks, notably its sole measurement of material hardness without deeper analysis. Extensive efforts have been dedicated to correlating Vickers hardness values with other material properties such as yield stress, ultimate tensile strength, and elastic modulus (Lucas et al., 1986). Some studies even attempt to assess properties such as strain hardening through pileup effects observed around conventional Vickers hardness tests (Cappia et al., 2016).

The question of mechanical property assessment of materials is also posed on actual in-service components. Aging LWR materials are facing the question of how much longer they can be in service and still fulfill licensing requirements [8]. Therefore, SSMT is relevant for actual in-service components and estimating the remaining life a component has. The techniques have also shown promise for evaluating complex components such as springs that are used in CANDU reactors where the component geometry makes it difficult to perform a macroscale test (Howard, Bhakhri, et al., 2019; Howard, Judge, et al., 2019).

Nevertheless, these correlations remain largely empirical in nature and require evaluation across diverse materials under consideration.

SSMT today goes far beyond vickers testing and includes a variety of testing techniques such as microcantilever, microcompression, microtensile, nanoindentation, and many more (Barnoush et al., 2019; Connolley et al., 2005; Dehm et al., 2018; Gianola et al., 2023; Islam et al., 2021; Jaya et al., 2022). This is not an exhaustive list of techniques but are the main ones used in the field today.

Nanoindentation, also known as instrumented indentation, was first developed in the 1980s where the main advantage of nanoindentation was that the hardness and reduced modulus of the material could be calculated without the need to measure the imprint of the indent (Oliver & Pharr, 1992). In addition, since the force and displacement data are constantly monitored during the indentation, it has allowed for a variety of different nanoindentation techniques to be developed enabling the measurement of such properties as creep (Fischer-Cripps, 2004).

Microcompression was the first of the techniques to be implemented in early 2000s with the development of the focused ion beam (FIB) microscopes that enable precise sputtering of material to manufacture the different geometries of the testing specimens (Uchic & Dimiduk, 2005). The growth of FIB has allowed for the development and implementation of all the different techniques as described earlier (Barnoush et al., 2019; Connolley et al., 2005; Dehm et al., 2018; Gianola et al., 2023; Islam et al., 2021; Jaya et al., 2022). During this time, there has been a large development of the different techniques for ion beam − irradiated materials and neutron-irradiated samples (Hosemann, 2018; Kiener et al., 2012).

In the following, we dive deeper into the assessment of mechanical properties using SSMT today.

20.2 Nanoindentation

Nanoindentation stands out as a straightforward technique for characterizing highly radioactive materials, owing to its simple sample preparation, which involves a polished flat surface and a small test volume (Fischer-Cripps, 2004). This simplicity makes it ideal for materials that typically require remote sample preparation. Furthermore, the ability to mount samples in epoxy provides some shielding for the radioactive material during subsequent handling.

Moreover, the automated nature of most nanoindentation instruments allows for the collection of tens to hundreds of data points while minimizing radiation exposure to researchers. The combination of small sample volumes and easy sample preparation renders nanoindentation an ideal characterization technique for nuclear materials, including both neutron-irradiated and ion beam − irradiated materials.

In neutron-irradiated materials, nanoindentation minimizes sample size, while in ion beam irradiations, it enables sampling of the thin irradiated region (Hardie et al., 2015; Hosemann et al., 2012). Nanoindentation instruments have the capability to measure various mechanical properties over a range of temperatures, including elastic modulus, hardness, creep, and wear, with high spatial resolution on the sample surface.

Moreover, nanoindentation systems offer high-temperature testing capabilities in inert, reducing, or vacuum environments to inhibit sample surface oxidation. Additionally, a wide range of attachments, such as cryogenic testing, humidity testing, fatigue and wear, scratch testing, liquid testing, and impact testing, further extend the versatility of nanoindentation instruments, enabling diverse testing configurations and evaluations under different operating conditions. These attributes have led to the widespread adoption of nanoindentation for measuring a broad range of mechanical properties.

However, the key challenge deploying nanoindentation for engineering challenges is the question on how to relate these values to engineering and materials design parameters. A main challenge with nanoindentation is sample size effects in which shallow indents appear harder (Huang et al., 2006; Nix & Gao, 1998). The size effect in nanoindentation is a well-known phenomenon. This size effect can make the interpretation of results for macroscale applications difficult. Additionally, while nanoindentation easily measures the hardness, it is a difficult parameter to relate back to the amount of elongation that a material can withstand in service (Rodriguez & Gutierrez, 2003). Being able to use nanoindentation to inform decisions about service materials has been the aim of the nuclear community and using nanoindentation elucidated material changes with irradiation.

20.2.1 Nanoindentation in ceramic materials and composites

Nanoindentation is highly relevant for both fuel forms (such as UO_2, USi_2, and UN) and ceramic structural materials such as SiC, graphite, and their composites. For the purpose of discussion, we will cover first the use of nanoindentation on structural ceramic materials and second, on fuel forms.

The advantages of modern nanoindenters' high spatial resolution are evident in studies on neutron-irradiated silicon carbide fiber−reinforced silicon carbide matrix (SiC−SiC) composites. Frazer et al. (2020) conducted elevated temperature nanoindentation on both control and neutron-irradiated SiC/SiC composites, separately measuring the mechanical properties of the matrix and fibers. Optical and SEM images showing the ability for indents to measure the matrix and fibers separately can be seen in Fig. 20.1. Their findings revealed distinct mechanical properties between the matrix and fibers, with the matrix exhibiting higher elastic modulus and hardness, possibly due to residual carbon from manufacturing (Frazer

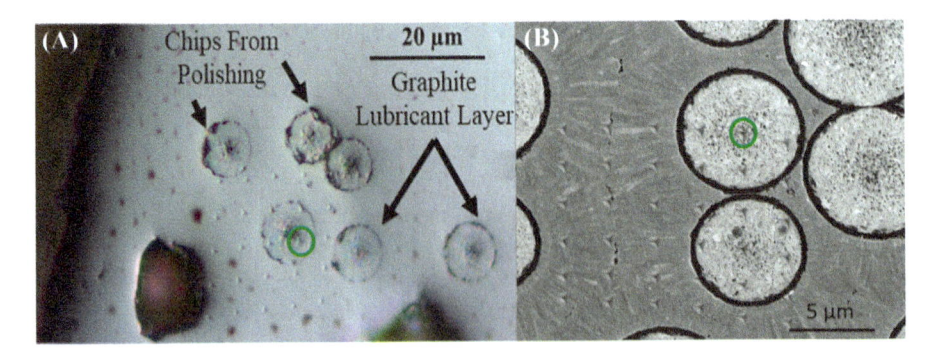

Figure 20.1 (A) An optical image of a field of indents illustrating the ability to place fields of indents with some of the indents landing in fibers as since with one example in the green circle. (B) Another field of indents with indents in both the matrix and fibers of the SiC−SiC composite. A green circle highlights one example of an indent in the fiber.

et al., 2015, 2020; Katoh et al., 2014). Nanoindentation accurately detected changes in hardness and Young's modulus with temperature and irradiation damage, aligning well with macroscale measurements (Snead et al., 2007). Furthermore, it differentiated the decrease in properties with temperature between the fibers and matrix, showing a larger relative decrease in the matrix over the same temperature range. This capability to separately characterize the matrix and fibers could inform modeling efforts for overall composite behavior.

Similarly, Rohbeck et al. (2015) conducted elevated temperature nanoindentation on the SiC layer of TRISO particles (~ 35 μm wide) with alumina kernels irradiated in PYCASSO I experiments. They observed irradiation hardening in the SiC layer, consistent with expected irradiation conditions. This work illustrates the benefits of nanoindentation's high spatial resolution, enabling the measurement of individual components of composite materials even after neutron irradiation. This capability could enhance the accuracy of modeling the interaction between different components in composite materials by incorporating their individual postirradiation properties into overall behavior models.

Nanoindentation is also performed on functional components such as graphite. Many nuclear applications consider graphite as a moderator to tailor the neutron spectrum to the desired shape. However, graphite is often exposed and used at elevated temperatures in harsh environments where some mechanical integrity and radiation damage resistance are needed. Graphite as a material is a rather exciting and complicated composite material containing multiple phases and often is not dense.

Poly granular graphite as it is used in nuclear applications exhibits a complex nature. While being a ceramic at room temperature, elastic properties demonstrate nonlinear behavior (Yoda et al., 1983). This phenomenon is attributed to various mechanisms, including microcracking (Liu et al., 2017; Marrow et al., 2016), which can lead to a decrease in the elastic modulus with increasing tensile strain (Marrow et al., 2016). Interestingly, the elastic modulus may however increase with rising temperature due to the closure of these microscopic pores caused by thermal expansion (Mason & Knibbs, 1960). The material's thermal expansion and the temperature dependence of its elastic modulus are influenced by fine-scale microstructure features, such as "Mrozovski" accommodation pores (Mrozowski, 1954), which are dependent on the raw materials and manufacturing processes. When subjected to interactions with fast neutrons, graphite crystals undergo dimensional changes, resulting in the closure of fine pores and other alterations in physical properties (Kelly, 1982). Particularly, irradiation eliminates the nonlinear behavior observed in the elastic deformation of poly granular graphite (Brocklehurst & Kelly, 1993) and impacts its fracture behavior (Jin et al., 2021). Studying graphite using nanoindentation is rare since it is difficult to conduct these experiments. Preparing good smooth samples of graphite is most certainly a challenge since it is difficult to ensure that the sample preparation method did not alter the microstructure. Especially, poly granular material with its many pores and features makes sample preparation difficult.

Therefore, fundamental studies are oftentimes carried out on highly oriented pyrolytic graphite (HOPG), which is smooth to start with. Fundamental studies using nanoindentation of single-crystal graphite materials (Barsoum et al., 2004) has shown that graphite can deform at room temperature with kink band formation. This mechanism may depend on mobile basal dislocations (Barsoum et al., 2004; Basu et al., 2009), and a universal "ripplocation" model has been proposed (Barsoum et al., 2019).

Recently, nanoindentation at elevated temperature (600°C) (Marrow et al., 2022) was applied to HOPG using a spherical diamond tip (~ 10 μm diameter) along the (Lucas et al., 1986) surface. There the combination of indenting and microscopy found that the graphite crystals deform inelastically by buckling with kink formation. Elevated temperature was found to increase the ease of buckling and its intensity likely caused by the effect of temperature on the shear modulus. However, how irradiation affects these behaviors is not understood today.

20.2.2 Nanoindentation on fuel forms

The mechanical properties of nuclear fuel play a critical role as they constitute the primary barrier against the release of fission products and radioactive materials. However, evaluating these properties has been challenging due to the radioactive nature of fuels containing uranium, plutonium, and thorium compounds. Postirradiation, the fuel becomes highly radioactive, making handling outside of a hot cell nearly impossible without sectioning it down into extremely small pieces. Typically, the fuel is fabricated into pellets for irradiation, but these pellets often crack during operation, complicating macroscale testing.

Nanoindentation has emerged as a valuable tool in recent years for evaluating the mechanical properties of nuclear fuels, both at room and elevated temperatures. Key parameters, such as Young's modulus and creep behavior of UO_2, are crucial for understanding and modeling pellet − clad mechanical interactions during operation. The use of vacuum systems and environmental control has proven beneficial in applying nanoindentation techniques to nuclear fuels, as uranium compounds readily react with oxygen.

Several studies have investigated nanoindentation of both fresh/nonirradiated fuels and irradiated/spent fuel. For instance, nanoindentation of spent nuclear fuel at room temperature has revealed variations in hardness and Young's modulus along the radius of the spent fuel. Additionally, the high temperature gradients and neutron self-shielding effects during operation lead to the development of diverse microstructures in UO_2 fuel, which can be examined using nanoindentation techniques (Rondinella & Wiss, 2010; Roostaii et al., 2021; Xiao et al., 2021).

Nanoindentation serves as a valuable tool for assessing the impact of varying grain sizes on the mechanical properties of materials. This is particularly pertinent in the field of nuclear engineering, where materials can exhibit a wide range of microstructures stemming from different fabrication processes.

In a study conducted by Gong et al. (2019), the researchers investigated several microstructures of UO_2 fuel pellets produced through spark plasma sintering to analyze the influence of grain size on material properties. Employing nanoindentation, they assessed the hardness and Young's modulus across different temperature ranges for each microstructure. The findings revealed that the specimen with the smallest grain size (125 nm) exhibited the highest hardness, indicative of a Hall − Petch-like effect. Conversely, the samples with grain sizes of 2 and 7 μm demonstrated lower hardness values.

Interestingly, the microstructure appeared to have no discernible effect on the Young's modulus, contrary to expectations. Additionally, consistent with theoretical predictions, both hardness and Young's modulus decreased with increasing temperature. This study underscores the utility of nanoindentation in elucidating the intricate relationship between grain size, mechanical properties, and temperature in nuclear materials.

The ability to perform elevated temperature testing is a great feature as nuclear fuels operate at high temperatures. In addition, the ability to perform creep testing is needed as pellet − clad mechanical interactions modeling need to be informed about the fuel's ability to creep to relieve stress on the cladding material as seen in the following studies. Expanding on these techniques for fresh fuel Frazer, Shaffer, et al. (2021) performed elevated nanoindentation and nanoindentation creep studies on fresh, plastically deformed and spark plasma sintered samples. In a comparison of the fresh and plastically deformed UO_2 (1200°C until 9% strain during a 6-hour creep test), it was observed that the plastically deformed sample had lower hardness values at all of the temperatures tested compared with the fresh UO_2 samples. In the study, both samples were taken from the same block of material initially. In comparing the creep exponents of the fresh and plastically deformed UO_2 at 300°C, it is observed that plastically deformed sample had a stress exponent less than 10 while the fresh sample had a value greater than 10 suggesting that the plastic deformation of the UO_2 might be hindered at lower temperatures by the need to nucleate dislocations.

Nanoindentation creep has also been to study properties of U-based metallic glasses by Ke et al. (2019) showing a difference in rigid structure of the glasses study. It was found that U60Fe27.5Al12.5 metallic glass had a more rigid state as compared with other glass tested U65Fe30Al5 by examining the secondary stage creep. This increase in rigid state could suggest a more compact atomic structure of metallic glass.

In expanding nanoindentation in testing nuclear fuels, Frazer et al. (2024) performed elevated temperature nanoindentation on $(U,Ce)O_2$ compounds. CeO_2 is used as a surrogate for PuO_2 in studying mixed oxide fuel as it is less hazardous and has very similar thermophysical and mechanical properties. Mixed oxide fuels are being examined for advanced and current reactors. In this work, the creep exponents at 800°C were measured, and it showed dislocation glide as the main creep mechanism. In addition, there has been other nanoindentation work on $(U,Ce)O_2$ compounds at room temperature measuring the hardness and Young's modulus. The work by Kurosaki et al. (2004) and Frazer et al. (2024) agree well with each other, illustrating the repeatability of nanoindentation experiments.

The small volume probed and the high spatial resolution of nanoindentation is great in evaluating materials development at the laboratory scale and the phases that form at interactions between two materials as utilized in the following studies of nuclear materials. The ability to measure the mechanical properties at the laboratory scale enables the quick and rapid screening of potential new fuel forms for nuclear reactors. In addition, understanding the individual phases that form in diffusion couple or between the fuel and cladding will inform modeling efforts of the behavior during operation. Utilizing the small volume probed, Kurosaki et al. (2004) were able to use nanoindentation to measure the phases at the interaction of AA6061 cladding with Zr diffusion barrier and uranium 10 wt.% molybdenum fuel for the development of monolithic uranium − molybdenum metallic fuels for research reactors.

On development of new uranium fuel forms for current and next-generation nuclear reactors, nanoindentation has been useful in getting the hardness and Young's modulus on laboratory quantities of material. In the work by Newell et al. (2017), nanoindentation was used to the mechanical properties of three forms of uranium silicides that are potential new nuclear fuel forms. These were being investigated after the Fukushima accident as accident tolerant fuel forms. In addition, the nanoindentation curves were fitted with finite element to measure the Young's modulus, hardness, and yield stress showing agreement with nanoindentation results.

Frazer et al. were able to expand on measuring new fuel forms using nanoindentation looking at UN, UB$_2$, and U$_3$Si$_2$ (Frazer, Maiorov, et al., 2021) over temperature and mixtures of UO$_2$/UB$_2$ and UO$_2$/UB$_4$ (Kardoulaki et al., 2020, 2021), see Fig. 20.2. UO$_2$, the current fuel, has a low thermal conductivity that is trying to be improved with different uranium compounds or mixtures of UO$_2$ and other uranium-bearing compounds. In the UN, UB$_2$, and U$_3$Si$_2$, elevated-temperature nanoindentation was applied to measure the Young's modulus and hardness on the laboratory scale (Frazer, Maiorov, et al., 2021). This can enable a quick and cost-effective manner to measure the mechanical properties of next generation nuclear

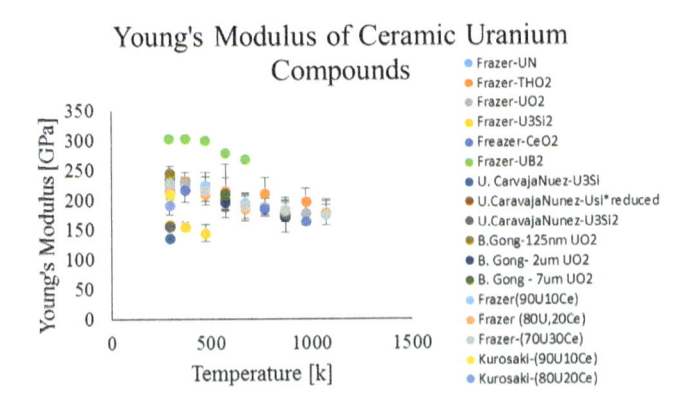

Figure 20.2 A plot of Young's modulus versus temperature for a variety of ceramic U compounds (Carvajal-Nunez et al., 2018; Frazer et al., 2024; Frazer, Shaffer, et al., 2021; Gong et al., 2019; Kurosaki et al., 2004; Kardoulaki et al., 2020, 2021).

fuels. In the UO$_2$/UB$_2$ and UO$_2$/UB$_4$, work nanoindentation was able to measure the difference in the material at room temperature, and it correlated well with the microscopy investigations as seen in Fig. 20.3 (Kardoulaki et al., 2021).

Continuing the use of nanoindentation to explore new uranium compounds, Li et al. (2018) used nanoindentation to measure the hardness and reduced modulus of U−Cu intermetallic compound that formed in an U and Cu diffusion couple. It showed that the intermetallic compounds had a higher hardness and Young's modulus as compared with U and Cu.

A challenge with fabrication of different uranium compounds is the inclusions in the material. In the work by Chen et al. (2017), the inclusions in alpha U and aged U-5.5 Nb are studied with nanoindentation to measure the hardness and Young' modulus. In the specimens, most of the inclusions were uranium nitrides and carbides or Nb$_2$C, which had higher hardnesses as compared with the metallic uranium compounds.

In addition, the expansion of nanoindentation for radioactive material enables fundamental studies of the material behavior to compare with modeling. To understand the anisotropic behavior of low-temperature alpha phase of uranium, nanoindentation was used by Li et al. to evaluate hardness and elastic properties to compare with theoretical values and first-principle calculations (Li et al., 2022).

With the success of measuring the mechanical properties of fresh nuclear fuel and the ease of the sample preparation, nanoindentation techniques have been applied to irradiated fuels to measure the change in the mechanical properties. The small volume probed as the advantage of being able to measure the different microstructural zones that develop in a fuel during operation.

The work by Terrani, et al. explores the hardness and Young's modulus of the irradiated fuel showing that hardness increases and Young's modulus decreases with irradiation damage as expected for ceramic material. In addition, they were able to look at how different regions of the pellets behaved based on burnup. It was observed that in the nanocrystalline region around the periphery of the pellet called the high burnup structure (HBS) that the Young's modulus decreased more than in the center. The HBS forms as more fissions and damages occur in this region due

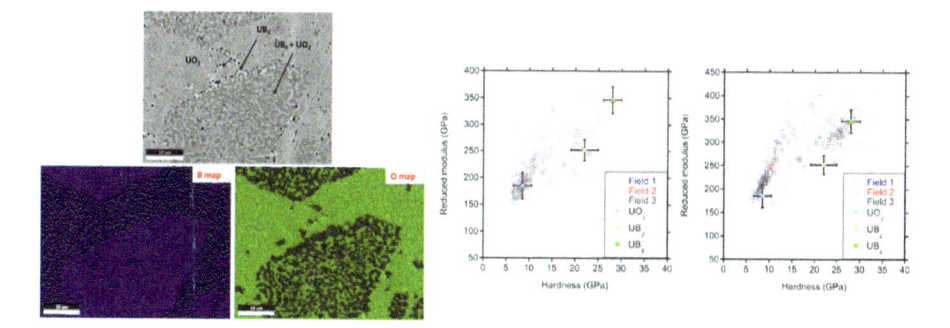

Figure 20.3 An example of the UO$_2$/UB$_4$ microstructure and plot of the reduced modulus versus hardness showing the nanoindentation results agreed well with the different phases identified with microstructural investigations (Kardoulaki et al., 2021).

to a self-shielding effect with neutrons in the reactor (Rondinella & Wiss, 2010; Roostaii et al., 2021; Xiao et al., 2021). The ability of nanoindentation to measure the volume of material independently of porosity or cracking allows the influence of these microstructure features to be removed. This is an important consideration for irradiated fuel ceramics due to the large amount of porosity and cracking during operation. In addition, due to the nanocrystalline grain size, there was a slight increase in the hardness in HBS as compared with the center of the pellet. This again demonstrates the power of being able to probe small volumes of material.

In the work by Elbakhshwan et al. (2016), nanoindentation was used to investigate the mechanical properties of ion beam − irradiated UO_2 thin films. The results showed an increase in the film hardness and yield strength with dose. The elastic modulus of the film initially decreased with irradiation and then subsequently increased with larger dose when the irradiation was performed at room temperature. When the irradiation was performed at 600°C, it resulted in a decrease in the hardness and elastic modulus up to a dose of 1E14 ions/cm^2.

In work by Schneider et al. (2022) on nanoindentation of the interface of high-burnup PWR fuel and cladding material, it was observed that there were four distinct zones related in the interaction layer between the fuel and cladding. These zones depended on the interactions with fission products in the fuel. These results were also able to measure the irradiated UO_2 results and showed good agreement with Terrani's work (Terrani et al., 2018).

Finally, on irradiated fuel for nanoindentation, Frazer et al. (2022) were able to use a liftout technique combined with a plasma FIB to remove small cubes of irradiated MOX fuel to measure the hardness and Young's modulus of the fuel. It was observed that Young's modulus decreased and the hardness increased with irradiation and that liftout technique enables additional microstructural investigations of the fuel pre and postmechanical testing.

20.2.3 Nanoindentation studies on metallic materials for nuclear applications

Similar to the aforementioned section on fuels and functional and structural ceramics, nanoindentation on metallic materials is conducted to minimize the volume required for assessing the material's mechanical performance. However, the significant advantage of metallic materials, particularly in structural applications, presents a major challenge when employing nanoindentation. Unlike most ceramic materials mentioned earlier, metallic materials can undergo substantial plastic deformation, complicating the assessment of mechanical properties, especially at small scales (Huang et al., 2006; Nix & Gao, 1998).

Three key practical reasons drive researchers to perform indentation on materials for nuclear structural applications. First, it allows for the reduction of volume and thus the amount of potentially highly radioactive material. Second, it may be the case that only ion beam − irradiated material is available (Hardie et al., 2015; Hosemann et al., 2012). Third, researchers may be interested in specific regions of interest such as welds, joints, or interfaces (Fischer-Cripps, 2004). In general, the objective is to extract bulk properties from these measurements to gain insights into the material's property changes due to exposure in the nuclear environment or to qualitatively trace a property change. While the latter is relatively straightforward, demonstrating a property change, the former can be challenging, especially considering the significant plasticity experienced by metallic materials, making it difficult to compare nanoindentation properties with macroscale behavior (Hosemann, 2018; Kiener et al., 2012). Many methodologies have been developed to translate nanoindentation techniques to bulk properties such as yield stress, ultimate tensile strength (UTS), and strain hardening.

Several approaches exist for performing nanoindentation on irradiated materials. Before the early 1990s, microhardness, where one evaluates the remaining imprint with a set force in the material, was mainly used to assess material properties, with many correlations and studies established relating microhardness to bulk properties (Cappia et al., 2016; Lucas et al., 1986). With the advent of more sophisticated instruments and instrumented nanoindentation, particularly following the establishment of the Oliver and Pharr methodology in 1992 (Oliver & Pharr, 1992), techniques using a Berkovich tip to evaluate changes in mechanical properties with radiation became more established. Subsequently, scientists utilized nanoindentation to extract qualitative and quantitative properties of irradiated materials (Hardie et al., 2015; Hosemann, 2018; Hosemann et al., 2012; Kiener et al., 2012). However, size effects remain a significant challenge when attempting to quantitatively extract mechanical bulk properties from nanoindentation (Huang et al., 2006; Nix & Gao, 1998). Various approaches, as outlined by Krumwiede, Hosemann, Zinkle, and Kasada using Berkovich indenters (Kasada et al., 2014; Krumwiede et al., 2018; Zhu et al., 2022), as well as spherical indentation approaches pursued by Pathak, Mara, Hardie (Bushby et al., 2012; Hardie et al., 2015; Pathak & Kalidindi, 2015; Pathak et al., 2016) and others, have been explored.

While challenges persist, the wealth of work performed has led to several viable approaches for extracting bulk properties from nanoindentation on irradiated materials (Bushby et al., 2012; Hardie et al., 2015; Krumwiede et al., 2018; Kasada et al., 2014; Pathak & Kalidindi, 2015; Pathak et al., 2016; Zhu et al., 2022). What largely remains unexplored is whether and how these approaches can be transferred to non-ambient conditions, which is where reactors primarily operate. Only a few papers have addressed this challenge to date, outlining why the scaling laws known at ambient conditions likely change at higher temperatures. Only Prasitthipayong et al. (2018), addressed this challenge today and outlined why the scaling laws known at ambient conditions likely change at higher temperatures. However, while not claiming comprehensiveness, we highlight a few specific studies on how nanoindentation is used in nuclear applications.

In the work by Kese et al. (2017), nanoindentation was used to measure the hardness and elastic modulus of Zircaloy-2 and the hydrides from fuel with >70 GWd/tU fuel. The nanoindentation was enable the measure of the hardness and elastic modulus up to 300°C accurately with 1% of the material it would have taken to perform for the axial tensile testing of material. In addition, the nanoindentation enabled the measurement of the hydride properties independently, which would not have been possible with macroscale mechanical testing.

Bridging the length scales from macroscale mechanical testing to microscale mechanical testing has been a topic of discussion in the SSMT community for decades and discussed earlier in this book (Hosemann, 2018). Due to the small size of the specimens, there is the opportunity to perform tests on undeformed regions of macroscaled tested material enabling direct comparison between macroscale data and microscale data. Several examples of deploying this experimental opportunity can be examined below enabling direct comparison of SSMT values and macroscale mechanical testing values for neutron-irradiated structural materials.

In the work by Krumwiede et al. (2018), nanoindentation was performed on tensile test grips sections of previously tested tensile bars of several neutron-irradiated structural materials. The ability to test the undeformed grip section of the tensile bar allowed for a direct comparison of the tensile macroscale tensile properties and the nanoindentation while being able to handle the samples outside of a hot cell. The studies evaluated the ability of nanoindentation data to predict bulk-scale tensile properties. It was found that the empirical correlations used and developed could predict the neutron-irradiated condition tensile yield stresses to within 15% and the flow stress to within 5% on average between the eight alloys tested. Developing an experimental protocol like this could increase the confidence of limited marcoscale test specimens from the hot cell without impinging on the space in a neutron irradiation facility.

Hosemann et al. (2010) performed nanoindentation and microcompression on the grips of pervious tested tensile bars of F82H, Fe-8Cr ODS, and Fe-8Cr-2W ODS irradiated to irradiated between 7.7–13 dap at temperatures ranging from 159°C to 382°C. The irradiation was performed in the Spallation Target Irradiation Program (STIP) and had average He/dpa ratio of 60 appm He/dpa. The results of the CSM nanoindentation showed a size effect on all of the materials; however, at larger depths, the values agreed well with Vicker's hardness indents performed on the same sample. The microcompression experiments were performed at room temperature, and the yield stress was evaluated and compared with the macroscale tensile tests. The microscale tests were performed at room temperature while the macroscale tensile tests were performed over temperature; however, similar trends in the data are seen.

Localized deformation of nanoindentation enables the ability to evaluate the plastic and elastic damage of the material using electron microscopy techniques to better understand the deformation behavior as elucidated with the following studies. Dolph et al. (2016) compared the plastic zone under a nanoindent in a Fe-9%Cr ODS alloy in both the control and 3 dpa at 500°C neutron irradiation material. The results of this work show that the hardness and strain hardening coefficient increased from irradiation and is believed to be consistent with irradiation-induced oxide nanocluster dissolution. In addition, it was shown that the plastic zone under the irradiated indent was more spherically shaped and had reduction in size compared with the plastic zone under the control indent. The plastic zone under the control indent was elongated in shape with it extending further below the indent than radially.

As described earlier, a benefit of nanoindentation is the ability to sample small volumes of material and allow investigating the properties of individual grains in a polycrystalline material. This advantage was exploited in a study by Mao et al. (2019), where the effect of the neutron irradiation on particular crystallographic grain orientations in 304L stainless steel was investigated. The results of this work showed that the {110} planes exhibited a slightly larger irradiation-induced increase in the hardness as compared with the {100} and {111} plants. Post nanoindentation TEM of the deformed areas showed that the control specimen deformed by twinning while there is an austenite-to-martensite phase transformation arising in the irradiated specimen.

In addition to measuring the hardness and Young's modulus, nanoindentation can be used to evaluate creep properties as seen in the work by Chen et al. (2020) of nanoindentation, and nanoindentation creep was used to investigate the mechanical properties of a neutron irradiated alloy D9 in three different conditions. Evaluating the creep properties of structural materials in a nuclear reactor is essential for predicting the lifetime of components in a nuclear reactor. It was found that samples irradiated at the highest temperature (683°C vs 430 and 448) exhibited the highest hardness values and the least amount of creep. The study also observed that the sample irradiated at 430°C had a higher hardness value as compared with the 448°C sample they had similar nanoindentation creep displacements.

20.3 Other small-scale mechanical test techniques used today

While Berkovich pyramidal indents have grown to be standard tip geometry for nanoindentation for hardness and Young's modulus measurements, there are several other options such as spherical tips that can be used to probe the mechanical response of materials. Spherical tips have been used to manufacture stress versus strain curves to examine the mechanical properties of materials including ion beam — irradiated materials. In Weaver et al. (2017), they used spherical indentation to produce stress versus strain curves looking at the increase in the yield strength because of irradiation. In addition, a comparison of a variety of different techniques that were used to evaluated the ion beam — irradiated 304 SS material is shown elucidating the differences between these techniques. A comparison of different techniques can be seen in Fig. 20.4.

The ability of spherical nanoindentation to enable the production of stress versus strain curves has encouraged its use in examining ion beam irradiations of materials. It has been seen in a variety of studies to enable the ability to evaluate the change in the mechanical properties with irradiation.

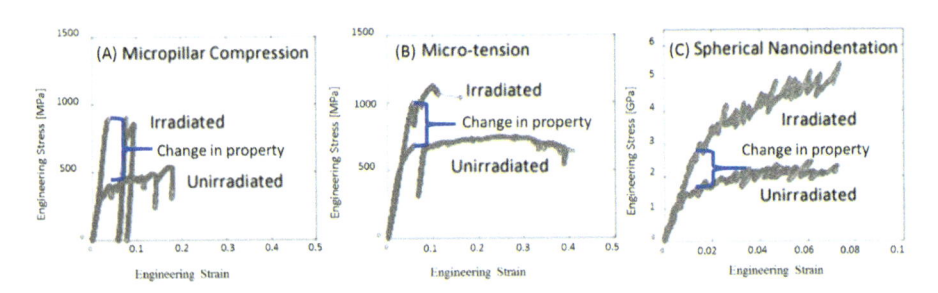

Figure 20.4 A variety of small-scale mechanical testing techniques showing the change in the unirradiated versus irradiated state in ion beam irradiated 304 SS.
Source: Adapted from Weaver, J. S., Pathak, S., Reichardt, A., Vo, H. T., Maloy, S. A., Hosemann, P., & Mara, N. A. (2017). Spherical nanoindentation of proton irradiated 304 stainless steel: A comparison of small scale mechanical test techniques for measuring irradiation hardening. *Journal of Nuclear Materials*, *493*, 368 −379, https://doi.org/10.1016/j.jnucmat.2017.06.031.

20.3.1 Focused ion beam-based techniques

While nanoindentation enables testing of easily prepared samples, it is limited in its ability to test constrained sample volumes enabling uniaxial stress tests versus the rather complex stress state under pyramidal indenter tips. The ability to constrain the volume is an important and interesting phenomenon to explore as Reichardt et al. (2017) did on an ion beam − irradiated 304SS specimen. In this work, the 304SS specimen was irradiated to 10 dpa in the plateau region of the dose profile with the stopping peak experiencing a higher dose. As seen in the Fig. 20.2, in a comparison of nanoindentation and microcompression experiments, the microcompression experiments were able to detect a change in the mechanical properties of plateau region versus the stopping peak while the nanoindentation had a more convoluted response. In addition, Vo et al. (2017) work on the same materials using microtensile testing shows the ability for microtensile testing to evaluate the reduction in elongation from irradiation damage as seen at the macroscale further demonstrating the need for the constrained volume for mechanical testing.

The ability to combine nanoindentation instruments with FIB-based manufacturing provides for a large variety of test configurations to be employed (Hosemann, 2018). FIB-based manufacturing techniques would enable microscale tensile, compression, and bending tests to evaluate these properties at a variety of length scales. In addition to constraining the volume of material, these techniques can allow for targeting individual microstructural features such as grain boundaries for better understanding of fundamental materials properties to improve material performance. These SSMT techniques can be combined with ion beam irradiation or radioactive materials to reduce the volume of material tested and measure the changes from irradiation damage as seen in Fig. 20.5. There have been numerous studies done with these techniques on radioactive material and while not a comprehensive list, a few are summarized.

Understanding the materials' properties in tension at the microscale would enable evaluating things such as grain boundary strength and strengths of other microstructural features. In addition, manufacturing microtensile specimens would allow for comparing the values measured with macroscale specimens and understanding the plastic deformation at the microscale. This direct comparison can be seen in Ando et al. (2018) on neutron-irradiated F82H to 5 dpa. In the work, FIB-based microtensile bars with gage size of $10 \times 1 \times 1$ μm were manufactured and tested at room temperature. The microscale samples were compared with macroscale samples of the SS3, and it was seen that the tensile properties of micrometer and millimeter size specimens were roughly in qualitative agreement. In addition, the change in tensile properties due to neutron irradiation agreed qualitatively for both micrometer and millimeter size F82H specimens. This illustrates that microscale samples can be used to measure the change in the mechanical properties of material due to irradiation.

In addition, understanding plastic deformation at the microscale, Vo et al. (2023) performed bicrystal microtensile testing on a push-to-pull device to evaluate the effect of void swelling on embrittlement in the material. In this work, samples of 304SS were irradiated to 33 dpa, which lead to 2% and 3.7% void swelling. The work was used in situ scanning electron microscope testing to provide direct observation on the effect of void-dominated microstructure on channel formation and intergranular cracking. It was found that voids fully suppress localization and promote necking within the microscale specimens.

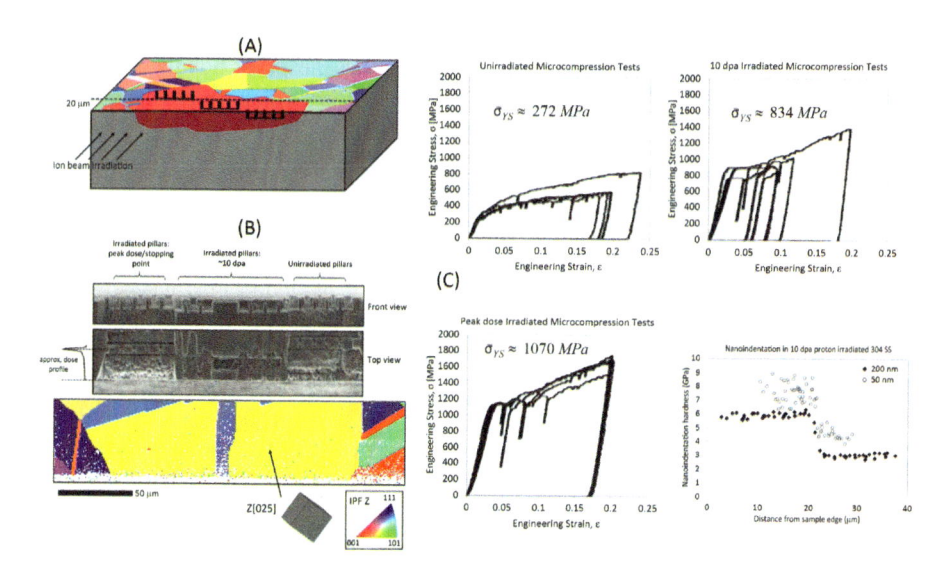

Figure 20.5 (A) A schematic of the location of the microcompression experiments in the large grain found. (B) SEM images of the microcompression experiments in the large grain and EBSD map showing the grain. (C) The stress versus strain curves for the three locations of the pillars (unirradiated, 10 dpa region and the stopping peak) showing the difference in yield stress, and finally, the nanoindentation at both 200 and 50 nm not showing an increase in hardness at the stopping peak.

Hosemann et al. (2011) investigated neutron-irradiated HT-9 using nanoindentation and microcompression on the grips on HT-9 tensile tests. The microcompression experiments were performed at room temperature while the macroscale tensile tests were performed at elevated temperature, which led to difficulty in comparing the data. However, similar trends in the data were observed. In addition, the article discusses the level of radioactivity produced by the samples that area of a pillar with a size of $8 \times 8 \times 16\ \mu m^3$ would still be below background levels.

Using SSMT to measure service components for nuclear reactors has been the aim of the community. Bend testing was used by Howard et al. (2018) on X750 spacer springs from a CANDU reactor to measure the mechanical properties. The X750 was from an inservice component, and the study was able to measure two regions on the spacer ring enabling the ability to elevate different regions in the space ring shows the benefits of SSMT. In this work, the service ring saw 58 dpa and 63 dap. The study found that the difference in yield strengths for the edge and center regions was ~ 740 MPa for nonirradiated material. After irradiation to a dose of 67 dpa, these differences were ~ 570 MPa for the lower irradiation temperature and ~ 710 MPa for higher irradiation temperature.

To measure the fracture properties of irradiated UO_2 microcantilevers, studies were conducted by Henery and coworkers (Henry, Zacharie-Aubrun, Blay, Chalal, et al., 2020; Henry, Zacharie-Aubrun, Blay, Tarisien, et al., 2020). In these studies, it was observed that the single crystal did not lose much strength from irradiation in comparison with the grain boundaries that had a significant reduction in strength with a burnup of 30 GWd/tHM. This is understood as the grain boundaries would be sinks for irradiation-produced defects in the material, which would lead to the reduction in the mechanical strength during a bending test. During this time, the internal grains of the UO_2 would have minimal increase in the defect concentration and retain the mechanical properties.

A 70 GWd/tHM microtensile testing was performed by Cappia et al. (2021), which was able to build off the previous microbending tests at lower burnup. In the microtensile test, it was able to show a continuation of reduction in strength of both grain boundaries and single crystals. However, it was observed that the strength of the grain boundaries only decreased slightly compared with the bending tests in Henry's work (Henry, Zacharie-Aubrun, Blay, Chalal, et al., 2020; Howard et al., 2018), and the values for the single crystals saw a significant reduction from 30 to 70 GWd/tHM. This can be used to hypothesize that the defect concentration at grain boundaries reaches saturation at relatively low burnups weakening the grain boundaries first. After saturation is reached, the defect concentration in the internal of the grain increases leading to decrease of the single crystal properties of subgrain properties. An understanding of how different microstructural features in the material degrade with operation can help develop improved models of the materials' performance during operation. This could enable new insights into the development of materials for different applications.

The interaction of the fuel with the cladding material is important for understanding the operation life of the reactor. In the work by Wang et al. (2022), microtensile testing was done to investigate this zone of interaction. The work showed hardening and embrittlement in the interaction zone with brittle fracture mainly attributed to the formation of nanocrystallized intermetallic σ-FeCr phases. In addition, there was also irradiation-induced mechanical softening by the disappearance of the martensitic lath structure in the unreacted HT9 cladding material.

The SSMT can be used to study the fundamental properties of materials such as work by Frazer and Hosemann (2019) on single-crystal UO_2 measuring the Peierls stress. In the work, the authors were able to use microcompression experiments at different temperature on UO_2 to calculate a Peierls stress of 3.65 GPa for UO_2 on the $\{100\}1/2 < 110>$ system.

A challenge to deploying SSMT on ceramic materials is in a heterogeneous microstructure with features on the same length scale. This can especially be seen in ceramic materials when the poresare roughly the size of the test specimens. In a work by Gong et al. (2021), microcantilevers in porous UO_2 were used, and the influence on the measured Young's modulus was examined. Results indicated that the presence of pore clusters near the substrate, that is, the clamp of the microcantilever beam, has the strongest effect on the load-deflection behavior, with the porosity leading to a reduction of stiffness that is the largest for any location of the pore clusters. Furthermore, it was also found that pore clusters located toward the middle of the span and close to the end of the beam have a comparatively small effect on the load-deflection behavior. Therefore, it is concluded that accurate estimates of Young's modulus can be obtained from microcantilever experiments after accounting for porosity on the one-third of the beam length close to the clamp. Adaptations of Gong's models can be seen in Fig. 20.2.

In addition, Doitrand et al. (2020) performed similar experiments on irradiated UO_2 investigating the influence of porosity on microcantilever testing. In this work, it was observed that premature fracture may occur in the presence of a pore, provided it is located close enough to the specimen surface undergoing tension, the resulting failure force decreasing with increasing pore size. A network of small porosities located on the fracture surface induces a decrease in the failure force, the magnitude of which mainly depends on the total surface fraction and on the porosity locations rather than on the porosity size as seen in Fig. 20.6. While there are challenges with performing microcantilever of ceramic U compounds as shown here, there are different ones with employing this testing method to metallic U compounds as seen in Frazer, Jadernas, Bolender, et al. (2019) on U-10 wt.% Mo fuels. In this work, the yield stress measure is higher than the macroscale values because of the size effects in materials testing, and after yield, the microcantilevers no longer follow the linear elastic assumptions used in the equations to calculate the stress and Young's modulus. It can however been seen in Frazer, Jadernas, Bolender, et al. (2019) that the calculation of the Young's modulus agrees quite well over temperature with macroscale techniques.

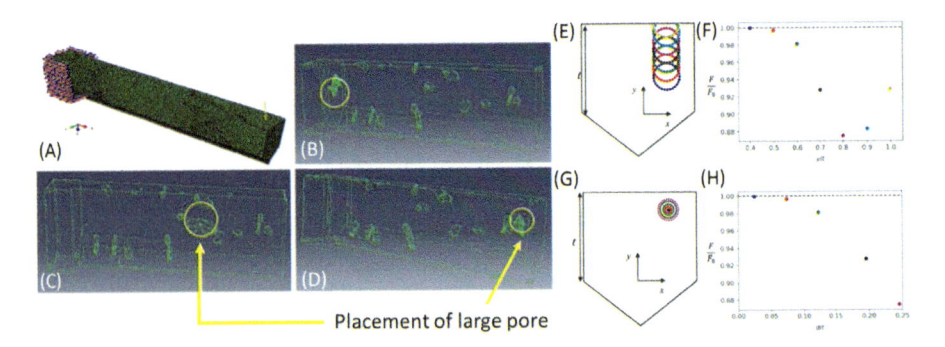

Figure 20.6 (A) The finite element model of the microcantilever with different grains. (B) A large pore placed near the substrate and clamped end of the cantilever. (C) A large pore placed near the middle of the span of the microcantilever. (D) A large pore place near the end of the span of the microcantilever (Gong et al., 2021). (E) A schematic showing the same size pore at different positions form the neutral fiber of the microcantilever shape. (F) The resulting change in the fracture strength because of the pore location as compared to fully dense material. (G) A schematic showing different size pores in the cantilever at the same location. (H) The resulting change in the fracture strength as a result of the pore size (Doitrand et al., 2020).

20.4 Future avenues for exploration

20.4.1 Femto-second laser and extreme environments

One of the challenges of SSMT is the reduced specimen volume, which significantly reduced the number of strength determining features in the test volume of material. One avenue to explore in manufacturing SSMT test specimens is femtosecond lasers that ablate material but still allow the manufacture of testing geometries (Dong et al., 2021; Maganosc et al., 2017; McCulloch et al., 2020; Pfeifenberger et al., 2017). One advantage of using the femtosecond lasers would be the ability to rapidly manufacture test specimens that are 10–100 s of microns in size. This would allow materials to have a sufficient number of strength determining features such as grain boundaries in the test volume to evaluate macroscale values of their mechanical properties. A combination of FIB and femtosecond laser manufacturing would allow an avenue to explore the region from size effect influence to macroscale values to understand the smallest size samples that enable measuring macroscale values. However, in the case of highly irradiated material, this would have to be balanced with the dose of the samples. This would provide the opportunity for extracting more information out of test material while also reducing the time and cost of PIE. In addition, as materials are placed in more extreme environments with multiple factors such as corrosion, high temperature, cryogenic, and liquid require the in situ environmental mechanical testing of materials. The need for these testing environments can be seen in the growth of the different testing options from vendors. Most vendors now offer options for high-temperature testing and low temperature with several offering options for humidity, liquid, scratch, fretting, wear, and several others. The growth of the in situ testing options illustrates the usefulness and acceptance of these techniques in the research and commercial communities with continued development in industry.

20.4.2 Lift out techniques

In addition, the levels of radioactivity of these samples are a limit for facilities that evaluate these samples. The ability to combine multiple investigations on the sample specimen of material would increase the knowledge of material being studied with the limited time available on the instruments. This could be seen with manufacturing of transmission electron microscopy foils from the plastically deformed material under an indent (Dolph et al., 2016). The ability to perform microstructural investigations before SSMT on the sample and then investigate the microstructure with other techniques such as TEM, APT, SEM, and EBSD on the deformed sample could provide a wealth of fundamental deformation behavior of the material. These techniques would be beneficial as they could lift out a large volume of material that could be mounted for both the SSMT and microstructure investigations. Examples of this can be seen in the studies already presented in this work with X750 alloy where the lift bending of the material enables EBSD and posttest TEM of the deformed region and undeformed regions (Howard et al., 2018). Other examples of this could be seen with microcompression experiments of pillars mounted on APT grids as shown in Fig. 20.7. The pillars could be crushed on the posts measuring the mechanical properties and then subsequently manufactured into APT needles for microstructural investigation about the deformation. One could even expand this to include deformation over temperature with subsequent APT to evaluate the temperature-dependent deformation mechanisms. The combination of these different techniques allows for detailed analysis and thorough analysis of the nuclear materials, which will accelerate the development and deployment of these new materials for the nuclear community.

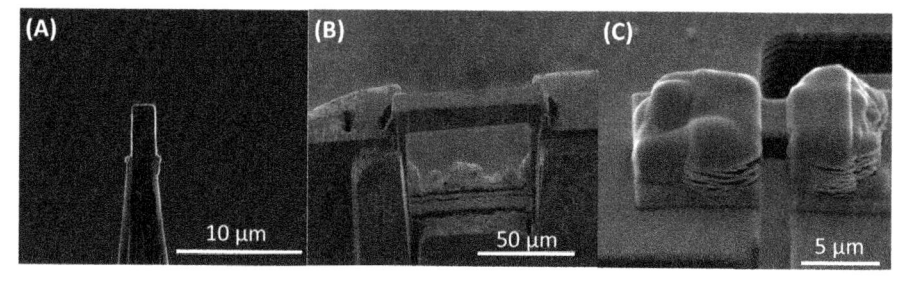

Figure 20.7 (A) An example of microcompression experiment on top of an APT post that could enable later manufacturing of the deformed material into an APT needle for microstructural investigations. (B) An example of 3-point bend test that can be thinned to electron transparency for microstructural investigation in the TEM after deformation. (C) A push-to-pull specimen that could be cleaned after testing for additional microstructural investigations.

20.5 Conclusions

In these examples, it was demonstrated how nanoindentation and SSMT testing have influenced the mechanical testing of nuclear materials. SSMT has enabled the mechanical properties of radioactive materials to be gathered, which would have been difficult or impossible previously. The use of the techniques have reduced the dose of test materials significantly enabling the testing of service components of reactors and spent nuclear fuel in what would have been pervious nearly impossible to gather data. In these cases, it has provided a wealth of information as compared with the rudimentary investigations available beforehand. In addition, it has advanced the field by enabling testing of materials at the laboratory-scale quantities to accelerate materials development for nuclear systems. Finally, it has also enabled fundamental studies into the deformation of nuclear materials to inform models for behavior to improve the operation and safety. SSMT will continue to enable the advanced and development of nuclear materials for Gen-IV reactors and the current fleet to keep nuclear power growing as a carbon-free source of electrical power.

References

Alam, M. E., Pal, S., Fields, K., Maloy, S. A., Hoelzer, D. T., & Odette, G. R. (2016). Tensile deformation and fracture properties of a 14YWT nanostructured ferritic alloy. *Materials Science and Engineering: A, 675*, 437–448. Available from https://doi.org/10.1016/j.msea.2016.08.051.

Ando, M., Tanigawa, H., Kurotaki, H., & Katoh, Y. (2018). Mechanical properties of neutron irradiated F82H using micro-tensile testing. *Nuclear Materials and Energy, 16*, 258–262. Available from https://doi.org/10.1016/j.nme.2018.07.008.

Barnoush, A., Hosemann, P., Molina-Aldareguia, J., & Wheeler, J. M. (2019). In situ small-scale mechanical testing under extreme environments. *MRS Bulletin, 44*(6), 471–477. Available from http://journals.cambridge.org/MRS, 10.1557/mrs.2019.126.

Barsoum, M. W., Murugaiah, A., Kalidindi, S. R., Zhen, T., & Gogotsi, Y. (2004). Kink bands, nonlinear elasticity and nanoindentations in graphite. *Carbon, 42*(8–9), 1435–1445. Available from https://doi.org/10.1016/j.carbon.2003.12.090.

Barsoum, M. W., Zhao, X., Shanazarov, S., Romanchuk, A., Koumlis, S., Pagano, S. J., Lamberson, L., & Tucker, G. J. (2019). Ripplocations: A universal deformation mechanism in layered solids. *Physical Review Materials, 3*(1). Available from https://doi.org/10.1103/physrevmaterials.3.013602.

Basu, S., Zhou, A., & Barsoum, M. W. (2009). On spherical nanoindentations, kinking nonlinear elasticity of mica single crystals and their geological implications. *Journal of Structural Geology, 31*(8), 791–801. Available from https://doi.org/10.1016/j.jsg.2009.05.008.

Brocklehurst, J. E., & Kelly, B. T. (1993). Analysis of the dimensional changes and structural changes in polycrystalline graphite under fast neutron irradiation. *Carbon, 31*(1), 155–178. Available from https://doi.org/10.1016/0008-6223(93)90169-b.

Bushby, A. J., Roberts, S. G., & Hardie, C. D. (2012). Nanoindentation investigation of ion-irradiated Fe–Cr alloys using spherical indenters. *Journal of Materials Research, 27*(1), 85–90. Available from https://doi.org/10.1557/jmr.2011.304.

Cappia, F., Frazer, D., Teng, F., Bawane, K., Jensen, C., Wachs, D., & Daum, R. (2021). Fracture properties and advanced characterization of high burnup fuel relevant to fragmentation during accident conditions. *Top Fuel 2022 Light Water Reactor Fuel Performance Conference proceedings*.

Cappia, F., Pizzocri, D., Marchetti, M., Schubert, A., Van Uffelen, P., Luzzi, L., Papaioannou, D., Macián-Juan, R., & Rondinella, V. V. (2016). Microhardness and Young's modulus of high burn-up UO_2 fuel. *Journal of Nuclear Materials*, *479*, 447−454. Available from https://doi.org/10.1016/j.jnucmat.2016.07.015.

Carvajal-Nunez, U., Elbakhshwan, M. S., Mara, N. A., White, J. T., & Nelson, A. T. (2018). Mechanical properties of uranium silicides by nanoindentation and finite elements modeling. *JOM*, *70*(2), 203−208. Available from https://doi.org/10.1007/s11837-017-2667-1.

Chen, D., Li, R., Lang, D., Wang, Z., Su, B., Zhang, X., & Meng. (2017). Determination of the mechanical properties of inclusions and matrices in α-U and ages U-5.5Nb alloy by nanoindentation measurements. *Materials Research Express*, *4*, 116516. Available from https://doi.org/10.1088/2053-1591/aa9864.

Chen, T., He, L., Cullison, M. H., Hay, C., Burns, J., Wu, Y., & Tan, L. (2020). The correlation between microstructure and nanoindentation property of neutron-irradiated austenitic alloy D9. *Acta Materialia Inc, United States Acta Materialia*, *195*, 433−445. Available from https://doi.org/10.1016/j.actamat.2020.05.020, http://www.journals.elsevier.com/acta-materialia/.

Connolley, T., Mchugh, P. E., & Bruzzi, M. (2005). A review of deformation and fatigue of metals at small size scales. *Fatigue & Fracture of Engineering Materials & Structures*, *28*(12), 1119−1152. Available from https://doi.org/10.1111/j.1460-2695.2005.00951.x.

Dehm, G., Jaya, B. N., Raghavan, R., & Kirchlechner, C. (2018). Overview on micro- and nanomechanical testing: New insights in interface plasticity and fracture at small length scales. *Acta Materialia*, *142*, 248−282. Available from https://doi.org/10.1016/j.actamat.2017.06.019.

Doitrand, A., Henry, R., Zacharie-Aubrun, I., Gatt, J. M., & Meille, S. (2020). UO_2 micron scale specimen fracture: Parameter identification and influence of porosities. *Theoretical and Applied Fracture Mechanics*, *108*. Available from https://doi.org/10.1016/j.tafmec.2020.102665, https://www.journals.elsevier.com/theoretical-and-applied-fracture-mechanics.

Dolph, C. K., da Silva, D. J., Swenson, M. J., & Wharry, J. P. (2016). Plastic zone size for nanoindentation of irradiated Fe—9%Cr ODS. *Journal of Nuclear Materials*, *481*, 33−45. Available from https://doi.org/10.1016/j.jnucmat.2016.08.033.

Dong, A., Duckering, J., Peterson, J., Lam, S., Routledge, D., & Hosemann, P. (2021). Femtosecond laser machining of micromechanical tensile test specimens. *JOM*, *73*(12), 4231−4239. Available from https://doi.org/10.1007/s11837-021-04971-w.

Elbakhshwan, M. S., Miao, Y., Stubbins, J. F., & Heuser, B. J. (2016). Mechanical properties of UO_2 thin films under heavy ion irradiation using nanoindentation and finite element modeling. *Journal of Nuclear Materials*, *479*, 548−558. Available from https://doi.org/10.1016/j.jnucmat.2016.07.047.

Fischer-Cripps, A. C. (2004). *Nanoindentation* (Third edition). Springer. Available from https://doi.org/10.1007/978-1-4419-9872-9.

Frazer, D., Abad, M. D., Krumwiede, D., Back, C. A., Khalifa, H. E., Deck, C. P., & Hosemann, P. (2015). Localized mechanical property assessment of SiC/SiC composite materials. *Composites Part A: Applied Science and Manufacturing*, *70*, 93−101. Available from https://doi.org/10.1016/j.compositesa.2014.11.008.

Frazer, D., Cappia, F., Miller, B., Murray, D., Winston, A., Pomo, A., & White, J. T. (2022). Nanoindentation testing of high burn-up fast reactor mixed oxide fuel. *Journal of Nuclear Materials, 564.* Available from https://doi.org/10.1016/j.jnucmat.2022.153668.

Frazer, D., Deck, C. P., & Hosemann, P. (2020). High-temperature nanoindentation of SiC/SiC composites. *JOM, 72*(1), 139−144. Available from https://doi.org/10.1007/s11837-019-03860-7.

Frazer, D., & Hosemann, P. (2019). Plasticity of UO_2 studied and quantified via elevated temperature micro compression testing. *Journal of Nuclear Materials, 525,* 140−144. Available from https://doi.org/10.1016/j.jnucmat.2019.07.022.

Frazer, D., Jadernas, D., Bolender, N., Madden, J., Giglio, J., & Hosemann, P. (2019). Elevated temperature microcantilever testing of fresh U-10Mo fuel. *Journal of Nuclear Materials, 526.* Available from https://doi.org/10.1016/j.jnucmat.2019.151746.

Frazer, D., Maiorov, B., Carvajal-Nuñez, U., Evans, J., Kardoulaki, E., Dunwoody, J., Saleh, T. A., & White, J. T. (2021). High temperature mechanical properties of fluorite crystal structured materials (CeO_2, ThO_2, and UO_2) and advanced accident tolerant fuels (U_3Si_2, UN, and UB_2. *Journal of Nuclear Materials, 554.* Available from https://doi.org/10.1016/j.jnucmat.2021.153035.

Frazer, D., Shaffer, B., Gong, B., Peralta, P., Lian, J., & Hosemann, P. (2021). Elevated temperature nanoindentation creep study of plastically deformed and spark plasma sintered UO_2. *Journal of Nuclear Materials, 545.* Available from https://doi.org/10.1016/j.jnucmat.2020.152605.

Frazer, D., Saleh, T. A., Matsumoto, T., Hirooka, S., Kato, M., McClellan, K., & White, J. T. (2024). High temperature nanoindentation of $(U,Ce)O_2$ compounds. *Nuclear Engineering and Design, 423.* Available from https://doi.org/10.1016/j.nucengdes.2024.113136.

Gianola, D. S., della Ventura, N. M., Balbus, G. H., Ziemke, P., Echlin, M. P., & Begley, M. R. (2023). Advances and opportunities in high-throughput small-scale mechanical testing. *Current Opinion in Solid State and Materials Science, 27*(4). Available from https://doi.org/10.1016/j.cossms.2023.101090, https://www.journals.elsevier.com/current-opinion-in-solid-state-and-materials-science.

Gong, B., Frazer, D., Shaffer, B., Lim, H. C., Hosemann, P., & Peralta, P. (2021). Microcantilever beam experiments and modeling in porous polycrystalline UO_2. *Journal of Nuclear Materials, 557.* Available from https://doi.org/10.1016/j.jnucmat.2021.153210, https://www.journals.elsevier.com/journal-of-nuclear-materials.

Gong, B., Frazer, D., Yao, T., Hosemann, P., Tonks, M., & Lian, J. (2019). Nano- and micro-indentation testing of sintered UO_2 fuel pellets with controlled microstructure and stoichiometry. *Journal of Nuclear Materials, 516,* 169−177. Available from https://doi.org/10.1016/j.jnucmat.2019.01.021.

Hardie, C. D., Roberts, S. G., & Bushby, A. J. (2015). Understanding the effects of ion irradiation using nanoindentation techniques. *Journal of Nuclear Materials, 462,* 391−401. Available from https://doi.org/10.1016/j.jnucmat.2014.11.066.

Henry, R., Zacharie-Aubrun, I., Blay, T., Chalal, S., Gatt, J. M., Langlois, C., & Meille, S. (2020). Fracture properties of an irradiated PWR UO_2 fuel evaluated by microcantilever bending tests. *Journal of Nuclear Materials, 538.* Available from https://doi.org/10.1016/j.jnucmat.2020.152209.

Henry, R., Zacharie-Aubrun, I., Blay, T., Tarisien, N., Chalal, S., Iltis, X., Gatt, J. M., Langlois, C., & Meille, S. (2020). Irradiation effects on the fracture properties of UO_2 fuels studied by micro-mechanical testing. *Journal of Nuclear Materials, 536.* Available from https://doi.org/10.1016/j.jnucmat.2020.152179.

Hosemann, P. (2018). Small-scale mechanical testing on nuclear materials: Bridging the experimental length-scale gap. *Scripta Materialia*, *143*, 161−168. Available from https://doi.org/10.1016/j.scriptamat.2017.04.026.

Hosemann, P., Dai, Y., Stergar, E., Leitner, H., Olivas, E., Nelson, A. T., & Maloy, S. A. (2011). Large and small scale materials testing of HT-9 irradiated in the STIP irradiation program. *Experimental Mechanics*, *51*(7), 1095−1102. Available from https://doi.org/10.1007/s11340-010-9419-2.

Hosemann, P., Kiener, D., Wang, Y., & Maloy, S. A. (2012). Issues to consider using nano indentation on shallow ion beam irradiated materials. *Journal of Nuclear Materials*, *425* (1-3), 136−139. Available from https://doi.org/10.1016/j.jnucmat.2011.11.070.

Hosemann, P., Pouchon, M., Dai, Y., & Maloy, S. (2010). Micro, macro scale mechanical testing and characterization on irradiated structural materials for nuclear applications. *Minerals, Metal and Materials Society/AIME , 420 Commonwealth Dr., Warrendale PA, USA.*.

Howard, C., Bhakhri, V., Dixon, C., Rajakumar, H., Mayhew, C., & Judge, C. D. (2019). Coupling multi-scale mechanical testing techniques reveals the existence of a transgranular channel fracture deformation mechanism in high dose Inconel X-750. *Journal of Nuclear Materials*, *517*, 17−34. Available from https://doi.org/10.1016/j.jnucmat.2019.01.051.

Howard, C., Judge, C. D., & Hosemann, P. (2019). Applying a new push-to-pull microtensile testing technique to evaluate the mechanical properties of high dose Inconel X-750. *Materials Science and Engineering: A*, *748*, 396−406. Available from https://doi.org/10.1016/j.msea.2019.01.113.

Howard, C., Judge, C. D., Poff, D., Parker, S., Griffiths, M., & Hosemann, P. (2018). A novel in-situ, lift-out, three-point bend technique to quantify the mechanical properties of an ex-service neutron irradiated inconel X-750 component. *Journal of Nuclear Materials*, *498*, 149−158. Available from https://doi.org/10.1016/j.jnucmat.2017.10.002.

Huang, Y., Zhang, F., Hwang, K. C., Nix, W. D., Pharr, G. M., & Feng, G. (2006). A model of size effects in nano-indentation. *Journal of the Mechanics and Physics of Solids*, *54* (8), 1668−1686. Available from https://doi.org/10.1016/j.jmps.2006.02.002.

Islam, M. M., Shakil, S. I., Shaheen, N. M., Bayati, P., & Haghshenas, M. (2021). An overview of microscale indentation fatigue: Composites, thin films, coatings and ceramics. *Micron*, *148*, 103110. Available from https://doi.org/10.1016/j.micron.2021.103110.

Jaya, B. N., Mathews, N. G., Mishra, A. K., Basu, S., & Jacob, K. (2022). Non-conventional small-scale mechanical testing of materials. *Journal of the Indian Institute of Science*, *102*(1), 139−171. Available from https://doi.org/10.1007/s41745-022-00302-3.

Jin, X., Wade-Zhu, J., Chen, Y., Mummery, P. M., Fan, X., & Marrow, T. J. (2021). Assessment of the fracture toughness of neutron-irradiated nuclear graphite by 3D analysis of the crack displacement field. *Carbon*, *171*, 882−893. Available from https://doi.org/10.1016/j.carbon.2020.09.072, http://www.journals.elsevier.com/carbon/.

Kardoulaki, E., Frazer, D. M., White, J. T., Carvajal, U., Nelson, A. T., Byler, D. D., Saleh, T. A., Gong, B., Yao, T., Lian, J., & McClellan, K. J. (2021). Fabrication and thermophysical properties of UO$_2$-UB$_2$ and UO$_2$-UB$_4$ composites sintered via spark plasma sintering. *Journal of Nuclear Materials*, *544*. Available from https://doi.org/10.1016/j.jnucmat.2020.152690.

Kardoulaki, E., White, J. T., Byler, D. D., Frazer, D. M., Shivprasad, A. P., Saleh, T. A., Gong, B., Yao, T., Lian, J., & McClellan, K. J. (2020). Thermophysical and mechanical property assessment of UB$_2$ and UB$_4$ sintered via spark plasma sintering. *Journal of*

Alloys and Compounds, 818. Available from https://doi.org/10.1016/j.jallcom.2019.153216.

Kasada, R., Konishi, S., Yabuuchi, K., Nogami, S., Ando, M., Hamaguchi, D., & Tanigawa, H. (2014). Depth-dependent nanoindentation hardness of reduced-activation ferritic steels after MeV Fe-ion irradiation. *Fusion Engineering and Design*, *89*(7-8), 1637–1641. Available from https://doi.org/10.1016/j.fusengdes.2014.03.068, http://www.journals.elsevier.com/fusion-engineering-and-design/.

Katoh, Y., Ozawa, K., Shih, C., Nozawa, T., Shinavski, R. J., Hasegawa, A., & Snead, L. L. (2014). Continuous SiC fiber, CVI SiC matrix composites for nuclear applications: Properties and irradiation effects. *Journal of Nuclear Materials*, *448*(1-3), 448–476. Available from https://doi.org/10.1016/j.jnucmat.2013.06.040.

Ke, H. B., Zhang, P., Sun, B. A., Zhang, P. G., Liu, T. W., Chen, P. H., Wu, M., & Huang, H. G. (2019). Dissimilar nanoscaled structural heterogeneity in U-based metallic glasses revealed by nanoindentation. *Journal of Alloys and Compounds*, *788*, 391–396. Available from https://doi.org/10.1016/j.jallcom.2019.02.256.

Kelly, B. T. (1982). Graphite—The most fascinating nuclear material. *Carbon*, *20*(1), 3–11. Available from https://doi.org/10.1016/0008-6223(82)90066-5.

Kese, K., Olsson, P. A. T., Alvarez Holston, A.-M., & Broitman, E. (2017). High temperature nanoindentation hardness and Young's modulus measurement in a neutron-irradiated fuel cladding material. *Journal of Nuclear Materials*, *487*, 113–120. Available from https://doi.org/10.1016/j.jnucmat.2017.02.014.

Kiener, D., Minor, A.M., Anderoglu, O., Wang, Y., Maloy, S.A., Hoseman, P., Naziris Versteylen, C., Frith, F., Bregman, M., & Kolluri, M. (2012). Development of mini-compact tension specimen fabrication and test methods in hot cell for post-irradiation examination of reactor pressure vessel steels. *Proceedings of the ASME 2021 Pressure Vessels & Piping Conference*, 27.

Krumwiede, D. L., Yamamoto, T., Saleh, T. A., Maloy, S. A., Odette, G. R., & Hosemann, P. (2018). Direct comparison of nanoindentation and tensile test results on reactor-irradiated materials. *Journal of Nuclear Materials*, *504*, 135–143. Available from https://doi.org/10.1016/j.jnucmat.2018.03.021.

Kurosaki, K., Saito, Y., Muta, H., Uno, M., & Yamanaka, S. (2004). Nanoindentation studies of UO_2 and $(U,Ce)O_2$. *Journal of Alloys and Compounds*, *381*(1-2), 240–244. Available from https://doi.org/10.1016/j.jallcom.2004.03.084.

Li, R., Liu, K., Luo, W., Su, B., & Zou, D. (2022). Anisotropy in the elasticity of α-U from first-principles calculations and nanoindentation αAnisotropy in the elasticity of α-U from first-principles calculations and nanoindentation. *Journal of Nuclear Materials*, *558*. Available from https://doi.org/10.1016/j.jnucmat.2021.153351.

Li, R., Mo, C., & Liao, Y. (2018). Mechanical properties of U-Cu intermetallic compound measured by nanoindentation. *Materials*, *11*(11). Available from https://doi.org/10.3390/ma11112215.

Liu, D., Mingard, K., Lord, O. T., & Flewitt, P. (2017). On the damage and fracture of nuclear graphite at multiple length-scales. *Journal of Nuclear Materials*, *493*, 246–254. Available from https://doi.org/10.1016/j.jnucmat.2017.06.021, https://www.journals.elsevier.com/journal-of-nuclear-materials.

Lucas, G.E., Odette, G.R., & Sheckherd, J.W. (1986). Shear punch and microhardness tests for strength and ductility measurements. *ASTM Special Technical Publication*. 112-140 ASTM undefined. https://doi.org/10.1520/stp32998s.

Maganosc, D. J., Ligda, J. P., Sano, T., & Schuster, B. E. (2017). Femtosecond laser machining of micro-tensile specimens for high throughput mechanical testing micro and nanomechanics. *Conference Proceesings of the Society for Experimental Mechanics Series*, 5, 7−9.

Maloy, S. A., Romero, T. J., Hosemann, P., Toloczko, M. B., & Dai, Y. (2011). Shear punch testing of candidate reactor materials after irradiation in fast reactors and spallation environments. *Journal of Nuclear Materials*, *417*(1-3), 1005−1008.

Mao, K. S., Sun, C., Huang, Y., Shiau, C. H., Garner, F. A., Freyer, P. D., & Wharry, J. P. (2019). Grain orientation dependence of nanoindentation and deformation-induced martensitic phase transformation in neutron irradiated AISI 304L stainless steel. *Materialia*, 5. Available from https://doi.org/10.1016/j.mtla.2019.100208, http://www.journals.elsevier.com/materialia.

Marrow, T. J., Liu, D., Barhli, S. M., Saucedo-Mora, L., Vertyagina, Y., Collins, D. M., et al. (2016). In situ measurement of the strains within a mechanically loaded polygranular graphite. *Carbon*, *96*, 285−302. Available from https://doi.org/10.1016/j.carbon.2015.09.058.

Marrow, T. J., Šulak, I., Li, B.-S., Vukšić, M., Williamson, M., & Armstrong, D. E. J. (2022). High temperature spherical nano-indentation of graphite crystals. *Carbon*, *191*, 236−242. Available from https://doi.org/10.1016/j.carbon.2022.01.067.

Mason, I. B., & Knibbs, R. H. (1960). Variation with temperature of Young's modulus of polycrystalline graphite. *Nature*, *188*(4744), 33−35. Available from https://doi.org/10.1038/188033a0.

McCulloch, Q., Gigax, J. G., & Hosemann, P. (2020). Femtosecond laser ablation for mesoscale specimen evaluation. *JOM*, *72*(4), 1694−1702. Available from https://doi.org/10.1007/s11837-020-04045-3.

Mrozowski, S. (1954). *Mechanical strength, thermal expansion and structure of cokes and carbons* (31, pp. 31−45). *Buffalo*, NY: *Carbon Waverly Press*.

Naziris, F., Versteylen, C., Frith, F., Bregman, M., & Kolluri, M. (2021). *Development of Mini-Compact Tension Specimen Fabrication and Test Methods in Hot Cell for Post-Irradiation Examination of Reactor Pressure Vessel Steels. Proceedings of the ASME 2021 Pressure Vessels & Piping Conference*. ASME V004T06A064. Available from https://doi.org/10.1115/PVP2021-61027.

Newell, R., Park, Y., Mehta, A., Keiser, D., & Sohn, Y. (2017). Mechanical properties examined by nanoindentation for selected phases relevant to the development of monolithic uranium-molybdenum metallic fuels. *Journal of Nuclear Materials*, *487*, 443−452. Available from https://doi.org/10.1016/j.jnucmat.2017.02.018.

Nix, W. D., & Gao, H. (1998). Indentation size effects in crystalline materials: A law for strain gradient plasticity. *Journal of the Mechanics and Physics of Solids*, *46*(3), 411−425. Available from https://doi.org/10.1016/S0022-5096(97)00086-0.

Oliver, W. C., & Pharr, G. M. (1992). An improved technique for determining hardness and elastic modulus using load and displacement sensing indentation experiments. *Journal of Materials Research*, 7, 1546−1583.

Pathak, S., & Kalidindi, S. R. (2015). Spherical nanoindentation stress-strain curves. *Materials Science and Engineering R: Reports*, *91*, 1−36. Available from https://doi.org/10.1016/j.mser.2015.02.001.

Pathak, S., Kalidindi, S. R., & Mara, N. A. (2016). Investigations of orientation and length scale effects on micromechanical responses in polycrystalline zirconium using spherical nanoindentation. *Scripta Materialia*, *113*, 241−245. Available from https://doi.org/10.1016/j.scriptamat.2015.10.035.

Pfeifenberger, M. J., Mangang, M., Wurster, S., Reiser, J., Hohenwarter, A., Pfleging, W., Kiener, D., & Pippan, R. (2017). The use of femtosecond laser ablation as a novel tool for rapid micro-mechanical sample preparation. *Materials and Design*, *121*, 109−118. Available from https://doi.org/10.1016/j.matdes.2017.02.012.

Prasitthipayong, A., Vachhani, S. J., Tumey, S. J., Minor, A. M., & Hosemann, P. (2018). Indentation size effect in unirradiated and ion-irradiated 800H steel at high temperatures. *Acta Materialia*, *144*, 896−904. Available from https://doi.org/10.1016/j.actamat.2017.11.001.

Reichardt, A., Lupinacci, A., Frazer, D., Bailey, N., Vo, H., Howard, C., Jiao, Z., Minor, A. M., Chou, P., & Hosemann, P. (2017). Nanoindentation and in situ microcompression in different dose regimes of proton beam irradiated 304 SS. *Journal of Nuclear Materials*, *486*, 323−331. Available from https://doi.org/10.1016/j.jnucmat.2017.01.036.

Rodriguez, R., & Gutierrez, I. (2003). Correlation between nanoindentation and tensile propertiesInfluence of the indentation size effect. *Materials Science and Engineering A*, *361*(1-2), 377−384. Available from https://doi.org/10.1016/s0921-5093(03)00563-x.

Rohbeck, N., Tsivoulas, D., Shapiro, I. P., Xiao, P., Knol, S., Escleine, J. M., & Perez, M. (2015). In-situ nanoindentation of irradiated silicon carbide in TRISO particle fuel up to 500°C. *Journal of Nuclear Materials*, *465*, 692−694. Available from https://doi.org/10.1016/j.jnucmat.2015.06.035.

Rondinella, V. V., & Wiss, T. (2010). The high burn-up structure in nuclear fuel. *Materials Today*, *13*(12), 24−32. Available from https://doi.org/10.1016/S1369-7021(10)70221-2, http://www.journals.elsevier.com/materials-today/.

Roostaii, B., Kazeminejad, H., & Khakshournia, S. (2021). Including high burnup structure effect in the UO_2 fuel thermal conductivity model. *Progress in Nuclear Energy*, *131*. Available from https://doi.org/10.1016/j.pnucene.2020.103561.

Schulthess, J. 2011. National Postirradiation Examination Workshop Report INL/EXT-11-21922. Available from https://inldigitallibrary.inl.gov/sites/sti/sti/5026014.pdf.

Schneider, C., Fayette, L., Zacharie-Aubrun, I., Blay, T., Sercombe, J., Favergeon, J., & Chevalier, S. (2022). Study of the hardness and Young's modulus at the fuel-cladding interface of a high-burnup PWR fuel rod by nanoindentation measurements. *Journal of Nuclear Materials*, *560*, 153511. Available from https://doi.org/10.1016/j.jnucmat.2022.153511.

Snead, L. L., Nozawa, T., Katoh, Y., Byun, T. S., Kondo, S., & Petti, D. A. (2007). Handbook of SiC properties for fuel performance modeling. *Journal of Nuclear Materials*, *371*(1-3), 329−377. Available from https://doi.org/10.1016/j.jnucmat.2007.05.016.

Terrani, K. A., Balooch, M., Burns, J. R., & Smith, Q. B. (2018). Young's modulus evaluation of high burnup structure in UO_7 with nanometer resolution. *Journal of Nuclear Materials*, *508*, 33−39. Available from https://doi.org/10.1016/j.jnucmat.2018.04.004.

Uchic, M. D., & Dimiduk, D. M. (2005). A methodology to investigate size scale effects in crystalline plasticity using uniaxial compression testing. *Materials Science and Engineering: A*, *400-401*(1-2), 268−278. Available from https://doi.org/10.1016/j.msea.2005.03.082, http://www.elsevier.com.

Vo, H. T., Frazer, D., Kohnert, A. A., Teysseyre, S., Fensin, S., & Hosemann, P. (2023). Role of low-level void swelling on plasticity and failure in a 33 dpa neutron-irradiated 304 stainless steel. *International Journal of Plasticity*, *164*. Available from https://doi.org/10.1016/j.ijplas.2023.103577.

Vo, H. T., Reichardt, A., Frazer, D., Bailey, N., Chou, P., & Hosemann, P. (2017). In situ microtensile testing on proton beam-irradiated stainless steel. *Journal of Nuclear Materials*, *493*, 336−342. Available from https://doi.org/10.1016/j.jnucmat.2017.06.026.

Wang, Y., Frazer, D. M., Cappia, F., Teng, F., Murray, D. J., Yao, T., Judge, C. D., Harp, J. M., & Capriotti, L. (2022). Small-scale mechanical testing and characterization of fuel cladding chemical interaction between HT9 cladding and advanced U-based metallic fuel alloy. *Journal of Nuclear Materials, 566.* Available from https://doi.org/10.1016/j.jnucmat.2022.153754, https://www.journals.elsevier.com/journal-of-nuclear-materials.

Weaver, J. S., Pathak, S., Reichardt, A., Vo, H. T., Maloy, S. A., Hosemann, P., & Mara, N. A. (2017). Spherical nanoindentation of proton irradiated 304 stainless steel: A comparison of small scale mechanical test techniques for measuring irradiation hardening. *Journal of Nuclear Materials, 493,* 368−379. Available from https://doi.org/10.1016/j.jnucmat.2017.06.031.

Xiao, H., Long, C., & Chen, H. (2021). The formation mechanisms of high burnup structure in UO_2 fuel. *Journal of Nuclear Materials, 556.* Available from https://doi.org/10.1016/j.jnucmat.2021.153151.

Yoda, S., Eto, M., & Oku, T. (1983). Change in dynamic Young's modulus of nuclear-grade isotropic graphite during tensile and compressive stressing. *Journal of Nuclear Materials, 119*(2-3), 278−283. Available from https://doi.org/10.1016/0022-3115(83)90204-0.

Zacharie-Aubrun, I., Henry, R., Blay, T. H., Brunaud, L., Gatt, J. M., Noirot, J., & Meille, S. (2021). Effects of irradiation on mechanical properties of nuclear UO_2 fuels evaluated by Vickers indentation at room temperature. *Journal of Nuclear Materials, 547.* Available from https://doi.org/10.1016/j.jnucmat.2021.152821.

Zhu, P., Zhao, Y., Agarwal, S., Henry, J., & Zinkle, S. J. (2022). Toward accurate evaluation of bulk hardness from nanoindentation testing at low indent depths. *Materials and Design, 213.* Available from https://doi.org/10.1016/j.matdes.2021.110317, https://www.journals.elsevier.com/materials-and-design.

Aerospace coatings - surface engineering for operation in extreme environments

21

John Rayment Nicholls, Christine Deborah Chalk and Luis Isern Arrom
Coating Technology, Surface Engineering and Precision Centre, Cranfield University, Cranfield, United Kingdom

21.1 Introduction

21.1.1 The gas turbine and extreme environments

The aero-gas turbine is a major driving force of the aerospace industry. It powers the sector and is a component like no other, in terms of power density, operating temperatures, and its extreme operating environments. A combination of cooling and a multilayer thermal barrier coating (TBC) system is needed to prevent the hottest metal parts from melting or oxidizing. Indeed, combustion temperatures can exceed 1800°C, with turbine inlet temperatures exceeding 1600°C at take-off, while ingesting runway dust, sea water spray (near the coast or over the sea), desert sand, and volcanic ash depending on the aeroplanes particular flight path. At such high service temperatures, these ingested particulates melt and form glassy deposits, known as CMAS (calcium, magnesium, alumino-silicate) that severely degrade engine performance through blocking cooling passages and chemically attacking the TBC systems applied to thermally protect the turbine hot gas path components (Morrell et al., 2010; Bold, 1999; Rolls-Royce Plc, 1986).

The aero-gas turbine sector consists of four major players, General Electric (GE) and Pratt and Whitney in the USA, and Rolls-Royce plc and Safran Aircraft Engines (formerly Snecma Moteurs) within Europe, supported by other smaller engine companies that develop niche markets or subcomponents of the engine, alongside these major players. Thus, Rolls-Royce bought out Allison in the USA to form Rolls-Royce Corporation, and MTU in Germany to form Rolls-Royce-Deutschland while GE and Safran/Snecma work collaboratively together through CFM International. Similarly, Pratt and Witney have clustered with MTU Aero-engines and Avio (formally Fiat) in Europe to support the short haul market sector needs (further complicated by GE's acquisition of Avio in 2013). These collaborations and clustering across the sector lead to long-term stability, supporting the needed to develop new, more efficient, engines and powertrains for all application sectors of the aero-gas turbine (Gao et al., 2024; Kim et al., 1993; Smialek et al., 1994).

Nanomechanics for Coatings and Engineering Surfaces. DOI: https://doi.org/10.1016/B978-0-443-13334-3.00021-8
Copyright © 2025 Elsevier Inc. All rights reserved, including those for text and data mining, AI training, and similar technologies. This chapter contributors own copyright for the Cranfields 'know-how' contents.

The cost of building a new engine is a closely guarded secret but is estimated to be 1 − 2 billion dollars, and once built, it has a life cycle of 30−50 years (Kim et al., 1993; Smialek et al., 1994). It is therefore imperative that the design is correct.

21.1.2 The role of surface engineering

Surface engineering is ubiquitous in the aerospace industry. Surface engineering describes the process of modifying the surface properties of a substrate material to enhance the performance of a component under operating conditions. These surface properties will be different from those of the bulk material, although both should be integrated and considered together as a system. Surface engineering covers the application of coatings (e.g., for oxidation or corrosion protection, wear resistance, or thermal protection), as well as the physical modification of a surface using mechanical or thermal processes to alter its properties without application of a coating (e.g., laser or shot peening where surfaces are modified through application of a compressive stress). While some coatings can be purely decorative, such as the application of gold plate to a cheaper base metal to give the appearance of a solid gold object, many coating systems today are multifunctional, comprising one or more layers, and offer protection against one or more than one degradation processes. For instance, erosion and corrosion often occur together in service and present a major concern in the surface science and engineering community.

In the aeroengine, TBCs for nickel superalloys are highly surface-engineered, multilayer systems applied to hot section components in the gas turbine, as seen in Fig. 21.1. TBC systems consist of a low thermal conductivity ceramic topcoat and a

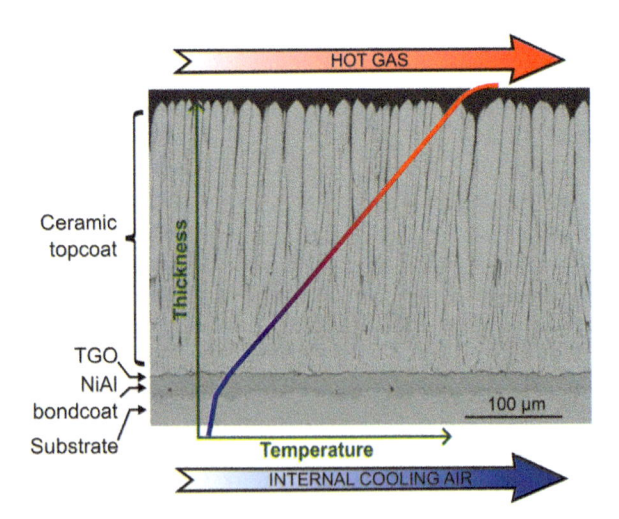

Figure 21.1 Diagram of a TBC. Columnar TBC system consisting of 7YSZ topcoat, TGO, and NiAl bondcoat on a nickel alloy substrate with thermal gradient superimposed from the gas-washed surface to the blade interior in arbitrary units.
Source: Courtesy Cranfield University.

metallic bondcoat providing oxidation protection to the underlying nickel superalloy substrate by forming a thin slow-growing oxide, usually alumina, known as a thermally grown oxide (TGO). The TGO also improves adhesion between the topcoat and the bondcoat. TBCs maximize operational lifetime under extreme operating conditions, which include high temperatures and high mechanical loads (with parts rotating at up to 12,000 rpm), thermal shock, and physical and chemical damage. The coating system itself, supported by highly engineered internal cooling of the substrate and film cooling of the surface coating, allows a temperature gradient of up to 170°C between the substrate and the coating/gas interface. The TBC system is now prime reliant: turbines are operating at temperatures exceeding the melting point of the underlying substrate alloy and failure of the TBC can lead to component failure.

The need to run engines hotter, a route to maximizing efficiency and reducing emissions, places increasing demand on the chemical, physical, and mechanical properties of coating materials. The industry standard TBC ceramic material is 7–8 wt.% yttria-stabilized zirconia (7YSZ). Electron beam physical vapor deposition (EB-PVD) is the preferred method for coating high-pressure, high-temperature turbine components; it produces columnar structures with improved aerodynamics that are strain-tolerant for operation under high load conditions. The resulting columnar microstructure accommodates differences in thermal expansion coefficient between the metallic substrate and bondcoat (typically $\sim 16 \times 10^{-6}\,\text{K}^{-1}$) and the ceramic topcoat (typically $10.5 \times 10^{-6}\,\text{K}^{-1}$). Currently, operating temperatures are limited to below 1200°C for two reasons. First, 7YSZ ceramic undergoes deleterious phase changes above this temperature leading to TBC loss. Second, chemical attack by ingested airborne dusts (CMAS) leads to TBC loss through the formation of glassy deposits and spallation. Recent volcanic eruptions in Iceland have led to grounding of aircraft in affected regions to avoid this failure mechanism. Higher operation temperatures mean that these ingested dusts are more likely to melt on the TBC surface, reducing its strain compliance. Hence, new TBC materials are required for the net-zero journey, and work is underway to design new TBC systems to meet these challenges. To aid this discussion, each layer of the TBC system is described further starting with the bondcoat.

21.2 Environmental protection coatings and bondcoats

Once it was recognized that first-stage turbine blades must be coated to provide additional surface protection, coating manufacturing methods developed in parallel with the evolution in blade alloy materials and blade design. In the early 1990s, aeroengine mean blade temperatures were around 1050°C, with peak temperatures circa 1150°C. TBCs were in their infancy, especially on aerofoil surfaces, but turbine components had to be protected against environmental degradation from oxidation and hot corrosion. As a result, two classes of alloy coating system were developed: intermetallic diffusion coatings using chemical vapor deposition (CVD), which evolved from earlier pack-cementation processes first used in turbines in 1957 (Goward & Cannon, 1988), and MCrAlY alloy overlay coatings using thermal spray methods and EB-PVD, where M is a combination of Ni and/or Co for aerospace applications (Goward, 1983). These environmental protection coatings formed the basis of bondcoats for the development of TBC technologies.

Since the early 1990s, one has seen turbine entry temperatures (TET) increase by a further 120°C, in pursuit of increased engine efficiency and lower specific fuel consumption. Thus, metal temperatures at the leading and trailing edges of a blade may exceed 1150°C. Today, environmental protection coatings are an integral part of turbine blade manufacture, to mitigate oxidation and high temperature corrosion attack. The turbine blades are single-crystal nickel superalloys, bondcoated with either a platinum aluminide intermetallic diffusion coating or an MCrAlY overlay coating, before overcoating with the thermal barrier.

21.2.1 Diffusion coatings

As the name suggests, diffusion coatings are formed by the diffusion of a depositing element into the surface of the superalloy. As a result, an intermetallic layer is formed that provides a reservoir of oxide scale forming element (Nicholls, 2000; Pomeroy, 2005). For high-temperature oxidation protection, aluminide coatings based on the formation of β-NiAl are the preferred choice (Pomeroy, 2005). This system is relatively low cost and provides a reservoir of aluminum that oxidizes in service to form a slow-growing, protective α-alumina scale, known as the thermally grown oxide (TGO) providing an environmental barrier at temperatures up to 1050°C. For higher-temperature service, at temperatures upward of 1100°C, modified aluminide coatings are used, with the most likely modification being the addition of platinum. This modification results in the formation of a platinum aluminide coating, (Pt,Ni)Al, where the platinum layer is first deposited by electroplating, then heat-treated to interdiffuse the Pt with Ni from the superalloy substrate, before aluminizing. Other bondcoat alloying additions can be added, including Cr, Si, Ta, Hf, the rare earths, and precious metals. Then, depending on the materials and parameters used during the coating deposition and its heat treatment (including temperature, aluminum activity, and process time), the thickness, microstructure, and properties of the resultant diffusion coating can be varied. For example, Fig. 21.2 (A) illustrates a micrograph of a fully processed pack aluminide coating (high activity) on IN738 (Nicholls, 2003), the formation of which is controlled by the inward diffusion of aluminum. By contrast, Fig. 21.2(B) illustrates a micrograph of a platinum aluminide (RT22LT) intermetallic coating (Nicholls, 2003) whose microstructure now depends on the outward diffusion of nickel.

Numerous studies have demonstrated the beneficial effects of adding various elements to a coating system. For example, Cr decreases the minimum Al content needed to form a slow-growing, continuous alumina scale (Goward & Cannon, 1988; Goward, 1983), while reactive elements (Y, Hf, etc.) aid adhesion and decrease the scale growth rate (Nicholls, 2000). In addition, Pt and the precious

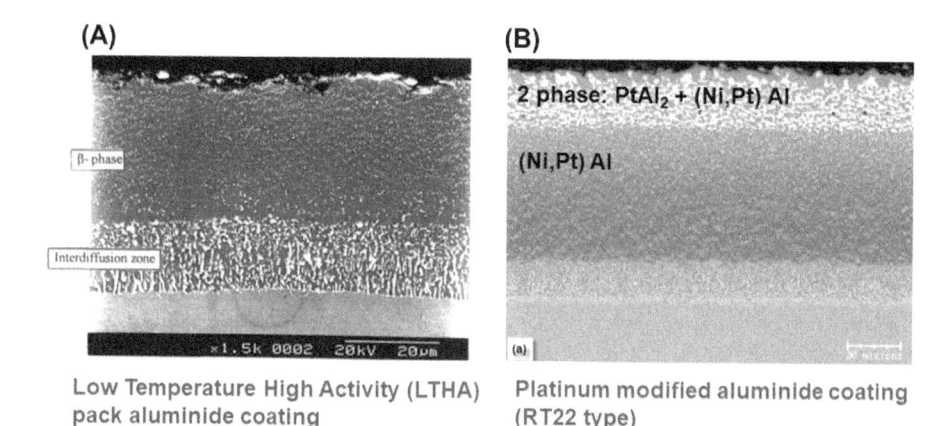

Aerospace coatings - surface engineering for operation in extreme environments 603

(A)

Low Temperature High Activity (LTHA) pack aluminide coating

(B)

Platinum modified aluminide coating (RT22 type)

Figure 21.2 Aluminide coatings. Comparison of bondcoats produced by (A) pack aluminide and (B) platinum aluminide.
Source: Adapted from Nicholls, J. R. (2003). Advances in coating design for high-performance gas turbines. *MRS Bulletin, 28*(9), 659–670. https://doi.org/10.1557/Mrs2003.194.

metals suppress spinel formation, decrease scale growth rates, and improve spallation resistance (Nicholls et al., 2010; Pomeroy, 2005).

In the NiAl system, the minimum concentration of Al to form an α-alumina scale is around 33 at.%, but with the addition of 5 at.% Cr, this limit for Al drops to 12 at.% (Goward & Cannon, 1988; Goward, 1983). This shows that Cr has a strong effect on the selective oxidation of Al and the formation of a continuous, slow-growing alumina scale. This effect is well known and actively used when designing MCrAlY overlay coatings, where M is Ni, Co, or Fe, or a combination of these elements, and forms the coating alloy base.

21.2.2 Overlay coatings

Overlay coatings are a custom-designed oxidation/corrosion-resistant alloy to provide environmental protection. Overlay coatings are deposited much thicker than diffusion coatings (approximately 150 μm compared with some 30–40 μm for a diffusion coating), so offer increased oxidation/corrosion resistance, when compared with diffusion coatings due to this increase in thickness. Overlay coatings are more generally mechanically bonded to the alloy substrate and are most often deposited by thermal spray methods. The surface to be coated is first grit-blasted, to mechanically clean the surface, while providing a mechanical key for the thermal sprayed coating (Goward & Cannon, 1988). Overlay coatings offer increased alloy design flexibility, as a prealloyed material is mechanically deposited onto the substrate at temperature. Limited interdiffusion may occur during the thermal spraying process and post spraying heat treatment supporting mechanical bonding to the substrate.

The coating composition and microstructure are decided at the prealloyed material powder manufacture stage (Goward & Cannon, 1988; Nicholls, 2000). Hence, overlay coatings are generally known as MCrAlY coatings, based on the original alloy specification containing Cr, Al, and Y, the latter an oxygen active element. "M" defines the alloy base, a possible blend of Ni, Co, or Fe. More recently, more complex alloy formulations have been proposed containing multiple active elements; thus alloys with blends containing Y, Ta, Hf, and Si have been proposed in an alloy base of Ni, plus Co (Bose, 2007; Herman, 1991; Khanna, 2002; Nicholls, 2003).

Early overlay coatings were custom-designed alloys for improved oxidation resistance. These were CoCrAlY alloy coatings and contained Cr in the range 20–40 wt.%, Al additions between 12 and 20 wt.%, and Y levels between 0.3 and 0.5 wt.%, with the most successful coating at that time being Co-25Cr-14Al-0.5Y (Tawancy et al., 1991). As well as thermal spraying, overlay coatings have been deposited by EB-PVD, sputter deposition, argon shrouded plasma spraying, and vacuum plasma spray (VPS) methods.

21.3 Thermal barrier coatings – design considerations

The outermost layer of a TBC system is a ceramic topcoat. While potential candidate topcoat materials are primarily selected by design engineers for their low thermal conductivity, other characteristics are also required, including:

- High thermal expansion coefficient.
- High-temperature sintering resistance and phase stability.
- Chemical resilience (against CMAS attack).
- Toughness for resistance to erosion/foreign object damage (FOD).

It can be difficult to obtain enhanced performance in all the properties with a simple ceramic chemical composition. Currently, compromises must be made, and further mitigation strategies sought. These may arise through use of more complex chemical compositions and/or modifications to microstructure brought about through variations in deposition conditions.

21.3.1 Low thermal conductivity

The thermal conductivity of the ceramic coating is influenced by both material selection and deposition method; the latter plays a significant role in determining the microstructure of the deposited coating. Heat is transferred in crystalline solids by three mechanisms: through electrons, lattice vibrations (phonons), and radiation (photons). In the case of TBCs, phonon scattering is most relevant, and several material features contribute to lowering thermal conductivity; these are the number of (oxygen) vacancies and atomic substitutions within the crystalline lattice and, to a lesser extent, the presence of grain boundaries (Nicholls et al., 2002). In the case of atomic substitutions, atoms and ions of different radii within the crystal structure distort the bond length and introduce elastic strains leading to lower thermal conductivity. More complex structures tend to have lower thermal conductivity. For 7YSZ, oxygen vacancies are introduced into the crystal lattice to compensate for the exchange of Y^{3+} for Zr^{4+} ions, and these account for its relatively low thermal conductivity (1.5–1.9 W mK^{-1} for EB-PVD coatings). Elements other than yttrium can be introduced into the crystal lattice, and the lanthanide series of elements, La to Yb, are often used in varying amounts to provide lower-conductivity doped ZrO2-based TBCs.

A wider group of TBC materials, including pyrochlores (rare earth zirconates), defect cluster materials, rare earth aluminates, and perovskites are being considered as candidate materials for next-generation TBCs based upon their low thermal conductivity. Fig. 21.3 maps thermal conductivity of different groups of materials alongside thermal expansion coefficient (Vaßen et al., 2022). A second consideration here is that ceramic materials with higher thermal expansion coefficients, closer to those of the bondcoat and substrate, are less likely to spall during service. It is noticeable that highly complex lanthanide zirconates feature in the low thermal conductivity, high thermal expansion coefficient corner of the chart.

In EB-PVD, an electron beam is used to melt a solid ceramic target (ingot) forming a vapor cloud. The parts to be coated are rotated in the vapor cloud above the evaporating ingot. Substrate rotation leads to a sunrise/sunset feature resulting in

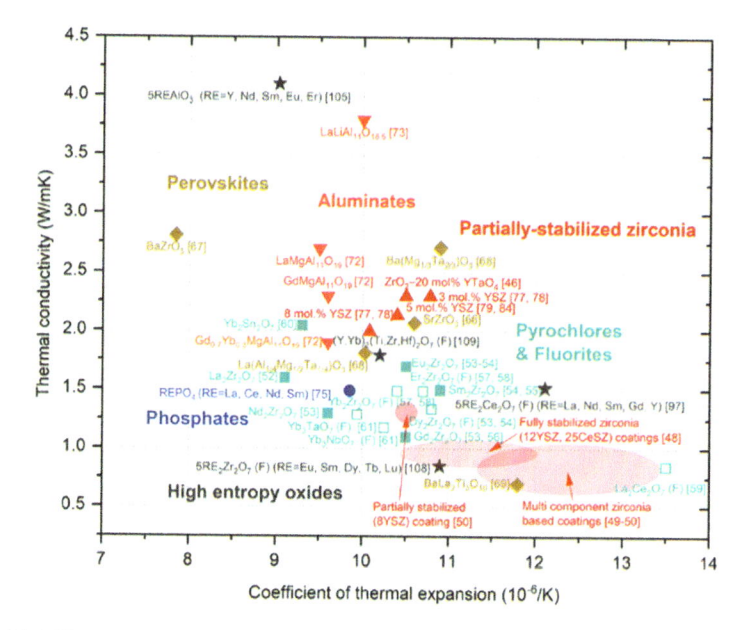

Figure 21.3 Thermal properties of ceramics. Thermal conductivity versus coefficient of thermal expansion for current and candidate TBC materials.
Source: From Vaßen, R., Bakan, E., Mack, D. E., & Guillon, O. (2022). A perspective on thermally sprayed thermal barrier coatings: Current status and trends. *Journal of Thermal Spray Technology*, *31*(4), 685–698. https://doi.org/10.1007/s11666-022-01330-2.

characteristic columnar microstructure. In the early stages of deposition, competitive growth occurs resulting in an equiaxed texture; with time, some grain orientations are favored, and other growth directions are lost from the structure giving rise to the classic appearance of an EB-PVD TBC. Coating growth can be explained by the classic Thornton structure zone model for thin film and coating growth with temperature and pressure and has been adapted by Barna and Adamik to include the effect of a second element on coating growth (Barna and Adamik, 1997). This deposition method results in intercolumnar, intracolumnar, and internal porosity all of which contribute to lowering thermal conductivity.

21.3.2 High-temperature stability

7YSZ, the current industry standard TBC, experiences ferroelastic toughening through formation of a metastable tetragonal phase, designated 't'. However, 7YSZ can undergo deleterious phase changes at operational temperatures above 1200°C. In such phase changes, the partially stabilized tetragonal polymorph transforms to cubic, tetragonal, and monoclinic forms upon cooling with volume changes, which can cause cracking (Witz et al., 2007). Newer TBC materials should always demonstrate higher-temperature stability without deleterious phase changes. Selection of rare earth zirconate materials is somewhat limited in this respect as some can undergo unwelcome phase changes at moderately high temperatures. For this reason, La, Sm, and Gd zirconates are more widely researched than the remaining rare earth elements in the lanthanide series. The latter has merited most attention, and indeed, gadolinium zirconate TBCs are already in service with some engine manufacturers.

Sintering is a further challenge for high-temperature porous ceramic materials. Under such conditions, atoms diffuse along grain boundaries, through pores, and move around lattice structures to minimize surface energy. The as-deposited, feathery features of EB-PVD columns become rounded and, with time, the rounded undulations eventually touch forming necks, locking columns together (Fig. 21.4) (Gao et al., 2024). This increases thermal conductivity as there are fewer defects to scatter phonons. In addition, this modified microstructure is more prone to physical damage; see the following section on erosion resistance.

Interaction of TBC elements with the TGO can also be accelerated with prolonged exposure at high temperatures. In the case of rare earth zirconates, some lanthanide elements interact with the TGO forming rare earth aluminates and disrupting the slow stable growth of the alumina scale accelerating spallation of the TBC under thermocyclic conditions. This is the case for gadolinium zirconate where a thin interlayer of 7YSZ TBC is first deposited on the TGO to limit such deleterious reactions.

21.3.3 Chemical resilience (calcium, magnesium, alumino-silicate attack)

Siliceous debris (sands, volcanic ash, and runway dusts) can form molten deposits when ingested into the engine, which adhere to the surfaces of high-temperature

(A) As-deposited feathery structure   **(B)** Surface smoothing   **(C)** Surface undulation to form necks   **(D)** CMAS infiltration

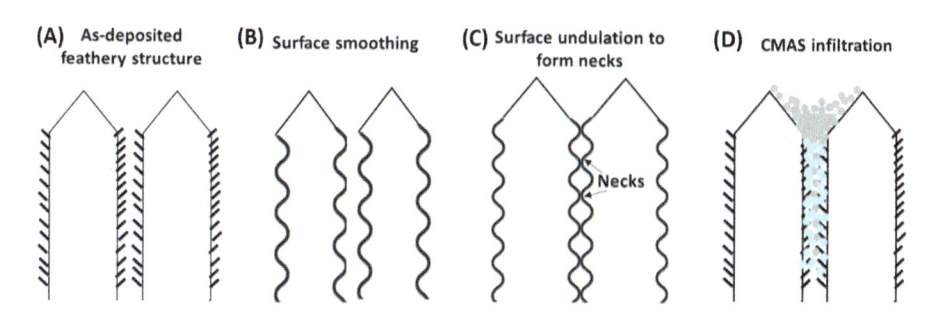

Figure 21.4 Sintering and CMAS attack of TBC columns. Microstructural evolution of TBCs upon thermal exposure and CMAS attack. (A–C) Feathery structure evolution of TBCs as function of TBCs as function of increasing temperature and duration; (D) CMAS infiltration of adjacent columnar gaps.
Source: From Gao, Z., Zhang, X., Chen, Y., Chalk, C., Nicholls, J., Brewster, G., & Xiao, P. (2024). Strain tolerance evolution of EB-PVD TBCs after thermal exposure or CMAS attack. *Journal of the European Ceramic Society*, 44(1), 426–434. https://doi.org/10.1016/j.jeurceramsoc.2023.08.047.

components. While the major constituents are CaO, MgO, Al_2O_3, and SiO_2, these dusts also can contain smaller amounts of oxides of Na, K, and Fe of geological origin and Ni and Ti oxides produced by wear in the engine. Most TBCs have inherent porosity, and the molten CMAS deposits can infiltrate the TBC and solidify under the imposition of a thermal gradient. This has the effect of reducing in-plane compliance of the columnar TBC and can lead to coating failure.

Although siliceous debris varies according to location (Levi et al., 2012), the nature of the molten deposits found on engine components is very similar, largely dictated by the lowest-melting-point compositions of debris. $C_{33}M_9A_{13}S_{45}$ is a typical deposit and is widely used in laboratory trials—here the abbreviated formula corresponds to 33 mol.% CaO; 9 mol.% MgO; 13 mol.% $AlO_{1.5}$, and 45 mol.% SiO_2. Liquid phases can form in this composition at temperatures as low as 1125°C. Deposits tend to accumulate on the hottest regions of the aerofoil, the pressure surface, and leading edge, due to local component temperatures, as well as the tendency for particles to impinge on these locations due to the angle and relative velocity of the airflow.

The nature of the multicomponent oxide mixtures is difficult to model, but ternary phase diagrams of the major constituents shed light on the complex phase chemistry of CMAS (Fig. 21.5) (Poerschke et al., 2016). In this diagram for CaO-$AlO_{1.5}$-SiO_2 in the absence of Mg and calculated for a temperature of 1300°C, the lowest-melting composition space is highlighted in dark blue. The light blue regions identify regions of the composition space containing some liquid phases.

It is important to note the viscosities at the lower end of the dark blue liquidus region in Fig. 21.5(A) are of the order of 1 Pa s^{-1}, and hence, it takes less than 4 s for a 7YSZ TBC, ~ 150 μm thick, to be fully infiltrated in the absence of a thermal gradient. At the top right of this dark blue region, the equivalent values for this

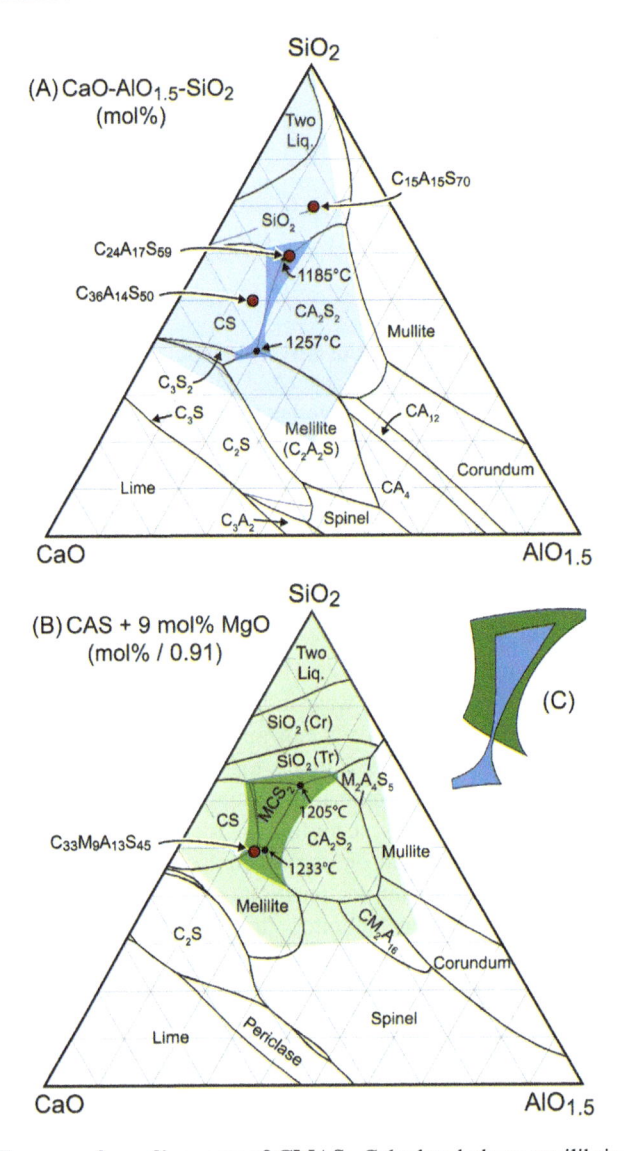

Figure 21.5 Ternary phase diagrams of CMAS. Calculated phase equilibria for the (A) CaO-SiO$_2$-AlO$_{1.5}$ and (B) section at 9 mol.% MgO at 1300°C. The liquid phase in each ternary is identified as dark shading, and phase fields containing liquid as light shading. The effect of the MgO addition on the shape of the phase field is shown in (C).
Source: From Poerschke, D. L., Barth, T. L., & Levi, C. G. (2016). Equilibrium relationships between thermal barrier oxides and silicate melts. *Acta Materialia*, *120*, 302–314. https://doi.org/10.1016/j.actamat.2016.08.077.

composition are 7000 Pa s^{-1} and 45 min (Poerschke et al., 2017). Hence, the higher viscosity means this composition is less likely to cause significant damage. Sensitivity to such low CMAS viscosity and short infiltration time led to the grounding of flights during volcanic eruptions and the need for shorter planned service intervals for aircraft operating in dusty environments.

The phases identified in Fig. 21.5 relate to potential intrinsic crystallization products, that is, those originating solely from CMAS constituents, in the absence of a reaction with TBC chemistry.

Other types of product can form in reactions between CMAS and the TBC, and these offer a route to identification of next-generation TBC materials. Reactive crystallization products involve elements from both CMAS and the TBC forming new materials and show most promise in acting to block further CMAS penetration of the TBC if they can form quickly enough (perhaps a timescale of seconds is needed, dependent upon viscosity of the CMAS melt). Reprecipitation products can also form; these also involve reaction between CMAS and the TBC, but CMAS modified (usually Ca) TBC phases result, for example, cubic (fluorite) or tetragonal ZrO_2, which are ineffective in limiting CMAS penetration.

Extensive studies in CMAS attack of $Gd_2Zr_2O_7$ type TBCs reveal that reactive crystallization of needle-like apatite $(Ca)_2(RE)_8(SiO_4)6O_2$ and fluorite $(Zr,RE,Ca)O_{1-x}$ phases is rapid (in a timescale of s). The reaction products, in the form of acicular crystals, block intercolumnar gaps in the TBC and prevent further penetration of low-viscosity melts. Of the RE zirconates considered for TBC materials, $Gd_2Zr_2O_7$ is the most widely reported. Levi showed that $Gd_2Zr_2O_7$ TBCs are more resistant to CMAS degradation than 7YSZ as demonstrated in Pratt and Whitney engine trials in 2011 (Levi et al., 2012).

However, while beneficial reactions can occur, interaction between CMAS and the TBC can bring about unwanted phase changes in TBC, leading to destruction of the strain-tolerant microstructure and premature failure. A further drawback is that these rare earth zirconates TBC structures are less tough than 7YSZ, and this limits their potential as next-generation TBCs.

This topic is discussed at length in a subsequent section as it is here that nanoindentation and nanoimpact tools are widely employed to shed light on the effect of materials and microstructures on this particular damage mechanism.

21.3.4 Resistance to erosion and foreign object damage

The previous section has shown that CMAS enters the gas stream of turbine engines during service, and it does so in the form of small, solid particles. The repeated impact of particles onto solid surfaces will lead to material loss of the engine components, which is defined as erosion (ASTM, 2022). Both static and rotating parts are affected by erosion, although the impact velocity will be much higher in the case of rotating parts. For instance, high-pressure turbine blades spin at up to 12,000 rpm and will experience relative impact velocities of up to 200 m s^{-1}. TBCs, as the most external layer of several engine components, are in direct line of fire and will experience erosion damage during their regular service. Therefore, the engineers in charge of TBC design must:

1. consider particle impact mechanics in their particular application,

2. understand how the properties and microstructure of TBCs affect erosion resistance,
3. be able to test their designs in conditions representative of their service environment.

The first step to improve erosion resistance is to consider the impact mechanics of erosion by solid particles. If we concentrate on the particles for a moment, the erosion damage depends mainly on their size, velocity, and shape. This aspect is covered in the work of Nicholls, Wellman and colleagues and summarized in two review papers (Nicholls et al., 2003; Wellman & Nicholls, 2007), which can be consulted for further detail. Fig. 21.6 is a visual summary of their findings. For small particle sizes and low velocities, the damage on the surface is all elastic up to a threshold at which permanent damage (plastic deformation) starts occurring, which appears as a negative slope on the velocity/particle diameter plot. This slope flattens out into a horizontal velocity threshold above which any impact from a sufficiently large particle will generate permanent damage. Cleverly, Fig. 21.6 considers different horizontal lines to account for particle shape. Sharper particles will induce plastic deformation at lower velocities than smoother particles of the same size, as the latter spreads the load of the impact over a larger area. Nicholls, Wellman and colleagues propose to quantify the particle shape using a single value, the r/R ratio, which divides the effective particle radius over the average particle radius of the particle. Therefore, r/R ratios closer to 1 correspond to smooth particles, whereas sharp particles have much lower values.

Another important factor is the angle of attack of the particles, but its consequences are very dependent on the material of the surface. Erosion damage can be initially divided into ductile and brittle modes, as detailed in depth by the works of

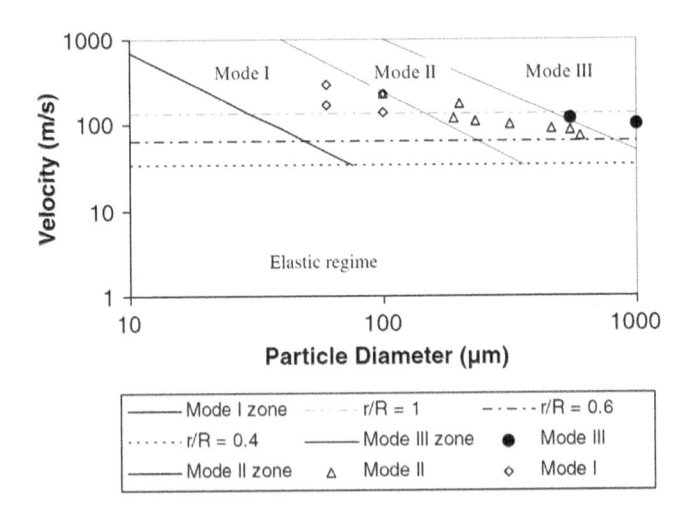

Figure 21.6 Diagram of erosion damage modes. Damage regimes on erosion/FOD depending on the velocity and diameter of the impinging particles.
Source: From Wellman, R. G. & Nicholls, J. R. (2007). A review of the erosion of thermal barrier coatings. *Journal of Physics D: Applied Physics*, *40*(16), R293–R305. https://doi.org/10.1088/0022-3727/40/16/R01.

(Aquaro & Fontani, 2001; Bousser et al., 2014). In summary, ductile materials such as metals experience material loss through plastic deformation or plastic flow (akin to cutting or plowing), and consequently, they experience maximum damage at shallow angles such as 20 degrees while being very resistant to an angle of attack of 90 degrees (surface perpendicular to the velocity vector of the particle). Brittle materials such as ceramics are the opposite, as the main damage mechanism is based on crack formation, propagation, and intersection. In this case, an angle of attack 90 degrees is the most damaging for the materials, which resist better shallower angles as supported by several experimental studies (Nicholls et al., 1998; Wellman & Nicholls, 2007; Wellman et al., 2005b).

The divergent behaviours to angle of attack hint at the importance of the properties and microstructure of the material being eroded. Let us forget about the particles now and focus on a surface that is covered by a TBC, a case that has been extensively studied by several research groups (Chen et al., 2004; Nicholls et al., 1999, 2003; Schmitt et al., 2017; Schubert et al., 2016; Steinberg et al., 2022; Wellman et al., 2005a; Nicholls et.al., 2003). The very first aspect to consider is chemical composition. Yttria-stabilized zirconia (YSZ), a traditional TBC material, is a tough ceramic. Other ceramics, such as gadolinium zirconate and other rare-earth zirconates, have attracted researchers due to their superior thermal and CMAS-resistant properties. However, their toughness and erosion resistance are much lower; for instance, stoichiometric gadolinium zirconate can have an erosion resistance up to an order of magnitude lower than YSZ, with in-between stoichiometries decreasing in erosion resistance as their Gd_2O_3 fraction increased (Schmitt et al., 2017); the same happens when rare-earth oxides are dopants in YSZ (Steenbakker et al., 2006).

Another critical aspect is the microstructure of the material, such as the differences derived by the method of manufacture. Studies on YSZ showed that TBCs deposited by electron-beam physical vapor deposition (EB-PVD) can have an erosion resistance between 2.5 and over 10 times larger than those deposited by air plasma spray (APS) (Nicholls et al., 1998, 1999). The damage mechanism in a characteristic EB-PVD YSZ microstructure is the generation of intracolumnar cracks 10–20 μm under the surface, as seen in Fig. 21.2(A); material removal happens when enough neighboring columns are affected, and the material above the crack is removed. On the other hand, the damage mechanism governing APS YSZ erosion is the failure of the splat boundaries, which results in the splat detaching from the rest of the coating. It appears that damaging the splat boundaries requires less energy than creating an intracolumnar crack. Moreover, the independent nature of the EB-PVD columns makes the intercolumnar space an effective crack-arresting feature, making it very difficult for cracks to jump between columns. Because of this, coatings with smaller column diameter (and thus more intercolumnar gaps) are more erosion resistant, up to a point. For instance, a YSZ TBC with an average column radius of 5 μm is five times more erosion-resistant than a TBC with an average column radius of 15 μm (Wellman et al., 2005b). Further reduction of the column below a certain threshold can result in the intercolumnar space not being able to stop crack propagation, reducing erosion resistance (Steenbakker et al., 2006):

cracks propagate between columns of 2.5 μm in radius (although it is worth mentioning that this particular TBC also had rare-earth oxides as dopants in the YSZ). Moreover, this crack-arresting key feature can be exported to other microstructures, for instance, the introduction of vertical cracks on APS YSZ reduced their erosion rates about five times compared with their regular microstructure (Wellman et al., 2005b). The benefits of the EB-PVD columnar structure hold true as long as the growth of the column is roughly perpendicular to the substrate surface. If the columns diverge more than 20 degrees from the perpendicular direction, the erosion resistance drops suddenly (Wellman et al., 2005b), which is something to consider in components with complex shapes. Another issue appears if the EB-PVD columns lose their independence, such as in the case of CMAS attack. If CMAS infiltrates between columns and solidifies, it acts as a bridge for crack propagation and severely reduces erosion resistance, as in a recent study written by (Steinberg et al., 2022).

The intention of the two previous paragraphs was to study the surface properties in isolation, but this is not completely possible, so their damage descriptions are only true for a certain range of particle sizes and velocities. The work of Nicholls, Wellman and colleagues covers a wider range of impact energies and distinguishes different mechanisms of permanent damage on columnar TBCs made of YSZ (Nicholls et al., 2003; Wellman & Nicholls, 2007). The least energetic impacts that cause permanent damage (due to smaller and/or slower particles) fit the previous description: the subsurface cracks described before and illustrated in Fig. 21.7(A); this is defined as Mode I erosion, and its range is defined in Fig. 21.6. In Mode I, a single particle affects only a few columns at once and temperature increases do not affect the fracture mechanisms. However, an increased temperature will increase the plasticity of the ceramic, which can lead to a change in the erosion rate. The experiments published in the literature show a counterintuitive, inconsistent trend of increased erosion resistance with increased temperature up to around 550°C, followed by a reduction of erosion resistance with a further increase in temperature (Nicholls et al., 1998, 2003). Several other studies that focus on the effects of high temperature on erosion, which the reader can consult, such as (Chen et al., 2004; Schubert et al., 2016).

On the opposite side of the plot from Fig. 21.6, one can find the most energetic impacts (due to larger and faster particles). This area is labeled as Mode III foreign object damage (FOD) and is characterized by gross plastic deformation, compaction, and the formation of kink bands. The latter are cracks or bends in the columns that originate from the center of the impact area and develop in a cone-shaped area at roughly 45 degrees from the vertical, as seen in Fig. 21.7(C). The plasticity induced by high temperatures is so significant that can lead to columns bending without cracking. Several FOD subtypes have been identified depending on the extent of the three failure mechanisms mentioned earlier can be found on the same samples; the reader is referred to review (Wellman & Nicholls, 2007) for further detail than goes beyond the scope of this chapter. In Mode III, a single particle affects dozens of columns as illustrated in Fig. 21.7(C), which can be caused by a very large particle traveling at moderate velocities or a medium particle traveling

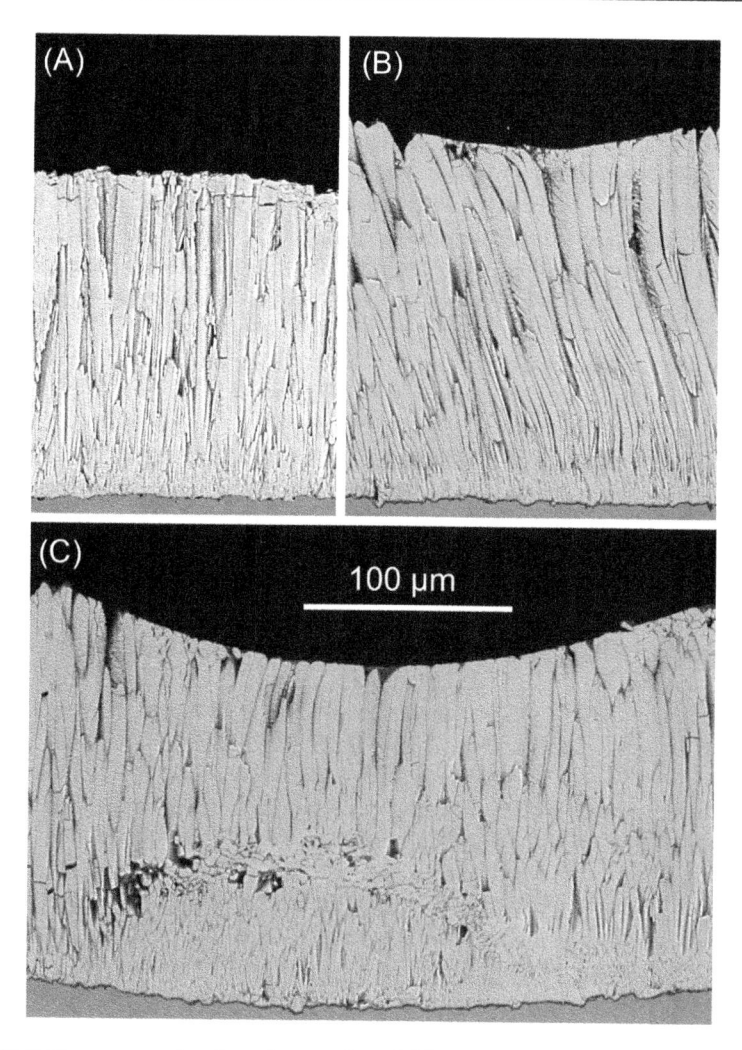

Figure 21.7 Damage mechanisms for erosion of TBCs. Characteristic damage mechanisms of different erosion regimes: (A) Mode I erosion, (B) Mode II compaction, (C) Mode III Foreign Object Damage. Images taken at Cranfield University on 7YSZ topcoats, NiAl bondcoats, and Ni-Cr alloy substrates.
Source: Courtesy Cranfield University.

very fast. To harmonize the effects of size and speed, the dimensionless parameter D/d or contact footprint parameter was introduced. Its value is the ratio of the contact footprint (D) divided by the average column diameter (d). For Mode I erosion, D/d is generally between 1 and 2 (on average), whereas for Mode III FOD, it is above 14 (Wellman & Nicholls, 2007).

Between Mode I and Mode III, a transition mechanism called Mode II compaction exists. This regime has been less studied, but some characteristics are known.

Mode II is characteristic of larger particles traveling at intermediate speeds and intermediate D/d ratios (3–12) and mainly displays compaction without major presence of subsurface cracking or kink band formation (see Fig. 21.7B), possibly enabled by the large porosity of the columns.

Finally, choosing the right erosion testing methodology is very important to evaluate the effectiveness of different erosion-resistance strategies. The testing regime should replicate the key aspects of real service conditions despite constraints and limitations inherent to any laboratory testing. For instance, the previous text highlights that the mechanism of failure in Mode I is not affected by temperature; therefore, many lessons can be learned from testing at room-temperature components that will see high temperatures and Mode I erosion in service. On the other hand, the size and velocity of the particles seem to determine fracture mechanisms; therefore, it would not be appropriate to rely on the intercolumnar space as a crack-arresting feature in Mode I conditions when the part will see service under Mode III FOD conditions, as Fig. 21.7(C) shows how kink bands jump from column to column. Because of this, some research groups have deviated from the main erosion standard, ASTM G76 (ASTM, 2018), especially when studying erosion in turbine engines.

ASTM G76 defines their standard testing conditions using: compressed gas, with small particles (angular alumina 50 μm) traveling at relatively low speeds (30 m s^{-1}) in a small accelerator tube (inner diameter 1.5 mm), at room temperature, at a perpendicular angle of attack, measures volume loss, etc. Some of these characteristics are rather limiting when studying TBCs in their representative working environment. Because of this, some researchers use modified testing rigs in their testing, such as Cranfield University's rig that can operate at high temperatures, with larger erodent size (up to 1 mm) that travels at much higher speeds (100–250 m s^{-1} is typical), with impact angles that can vary from 20 to 90 degrees and base erosion rate of mass loss instead of volume loss (which is difficult for thin layers of porous materials). However, the spirit of the standard is followed, especially on the distinction between transient and steady-state erosion rate (the reported value is from the testing regime at which mass loss is linear), the measuring intervals and general test procedure, and the use of calibration material (as many parts of the erosion rig are treated as consumables because they erode over time, subtly changing particle speeds and paths, etc.).

21.4 Design tools for developing enhanced TBCs — probing sintering resistance, chemical resilience, and erosion resistance using nano- and microindentation

We have seen how the TBC chemistry and the porosity inherent in the TBC microstructure play a significant role in its thermal, physical, and chemical properties. The experimentalist needs to select characterization tools carefully to probe relevant properties on representative sample materials. Nano- and micromechanical test tools have a useful role to play here.

21.4.1 Nanoindentation

The extreme environment in the gas turbine makes the TBC prone to sintering, CMAS attack, and erosion; such damage mechanisms are often studied using nanoindentation. Knowing the contact area of indenter, the resulting load versus displacement unloading curves can be analyzed to generate hardness (H) and reduced elastic modulus (E_R), values useful for those designing and modeling new TBC systems. Hardness measurements can give a proxy guide to fracture toughness while the reduced modulus, a measure of stiffening, can be directly correlated with sintering behavior and CMAS damage. Moreover, the ratios, H/E_R and H^3/E_R^2 are sometimes used to equate to erosion resistance; the latter is said to reflect resistance to plastic deformation and is commonly known as the plasticity index (Chen et al., 2012). Bull also found that high H/E_R materials showed better erosion performance as this ratio is shown to be linked to indentation energy (Bull, 2005). It is important to note caveats about the ratio of indenter depth to coating thickness and substrate behavior. As a rule of thumb, indenter displacement should be less than one-tenth of the coating thickness to reflect coating properties rather than those of the substrate, but this is seldom a problem for ~200 μm thick TBCs.

In the case of TBCs, nanoindentation can probe the mechanical properties of an individual column, but it is just as likely to indent several columns or part of a column with some porosity. This porosity allows the column to move sideways under indentation (Wellman, Dyer, & Nicholls, 2004; Wellman et al., 2004). These authors showed that the hardness of an individual TBC column (14 GPa) measured on polished coating cross sections was over 10 times higher than the hardness of the coating (2.4 GPa) measured by microindentation using increasing loads, measured on top of the coating, and polished to less than 1 μm R_a.

Hence, the scale of the method needs to be taken into consideration. However, sometimes the design engineer is concerned with macroscale properties and other techniques must be used. In the case of probing sintering of TBC columns through measurement of Young's modulus as shown in Fig. 21.8 nano-/microindentation, beam bending, and resonant frequency damping analysis have been used to compare different length scales of measurements (Gao et al., 2024). The measurements span those from a few columns to thousands of columns and finally to those for a coated flat beam. Nano- and microindentation measurements were made on polished cross sections some 20–80 μm from the top surface. The absolute values of the moduli measured vary widely with technique, with nanoindentation yielding the highest values, but the overall trends with thermal aging are very similar. Nanoindentation values show the modulus increases from 87.3 GPa as deposited to 198 GPa after sintering at 1400°C for 100 hours, this latter value is closer to the value of 200–210 GPa reported for bulk 7YSZ. Both nano- and microindentation show an increase in Young's modulus of ×2.3 for this thermal exposure, whereas beam bending and RFDA show much larger increases, ×20.6 in the case of beam bending. In any of these measurements, the influence of the substrate needs to be

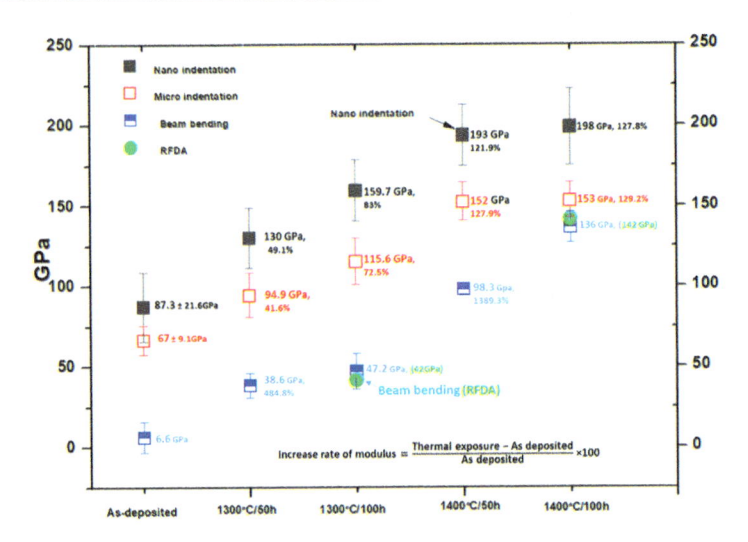

Figure 21.8 Elasticity measurements of TBCs. Young's moduli of TBCs at different length scales upon different thermal exposure.
Source: From Gao, Z., Zhang, X., Chen, Y., Chalk, C., Nicholls, J., Brewster, G., & Xiao, P. (2024). Strain tolerance evolution of EB-PVD TBCs after thermal exposure or CMAS attack. *Journal of the European Ceramic Society*, 44(1), 426–434. https://doi.org/10.1016/j.jeurceramsoc.2023.08.047.

considered, especially for macroscale measurements and elastic modulus values. Often, as here, alumina substrates are used and interdiffusion of Al with 7YSZ does occur at high temperatures, chemically interfering with the progress of sintering through the formation of yttrium aluminates (Schmitt et al., 2019).

Fracture toughness of materials is an important property to quantify. This has been carried out on TBCs by studies of crack patterns resulting from nanoindentation. This method has some limitations for comparing a wide range of materials, as set out by Quinn (2006). More recently, researchers have come with strategies to minimize these concerns.

GZO, a TBC coating already in commercial use, has a much lower fracture toughness than 7YSZ (~ 1 vs. 10–15 MPa m$^{1/2}$ respectively), which impacts on its erosion resistance. Schmitt et al. used this crack pattern indentation method to investigate the toughening of gadolinium zirconate dense compacts with 10–50 wt. % additions of gadolinia alumina perovskite GdAlO$_3$ (GAP) (Schmitt et al., 2019). They compared their results with erosion rates for the same material compositions. They used dense pellets rather than TBCs to avoid microstructural variations. They calculate the indentation fracture resistance, K_{IRF}, as in the equation below:

$$K_{IRF} = \frac{0.018 F \sqrt{\frac{E}{H}}}{c^{1.5}}$$

Here F is the applied load (N), E the elastic modulus (GPa), H the hardness (GPa), and c the average crack length (μm).

The authors showed that a 50 wt.% GAP addition reduced the erosion rate by ∼ 60% with respect to GZO, and indeed, this erosion rate was slightly lower than that of GAP alone. In contrast, the authors found that K_{IRF} increases slightly with 10 wt.% addition of GAP but is almost insensitive to further increases. This was driven by the hardness values, which are somewhat reduced at 10 wt.% GAP addition but then increase slightly and remain constant with further GAP additions. K_{IRF} shows very little sensitivity to Young's modulus, which falls sharply with GAP content.

This difference in behavior led Schmitt et al. to conclude that the mechanism that operates in erosion (a dynamic process with high strain rates) resulting in material loss is different from that operating in the indentation test, and hence, alternative test methods are required.

Nano-/microimpact systems offer a rapid test method to speed up the pace of development for new TBC systems. They also allow a more detailed understanding of microstructural behavior and crack propagation under erosive conditions.

Cyclic nanoimpact offers a higher strain rate (10^5 s^{-1}) when compared with nanohardness measurements (where the indentation is quasistatic). In the impact test, impact energy is controlled by the static force and acceleration distance. The test allows for cyclic repetitive contact at the same point and provides a comparative method for determining the number of impacts to failure in brittle materials.

Allowing load-controlled impacts at different locations (employing precision-controlled stage movement) enables us to understand coating response and damage propagation at a single impact level. A further development is the introduction of random distributions of impacts within a defined area, which allows for a more realistic erosion scenario where impacts can and do overlap, a technique introduced in a recent paper (Beake et al., 2024). This test method has a distinct advantage in that the same "random distribution" pattern can be carried out on multiple samples enabling more precise comparisons between coating material and microstructure systems. Erosion in the engine and, indeed, in erosion testing, is a stochastic event; there will always be variation in the number and the momentum of particles involved during each erosive episode. It is therefore difficult to test a range of coating variants under identical erosive conditions.

Further advantages are that the position of each impact is identified in order of impact allowing damage mechanisms to be tracked for further analysis. The instrument provides the maximum penetration depth, residual penetration depth, impact velocity, and rebound velocity for each impact. Knowing the instrument characteristics (the effective mass of the loading head), the coefficient of restitution and the kinetic energy absorbed can be calculated. The former allows us to investigate whether any changes in mechanism occur with continued impacting or not. The complementary value, or percentage of dissipated energy, is a useful measure of whether the coating is work-hardening the surface (when it goes down slightly) or whether there is suddenly more cracking when the response is more variable.

Scaling up to microimpact tests means that the influence of intrinsic TBC surface roughness reduced as both the contact area and the impact depth are larger but still within the $D/d < 2$ range and representative of the erosion regime. Chen et al. (2012) indicated that erosive behavior could be better reproduced by microimpact tests.

A combination of scaled-up microimpacts with the random impact testing proposed earlier took place recently in an Innovate UK-funded program, the results of which will be soon submitted for peer-review publication. This joint research between Micro Materials Ltd, Cranfield University, and the National Physics Laboratory compared a combination of random and repetitive impact distributions using sphero-conical ($R = 25$ μm) and sharp ($R = 100$ nm) cube corner diamond probes with erosion tests with (112 μm) spherical and (90–125 μm) angular alumina erodent. Several TBC compositions deposited by EB-PVD were tested, including 7YSZ and gadolinium zirconate (GZO) among others. Cyclic and specially random impact testing has been shown to replicate some of the erosion mechanisms observed in erosion testing, as exemplified below with 7YSZ versus GZO impact and erosion.

For 7YSZ, the damage mechanisms induced by cyclic microimpact tests using a sphero-conical probe and 500–3000 mN load are dominated by compaction damage on the column tips, identified as the light band in Fig. 21.9(A) and (B), followed by a network of cracks. By contrast, typical erosion test damage on the same TBC system is dominated by subsurface cracking especially when using angular erodent (Fig. 21.9C), with some degree of compaction present on a thin top layer

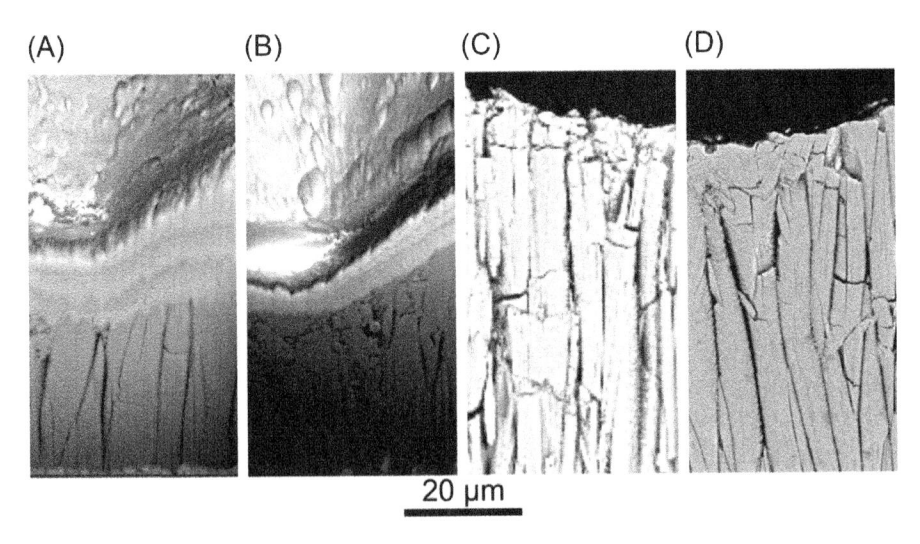

Figure 21.9 Comparison between micro-impact and erosion damage on a 7YSZ TBC. SEM backscattered images of 7YSZ TBC after (A) cyclic impact damage with spherical indenter at 500 mN and FIB milling, (B) cyclic impact damage with spherical indenter at 3000 mN and FIB milling, (C) erosion testing with angular alumina 90–125 μm, (D) erosion testing with spherical alumina ∼112 μm.
Source: Courtesy Cranfield University.

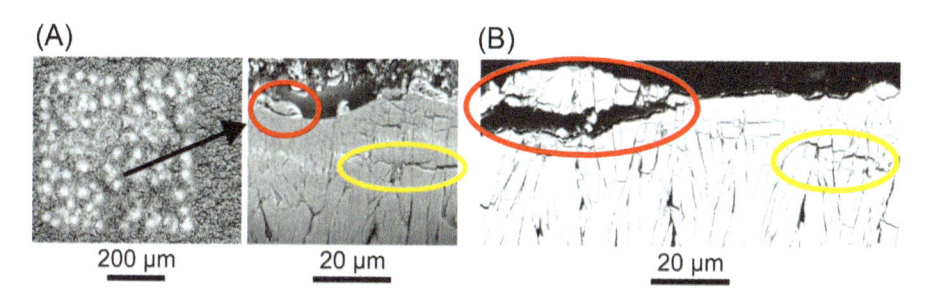

Figure 21.10 Comparison between microimpact and erosion damage on a GZO TBC. SEM images of Gadolinium Zirconate TBC after (A) random impact damage with sphero-conical indenter on a 0.5 × 0.5 mm² area, (B) erosion testing with angular alumina erodent (112 μm).
Source: Courtesy Cranfield University.

when using spherical erodent (Fig. 21.9D)—albeit in a much smaller scale than in microimpact.

For GZO, the random impact test with sphero-conical indenter produces a significant compaction zone (Fig. 21.10A) like in 7YSZ, plus some detachment at the tips (red circle). Subsurface cracking under the compaction layer is again prominent, but some of the cracks generated in this region propagate across multiple columns (yellow circle). In addition, there may be some column plasticity evidenced by column bending fanning out on both sides from the site of a multiple impact crater. Erosion testing on GZO generates similar damage mechanisms, specially when using spherical erodent (Fig. 21.10B); a compaction layer (although smaller) is generated with tip detachment (red circle), as well as the network of subsurface cracking with cracks spanning multiple columns (yellow circle), but there is no evidence of plasticity.

This joint research program concluded with a set of ranking experiments comparing behavior of reference and novel TBC systems (with material and microstructure variations) under erosion testing and cyclical impact testing. Results, presented in Table 21.1, show that all TBC systems tested were correctly ranked except for TBC variant 2. Its anomalous but interesting behavior merits further investigation and will support studies in the development of tougher new TBC materials and microstructures. It should be noted that the distinction between erosion rates for materials ranked 2, 3, and 4 is very small, and fine-tuning of the impact test conditions can be adapted to match the damage features expected for each material.

21.5 Modeling erosion of thermal barrier coatings

Modeling the damage mechanisms of solid particle erosion has been an ongoing task since at least the 1950s, a great summary of which can be found in (Aquaro & Fontani, 2001; Bousser et al., 2014). The two papers present the initial theoretical

Table 21.1 Comparison of the relative performance of different TBC systems for erosion testing and rmicroimpact testing using the random distribution pattern. Data courtesy of Micro Materials Ltd and Cranfield University.

TBC variant	Relative Erosion rate	Erosion rank	Random impact rank
TBC1	1.0	1	1
TBC2	6.3	2	4
TBC3	6.9	3	2
TBC4	20.0	5	5
TBC5	8.8	4	3

Source: Courtesy Micro Materials Ltd. and Cranfield University.

and mathematical models, which are very helpful in understanding the underlying mechanisms of erosion damage. However, the assumptions and simplifications of those models limit their practical applications and were soon complemented with computer-based numerical models, which improved their predictive power (Aquaro & Fontani, 2001). The unique microstructure of TBCs further limits the application of the previous theoretical and numerical models for generic erosion, and thus scholars in the field have developed independent computer models. Ranging from Monte-Carlo models to fully fledged finite-element models, these simulation tools provide a fundamental understanding of TBC fracture mechanics and are an effective way to screen different compositions and microstructures for TBC design.

Two main theoretical models provide a fundamental understanding of TBC fracture mechanics during erosion, both complementary and in agreement with the basic principles of material loss. The first is the work of Wellman, Nicholls and coworkers in Cranfield, UK (Nicholls et al., 2003; Wellman & Nicholls, 2007), which is covered in a previous subsection. The other main example is the work of Chen, Evans and coworkers in Santa Barbara and other US universities together with Fleck and coworkers in Cambridge, UK (Bousser et al., 2014). This publication follows a similar approach to the former to distinguishing failure mechanisms depending on particle size and velocity, explaining subsurface cracking of columns by way of generation and interaction of elastic stress waves, and focusing on elastoplastic interactions of high-temperature FOD. Both models have used computer-based simulations to understand the erosion effects observed in laboratory experiments and service.

The Monte Carlo method has been successfully applied to estimate material loss in Mode I erosion of TBCs (Wellman & Nicholls, 2004). Proposed in the 1940s for the very first computers by Stanislaw Ulam and Nicholas Metropolis, the Monte Carlo method is based on repeating a deterministic computation on random inputs from a defined range to obtain a group of results that can be analyzed together. Wellman and Nicholls based their model on earlier modeling erosion work on alloys and oxide scales (Stephenson & Nicholls, 1993). Their proposed algorithm considers several variables such as erodent used, particle size, and impact location. Then, an individual impact is generated by randomizing the value of the variables and calculating the potential damage generated. This process is repeated for many impacts, accumulating the calculated damage of subsequent impacts, which finally produces an erosion rate. The predictions marry well with experimental results within ± 10%−30% on a wide range of operating conditions.

Finite element analysis has been the preferred modeling technique to simulate erosion Mode III (FOD) at different temperatures. Different degrees of simplifications and assumptions have been used, from general models for FOD on brittle materials (Aquaro & Fontani, 2001), to full TBC systems will all layers represented by uniform bodies (Chen et al., 2003), to finally accounting for idealized, uniform-diameter, individual columns in more recent models (Geng et al., 2023). In all cases, the models replicate and explain key features observed in experimental testing such as densification areas, kink-band locations, and the effect of temperature. The models are also useful prediction tools (e.g., for impact depth) and can be used to measure unknown characteristics and properties based on iterative approaches around a known experimental measurement. Examples of the latter are the estimation of the impact velocity of the particles (Chen et al., 2003) and the fracture toughness of the TBC material (Geng et al., 2023) based on known penetration depths and crack lengths of experimental data.

Aerodynamic models are also worth mentioning even if they are not directly related to the impact mechanics of TBCs. Early models of aerodynamic effects, flow fields, and particle trajectories (Fackrell, 1989; Tabakoff, 1989) can be used to understand where erosion will be more severe in a real component, or what impact velocities and angles are most likely to occur on a given geometry and environment.

Regardless of the technique used, any model will require reliable information about the mechanical properties of the TBC (elastic modulus, yield strength, density, Poisson's ratio, etc.) at a relevant temperature and perhaps considering the strain rate for high-velocity impacts. Small-scale tools such as nanoindentation and microtensile testing are the tool of choice for many works given their ability to measure the properties of individual columns, which can be quite different from the values for bulk material. Complementary data can be obtained by analyzing experimental impacts with scanning electron microscopy (SEM) imaging, perhaps in combination with focused-ion beam (FIB) milling, to characterize the fracture mechanisms present and measure key features such as column size, crack length and location, and evaluate the impact damage.

21.6 Conclusions

We have seen that TBCs are critical materials in aviation gas turbines. Material and microstructural developments will facilitate the journey toward net-zero enabling manufacturers to meet 2050 targets.

This journey is assisted by concomitant developments in superalloys (alloy development and cooling technologies), bondcoats (chemistry, physical, and mechanical properties), and TBCs (chemistry and microstructure). The system must be viewed as a whole if significant improvements are to be made, as the operating conditions mean that elements migrate from one system to the next driven by diffusion and microstructures evolve leading to degradation in properties. The design and research engineers need access to many test procedures and analytical techniques.

References

Aquaro, D., & Fontani, E. (2001). Erosion of ductile and brittle materials. *Meccanica, 36*(6), 651−661. Available from https://doi.org/10.1023/A:1016396719711.

ASTM. (2018). ASTM G76 − 18 Standard test method for conducting erosion tests by solid particle impingement using gas jets 1. https://doi.org/10.1520/G0076-18.

ASTM. (2022). ASTM G40 − 22 Standard terminology relating to wear and erosion. https://doi.org/10.1520/G0040-22.

Barna, P. B., & Adamik, M. (1997). *Formation and characterisation of the structure of surface coatings* (pp. 279−297). Springer Nature. Available from http://doi.org/10.1007/978-94-011-5644-8_21.

Beake, B. D., Goodes, S. R., Zhang, H., Isern, L., Chalk, C., Nicholls, J. R., & Gee, M. G. (2024). Randomised nano-/micro- impact testing − A novel experimental test method to simulate erosive damage caused by solid particle impacts. *Tribology International, 195*. Available from https://doi.org/10.1016/j.triboint.2024.109647, https://www.sciencedirect.com/science/journal/0301679X.

Bold S.E. (1999). Technology development for aero-engines gas turbines: Materials make a Difference. *Proc. DERA Symposium.*

Bose, S. (2007). *High temperature coatings high temperature coatings*. United States: Elsevier Inc.. Available from https://doi.org/10.1016/B978-0-7506-8252-7.X5000-8, http://www.sciencedirect.com/science/book/9780750682527.

Bousser, E., Martinu, L., & Klemberg-Sapieha, J. E. (2014). Solid particle erosion mechanisms of protective coatings for aerospace applications. *Surface and Coatings Technology, 257*, 165−181. Available from https://doi.org/10.1016/j.surfcoat.2014.08.037.

Bull, S. J. (2005). Nanoindentation of coatings. *Journal of Physics D: Applied Physics, 38* (24), R393−R413. Available from https://doi.org/10.1088/0022-3727/38/24/r01.

Chen, J., Beake, B. D., Wellman, R. G., Nicholls, J. R., & Dong, H. (2012). An investigation into the correlation between nano-impact resistance and erosion performance of EB-PVD thermal barrier coatings on thermal ageing. *Surface and Coatings Technology, 206* (23), 4992−4998. Available from https://doi.org/10.1016/j.surfcoat.2012.06.011.

Chen, X., He, M. Y., Spitsberg, I., Fleck, N. A., Hutchinson, J. W., & Evans, A. G. (2004). Mechanisms governing the high temperature erosion of thermal barrier coatings. *Wear, 256*(7−8), 735−746. Available from https://doi.org/10.1016/s0043-1648(03)00446-0.

Chen, X., Wang, R., Yao, N., Evans, A. G., Hutchinson, J. W., & Bruce, R. W. (2003). Foreign object damage in a thermal barrier system: Mechanisms and simulations. *Materials Science and Engineering: A, 352*(1−2), 221−231. Available from https://doi.org/10.1016/s0921-5093(02)00905-x.

Fackrell, J. E. (1989). Aerodynamic effects on PFBC erosion target experiments. *Wear, 134* (2), 237−252. Available from https://doi.org/10.1016/0043-1648(89)90128-2.

Gao, Z., Zhang, X., Chen, Y., Chalk, C., Nicholls, J., Brewster, G., & Xiao, P. (2024). Strain tolerance evolution of EB-PVD TBCs after thermal exposure or CMAS attack. *Journal of the European Ceramic Society*, *44*(1), 426−434. Available from https://doi.org/10.1016/j.jeurceramsoc.2023.08.047.

Geng, X., Wellman, R., Arrom, L. I., Chalk, C., & Castelluccio, G. M. (2023). Estimation of thermal barrier coating fracture toughness using integrated computational materials engineering. *Ceramics International*, *49*(15), 25788−25794. Available from https://doi.org/10.1016/j.ceramint.2023.05.124, https://www.journals.elsevier.com/ceramics-international.

Goward, G. W. (1983). Recent developments in high temperature coatings for gas turbine airfoils. *International Corrosion Conference Series NACE*, *12*, 553−560.

Goward, G. W., & Cannon, L. W. (1988). Pack cementation coatings for superalloys: A review of history, theory, and practice. *Journal of Engineering for Gas Turbines and Power*, *110*(1), 150−154. Available from https://doi.org/10.1115/1.3240078.

Herman, H. (1991). Powders for thermal spray technology. *KONA Powder and Particle Journal*, *9*, 187−199. Available from https://doi.org/10.14356/kona.1991024.

Khanna, A. S. (2002). *Introduction to high temperature oxidation* (1st ed.). ASM International.

Kim, J., Dunn, M. G., Baran, A. J., Wade, D. P., & Tremba, E. L. (1993). Deposition of volcanic materials in the hot sections of two gas turbine engines. *Journal of Engineering for Gas Turbines and Power*, *115*(3), 641−651. Available from https://doi.org/10.1115/1.2906754.

Levi, C. G., Hutchinson, J. W., Vidal-Sétif, M. H., & Johnson, C. A. (2012). Environmental degradation of thermal-barrier coatings by molten deposits. *MRS Bulletin*, *37*(10), 932−941. Available from https://doi.org/10.1557/mrs.2012.230.

Morrell, P., Rickerby, D. S., Brewster, G., Lee, K., & Nicholls, J. R. (2010). *High-temperature coatings in turbines in extreme military environments. NATO Science and Technology Organization Meeting Proceedings AVT-187* (pp. 10−28).

Nicholls, J. R. (2000). Designing oxidation-resistant coatings. *JOM*, *52*(1), 28−35. Available from https://doi.org/10.1007/s11837-000-0112-2.

Nicholls, J. R. (2003). Advances in coating design for high-performance gas turbines. *MRS Bulletin*, *28*(9), 659−670. Available from https://doi.org/10.1557/mrs2003.194.

Nicholls, J. R., Deakin, M. J., & Rickerby, D. S. (1999). A comparison between the erosion behaviour of thermal spray and electron beam physical vapour deposition thermal barrier coatings. *Wear*, *233−235*, 352−361. Available from https://doi.org/10.1016/S0043-1648(99)00214-8.

Nicholls, J. R., Jaslier, Y., & Rickerby, D. S. (1998). Erosion of EB-PVD thermal barrier coatings. *Materials at High Temperatures*, *15*(1), 15−22. Available from https://doi.org/10.1080/09603409.1998.11689572, http://www.google.co.in/url?sa = t&rct = j&q = &esrc = s&source = web&cd = 2&cad = rja&uact = 8&ved = 0ahUKEwiz2P3SgfTLAhWBVhoKHRuoCDU-QFgghMAE&url = http%3A%2F%2Fhttp://www.tandfonline.com%2Floi%2Fymht20&usg = AFQjCNGZDXW4jh1adNtw1x2xb_Pp2YZp9A&bvm = bv.118443451,d.amc.

Nicholls, J. R., Lawson, K. J., Johnstone, A., & Rickerby, D. S. (2002). Methods to reduce the thermal conductivity of EB-PVD TBCs. *Surface and Coatings Technology*, *151−152*, 383−391. Available from https://doi.org/10.1016/s0257-8972(01)01651-6.

Nicholls, J. R., Long, K. A., & Simms, N. J. (2010). *Diffusion coatings Shreir's corrosion* (pp. 2532−2555). United Kingdom: Elsevier. Available from https://doi.org/10.1016/B978-044452787-5.00176-1, http://www.sciencedirect.com/science/referenceworks/9780444527875.

Nicholls, J. R., Wellman, R. G., & Deakin, M. J. (2003). Erosion of thermal barrier coatings. *Materials at High Temperatures*, *20*(2), 207–218. Available from https://doi.org/10.3184/096034003782749008.

Poerschke, D. L., Barth, T. L., & Levi, C. G. (2016). Equilibrium relationships between thermal barrier oxides and silicate melts. *Acta Materialia*, *120*, 302–314. Available from https://doi.org/10.1016/j.actamat.2016.08.077, http://www.journals.elsevier.com/acta-materialia/.

Poerschke, D. L., Jackson, R. W., & Levi, C. G. (2017). Silicate deposit degradation of engineered coatings in gas turbines: Progress toward models and materials solutions. *Annual Review of Materials Research*, *47*, 297–330. Available from https://doi.org/10.1146/annurev-matsci-010917-105000, http://arjournals.annualreviews.org/loi/matsci.

Pomeroy, M. J. (2005). Coatings for gas turbine materials and long term stability issues. *Materials and Design*, *26*(3), 223–231. Available from https://doi.org/10.1016/j.matdes.2004.02.005, https://www.journals.elsevier.com/materials-and-design.

Quinn, G. D. (2006). Fracture toughness of ceramics by the Vickers indentation crack length method: A critical review. *Ceramic Engineering and Science Proceedings*, *27*(2), 45–62.

Rolls-Royce Plc. (1986). The jet engine. Rolls-Royce Plc.

Schmitt, M. P., Stokes, J. L., Gorin, B. L., Rai, A. K., Zhu, D., Eden, T. J., & Wolfe, D. E. (2017). Effect of Gd content on mechanical properties and erosion durability of substoichiometric $Gd_2Zr_2O_7$. *Surface and Coatings Technology*, *313*, 177–183. Available from https://doi.org/10.1016/j.surfcoat.2016.12.045, http://www.journals.elsevier.com/surface-and-coatings-technology/.

Schmitt, M. P., Stokes, J. L., Rai, A. K., Schwartz, A. J., & Wolfe, D. E. (2019). Durable aluminate toughened zirconate composite thermal barrier coating (TBC) materials for high temperature operation. *Journal of the American Ceramic Society*, *102*(8), 4781–4793. Available from https://doi.org/10.1111/jace.16317, http://onlinelibrary.wiley.com/journal/10.1111/(ISSN)1551-2916.

Schubert, A. B., Wellman, R., Nicholls, J., & Gentleman, M. M. (2016). Direct observations of erosion-induced ferroelasticity in EB-PVD thermal barrier coatings. *Journal of Materials Science*, *51*(6), 3136–3145. Available from https://doi.org/10.1007/s10853-015-9623-7.

Smialek, J. L., Archer, F. A., & Garlick, R. G. (1994). Turbine airfoil degradation in the persian gulf war. *JOM*, *46*(12), 39–41. Available from https://doi.org/10.1007/BF03222663.

Steenbakker, R. J. L., Wellman, R. G., & Nicholls, J. R. (2006). Erosion of gadolinia doped EB-PVD TBCs. *Surface and Coatings Technology*, *201*(6), 2140–2146. Available from https://doi.org/10.1016/j.surfcoat.2006.03.022.

Steinberg, L., Mikulla, C., Naraparaju, R., Pavlov, P., Löffler, M., Schulz, U., & Leyens, C. (2022). Erosion behavior of CMAS/VA infiltrated EB-PVD Gd2Zr2O7 TBCs: Special emphasis on the effect of mechanical properties of the reaction products. *Wear*, *506–507*. Available from https://doi.org/10.1016/j.wear.2022.204450.

Stephenson, D. J., & Nicholls, J. R. (1993). Modelling erosive wear. *Corrosion Science*, *35*(5–8), 1015–1026. Available from https://doi.org/10.1016/0010-938x(93)90320-g.

Tabakoff, W. (1989). Investigation of coatings at high temperature for use in turbomachinery. *Surface and Coatings Technology*, *39–40*(C), 97–115. Available from https://doi.org/10.1016/0257-8972(89)90045-5.

Tawancy, H. M., Abbas, N. M., & Rhys-Jones, T. N. (1991). Role of platinum in aluminide coatings. *Surface and Coatings Technology*, *49*(1–3), 1–7. Available from https://doi.org/10.1016/0257-8972(91)90022-o.

Vaßen, R., Bakan, E., Mack, D. E., & Guillon, O. (2022). A perspective on thermally sprayed thermal barrier coatings: Current status and trends. *Journal of Thermal Spray Technology*, *31*(4), 685−698. Available from https://doi.org/10.1007/s11666-022-01330-2, http://www.springer.com/journal/11666.

Wellman, R. G., Deakin, M. J., & Nicholls, J. R. (2005a). The effect of TBC morphology and aging on the erosion rate of EB-PVD TBCs. *Tribology International*, *38*(9), 798−804. Available from https://doi.org/10.1016/j.triboint.2005.02.008.

Wellman, R. G., Deakin, M. J., & Nicholls, J. R. (2005b). The effect of TBC morphology on the erosion rate of EB PVD TBCs. *Wear*, *258*(1−4), 349−356. Available from https://doi.org/10.1016/j.wear.2004.04.011.

Wellman, R. G., Dyer, A., & Nicholls, J. R. (2004). Nano and micro indentation studies of bulk zirconia and EB PVD TBCs. *Surface and Coatings Technology*, *176*(2), 253−260. Available from https://doi.org/10.1016/s0257-8972(03)00737-0.

Wellman, R. G., & Nicholls, J. R. (2004). A Monte Carlo model for predicting the erosion rate of EB PVD TBCs. *Wear*, *256*(9−10), 889−899. Available from https://doi.org/10.1016/j.wear.2003.09.001.

Wellman, R. G., & Nicholls, J. R. (2007). A review of the erosion of thermal barrier coatings. *Journal of Physics D: Applied Physics*, *40*(16), R293−R305. Available from https://doi.org/10.1088/0022-3727/40/16/R01.

Wellman, R. G., Tourmente, H., Impey, S., & Nicholls, J. R. (2004). Nano and microhardness testing of aged EB PVD TBCs. *Surface and Coatings Technology*, *188−189*(1−3), 79−84. Available from https://doi.org/10.1016/j.surfcoat.2004.08.018.

Witz, G., Shklover, V., Steurer, W., Bachegowda, S., & Bossmann, H. P. (2007). Phase evolution in yttria-stabilized zirconia thermal barrier coatings studied by rietveld refinement of X-ray powder diffraction patterns. *Journal of the American Ceramic Society*, *90*(9), 2935−2940. Available from https://doi.org/10.1111/j.1551-2916.2007.01785.x.

Advanced solid lubricants

22

Diana Berman
Department of Materials Science and Engineering, University of North Texas, Denton, TX, United States

22.1 Introduction

Lubrication is a vital aspect of mechanical engineering, essential for minimizing friction and wear in various mechanical systems. While liquid lubricants such as oils and greases have long been the go-to solutions for reducing friction and ensuring smooth operation, solid lubrication has emerged as an equally important and, in some cases, superior alternative. Solid lubrication offers distinct advantages in specific applications, where the use of traditional liquid lubricants is restricted or prohibited (Kumar et al., 2022).

Solid lubricants are in various forms, such as powders, coatings, or bulk materials. Depending on the desired properties and compatibility requirements, solid lubricants can be used as single-component materials or fillers in composite structures. The most common powdery solid lubricant solutions include graphite, molybdenum disulfide (MoS_2), and boron nitride (BN), traditionally used in seals and gaskets as well as in high-temperature sintering and powder metallurgy processes. The popular coating materials that are widely known and commonly used in tribological applications are diamond-like carbon (DLC) and polytetrafluoroethylene (PTFE, often known as Teflon). DLC coatings are commonly used in engine parts, piston rings, and camshafts, while PTFE is used as a solid lubricant in bearings and bushings. In the case of composites, which combine the beneficial properties of the solid lubricant with the structural support and stability of the matrix material, the most common are carbon-based and PTFE-based ones, widely used in seals, bearings, and breaks.

Solid lubricants can significantly enhance the efficiency, reliability, and durability of mechanical systems in challenging environments, high-temperature conditions, or vacuum settings where conventional liquid lubricants may fail to perform adequately. Consequently, they have found applications in diverse industries, ranging from aerospace and automotive to manufacturing and space exploration.

In this chapter, we review the general mechanisms of solid lubrication and delve into the novel concepts of adaptive and self-repairing solid lubricant solutions.

22.2 Mechanism of lubrication in traditional solid lubricants and approaches for application of solid lubricants

22.2.1 Basics of solid lubrication

Before going into the details on the possible material candidates for solid lubricants, we should review the basic mechanisms of their lubrication. Since generally solid lubricants are added on the surfaces, they play the role of a third body in the tribological systems. Because of these, the important parameters that affect their lubrication efficiency are adhesion, shearing, and conformity.

Solid lubricants interact through van der Waals forces, the weak forces between atoms and molecules that affect the adhesion of solid lubricants to the substrates, thus contributing to the formation of a continuous and uniform layer. The van der Waals forces can be modified further by the formation of chemical bonds with the surface atoms strengthening the attachment of the lubricant to the surface and allowing it to resist removal by frictional forces.

During the relative movement of surfaces with the solid lubricant in between, the shear mechanism comes into play. The shear mechanism induced by the sliding of the lubricant layer over the surface affects the layer shearing and deformation. The shear forces break the bonds between the layers of the lubricant and at the lubricant/surface interface, allowing the lubricant to adapt and transform. Therefore the shear strength largely affects the lubrication performance of the material. For example, a solid lubricant with a low shear strength creates favorable friction performance but may wear away quickly, thus limiting the duration of the protection. Conversely, a solid lubricant with a high shear strength will provide better protection but may cause increased friction and wear due to the reduction in the shearing abilities and adaptation.

From the adhesive friction model, the coefficient of friction between two sliding surfaces is proportional to the shear strength of the shearing material and inversely proportional to the hardness supporting the applied contact loading. This major concept is used for the design of solid lubricant solutions. Specifically, the lubricant should be thin enough to ensure the hardness support of the substrate while having low shear strength for easier shearing during sliding:

$$\mu = \frac{\tau}{H} \tag{22.1}$$

Notably, the properties of the substrate material largely affect the effectiveness of the solid lubricant. For example, substrate hardness defines the contact geometry, especially in the systems implying thin solid lubrication layers. At the same time, the surface roughness affects the conformality of the solid lubricant layer, providing the lubricant storing capability.

22.2.2 Approaches for application of solid lubricants

The overall approaches to using solid lubricants vary depending on the final requirements of the applications. Their application can be as simple as dispersions in the carrier solutions drop-casted or spray-coated on the substrate followed by the evaporation of the solvent or as complex as plasma and high-temperature-assisted deposition in a vacuum chamber. Table 22.1 summarizes standard approaches for the deposition of solid lubricants, the characteristics of the deposition techniques, and the properties of the resulting materials.

Solid lubricant can be also introduced in the sliding interfaces as components of composite structures that are formed by embedding solid lubricant particles into a bulk material. The matrix of the bulk material can be presented in the form of metals, polymers, or ceramics, depending on the application needs. In this case, the matrix material provides mechanical strength and support to the solid lubricant particles, allowing them to maintain their position and orientation within the composite

Table 22.1 Summary of common methods used for deposition of solid lubricants.

Process	Physical vapor deposition	Chemical vapor deposition	From solution	Electro-chemical
Examples	Sputtering, thermal evaporation, electron beam, arc discharge, pulsed laser	Atomic layer deposition, hot filament, sequential infiltration synthesis	Spray-coating, drop-casting, spin-coating	Electroless deposition, electroplating
Compatibility	Ferrous and nonferrous	Heat-resistant steels and alloys	Ferrous and nonferrous	Conductive metals
Treatment media	Mild vacuum or gas	Mild to atm. reactive gas	Air, dry nitrogen box	Diluted salts
Treatment temperature	Up to 500°C	Up to 1000°C	RT to 100°C	RT to 100°C
Treatment duration	120–360 min	600–900 min	10–30 min	15–300 min
Thickness	1–10 μm μ1–10 μm	0.1–10 μm μ0.1–10 μm	1–30 μm μ1–30 μm	1–50 μm μ1–50 μm
Hardness	15–40 GPa	15–35 GPa	1–5 GPa	10–20 GPa
Adhesion to substrate	Good	Good	Moderate	Good

structure. The matrix also helps to prevent the solid lubricant particles from being removed from the composite during use. In this case, the lubricant is exposed to the sliding interfaces on demand, upon gradual wear of the composites.

22.2.3 Traditional solid lubricants

Selection of the materials being used as solid lubricants is made based on the operation requirements. In traditional solid lubrication approaches, the major mechanism relies on the mechanically, thermally, or chemically produced easy shearing of the planes or platelets of atoms to minimize friction energy dissipation. Table 22.2 summarizes the major solid lubrication classes and their relevant characteristics as well as benefits and challenges associated with each of them.

Thus hexagonal solids such as graphite or MoS_2 powders have been widely used in various moving systems due to the relatively low cost of production and easy application. However, their susceptibility to oxidation limits the application conditions of such coatings to temperatures below 300°C. To expand the application regime of solid lubrication solutions to elevated temperatures, soft materials such as silver or gold can be used. In this case, the lubrication mechanism relies on the melting of the metals encapsulated in the supporting ceramic matrix and its diffusion to the contact interface. In this case, however, the diffusion to the surface quickly depletes metal lubricant reservoirs significantly limiting the lifetime of the coating materials. In this case, the temperature is usually limited to 500°C. With further increases in the operation temperature, the lubrication approach can rely on the formation of lubricious phases of oxides, such as Magneli phases, which exhibit low shearing strength at high temperatures. For example, during high-temperature sliding, TiO_2 has been reported to transform into gamma-Ti_3O_5, Ti_5O_9, Ti_9O_{17} ($TinO_{2n-1}$), and $TiO_{1.93-1.98}$ sliding; resulting in a noticeable reduction in the shear strength on the surface as well as in the bulk materials (Gardos, 2000; Lu et al., 2012). For all three classes of materials, still, the fundamental mechanism of lubrication remains the same; it involves easy shearing of the planes in the materials during sliding.

Even with an extensive library of material choices satisfying these traditional approaches of lubrication, they still experience issues such as oxidation, incompatibility with the changes in environment and temperature conditions, and fast removal of material from the contact. To address the existing challenges with traditional solid lubricants, new material solutions are needed. In the following chapters, we overview alternative methods of solid lubrication, using advanced solutions as the remedy to the existing challenges.

22.3 Advanced solid lubricants

In recent years, lots of progress has been made to improve the area of solid lubrication through the application of elaborated material solutions, thus solving the long-lasting quest of reducing friction to unmeasurable values (Ayyagari et al., 2022).

Table 22.2 Traditional lubrication approaches.

Lubrication approach	Material examples	Benefits	Temperature range (in air)	Challenges
Hexagonal solids	MoS_2, WS_2, Graphite	– Easy application – Low cost	<300°C	Oxidation at higher temperatures
Diffusion of soft metals	Ag and Au encapsulated in oxide and nitride ceramic matrices	– Oxidation stable; – Temperature self-regulated	300°C–500°C	Fast diffusion to the surface depletes metal lubricant reservoirs
Thermally-activated formation of lubricious oxides	Magnéli phases: – V_2O_5, MoO_3, TiO_2, – WO_3, PbO, ZnO	– Provides liquid lubrication at high temperatures – Environment supplies oxygen	500°C–1000°C	– Abrasion at low temperatures – Lubricant extrusion from contact by the load

Among those are 2D materials, adaptive/chameleon coatings, self-healing materials, and tribocatalytic materials. Each of these advanced approaches provides solutions to address the potential challenges of friction and wear relevant to the application conditions. For example, 2D materials, which are often deposited using dropcasting or spray-coating techniques, eliminate the need for high-temperature processing and ensure good conformality to the substrate, which is essential for complex contact geometries. Adaptive/chameleon coatings resolve the issue of tribological inconsistency upon transitioning from one environment to another. Self-healing materials enable the repair of the damaged surfaces. And finally, tribocatalytic materials provide a solution for coating replenishment directly from the environment during sliding. Further, we review each of the advanced solutions and provide representative examples of their performances.

22.3.1 Two-dimensional materials

While hexagonal materials, such as graphite and transition metal dichalcogenides, were used for lubrication purposes for decades, their application usually involved applying powders of their bulk forms. As a result, low adhesion to the substrates posed a major challenge since such powders could be easily pushed away from the contact interface during sliding. With the discovery of two-dimensional (2D) materials, such as graphene, layered MoS_2, and hexagonal BN, their tribological properties have become a topic of interest for researchers (Berman et al., 2018; Marian, Berman, Nečas, et al., 2022; Marian, Berman, Rota, et al., 2022).

Graphene, for example, has exceptional mechanical properties, including high strength and stiffness, which make it a promising material for tribological applications. Studies have shown that graphene can reduce friction and wear between sliding surfaces due to its low shear strength and high surface area (Berman, Erdemir, et al., 2014). Prior studies demonstrated the unique effectiveness of graphene as a solid lubricant solution for a wide range of environmental conditions and substrates (Berman, Deshmukh, et al., 2014; Berman et al., 2013a, 2013b). While traditionally materials are highly sensitive to the water presence, graphene demonstrated a stable coefficient of friction behavior upon transitioning between humid and dry environments (Fig. 22.1) (Berman, Erdemir, et al., 2014). Such exceptional behavior has been attributed to the nonpermeability of graphene to liquids, suppressing the environment-induced corrosion of steel substrate while supporting easy shearing (Berman & Krim, 2013; Berman et al., 2013b). However, though the friction stayed low, it was still above the threshold for the superlubricity, or near-zero friction, regime. Superlubricity became possible with the incorporation of a combination of solid lubricants, all made from carbon: graphene, diamond nanoparticles, and amorphous carbon or DLC coating (Berman et al., 2015).

A widely known alternative to graphene, MoS_2, is also used in various studies and applications for friction and wear reduction. In contrast to graphene, though, the structure of MoS_2 enables further friction decrease to very low regimes once exposed to dry sliding conditions (Cairns et al., 2023). MoS_2 has sulfur atoms that form iono-covalent bonds with the molybdenum atoms (positively charged) within

Advanced solid lubricants 633

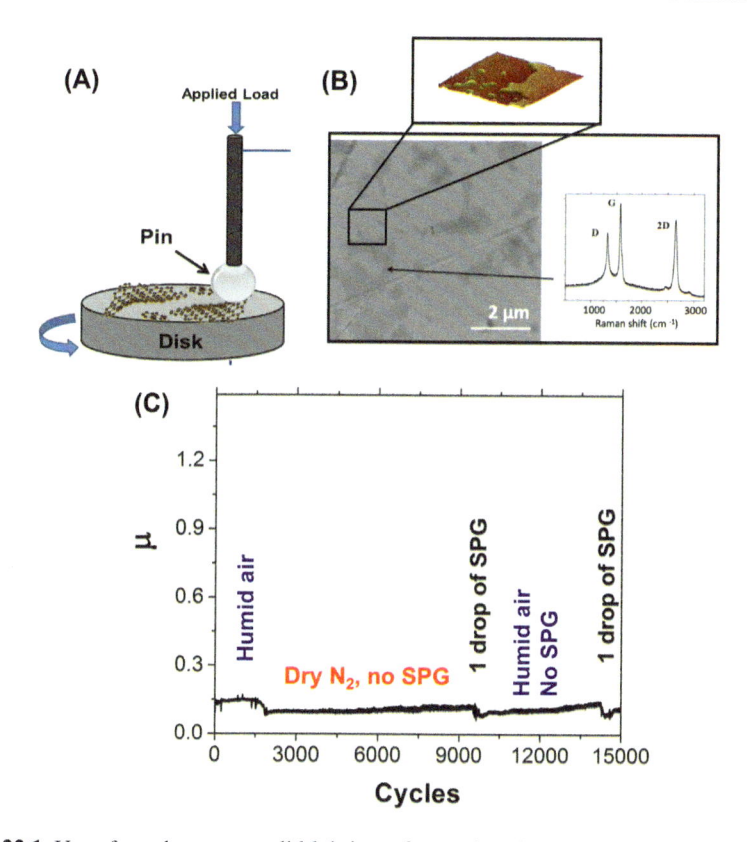

Figure 22.1 Use of graphene as a solid lubricant for steel surfaces. (A) Schematic of the tested system and (B) the corresponding analysis of the coated steel surfaces. (C)The COF stays stable upon changing between humid and dry environments.
Source: Adapted with permission from Berman, D., Erdemir, A., & Sumant, A. V. (2014). Graphene: A new emerging lubricant. *Materials Today*, 17(1), 31−42. Available from: http://www.journals.elsevier.com/materials-today/. https://doi.org/10.1016/j.mattod.2013.12.003.

the layer and with other sulfur atoms (negatively charged) above and below the layer. The shearing occurring between sulfur layers supported by the electric repulsion of sulfur atoms has the ability to further decrease the friction to very low regimes. Boron nitride, another 2D material, has similar properties to MoS_2, with low friction and wear due to its layered structure. However, it also has a high thermal conductivity and can withstand high temperatures, making it suitable for high-temperature tribological applications.

Recently, the tribological characteristics of MoS_2 were further improved by incorporating a new class of 2D materials, MXenes (Macknojia, Ayyagari, Shevchenko, et al., 2023; Macknojia, Ayyagari, Zambrano, et al., 2023). In this case, the mixture of MoS_2 and MXene, in the form of $Ti_3C_2T_x$, has been spray-coated on rough steel surfaces. During sliding, the friction values dropped to the superlubricity regime (the coefficient of friction below 0.01) under high load and sliding velocity conditions (Fig. 22.2). The observed dramatic reduction in friction and wear is attributed to the reorientation of MoS_2 and MXene basal planes creating an incommensurate contact (Macknojia, Ayyagari, Zambrano, et al., 2023). These results promise an easy and robust lubrication approach for various mechanical systems.

22.3.2 Chameleon coatings

Chameleon, or adaptive, tribological coatings are a class of coatings designed to respond to changes in the local tribological environment in real time. These coatings

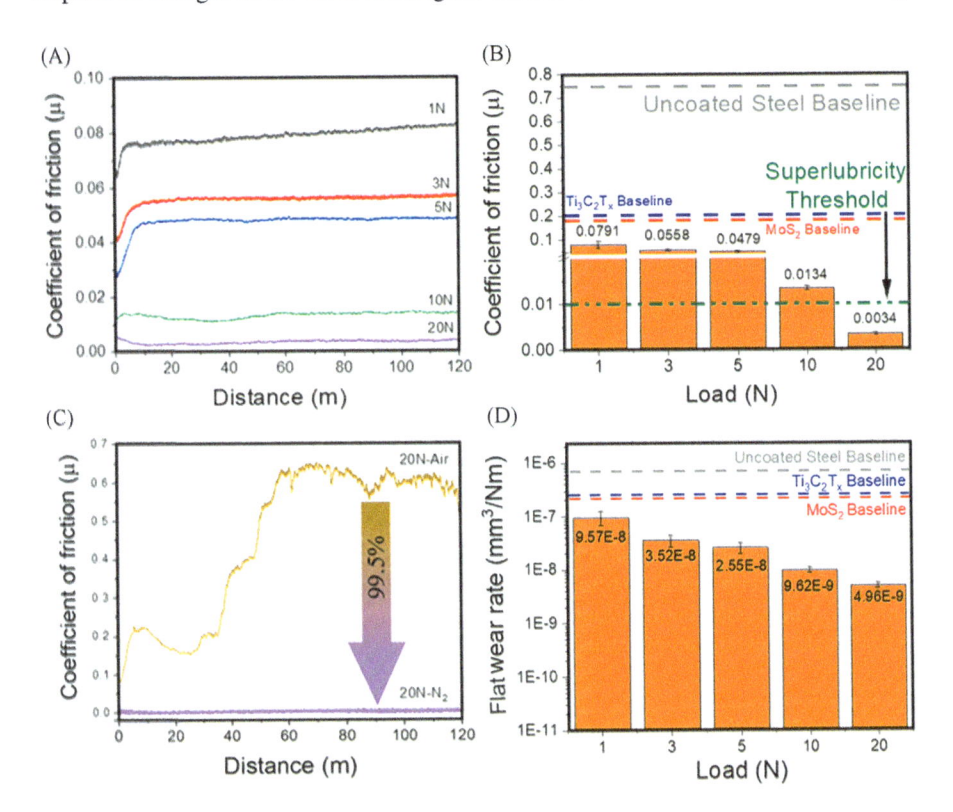

Figure 22.2 Demonstration of the superlubricity regime achieved for MXene + MoS_2 combination deposited on rough steel surfaces. (A) COF evolution at different applied loads with (B) the corresponding summary of the final COF values. (C) Comparison of the COF for dry and humid testing conditions. (D) Summary of the wear rates observed for different loads.
Source: Reproduced with permission from Macknojia, A., Ayyagari, A., Zambrano, D., Rosenkranz, A., Shevchenko, E. V., & Berman, D. (2023). Macroscale superlubricity induced by MXene/MoS_2 nanocomposites on rough steel surfaces under high contact stresses. *ACS Nano*, 17(3), 2421–2430. Available from: http://pubs.acs.org/journal/ancac3. https://doi.org/10.1021/acsnano.2c09640.

can change their properties, such as friction coefficient, wear resistance, or lubricity, in response to changes in the operating conditions. This allows the coatings to adapt to the changing demands of the application, optimizing performance and extending the lifespan of the components. Applicability of chameleon coatings has been successfully demonstrated for lightweight metals, such as aluminum and titanium alloys, often used in aviation, automobile, and rail transportation applications due to their low density and high strength-to-weight ratio, which helps to reduce fuel consumption (Dong et al., 2010; Peters et al., 2003).

Without coating, due to their low hardness (Molinari et al., 1997; Wilson & Alpas, 1996) and absence of a low cycle fatigue limit, light alloys exhibit poor sliding, rolling, and fretting wear performance. Surface treatments are becoming more important in enhancing the surface performance of light alloys used in aeronautical components. Surface engineering of light alloys remains challenging because they cannot be effectively hardened, and surface oxides can impair bonding between applied hard coatings and the substrate (Dong et al., 2010). To address this challenge, plasma electrolytic oxidation (PEO) has been used to improve the wear, heat, and corrosion resistance of aluminum and titanium alloys (Liu et al., 2018; Yerokhin et al., 1999). PEO creates a hard, well-adhered aluminum oxide or titanium oxide ceramic coating with a graded morphology that transitions from a dense zone near the substrate interface to a porous outer region. As a result, PEO can be a good underlying layer for the application of solid lubricants, which can be encapsulated in external pores and serve as reservoirs for tribological contact lubrication, due to these properties (Liu et al., 2018; Zabinski et al., 2006).

Indeed, Liu et al. explored the fretting wear properties and adaptation processes for a hard PEO-produced aluminum oxide surface with a top layer of an $MoS_2/Sb_2O_3/C$ chameleon coating designed to self-adapt in multiple humidity conditions for friction and wear control. The experiments demonstrated low friction coefficients, a considerable reduction in critical amplitude for the stick-slip transition, and self-adaptive tribological behavior in a cyclic, humid/dry, environment. Zabinski et al. (2006) further investigated the adaptability mechanism of the chameleon coating by performing a Raman analysis of the surfaces during transitions. In the dry cycle, Raman activity was seen in the MoS_2 area, with essentially negligible carbon-related scattering identified. In the Raman spectrum for the humid cycle, meanwhile, the signal from graphite-like carbon was detected.

These results suggested that in a dry environment, which is preferable for MoS_2 but not for graphite, MoS_2 prevails at the contacting surfaces, creating easy-shearing interfaces. When the humidity was raised, the MoS_2 layers on the counterfaces were replaced by basal-oriented graphite, with the basal surface remaining the most inert and slippery (Savage, 1948). Upon further transition to a dry environment, the desorption of water molecules promotes fast wear-off of graphite sheets, revealing again fresh MoS_2 layers that recoat the surface. In all these cases, Sb_2O_3 creates a sublayer for the active lubricant that could transfer proper/dominant lubricant toward the contact area while helping to limit fracture formation and propagation (Zabinski et al., 1993).

The performance of the chameleon coating applied on an aluminum oxide substrate has been further evaluated under varied temperature conditions (Fig. 22.3) (Shirani et al., 2020). The increased temperature had a significant deteriorating effect on the frictional behavior of the bare sample (Fig. 22.3B). Above 200°C, aluminum started to develop scuffing behavior, which led to the formation of a large amount of debris and material transformation from the substrate to the Si_3N_4 counter body. In contrast, for the PEO-Chameleon coatings, the temperature increase from 25°C to 300°C promoted the COF to decrease from 0.08 to 0.02, as a result of water removal and MoS_2 exposure to the sliding interfaces (Fig. 22.3C). High temperature also helped to remove microcracks from the coating making them more uniform and aligned.

The in situ Raman analysis indicated no signs of the oxidation of the MoS_2 compound as confirmed by the intensity of MoS_2 peaks at 380 and 410 cm^{-1} remaining

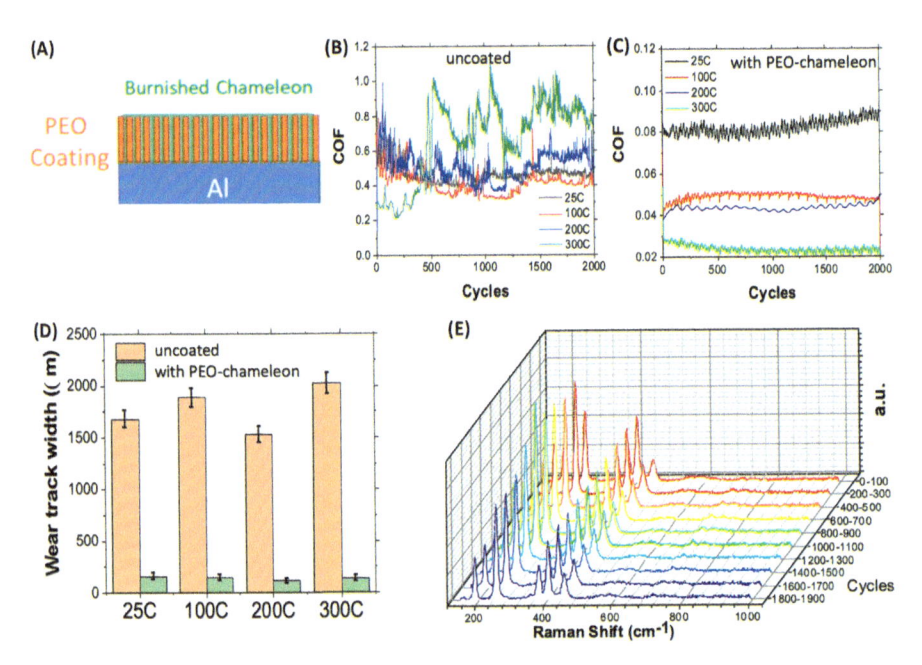

Figure 22.3 Adaptive behavior of chameleon coatings at different temperatures. (A) Schematic of the PEO-chameleon coating. COF for (B) uncoated and (C) coated samples tested at different temperatures. (D) Summary of the resulting wear track width. (E) Evolution of the Raman spectrum for the coated samples during testing at 300°C. *Source*: Adapted with permission from Shirani, A., Joy, T., Rogov, A., Lin, M., Yerokhin, A., Mogonye, J. E., Korenyi-Both, A., Aouadi, S. M., Voevodin, A. A., & Berman, D., (2020). PEO-Chameleon as a potential protective coating on cast aluminum alloys for high-temperature applications. *Surface and Coatings Technology*, *397*, 126016. Available from: http://www.journals.elsevier.com/surface-and-coatings-technology/. https://doi.org/10.1016/j.surfcoat.2020.126016.

strong with no shift during the tribotests at 300°C (Fig. 22.3E). This chemical stability of MoS_2 is crucial for maintaining a low COF over large sliding distances.

22.3.3 Self-healing materials

Self-healing tribological materials are a class of materials that have the ability to repair the caused by wear and friction damage without external intervention by triggering a chemical or physical reaction, restoring their original properties and functionality (Voevodin et al., 2014). Examples of self-healing materials include those that contain microcapsules or microchannels filled with a healing agent, such as a polymer or a lubricant that is released upon rupturing of the microcapsules, or material combinations undergoing reversible oxidation or reduction reactions that can repair surface damage by restoring the surface chemistry.

Related to the tribological contacts, the proposed systems with the first self-healing approach apply noble metal diffusion. All noble metals distributed in a hard ceramic matrix can be used for such systems, though silver is the most explored one due to its relatively lower cost than others and high mobility at elevated temperatures ($\sim 300°C-500°C$) (Voevodin et al., 2014). Examples of systems that were investigated include TiC-Ag (Endrino et al., 2002), YSZ-Ag (Muratore et al., 2006), Cr_2AlC-Ag (Gupta et al., 2007), CrN-Ag (Mulligan et al., 2010), CrAlN-Ag (Basnyat et al., 2007), Mo_2N-Ag (Shtansky et al., 2013), and MoCN-Ag (Voevodin & Zabinski, 2000). In all of these cases, solid lubricant is introduced in low quantities ($< \sim 20$ at%) as amorphous or poorly crystalline inclusions to maintain the relatively large hardness and elastic modulus values of the ceramic and enhance toughness through stress minimization, crack deflection, and ductility (Aouadi et al., 2009; Voevodin & Zabinski, 2000).

In their study, Shirani et al. (2019) proposed to use of friction-induced heating for regulating the self-healing process. They conducted macroscale pin-on-disk tests on a material system comprising niobium oxide and silver (Fig. 22.4). The selection of this specific sliding chemistry resulted in the formation of a lubricious ternary $AgNbO_3$ phase at the sliding interfaces when tested at 600°C, in contrast to $\sim 945°C$ required under static conditions. The formation of the ternary phase was accompanied by a significant reduction in the friction coefficient value, about three times lower than the niobium oxide/silicon nitride system. This improved tribological performance was attributed to a tribologically activated surface reconstruction process inside the wear track. This novel solid lubrication approach enabled enhancement in the wear and crack resistance characteristics of ceramic components, which can be adjusted on demand to achieve the desired frictional response.

22.3.4 Tribocatalytic materials

The use of solid lubricants also enabled the development of the field of tribochemistry and tribocatalysis. Though the concept originated from the idea of enhancing the liquid lubrication approaches by facilitating the formation of the protective coatings on the sliding surfaces from the additives in the liquid lubricants, recent efforts

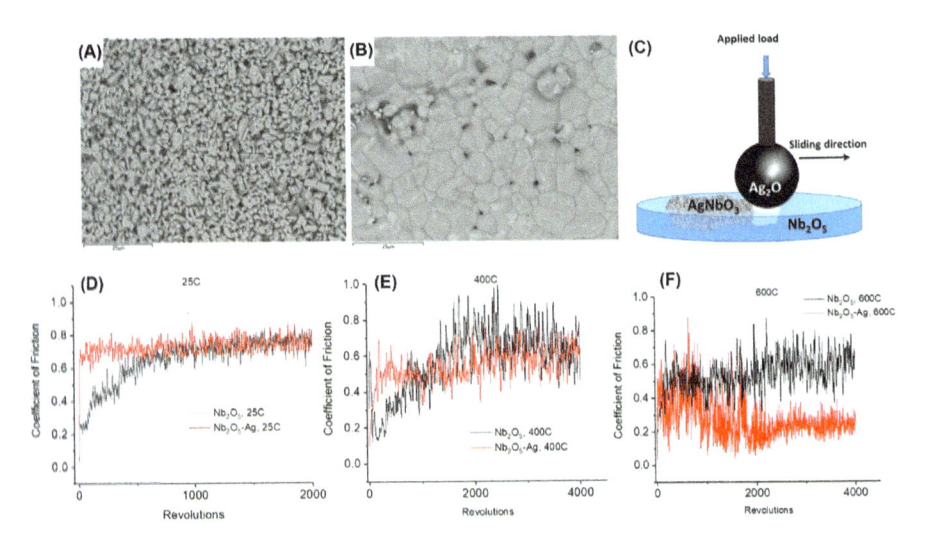

Figure 22.4 Self-healing for in situ repair of coatings. Preparation of the bulk niobium oxide sample by (A) pressing and (B) sintering the pellet. The sample was further tested for tribological performance (C). Tribology test of Nb_2O_5 with and without the presence of Ag at (D) 25°C, (E) 400°C, and (F) 600°C. Results indicate a reduction in the coefficient of friction in the case of silver presence at 600°C (F).
Source: Reproduced with permission from Shirani, A., Gu, J., Wei, B., Lee, J., Aouadi, S. M., & Berman, D. (2019). Tribologically enhanced self-healing of niobium oxide surfaces. *Surface and Coatings Technology*, *364*, 273−278. Available from: http://www.journals.elsevier.com/surface-and-coatings-technology/. https://doi.org/10.1016/j.surfcoat.2019.03.002.

demonstrated the adaptation of the approach to solid and gas sources (Fig. 22.5). In this case, in contrast to the traditional solid lubricant use, the mechanism of lubrication relies on the in situ generation of solid lubricants at the sliding interfaces. Particular interest has been dedicated to the generation of carbon films of different structures. These carbon films, in turn, provide the benefit of solid lubrication delivery directly to the sliding interface playing the role of both friction-reducing and wear-repairing materials.

Within sliding systems, the intentional inclusion of catalytic species can be utilized to decrease the activation energy to initiate lubricity-aiding reaction processes on contacting surfaces. One such alternative is the addition of magnesium silicate hydroxide (MSH) nanoparticles (NPs) into oils that facilitates the formation of a protective tribofilm directly on steel surfaces during sliding (Chang et al., 2017). Later, MSH NPs were also suggested as reactive components in solid lubricant powders (Gao et al., 2021; Wang et al., 2021). While these lubricant additives are effective in protecting steel surfaces under the lubrication of various oils, for certain applications, the modifications of the lubricant chemistry can be unacceptable. In such cases, the tribocatalytic materials can be incorporated in a solid material form,

Advanced solid lubricants 639

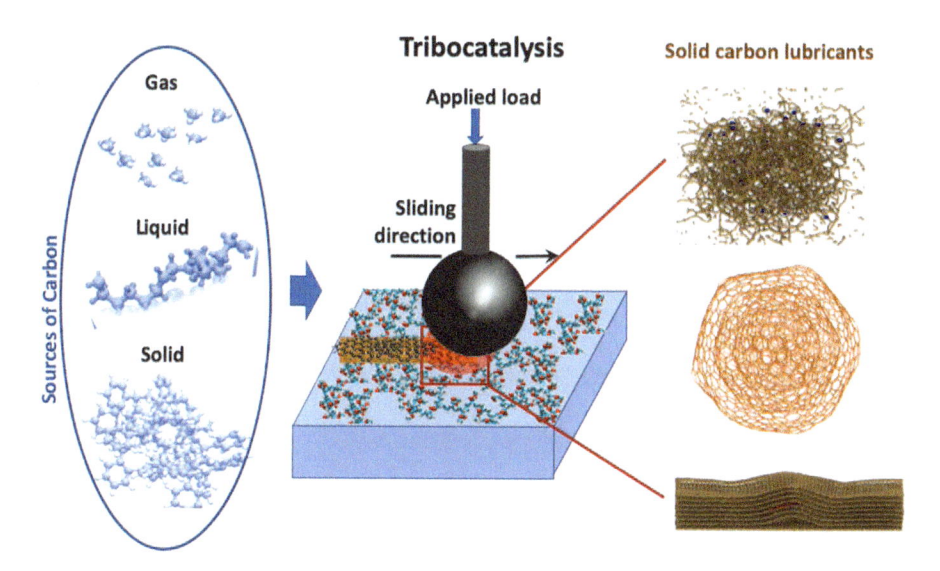

Figure 22.5 Schematic of the tribocatalysis concept for in situ formation of lubricating carbon films inside the contact. Formation of carbon-based tribofilms is facilitated during sliding from the hydrocarbon sources.
Source: Reproduced with permission from Berman, D., & Erdemir, A. (2021). Achieving ultralow friction and wear by tribocatalysis: Enabled by in-operando formation of nanocarbon films. ACS Nano, 15(12), 18865−18879, Available from: http://pubs.acs.org/journal/ancac3. https://doi.org/10.1021/acsnano.1c08170.

as components of protective coatings. These tribocatalytic coatings with catalytically active inclusions can replenish the worn-away solid material with carbon-rich transfer films formed with the aid of a hydrocarbon environment, in the forms of gas, liquid, or solid (Berman & Erdemir, 2021; Hiratsuka et al., 2018; Kajdas et al., 2017). As a result, the lifetime of the protection is significantly extended and does not require high maintenance costs.

The concept of the tribocatalysis implies in situ formation and repair of protective carbon films being facilitated by the presence of catalytically reactive metals, such as Pt (Argibay et al., 2018; Curry et al., 2018; Shirani et al., 2022), Ni (Ramirez et al., 2020), Fe (Berman et al., 2019), and Mg (Gao et al., 2021). Erdermir et al. (Erdemir et al., 2016) synthesized nitride-copper-based coatings that enabled the formation of protective carbon-rich tribofilms when tested in an oil-lubricant environment. This formation of the protective tribofilm has been promoted by the inclusion of catalytically reactive copper grains, which interact with the hydrocarbon molecules of the oil during tribological wear. More recently, similar coatings (with the inclusion of copper in various metal nitride matrices) were tested in different alkanes demonstrating improved friction and wear characteristics depending on the chain length of the alkane molecules (Shirani et al., 2021). In general, metal nitride − based hard coatings are known for their high toughness, wear

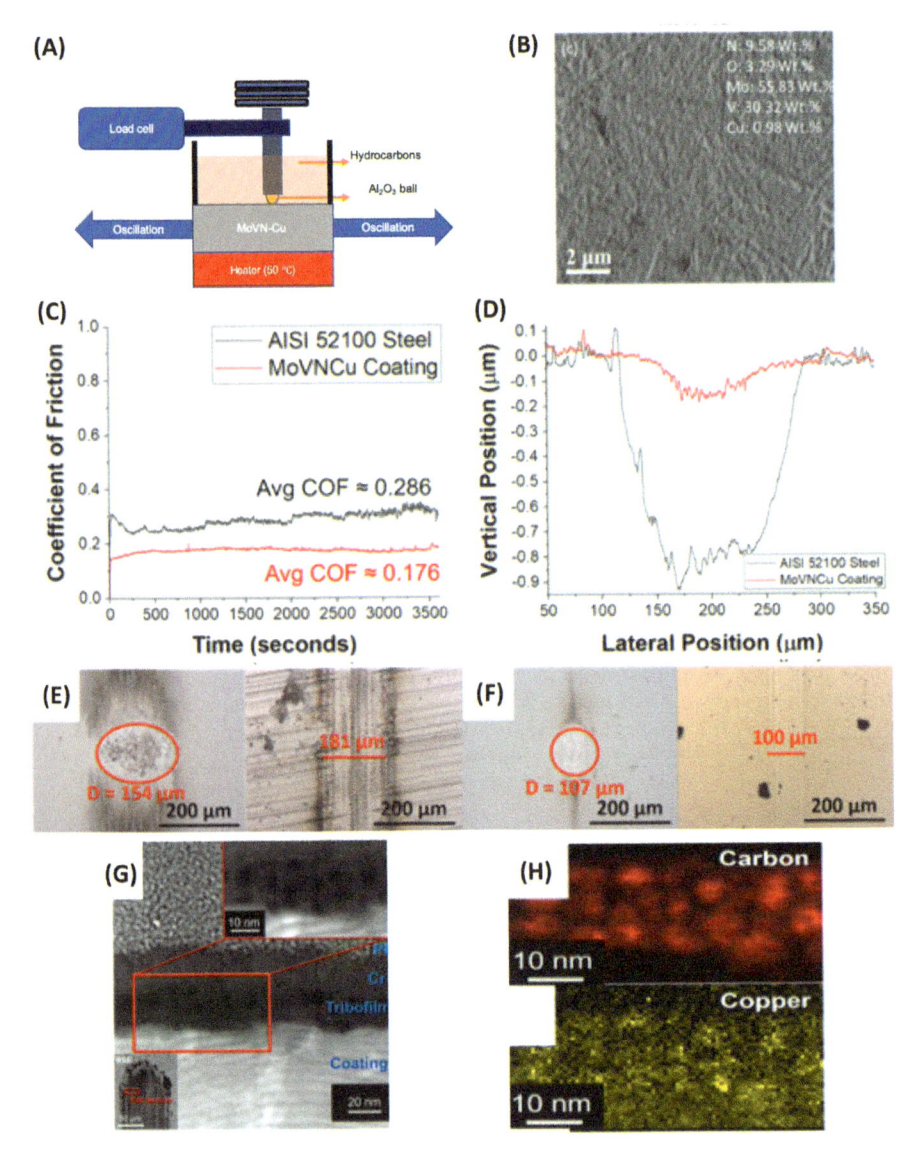

Figure 22.6 Tribocatalytically driven formation of carbon films from hydrocarbon-rich dodecane environment. (A) schematic of the experimental setup for testing (B) tribocatalytically active MoVN-Cu coating. (C) The COF values for uncoated and coated 52100 steel samples. (D) The corresponding profilometry analysis of the wear produced after the tests. The optical micrographs of the ball and flat wear for (E) uncoated and (F) coated samples. (G) TEM analysis of the carbon-based tribofilm tribocatalytically grown on the MooVN-Cu coating from the hydrocarbon environment. (H) The elemental mapping of carbon and copper in the tribofilm.
Source: Adapted with permission from Jacques, K., Shirani, A., Smith, J., Walck, S., Berkebile, S., Aouadi, S., Voevodin, A., Berman, D. (2023). MoVN-Cu coatings for in situ tribocatalytic formation of carbon-rich tribofilms in low-viscosity fuels. ACS Applied Materials Interfaces 15, 30070.

resistivity, and stability in corrosive and reactive environments (Aouadi et al., 2009; Subramanian & Strafford, 1993; Walsh & Ponce de Leon, 2014), and thus their use as matrices for tribocatalytic element inclusion can provide a beneficial approach for surface engineering of fuel components to operate with low-viscosity hydrocarbons. The presence of catalytically reactive metal, such as copper, distributed in the nitride matrix facilitated the transformation of hydrocarbons into protective amorphous carbon film (Fig. 22.6), thus reducing friction and wear of the surfaces in comparison to the uncoated steel substrates (Jacques et al., 2023). In situ formation of the protective films not only allows to repair the wear on demand but also creates a protective tribofilm that can extend the lifetime of the sliding interface in case the lubricant is removed.

22.4 Conclusions

In conclusion, this chapter has provided a comprehensive overview of solid lubrication approaches, highlighting their importance for solving existing tribological challenges. Solid lubricants already play a pivotal role in modern engineering and manufacturing, offering crucial solutions for reducing friction, minimizing wear and tear, and enhancing the overall efficiency and lifespan of machinery and equipment across a wide range of industries.

With the novel solutions, such as the use of 2D materials, self-adaptive, self-healing, and self-generating materials, the use of the solid lubrication approach becomes even more robust and efficient, thus addressing challenges associated with extreme temperatures, high loads, and harsh operating conditions, where traditional liquid lubricants may fall short.

Acknowledgments

The author acknowledges the support of this work by the National Science Foundation (NSF) (Award No. 2323452).

References

Aouadi, S. M., Luster, B., Kohli, P., Muratore, C., & Voevodin, A. A. (2009). Progress in the development of adaptive nitride-based coatings for high temperature tribological applications. *Surface and Coatings Technology*, *204*(6−7), 962−968. Available from https://doi.org/10.1016/j.surfcoat.2009.04.010.

Argibay, N., Babuska, T. F., Curry, J. F., Dugger, M. T., Lu, P., Adams, D. P., Nation, B. L., Doyle, B. L., Pham, M., Pimentel, A., Mowry, C., Hinkle, A. R., & Chandross, M. (2018). In-situ tribochemical formation of self-lubricating diamond-like carbon films. *Carbon*, *138*, 61−68. Available from https://doi.org/10.1016/j.carbon.2018.06.006.

Ayyagari, A., Alam, K. I., Berman, D., & Erdemir, A. (2022). Progress in superlubricity across different media and material systems—A review. *Frontiers in Mechanical Engineering*, *8*. Available from https://doi.org/10.3389/fmech.2022.908497, http://www.frontiersin.org/journals/mechanical-engineering#.

Basnyat, P., Luster, B., Kertzman, Z., Stadler, S., Kohli, P., Samir Aouadi, J., Xu, S. R., Mishra, O. L., Eryilmaz, A., & Erdemir. (2007). Mechanical and tribological properties of CrAlN-Ag self-lubricating films. *Surface and Coatings Technology*, *202*(4−7), 1011−1016. Available from https://doi.org/10.1016/j.surfcoat.2007.05.088.

Berman, D., Deshmukh, S. A., Sankaranarayanan, S. K. R. S., Erdemir, A., & Sumant, A. V. (2014). Extraordinary macroscale wear resistance of one atom thick graphene layer. *Advanced Functional Materials*, *24*(42), 6640−6646. Available from https://doi.org/10.1002/adfm.201401755, http://onlinelibrary.wiley.com/journal/10.1002/(ISSN)1616-3028.

Berman, D., Deshmukh, S. A., Sankaranarayanan, S. K. R. S., Erdemir, A., & Sumant, A. V. (2015). Macroscale superlubricity enabled by graphene nanoscroll formation. *Science (New York, N.Y.)*, *348*(6239), 1118−1122. Available from https://doi.org/10.1126/science.1262024, http://www.sciencemag.org/content/348/6239/1118.full.pdf.

Berman, D., & Erdemir, A. (2021). Achieving ultralow friction and wear by tribocatalysis: Enabled by in-operando formation of nanocarbon films. *ACS Nano*, *15*(12), 18865−18879. Available from https://doi.org/10.1021/acsnano.1c08170, http://pubs.acs.org/journal/ancac3.

Berman, D., Erdemir, A., & Sumant, A. V. (2013a). Few layer graphene to reduce wear and friction on sliding steel surfaces. *Carbon*, *54*, 454−459. Available from https://doi.org/10.1016/j.carbon.2012.11.061.

Berman, D., Erdemir, A., & Sumant, A. V. (2013b). Reduced wear and friction enabled by graphene layers on sliding steel surfaces in dry nitrogen. *Carbon*, *59*, 167−175. Available from https://doi.org/10.1016/j.carbon.2013.03.006.

Berman, D., Erdemir, A., & Sumant, A. V. (2014). Graphene: A new emerging lubricant. *Materials Today*, *17*(1), 31−42. Available from https://doi.org/10.1016/j.mattod.2013.12.003, http://www.journals.elsevier.com/materials-today/.

Berman, D., Erdemir, A., & Sumant, A. V. (2018). Approaches for achieving superlubricity in two-dimensional materials. *ACS Nano*, *12*(3), 2122−2137. Available from https://doi.org/10.1021/acsnano.7b09046, http://pubs.acs.org/journal/ancac3.

Berman, D., & Krim, J. (2013). Surface science, MEMS and NEMS: Progress and opportunities for surface science research performed on, or by, microdevices. *Progress in Surface Science*, *88*(2), 171−211. Available from https://doi.org/10.1016/j.progsurf.2013.03.001.

Berman, D., Mutyala, K. C., Srinivasan, S., Sankaranarayanan, S. K. R. S., Erdemir, A., Shevchenko, E. V., & Sumant, A. V. (2019). Iron-nanoparticle driven tribochemistry leading to superlubric sliding interfaces. *Advanced Materials Interfaces*, *6*(23). Available from https://doi.org/10.1002/admi.201901416, http://onlinelibrary.wiley.com/journal/10.1002/(ISSN)2196-7350.

Cairns, E., Ayyagari, A., McCoy, C., Berkebile, S., Berman, D., Aouadi, S. M., & Voevodin, A. A. (2023). Tribological behavior of molybdenum disulfide and tungsten disulfide sprayed coatings in low viscosity hydrocarbon environments. *Tribology International*, *179*. Available from https://doi.org/10.1016/j.triboint.2022.108206.

Chang, Q., Rudenko, P., Miller, D. J., Wen, J., Berman, D., Zhang, Y., Arey, B., Zhu, Z., & Erdemir, A. (2017). Operando formation of an ultra-low friction boundary film from synthetic magnesium silicon hydroxide additive. *Tribology International*, *110*, 35−40. Available from https://doi.org/10.1016/j.triboint.2017.02.003, http://www.elsevier.com/inca/publications/store/3/0/4/7/4.

Curry, J. F., Babuska, T. F., Furnish, T. A., Lu, P., Adams, D. P., Kustas, A. B., Nation, B. L., Dugger, M. T., Chandross, M., Clark, B. G., Boyce, B. L., Schuh, C. A., & Argibay, N. (2018). Achieving ultralow wear with stable nanocrystalline metals. *Advanced Materials*, *30*(32). Available from https://doi.org/10.1002/adma.201802026, http://onlinelibrary.wiley.com/journal/10.1002/(ISSN)1521-4095.

Dong P. Qingyan X. Baicheng L. Jiarong L. Llailong Y. Iiaipeng J. 2010 Modeling of grain selection during directional solidification of single crystal superalloy turbine blade castings. *TMS annual meeting* (Vol. 1, pp. 243−247). China.

Endrino, J. L., Nainaparampil, J. J., & Krzanowski, J. E. (2002). Microstructure and vacuum tribology of TiC-Ag composite coatings deposited by magnetron sputtering-pulsed laser deposition. *Surface and Coatings Technology*, *157*(1), 95−101. Available from https://doi.org/10.1016/S0257-8972(02)00138-X.

Erdemir, A., Ramirez, G., Eryilmaz, O. L., Narayanan, B., Liao, Y., Kamath, G., & Sankaranarayanan, S. K. R. S. (2016). Carbon-based tribofilms from lubricating oils. *Nature*, *536*(7614), 67−71. Available from https://doi.org/10.1038/nature18948, http://www.nature.com/nature/index.html.

Gao, K., Wang, B., Shirani, A., Chang, Q., & Berman, D. (2021). Macroscale superlubricity accomplished by Sb_2O_3-MSH/C under high temperature. *Frontiers in Chemistry*, *9*. Available from https://doi.org/10.3389/fchem.2021.667878, http://journal.frontiersin.org/journal/chemistry.

Gardos, M. N. (2000). Magnéli phases of anion-deficient rutile as lubricious oxides. Part I. Tribological behavior of single-crystal and polycrystalline rutile ($TinO_{2n-1}$) éMagnéli phases of anion-deficient rutile as lubricious oxides. Part I. Tribological behavior of single-crystal and polycrystalline rutile ($TinO_{2n-1}$). *Tribology Letters*, *8*(2-3), 65−78. Available from https://doi.org/10.1023/a:1019122915441, http://www.springerlink.com/(snpxut45gxflnr45vb2gia45)/app/home/journal.asp?
referrer = parent&backto = searchpublicationsresults,1,2.

Gupta, S., Filimonov, D., Palanisamy, T., El-Raghy, T., & Barsoum, M. W. (2007). Ta_2AlC and Cr_2AlC Ag-based composites—New solid lubricant materials for use over a wide temperature range against Ni-based superalloys and alumina. *Wear*, *262*(11−12), 1479−1489. Available from https://doi.org/10.1016/j.wear.2007.01.028.

Hiratsuka, K., Kajdas, C., Kulczycki, A., & Dante, R. C. (2018). *Tribocatalysis, tribochemistry tribocatalysis, tribocorrosion* (pp. 163−255). Jenny Stanford Publishing.

Jacques, K., Shirani, A., Smith, J., Walck, S., Berkebile, S., Aouadi, S., Voevodin, A., & Berman, D. (2023). MoVN-Cu coatings for in situ tribocatalytic formation of carbbon-rich tribofilms in low-viscosity fuels. *ACS Applied Materials Interfaces*, *15*(25), 30070. Available from https://doi.org/10.1021/acsami.3c01953.

Kajdas, C., Kulczycki, A., & Ozimina, D. (2017). A new concept of the mechanism of tribocatalytic reactions induced by mechanical forces. *Tribology International*, *107*, 144−151. Available from https://doi.org/10.1016/j.triboint.2016.08.022, http://www.elsevier.com/inca/publications/store/3/0/4/7/4.

Kumar, R., Hussainova, I., Rahmani, R., & Antonov, M. (2022). Solid lubrication at high-temperatures—A review. *Materials*, *15*.

Liu, Y. F., Liskiewicz, T., Yerokhin, A., Korenyi-Both, A., Zabinski, J., Lin, M., Matthews, A., & Voevodin, A. A. (2018). Fretting wear behavior of duplex PEO/chameleon coating on Al alloy. *Surface and Coatings Technology*, *352*, 238−246. Available from https://doi.org/10.1016/j.surfcoat.2018.07.100, http://www.journals.elsevier.com/surface-and-coatings-technology/.

Lu, Y., Matsuda, Y., Sagara, K., Hao, L., Otomitsu, T., & Yoshida, H. (2012). Fabrication and thermoelectric properties of Magneli phases by adding Ti into TiO$_2$. *Advanced Materials Research, 415-417*, 1291−1296. Available from https://doi.org/10.4028/http://www.scientific.net/AMR.415-417.1291, 10226680. Japan.

Macknojia, A., Ayyagari, A., Zambrano, D., Rosenkranz, A., Shevchenko, E. V., & Berman, D. (2023). Macroscale superlubricity induced by MXene/MoS$_2$ nanocomposites on rough steel surfaces under high contact stresses. *ACS Nano, 17*(3), 2421−2430. Available from https://doi.org/10.1021/acsnano.2c09640, http://pubs.acs.org/journal/ancac3.

Macknojia, A. Z., Ayyagari, A., Shevchenko, E., & Berman, D. (2023). MXene/graphene oxide nanocomposites for friction and wear reduction of rough steel surfaces. *Scientific Reports, 13*(1). Available from https://doi.org/10.1038/s41598-023-37844-0, https://www.nature.com/srep/.

Marian, M., Berman, D., Nečas, D., Emami, N., Ruggiero, A., & Rosenkranz, A. (2022). Roadmap for 2D materials in biotribological/biomedical applications − A review. *Advances in Colloid and Interface Science, 307*. Available from https://doi.org/10.1016/j.cis.2022.102747, https://www.journals.elsevier.com/advances-in-colloid-and-interface-science.

Marian, M., Berman, D., Rota, A., Jackson, R. L., & Rosenkranz, A. (2022). Layered 2D nanomaterials to tailor friction and wear in machine elements—A review. *Advanced Materials Interfaces, 9*(3). Available from https://doi.org/10.1002/admi.202101622, http://onlinelibrary.wiley.com/journal/10.1002/(ISSN)2196-7350.

Molinari, A., Straffelini, G., Tesi, B., & Bacci, T. (1997). Dry sliding wear mechanisms of the Ti$_6$Al$_4$V alloy. *Wear, 208*(1-2), 105−112. Available from https://doi.org/10.1016/s0043-1648(96)07454-6.

Mulligan, C. P., Blanchet, T. A., & Gall, D. (2010). CrN−Ag nanocomposite coatings: High-temperature tribological response. *Wear, 269*(1-2), 125−131. Available from https://doi.org/10.1016/j.wear.2010.03.015.

Muratore, C., Voevodin, A. A., Hu, J. J., & Zabinski, J. S. (2006). Tribology of adaptive nanocomposite yttria-stabilized zirconia coatings containing silver and molybdenum from 25 to 700°C. *Wear, 261*(7-8), 797−805. Available from https://doi.org/10.1016/j.wear.2006.01.029.

Peters, M., Kumpfert, J., Ward, C. H., & Leyens, C. (2003). Titanium alloys for aerospace applications. *Advanced Engineering Materials, 5*(6), 419−427. Available from https://doi.org/10.1002/adem.200310095.

Ramirez, G., Eryilmaz, O. L., Fatti, G., Righi, M. C., Wen, J., & Erdemir, A. (2020). Tribochemical conversion of methane to graphene and other carbon nanostructures: Implications for friction and wear. *ACS Applied Nano Materials, 3*(8), 8060−8067. Available from https://doi.org/10.1021/acsanm.0c01527, https://pubs.acs.org/journal/aanmf6.

Savage, R. H. (1948). Graphite lubrication. *Journal of Applied Physics, 19*(1), 1−10. Available from https://doi.org/10.1063/1.1697867.

Shirani, A., Gu, J., Wei, B., Lee, J., Aouadi, S. M., & Berman, D. (2019). Tribologically enhanced self-healing of niobium oxide surfaces. *Surface and Coatings Technology, 364*, 273−278. Available from https://doi.org/10.1016/j.surfcoat.2019.03.002, http://www.journals.elsevier.com/surface-and-coatings-technology/.

Shirani, A., Joy, T., Rogov, A., Lin, M., Yerokhin, A., Mogonye, J. E., Korenyi-Both, A., Aouadi, S. M., Voevodin, A. A., & Berman, D. (2020). PEO-Chameleon as a potential protective coating on cast aluminum alloys for high-temperature applications. *Surface and Coatings Technology*, *397*. Available from https://doi.org/10.1016/j.surfcoat.2020. 126016, http://www.journals.elsevier.com/surface-and-coatings-technology/.

Shirani, A., Li, Y., Eryilmaz, O. L., & Berman, D. (2021). Tribocatalytically-activated formation of protective friction and wear reducing carbon coatings from alkane environment. *Scientific Reports*, *11*(1). Available from https://doi.org/10.1038/s41598-021-00044-9, http://www.nature.com/srep/index.html.

Shirani, A., Li, Y., Smith, J., Curry, J. F., Lu, P., Wilson, M., Chandross, M., Argibay, N., & Berman, D. (2022). Mechanochemically driven formation of protective carbon films from ethanol environment. *Materials Today Chemistry*, *26*. Available from https://doi.org/10.1016/j.mtchem.2022.101112.

Shtansky, D. V., Bondarev, A. V., Kiryukhantsev-Korneev, P. V., Rojas, T. C., Godinho, V., & Fernández, A. (2013). Structure and tribological properties of MoCN-Ag coatings in the temperature range of 25-700 °C. *Applied Surface Science*, *273*, 408–414. Available from https://doi.org/10.1016/j.apsusc.2013.02.055, http://www.journals.elsevier.com/applied-surface-science/.

Subramanian, C., & Strafford, K. N. (1993). Review of multicomponent and multilayer coatings for tribological applications. *Wear*, *165*(1), 85–95. Available from https://doi.org/10.1016/0043-1648(93)90376-w.

Voevodin, A. A., Muratore, C., & Aouadi, S. M. (2014). Hard coatings with high temperature adaptive lubrication and contact thermal management: Review. *Surface and Coatings Technology*, *257*, 247–265. Available from https://doi.org/10.1016/j.surfcoat.2014.04.046.

Voevodin, A. A., & Zabinski, J. S. (2000). Supertough wear-resistant coatings with 'chameleon' surface adaptation. *Thin Solid Films*, *370*(1-2), 223–231. Available from https://doi.org/10.1016/s0040-6090(00)00917-2.

Walsh, F. C., & Ponce de Leon, C. (2014). A review of the electrodeposition of metal matrix composite coatings by inclusion of particles in a metal layer: An established and diversifying technology. *Transactions of the IMF*, *92*(2), 83–98. Available from https://doi.org/10.1179/0020296713z.000000000161.

Wang, B., Gao, K., Chang, Q., Berman, D., & Tian, Y. (2021). Magnesium Silicate Hydroxide-MoS_2-Sb_2O_3 Coating Nanomaterials for High-Temperature Superlubricity. *ACS Appl Nano Mater*, *4*, 7097. Available from https://doi.org/10.1021/acsanm.1c01104.

Wilson, S., & Alpas, A. T. (1996). Effect of temperature on the sliding wear performance of Al alloys and Al matrix composites. *Wear*, *196*(1-2), 270–278. Available from https://doi.org/10.1016/0043-1648(96)06923-2.

Yerokhin, A. L., Nie, X., Leyland, A., Matthews, A., & Dowey, S. J. (1999). Plasma electrolysis for surface engineering. *Surface and Coatings Technology*, *122*(2-3), 73–93. Available from https://doi.org/10.1016/s0257-8972(99)00441-7.

Zabinski, J. S., Bultman, J. E., Sanders, J. H., & Hu, J. J. (2006). Multi-environmental lubrication performance and lubrication mechanism of MoS_2/Sb_2O_3/C composite films. *Tribology Letters*, *23*(2), 155–163. Available from https://doi.org/10.1007/s11249-006-9057-0.

Zabinski, J. S., Donley, M. S., & McDevitt, N. T. (1993). Mechanistic study of the synergism between Sb_2O_3 and MoS_2 lubricant systems using Raman spectroscopy. *Wear*, *165*(1), 103–108. Available from https://doi.org/10.1016/0043-1648(93)90378-y.

Nanomechanics of tribologically transformed surfaces

23

Guillaume Kermouche[1] and Gaylord Guillonneau[2]
[1]Mines Saint-Etienne, LGF UMR5307 CNRS, Saint-Etienne, France, [2]Ecole Centrale de Lyon, CNRS, ENTPE, LTDS, UMR5513, Ecully, France

23.1 Introduction

When a material is subjected to intense contact or friction loading, its near surface tends to undergo changes to protect its bulk. In some cases, a new material is formed, referred to as tribologically transformed structures or tribologically transformed surfaces (TTSs), especially when there is no mechanical alloying between the two contacting bodies. TTSs typically have a thickness ranging from a few hundred nanometers to a few tens of micrometers (see Fig. 23.1) and are known to significantly enhance the material's resistance to wear and surface damage.

TTS formation often results from cyclic plastic deformation observed in fretting or severe plastic deformation occurring in sliding friction or impact loadings, such

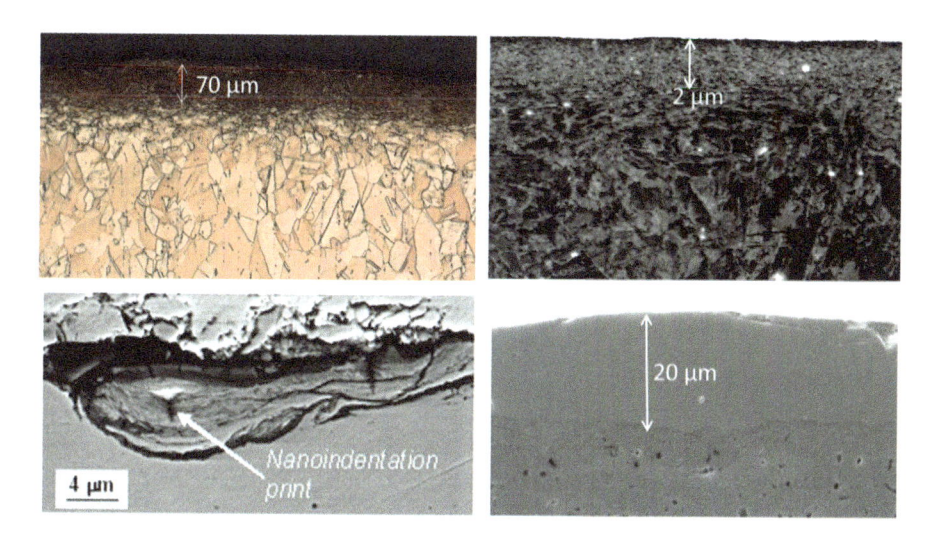

Figure 23.1 Some examples of tribologically transformed surfaces. Top-left: after severe shot peening of a pure copper. Top-right: White Etched Layer after turning of a 15–5PH martensitic stainless steel (Dumas et al., 2021). Bottom-left: after impact-sliding tests on a AISI304L austenitic stainless steel (Kermouche et al., 2007). Bottom-right: Glaze Layer resulting of high-temperature fretting applied to a cobalt-based alloy (Viat et al., 2017).

as those encountered in certain manufacturing processes such as peening and grinding. High strain rates and temperature increases can also facilitate structural transformations, such as phase changes or recrystallization, as observed in machining processes. Due to their thin nature, the mechanical properties of TTS can only be accurately probed using mechanical tests conducted at the micron scale.

Thanks to significant advancements in this field over the past two decades, it has become possible to perform precise measurements of TTS mechanical properties under both standard conditions and what are commonly referred to as extreme conditions, including high temperature, high strain rate, and aggressive environments. This chapter is dedicated to the utilization of micromechanical tests to investigate the mechanical properties of tribologically transformed surfaces.

In this chapter, the concept of TTS is elaborated upon in detail. A brief overview of micromechanical tests, such as nanoindentation and micropillar compression, along with microstructural characterization methods such as electron backscattering diffraction and transmission electron microscopy, is provided. The chapter also covers two main applications: TTS resulting from manufacturing processes and glaze layers—metallic glass-like protective layers generated by high-temperature fretting loads on certain metals. Although the latter application is more related to the mechanical characterization of micron-thick tribofilms rather than TTS, the same micromechanical testing methodology can be applied, warranting its inclusion in a specific subchapter.

Finally, the chapter concludes with a discussion on upcoming developments in this field.

23.2 About tribologically transformed surfaces

Upon initial examination, defining a tribologically transformed surface (TTS) appears to be as challenging as characterizing its microstructure or mechanical properties. Typically, TTSs are distinguished from the mechanically mixed layer (MML) introduced by Rigney (2000) in analogy with mechanical alloying. TTSs are formed from one of the contacting bodies without materials transferring from the counterpart. In contrast, in the MML process, mechanical mixing involves both contact bodies and the reaction of the environment, resulting in a different chemical composition for the MML. Sometimes, the term "tribofilms" is used interchangeably.

Extensive research on the formation of TTS has been conducted in the field of fretting wear. It is considered the preliminary stage before the creation of wear particles. Sauger et al. (2000) highlighted its nanocrystalline nature and chemical composition close to the initial material. They concluded that TTS formed once a cumulative plastic strain threshold was reached, which was macroscopically related to a threshold in energy dissipation. They even proposed that the formation mechanism is based on recrystallization. More recently, utilizing advanced microstructural characterization techniques such as focused ion beam and transmission electron microscopy (Lefranc et al., 2023) revealed that fretting-induced TTS in titanium alloys consisted of nanosized grains surrounded by a highly deformed zone (referred to as the general deformed layer in their paper). Under harsh conditions, a sandwich-like structure was observed. Similar sandwich-like features were also noted by Kaiser et al. (2006) and Kermouche et al. (2007) in TTS resulting from impact-sliding loadings in an austenitic stainless steel. Sekkal, Langlade and Vannes (2005) observed the formation of TTS under repeated normal impact conditions in a titanium alloy using a micropercussion device, with the TTS thickness being much larger than in fretting, emphasizing the importance of contact size and plastic deformation. Through a combined experimental/numerical approach, Kermouche et al. (2011) observed and modeled the formation of a thick TTS during micropercussion of an AISI1045 steel grade, where cumulative plastic deformation was confirmed as the governing parameter. The structure shakedowned macroscopically to an elastic response but exhibited cyclic plastic deformation. Hydrostatic pressure likely functioned as a damage preventer rather than promoting metallurgical transformations. Tumbajoy-Spinel et al. (2017) experimentally investigated the formation and spreading of a TTS under repeated oblique impacts on pure iron. The TTS thickness was more than twice the one obtained under normal impact conditions, pointing out the effect of shear in the process.

These studies pointed out the primary role of plastic deformation in TTS formation. However, the influence of temperature should not be overlooked in the process. For example, high-speed friction experiments on pure copper conducted by Kermouche, Jacquet, et al. (2016) resulted in a depth-graded TTS. The near-surface high-temperature rise, combined with shear-induced severe plastic deformation, activated dynamic recrystallization/grain refinement and strain-hardening mechanisms. An architectured-like TTS was obtained with graded mechanical properties (see Fig. 23.2).

Tribologically transformed surfaces (TTS) are sometimes referred to as white etching layers (WEL), a phenomenon encountered in railway steel grades. Their formation has been attributed to solid − solid-phase transformations induced by rolling contact fatigue. According to Thiercelin et al. (2020), WEL consists of squats with mechanical properties significantly different from those of the neighboring bulk material. The resulting deformation incompatibilities lead to crack propagation in the surrounding areas. The morphology of these squats is similar to those obtained in fretting and impact sliding. It's worth mentioning that "White Etching Layer" is also used in the field of machining to describe TTS induced by the contact between the tool and the workpiece (Brown et al., 2018). Here, TTS is created due to high strain rates, high temperature rise, and severe plastic deformation resulting from high-speed contact. It appears as a continuous layer a few micrometers thick rather than localized squats. Mondelin et al. (2013) demonstrated that the WEL resulting from the turning of a 15−5PH steel was a consequence of grain refinement induced by dynamic recrystallization.

Glaze layers, on the other hand, are tribofilms formed by tribological loadings (Viat et al., 2016), usually fretting or impact/sliding, at high temperatures (see Fig. 23.1). They differ from TTS as the formation process involves the

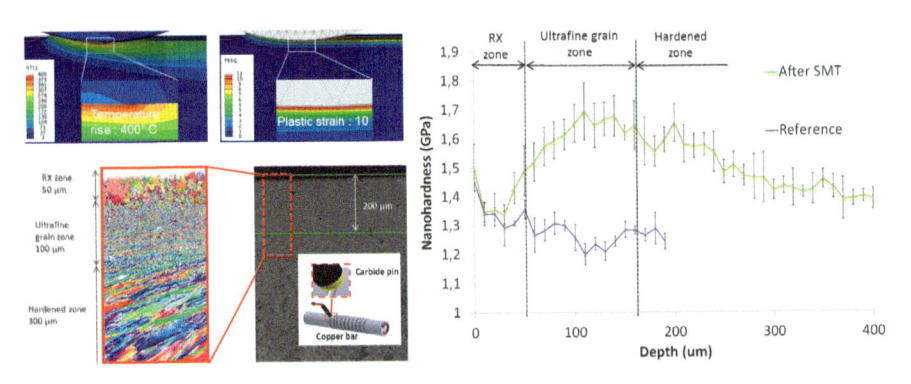

Figure 23.2 TTS induced on a copper bar by high sliding friction. A dynamically recrystallized zone in the near surface is followed by a ultrafine grain region. A transition zone composed of a microstructure oriented in the sliding direction is observed below. The microstructure evolution results from the high temperature rise and the large plastic deformation in the near surface (Jacquet et al., 2014). Hardness is lower in the recrystallized region than in ultrafine grains and hardened regions as expected (Kermouche, Jacquet, et al., 2016).

environment, including factors such as temperature and oxygen, leading to a mixing of chemical elements from the two surfaces in contact. An important consideration is that these layers form only when a threshold temperature is reached (Fig. 23.1). Below this temperature, no glaze layer is formed, and wear by oxidation/abrasion occurs. Glaze layers can be created from various metals, especially Ni, Co, and steel alloys. The formation process has been extensively studied for 50 years (Stott et al., 1976). It is argued that this layer is formed by debris detachment due to adhesion, followed by sintering/compaction, and usually adheres the two parts in contact (Jiang et al., 2004). High temperature and surface oxidation play crucial roles in glaze layer formation. The structure of this layer can be either homogeneous or heterogeneous with multilayer structures, and its thickness typically ranges from 3 to 20 μm (Dreano et al., 2020). It can be surrounded by a highly deformed structure. Generally, the glaze layer structure is nanocrystalline, with the grain size dependent on the material couples and the degree of heterogeneity of the layer as shown by Viat et al. (2017).

23.3 About micromechanical tests used to characterize tribologically transformed surface mechanical properties

One significant objective in the mechanical characterization of tribologically transformed surfaces (TTSs) is to obtain their stress − strain curves. This enables a comparison of their mechanical properties with those of the bulk material (Tumbajoy-Spinel et al., 2018) and is crucial for incorporating them into numerical models aimed at predicting their lifetime (Kermouche et al., 2007). Over the past two decades, substantial efforts have been dedicated to characterizing materials' mechanical properties on a micron scale, which is precisely the scale of interest for TTS.

One of the most prominent techniques, both in terms of age and practical relevance, is the instrumented nanoindentation test, extensively described in other chapters of this book. This method involves pressing a hard tip into the material, continuously measuring the load and displacement on the surface. Some devices even allow for load oscillation, resulting in continuous loading/unloading during testing. The initial slope of the unloading part, often referred to as stiffness, is valuable as it relates to the elastic modulus and the contact area under load. If the material's elastic properties are known, it is possible to compute the mean contact pressure, typically named (nano)hardness, as a function of the tip's penetration depth. If the elastic properties are unknown, various contact models, such as the one proposed by Oliver and Pharr (1992), can be utilized. This approach allows probing material hardness at the scale allowed by the tip sharpness, making it particularly suitable for characterizing TTS. As an example, a nanoindentation impression a few micrometers in size is clearly visible over the TTS created by impact-sliding tests on AISI304L steel (see Fig. 23.1).

To complete the characterization, it is necessary to transform nanoindentation measurements into stress − strain curves. Numerous articles have been written on this topic, especially employing inverse identification methodologies based on finite element calculations. However, despite significant efforts, this remains a challenge. This difficulty can be traced back to the pioneering work of Tabor in the 1950s, which demonstrated that sharp indentation could lead to the measurement of only one point on the stress − strain curve due to its self-similarity (see Fig. 23.1). For instance, when a Berkovich tip indents a rigid plastic material, the yield stress (σ_r) for an 8% plastic strain can be estimated by dividing the hardness (H) by a factor of 3. This factor is often referred to as a constraint factor, with its value depending on the ratio of hardness (H) to Young's modulus (E). The Tabor relation has been extended with varying degrees of success over the years and has been applied successfully to polymers (Kermouche et al., 2006), glasses (Barthel et al., 2020), and metals (Tiphéne et al., 2023):

$$\sigma_r = \left(\frac{0.087}{0.243 - 0.783\frac{H}{E}} \right) H$$

Indeed, deriving a representative uniaxial yield stress σ_r from hardness might seem nonsensical from purely mechanical considerations. The stress state beneath the indenter is not uniaxial, particularly directly under the tip. Therefore, this relation should be regarded more as a qualitative indicator than a quantitative measurement. Additionally, sharp indentation is not suitable for extracting the onset of plastic flow, a crucial aspect when investigating the relationship between TTS microstructure and mechanical properties in a quantitative manner.

Spherical indentation theoretically provides a portion of the stress − strain curves, but it necessitates meticulous attention to the indenter geometry. Moreover, assessing size effects with spherical indentation is challenging, making it a significant drawback in the mechanical characterization of TTS. Consequently, for TTS characterization, sharp indentation is often preferred. If obtaining stress − strain curves for TTS is the ultimate objective, it is advisable to explore alternative micromechanical testing methods such as microtraction or microcompression tests, such as micropillar compression. These techniques offer more accurate insights into the mechanical behavior of TTS, ensuring a more comprehensive and reliable analysis of their properties.

The micropillar compression technique, introduced by Uchic et al. (2004) in the early 2000s, has been extensively utilized since then. This method involves machining or etching micron-sized pillars (see Fig. 23.3) and utilizing a nanoindentation setup equipped with a flat punch to apply the loading. To ensure precise positioning, these tests are often conducted inside a scanning electron microscope (SEM). This approach allows for in-situ deformation observation, providing valuable insights into the plasticity mechanisms at play in TTS.

However, there are several drawbacks associated with micropillar compression experiments. One limitation arises from the continuity between the micropillar and its substrate. Consequently, the stress state is not purely uniaxial due to constraints at the pillar's bottom and top, leading to a more complex mechanical behavior. Additionally, aligning the pillars and punch axes precisely is challenging, potentially affecting the measurement of the elastic modulus (Zhang et al., 2006) and occasionally causing buckling phenomena (Lacroix et al., 2012). Despite these

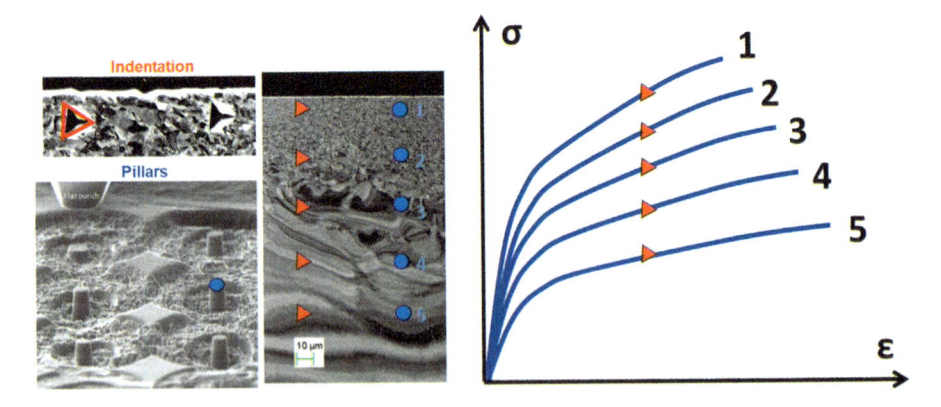

Figure 23.3 Illustration of indentation and micropillar compression tests applied on the cross-section of a tribologically transformed surface (pure iron submitted to sever shot peening). The micropillar compression can be used to measure the stress − strain curve in the plastic regime whereas the nanoindentation test helps at determining the elastic properties and one point of the stress − strain curve.

challenges, micropillar compression experiments are well-suited for characterizing the plastic part of the stress − strain curves (Kermouche, Guillonneau, et al., 2016), complementing the efficient elastic modulus measurement provided by nanoindentation tests.

Another significant challenge lies in the cost associated with micropillar fabrication, particularly when using techniques such as focus ion beam (FIB) machining. This cost limitation often hinders the acquisition of a large dataset for cross-correlation studies between microstructure and mechanical properties. Fast nanoindentation mapping, with indentation cycles lasting less than 1 second, has emerged as an alternative, enabling efficient data collection. The combination of micropillar compression and nanoindentation tests proves to be highly complementary in overcoming these challenges.

Nevertheless, accurately characterizing the microstructure of TTS is essential for meaningful interpretation of mechanical properties obtained through advanced micromechanical methods. This task is challenging due to the microstructural gradient inside and around the TTS zone. Electron back scattering diffraction (EBSD) is a promising technique, particularly when the grain size remains sufficiently large (>200 nm) and plastic deformation is not excessively severe. EBSD allows measurement of crystal orientations at each point with lateral resolution down to a few tens of nanometers. This technique enables determination of grain size and intragrain misorientation (see Fig. 23.4), theoretically linked to geometric dislocation density (Breumier et al., 2021). In cases where TTS microstructure is heavily deformed or the grain size is too small, transmission electron microscopy (TEM) can be employed. Thin slices within the TTS zone can be cut and analyzed using these characterization methods (Lefranc et al., 2023).

23.4 Application to tribologically transformed surface resulting of manufacturing processes

TTSs are formed during tribological contacts, as indicated by Chromik and Zhang (2018). This transformation can occur due to in-service loading, such as fretting, impact, and sliding wear, and is created and eliminated during the mechanical part's lifetime. Additionally, TTS can result from manufacturing processes involving thermomechanical contacts, encompassing a wide range of mechanical processes. These processes can lead to the formation of a white layer, as observed in machining and grinding, or a severely deformed layer, as seen in shot peening and hammering. The unique mechanical strength of TTS created in this manner can have intriguing implications for the component's lifetime. Specific treatments such as surface mechanical attrition treatments (Roland et al., 2006), nanopeening treatments (Tumbajoy-Spinel et al., 2016), and surface mechanical grinding treatments (Li et al., 2008) have been developed for that purpose. However, the impact of machining-induced white layers on the component's lifetime remains a topic of debate (La Monaca et al., 2021). Consequently, there is significant interest in

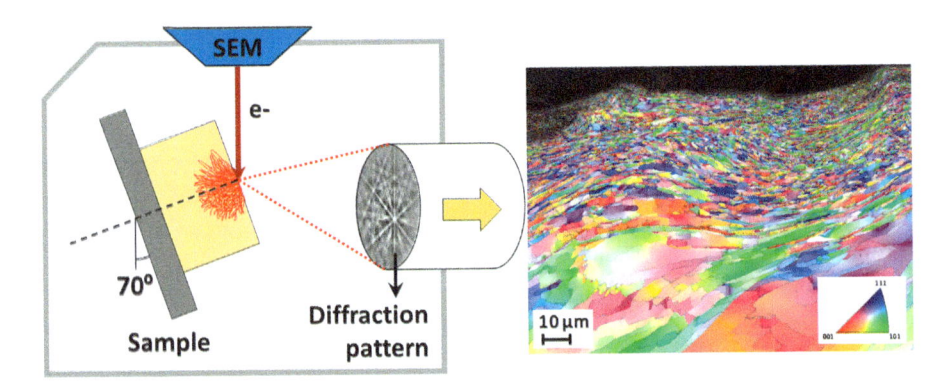

Figure 23.4 Principle of EBSD characterization and application on the cross-section of a tribologically transformed surface resulting of severe shot peening of a pure iron (Inverse Pole Figure along X). Grain size and morphology change along the depth. The intragrain misorientation in the grain below the TTS is related to the presence of dislocations in the subsurface.

characterizing TTS resulting from various manufacturing processes to gain a deeper understanding of their properties and their potential implications for the mechanical part's durability and performance.

23.4.1 Strengthening mechanisms in tribologically transformed surface

The strength of TTS is directly correlated to the strengthening mechanisms at play in metals. The yield stress of a crystal is usually the sum of several contributions: the crystalline lattice resistance σ_c, the solid solution hardening $\Delta\sigma_s$, the precipitates hardening $\Delta\sigma_p$, the dislocation strain hardening $\Delta\sigma_d$, and the grain boundary hardening $\Delta\sigma_g$:

$$\sigma = \sigma_c + \Delta\sigma_s + \Delta\sigma_p + \Delta\sigma_d + \Delta\sigma_g$$

TTS resulting from manufacturing processes typically emerge due to severe plastic deformation or cyclic plastic deformation, often assisted by high strain rates and/or low/high temperatures. In specific processes such as surface mechanical grinding treatment (SMGT) developed by Li et al. (2008), conducting treatments under cryogenic conditions is recommended to enhance the grain refinement process. This process is expected to significantly strengthen the material due to mechanisms such as dislocation strain hardening and grain boundary hardening.

In situations where high temperatures are involved, the hardening from precipitates can be affected by modifications in precipitate size through processes such as dissolution, nucleation, or coalescence. However, the timespan in these processes is usually too short to allow for significant changes in precipitate kinetics. Instead, hardening is more likely to occur through displacive transformations, such as martensitic transformations or mechanical twinning as shown by Dumas et al. (2021).

In the subsequent sections, the nanomechanical methodology is applied to metals not prone to such transformations, specifically metals with a sufficiently high stacking fault energy. Therefore, TTS strengthening mechanisms can be modeled using the Taylor's law regarding the dislocation strain hardening: $\Delta\sigma_d \propto \sqrt{\rho}$, where ρ is the dislocation density, the Hall − Petch regarding the grain boundary hardening: $\Delta\sigma_g \propto 1/\sqrt{\varnothing}$, where $\varnothing$ is the grain size. The other contributions are considered as constant and are included into an initial yield stress $\sigma_y = \sigma_c + \Delta\sigma_s + \Delta\sigma_p$.

23.4.2 Application to impact-induced tribologically transformed surface in pure iron

The following paragraph is a summary of the PhD thesis of David Tumbajoy-Spinel. He investigated the strength of tribologically transformed surfaces (TTSs) in low-carbon steels generated through severe shot peening, as documented in his three published papers (Tumbajoy-Spinel et al., 2016, 2017, 2018). For this study, high-purity pure iron, prepared using a cold-crucible method, was chosen as the base material. The material underwent thermomechanical treatment and recrystallization at a temperature of approximately 650°C, resulting in equiaxed grains with an average size of about 250 μm.

A specific set of samples underwent intense shot peening to create a 60 μm thick ultrafine-grained surface layer, identified as TTS (see Fig. 23.1). Microstructural and mechanical assessments were conducted on cross sections of these samples. EBSD analysis revealed an internal gradient in grain size within the TTS. Nanoindentation measurements exhibited a depth-dependent nanohardness gradient, primarily within the TTS region. This increase in mechanical properties was attributed to grain boundary hardening. This assertion was supported by a Hall − Petch plot (see Fig. 23.5), demonstrating a linear decrease in nanohardness corresponding to the logarithm of the average grain size. The Hall − Petch coefficient measured closely matched those reported by Fu et al. (2001) and by Zhao et al. (2003) for pure iron.

Within the TTS zone, a micropillar approximately 10 μm in diameter was fabricated. Microcompression experiments revealed a yield stress of approximately 700 MPa for the TTS, with minimal strain hardening observed. As the micropillar was positioned midway through the thickness and had sufficiently large dimensions, its yield stress could be compared with the average hardness of the TTS, which is measured around 2000 MPa. Notably, the hardness-to-compression yield stress ratio was approximately 3, consistent with the constraint factor value proposed by Tabor.

This study demonstrates the potential of pure iron for generating significant tribologically transformed surfaces (TTS) layers through intense shot peening. Nanoindentation tests provide a valuable means of quantifying the Hall − Petch strengthening mechanisms within these layers. However, when considering microcompression testing to quantify the stress − strain curve in the TTS zone, it is

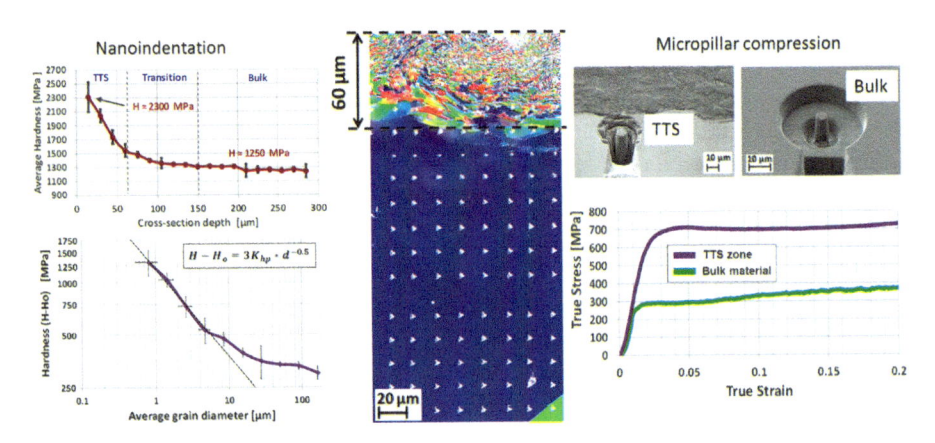

Figure 23.5 TTS resulting of severe shot peening of a pure iron. Left-hand side: nanoindentation hardness gradient reveals the Hall − Petch strengthening mechanisms in the TTS zone (Tumbajoy-Spinel et al., 2016). Middle: EBSD IPF-Z cross-section map reveals the in-depth microstructure gradient in the TTS zone. The white triangles are the signature of indent prints. Right-hand side: micropillar compression reveals the significant increase in mechanical properties in the TTS zone.

crucial to recognize that it provides an average stress − strain curve over a representative elementary volume, roughly equivalent to the pillar's diameter.

To emphasize this point, consider the work of Zhang et al. (2020), who conducted unconstrained high-pressure torsion tests on pure iron samples prepared using a similar method. Through a grain fragmentation process primarily based on continuous and geometric dynamic recrystallization (Sakai et al., 2014), they produced ultrafine grains with an average size of about 200 nm, significantly smaller than the average grain size within the TTS (~ 500 nm). Despite the similar shape of the stress − strain curve, their study yielded a yield stress of about 1.2 GPa.

In conclusion, if the TTS size is sufficiently large and the gradient not excessively steep, it becomes feasible to measure the in-depth gradient of mechanical properties through micropillar compression testing. This approach should enable a more precise quantification of the strengthening mechanisms compared with nanoindentation testing.

For that purpose, it was decided to apply this methodology to very thick TTS generated on a pure iron using a micropercussion device (Tumbajoy-Spinel et al., 2017). Oblique impacts were repeated 10,000 times in the same location, resulting in a TTS zone approximately 100 microns thick. A distinct transition zone was observed beneath the TTS, transitioning to the bulk material, which, at this size, can be considered a single crystal. The transition zone displayed severe deformation close to the TTS and exhibited a high crystal orientation gradient, as evident from the color gradient in the electron back scattering diffraction (EBSD) map. The kernel average misorientation (KAM) map, providing the average misorientation angle for each pixel, was employed to estimate the dislocation density (Kubin &

Mortensen, 2003), although more advanced techniques such as the Nye tensor could be utilized for more refined calculations (Breumier et al., 2021).

Multiple pillars were machined in different zones of interest: the TTS, the transition zone, and the single crystal (bulk). For each pillar, the average dislocation density and average grain size were determined to estimate dislocation strain hardening and grain boundary hardening contributions. Successful compression tests were conducted on each micropillar (see Fig. 23.6), revealing a clear gradient of mechanical properties between the bulk material and the TTS zone. Within the TTS zone, results exhibited considerable variability, reflecting the localized but controlled impacts that created the TTS, contrasting with the continuous TTS layer induced by shot peening (as shown in Fig. 23.5). In micropercussion-induced TTS, there was a microstructure gradient in every direction, as illustrated in Fig. 23.6. It is worth noting that stress − strain curves measured after micropercussion or shot peening were in the same order of magnitude, indicating similar types of microstructural modifications. The notable difference lay in the presence of strain hardening in the

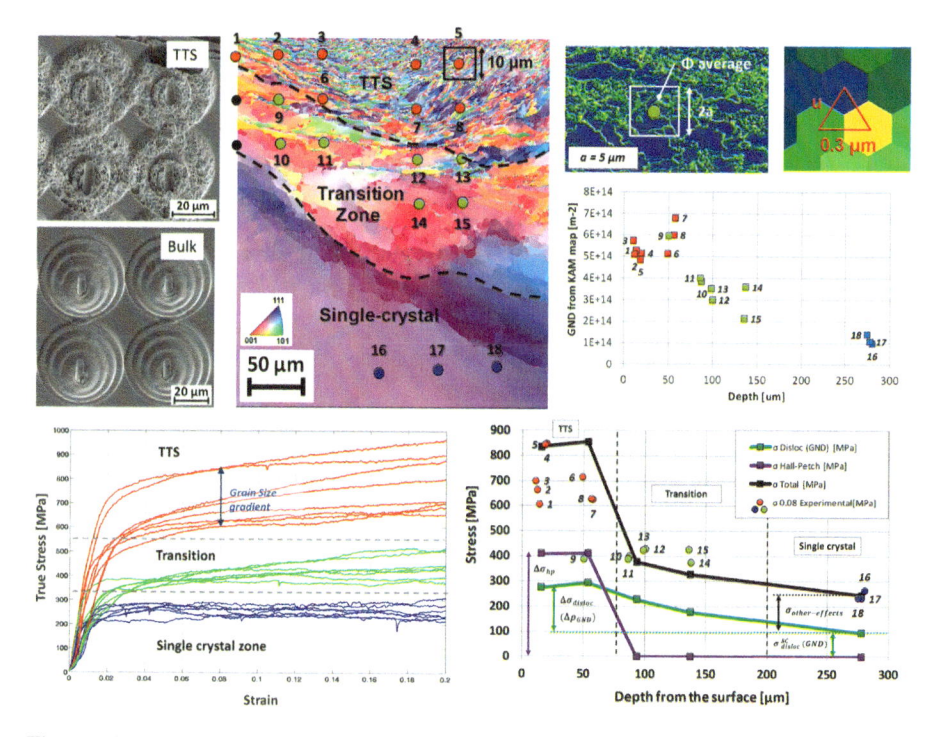

Figure 23.6 Thick TTS generated by oblique micro-percussion. Micropillars were machined in various locations in the TTS zone and its neighborhood (Tumbajoy-Spinel et al., 2018). The resulting stress − strain curves highlighted a significant gradient of mechanical properties along the distance to the surface. For each pillar the average geometric dislocation density was measured from KAM map and then plotted as a function of the depth (distance to the surface). The stress computed for 8% plastic strain is in good agreement with the expected microstructural contributions.

micropercussion-induced TTS, likely due to the microstructural gradient inside the TTS zone.

Comparisons between the measured yield stress and theoretical predictions based on dislocation strain hardening and grain boundary hardening revealed a fairly good agreement. The primary contribution in the TTS zone was grain boundary strengthening, while in the transition zone, dislocation strain hardening played a more significant role. This investigation provided valuable insights, outlining the contributions of these mechanisms in the impact-induced mechanical strengthening of ferritic steels.

23.5 Application to glaze layers

This section focuses on the research conducted Ariane Viat and Alixe Dreano during their PhD thesis, investigating the formation process and mechanical properties of glaze layers. Viat et al., (2016, 2017) studied the behavior of this layer with a HS25/silicate ceramics contact (punch/plate, fretting), while Dreano et al. (2019, 2020) studied the behavior with a HS25/Al_2O_3 contact (cross cylinders, fretting). As mentioned earlier, the glaze layer forms only when a specific threshold temperature is reached, as illustrated in Fig. 23.1, where the glaze layer appears only for temperatures higher than 400°C. It becomes detectable when the wear volume approaches zero. Depending on the materials involved, the friction coefficient can significantly decrease when the glaze layer is formed, leading to improved wear resistance. Typically, the upper surface of the glaze layer appears bright and is small in comparison to wear tracks obtained at lower temperatures, often displaying some scratches inside the track.

Before becoming wear-resistant, the glaze layer must undergo a formation process. Fig. 23.7 illustrates that the formation of the glaze layer is not immediate; a certain number of cycles are required before achieving optimal glaze layer performance. During this initial period, the friction coefficient reaches a maximum value, and the wear volume increases with the number of cycles. The fretting action causes debris to be torn off, which are then compacted and sintered at high temperatures (Kato & Komai, 2007). At a specific threshold cycle (NGL), the friction coefficient decreases and stabilizes at a constant value. Simultaneously, the wear volume remains constant, indicating that the glaze layer is now performing effectively. At this point, almost no wear occurs in the contact. Fig. 23.7 provides an example of a formed glaze layer. Depending on the materials tested and the contact configuration, the glaze layer can have a thickness ranging from 3 to 20 μm. Additionally, it may be homogeneous or composed of different layers. For instance, Fig. 23.7 displays three layers, all oxidized, within the glaze layer. The layer in contact with the bulk is rich in chromium (CRL). The thickest layer, with a nanocrystalline structure approximately 10 nm in size, consists of cobalt and chromium oxides (MOL). Porosities might be present in this layer. Finally, the surface layer, less than a micron thick, possesses a larger grain size (around 50 nm) compared

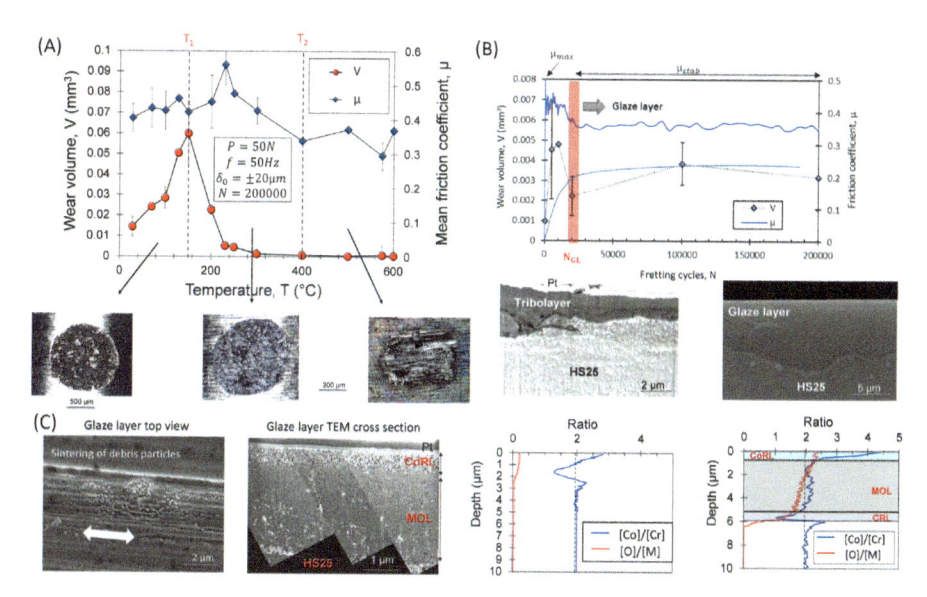

Nanomechanics of tribologically transformed surfaces 659

Figure 23.7 Glaze layer formation mechanisms in a HS25/silicate ceramics contact and relation to wear resistance. (A) Wear volume and friction coefficient versus temperature measured by fretting (Al$_2$O$_3$ vs HS25), and wear tracks observed on HS25 in the three domains. A glaze layer is formed at high temperature (here at temperatures above 400°C), reducing the wear and the friction. (B) Wear volume and friction coefficient measured as a function of fretting cycles at 575°C, along with cross sections obtained before and after glaze layer formation and their Cr/Co/O contents in the cross section. (C) SEM top view and TEM cross section of the glaze layer (Dreano et al., 2020).

with the MOL layer and is primarily composed of cobalt oxides (CoRL). It is presumed that this surface layer acts as a lubricant, as it is not observed during the formation process and is directly in contact with the counterpart.

In the high-temperature micromechanical characterization of glaze layers, a specific example is presented in Fig. 23.8. This particular glaze layer was formed through fretting at 700°C, using a punch/flat contact between HS25/silicate ceramics. The glaze layer has a thickness ranging from 10 to 20 μm and is presumed to be homogeneous in terms of microstructure, exhibiting a grain size along the thickness of 10 nm with more or less amorphous zones around the grains (Viat et al., 2016). To mechanically characterize this layer, micropillar compression tests were performed, allowing the fabrication of 4 μm diameter pillars within this third body. The tests were conducted in the cross section of the glaze layer to ensure that 100% of the micropillars were made of the glaze layer, minimizing the impact of surface irregularities and ensuring a uniform composition.

A critical consideration was the characterization of this glaze layer at elevated temperatures (between 25°C and 500°C) to capture its performance range, as the glaze layer is effective only at temperatures higher than 500°C. Micropillar compression was chosen due to its reduced sensitivity to thermal drift compared with

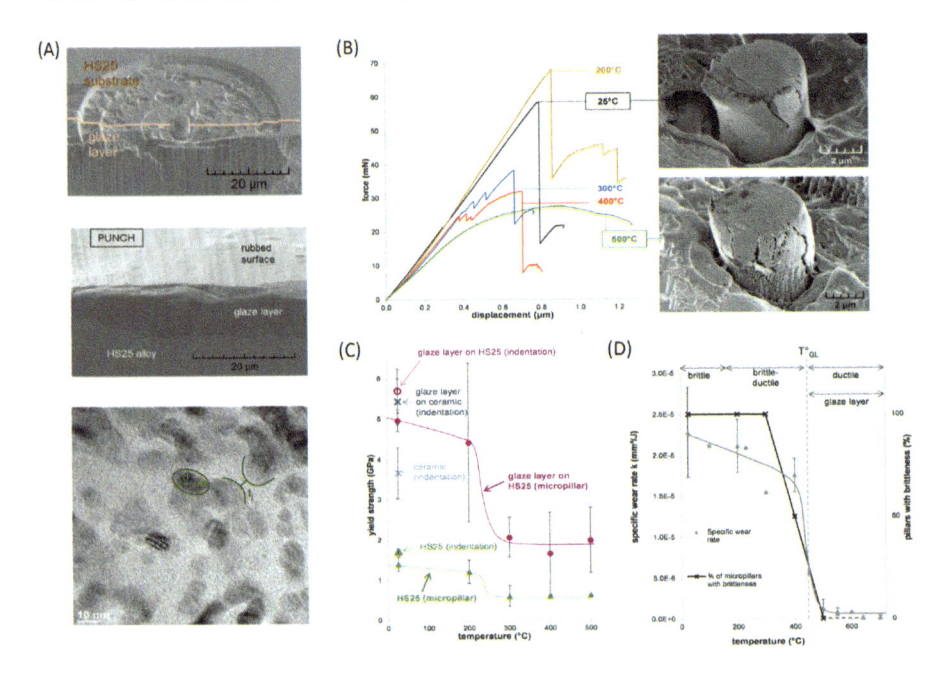

Figure 23.8 High-temperature microcompression testing of a glaze layer and relation to wear resistance. (A) Glaze layer formed by high-temperature fretting with an example of micropillar machined in the glaze layer cross section, and TEM image of the glaze layer showing its nanostructure. (B) Load displacement curves measured by micropillar compression as a function of temperature, with deformed micropillar images obtained after compression at 200°C (brittle behavior) and 500°C (ductile behavior). (C) Yield stress measured on the glaze layer and the HS25 by microcompression as a function of temperature, along with properties measured at room temperature glaze layer, HS25, and the silicate ceramics, by nanoindentation. (D) Wear rate versus temperature compared with the percentage of brittle pillars (Viat et al., 2017).

other techniques, making it suitable for high-temperature testing. Additionally, micropillar compression is less dependent on tip wear than indentation methods. Fig. 23.8 shows the force − displacement curves obtained from micropillar compression tests conducted at different temperatures. The mechanical behavior of the glaze layer exhibited drastic changes with temperature. At low temperatures (below 200°C), the curves indicated an elastic response followed by fracture, indicating a brittle behavior of the glaze layer. At 500°C, the curves exhibited both elasticity and plasticity, revealing a completely ductile behavior. Intermediate behaviors with load drops and plasticity were observed between these extremes (Viat et al., 2017).

The difference in behavior was further illustrated by the deformation patterns of the pillars. Pillars deformed at low temperatures displayed elastic deformation with vertical cracks along their length, whereas those deformed at 500°C exhibited plastic deformation, resulting in a barrel shape with cracks along the edges. To understand the link between mechanical properties and tribological behavior, the yield strengths of the glaze layer and HS25 were compared as a function of temperature, using both microcompression testing and nanoindentation at room temperature (see Fig. 23.8). The glaze layer consistently showed higher yield strength than HS25 as a function of temperature and silicate ceramics at room temperature. However, the glaze layer was not effective at temperatures below 500°C. Interestingly, both HS25 and the glaze layer exhibited a drop in yield strength at temperatures lower than those at which the glaze layer demonstrated its performance. This drop in yield strength coincided with the observed ductility from 300°C (see Fig. 23.8).

To elucidate the tribological behavior, a ductility/brittleness criterion was developed, considering a behavior as brittle when a sudden drop in force was observed on the force − displacement curve. Using this criterion, the percentage of brittle micropillars was determined. This criterion closely followed the wear rate curve as a function of temperature, indicating that the lubricating properties of the glaze layer were primarily attributed to its lack of brittleness rather than its high hardness (see Fig. 23.8). This research highlighted the importance of microcompression testing in the mechanical study of glaze layers, as it provided access to ductility/brittleness properties that are challenging to assess using nanoindentation.

23.6 Thermal stability of tribologically transformed surface

The high energy stored in terms of substructures, dislocation density in tribologically transformed surfaces makes their microstructure prone to change upon a thermal loading. The main mechanisms at stake are restoration mechanisms such as static recovery, static recrystallization, and grain growth (Rollett et al., 2017). All of these mechanisms result in a decrease of the yield stress/hardness with time and temperature. Static recovery consists of the reorganization/annihilation of dislocations and thus lowers the dislocation strain hardening component of the TTS strength. The grain growth mechanism results in the increase of the grain size and thus lowers the grain boundary hardening component. Static recrystallization consists of the nucleation of new grains in zone of high stored energy such as grain boundary. These new grains, of low stored energy, then grow at the extent of the former microstructure. Upon completion, static recrystallization leads to a new microstructure of very low stored energy. Usually, the only remaining component in the TTS strength is the grain boundary strentgthening component, but which can be lowered by subsequent grain growth. A large decrease of TTS hardness can thus be expected. To quantify recrystallization evolution, it is often referred to the definition of a fraction recrystallized X, such as $H = XH_{Rex} + (1 - X)H_{init}$, where H is the TTS hardness after heat treatment, H_{init} is the TTS hardness before heat tretament, and H_{Rex} is the TTS hardness once fully recrystallized. The latter is usually close to the hardness of the bulk once recrystallized. The evolution of the fraction recrystallized is usually modeled through Avrami's evolution law usually referred as JMAK (Johnson − Mel − Avrami − Kolmogorov). It is written as $X = 1 - \exp(-Ct^n)$, where n is the JMAK exponent, t is the time, and C is a temperature-dependent parameter. This equation can also include a nucleation time, which is a temperature-dependent parameter too. One important feature of static recrystallization is that its rate highly depends on the stored energy. The higher the stored energy, the faster the recrystallization kinetics. Consequently, the JMAK parameters should also depend on the stored energy. This is of primary importance regarding TTS since their in-depth microstructure stored energy may vary significantly as shown in Figs. 23.1 and 23.6.

In the study illustrated in Fig. 23.9, the tribologically transformed surface (TTS) induced by sliding friction on a copper (Cu) bar was subjected to isothermal heat treatments at two different temperatures: 300°C and 450°C, with the latter chosen to achieve a fully recrystallized state. The resulting in-depth nanohardness gradient was measured using conventional nanoindentation testing (Kermouche, Jacquet, et al., 2016). The results revealed interesting insights into the recrystallization process and its effects on the material's mechanical properties.

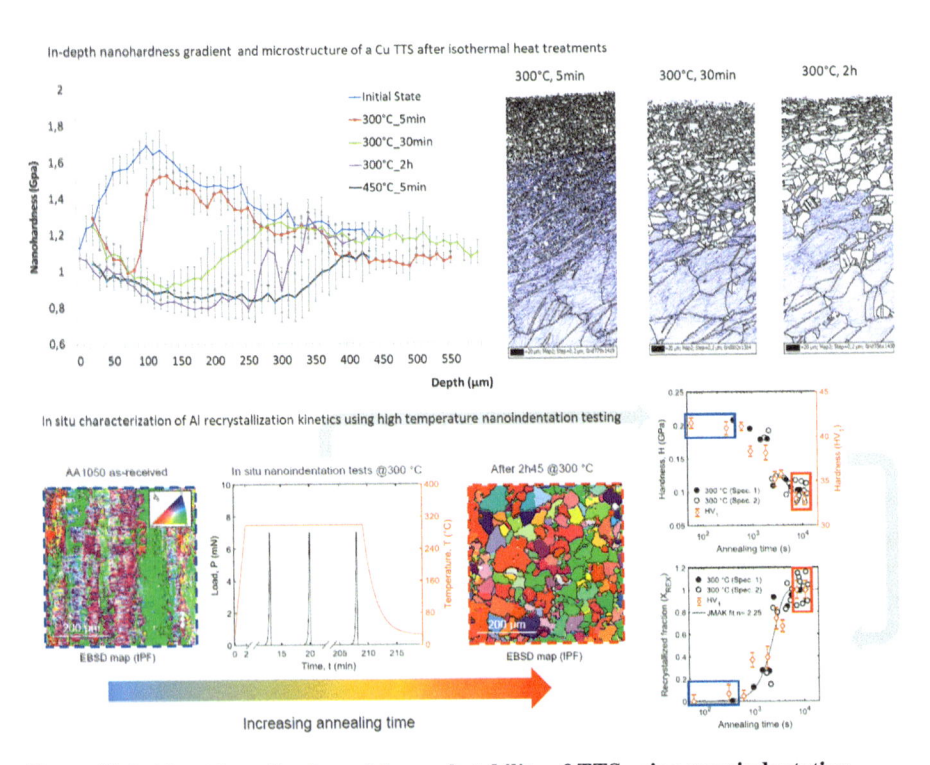

Figure 23.9 About investigation of thermal stability of TTS using nanoindentation testing. Top: resulting nanohardness gradient after a heat treatment applied on a Cu TTS (Kermouche, Jacquet, et al., 2016). Bottom: proof of concept of high temperature nanoindentation testing to quantify in situ recrystallization kinetics of cold-rolled Al (Baral et al., 2018).

The nanohardness in the extreme surface did not vary significantly with annealing time or temperature, indicating that this region had already undergone recrystallization during the sliding friction process (as observed in Fig. 23.9). The effect of grain boundary softening through grain growth seemed to be negligible. Upon heat treatment, the nanohardness decreased with increasing depth until reaching a certain value before increasing again to approach the initial nanohardness gradient. This phenomenon was observed for various heat treatment parameters, with the depth at which nanohardness started increasing being influenced by annealing temperature and time. This behavior indicated an ongoing recrystallization process, forming a recrystallization wave that spread through the bulk of the material. The in-depth variation in recrystallization kinetics was attributed to the gradient in stored energy within the material, suggesting that the microstructure's stored energy influenced the recrystallization process.

Notably, even for the highest temperature or longest annealing time, the apparent bulk nanohardness remained higher than the fully recrystallized hardness. This indicated that the material in this zone had undergone deformation during the sliding friction process but did not possess enough stored energy for static recrystallization to occur. EBSD maps revealed a slight decrease in stored energy in this zone, indicating a slower static recovery process compared with the recrystallization process closer to the surface. The variation in final recrystallized microstructure along the depth led to differences in nanohardness within the recrystallized zone, highlighting the depth-dependent nucleation rate and recrystallization kinetics influenced by stored energy.

This study demonstrated the relations between microstructural stored energy, recrystallization kinetics, and resulting mechanical properties. It shed light on the complex behavior of materials undergoing tribological transformations and subsequent heat treatments.

The recent advancements in high-temperature nanoindentation testing have opened up new avenues for studying thermal stability of tribologically transformed surfaces (TTS) and other materials with complex microstructures. As shown earlier, traditional methods for investigating restoration mechanisms, involving heat treatments followed by postmortem mechanical properties characterization, are very efficient, but they are time-consuming and costly.

High-temperature nanoindentation setups now incorporate heating stages that allow for in-situ annealing treatments within the nanoindenter chamber. This advancement can be advantageous for studying materials undergoing thermally activated transitions, such as recrystallization or even phase transformations. Baral et al. (2018) demonstrated the feasibility of in-situ measurements of hot nanohardness versus time using high-temperature nanoindentation on a cold-rolled pure aluminum specimen (see Fig. 23.9), providing results consistent with macroscopic methods. However, challenges related to thermal stability, particularly the critical stabilization time, were observed. A critical stabilization time was necessary to minimize thermal drift related to the contact zone's isothermal stabilization. This limitation constrained the exploration of recrystallization kinetics with durations longer than the critical time.

To address these challenges, the concept of high temperature scanning indentation (HTSI) was introduced by Tiphéne et al. (2021). HTSI combines the capabilities of high-speed nanoindentation and high-temperature nanoindentation to overcome thermal drift issues. By running experiments at a high speed, thermal drift becomes negligible, allowing for rapid and accurate measurements. The HTSI method, with an indentation cycle duration of less than 1 second, enables measurements during the heating and cooling stages. This short cycle time facilitates high-data acquisition along a temperature ramp, aiding in the quantification of thermally activated mechanisms in materials.

Tiphéne et al. (2023) recently proposed an inverse methodology based on HTSI measurements, allowing the quantification of restoration kinetics in materials such as cold-rolled copper and cold-rolled aluminum. This approach discriminates between static recovery and static recrystallization and highlights the competition between these mechanisms concerning the final microstructure. The HTSI technique has also been successfully applied to investigate the thermal stability of metallic glasses (Comby-Dassonneville et al., 2021), revealing intriguing features such as the activation of a superplastic flow mechanism just before the crystallization stage.

Applying HTSI to the study of thermal stability in TTS and tribofilms presents promising opportunities regarding the wear resistance at high temperature. The ability to perform rapid and precise measurements during temperature variations offers valuable insights into the behavior of these complex materials, leading to a deeper understanding of their mechanical properties and restoration mechanisms.

23.7 Upcoming developments

Certainly, the developments in characterizing the mechanical properties of tribologically transformed structures (TTSs) and tribofilms have made significant progresses in recent years. While techniques such as nanoindentation and microcompression testing have been invaluable in understanding the mechanical behavior of these structures under compression loadings and deriving stress − strain curves at different temperatures and strain rates, there are still aspects of TTS mechanical properties that remain unexplored. Direct measurements of TTS shear flow, toughness, and fatigue resistance are currently beyond the reach of these techniques. However, ongoing research and advancements in nanomechanical testing are likely to pave the way for new developments in this field. Here are a few potential areas of progress in the characterization of TTS mechanical properties in the coming years:

23.7.1 The microshear compression test

The development of microshear tests, specifically the micro-shear compression test introduced by Guillonneau et al. (2022) derived from the macroscopic "shear compression test" designed by Dorogoy and Rittel (2005), may represent a significant advancement in measuring the shear flow resistance of tribologically transformed structures (TTSs) and tribofilms. These tests are essential because TTS and tribofilms often act as protective barriers against wear or accommodate relative displacement between sliding bodies. The microshear compression test offers several advantages in this regard.

Indeed, by compressing a micropillar-shaped specimen with notches inclined at 45 degrees, the microshear compression test applies a shear compression stress state to the material (see Fig. 23.10). This loading configuration enables the measurement of shear flow resistance without introducing bending torques, providing a more accurate representation of shear behavior in TTS. Finite element calculations validate the application of shear stress and demonstrate the localization of shear flow inside the gage. Comparisons between experimental results and FEA confirm the reliability of the test for the elastic part of the material's behavior up to the yield point on fused silica.

Unlike previous microshear tests that involved the machining of two symmetrical gages linked to a cube (Heyer et al., 2014), the microshear compression test focuses on a single sheared zone. This design avoids complications related to machining accuracy and loading, making it more suitable for TTS testing.

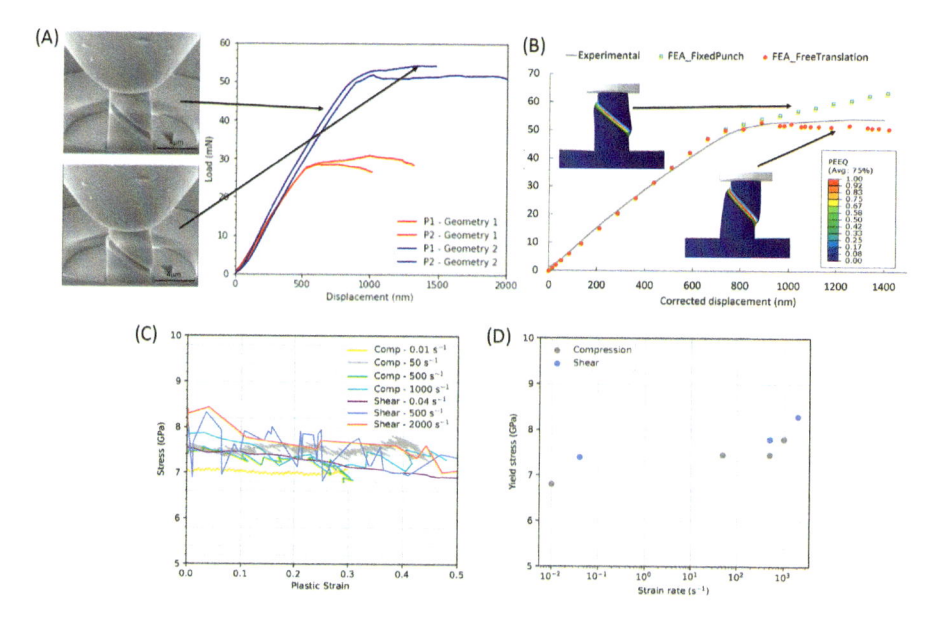

Figure 23.10 The microshear-compression test. (A) Load displacement curves measured by microshear, along with SEM images showing the deformation behavior in the elastic and plastic domains. (B) Comparison between experimental an numerical load-displacement curves, obtained by microshear. The numerical data letting the punch free laterally correlates better the experimental data. (C) Stress-strain curves calculated in the plastic domain, by microcompression and microshear, at low and high strain rates. Higher strain rates are reachable using microshear. (D) Yield stress measured by microcompresion and microshear as a function of the strain rate. Slow strain rate dependence is observed (Guillonneau et al., 2022). All experimental tests were performed on fused silica.

The microshear compression test allows for the achievement of very high strain rates by manipulating the gage geometry. The localized deformation within the gage results in geometrically increased strain rates. An example is shown in Fig. 23.10. As shown here, the maximum strain rate achieved by microcompression was 1000 s^{-1}, whereas by microshear, the maximum strain rate achieved was 2000 s^{-1}. Strain rates of 10,000 s^{-1} are clearly achievable using the microshear compression test. This capability is of high interest for studying TTS under high strain rate conditions, which may happen when submitted to high-speed tribological contacts, especially in the framework of manufacturing process as described earlier. Fused silica, used in these tests, demonstrates low dependence of yield strength on strain rate as observed by Widmer et al. (2022). Let us note that this characteristic makes fused silica here again an excellent candidate for calibration purposes in nanomechanical testing at high strain rates.

Ongoing efforts are directed toward improving the reliability of microshear tests, particularly in the plastic deformation region. Addressing issues related to lateral stiffness and friction between the punch and specimen will enhance the accuracy of experimental results, making the microshear compression test an even more robust technique.

In summary, the microshear compression test offers a promising avenue for accurately measuring shear flow resistance in TTS and tribofilms. Its innovative design, validated through finite element analysis and demonstrated high strain rate measurement capabilities, positions it as a valuable tool for research in the field of materials science and tribology.

23.7.2 Microtensile testing

At the millimeter scale, tensile tests are often preferred to compression experiments for their ability to measure ductility, strain to failure, and ultimate tensile strength (UTS). However, conducting microtensile tests at the micron scale presents significant challenges. One major challenge is the time-consuming process of machining microtensile samples using focused ion beam (FIB) milling. Additionally, specialized gripper designs (Della Ventura et al., 2021) and precise positioning procedures are essential for applying tensile loads to these small specimens. While MEMS-based push-to-pull devices have been developed to simplify the testing process (Idrissi et al., 2016), issues related to ensuring proper uniaxial tensile loading still need to be addressed.

In the case of tribologically transformed structure (TTS) zones, these challenges are further compounded due to their inherent microstructural gradients and small size. To date, microtensile testing of TTS has not been widely explored, likely due to these complexities. However, microtensile testing offers a unique advantage over compression and nanoindentation loadings—it allows for full-field measurements during straining. Techniques such as high-resolution digital image correlation (in a scanning electron microscope) and electron back scattering diffraction can be employed to measure strain fields (Arnaud et al., 2021) and track microstructural evolution (Della Ventura et al., 2021), providing valuable insights into the mechanical properties of TTS.

Despite the challenges, the ongoing progress in FIB machining and nanomechanical testing techniques is expected to facilitate microtensile testing of TTS in the coming years. As these advancements continue, it is likely that microtensile testing will become a valuable tool for researchers studying the mechanical behavior of TTS, offering a deeper understanding of their properties and behavior under tensile loading conditions.

23.7.3 Fracture toughness at the micron scale

Evaluating the toughness of tribologically transformed structures (TTS) is crucial given their small grain size, the presence of nanocracks, and potential chemical contamination, which can influence the nucleation and propagation of surface cracks. To address this question, micromechanical tests are essential to quantify the material's toughness at the TTS scale. While microtensile testing is a possibility, it has limitations, including the challenges related to sample preparation and the small size of TTS zones. However, several methodologies developed for evaluating the toughness of thin films could be adapted for TTS characterization, taking advantage of the precision of focused ion beam (FIB) milling to create accurate microgeometries.

Among the various tests used for thin films, the microcantilever bending test (Ast et al., 2019) and the double-cantilever beam test (Sebastiani et al., 2015) are promising candidates for TTS characterization. These tests are embedded in linear elastic fracture mechanics (LEFM) and elastic − plastic fracture mechanics (EPFM) frameworks, enabling reliable posttreatment of data to evaluate fracture toughness. High-resolution electron back scatter diffraction (HR-EBSD) or high-resolution digital image correlation (HR-DIC) techniques can also aid in identifying the plastic zone near the notch edge. While these tests are widely used in nanomechanics, it is essential to address geometrical imperfections, such as ensuring a sharp initial notch and homogeneity in notch length across the specimen width, for accurate results.

The micropillar splitting test is another micromechanical test able to quantify materials toughness at a very small scale. It is based on the measurement of the critical load required to nucleate and propagate cracks in a micropillar using pyramidal nanoindentation. This test is suitable for quasibrittle materials, where crack propagation occurs rather than plastic flow under the loading conditions. One advantage of this test is its insensitivity to FIB milling since no notch needs to be machined. However, quantifying fracture toughness in this test relies on a conversion parameter, which is material- and tip-geometry-dependent and must be known or calculated through finite element simulations (Sebastiani et al., 2015).

23.7.4 Fatigue testing at the micron scale

The role of tribologically transformed structures (TTSs) in the fatigue resistance of engineering parts is an intriguing question and understanding their impact on fatigue crack nucleation and propagation are essential. While it is challenging to predict whether TTS will have a positive or detrimental effect on fatigue crack nucleation during cyclic loading, investigating the fatigue resistance of TTS through micromechanical tests is of primary importance.

Micromechanical fatigue testing at the micron scale is a relatively recent field, experiencing significant growth due to advancements in nanoindentation setups capable of high-frequency loading. Researchers have been focusing on solving instrumentation and methodology issues, primarily using model materials such as ultrafine-grained metals or single crystals. For instance, Merle and Höppel (2018) used the continuous stiffness measurement mode to conduct high cycle fatigue test on micropillars. Gabel et al., (2020) extended the methodology to microcantilevers specimens to simulate tensile cyclic loading. Lavenstein et al. (2018) glued the tip to the cantilever to simulate fully reversed high cycle fatigue. The same idea was used by Huang et al. (2020) on microtensile experiments where the gripper was glued to the specimen head. MEMS microresonator has also been successfully applied to the quantification of very high cycle fatigue tests of microcantilevers as shown by Barrios et al. (2018). This field is growing significantly in terms of experimental maturity, and it is very likely that these developments would be applied to TTS in a very near future.

For instance, given the expected nanocrystalline nature of TTS, the paper of Gabel et al. (2020) on microfatigue of ultrafine-grained copper is very instructive. They have shown that microstructural coarsening occurs on the top and bottom regions of the microcantilevers that result in a cyclic softening up to the point that some grains grew in size by an order of magnitude. Grain coarsening is then replaced by slip band formations and surface extrusions, which grow to small cracks. The ultrafine grained microstructure delayed the formation of fatigue crack compared with a classical coarse grain sample. One might therefore expect that TTS could delay the nucleation of surface fatigue cracks, similar to the observed delay in ultrafine-grained materials. However, this hypothesis remains to be tested and validated through systematic micromechanical fatigue testing on TTS samples.

As the field of micromechanical fatigue testing continues to mature, applying these techniques to investigate the fatigue behavior of TTS is a promising avenue for future research. By conducting such experiments, researchers can gain valuable insights into how TTS microstructure influences fatigue resistance, contributing to a better understanding of their mechanical properties under cyclic loading conditions.

23.8 Conclusion

In summary, the recent advancements in nanomechanical testing techniques, particularly nanoindentation and microcompression testing, have significantly contributed to our understanding of the mechanical properties of tribologically transformed surfaces (TTSs). These methods, when coupled with microstructural characterization, have enabled the discrimination of various strengthening mechanisms within TTS, including dislocation strain hardening and grain boundary hardening. Additionally, when applied postmortem after annealing treatments, nanomechanical tests have provided valuable insights into the thermally activated restoration mechanisms at play in TTS after annealing treatments. The recent development of the high-temperature scanning indentation (HTSI) by Tiphéne et al. (2021) should also help to speed up such investigations. However, challenges still exist, especially in quantifying yield stress and thermal stability of TTS, which depend on their size and microstructure.

Future developments are expected to focus on developing micromechanical tests that can extract other components of TTS mechanical behavior. One crucial aspect is understanding the shear resistance of TTS, considering their role in sustaining sliding friction. The recently developed microshear compression test by Guillonneau et al. (2022) shows promise in this regard. Additionally, investigating the fracture toughness and ultimate tensile strength of TTS, particularly in relation to their role as wear debris sources, remains a challenge that requires further exploration. Full-field measurements, combined with techniques such as micro-cantilever bending, could provide valuable insights into the failure mechanisms of TTS.

Furthermore, the progress in small-scale fatigue tests offers an opportunity to unravel the impact of TTS on the fatigue resistance of structural materials, especially those generated by manufacturing processes such as severe shot peening or machining operations. Understanding the role of TTS in fatigue resistance is essential for ensuring the durability and reliability of engineered structural components.

In conclusion, the application of nanomechanical testing to tribologically transformed surfaces represents a rapidly growing research field with significant potential. Continued advancements in this area are expected to provide deeper insights into the mechanical behavior of TTS, leading to improved materials design and enhanced performance of engineered parts.

Acknowledgments

This work was supported by the RATES project (ANR-20-CE08−0022) and by the LABEX MANUTECH- SISE (ANR-10-LABX-0075) operated by French National Research Agency (ANR). The authors are very grateful to D. Tumbajoy-Spinel, S. Breumier, P. Baral, G. Tiphène, A. Viat, A. Dréano,.L. Loubet, S. Fouvry. S. Descartes. S. Sao-Joao, and S. Kalacska for their contributions to the results presented in this chapter. The authors would like to thank ChatGPT 3.5 for its assistance in language corrections.

References

Arnaud, P., Heripre, E., Douit, F., Aubin, V., Fouvry, S., Guiheux, R., Branger, V., & Michel, G. (2021). Micromechanical tensile test investigation to identify elastic and toughness properties of thin nitride compound layers. *Surface and Coatings Technology*, *421*, 127303. Available from https://doi.org/10.1016/j.surfcoat.2021.127303.

Ast, J., Ghidelli, M., Durst, K., Göken, M., Sebastiani, M., & Korsunsky, A. M. (2019). A review of experimental approaches to fracture toughness evaluation at the micro-scale. *Materials & Design, 173*, 107762. Available from https://doi.org/10.1016/j.matdes.2019.107762.

Baral, P., Laurent-Brocq, M., Guillonneau, G., Bergheau, J. M., Loubet, J. L., & Kermouche, G. (2018). In situ characterization of AA1050 recrystallization kinetics using high temperature nanoindentation testing. *Materials and Design, 152*, 22−29. Available from https://doi.org/10.1016/j.matdes.2018.04.053.

Barrios, A., Gupta, S., Castelluccio, G. M., & Pierron, O. N. (2018). Quantitative in situ SEM high cycle fatigue: The critical role of oxygen on nanoscale-void-controlled nucleation and propagation of small cracks in Ni microbeams. *Nano Letters, 18*(4), 2595−2602. Available from https://doi.org/10.1021/acs.nanolett.8b00343, http://pubs.acs.org/journal/nalefd.

Barthel, E., Keryvin, V., Rosales-Sosa, G., & Kermouche, G. (2020). Indentation cracking in silicate glasses is directed by shear flow, not by densification. *Acta Materialia, 194*, 473−481. Available from https://doi.org/10.1016/j.actamat.2020.05.011, http://www.journals.elsevier.com/acta-materialia/.

Breumier, S., Adamski, F., Badreddine, J., Lévesque, M., & Kermouche, G. (2021). Microstructural and mechanical characterization of a shot peening induced rolled edge on direct aged Inconel 718 alloy. *Materials Science and Engineering: A, 816*, 141318. Available from https://doi.org/10.1016/j.msea.2021.141318.

Brown, M., Wright, D., M'Saoubi, R., McGourlay, J., Wallis, M., Mantle, A., Crawforth, P., & Ghadbeigi, H. (2018). Destructive and non-destructive testing methods for characterization and detection of machining-induced white layer: A review paper. *CIRP Journal of Manufacturing Science and Technology, 23*, 39−53. Available from https://doi.org/10.1016/j.cirpj.2018.10.001, http://www.elsevier.com/wps/find/journaldescription.cws_home/714185/description#description.

Chromik, R. R., & Zhang, Y. (2018). Nanomechanical testing of third bodies. *Current Opinion in Solid State and Materials Science, 22*(4), 142−155. Available from https://doi.org/10.1016/j.cossms.2018.05.001.

Comby-Dassonneville, S., Tiphéne, G., Borroto, A., Guillonneau, G., Roiban, L., Kermouche, G., Pierson, J. F., Loubet, J. L., & Steyer, P. (2021). Real-time high-temperature scanning indentation: Probing physical changes in thin-film metallic glasses. *Applied Materials Today, 24*. Available from https://doi.org/10.1016/j.apmt.2021.101126, http://www.journals.elsevier.com/applied-materials-today/.

Della Ventura, N. M., Kalácska, S., Casari, D., Edwards, T. E. J., Sharma, A., Michler, J., Logé, R., & Maeder, X. (2021). {10$\bar{1}$2} twinning mechanism during in situ microtensile loading of pure Mg: Role of basal slip and twin-twin interactions. *Materials and Design, 197*. Available from https://doi.org/10.1016/j.matdes.2020.109206, https://www.journals.elsevier.com/materials-and-design.

Dorogoy, A., & Rittel, D. (2005). Numerical validation of the shear compression specimen. Part I: Quasi-static large strain testing. *Experimental Mechanics, 45*(2), 167−177. Available from https://doi.org/10.1007/bf02428190.

Dreano, A., Fouvry, S., & Guillonneau, G. (2020). Understanding and formalization of the fretting-wear behavior of a cobalt-based alloy at high temperature. *Wear, 452−453*, 203297. Available from https://doi.org/10.1016/j.wear.2020.203297.

Dreano, A., Fouvry, S., Sao-Joao, S., Galipaud, J., & Guillonneau, G. (2019). The formation of a cobalt-based glaze layer at high temperature: A layered structure. *Wear, 440−441*, 203101. Available from https://doi.org/10.1016/j.wear.2019.203101.

Dumas, M., Kermouche, G., Valiorgue, F., Van Robaeys, A., Lefebvre, F., Brosse, A., Karaouni, H., & Rech, J. (2021). Turning-induced surface integrity for a fillet radius in a 316L austenitic stainless steel. *Journal of Manufacturing Processes, 68*, 222−230. Available from https://doi.org/10.1016/j.jmapro.2021.05.031, http://www.elsevier.com/wps/find/journaldescription.cws_home/620379/description#description.

Fu, H. H., Benson, D. J., & Meyers, M. A. (2001). Analytical and computational description of effect of grain size on yield stress of metals. *Acta Materialia, 49*(13), 2567−2582. Available from https://doi.org/10.1016/S1359-6454(01)00062-3.

Gabel, S., Merle, B., & Merle, B. (2020). Small-scale high-cycle fatigue testing by dynamic microcantilever bending. *MRS Communications, 10*(2), 332−337. Available from https://doi.org/10.1557/mrc.2020.31, http://journals.cambridge.org/action/displayJournal?jid = MRC.

Guillonneau, G., Sao Joao, S., Adogou, B., Breumier, S., & Kermouche, G. (2022). Plastic flow under shear-compression at the micron scale-application on amorphous silica at high strain rate. *JOM, 74*(6), 2231−2237. Available from https://doi.org/10.1007/s11837-021-05142-7, http://www.springer.com/materials/journal/11837.

Heyer, J. K., Brinckmann, S., Pfetzing-Micklich, J., & Eggeler, G. (2014). Microshear deformation of gold single crystals. *Acta Materialia, 62*(1), 225−238. Available from https://doi.org/10.1016/j.actamat.2013.10.002, http://www.journals.elsevier.com/acta-materialia/.

Huang, K., Sumigawa, T., & Kitamura, T. (2020). Load-dependency of damage process in tension-compression fatigue of microscale single-crystal copper. *International Journal of Fatigue, 133*, 105415. Available from https://doi.org/10.1016/j.ijfatigue.2019.105415.

Idrissi, H., Bollinger, C., Boioli, F., Schryvers, D., & Cordier, P. (2016). Low-temperature plasticity of olivine revisited with in situ TEM nanomechanical testing. *Science Advances, 2*(3). Available from https://doi.org/10.1126/sciadv.1501671, http://advances.sciencemag.org/content/advances/2/3/e1501671.full.pdf.

Jacquet, G., Kermouche, G., Courbon, C., Tumbajoy, D., & Rech, J. (2014). Effect of sliding velocity on friction-induced microstructural evolution in Copper. *IOP Conference Series: Materials Science and Engineering, 63*(1), 012039. Available from https://doi.org/10.1088/1757-899x/63/1/012039.

Jiang, J., Stott, F. H., & Stack, M. M. (2004). A generic model for dry sliding wear of metals at elevated temperatures. *Wear, 256*(9−10), 973−985. Available from https://doi.org/10.1016/j.wear.2003.09.005.

Kaiser, A. L., Bec, S., Vernot, J. P., & Langlade, C. (2006). Wear damage resulting from sliding impact kinematics in pressurized high temperature water: Energetical and statistical approaches. *Journal of Physics D: Applied Physics, 39*(15), 3193−3199. Available from https://doi.org/10.1088/0022-3727/39/15/S09.

Kato, H., & Komai, K. (2007). Tribofilm formation and mild wear by tribo-sintering of nanometer-sized oxide particles on rubbing steel surfaces. *Wear, 262*(1−2), 36−41. Available from https://doi.org/10.1016/j.wear.2006.03.046.

Kermouche, G., Loubet, J. L., & Bergheau, J. M. (2006). A new index to estimate the strain rate sensitivity of glassy polymers using conical/pyramidal indentation. *Philosophical Magazine, 86*(33−35), 5667−5677. Available from https://doi.org/10.1080/14786430600778682.

Kermouche, G., Guillonneau, G., Michler, J., Teisseire, J., & Barthel, E. (2016). Perfectly plastic flow in silica glass. *Acta Materialia, 114*, 146−153. Available from https://doi.org/10.1016/j.actamat.2016.05.027, http://www.journals.elsevier.com/acta-materialia/.

Kermouche, G., Jacquet, G., Courbon, C., Rech, J., Zhang, Y. Y., & Chromik, R. (2016). Microstructure evolution induced by sliding-based surface thermomechanical treatments

- Application to pure copper. *Materials Science Forum*, *879*, 915−920. Available from https://doi.org/10.4028/http://www.scientific.net/msf.879.915.
Kermouche, G., Kaiser, A. L., Gilles, P., & Bergheau, J. M. (2007). Combined numerical and experimental approach of the impact-sliding wear of a stainless steel in a nuclear reactor. *Wear*, *263*(7−12), 1551−1555. Available from https://doi.org/10.1016/j.wear.2007.02.015.
Kermouche, G., Pacquaut, G., Langlade, C., & Bergheau, J. M. (2011). Investigation of mechanically attrited structures induced by repeated impacts on an AISI1045 steel. *France Comptes Rendus − Mecanique*, *339*(7−8), 552−562. Available from https://doi.org/10.1016/j.crme.2011.05.012, http://www.elsevier.com/journals/comptes-rendus-mecanique/1631-0721.
Kubin, L. P., & Mortensen, A. (2003). Geometrically necessary dislocations and strain-gradient plasticity: A few critical issues. *Scripta Materialia*, *48*(2), 119−125. Available from https://doi.org/10.1016/S1359-6462(02)00335-4.
La Monaca, A., Murray, J. W., Liao, Z., Speidel, A., Robles-Linares, J. A., Axinte, D. A., Hardy, M. C., & Clare, A. T. (2021). Surface integrity in metal machining − Part II: Functional performance. *International Journal of Machine Tools and Manufacture*, *164*. Available from https://doi.org/10.1016/j.ijmachtools.2021.103718, http://www.journals.elsevier.com/international-journal-of-machine-tools-and-manufacture/.
Lacroix, R., Chomienne, V., Kermouche, G., Teisseire, J., Barthel, E., & Queste, S. (2012). Micropillar testing of amorphous silica. *International Journal of Applied Glass Science*, *3*(1), 36−43. Available from https://doi.org/10.1111/j.2041-1294.2011.00075.x.
Lavenstein, S., Crawford, B., Sim, G. D., Shade, P. A., Woodward, C., & El-Awady, J. A. (2018). High frequency in situ fatigue response of Ni-base superalloy René-N5 microcrystals. *Acta Materialia*, *144*, 154−163. Available from https://doi.org/10.1016/j.actamat.2017.10.049, http://www.journals.elsevier.com/acta-materialia/.
Lefranc, V., Baydoun, S., Gandiollé, C., Héripré, E., Vallet, M., Fouvry, S., & Aubin, V. (2023). Heterogeneity in tribologically transformed structure (TTS) of Ti−6Al−4V under fretting. *Wear*, *522*, 204680. Available from https://doi.org/10.1016/j.wear.2023.204680.
Li, W. L., Tao, N. R., & Lu, K. (2008). Fabrication of a gradient nano-micro-structured surface layer on bulk copper by means of a surface mechanical grinding treatment. *Scripta Materialia*, *59*(5), 546−549. Available from https://doi.org/10.1016/j.scriptamat.2008.05.003.
Merle, B., & Höppel, H. W. (2018). Microscale high-cycle fatigue testing by dynamic micropillar compression using continuous stiffness measurement. *Experimental Mechanics*, *58*(3), 465−474. Available from https://doi.org/10.1007/s11340-017-0362-3, http://www.springer.com/dal/home/generic/search/results?SGWID = 1-40109-70-36417762-0.
Mondelin, A., Valiorgue, F., Rech, J., Coret, M., & Feulvarch, E. (2013). Modeling of surface dynamic recrystallisation during the finish turning of the 15-5PH steel. *Procedia CIRP*, *8*, 311−315. Available from https://doi.org/10.1016/j.procir.2013.06.108, http://www.sciencedirect.com/science/journal/22128271.
Oliver, W. C., & Pharr, G. M. (1992). An improved technique for determining hardness and elastic modulus using load and displacement sensing indentation experiments. *Journal of Materials Research*, *7*(6), 1564−1583. Available from https://doi.org/10.1557/jmr.1992.1564.
Rigney, D. A. (2000). Transfer, mixing and associated chemical and mechanical processes during the sliding of ductile materials. *Wear*, *245*(1-2), 1−9. Available from https://doi.org/10.1016/S0043-1648(00)00460-9.

Roland, T., Retraint, D., Lu, K., & Lu, J. (2006). Fatigue life improvement through surface nanostructuring of stainless steel by means of surface mechanical attrition treatment. *Scripta Materialia*, *54*(11), 1949−1954. Available from https://doi.org/10.1016/j.scriptamat.2006.01.049.

Rollett A., Rohrer G.S., Humphreys J., (2017). Recrystallization and related annealing phenomena. Elsevier, 1−704, https://www.elsevier.com/books/recrystallization-and-related-annealing-phenomena/rollett/978-0-08-098235-9. doi:10.1016/j.matchar.2020.110382.

Sakai, T., Belyakov, A., Kaibyshev, R., Miura, H., & Jonas, J. J. (2014). Dynamic and post-dynamic recrystallization under hot, cold and severe plastic deformation conditions. *Progress in Materials Science*, *60*(1), 130−207. Available from https://doi.org/10.1016/j.pmatsci.2013.09.002.

Sauger, E., Fouvry, S., Ponsonnet, L., Kapsa, P., Martin, J. M., & Vincent, L. (2000). Tribologically transformed structure in fretting. *Wear*, *245*(1-2), 39−52. Available from https://doi.org/10.1016/S0043-1648(00)00464-6.

Sebastiani, M., Johanns, K. E., Herbert, E. G., & Pharr, G. M. (2015). Measurement of fracture toughness by nanoindentation methods: Recent advances and future challenges. *Current Opinion in Solid State and Materials Science*, *19*(6), 324−333. Available from https://doi.org/10.1016/j.cossms.2015.04.003.

Sekkal, A. C., Langlade, C., & Vannes, A. B. (2005). Tribologically transformed structure of titanium alloy (TiAl6V4) in surface fatigue induced by repeated impacts. *Materials Science and Engineering: A*, *393*(1-2), 140−146. Available from https://doi.org/10.1016/j.msea.2004.10.008, http://www.elsevier.com.

Stott, F. H., Lin, D. S., Wood, G. C., & Stevenson, C. W. (1976). The tribological behaviour of nickel and nickel-chromium alloys at temperatures from 20° to 800°C. *Wear*, *36*(2), 147−174. Available from https://doi.org/10.1016/0043-1648(76)90002-8.

Thiercelin, L., Saint-Aimé, L., Lebon, F., & Saulot, A. (2020). Thermomechanical modelling of the tribological surface transformations in the railroad network (white etching layer). *Mechanics of Materials*, *151*, 103636. Available from https://doi.org/10.1016/j.mechmat.2020.103636.

Tiphéne, G., Baral, P., Comby-Dassonneville, S., Guillonneau, G., Kermouche, G., Bergheau, J. M., Oliver, W., & Loubet, J. L. (2021). High-temperature scanning indentation: A new method to investigate in situ metallurgical evolution along temperature ramps. *Journal of Materials Research*, *36*(12), 2383−2396. Available from https://doi.org/10.1557/s43578-021-00107-7, https://www.springer.com/journal/43578.

Tiphéne, G., Kermouche, G., Baral, P., Maurice, C., Guillonneau, G., Bergheau, J. M., Oliver, W. C., & Loubet, J. L. (2023). Quantification of softening kinetics in cold-rolled pure aluminum and copper using high-temperature scanning indentation. *Materials and Design*, *233*. Available from https://doi.org/10.1016/j.matdes.2023.112171, https://www.journals.elsevier.com/materials-and-design.

Tumbajoy-Spinel, D., Descartes, S., Bergheau, J.-M., Al-Baida, H., CLanglade, C., & Kermouche, G. (2017). Investigation of graded strengthened hyper-deformed surfaces by impact treatment: Micro-percussion testing. *IOP Conference Series: Materials Science and Engineering*, *194*(1), 012024. Available from https://doi.org/10.1088/1757-899x/194/1/012024.

Tumbajoy-Spinel, D., Descartes, S., Bergheau, J. M., Lacaille, V., Guillonneau, G., Michler, J., & Kermouche, G. (2016). Assessment of mechanical property gradients after impact-based surface treatment: Application to pure α-iron. *Materials Science and Engineering: A*, *667*, 189−198. Available from https://doi.org/10.1016/j.msea.2016.04.059, http://www.elsevier.com.

Tumbajoy-Spinel, D., Maeder, X., Guillonneau, G., Sao-Joao, S., Descartes, S., Bergheau, J. M., Langlade, C., Michler, J., & Kermouche, G. (2018). Microstructural and micromechanical investigations of surface strengthening mechanisms induced by repeated impacts on pure iron. *Materials and Design, 147*, 56−64. Available from https://doi.org/10.1016/j.matdes.2018.03.014.

Uchic, M. D., Dimiduk, D. M., Florando, J. N., & Nix, W. D. (2004). Sample dimensions influence strength and crystal plasticity. *Science (New York, N.Y.), 305*(5686), 986−989. Available from https://doi.org/10.1126/science.1098993.

Viat, A., De Barros Bouchet, M. I., Vacher, B., Le Mogne, T., Fouvry, S., & Henne, J. F. (2016). Nanocrystalline glaze layer in ceramic-metallic interface under fretting wear. *Surface and Coatings Technology, 308*, 307−315. Available from https://doi.org/10.1016/j.surfcoat.2016.07.100, http://www.journals.elsevier.com/surface-and-coatings-technology/.

Viat, A., Guillonneau, G., Fouvry, S., Kermouche, G., Sao Joao, S., Wehrs, J., Michler, J., & Henne, J. F. (2017). Brittle to ductile transition of tribomaterial in relation to wear response at high temperatures. *Wear, 392-393*, 60−68. Available from https://doi.org/10.1016/j.wear.2017.09.015.

Widmer, R. N., Groetsch, A., Kermouche, G., Diaz, A., Pillonel, G., Jain, M., Ramachandramoorthy, R., Pethö, L., Schwiedrzik, J., & Michler, J. (2022). Temperature−dependent dynamic plasticity of micro-scale fused silica. *Materials & Design, 215*, 110503. Available from https://doi.org/10.1016/j.matdes.2022.110503.

Zhang, H., Schuster, B. E., Wei, Q., & Ramesh, K. T. (2006). The design of accurate microcompression experiments. *Scripta Materialia, 54*(2), 181−186. Available from https://doi.org/10.1016/j.scriptamat.2005.06.043.

Zhang, Y., Sao-Joao, S., Descartes, S., Kermouche, G., Montheillet, F., & Desrayaud, C. (2020). Microstructural evolution and mechanical properties of ultrafine-grained pure α-iron and Fe-0.02%C steel processed by high-pressure torsion: Influence of second-phase particles. *Materials Science and Engineering: A, 795*, 139915. Available from https://doi.org/10.1016/j.msea.2020.139915.

Zhao, M., Li, J. C., & Jiang, Q. (2003). Hall−Petch relationship in nanometer size range. *Journal of Alloys and Compounds, 361*(1-2), 160−164. Available from https://doi.org/10.1016/S0925-8388(03)00415-8.

Microtribology experiments for hardmetals

Mark G. Gee
Department of Materials and Mechanical Metrology, National Physical Laboratory, Teddington, United Kingdom

24.1 Introduction

WC/Co hard metals are widely used in tool applications because of their excellent wear resistance properties. The dependence of their wear resistance on microstructure has been studied over many years (Gee et al., 2014). Recent advances in testing capability including computer control with data acquisition has enabled step changes in quality of the data achieved. When combined with modern microstructural characterization, this allows a much better understanding of wear mechanisms to be obtained.

One development has been the use of microtribology tests where tribological damage is caused by a diamond indenter pressed into and traversed across the surface of the sample (Csanádi et al., 2015; Gant et al., 2017; Gee, 2010; Gee & Nimishakavi, 2011; Gee et al., 2011; Pignie et al., 2013). The surface of the sample is examined before and after tribological damage to determine the mechanisms of wear. In the latest tests, tribological exposure has been carried out in situ in the SEM with incremental damage being caused to the sample with high-resolution images taken after every increment of damage so that a sequence of images can be created to give a time history of how damage increases with additional tribological exposure (Gee et al., 2017, 2022).

24.2 Microtribology experiments

Experiments were carried out with the National Physical Laboratory (NPL) microtribometer (Fig. 24.1) (Pignie et al., 2013). Initial experiments carried out on a 4 μm grain size WC/Co hardmetal show how damage accumulates quickly with number of repeated passes of the indenter over the surface of the sample (Fig. 24.2). With 50 passes, there was already considerable damage to the structure of the hardmetal with plastic deformation in the tungsten carbide (WC) grains as evidenced by slip line traces with cracking and fragmentation of some WC grains (Fig. 24.2A). After 100 passes, there is much more damage to the structure with very considerable fragmentation of the WC grains (Fig. 24.2B). The fragments of WC reembed into the cobalt to form a layer on the surface.

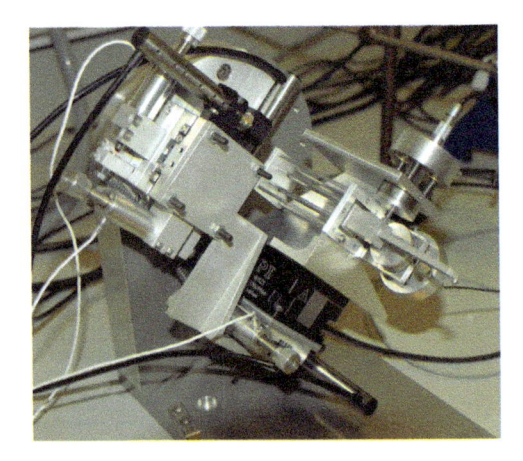

Figure 24.1 NPL microtribometer photograph of test system.
Source: From Gee, M. G., Nunn, J. W., Muniz-Piniella, A., & Orkney, L. P. (2011). Microtribology experiments on engineering coatings. *Wear*, *271*(9−10), 2673−2680. https://doi.org/10.1016/j.wear.2011.02.031.

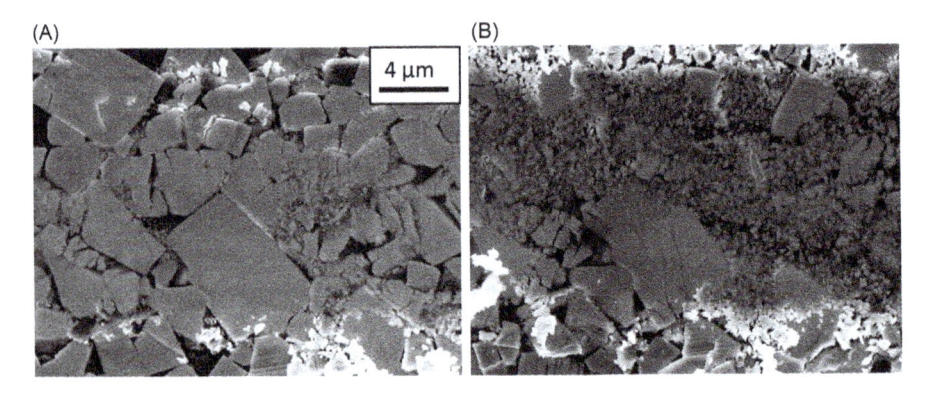

Figure 24.2 Surface of 4 μm grin size 11% binder WC/Co hardmetal after multiple pass scratching using a 30 μm radius diamond indenter, (A) 50 passes, (B) 100 passes of same area as (A). Surfaces of hardmetal scratched under specified conditions.

As the NPL microtribometer is fully controlled by computer, it can also be used to carry out abrasion simulation experiments (Gee, et al., 2011; Pignie et al., 2013). In this mode, the multiple scratches are made in a random controlled pattern over an area. Figs. 24.3 and 24.4 shows micrographs of one of these damaged areas. A ramping load has been used with a minimum load of 50 mN at one end of the area with a maximum of 250 mN at the other end. The micrographs show that the damage area is rectangular, with deeper damage as the load is increased. The roughness of the damaged area also becomes more pronounced as the load is increased.

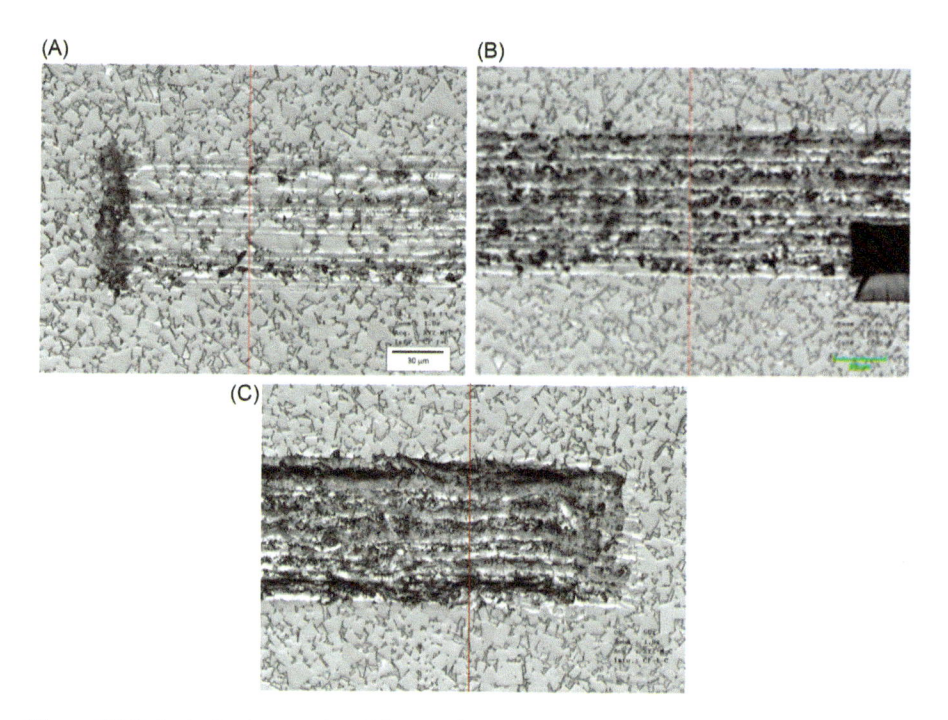

Figure 24.3 Optical micrographs of damaged area produced in abrasion simulation study of 4 μm grin size 11% binder WC/Co hardmetal using a 10 μm radius diamond indenter (A) at 50 mN end of damage zone, (B) 150 mN position (C) 250 mN end. Note that red lines show positions of profiles shown in Fig. 24.5.

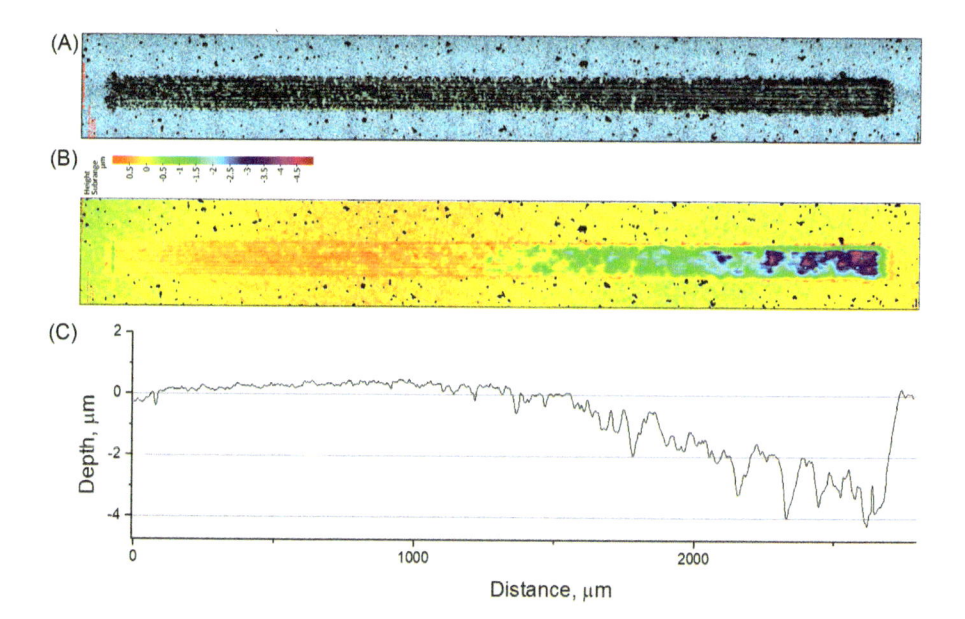

Figure 24.4 3D optical microscopy on 4 μm grain size 11% binder hardmetal subjected to 2000 repeat scratches (A) optical image, (B) height map, (C) profile along length. Low applied load end is at left, high load is at right.

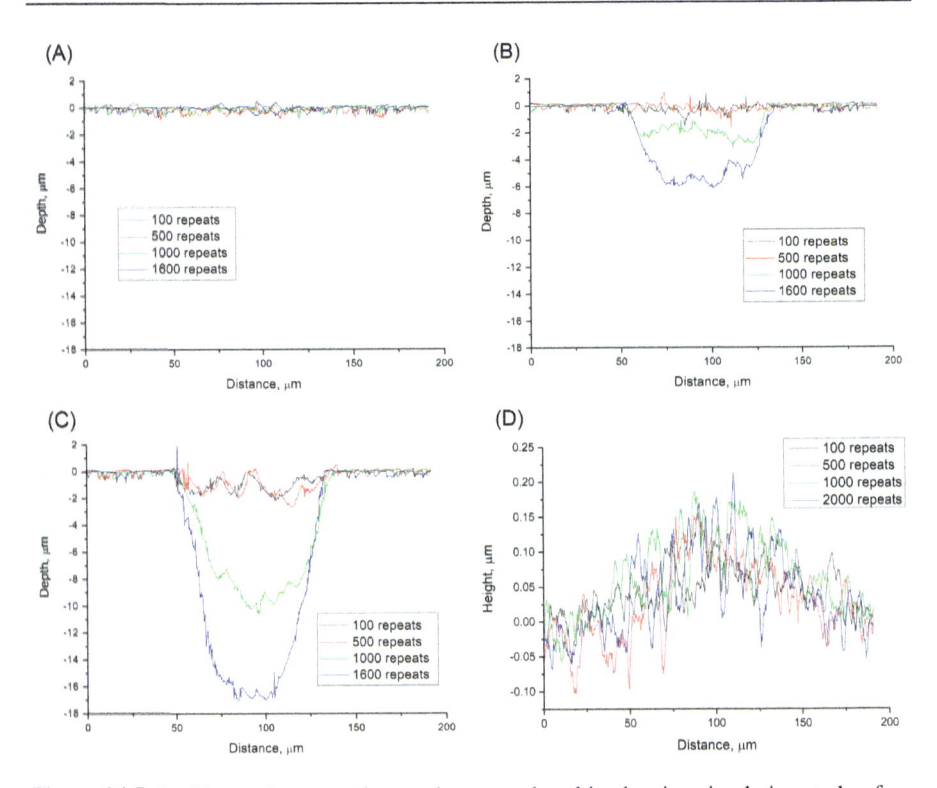

Figure 24.5 Profiles made across damaged area produced in abrasion simulation study of 4 μm grin size 11% binder hardmetal, (A) at 50 mN end of damage zone using a 10 μm radius diamond indenter, (B) 150 mN position using a 10 μm radius diamond indenter, (C) 250 mN using a 10 μm radius diamond indenter, (D) 50 mN using a 20 μm radius diamond indenter.

Fig. 24.4 also shows a height map of the damaged area with a profile taken along the length of the damaged area. Both the height map and the profile show that unexpectedly, there is a small height gain in the low load region until a load of about 150 mN where reduction in height starts to occur with removal of material from the surface.

These results are confirmed by profiles made across the damaged area Fig. 24.5. Figs. 24.5A–C show the progression of damage as the number of passes increases. At low load, there is little evident damage at first view, although when the vertical scale is expanded, there is a small swelling of the surface (Fig. 24.5D). At a load of 150 mN, there is initially little damage to the surface until damage starts to take place as the number of passes increases (more than 1000 passes, Fig. 24.5B). At the highest load, damage starts to take place much earlier in the test (Fig. 24.5C). Table 24.1 gives a summary of these results and shows that there is a threshold load and number of passes below which significant damage to the WC/Co does not occur.

Table 24.1 Threshold load for significant damage (loss of material) and maximum depth of damage for 6A (6% Co and 1 μm grain size) and 11E (11% Co and 4 μm grain size) samples tested in abrasion simulation experiments with 20 μm radius diamond indenter μ.

Number of passes	Threshold load for damage, mN		Maximum depth of damage, μm	
	11E	6A	11E	6A
100	–	–	–	–
500	170	–	– 0.5	–
1000	147	190	– 2.5	– 0.23
2000	101	161	– 3.6	– 0.5

Source: From Pignie, C., Gee, M. G., Nunn, J. W., Jones, H., & Gant, A. J. (2013). Simulation of abrasion to WC/Co hardmetals using a microtribology test system. *Wear, 302*(1–2), 1050–1057. https://doi.org/10.1016/j.wear.2012.11.057.

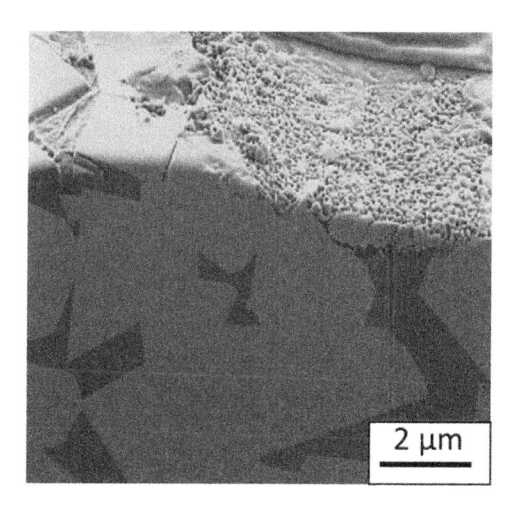

Figure 24.6 FIB section through scratch made on 4 μm grain size WC/Co material. Damage to the surface is less than a carbide grain in depth. Red highlights area of subsurface cracking.

The reason for the slight swelling in the surface seen at low loads was found to be due to oxidation of the Co binder phase as demonstrated by EDX analysis of the surface (Gant et al., 2017).

Fig. 24.6 shows an field ion bombardment (FIB) cross section of the scratch, which shows that the damage that was caused was quite shallow and was confined to one or two grains of the surface. Considerable fragmentation of these surface grains took place. A video (video at https://youtu.be/9V5_aG7YVQ0) shows a full 3D slice and view analysis, which shows the depth of damage that is produced.

Operation of the microtribometer in the SEM allows the progression of damage to be followed with repeated passes. In these experiments, high-resolution electron images of the damage are captured after every pass. These images are registered with respect to each other in ImageJ to allow a video of the build up of damage to be built up with respect to number of passes. These videos are given in the supplementary information (videos at https://youtu.be/s7-mPyTgFX0 and https://youtu.be/aFZgzWP-Ij0) and are worth examining. They show the dynamics of the damage processes and the flow of material from the damaged area to form debris and eventual material removal processes.

Figs. 24.7 and 24.8 show sequences of single images from these videos to demonstrate some of the features of the damage processes. In the larger grain size WC/Co, there was considerable damage to the WC grains even after a single pass with considerable fracture to the WC grains (Fig. 24.7B). It is also interesting that the scratch is wider at the large binder phase zone at the center of the scratch, presumably due to the reduced hardness of the binder relative to the WC. As the number of passes increases, the fracture to the WC grains increases with fragmentation of the WC (Fig. 24.7D). By 50 passes, the large binder phase region at the center of the scratch has disappeared revealing an underlying WC grain (Fig. 24.7F). Reembedment of WC fragments also occurs to form composite films over large fractions of the scratch surface.

Similar mechanisms of damage occur for the finer grain sized WC/Co material with fracture and fragmentation of WC grains and reembedment of WC fragments in binder to form a composite surface layer (Fig. 24.8). However, damage took much longer to accumulate with this material. This is possibly due to the difficulty in generating and moving dislocations with the smaller grains of this material.

The damage can be quantified by measuring the width of the scratch (Fig. 24.9). This shows an initial sharp increase in scratch width in the initial stages of the experiment with a much lower increase in width from this point (Fig. 24.9A). When converted to scratch pressure, the initial pressures are extremely high before they reduce to a lower value.

24.3 Wear-corrosion synergy

Microtribology experiments can also be used to explore how corrosion can act synergistically with wear. Figs. 24.2 and 24.10 show micrographs from experiments on a 4 μm grain size hardmetal. The figure shows how you get different progression of damage when you have corrosion as well as wear damage. Fig. 24.2A and B show the results of scratch tests in dry conditions. Here, although there is considerable fragmentation and reembedment of WC into the surface of the scratch, the microstructure is still recognizable. In the equivalent case where the scratching was carried out under a corrosive liquid (HCl) placed on the area being scratched, after 100 passes, the surface structure has been completely removed (Fig. 24.10A—C).

Similar scratches performed in dry conditions and under HCl were sectioned with an FIB. Figs. 24.11A and B show two sections from this analysis showing how the cobalt binder phase has been removed from the surface region under the

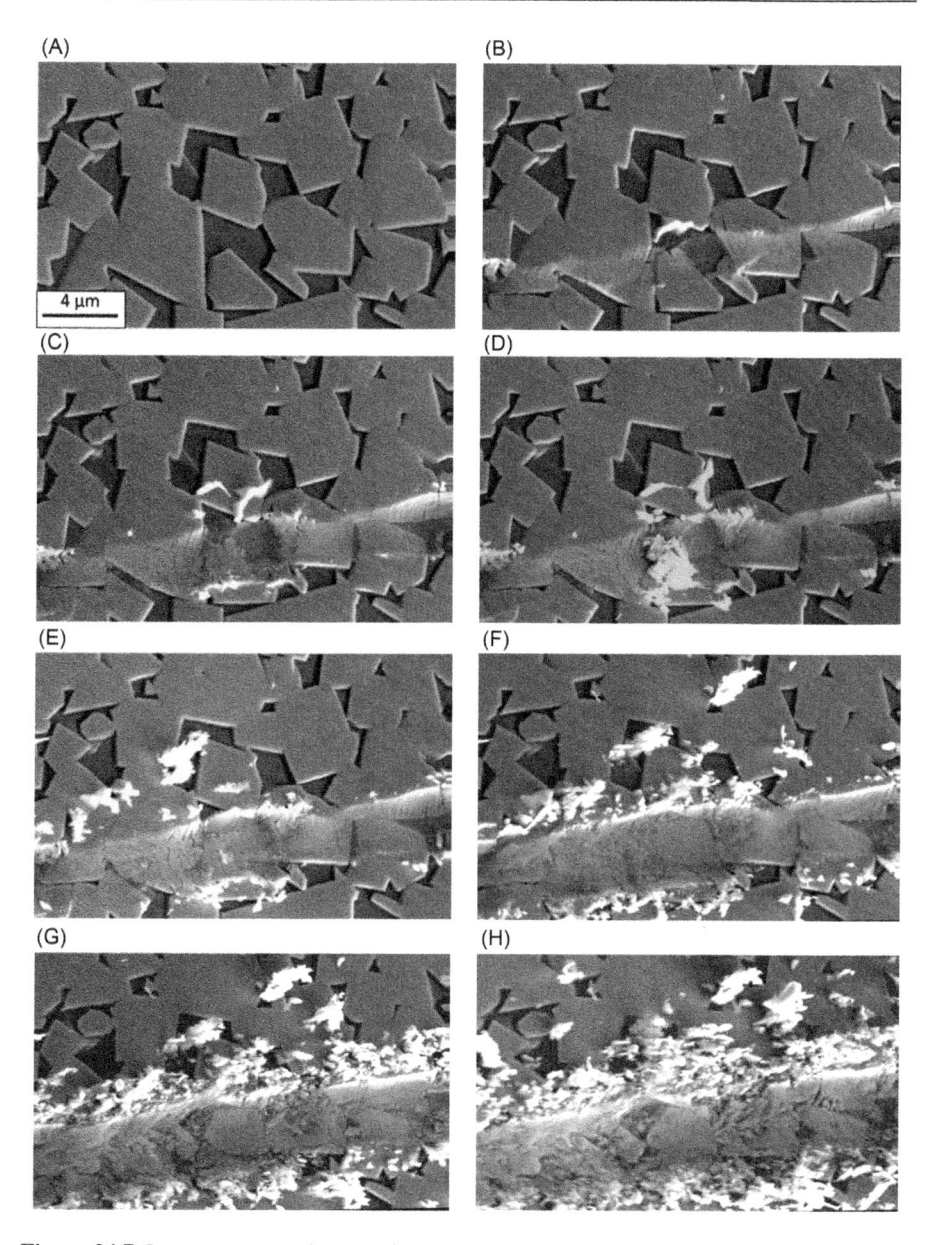

Figure 24.7 Image sequence from in situ experiment at 55 mN with a 1 μm radius indenter on 11% binder 4 μm grain size hardmetal, (A) initial surface, (B) pass 1, (C) pass 3, (D) pass 6, (E) pass 10, (F) pass 50, (G) pass 124, (H) pass 243.

scratch. This has weakened the surface structure of the hardmetal and has resulted in considerable fracture and removal of material from the surface structure. The magnitude of damage was evaluated by using the width of the scratch to calculate

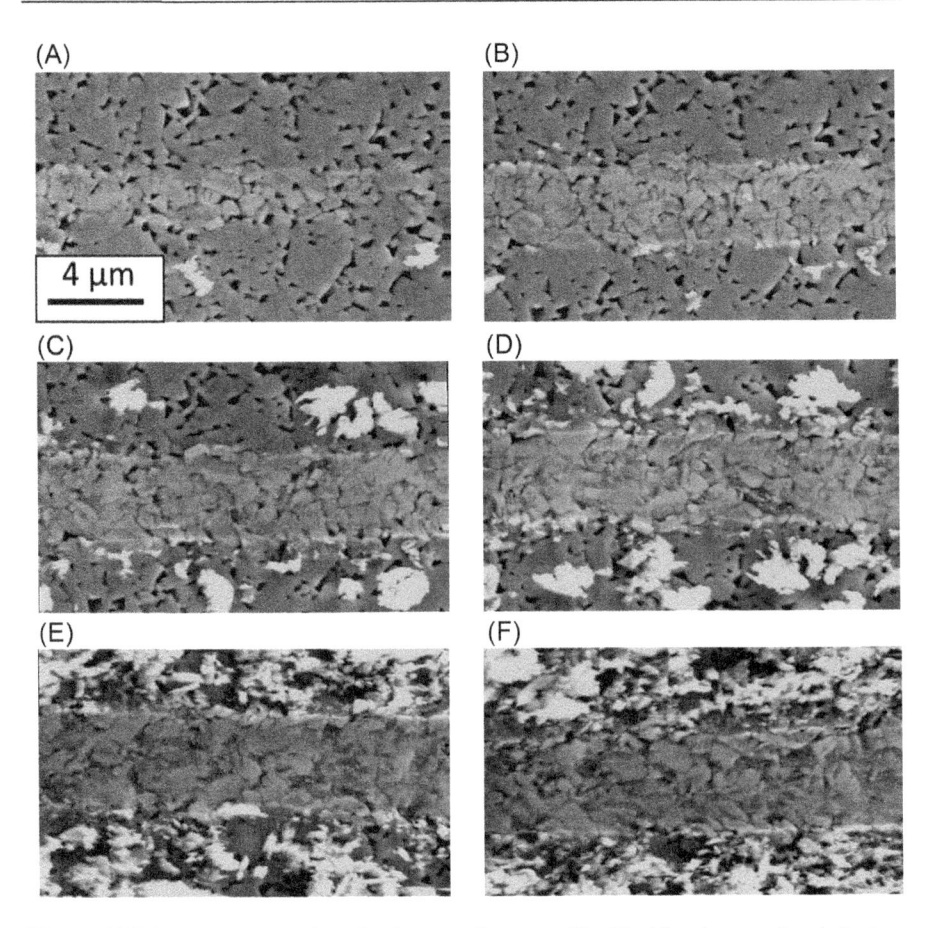

Figure 24.8 Image sequence from in situ experiment at 55 mN with a 1 μm radius indenter on 6% binder 1 μm grain size hardmetal (A) pass 1, (B) pass 6, (C) pass 21, (D) pass33, (E) pass 70, (F) pass 108.

the volume of wear for the two cases. Complementary experiments were carried out to look at the corrosion performance of this hard metal in the same acid that was used for the scratch test. It was found (Fig. 24.11C) that the mass loss for the dry scratch or the corroded hardmetal samples were both lower by at least an order of magnitude showing that there was a considerable synergistic effect from the damage that took place.

24.4 Electron back scattered diffraction analysis

Electron back scattered diffraction (EBSD) analysis can be combined with microscratch testing of hard metals to reveal more information about how the material

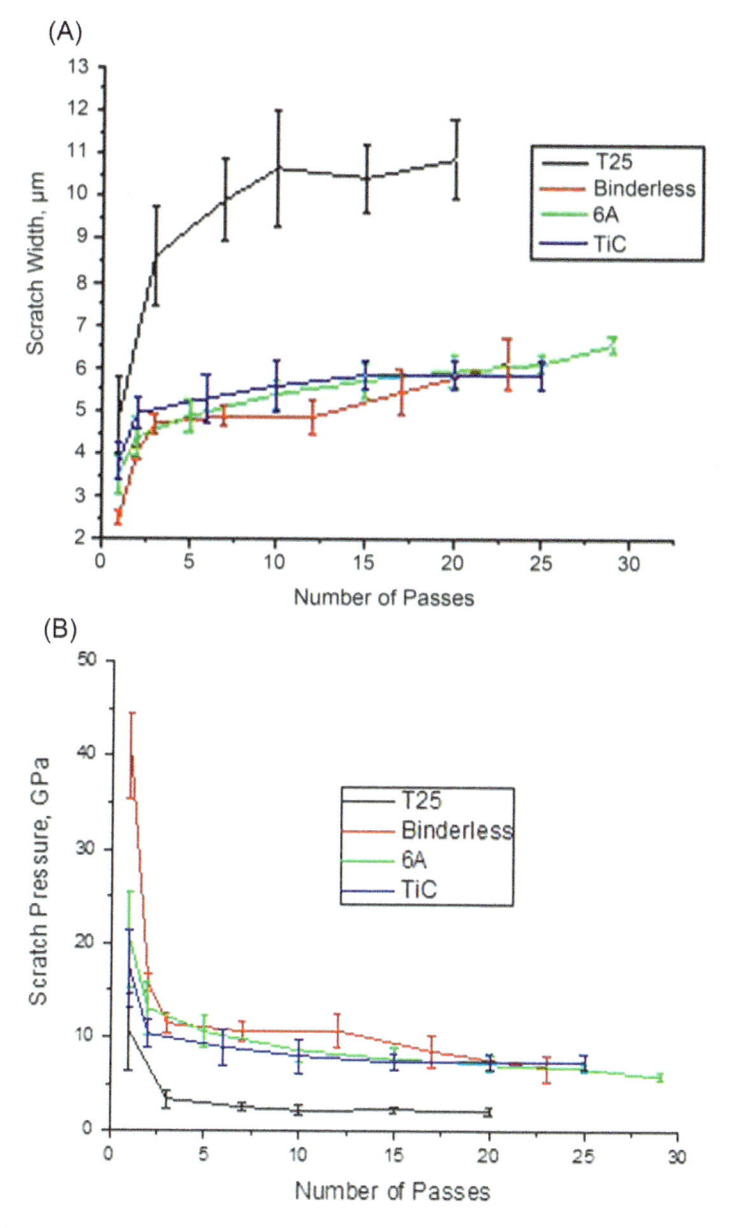

Figure 24.9 Scratch width and scratch pressure for a number of different hardmetals (A) Scratch width (B) scratch pressure.

deforms under scratching. EBSD reveals the crystallographic orientation of grains, with any deformation revealed by local changes in this orientation. Fig. 24.12 shows an EBSD analysis of a hard metal region that has been scratched in

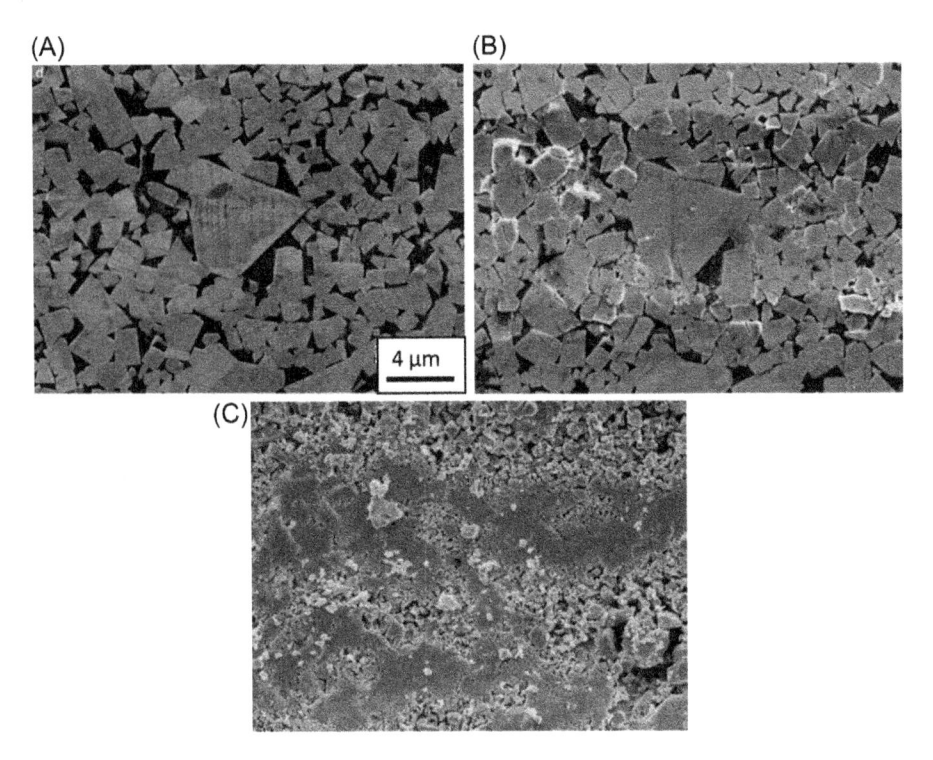

Figure 24.10 Scratch tests on 4 μm grain size 11% binder hardmetal tested with a 20 μm radius diamond indenter under a 400 mN applied load immersed in HCl (A) 2 pass HCl, (B) 50 pass HCl, (C) 100 pass HCl.

single-pass experiments under a variety of applied loads. The scratches are mainly revealed by the loss of information giving black bands as the deformation disrupts the diffraction processes. Some color changes are nevertheless seen near the edges of the scratches, indication that deformation has occurred in those positions through rotation of the lattice at these points.

More interesting is that the degree of scratch damage as evidenced by the degree of diffraction information is linked to the orientation of the WC crystals. This is shown in Fig. 24.12B where the orientation of the hexagonal WC crystals is labeled showing a correlation of grain orientation with degree of damage. This suggests that it may be possible to improve the resistance of WC to scratching damage by processing routes that control the orientation of grains in the surface layers.

Figs. 24.13 and 24.14 show how the use of misorientation maps can more clearly show and to some extent quantify the degree of deformation in the scratches (Gee et al., 2023). Fig. 24.13 shows a sequence of misorientation maps for scratches made under increasing loads for a 1 μm grain size 6% binder phase WC/Co material. Here blue indicates no deformation, and green through red signifies increasing amounts of deformation as determined by the change in orientation from one pixel

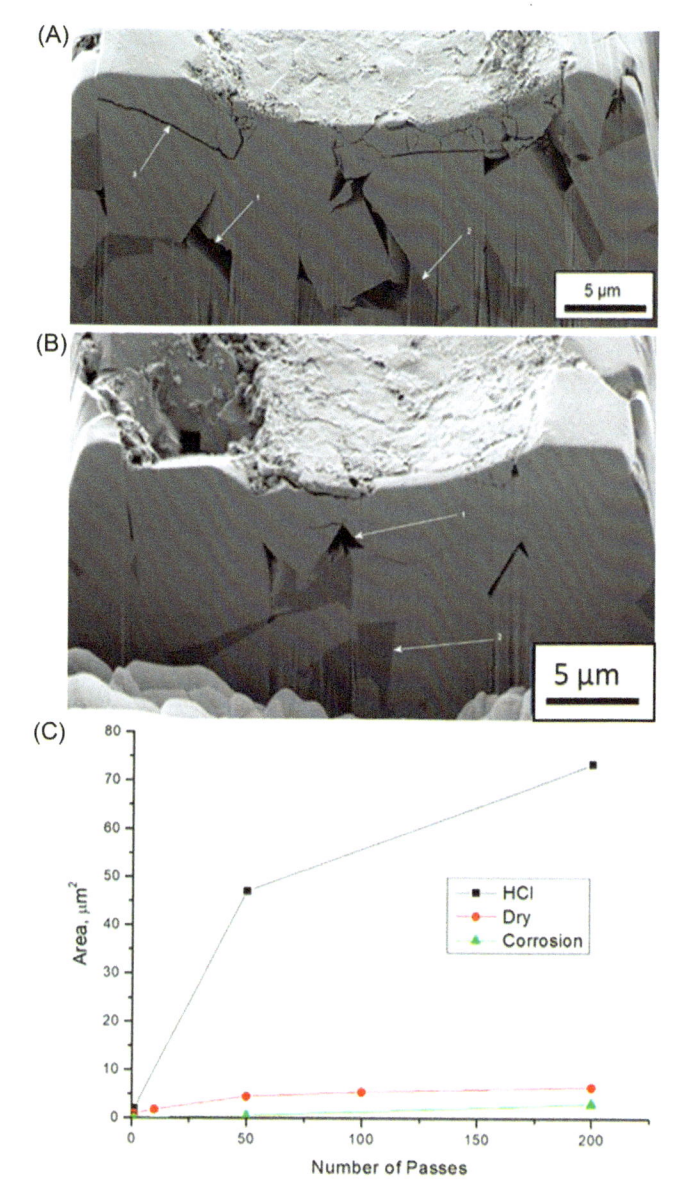

Figure 24.11 Results from tribocorrosion scratch tests on 4 μm grain size 11% binder hardmetal. (A) and (B) FIB cross sections through surface of scratch showing removal of binder phase (black areas), (C) comparison between removal rates (expressed as cross sectional area) for dry tests, tests in HCl and HCl corrosion tests.

to another local pixel (Gee et al., 2009). The degree of misorientation increases as the load increases. The black area at the center of the scratch where there is a lack of diffraction due to high deformation also increases.

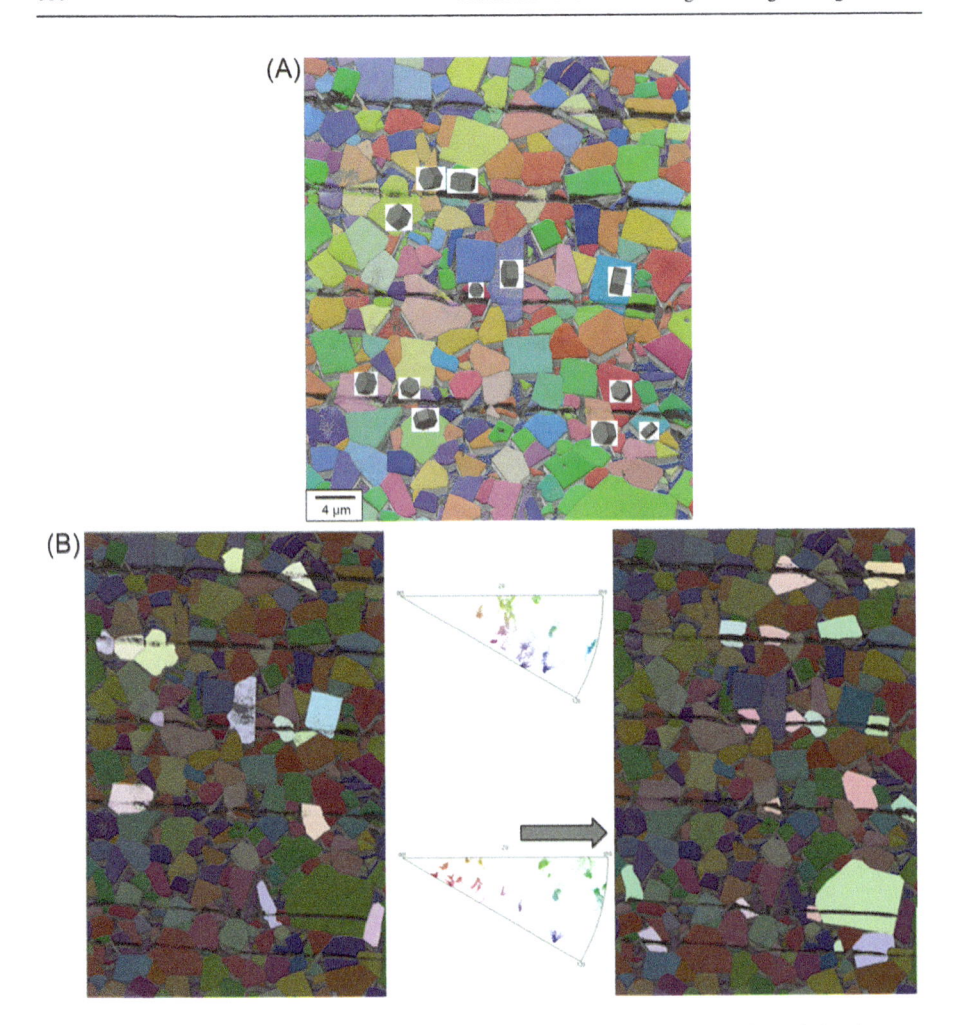

Figure 24.12 Inverse pole figure maps of 4 μm grain size 11% binder hardmetal that has been scratched by an array of scratches (A) revealing difference in damage with orientation of crystal, (B) illustrating the same point identifying specific grains and their orientation. Note that applied load for scratches from top of map is 50, 45, 35, 30, 25, 20 mN.

As it was expected that the surface had higher deformation, the surface of a scratched sample was broad ion beam polished to remove a few 10 s of nm of material from the surface. Fig. 24.14A shows the inverse pole figure (IPF) map of this scratch showing color changes in and around the scratched region. The misorientation map (Fig. 24.14B) reveals many linear features associated with these changes in color. Although these are not dislocations themselves, they are related to the network of dislocations that are created by the deformation from the scratch.

Microtribology experiments for hardmetals 687

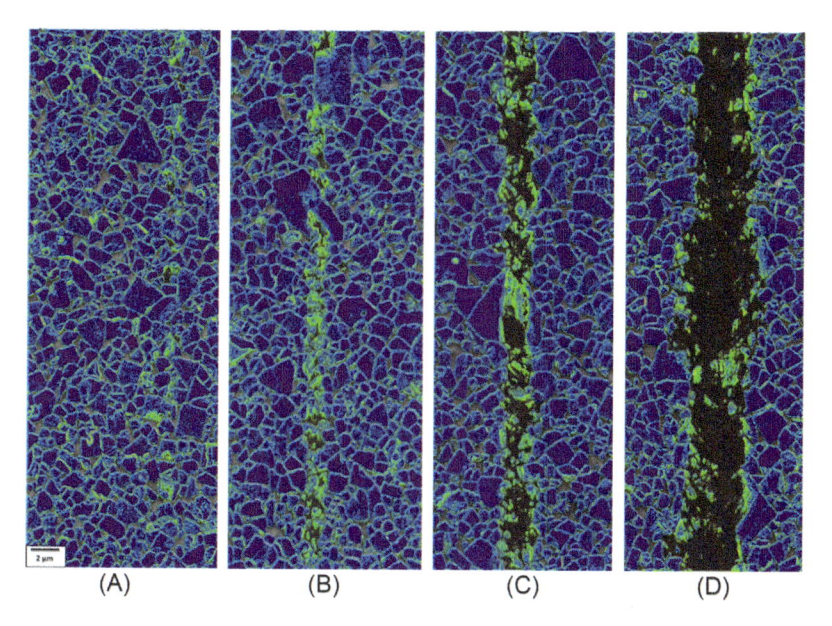

Figure 24.13 WC misorientation maps for scratches on 1 μm grain size 6% binder hardmetal (A) 2 mN, (B) 10 mN, (C) 40 mN, (D) 100 mN. Magnification in all images is the same.

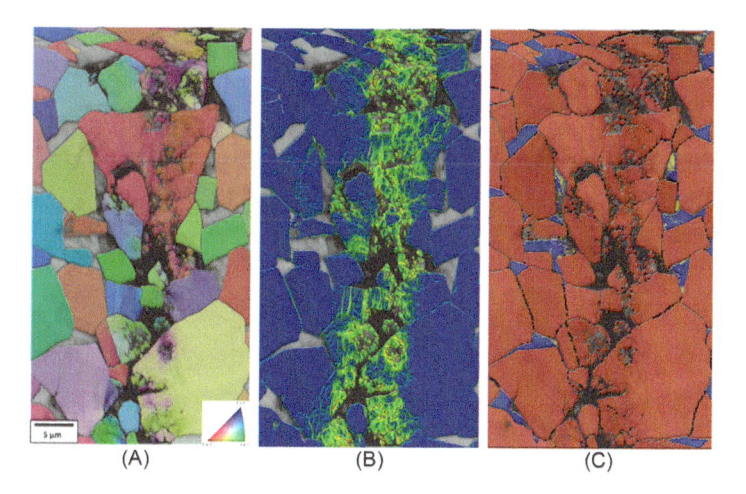

Figure 24.14 Electron back scattered diffraction analysis of microscratches on 4 μm grain size 11% binder hardmetal sample following broad ion beam (BIB) polishing. (A) Inverse pole figure map, (B) WC misorientation, (C) phase map with WC in red, fcc Co in blue, and hcp Co in yellow. All images are the same magnification.

The phase of the material near to the scratch can also be indexed by EBSD. Fig. 24.14C shows an example of this with some of the fcc Co binder phase transforming to hcp form of cobalt.

24.5 Summary and conclusions

The results presented in this part of the chapter illustrate how microtribology experiments can provide a much better understanding of the mechanisms of wear and damage for WC-Co hardmetals. The results confirm that wear in WC-Co from abrasive processes takes place through a progressive process of:

- Accumulation of plastic deformation leading to fracture of the WC grains
- Fragmentation of the WC grains
- Reembedment of WC fragments in binder phase to form composite surface layers
- Removal of material from the surface to form wear debris

Work has also shown that plastic deformation also occurs in the binder phase leading in some cases to fracture of the cobalt (Gee et al., 2023). EBSD analysis has also shown that there is likely to be a dependence of the degree of damage on the orientation of the WC grains. It also shows that transformation can occur between the normal fcc form of Co binder phase to hcp form.

24.5.1 Future work

Future work could include the further development of the in situ technology to enable experiments between WC-Co hardmetal and other materials such as the rocks and other abrasives that will be encountered in applications such as rock drilling and mineral extraction. The development of a high-temperature in situ system would also enable experiments to be conducted, which are relevant to the metalworking industry. It would also be beneficial to install the in situ tribometer into an environmental SEM so that the effect of the atmosphere of the test on the mechanisms of damage could be determined. It will also be important to combine the images of damage with techniques such as electron back scattered diffraction (EBSD) and electron channeling contrast imaging (ECCI), preferably in situ simultaneously with the testing. This has the potential to provide qualitative and quantitative information on the phase structure of the hard metal and its plastic deformation state, and some steps in this direction have recently been published (Gee et al., 2023).

Acknowlegments

The author would like to acknowledge the contribution of colleagues at the National Physical Laboratory to the work presented in this chapter and also acknowledges funding from the Department for Science, Innovation and Technology under the NMS program.

References

Csanádi, T., Bl'Anda, M., Chinh, N. Q., Hvizdoš, P., & Dusza, J. (2015). Orientation-dependent hardness and nanoindentation-induced deformation mechanisms of WC crystals. *Acta Materialia, 83*, 397−407. Available from https://doi.org/10.1016/j.actamat.2014.09.048, http://www.journals.elsevier.com/acta-materialia/.

Gant, A. J., Nunn, J. W., Gee, M. G., Gorman, D., Gohil, D. D., & Orkney, L. P. (2017). New perspectives in hardmetal abrasion simulation. *Wear, 376−377*, 2−14. Available from https://doi.org/10.1016/j.wear.2017.01.038.

Gee, M., Kamps, T., Woolliams, P., Nunn, J., & Mingard, K. (2022). In situ real time observation of tribological behaviour of coatings. *Surface and Coatings Technology, 442*. Available from https://doi.org/10.1016/j.surfcoat.2022.128233.

Gee, M., Mingard, K., Nunn, J., Roebuck, B., & Gant, A. (2017). In situ scratch testing and abrasion simulation of WC/Co. *International Journal of Refractory Metals and Hard Materials, 62*, 192−201. Available from https://doi.org/10.1016/j.ijrmhm.2016.06.004, http://www.journals.elsevier.com/international-journal-of-refractory-metals-and-hard-materials/.

Gee, M., Mingard, K., & Roebuck, B. (2009). Application of EBSD to the evaluation of plastic deformation in the mechanical testing of WC/Co hardmetal. *International Journal of Refractory Metals and Hard Materials, 27*(2), 300−312. Available from https://doi.org/10.1016/j.ijrmhm.2008.09.003.

Gee, M. G. (2010). Model scratch corrosion studies for WC/Co hardmetals. *Wear, 268*(9−10), 1170−1177. Available from https://doi.org/10.1016/j.wear.2010.01.004.

Gee, M. G., Gant, A. J., Roebuck, B., & Mingard, K. P. (2014). Wear of hardmetals. *Comprehensive Hard Materials, 1*. Available from https://doi.org/10.1016/B978-0-08-096527-7.00012-X, http://www.sciencedirect.com/science/referenceworks/9780080965284.

Gee, M. G., Mingard, K., & Nunn, J. W. (2023). EBSD evaluation of damage in microtribology experiments on WC/Co hardmetals. *Wear, 524−525*. Available from https://doi.org/10.1016/j.wear.2023.204784.

Gee, M. G., & Nimishakavi, L. (2011). Model single point abrasion experiments on WC/Co hardmetals. *International Journal of Refractory Metals and Hard Materials, 29*(1), 1−9. Available from https://doi.org/10.1016/j.ijrmhm.2010.04.009.

Gee, M. G., Nunn, J. W., Muniz-Piniella, A., & Orkney, L. P. (2011). Micro-tribology experiments on engineering coatings. *Wear, 271*(9−10), 2673−2680. Available from https://doi.org/10.1016/j.wear.2011.02.031.

Pignie, C., Gee, M. G., Nunn, J. W., Jones, H., & Gant, A. J. (2013). Simulation of abrasion to WC/Co hardmetals using a micro-tribology test system. *Wear, 302*(1−2), 1050−1057. Available from https://doi.org/10.1016/j.wear.2012.11.057.

Section V

Summary

Trends and future directions

25

Ben Beake[1] and Tomasz Liskiewicz[2]
[1]Micro Materials Ltd., Wrexham, United Kingdom, [2]Department of Engineering, Manchester Metropolitan University, Manchester, United Kingdom

25.1 More extreme temperatures

To enable characterization of the mechanical properties of materials at the microstructural level under increasingly severe operating conditions ("in operando") in gas turbines, very-high-temperature nuclear reactors, ultrahigh-speed cutting, cryomachining, space applications, etc., there is a strong drive for instrumentation manufacturers to develop their instruments to be able to test hotter and colder (e.g., between $-180°C$ and $1200°C$). To test reliability at $1200°C$, this will be challenging requiring (1) high vacuum, (2) control of chemical reactivity between indenter and sample, (3) development of a suitable test methodology and analysis protocol to mitigate increased time-dependent deformation, and (4) innovations in thermocouple and heater design. Best practice in elevated temperature nanomechanics will be addressed in a new part of the ISO 14577 standard for nanoindentation: Metallic materials — Instrumented indentation test for hardness and materials parameters — Part 6: Instrumented indentation test at elevated temperature. To test reliability at $-180°C$ will require (1) high vacuum to eliminate traces of water vapor that can otherwise condense on the sample (icing) and (2) matching temperature of tip and sample analogously to high temperature, by using separate tip and sample cooling. Reliable nanomechanical testing at these low temperatures will enable brittle — ductile transitions to be studied at the microstructural level. To simulate specific working environments, nanomechanical setups could be developed where tests could be performed under conditions with active control of the moisture level and test temperature, for example, hot and humid conditions, or low temperatures in liquid.

25.2 More complex experiments — multisensing, big data, and AI

The benefits in multisensing where, for example, the signature of a change in deformation may not be observed in one signal but be clearly shown in another will be another active area of development, aided by improvements to electronics (signal synchronisation) and artificial intelligence/machine learning methods for data analysis. Developments in computational power to handle large and complex datasets will help with high-resolution multisensing in microtribological tests that generate significant amounts of data. Too much data will no longer be a problem.

An open challenge is to what extent the multisensing approaches could be developed to have predictive power, for example, to provide early warning of a change to a more severe deformation mechanism or failure of a coated component operating in a mechanically loaded contact. Variability in individual signals that is not related to failure (i.e., noise) could be mitigated by using multiple signals at once. Precursors to these changes could be detected using parameters, or changes in parameters, derived from combinations of multiple signals (e.g., a combined "figure of merit" parameter using friction, depth, and electrical resistance could not only show a change in mechanism but also indicate what type of change). Machine learning would be highly useful here.

25.3 More complex experiments to simulate abrasion and erosion

Currently popular model tests used to simulate abrasion and erosion—scratch testing and impact testing respectively—simplify the real complex tribocontact situation to study the damage created by a single isolated asperity in a controlled environment so that the fundamental mechanisms involved can be isolated and determined. However, in abrasion, the production of a single wear groove on a flat surface (as happens in a repetitive scratch test) is an oversimplification of what actually happens in practice when individual wear events are superimposed on each other and interact to produce the worn surface. In erosion, the individual impacts occur with a statistical distribution of contacts over a surface, and the interaction between these, for example, of lateral crack systems in brittle materials, determines how the surface erodes. In the cyclic impact test, the repetitive contacts occur at the same position rather than with a statistical distribution of contacts over a surface. Nevertheless, despite these limitations, in many cases behavior in these tests has shown excellent correlation to performance in actual applications. To be better integrated in coatings/surface optimization campaign, it is desirable that the tests can even more closely simulate the actual contact conditions. With improvements to modeling (both analysis approach and speed of calculation) and understanding, we can start to design more complex experiments, which capture the real situations more faithfully. There has been recent progress in this direction through development of the statistically distributed nanoscratch test, which involves parallel scratches with statistical distribution, which is capable of more closely simulating the damage progression in abrasion where material removal can be influenced by the interaction between damage produced by previous scratches in close proximity. Similarly, the randomized impact test has been developed as more direct way to experimentally simulate the stochastic multiple contact nature of erosion by performing repetitive controlled impacts at different locations on the sample surface. Further technique development could include even closer simulation through varying angle (of incidence for impacts, or between scratches), introducing statistical distribution of loads, and testing at elevated temperatures to simulate high-temperature erosion and high-temperature abrasion. There may be a trend toward using other probe materials than diamond. For example, (1) changing indenter hardness for impact testing, (2) changing to steel for microtribology, and (3) switching to cBN or WC for testing in air above the oxidation onset temperature of diamond. These might require development of integrated protocols for more regularly checking indenter wear, either determining tip shape from indentation or may require imaging if directional tip wear is suspected. Repetitive nanoscratch tests and cyclic nano or microimpact tests are typically accelerated tests that provide a severe environment so that the surface is subjected to high contact stresses (including high bending stresses for coatings, through high loads and small contacts). These more complex abrasion and erosion resistance tests could be performed (as longer experiments) with lower stresses.

25.4 Mechanical property mapping

Undoubtably, an important growth area in recent years has been in the development of microstructure − mechanical property correlations made possible by the larger datasets obtained from faster measurements together with statistical treatments for data analysis. Provided that any surface polishing to obtain sufficiently flat and smooth surfaces for large-area mapping does not unduly modify (e.g., work-harden) the surface, it has been shown that small indentations can be placed close together and extremely detailed 2D surface property maps obtained.

However, an open question is how to develop the methodology to handle 3D information required for characterising microstructural influences in coatings, hierarchical structures, and mechanically graded materials given that with commonly used indenter types the plastic and elastic stress fields widen (and at different degrees) as indentations become deeper. This puts a limitation on how close these indentations could be placed as stress fields will overlap.

A goal would be predictive models to correlate material structure with 3D mechanical properties, perhaps aided by AI and 3D microstructural information from, for example, X-ray tomography or FIB milling and imaging. Biomimetics, or nature-inspired design, would benefit from modeling seamlessly integrating information at different length scales.

25.5 Tribological and impact mapping

In contrast to mechanical property mapping by nanoindentation, as yet mapping approaches with scratch and impact tests are almost completely unexplored. Mapping varying impact or scratch resistance across the surface could represent an opportunity to more directly develop microstructure − properties − performance relationships. To do this effectively it will be necessary to first develop understanding of the test condition dependent interaction thresholds, to choose a suitable spacing between scratches or impacts.

25.6 Novel experiments

Tests on specialized test piece geometries such as pillars and cantilevers to simplify stress states will continue to be popular, particularly when performed in an SEM to observe deformation in real-time. With sufficiently high instrumental stability, multicycling tests can be performed over extended periods to study slow crack growth. Constant stress indentation creep tests, where the load is adjusted through feedback to provide constant stress creep, could be performed at elevated temperatures.

25.7 A tool for surface optimization

Through improved understanding of the location and magnitude of the stresses developed in contact, there is a growing realization of the requirements *at different length scales/contact sizes* for surface/coating wear resistance/damage tolerance and how the properties of the system rather than the coating alone provide these. Advanced nanomechanics/tribology is predicted to play a more central role in surface optimization rather than simply "characterisation under lab conditions" through closer simulation of actual conditions.

25.8 Summary

These emerging capabilities in nanomechanics instrumentation and experimental design are shaping the way how we study and optimize material surfaces and coatings. A common theme linking several of these developments is closer simulation of real conditions so that the nanomechanical data obtained are more useful, which is being achieved through advances due to greater test control, by pushing more extreme conditions, and in combination with other advances (e.g., in microscopy and simulation). By enabling sophisticated multisensory measurements combined with AI/machine learning for data analysis, we can gain unprecedented insights into deformation mechanisms and damage precursors. Innovations in simulating complex contact conditions such as abrasion and erosion through statistically distributed scratches/impacts will allow more realistic modeling of real-world performance. Mapping approaches that correlate 3D microstructural features with spatially resolved mechanical properties aided by techniques such as X-ray tomography promise to expand our understanding of structure − property relationships. Novel geometries such as micropillars combined with in-situ microscopy can reveal deformation dynamics. Ultimately, these advances will transform nanomechanics from a technique for characterising lab samples into a powerful tool for rationally designing optimized surfaces tailored to demanding applications through virtual simulation of operating environments.

Index

Note: Page numbers followed by "*f*" and "*t*" refer to figures and tables, respectively.

A

Abrasion, 694–695
Accelerated wear tests, 39
Acetylene (C_2H_2), 521
Achard wear model, 383
Acoustic emission (AE), 301–304, 306, 532
  advanced evaluation during scratch test, 310–313, 312*f*
  in bulk analysis, 306, 308*f*
  in impact testing, 306–309, 310*t*
  monitoring, 26, 306–307
  multisensing approach using simultaneous record of, 301–313, 303*f*
  process, 310–311
  recording, 306
  in scratch test, 303–306
  sensitivity, 306
Actuators of multisensing approaches, 320–321
Adaptation processes, 635
Adaptive coatings, 630–632
Adhesion energy, models for determining, 169
Adhesion testing of diamond-like carbons, scratch testing for, 532–535
Adhesive, simulation of, 378–379
Advanced engineering materials, 431
AE. *See* Acoustic emission (AE)
AEI Ltd. *See* Associated Electrical Industries Ltd (AEI Ltd)
Aero-gas turbine sector, 599
Aerodynamic models, 621
Aerospace coatings
  design tools for developing enhanced TBCs, 614–619
  environmental protection coatings and bondcoats, 601–604
  gas turbine and extreme environments, 599–600
  modeling erosion of thermal barrier coatings, 619–621
  role of surface engineering, 600–601, 600*f*
  thermal barrier coatings—design considerations, 604–614
AFM. *See* Atomic force microscopy (AFM)
AFM-IR. *See* Atomic force microscopy coupled with infrared spectroscopy (AFM-IR)
AI algorithms. *See* Artificial intelligence algorithms (AI algorithms)
Air plasma spray (APS), 252, 611–612
ALD. *See* Atomic layer deposition (ALD)
Algebraic models, coupling of, 383
Algebraical equation, 280–281
α-alumina scale, 602
Alumina, 306–307
Alumina grit, 457
Alumino-silicate attack, 606–609
Aluminum, 9–10, 379
Amorphous carbon compounds, 510, 516–517
Amorphous coatings, 538
Amorphous phases, 446–447
Analytical digital twin simulation approach, 365
Analytical methods, 343–345, 365, 370–371
  analytical modeling of coated systems, 343–346
  thought experiments, 361–363
    graphene simulation, 363
    impact simulation-mixed droplet and particle impact, 361–363

Analytical methods (*Continued*)
  working through practical examples, 346–361
    comparing stress and modulus profile tuning for same cathodic arc deposited coating system, 353–355
    from geological to nanoscale, 349–351
    impact simulation on fused silica, 359–361
    representation of different indentation scales of measurement, 348f
    scratch testing of zirconia coatings, 351–353
    stress profiling of DLC coating on A2 tool steel, 356–359
Anisotropic mechanisms during nanoindentation, 411–419
Annealing, 76
APS. *See* Air plasma spray (APS)
Archard approach, 456
Archard equation, 457–458
Archard wear equation, 447–453, 456–458
Archard wear model, 67
Area function determination, 9–10
Artificial intelligence algorithms (AI algorithms), 313, 693–694
Associated Electrical Industries Ltd (AEI Ltd), 453–454
Atomic force microscopy (AFM), 38–39, 143, 217–218
  nanowear
    AFM-based nanoindentation tests, 223
    friction and wear of ultrathin films, 225–229
    limitations and strengths of atomic force microscopy, 223–224
    line and area scanning comparison, 230–232
    single-asperity nanotribology by atomic force microscopy, 217–222
  scratch and wear, 48–49
Atomic force microscopy coupled with infrared spectroscopy (AFM-IR), 223–224
Atomic layer deposition (ALD), 129–130
Atomic stress, 404–405
Atomic substitutions, 604–605
Atomistic simulations, 48–49
Atomistic vaporization technology, 519

Attack angle, 48–49
Austenitic stainless steels, 560

B
Ball-on-plate impact test, 51–52
BE. *See* Boundary element (BE)
Berkovich indenters, 6, 9, 21, 115, 488–489
Berkovich pyramidal indents, 583
Berkovich tip, 581
Biaxial residual stress, 279
Biaxial stress, 354–355
Big data, 693–694
Bilinear traction-separation-based cohesive zone modeling, 289–290
Biomedical materials
  case studies on, 194–202
  nanofretting, 195–198
  reciprocating microscale tests on 316L stainless steel and Ti6Al4V, 198–202
BMG. *See* Bulk metallic glass (BMG)
Boron nitride (BN), 627, 632–633
Boundary element (BE), 366
Bowden and Tabor model, 166
Brinell indentation hardness, 454–455
Brittle ceramic coating, 459–460
Brittle chipping, 484
Brittle semiconductive material, 428
Brittleness index, 18
Buckling failure, 41–42
Bulk metallic glass (BMG), 223
Burgers vector, 411
Burris and Sawyer approach, 193

C
CAFM. *See* Conductive atomic force microscopy (CAFM)
Calcium attack, 606–609
Calcium–magnesium–aluminum–silicate erosion (CMAS), 252, 599
Calibrated contact load, 10–11
Calibrated weights, 113
Calibration sample, 333–334
Calo wear scar, 357
Calotte ball crater method, 347–349
Carbon, 396–397, 510
Carbonaceous transfer layer formation, 515
CAT. *See* Crystal analysis tool (CAT)
Catalytically reactive metals, 639–641
Cathodic arc deposition, 353

Index 699

comparing stress and modulus profile tuning for same, 353−355
indent positions with Calotte ball crater wear scar, 355f
Cathodic arc evaporation processes, 380−381, 561−562
Ceramics, 451−452, 610−611
  coatings, 604−605
  components, 637
  interlayers, 520
  materials, 447, 453−454, 580, 587
  nanoindentation in, 574−576
  thin films, 292−293
Cermet materials, 453−454
CFD. *See* Computation fluid dynamics (CFD)
CFUBMS. *See* Closed field unbalanced magnetron sputtering (CFUBMS)
Chameleon coatings, 630−632, 634−637
Chemical vapor deposition (CVD), 228−229, 370, 519, 601−602
Chip formation process, 557
Chipping, 41−42
Chromium, 515−516
  chromium-doped hydrogenated amorphous carbon, 519
  nitride coated systems, 379−380
Clamped microcantilevers, 83−84
Closed field unbalanced magnetron sputtering (CFUBMS), 495
CMAS. *See* Calcium−magnesium−aluminum−silicate erosion (CMAS)
Coated aircraft engine blades, 239
Coated systems
  analytical modeling of, 343−346, 344f
  digital twin, 345−346
Coatings, 54, 369, 512, 600
  adhesion, 533
  anisotropy, 347
  architecture, 563
  composition, 604
  in contact situations, 71−96
    breaking mutual exclusivity between hardness and toughness, 82−84
    coating behavior in nano-and microscale scratch tests, 72−80
    elevated temperature microtribology and impact testing, 93−96
    H^3/E^2

load-carrying capacity, 80−82
high-temperature testing, 90−93
impact resistance, 84−89
influence of elastic mismatch on coating behavior in nanoscratch testing, 73−77
nano-and microscratch testing of diamond-like carbon coatings on hardened tool steel, 77−80
weakening effects/high stiffness limitations, 72
deposition, 271−272, 521
design, 463−464
designers, 365
detachment during loading of indenter, 155f
engineering, 271−272
examples for measurement of, 134−139
hardness, 69−71
indentation and scratch analyses for coating simulation, 381−383
layer, 371−372
to protect silicon−thin ta-C films, 488−501
  challenges in measuring film hardness in ultrathin films, 488−490
  impact, 499−501
  nanofretting of ta-C films, 496−499
  nanoscratch of ta-C films, 491−496
  spherical indentation of ta-C films, 490−491
strategies for robust high-temperature measurements on, 332f, 333−334
effect of substrate anisotropy on coating behaviour, 376−377
systems, 67, 345−346, 353, 494, 528, 600−601
technology, 239, 557−558
theoretical considerations for measurement of, 123−127
toughness, 150
CoCrMo alloys, 194−202
Coefficient of friction (COF), 372−373, 518, 533, 537
Coefficient of variation (CV), 420−422
COF. *See* Coefficient of friction (COF)
Cohesive + adhesive failure, simulation of adhesive and, 378−379

Cohesive coating failure, simulation of, 377–378
Cohesive zone modeling (CZM), 378
Compaction, 252
Composites, 451–452
  modulus, 129
  nanoindentation in, 574–576
  properties, 67–68
  response, 333
Computation fluid dynamics (CFD), 394–395
Computational materials, 449–450
Conductive atomic force microscopy (CAFM), 223–224
Cone cracks, 14–15
Constant load scratch tests, 170
Constraint factor, 19
Contact area in unloaded state, 118–119
Contact depth, 118, 148
  in spherical indentation contact, 164
Contact detection accuracy, 23
Contact mechanics analysis, 69, 163
Contact pressure, 164
  calculation, 164–165
Contact radius, 164
Contact size effects on deformation versus fracture, 18
Contact stresses, 370–371, 372f
Contact-based FEM models, 383
Continuous stiffness measurement (CSM), 132, 331–332
Continuous stiffness technique, 23–24
Continuously recorded indentation tests, 143
Controllable parameters, 244
Conventional indentation toughness methods, 149
Conventional nanoindentation assessment, 144–147
Copper (Cu), 7, 662
Correction factor, 4
Crack closure due to high compressive stresses in the hard layer, 83–84
Crack deflection along weak interfaces, 83–84
Crack tip blunting and bridging due to plasticity in the metal layer, 83–84
Cracking, 482–488
Creep, 13–14
Creep rate, 13–14

Critical loads, 73, 162
Critical splitting load, 287
Crystal analysis tool (CAT), 406–407
Crystal defects, 406–411
Crystallization products, 609
CSM. *See* Continuous stiffness measurement (CSM)
Cube corner indenters, 490
Cube corner probes, 6
Cumulative strain relief function, 281
Cutting performance, 88–89
Cutting processes, hard coatings for, 557–560
Cutting tests, 558
Cutting tools, self-organization processes during wear of, 562–563
CVD. *See* Chemical vapor deposition (CVD)
Cyclic impact tests, 37–38, 84–85, 239–240
Cyclic nano tests, 566–567
Cyclic nanoindentation, 424–428
Cyclic wear evolution, 383
CZM. *See* Cohesive zone modeling (CZM)

D

DC. *See* Direct-current (DC)
Deformation
  estimation of stresses causing deformation during nanoindentation, 404–406
  processes, 302, 327, 461–462
  responses, 73
Degree of freedom, 349–351
Deposition, 358, 562
  of diamond-like carbon coatings, 519–521
  methods, 519–520, 523t
  technology, 359
Depth-dependent indentation modulus, 135f
Depth-sensing indentation, 3–4
Depth-sensing nanoindenter, 3–4
Detection threshold, 302
Diamond-like carbon (DLC) coatings, 84–85, 202–203, 227, 239, 294, 509–511, 627
  case study on diamond-like carbon coatings on steel, 202–209
  deposition of diamond-like carbon coatings, 519–521

Index   701

fretting testing of diamond-like carbons, 535–537
history of diamond-like carbons and types of diamond-like carbons, 509–511, 524t
impact for fatigue determination, 537–542
mechanical and structural characterization, 521–532
  multilayer structures and mechanical properties, 527–530, 531t
  partial-load indentation, 525–527
  structural characterization of doped diamond-like carbons, 530–532
performance of diamond-like carbons, 511–519
  common automotive applications for diamond-like carbons, 512–513
  lubricant interactions of diamond-like carbons, 516–519
  tribology of diamond-like carbon, 513–516
scratch testing for adhesion testing of diamond-like carbons, 532–535
stress profiling of DLC coating on A2 tool steel, 356–359
DIC analysis. *See* Digital image correlation analysis (DIC analysis)
Diffusion coatings, 602–603, 603f
Diffusion treatment, 466
Diffusion zone, 468–469
Digital image correlation analysis (DIC analysis), 274, 276–277
Digital model, 345
Digital twin, 345–346
  data, 346
  digital model, 345
  material data, 345–346
DIIT. *See* Dynamic instrumented indentation testing (DIIT)
Dimensional wear coefficient, 450
Dimensioning, 44, 159
Dimensionless index, 248–249
Direct linear relationship, 449
Direct measurement method, 117
Direct-current (DC), 519
Directional roughness, 170–171
Dislocation extraction algorithm (DXA), 406–407

Dislocations, 406–411
Doerner–Nix approach, 4
Doped diamond-like carbons
  electron energy loss spectroscopy, 530–532
  Raman spectroscopy, 530
  structural characterization of, 530–532
Dry sliding, 484
Ductile chipping, 484
Ductile materials, 610–611
DXA. *See* Dislocation extraction algorithm (DXA)
Dynamic hardness, 243
  of graphite-like carbon films, 258–261
  testing, 336
Dynamic instrumented indentation testing (DIIT), 132
Dynamic test methods, 125
  for measurement of hardness and modulus, 132–134

E
EB. *See* Electron-beam (EB)
EB-PVD. *See* Electron beam physical vapor deposition (EB-PVD)
EBPVD. *See* Electron-beam physical vapor deposition (EBPVD)
EBSD. *See* Electron back scattering diffraction (EBSD)
ECR. *See* Electrical contact resistance (ECR)
Edge cracking, 484
Edge forward scratch direction, 335–336
EDM. *See* Electric discharge machining (EDM)
EDX-ray spectroscopy. *See* Electron diffraction X-ray spectroscopy (EDX-ray spectroscopy)
EELS. *See* Electron energy loss spectroscopy (EELS)
Effective indenter concept, 5–6
Egg shell effect, 125–126
Eigenstrains, 272–273, 277
Elastic anisotropy of copper single crystals, 7
Elastic considerations, 448
Elastic contacts, 370, 452
Elastic mismatch, 468–469

Elastic mismatch (*Continued*)
    on coating behavior in nanoscratch testing, 73–77
Elastic modulus, 71, 332, 370, 400–402, 443, 445, 447, 460–461, 489–490, 525, 527–528, 580
    importance of in tribology and surface engineering, 443–447
Elastic radial displacement, 118–119
Elastic stress fields, 70–71, 70*f*
Elastica software, 72
Elastic–plastic deformation, 151–152
Elastic–plastic fracture mechanics (EPFM), 461–462, 667
Electric discharge machining (EDM), 334–335
Electric vehicles, 516
Electrical contact resistance (ECR), 190
    boron-doped diamond Berkovich unloading, 315*f*
    measurement, 313–317, 314*f*, 316*f*
    principle of, 314*f*
Electrification, 516
Electron back scattering diffraction (EBSD), 653, 656–657, 682–687, 687*f*
Electron beam physical vapor deposition (EB-PVD), 601, 605–606
Electron diffraction X-ray spectroscopy (EDX-ray spectroscopy), 521
Electron energy loss spectroscopy (EELS), 530–532
Electron microscopy techniques, 582–583
Electron-beam (EB), 465–466
Electron-beam physical vapor deposition (EBPVD), 252, 611–612
Electroplastic effect, 321–322
Electroplastic forming, 321–322
Electroplated multilayer sliding interconnectors, 190
Elevated temperature microtribology and impact testing, 93–96
Empirical power law dependence, 16–17
Energy balance, 449–450
Energy density, 542
Energy ratio, 487–488
Engineered surfaces
    appropriate use of hardness-to-modulus ratio as design tool, 463–464
    Archard wear equation, 447–453
    brief historical perspective on hardness, hardness-to-modulus ratio, and wear-rate determination, 453–458
    case study, 464–469
        triode-plasma nitrided surface layers, 467*f*
    importance of hardness and elastic modulus in tribology and surface engineering, 443–447
    practical use of hardness-to-modulus ratio-based parameters, 459–462
Engineering components, 441, 455–456
Engineering metals, 445
Engineering surfaces, 369
Environmental protection coatings and bondcoats, 601–604
    diffusion coatings, 602–603, 603*f*
    overlay coatings, 603–604
EP. *See* Extreme pressure (EP)
EPFM. *See* Elastic–plastic fracture mechanics (EPFM)
Epsilon factor, 118–119
Erosion, 252, 694–695
EU NANOINDENT projects, 159
Experimental techniques for nano-/microscale fretting and reciprocating wear testing, 49–51
Exploration
    femto-second laser and extreme environments, 588–589, 588*f*
    future avenues for, 588–589
    lift out techniques, 589
eXtended Finite Element Method (XFEM), 377–378
Extended Hertzian contact module, 129–130
Extrapolation method, 528
Extreme environments, 588–589
    of aerospace coatings, 599–600
Extreme pressure (EP), 469
Extreme temperatures, 693

F
Face forward scratch direction, 335–336
Fatigue determination, impact for, 537–542
Fatigue process, 499–500
Fatigue testing at micron scale, 667–668
Fatigue wear process, 498

Index

FCVA technology. *See* Filtered cathodic vacuum arc technology (FCVA technology)
FEA. *See* Finite element analysis (FEA)
FEM. *See* Finite element method (FEM)
Femto-second laser, 588–589
FIB. *See* Focused ion beam (FIB)
Film hardness in ultrathin films, challenges in measuring, 488–490, 489*f*
Filtered cathodic vacuum arc technology (FCVA technology), 481
Final impact depth, 247–248
Finite element (FE) methods, 19–21, 172, 366
Finite element analysis (FEA), 381, 394–395, 621
  contact stresses, 370–371
  indentation and scratch analyses for coating/substrate simulation, 381–383
  simulation of normal loads and lateral indenter displacement, 380–381
  thin film mechanics and substrate effects during indentation, 371–380
  indentation analyses, 379–380
  initial analyses of indentation problems, 371–376
  simulation of adhesive and cohesive + adhesive failure, 378–379
  simulation of cohesive coating failure, 377–378
  effect of substrate anisotropy on coating behaviour, 376–377
Finite element method (FEM), 279, 343–344, 369, 371, 380–382, 394–395, 499
Finite element simulations, 380
First impact depth, 247–248
First-stage turbine blades, 601–602
Flat punch approximation, 4
Fluorinated graphene nanosheets, 499
Focused ion beam (FIB), 275–277, 286, 306, 573, 621, 666–667
  FIB-based manufacturing techniques, 584
  FIB-based microtensile bars, 585
  FIB-DIC ring-core method, 277–279, 283
  focused ion beam-based techniques, 584–587, 584*f*
    microcompression experiments, 585*f*
    machining, 653
    milling techniques, 347–349
  ring-core method, 275–287, 295
    beam-digital image correlation ring-core measurement workflow, 276*f*
    eigenstrain approach to quantify focused ion beam-induced residual stress, 278*f*
    FEM analysis for average stress and stress gradient analysis, 277–279
    localized, position-resolved residual stress measurements, 285–286
    measurement of mean residual stress value at certain depth, 279–280
    measurement of residual stress depth profile, 280–283
    recent advances and applications, 283–287
    spatially resolved stress mapping, 286–287
    ultrathin coatings with nonequibiaxial stress states, 283–284
FOD. *See* Foreign object damage (FOD)
Force calibration, 114, 114*f*
Force field, 396–398
Foreign object damage (FOD), 252, 604, 612–613
Fourier transformation, 313
Fouvry and Liskiewicz approach, 193
Fracture energy, 148–156
Fracture events associated with coating/substrate system, 150*t*
Fracture probability, 247
Fracture resistance, 88–89
Fracture toughness, 15, 148–156
  of hard CrN coating, 261–264
  at micron scale, 667
Frame compliance correction, 7–8
Frank dislocation, 407
Fretting, 192–194
  testing of diamond-like carbons, 535–537
  wear, 487, 497, 535–537
Friction, 165–169, 514–515, 517–518
  coefficients, 162, 193, 451–452, 484, 487, 492, 495–496, 512–513
  of TiN coatings, 165–166
  force, 317–318
  loops, 192–194
  measurement, 317–319, 317*f*, 319*f*

Friction (*Continued*)
   friction coefficient versus load in nano-scratch tests, 318*f*
   model, 628
   and wear of ultrathin films, 225−229
Frictional heating, 463
Fuel forms
   nanoindentation on, 576−580, 579*f*
   Young's modulus *vs.* temperature for ceramic U compounds, 578*f*
Fukushima accident, 578
Fused silica, 8−9
   impact simulation on, 359−361

G

Gadolinia alumina perovskite (GdAlO$_3$), 616−617
Gadolinium zirconate (GZO), 618−619
Gallium (Ga), 521
   gallium-based focussed ion beam, 334−335
   ions, 521
Gas turbine of aerospace coatings, 599−600
GE. *See* General Electric (GE)
General Electric (GE), 599
Geological scales, examples, 349−351, 350*f*
Geometric estimation approach, 401−402
Geometrically necessary dislocations (GNDs), 16
Glaze layers, 649−650, 658−661, 659*f*
GLC. *See* Graphitic-like carbon (GLC)
Glycerol mono-oleate (GMO), 518
GMO. *See* Glycerol mono-oleate (GMO)
GNDs. *See* Geometrically necessary dislocations (GNDs)
GO. *See* Graphene oxide (GO)
Grain fragmentation process, 656
Grain refinement, 83
Graphene, 228−229, 365, 511, 632
   simulation, 363
Graphene oxide (GO), 229
Graphite, 510, 575
Graphitic-like carbon (GLC), 511
   determination of dynamic hardness of, 258−261
Graphitization, 515, 518
Gross slip, 192−194
GZO. *See* Gadolinium zirconate (GZO)

H

H/E optimization, 71
H^3/E^2
   load-carrying capacity, 80−82
Half-penny radial cracks, 14−15
Hard coatings, 83, 239, 557
   for cutting processes, 557−560, 559*f*
Hard nitride coatings for machining ductile materials, 560−562, 561*f*
Hardenend M2 tool steel (HTS), 520−521
Hardmetals, 680−682
   electron back scattered diffraction analysis, 682−687, 687*f*
   microtribology experiments, 675−680, 676*f*
      FIB section through scratch made on grain size WC/Co material, 679*f*
      threshold load for significant damage, 679*t*
   wear-corrosion synergy, 680−682
Hardness (H), 67−68, 143−148, 370, 402, 443, 445, 489−490, 525, 527−528, 615, 651
   brief historical perspective on, 453−458
   importance of hardness in tribology and surface engineering, 443−447
   uncertainty, 459−460, 462
Hardness-to-modulus ratios, 459
   brief historical perspective on, 453−458
   as design tool, 463−464
   practical use of hardness-to-modulus ratio-based parameters, 459−462
Hardness/toughness ratio, 18
HBS. *See* High burnup structure (HBS)
Hertz's theory, 370, 490
Hertzian contact model, 128
Hertzian contact theory, 399, 417
Hertzian correction, 12
Hertzian method, 494
Hertzian pressure distribution, 128
Heterogeneous coatings, 286
Hexagonal materials, 632
Hexagonal solids, 630
Hexagonal WC crystals, 684
Hexamethyldisiloxane (HMDSO), 521
High burnup structure (HBS), 579−580
High contact pressures and susceptibility to probe wear, 48−49

Index

High data acquisition single nano-impact technology, 256
High strain rate testing, 24–26
High surface roughness, 161
High temperature scanning indentation (HTSI), 663–664
High-order postprocessing analysis of molecular dynamics simulation files anisotropic mechanisms during nanoindentation, 411–419
  to correlate with experiments, 411–428
  cyclic nanoindentation, 424–428
  influence of indentation velocity, 420–423
High-performance cutting (HPC), 557–558
High-power impulse magnetron sputtering (HiPIMS), 284
High-precision single nano-impact testing, 254–264. See also Multiple nano/microimpact testing
  applications of single nano-impact testing, 258–264
    determination of dynamic hardness of graphite-like carbon films, 258–261
    evaluation of fracture toughness of hard CrN coating, 261–264
    physical mode of single nano-impact, 256–258
    research progress in single nano-impact testing, 254–256
High-resolution digital image correlation techniques (HR-DIC techniques), 667
High-resolution electron back scatter diffraction (HR-EBSD), 667
High-resolution electron microscopy, 277
High-speed cutting (HSC), 557–558
High-speed indentation, 336
High-speed machining
  hard coatings for cutting processes, 557–560
  hard nitride coatings for machining ductile materials, 560–562
  influence of fracture resistance and high-temperature mechanical properties in, 563–566
  self-organization processes during wear of cutting tools, 562–563
High-speed mapping, 26–27
High-speed nanoindentation mapping, 286

High-temperature testing, 90–93
  high-temperature pillar compression and microcantilever bending, 334–335
  high-temperature scratch and impact/high strain rate testing, 335–336
  strategies for reliable high-temperature nanomechanical test measurements, 327–334, 330f
  strategies for robust high-temperature measurements on coatings, 333–334
Higher-order polynomials, 9–10
Highly oriented polycrystalline/pyrolytic graphite (HOPG), 530–531, 576
HiPIMS. See High-power impulse magnetron sputtering (HiPIMS)
Hirth dislocation, 407
Hirth locks. See Hirth dislocation
Hit energy, 312
Hit spectrum, 313
Hits, 312
HMDSO. See Hexamethyldisiloxane (HMDSO)
Hold periods, 26
HOPG. See Highly oriented polycrystalline/pyrolytic graphite (HOPG)
HPC. See High-performance cutting (HPC)
HR-DIC techniques. See High-resolution digital image correlation techniques (HR-DIC techniques)
HR-EBSD. See High-resolution electron back scatter diffraction (HR-EBSD)
HSC. See High-speed cutting (HSC)
HTS. See Hardenend M2 tool steel (HTS)
HTSI. See High temperature scanning indentation (HTSI)
Hybridization, 510
Hydrocarbon precursors, 519
Hydrogen, 520
  hydrogen-free DLC, 515
Hydrogenated amorphous carbon film (a-C: H film), 381
Hydrostatic stresses, 422

I

IIT. See Instrumented indentation testing (IIT)
Impact fatigue mechanisms, 52, 85
Impact process, 321
Impact resistance, 84–89

Impact resistance (*Continued*)
  evaluation of impact resistance of hard diamond-like carbon coatings, 249–252
Impact simulation
  on fused silica, 359–361
  impact simulation-mixed droplet, 361–363
Impact testing
  acoustic emission in, 306–309, 310*t*
  depth and acoustic emission from nano-impact tests on
    alumina, 310*f*
    partially stabilized zirconia, 311*f*
In situ electrical contact resistance measurements, 483
In situ formation, 639–641
In situ Raman analysis, 636–637
In situ Raman spectroscopic measurements, 483–484
In situ scanning electron microscope, 585
In-situ systems, 334
In-wheel motor systems, 516
Indentation, 3–4, 372–373, 482–483, 485
  analyses, 379–380
    for coating/substrate simulation, 381–383
  curve data, 353–354
  fracture, 14–16
  influence of indentation velocity, 420–423
  initial analyses of indentation problems, 371–376, 373*f*
  methods, 445, 459
  to sliding transition, 170
  thin film mechanics and substrate effects during, 371–380, 375*f*
  work, hardness, and fracture, 22–23
Indentation energy, 19–23
  indentation energy-based analysis methods, 143
    conventional nanoindentation assessment, 144–147
    energy release rates and toughness, 153*t*
    fracture events associated with coating/substrate system, 150*t*
    fracture toughness and fracture energy, 148–156
    hardness and Young's modulus, 143–148
    load-displacement curve, 144*f*
  work of indentation approach, 147–148
  indentation work, hardness, and fracture, 22–23
  relationship between H/E and plasticity index, 19–22
Indentation size effects (ISEs), 16–19
  in coatings, 18–19
  contact size effects on deformation versus fracture, 18
  size effects in plasticity, 16–18
Indenter choice, 6
Indirect measurement method, 117
Industrial diamond, 511
Infrared (IR), 223–224
Instrumental effects, 13
Instrumented indentation. *See* Nanoindentation
Instrumented indentation methods, 442–443, 445
Instrumented indentation testing (IIT), 3–4, 113
Instrumented laboratory wear tests, 450–451
Interfacial fracture, 148
  energy and adhesion, 153–156
Interfacial friction, 167–168
Interfacial toughness, models for determining, 169
Interlayers, 519–520
Intermetallic compounds, 579
Intrinsic changes in strength due to the oscillation, 23
Intrinsic stacking faults (ISF), 407–409
Inverse pole figure (IPF), 686–687
Inverse problem, 273
Ion milling, 276
Ionic liquids, 518–519
Ions, 604–605
IPF. *See* Inverse pole figure (IPF)
IR. *See* Infrared (IR)
Irradiation, 571, 582, 586
ISEs. *See* Indentation size effects (ISEs)
ISF. *See* Intrinsic stacking faults (ISF)
ISO 14577, 7

J
J-integral theory, 461–462
JMAK. *See* Johnson–Mel–Avrami–Kolmogorov (JMAK)

Johnson−Mel−Avrami−Kolmogorov (JMAK), 661−662

K
KAM map. *See* Kernel average misorientation map (KAM map)
Kelvin probe force microscopy (KPFM), 223−224
Kernel average misorientation map (KAM map), 656−657
Kink pop-outs, 482−483
KPFM. *See* Kelvin probe force microscopy (KPFM)

L
Lab-scale test, 512
Laboratory-based accelerated tests, 84−85
Laser interferometer, 114
Lateral cracks, 14−15
Lateral dilation, 10
Lateral indenter displacement, simulation of, 380−381
LC locks. *See* Lomer−Cottrell locks (LC locks)
LEFM. *See* Linear elastic fracture mechanics (LEFM)
Lift out techniques, 589, 589f
Light water reactors (LWRs), 572
Lightweight metals, 634−635
Line and area scanning comparison, 230−232
Linear back-extrapolation, 12
Linear elastic fracture mechanics (LEFM), 44−45, 275, 667
Liquid lubricants, 627
Load support, 165
Load-dependent surface roughness parameter, 10−11
Load − displacement ($P-h$) curve, 127−128, 151
  estimation of, 399, 401f
  from molecular dynamics simulation, 400f
Lomer−Cottrell locks (LC locks), 407
Low friction, 84
Low sliding velocity, 48−49
Lubricants, 518
  interactions of diamond-like carbons, 516−519
Lubrication, 627
  approaches for application of solid lubricants, 629−630, 629t
  basics of solid lubrication, 628
  traditional solid lubricants, 630, 631t
  in traditional solid lubricants and approaches for application of solid lubricants, 628−630
LWRs. *See* Light water reactors (LWRs)

M
Machining ductile materials, hard nitride coatings for, 560−562
Machining process, 560
Macroscale data, 582
Macroscale methods, 346−347
Macroscale scratch tests, 162
Macroscopic residual stress evaluation methods, 275−276
Macroscopic Vickers, 124
Macroscratch testing, 41−44, 532−533
Magnesium alloys, 327
Magnesium attack, 606−609
Magnesium silicate hydroxide (MSH), 638−639
Manufacturing process, 344, 358, 647−648, 653−654, 666
  application to impact-induced tribologically transformed surface in pure iron, 655−658, 656f
  application to tribologically transformed surface resulting of, 653−658
  strengthening mechanisms in tribologically transformed surface, 654−655
Marked frictional oscillations, 74
Materials science, 441−442
Materials-related effects, 13
MCrAlY alloy, 601−602, 604
MD. *See* Molecular dynamics (MD)
MDS. *See* Molecular dynamics simulations (MDS)
Mechanical engineering, 441−442
Mechanical performance modulation, 272
Mechanical property mapping, 695
Mechanically mixed layer (MML), 648
Median cracks, 14−15
MEMS. *See* Microelectromechanical systems (MEMS)
Metal, 459−460
Metal cutting scenarios, 557

Metal nitride–based hard coatings, 639–641
Metallic elements, 511
Metallic interlayers, 520
Metallic materials, 7, 13–14
 nanoindentation studies on metallic materials for nuclear applications, 580–583
Metallic phases, 446–447
Metallic–intermetallic composites, 321–322
Micro Materials Vantage system, 538
Micro-electric discharge machining, 334–335
Micro-Raman spectroscopy, 274
Micro/nano-impact testing
 controllable parameters, 244
 experimental setup of, 242–243
Microcantilevers, 572–573, 587
 bending, 334–335
 tests, 16
Microchipping damage, 487
Microcompression, 572–573
 experiments, 584, 586, 589
 testing, 15–16
Microcutting transition, 172
Microelectromechanical systems (MEMS), 249–250, 481
 coatings to protect silicon–thin ta-C films, 488–501
 silicon (100), 482–488
Microhardness testing, 572
Microimpact, 541
 systems, 617
 tests, 54, 538, 558, 566–567
Microindentation, probing sintering resistance, chemical resilience, and erosion resistance using, 614–619
Micromechanical fatigue testing, 668
Micromechanical testing methods, 652
Micromechanical tests, 648
 used to characterize tribologically transformed surface mechanical properties, 650–653, 652f
Micron scale, 650–651
 fatigue testing at, 667–668
 fracture toughness, 667
 analysis, 287–295
 recent advances in residual stress measurement at, 275–287

Micropillar compression technique, 652
Micropillar splitting test, 667
Microresidual stresses, 272
Microring core method, 15–16
Microscale, 301–302
 data, 582
 mechanical testing, 582
 samples, 585
 scratch tests, 566–567
  coating behavior in, 72–80
 tests, 582
Microscratch testing, 44–46, 535
Microseconds, 312
Microshear compression test, 664–666, 665f
Microshear tests, 664–665
Microstructural changes, 83
Microstructural characterization methods, 648–649
Microtensile testing, 584, 586, 666–667
Microtribological tests, 189
Microtribology
 experiments, 680
 tests, 675
Microtribometer, 679–680
Milling process, 277–279
Misorientation maps, 684–687
Mixed oxide fuels, 577
MML. See Mechanically mixed layer (MML)
Modeling, 619–620
 erosion of thermal barrier coatings, 619–621
Modell Factor, 455
Modern coatings tribology, 450–451
Modern Surface Engineering deposition processes, 464
Modulus, 429
 measurements by fully elastic indentations, 127–131
 profile tuning for same cathodic arc deposited coating system, comparing stress and, 353–355
Molecular dynamics (MD), 343–344, 366, 369
 high-order postprocessing analysis of, 411–428
 postprocessing of molecular dynamics data to extract nanomechanical properties, 399–411

Molecular dynamics simulations (MDS), 394–395, 491–492, 497
  ingredients of trustworthy, 395–398
    boundary conditions and ensemble, 395–396, 397f
    considerations for simulation of nanoindentation process, 395
    potential energy function or force field, 396–398, 398t
    schematic diagram molecular dynamics simulation model of nanoindentation, 396f
  means to validate molecular dynamics simulation studies, 428–429
  restrictions and limitations of, 429–431
  scale differences between nanoindentation experiments and, 428–431
Molybdenum, 333–334
  carbide compound, 516–517
Molybdenum disulfide (MoS_2), 627, 632–633
Monolayered columnar coatings, 566
Monte-Carlo models, 619–621
MountainsMap software, 537
Mounting method, 302–303
"Mrozovski" accommodation pores, 575
Multiasperity testing, 37–38
Multifunctional nanomechanical NanoTest systems, 481–482
Multilayered systems, 292, 294–295
Multipass scratch testing, 174–179
Multiple impact depth–time curves, analysis method of, 246–249
Multiple impulse technique, 336
Multiple nano/microimpact testing, 244–254. *See also* High-precision single nanoi-mpact testing
  analysis method of multiple impact depth–time curves, 246–249
  applications of, 249–254
    evaluation of impact resistance of hard diamond-like carbon coatings, 249–252
    simulation of erosion failure of thermal barrier coating, 252–254
  research progress in multiple-impact testing, 244–246
Multiple shear loops, 407
Multiple-scratch test experiments, 159–160

Multisensing approaches, 301, 693–694
  electrical contact resistance measurement, 313–317, 314f
  electroplastic effect, 321–322
  friction measurement, 317–319, 317f
  multisensing in high strain rate contact, 321
  nanomechanical Raman spectroscopy, 319–320
  sensors and actuators, 320–321
  using simultaneous record of acoustic emission, 301–313, 303f
Multispeed transmission systems, 516

N

Nano-/microscale fretting, experimental techniques for, 49–51
Nano-and microscratch testing of diamond-like carbon coatings on hardened tool steel, 77–80
  critical loads in, 78t
  mechanical properties, 77t
  stress distributions in nanoscratch test, 78f
Nano/microtribological test techniques, 3
Nanocomposite coatings, 67, 292
Nanocomposite design strategy, 82–83
Nanoelectromechanical systems (NEMS), 217
Nanofretting, 187–189, 195–198
  of Si(100), 485–486
  of ta-C films, 496–499, 497f
  tests, 193, 485–486, 488, 497–498
Nanohardness, 663
Nano-impact testing, 25, 53, 240, 488, 499–500, 541, 558
  calculated fracture dissipated energy and fracture toughness for CrN coating, 263t
  experimental setup of micro/nanoi-mpact testing, 242–244
  high-precision single nano-impact testing, 254–264
  multiple nano/microimpact testing, 244–254
NANOINDENT-PLUS projects, 159
Nanoindentation, 3–4, 328, 394–395, 417–419, 445, 513, 572–573, 580, 615–619, 618f, 620t, 655
  anisotropic mechanisms during, 411–419

Nanoindentation (*Continued*)
  application of radial displacement correction, 116–118
  combined calibration of area function and instrument compliance, 118–120
  considerations for simulation of, 395
  creep, 577
  curves, 5*f*, 482–483
  dynamic test methods for measurement of hardness and modulus, 132–134
  estimation of stresses causing deformation during, 404–406
  examples for measurement of coatings, 134–139
  fatigue, 54
  hardness, 400–402
  influence of tip rounding, 120–123
  instruments, 573
  load–displacement curve, 150
  mapping, 515
  methods, 302–303
  model, 396
  modulus measurements by fully elastic indentations, 127–131
  nanoindentation-based techniques, 573–583
    future avenues for exploration, 588–589
    nanoindentation in ceramic materials and composites, 574–576, 574*f*
    nanoindentation on fuel forms, 576–580
    nanoindentation studies on metallic materials for nuclear applications, 580–583
    small-scale mechanical test techniques used today, 583–587
  practical factors influencing accuracy of nanoindentation data and reliability of properties, 6–14
    area function determination, 9–10
    creep, time dependency, and thermal drift correction, 13–14
    frame compliance correction, 7–8
    indenter choice, 6
    ISO 14577, 7
    pile-up and sink-in, 11–12
    reference materials, 8–9
    surface roughness, 10–11
    zero point correction, 12
  preconditions for correct and reliable measurements, 113–116
  probing sintering resistance, chemical resilience, and erosion resistance using, 614–619
  scale differences between molecular dynamics simulations and, 428–431
  systems, 573–574
  tests, 306, 428–429, 485–486, 488, 491, 655–656
  theoretical considerations for measurement of coatings, 123–127
Nanojackhammer effect, 24
Nanomechanical measurements, 564–566
Nanomechanical methodology, 655
Nanomechanical Raman spectroscopy, 319–320
Nanomechanical testing, 3
  constraint factor, 19
  continuous stiffness technique, 23–24
  high strain rate testing, 24–26
  high-speed mapping, 26–27
  historical background and development of methods for unloading curve analysis, 3–6
  indentation energy, 19–23
  indentation fracture, 14–16
  indentation size effects, 16–19
  methods
    concept of eigenstrain as invariant source of residual stresses, 273*f*
    micron-scale fracture toughness analysis, 287–295
    recent advances in residual stress measurement at micron scale, 275–287
  practical factors influencing accuracy of nanoindentation data and reliability of properties, 6–14
  testing under environmentally relevant conditions, 27
Nanomultilayer coating, 563–564
Nanomultilayered CrAlTiN coatings, 89
Nanoparticles (NPs), 638–639
Nanoscale, 346–347, 349
  coating behavior in nanoscale scratch tests, 72–80
  examples, 349–351, 350*f*

wear test techniques
  AFM scratch and wear, 48–49
  experimental techniques for nano-/microscale fretting and reciprocating wear testing, 49–51
  main features and differences between macro-, micro-, and nanoscale impact tests, 52t
  repetitive contact, 51–54
  scratch and wear testing, 40–48
  single-asperity tribology, 37–40
Nanoscratch, 44–46, 533–535
  behavior of CoCrMo alloys, 166
  data, 492
  module, 533
  resistance, 492
  of Si(100), 484–485
  of ta-C films, 491–496, 492f
    critical loads in nanoscratch experiment, 493f
    SEM image of nano-scratch, 493f
Nanoscratch testing, 159, 484–486, 488, 491–492, 495, 533
  experimental considerations, 160–162
  general features, 162–174
    constant load scratch tests, 170
    contact pressure calculation, 164–165
    friction, scratch recovery, and plowing, 165–169
    load, 165
    loading curves from indentation and scratch tests, 163f
    models for determining adhesion energy and interfacial toughness, 169
    probe geometry and size effects, 172–174
    probe radius dependence, 163–164
    scratch hardness, 170
    scratch size effects, 171–172
    surface roughness and scratch orientation relative to grinding marks, 170–171
    3-scan test, 162–163
  influence of test temperature, 179–180
  multipass scratch testing, 174–179
NanoTest system, 320–321, 558
NanoTest Vantage, 204
NanoTriboTest, 187–189
Nanowear resistance, 492

Negative thermal drift, 329
NEMS. See Nanoelectromechanical systems (NEMS)
Neutron diffraction, 274
Neutron irradiation, 583
  facilities, 571
  neutron-irradiated materials, 573
Nitrogen doping, 76
Nix-Gao model, 16–17
Noble metals, 637
Nonconventional environments, 519
Nonequibiaxial residual stress state, 279–280
Nonequibiaxial stress states, ultrathin coatings with, 283–284
Nonmetals, 511
Normal loads, simulation of, 380–381
Novel alloys, 315–317
Novel composite coatings, 292
Novel experiments, 696
NPs. See Nanoparticles (NPs)
Nuclear fuel, 576
Nuclear fusion device, 361–362
Nuclear materials, 589
Nuclear reactor, 583
Numerical analysis, 37
Numerical modeling, 377–378
Numerical simulation
  contact stresses, 370–371
  indentation and scratch analyses for coating/substrate simulation, 381–383
  simulation of normal loads and lateral indenter displacement, 380–381
  thin film mechanics and substrate effects during indentation, 371–380
  indentation analyses, 379–380
  initial analyses of indentation problems, 371–376
  simulation of adhesive and cohesive + adhesive failure, 378–379
  simulation of cohesive coating failure, 377–378
  effect of substrate anisotropy on coating behaviour, 376–377

O

Octahedral shear stress, 411–417
OLC. See Onion-like carbon (OLC)

Oliver and Pharr method, 145–146, 527
1/10 rule, 70–71
Onion-like carbon (OLC), 511
Oscillation-based technique, 23
Overlay coatings, 603–604
"Overload" tests, 39
Oxidation, 252, 336
Oxide tribofilm formation, 453–454
Oxygen, 320–321

P
PACVD system. *See* Plasma-assisted chemical vapor deposition system (PACVD system)
Palmqvist radial cracks, 14–15
Partial load–unload indentation, 527–528
Partial-load indentation, 525–527
Particle impact, 361–363
PECVD. *See* Plasma enhanced chemical vapor deposited (PECVD)
PEO. *See* Plasma electrolytic oxidation (PEO)
Periodic stiffness variation, 83–84
Phase transformation, 313–314, 482–488, 484*f*
Phosphate laser (PL), 221–222
Physical mode of single nano-impact, 256–258
Physical vapor deposition (PVD), 239, 370, 519, 521, 557
Physical-based analytical methodology, 44–45
PI. *See* Plasticity index (PI)
PIE. *See* Postirradiation examination (PIE)
Piezo stack, 187–188
Pile-up, 11–12
Pillar splitting method, 15, 275, 287–295
  experimental considerations, 290–291
  experiments, 290–291
  recent advances and applications, 291–295
    ceramic thin films, 292–293
    multilayered systems, 294–295
    nanocomposite coatings, 292
    thermal barrier coatings, 293–294
Pin-on-disk tests of Ti-Si-C-H coatings sliding, 84
PL. *See* Phosphate laser (PL)
Plasma electrolytic oxidation (PEO), 635

Plasma enhanced chemical vapor deposited (PECVD), 495
Plasma-assisted chemical vapor deposition system (PACVD system), 521
Plastic deformation, 331, 335, 377
Plastic stress fields, 70–71, 70*f*
Plastic work of indentation (Wp), 143
Plastic zone, 582–583
Plasticity, 6, 612–613
  errors, 23–24, 132
  size effects in, 16–18
Plasticity index (PI), 67–69, 445, 452, 513–514, 615
Platinum aluminide coating, 602
Plowing, 165–169
Poisson's ratio, 4–5, 115, 364, 445
Polishing techniques, 353
Poly granular graphite, 575
Polymers, 451–452, 459–460
Polytetrafluoroethylene (PTFE), 627
Pop-outs, 482–483
Post running-in stage, 562
Postirradiation, 576
Postirradiation examination (PIE), 571
Postprocessing of molecular dynamics data to extract nanomechanical properties, 399–411
  estimation of load – displacement curve, 399
  estimation of strain rate sensitivity, 402–404
  estimation of stresses causing deformation during nanoindentation, 404–406
  nanoindentation hardness and elastic modulus, 400–402, 402*f*
  subsurface deformation mechanisms, dislocations, and crystal defects, 406–411, 408*f*
Potential energy function, 396–398
Power law, 313–314
Precession electron diffraction, 16–17
Prismatic dislocation loops, 411
Probes, 40–41
  geometry, 172–174
  influence of, 499–501
  radius dependence, 163–164
Progressive load nanoscratch tests, 159
Progressive load scratch tests. *See* Ramped load scratch tests

Index

Pure iron, application to impact-induced tribologically transformed surface in, 655–658, 656f
PVD. *See* Physical vapor deposition (PVD)
Pyramidal indenters, 9, 395

Q

q parameter, 167–168
QCM. *See* Quartz crystal microbalance (QCM)
QCSM. *See* Quasicontinuous stiffness measurement (QCSM)
Quartz crystal microbalance (QCM), 223–224
Quasicontinuous stiffness measurement (QCSM), 133
Quasielastic Hertzian analytical approach, 344

R

Radial cracks, 15, 148–149
Radial displacement correction, 134
  application of, 116–118
Radiative heating, 328
Radio frequency (RF), 519
Raman activity, 635
Raman spectroscopy, 530
Ramped load scratch tests, 41
Reciprocating contacts, 37–38, 187
Reciprocating line scanning, 230
Reciprocating microscale sliding, 189–192
  electrical contact resistance and COF data from typical test, 191f
  friction and resistance for Ag/Ag tribocouple, 192f
Reciprocating microscale tests on 316L stainless steel and Ti6Al4V, 198–202
Reciprocating motion, 51
Reciprocating nano-and microscale wear testing, 187
  case studies on biomedical materials, 194–202
  case study on diamond-like carbon coatings on steel, 202–209
  friction loops, 192–194
  nanofretting, 188–189
  reciprocating microscale sliding, 189–192
  test conditions in nano-and microtribological tests, 188t

Reciprocating sliding tests, 469
Reciprocating wear testing, experimental techniques for, 49–51
Recrystallization process, 663
Reduced elastic modulus (E_R), 615
Reduced modulus, 4–5, 443–444
Reference materials, 8–9
Relative indentation depth (RID), 69–70
Repetitive (multipass) scratch testing, 46–48
Repetitive constant load tests, 41, 159
Repetitive contact, 51–54
Repetitive impact tests, 37–38
Repetitive scratch tests, 174
Reprecipitation products, 609
Residual creep, 332
Residual stresses, 271–272, 380
  measurement of residual stress depth profile, 280–283
  recent advances in residual stress measurement at micron scale, 275–287
Restoration mechanisms, 661–662
RF. *See* Radio frequency (RF)
RID. *See* Relative indentation depth (RID)
Rise time, 312
RMS. *See* Root mean square (RMS)
Rockwell C adhesion test, 16
Root mean square (RMS), 312
Rosette strain gage equations, 282
Round-robin intercomparison study scratch testing, 42
Round-robin interlaboratory intercomparisons, 159

S

Scale differences between molecular dynamics simulations and nanoindentation experiments, 428–431
  means to validate molecular dynamics simulation studies, 428–429
  restrictions and limitations of molecular dynamics simulations, 429–431
Scale factor (SF), 283
Scanning electron microscopy imaging (SEM imaging), 621, 652
Scanning near-field ellipsometry microscopy (SNEM), 223–224
Scanning probe techniques, 38–39
Scanning thermal microscopy (SThM), 223–224

Scratch analysis
 for coating/substrate simulation, 381–383
 software, 353
Scratch analyzer, 351
Scratch and wear testing, 40–48
 macroscratch testing, 41–44
 nano-and microscratch, 44–46
 repetitive (multipass) scratch testing, 46–48
Scratch crack propagation resistance, 532
Scratch damage, 684
Scratch hardness, 170
Scratch orientation relative to grinding marks, 170–171
Scratch recovery, 165–169
Scratch size effects, 171–172
 in coated systems, 173–174
Scratch Stress Analyzer, 77–79, 346
Scratch test(ing), 159, 160f, 335, 346, 380, 513
 acoustic emission in, 303–306
 for adhesion testing of diamond-like carbons, 532–535
  macroscratch, 532–533
  nanoscratch, 533–535
 advanced evaluation of acoustic emission during, 310–313, 312f
 data, 351
 evaluation, 303–304
 modeling, 458
 of zirconia coatings, 351–353, 351f, 352f
Secondary ion mass spectroscopy analysis, 518
Self-healing materials, 637
Self-healing tribological materials, 637
Self-organization processes during wear of cutting tools, 562–563, 563f
SEM imaging. *See* Scanning electron microscopy imaging (SEM imaging)
Sensors of multisensing approaches, 320–321
7 yttria-stabilized zirconia (7YSZ), 601, 606, 618–619
SF. *See* Scale factor (SF)
Sharp cube corner indenters, 53–54
Shear forces, 628
Siliceous debris, 606–607
Silicon, 396–397, 428, 481, 499, 509, 520
 coatings to protect silicon—thin ta-C films, 488–501
 silicon-based nanoelectromechanical systems, 217
 substrate, 317, 501
 surface, 488
 wafers, 11
Silicon (100), 482–488
 indentation, 482–483
 nanofretting of, 485–486
 nanoindentation, nanoscratch, and nanofretting tests of silicon, 486f
 nanoscratch of, 484–485
Silicon carbide (SiC), 379
Silicon carbide fiber—reinforced silicon carbide matrix composites (SiC–SiC composites), 574–575
Silver, 518–519
Simulation
 of adhesive and cohesive + adhesive failure, 378–379
 of cohesive coating failure, 377–378
 considerations for simulation of nanoindentation process, 395
 of erosion failure of thermal barrier coating, 252–254
 of normal loads and lateral indenter displacement, 380–381
Single nano-impact testing
 applications, 258–264
 physical mode, 256–258
 research progress in, 254–256
Single-asperity nanotribology by atomic force microscopy, 217–222
 atomic force microscopy topography images and average depth of wear mark, 219f
 nanoscratch (wear) tests on a polymer sample, 218f
Single-asperity tribology, 37–40
Single-layer graphene, 363
Sink-in, 11–12
Sintering, 252
Size effects, 172–174
 in plasticity, 16–18
 in yield, 172–173
Sliding ratio, 537
 in fretting, 192, 193f
Sliding systems, 638–639

Index   715

Slip systems, 407–409
Small track length, 188
Small-scale mechanical testing (SSMT), 571–572
  focused ion beam-based techniques, 584–587
  techniques, 583–587
Small-scale testing, 54, 275, 327
Smaller-scale nanofretting tests, 487–488
SMGT. See Surface mechanical grinding treatment (SMGT)
Sneddon's flat punch equation, 145
SNEM. See Scanning near-field ellipsometry microscopy (SNEM)
SOFC. See Solid oxide fuel cell (SOFC)
Soft materials, 630
Solenoid, 538
Solid lubricants, 627, 629–630, 635, 637–638
  advanced solid lubricants, 630–641
    chameleon coatings, 634–637
    self-healing materials, 637
    tribocatalytic materials, 637–641
    two-dimensional materials, 632–634, 633f
  approaches for application of, 629–630, 629t
  mechanism of lubrication in traditional solid lubricants and approaches for application of, 628–630
Solid lubrication, 627
  basics of, 628
Solid oxide fuel cell (SOFC), 320–321
Spallation, 532–533
Spallation Target Irradiation Program (STIP), 582
Spectroscopic analysis techniques, 442–443
Sphere-on-flat test geometry, 37–38
Spherical indentation, 652
  of ta-C films, 490–491, 491f
  tests, 490–491
Spherical indenters, 17–18
Spherical nanoindentation, 583
Spherical tips, 161, 583
Springs, 572
SS. See Stainless steel (SS)
SSMT. See Small-scale mechanical testing (SSMT)

Stainless steel (SS), 520–521, 560
Star-crack, 349
Static contacts, 452
Static friction coefficient, 193–194
Static indentation depth, 247–248
Static recrystallization, 661–662
Statistical distributed nanoscratch test, 159–160
Steel-on-steel flats, 452
SThM. See Scanning thermal microscopy (SThM)
STIP. See Spallation Target Irradiation Program (STIP)
Strain rate sensitivity, 403–404
  estimation of, 402–404, 403f, 404f
Strengthening mechanisms in tribologically transformed surface, 654–655
Stress and modulus profile tuning for same cathodic arc deposited coating system, 353–355
Stress distributions, 4
  in nanoscratch test, 78f
  in sliding contact, 40, 40f
Stress gradient analysis, FEM analysis for average stress and, 277–279
Stress profile, 358
  of DLC coating on A2 tool steel, 356–359
  measurement method, 364
Stress tensor, 404–405, 425–428
Stresses causing deformation during nanoindentation, estimation of, 399–411
Submicron scale, 275
Submicroscopic stresses, 272
Substrate anisotropy on coating behaviour, effect of, 375f, 376–377
Substrate deformation, 54
Substrate effects during indentation, 371–380
Substrate hardness, 628
Substrate simulation, indentation and scratch analyses for, 381–383
Subsurface deformation mechanisms, 406–411
Superhard coatings, 67
Superlattices, 83
Surface engineering, 441–442, 446, 600–601, 600f, 635
Surface influence, 277

Surface mechanical grinding treatment (SMGT), 654
Surface optimization, tool for, 696
Surface roughness, 10–11, 170–171, 353
Surface treatments, 635
Surfaces, 603, 628
Synchrotron radiation techniques, 274

T
ta-C films. *See* Tetrahedral amorphous carbon films (ta-C films)
Tabor relation, 651
Tangential force, 537
Tangential loading, 486–488
Taylor's law, 655
TBC system. *See* Thermal barrier coating system (TBC system)
Teflon, 627
TEM. *See* Transmission electron microscopy (TEM)
Tensile cracking, 494
Tensile stresses, 79, 372–373
Tensile test, 459
TERS. *See* Tip-enhanced Raman spectroscopy (TERS)
Test temperature, influence of, 179–180
Testing environments, 588–589
TET. *See* Turbine entry temperatures (TET)
Tetrahedral amorphous carbon films (ta-C films), 481, 488, 512–513
  coated systems, 501
  nanofretting of, 496–499
  nanoscratch of, 491–496
  spherical indentation of, 490–491
Thermal barrier coating system (TBC system), 241–242, 292–294, 599–601
  compositions, 618
  design considerations, 604–614
    chemical resilience, 606–609, 608*f*
    high-temperature stability, 606
    low thermal conductivity, 604–606
    resistance to erosion and foreign object damage, 609–614, 610*f*
  design tools for developing enhanced, 614–619
  elements, 606
  materials, 601, 605
  modeling erosion of, 619–621
  nanoindentation, 615–619

Thermal conductivity, 557–558
Thermal drift correction, 13–14, 525–527
Thermal endurance dynamics, 272
Thermal spraying process, 603
Thermal stability, 513
  of tribologically transformed surface, 661–664
Thermally grown oxide (TGO), 293–294, 600–602
Thermocouples, 328
Thin films, 161
  mechanics during indentation, 371–380
  systems, 481–482
Three-body abrasion, 458
3D
  3D-imaging digital microscopy, 453
  FEM model, 282
  mathematical complexities, 462
"Three-scan" tests, 161–163
316L stainless steel, 194–202, 520–521
Through-thickness cracking, 41–42
Through-thickness fracture, 148
Ti6Al4V, 194–202
TiAlCrSiYN-based coatings
  influence of fracture resistance and high-temperature mechanical properties in self adaptive behavior of, 563–566
  temperature dependency of mechanical properties of selected hard coatings, 565*f*
Time dependency, 13–14
Tip rounding, influence of, 120–123
Tip wear, 334
Tip-enhanced Raman spectroscopy (TERS), 223–224
Titanium (Ti), 284, 469, 515–516
  titanium-based bearing, 465
Titanium nitride (TiN), 464–467
TMP. *See* Trimethylolpropane (TMP)
Topography–scratch–topography, 533
Total penetration depth, 144
Total track length, 193–194
Toughness, 541–542
TPN. *See* Triode plasma nitriding (TPN)
Track length and small sliding distances, 48–49
Traditional energy-based methods, 321
Traditional solid lubricants, 630, 631*t*

Index

mechanism of lubrication in traditional solid lubricants and approaches for application of solid lubricants, 628–630
Transition zone, 656–657
Transmission electron microscopy (TEM), 530, 589, 653
  nanoindenter, 48–49
Tribocatalytic coatings, 638–639
Tribocatalytic materials, 637–641
Tribofilms, 516–517
Tribological and impact mapping, 695
Tribological coatings, 634–635
Tribological system, 459
Tribologically transformed surfaces (TTSs), 46, 189, 200, 647–650, 655–656, 662–664
  application to glaze layers, 658–661, 659f
  application to tribologically transformed surface resulting of manufacturing processes, 653–658
  developments, 664–668
    fatigue testing at micron scale, 667–668
    fracture toughness at micron scale, 667
    microshear compression test, 664–666, 665f
    microtensile testing, 666–667
  examples of, 647f
  formation, 647–648
  induced on copper bar by high sliding friction, 650f
  micromechanical tests used to characterize tribologically transformed surface mechanical properties, 650–653, 652f
  strengthening mechanisms in, 654–655
  thermal stability of, 661–664
  zones, 666–667
Tribology, 446
  of diamond-like carbon, 513–516
  importance of hardness and elastic modulus in, 443–447
Tribooxidation analysis, 562–563
Tribosystem, 192
Trimethylolpropane (TMP), 518
Triode plasma nitriding (TPN), 465–466
TRISO particles. See TRIstructural-ISOtropic particles (TRISO particles)

TRIstructural-ISOtropic particles (TRISO particles), 285–286
TTSs. See Tribologically transformed surfaces (TTSs)
Tungsten, 7, 328, 333–334, 518–519
Tungsten carbide, 328
Turbine blades, 602
Turbine entry temperatures (TET), 602
Twin boundaries, 411
Two dimensional materials (2D materials), 630–634, 633f

U

UHV. See Ultrahigh vacuum (UHV)
Ultimate tensile strength (UTS), 454–455, 580–581, 666
Ultrahigh vacuum (UHV), 218
Ultralow friction, 515–516
Ultrananocrystalline diamond (UNCD), 224
Ultrathin coatings with nonequibiaxial stress states, 283–284
Ultrathin films, challenges in measuring film hardness in, 488–490
Ultrathin hard carbon films, 488
Ultrathin ta-C film, 489
UNCD. See Ultrananocrystalline diamond (UNCD)
Unidirectional sliding tests, 37–38
Unit element cell, 378
Unloading curve analysis, historical background and development of methods for, 3–6
Uranium
  compounds, 579
  fuel forms, 578
UTS. See Ultimate tensile strength (UTS)

V

Vacuum plasma spray methods (VPS methods), 604
Van der Waals forces, 628
Vapor deposition process, 83, 465–466
Vickers hardness, 117–118
Vickers indentation, 14–15
Vickers indenter, 6
Vickers testing, 572
Viscoelastic correction, 13–14
Von Mises equation, 376
Von Mises stresses, 48, 125–126

VPS methods. *See* Vacuum plasma spray methods (VPS methods)

W
Water-based lubricants, 518–519
Water-droplet erosion (WDE), 460–461
Wavelength, 10–11
WDE. *See* Water-droplet erosion (WDE)
Weakening effects/high stiffness limitations, 72
Wear acceleration, 518
Wear mechanisms, tangential loading influence on, 486–488
Wear of cutting tools, self-organization processes during, 562–563
Wear processes, 461–462, 562
Wear resistance, 84
Wear volume, 67
Wear-corrosion synergy, 680–682
Wear-rate determination, brief historical perspective on, 453–458
Wear-resistant coatings, 67, 336
Wedge spallation, 41–42
WEL. *See* White etching layers (WEL)
White etching layers (WEL), 649
Work of indentation approach, 147–148

X
X-ray absorption spectroscopy (XAS), 562–563
X-ray diffraction (XRD), 273–274, 442–443, 510
X-ray emission (XES), 562–563
X-ray photoelectron spectroscopy (XPS), 530
XAS. *See* X-ray absorption spectroscopy (XAS)
XES. *See* X-ray emission (XES)
XFEM. *See* eXtended Finite Element Method (XFEM)
XPS. *See* X-ray photoelectron spectroscopy (XPS)
XRD. *See* X-ray diffraction (XRD)

Y
Yield
  size effects in, 172–173
  stress, 17
  tangential loading influence on, 486–488
Young's modulus, 115, 143–148, 345–346, 429, 576, 579–580, 615–617, 651
YSZ coatings. *See* Yttria-stabilized zirconia coatings (YSZ coatings)
Yttria-stabilized zirconia coatings (YSZ coatings), 293–294, 320–321, 611

Z
Zero point correction, 12, 134
Zirconia coatings, scratch testing of, 351–353